Applications of Quality Control in the Service Industries

QUALITY AND RELIABILITY

A Series Edited by

Edward G. Schilling
Center for Quality and Applied Statistics
Rochester Institute of Technology
Rochester, New York

1. Designing for Minimal Maintenance Expense: The Practical Application of Reliability and Maintainability, *Marvin A. Moss*
2. Quality Control for Profit, Second Edition, Revised and Expanded, *Ronald H. Lester, Norbert L. Enrick, and Harry E. Mottley, Jr.*
3. QCPAC: Statistical Quality Control on the IBM PC *Steven M. Zimmerman and Leo M. Conrad*
4. Quality by Experimental Design, *Thomas B. Barker*
5. Applications of Quality Control in the Service Industries, *A. C. Rosander*

In preparation

6. Integrated Product Testing and Evaluation, *Harold L. Gilmore and Herbert C. Schwartz*
7. Quality Management Handbook, *edited by Loren Walsh, Ralph Wurster, and Raymond J. Kimber*

Applications of Quality Control in the Service Industries

A. C. Rosander

Consultant
Loveland, Colorado

MARCEL DEKKER, INC. New York and Basel
ASQC Quality Press Milwaukee

Library of Congress Cataloging-in-Publication Data

Rosander, A. C. (Arlyn Custer), [date]
Applications of quality control in the service industries.

Includes index.
1. Service industries—Quality control. I. Title.
HD9980.5.R67 1985 658.5′62 85-25311
ISBN 0-8247-7466-3

MARCEL DEKKER, INC.
270 Madison Avenue, New York, New York 10016

Current printing (last digit):
10 9 8 7 6 5 4 3 2 1

PRINTED IN THE UNITED STATES OF AMERICA

To the memory of my mother and father
Nellie Palmer Rosander and John Carl Rosander
sturdy pioneer farmers of the Middle West

Preface

Quality control started in the factory to control the dimensions of manufactured products. While the concept and techniques have been successfully applied outside the factory to service industries and service operations, quality control is still associated primarily with factories and products. It is just as important to apply quality control to services as it is to apply quality control to products.

Walter Shewhart, originator of quality control, showed how variation in quantitative measurements made on a product could be separated into that due to chance and that due to assignable causes which could be found and eliminated. He did not try to translate human wants into factory products with physical characteristics which satisfy these wants, a general problem encountered in the service industries.

To apply quality control most effectively to service industries requires an analysis of the unique characteristics of service operations, the role of the buyer as well as that of the seller, and the application of appropriate quality concepts and techniques. A broader approach is needed than the approach implied in Shewhartian charts, even though the latter are often applicable. In services, errors, delays, failures, and dissatisfaction are top priority quality concerns. A wide variety of techniques are available that meet the basic quality concept of Shewhart but are not used in the factory.

In many services tens of millions of people buy services directly from the seller daily—food, housing, electricity, transportation, communication. Furthermore, the buyer of services does not write specifications, as in manufacturing. The buyer purchases in accordance with what laws, legislative acts, regulations, and established procedures specify. Examples of such service industries are banks, insurance, real estate, public utilities, health services, and communication. These characteristics alone make service industries radically different from manufacturing.

Successful applications of quality control techniques to service operations are not new, since several early applications were made in the 1940's by Deming, Ballowe, Rosander, Halbert, and Jones. Since then numerous applications have been made in a wide variety of service industries and operations including government, especially the Federal government. Despite this experience, quality concepts and techniques and management have not been accepted and applied to the extent they are in manufacturing. A major reason for this book is to help close the gap between proven principles and successful applications on the one hand, and countless potential problems and situations still untapped on the other.

This book is organized in four parts—Service Industry Applications, Statistics, Techniques, and an Appendix—with the purpose of making the book largely self-contained. While a considerable portion of the book is devoted to a discussion of basic concepts, principles, and quality characteristics in the service industries, an understanding of the techniques and their applications requires some knowledge of probability and statistics, which are discussed in Part 2.

Acknowledgments to individuals and other sources for materials used appear in the text. So many persons have helped that acknowledgment of the debt owed to each one is out of the question. I am very grateful to them for their materials, for without them this book would not have been possible. Organizations, agencies, and publishers that gave permission to use materials are:

American Society for Quality Control
American Association for the Advancement of Science
American Chemical Society
American Association of Textile Chemists and Colorists
Operations Research Society of America
Federal Department of Health and Human Services
Colorado Department of Health
Colorado Division of Insurance
American Society for Testing Materials: constants for quality control chart
Marcel Dekker, Inc.: table of random times
Iowa State University Press: F table
Longman Group, Ltd.: tables for t, chi-squared, and random numbers
Brooks Cole Publishing Co.: normal distribution table

A. C. ROSANDER

Contents

Applications of Quality Control in the Service Industries

Part 1
Service Industry Applications

1
Introduction

Origin of modern quality control. In 1924 officials at the Western Electric Company, manufacturer for the Bell Telephone System, had a problem at their Hawthorne plant in the Chicago area. They sent it to the Bell Telephone Laboratories, where Walter A. Shewhart, a physicist, originated and developed the quality control chart based on the principles of probability and statistics. Shewhart expanded the concept and applicable techniques, and described them in detail in a definitive work.[1] At about the same time and in the same organization, Dodge and Romig developed another approach to the control of quality. While Shewhart's techniques and charts were for process control using samples from the production line, Dodge and Romig used sampling inspection for accepting or rejecting lots of manufactured products.[2]

These techniques were applied in a few factories during the 1930s, but it was the huge military program of World War II that gave quality control its opportunity. The War Production Board's Office of Production Research and Development sponsored classes in quality control all over the country which were attended by thousands of inspectors, foremen, engineers, and others. At the War's end it was clear that quality control could be applied just as easily to peace-time production as it had been applied to war-time production. The rest is history.

Right after the War, in 1946, the American Society for Quality Control was founded, and has grown so rapidly that today it has over 43,000 members, mostly in the manufacturing industries. After the War, quality control experts Deming and Juran were invited to Japan to instruct managers, scientists, and engineers on how to apply quality control principles and techniques. How well they learned is reflected in the strong economic position Japan now has both at home and abroad.

From the beginning quality control was applied to the characteristics of manufactured products; hence, it was factory-engineering-product oriented, a view and practice that still prevails. Despite the demonstrations that the concepts and principles of quality control can be successfully applied to service industries such as banking, communication, insurance, transportation, and government, and to service operations such as accounting, auditing, clerical work, and marketing, quality control has not yet been widely applied in these industries and operations.

Employment in the service industries. The employment figures in the service industries show that a very large and important field has been overlooked. As of July 1975, the service industries employed *three times* as many persons as the manufacturing industries.[3] Manufacturing employed 18 million persons, but the service industries employed 54 million divided as follows: wholesale and retail trade, 16.9 million; local, State, and Federal governments, 14.3 million; transportation and public utilities, 4.5 million; finance, insurance, and real estate, 4.2 million; and 13.9 million in other services such as medical, legal, dental, recreation, entertainment, lodging, personal services, and business services. There were *four times* as many employees in local and State governments as there were in the Federal government: 11.6 million compared with 2.9 million, respectively.

Since quality is closely related to, and determined by, human performance, it is easy to see from these figures that quality is a tremendous challenge to everyone working in the service fields, including those employed in the local, State, and Federal governments.

Major differences between service industries and manufacturing. Several extremely important differences between service industries and manufacturing industries have a highly significant bearing on the type, nature, and extent of quality control applications. The major differences are the following:

1. Direct transactions with masses of people: customer, householder, depositor, insured, taxpayer, borrower, consumer, shipper, passenger, claimant, patient, client.
2. Large volume of transactions: sales, loans, payments, premiums, deposits, taxes, charges, interest.
3. Large volume of paper: sales slips, bills, checks, tickets, credit cards, charge accounts, claims, tax returns, mail, prescriptions, freight bills, waybills.
4. Large amount of paperwork movement: receiving, controlling, processing, mailing, filing, accounting.
5. Relatively small amounts of money per transaction.
6. An extremely large number of ways of making errors.
7. Machine and instrumental controls, such as exist in a factory, are missing or exist to a very limited degree. Computers are the exception.
8. Extremely large scale operations may be involved, the largest being the Postal Service and the Federal tax system; also large States, cities, banks, insurance companies, railroads, airlines, telephone companies, public utilities.
9. Usually the customer does not prepare a formal specification of the type and amount and quality of the service required, nor enter into a contract with the supplier of the service. In many cases, the customer has little or no choice, as in monopolies such as electricity, gas, telephone, water, garbage collection, and in banking/insurance where State and Federal laws, regulations, and decisions determine the nature of the service received.
10. Buying products is much more competitive than buying services, especially for the household buyer and consumer who constitute the great mass of buyers of services.

Implications for quality control in the service industries. These characteristics show that the situation faced by a service industry is quite different from that faced by a manufacturer. Several of these implications are listed and described below:

1. Immediate human needs, human performance, and large masses of paper predominate. Customers, employees, and managers are involved. Therefore, quality control must concentrate on the quality of human performance of employees and managers, on the quality of the large masses of data involved, on the quality of decisions made by employees at all levels including top managers, and on the quality of the responses made by customers.

2. The major quality characteristics are error rates, time, cost, and buyer satisfaction.

3. The exposure to human error is tremendous since errors can be made by employees, managers, and customers. Control charts similar to those used in a factory can be used, but they are not enough.

4. Quality of service is related to various time components required to perform the service. For the customer, these include arrangement time, immediate waiting time, service time, and post service time, if any. The customer is concerned with delay time and the total time to obtain satisfactory service. Examples are: time required for a store delivery, mail delivery, an installation, a repair, household goods moving, an insurance claim, transportation, medical service, dental service, a mail order to be filled. A company is interested in appointments, scheduling, billing, processing, controlling, delinquent accounts, auditing, complaints, costs, profit.

5. Quality of service is related to cost. The customer wants acceptable quality service at an affordable expenditure. The company wants to operate so as to make a profit on its investment. The basic problem of business and industry is to operate so that these two goals are met: customer satisfaction and company profit.

6. Management must stress human factors. This means appropriate training, supervision, instructions, procedural manuals, conditions of work, equipment, personal assistance, inter-employee relations, and other factors of a personal or psychological nature which affect the quality of the service. For the customer, it means receiving adequate instructions as to how to use a product including warnings, it means receiving polite and efficient service in connection with maintenance and repairs, and it means receiving adequate and satisfactory attention in connection with complaints. The goal is competent error-free performance.

7. Management itself must be subject to human factors. This means management must accept responsibility for quality, for determining quality policy and objectives; management must insure that employees are properly trained and oriented in quality, that they are provided all the facilities and conditions required to enable them to perform at an acceptable quality level. Policy and practice both must aim at building quality into performance, data, and decisions. Error-free performance is the goal.

8. Management must know how to run large operations effectively. This ability, however, is very hard to find. This failure to be able to run large projects with some degree of efficiency is very apparent in the Federal government, in State governments, and in many large cities. There are also many examples in private industry and business. Experience and observation show that our greatest problem in government, business, and industry is to train and develop managers and executives who can run effectively large scale projects.

9. Management must know how to use the new management tools such as applied probability and statistics on the one hand and computer technology on the other. This does not mean managers have to be experts in these two fields, but they have to know how each can be applied to management problems and situations, and how to hire professionals who are qualified by training and experience to use each field to aid management in solving its problems.

10. Efficient techniques must be used to handle large masses of paper, such as probability sampling, computer technology, and related techniques. There must be technical professionals and managers who know how to organize, design, plan, and implement these techniques for an effective operation.

11. Customer complaints must be handled in an understanding, expeditious, and polite manner, and used as a basis for corrective actions where needed. Service operations are customer intensive and should be handled with this in view. A manufacturer may have only very few customers he sells to directly. But in most service industries, customers may easily number hundreds of thousands if not millions. Customer satisfaction becomes very important; customer complaints may easily be a major factor and influence in service operations. The role of customers is so important that periodic customer surveys are necessary.

Dangers of imitating factory practices. There is a strong tendency to apply quality control techniques to service operations without a careful analysis of the quality aspects and requirements of the latter. The rendering of services is vastly different from the production of goods in a factory. The dangers of imitating factory practices are several:

- The application may be inefficient, if not inappropriate.
- Some very important quality aspects and problems are ignored.
- The benefits of a wider approach to quality control are lost because many areas of application are overlooked. Applications may be limited to a single instance.
- Techniques applicable to the factory may be misapplied. An example is the use of MIL-STD-105-D to obtain sample sizes.
- Those who pay the bills, the buyer or customer, may be ignored because all of the emphasis is placed on the internal operations of the company, agency, or firm.
- Customer satisfaction is overlooked because of the overwhelming emphasis on techniques such as systems analysis, cost-benefit analysis, statistical methods, and the like.

An example of this latter practice is the case where a hospital places more emphasis on systems analysis, the purchasing agent, and the quality of the incoming purchases than on what the real goal is: delivering acceptable quality health services to patients at a price they can afford. Indeed, it may be assumed that the patient cannot appraise with any degree of competence the services he or she is paying for. For some medical services this may be true, but for other medical services it certainly is not true.[4] Quality control over purchases is necessary, but so is quality applied to the medical and non-medical services the patient is paying for.

Service operations in a factory. In service industries quality control is applied to all of the service operations and their components. This differs from manufacturing where quality control may be applied solely to direct manufacturing operations and their components. Of the various service operations, only purchasing of materials and component parts are directly related to factory production, so this operation may receive considerable attention.

Quality control, however, is not applied to all of the service operations which are usually considered overhead costs or indirect costs: accounting, sales, personnel, finance, data collection, data processing, marketing research, clerical work, laboratory testing, computer operations, payroll, and inventory — although these operations may account for as much as 15 to 20 percent of the total cost of making the product.[5]

It is only fair to state that some of the early applications of quality control were in the service areas of a manufacturing company; examples are those made by Carl Noble at Kimberly Clark, those made by Charles Bicking, and those made in sales at 3M.

In other words, manufacturers are overlooking a very substantial area where cost reductions can be made, by not applying quality control as extensively to their service operations as they are applying it to their direct production operations.

The buyer-seller (vendee-vendor) relationship is very different. A very important area where the situation in the service industries is entirely different from that in manufacturing is the buyer-seller (vendee-vendor) relationship. It is so important that it needs special attention and treatment. The manufacturer (vendee) wants the supplier (vendor) to produce materials and component parts that meet the manufacturer's specifications and are produced under quality control so that little or no receiving inspection is required before immediate use in the factory. If vendor A cannot do a satisfactory job in meeting specifications, the manufacturer turns to vendor B, or to vendor C, or to some other vendor.

This practice does not exist in the service industries. The individual buyer does not buy according to specifications he or she prepares. In certain instances the company buyer may. Actually, the individual buyer must buy subject to the rules and prescriptions under which the seller operates. Examples are:

buyer (vendee)	**seller (vendor)**
depositor	bank
patient	hospital, nursing home, doctor's office
insurer	insurance company
passenger	airline, railroad, busline, shipline
home buyer	mortgage company
customer	retail store
customer	public utility

In a large number of service industries the basic specifications, or "rules of the game", are set for the seller or vendor by legislatures, regulatory agencies, and laws. The seller may have some flexibility to operate within these rules. But the buyer has little or no freedom to choose; he or she must buy within this framework or not at all. They must accept what is offered.

Examples of regulated industries are banks, insurance companies, real estate firms, public utilities, hospitals, nursing homes, and rental leases. Examples of where the buyer may be forced to buy some unnecessary service are the homeowner's insurance package which forces those with guaranteed income to pay premiums for loss of income which never will arise, forced increases in premiums due to arbitrary increases in value of properties even though they are rarely completely destroyed, and paying premiums on "adjoining buildings" which do not exist.

Three parties to service quality. There are three parties involved when the quality of any service operation is under consideration:

- The seller of the service.
- The buyer of the service who is an *individual* (householder, user).
- The buyer of the service which is another *company* (business, agency).

Each one of these has a different approach to quality and a different interpretation of what quality means. If the seller, however, is going to perform quality services for the buyers, then the seller must take into consideration what quality means in specific terms to each kind of buyer. Each one raises a question about quality:

1. What does quality mean to the seller of the service or to the provider of the service?

In private industry, the quality goal of a business is to provide a type of service that is generally satisfactory to the customers or buyers and which earns them a profit from one year to the next. The quality picture is very mixed: some firms stress customer satisfaction so as to build up and hold the trade of buyers. Others do not pay much attention, if any attention at all, to customer satisfaction as long as they are earning a steady profit.

In non-profit agencies such as government, quality of service is determined in large measure by past performance or by the rules laid down by the immediate supervisors and managers.

2. What does quality mean to the individual buyer?

The quality goal is to buy an acceptable quality service at an affordable price. Factors determining acceptability include errors, delay time, failures, and dissatisfaction. There are many different mass markets, not one market, and the quality goals of individuals in these various groups differ. Affordable price is a real factor for those with low and middle incomes, but not for the rich.

3. What does quality mean to the *company* buyer of services?

The company buyer, like the company seller, wants to make a profit. Therefore, it wants to buy services of acceptable quality at the lowest price, to keep costs down. Many of these companies can bargain or negotiate with the seller to obtain the most favorable price. Even under monopolies and under regulation, these companies obtain special rates if they are similar to a public utility, which are denied to the individual.

Quality requirements of the buyer. Several quality requirements can be listed as including the major desires of the buyer with regard to the quality of the services bought. These desires or requirements are of special concern to the seller of services, especially those of a competitive nature, because they indicate the nature of the services desired.

1. Acceptable quality products and services at an affordable price.
2. Minimum delay time in arranging for and obtaining service.
3. Service performed according to scheduled or promised times.
4. Error-free performance of people and very low failure rate of products and repairs.

5. They do not want to pay for the errors, mistakes, and goofs made by the service company or agency.

6. They want service free of trouble: they want satisfaction, courtesy, attention, and concern. They do not want long delays, especially on something important, e.g., auto repair. They want sellers of services to do what they promise or advertise.

7. They want a reliable product with a low life-cycle cost: low initial, low operating, low maintenance, and low repair cost.

8. They do not like prices going up while quality is going down. Of course many sellers of services recognize the importance of prices, of reducing if not eliminating errors, of reliable repair services, of honoring promises, of customer satisfaction, and of avoiding unreasonable delays.

Types of services. Service industries engage in a wide variety of activities as the following list shows:

1. Rentals: housing, automobile, truck, equipment, offices, buildings, farms, other.
2. Use of facilities: transportation, telephone, mail, streets, highways, parking lots, telegraph.
3. Safety and protection (private): banks, insurance companies, security.
4. Safety and protection (public): police, firemen, health departments, sanitation, pollution.
5. Energy and water: natural gas, electricity, water.
6. Health: doctors, hospitals, laboratories, nursing homes, nurses, health clubs, health research, government health departments, dentists.
7. Drainage and waste removal: sewers, garbage removal.
8. Keeping products fit for use: repairs, maintenance, cleaning.
9. Personal service: restaurants, hotels, motels, barbers, beauticians, laundries, legal services, government assistance.
10. Recreation and entertainment: movies, theaters, parks, amusement parks, sports, athletics, private clubs, vacation centers.
11. Business services: printing, duplicating, accounting, tax service, legal service, laboratories, computer service, data processing, personnel service.
12. Distribution of goods: wholesale and retail trade, mail order firms, transportation.
13. Financial: banks, mortgage companies, credit unions, finance companies, government.
14. Education: schools, colleges, universities, institutes, seminars, libraries.
15. Environmental quality: air, water, soil, radiation pollution control.

It should not be overlooked that local, State, and Federal governments engage in a wide variety of service operations, some of which are similar to those in private industry and some of which are peculiar to governments. Savings by government quality control programs can be just as large, if not larger, than the savings obtained in private industry, since one out of every four service employees works for some government.

A wider approach to quality control is required. This wide variety of services as well as experience has shown that a more comprehensive approach to quality control is needed in service operations than in manufacturing. Furthermore, in a more general approach the basic logic and aspects of Shewhartian quality control are retained. Quality control is applied not only to physical products, but to data, to human performance, to decisions, and to the environment. To the standard Shewhartian charts and acceptance-rejection sample plans are added a wide repertoire of other techniques: sampling for discovery, estimation, comparisons, testing effectiveness, random time sampling for work and other activity analysis, input-output analysis, learning curve analysis, written procedures and specifications, waiting or delay time analysis, and field testing and experimental design.

Control charts and acceptance sampling use the principles of probability and statistics to indicate trouble, that is, the existence of unacceptable quality products. The Shewhartian chart indicates possible trouble when a sample point falls outside the limits of random variation because an assignable cause may be acting that needs to be eliminated. Put another way, a point "out-of-control" may mean a significant shift in the value of the characteristic of the item being produced. Rejecting a lot of product on the basis of a random sample means that the quality as measured by the average quality level from a random sample may be departing from specifications.

These are effective ways of discovering trouble and taking effective action. But there are other objective indicators of trouble, such as the following:

1. Error rate is too high.
2. Idle time is excessive.
3. Delay is too long.
4. One massive error is found.
5. A violation of a basic procedure is discovered.
6. Length of life is too short.
7. Failure rate is too high.
8. Cost is too high.
9. Too many complaints.
10. Critical or control level is exceeded.
11. Illegal action is discovered.

Some examples are given below:

1. Work sampling: A random time sample for five successive days shows that idle time in a graphic arts section is 15 percent. The director says this level is much too high and requests that the supervisor re-schedule the work to reduce it. This was done.
2. Tax returns: An audit sample shows that taxpayers are erroneously claiming aliens as exemptions, each exemption representing a tax error of over $100. As a result of discovering a high tax error source, a new schedule was added to the tax return, and the instructions were elaborated to explain who could be claimed as an exemption and who could not.
3. Tax returns: A nationwide audit sample shows under-reported and unreported income, deductions, exemptions, and arithmetic, in that order, are major sources of error. Identifying the major sources of tax error and their magnitudes enabled management to begin to allocate audit resources more effectively.
4. Tax returns: A nationwide sample survey for the purposes of estimation showed that the amount of tax error on individual income tax returns was $1.5 billion. This was a big step forward because it defined in quantitative terms the size of the non-compliance problem. It also identified the problem with regard to various classes of taxpayers so audit resources could be allocated more effectively.
5. Rail transport: A probability sample survey shows a shortage of 10,000 box cars in a grain area. This was a serious shortage calling for immediate action.

6. Interviewing and data collection: Check interviews (sample audits) show family income in a study is underreported by 10 percent in one city and 20 percent in another. This error rate was too high and called for immediate instructions to interviewers to prevent the types of errors being made.

7. The Bureau of the Census sample audit found that about two million persons were omitted, for one reason or another, from the 1970 census. This was about one percent. Attempts are being made in the 1980 census to reduce this figure.

The wider approach follows a general quality control format consisting of the following steps:

1. An operation is identified and isolated.
2. Data are collected continuously on key characteristics of this operation.
3. Data are analyzed to discover levels, variations, relationships.
4. If trouble does not exist, leave alone.
5. If trouble exists, feedback key information to source immediately.
6. Investigate operation for sources of trouble.
7. Eliminate source of trouble or take steps to negate its effects.

Data can be collected and trouble indicated in many different ways, and by many different methods, not only by Shewhartian charts.

Small errors versus massive errors. In attempting to control and eliminate the smaller errors by various kinds of control charts, lot sampling, or by other means, the quality control specialist and quality control manager should not overlook likely sources of massive errors. Four actual examples are cited.

Due to the inclusion of a class of persons in a computer program that was not eligible for aid under a Federal financial help program, one large city paid out $18 million before the error was discovered. With proper procedural and supervisory control over computer programming, this error would never have been made.

In a large nationwide sample program using shipping documents as the data source and sampling units, the number of such units totalled about 230,000 based upon past counts obtained by a sample receipts control system. This manual system, independent of the computer, insured that companies sent in the prescribed number of sampling units monthly. At the end of one year, the computer unit submitted a count of 212,000 as the total in the final tabulation. The count was questioned, and both the computer unit and the sample receipts control desk were asked to check their records for possible errors. As a result, the computer unit reported that four reels of tape, amounting to about 17,000 sampling units, had been omitted from the tabulation. Control by the computer library had failed. Only the independent sample receipts control system prevented a disaster since the tabulation with more than seven percent of the sample missing would have been practically worthless.

A large information system firm had to run a tabulation of a large nationwide sample which required the calculation of ratio estimates and their sampling errors. Apparently the computer programmer thought he could pick up the required formulas in a statistics textbook. What he did is a mystery since in a 260 page tabulation about 26,000 variances were calculated, but not a single one of them was correct. A statistician familiar with these equations was readily available, but the programmer ignored him. There were no procedural or supervisory controls over this programmer, and as a result, an extremely serious defect existed in the tabulation. Under proper management, the error would have been caught at the programming stage. It certainly would never have gotten beyond a test check of the tabulation.

The press reported results of tests made by the United States Environmental Protectional Agency (EPA) to determine if differences in plutonium exposure existed between the Rocky Flats area near Denver and a test site in Nevada.[6] In a herd of 10 cattle near Rocky Flats the level of plutonium in lung tissue was 5.1 picocuries per kilogram (pc/kg) while a herd of 10 cattle in Nevada had a value of 1.34 pc/kg. The newspaper headline stated that plutonium was four times as strong near Rocky Flats as in Nevada (5.1 divided by 1.34).

These values were measurements on a scale, not a count of objects, and had to be treated as such. The variation between animals at each site was very large, showing that this ratio was unstable, and that in another test the magnitude might be reversed. This is precisely what happened; a second test of eight animals in Nevada gave 12 pc/kg, or more than twice the Rocky Flats figure of 5.1. The usual practice is to use the difference between 5.1 and 1.34, not the ratio, but since the within-herd variation was so large neither the difference nor the ratio was significant.

The test was not based on well-known principles of experimental design since animals were not randomly assigned to sites; lengths of exposure to radioactivity which varied widely among the cattle were not balanced nor adjusted for; and the data were not properly analyzed because variation among cattle at each site was ignored.

As a result of the press report, people became alarmed over a difference that did not exist; a federal senator called for a $100,000 study by EPA to research the situation further. This is what can happen when an experiment or test is not properly designed, when the resulting data are not properly analyzed, when the results are not verified by an expert, and when the findings are not clearly and accurately explained in writing to newspapers.

This was a question of vital public interest because it involved health, whether houses should be built in the vicinity, public funds, and the very existence of the Rocky Flats facility. To obtain accurate and valid information required proper application of well-known technical knowledge, but the situation was badly handled from beginning to end.

Increased human reliability as the goal. While reference is made to large and small errors of various kinds which should be detected and corrected, on the tacit assumption that they are bound to occur, the goal is to prevent them from occurring in the first place. *A realistic quality program is one aimed at the progressive reduction of errors at all levels.* In positive terms this means increased human reliability.

The view that certain error levels and certain delay levels are inevitable and fixed should be rejected. We need to change drastically the practice that sample plans resulting in lots of, say, two or five or eight percent defective items are acceptable. These error rates are not only costly to both the service company and the customer-buyer, but they do **not** represent the lowest error rate at which persons can reasonably operate. They do not represent the best we can do.

Since service companies are people oriented, the prevention and reduction of human errors at all levels are of vital importance. This means top priority must be given to human reliability with a set goal of 100 percent. Progressive movement toward this goal is characterized by emphasis on the two Zs:

- Zero Errors which means emphasis on *Accuracy*
- Zero Delay which means emphasis on *Urgency*

To attain, improve, and maintain *accuracy* means intensive training in appropriate knowledge, abilities, and skills; ability to apply knowledge, abilities, and skills to solve problems of the workplace; and a desire to use knowledge, abilities, and skills to improve performance.

To attain, improve, and maintain *urgency* means attention to, and development of, attitudes such as care, concern, alertness, and awareness; a desire to act in a timely fashion; a desire to act promptly as needed; and a desire to use the most effective techniques and methods regardless of the nature of the job or project.

In summary, human reliability means 100 percent reliance on persons to: make no errors, do the job right the first time, use the most efficient methods, perform in a timely fashion, act promptly as needed, keep promises and commitments, be ready when needed, and be available when needed.

Quantitative and non-quantitative methods of improving quality. There are two basic approaches to the improvement of quality: preventing trouble, identifying trouble, detecting problems, resolving problems, reducing errors, eliminating trouble, and eliminating errors.

One method is the quantitative or statistical approach which is described in detail in later chapters. The other is the non-quantitative approach of which four examples are described below: the use of written procedural manuals; the use of a budget; the use of rules, standards, and specifications; and the use of time tables and schedules.

Written procedures or procedural manuals. Service industries rely heavily upon a wide variety of procedural controls to keep human errors and human variation within acceptable limits; indeed, the real purpose is to prevent errors in the first place. These procedures take many forms, but they should be in written form. They include specific steps in an operation, a description of the order in which tasks are to be performed, various kinds of instructions, explanations, examples of operations correctly performed, examples of forms properly filled out, examples of problems correctly solved, special problem situations to watch for and how to handle them, and examples of errors to look for as well as to avoid.

All of this material is compiled and issued as procedural manuals covering such operations as the following:

manual editing	computer editing	reviewing	interviewing
coding	sample selection	key punching	typing
tabulating	transcribing	verifying	testing
calculating	analysis	inspecting	making observations
preparing charts	correspondence	filing	sample receipts control

Procedural manuals translate objectives and goals into specific jobs and tasks to be performed. They show the worker what has to be done; they show the types of skills and abilities and knowledge needed to do a job or perform an operation; and they show the worker as well as the supervisor the quality as well as the kind of work expected. They reduce, if not eliminate, uncertainties, ambiguities, conflicts, and misunderstandings; and by so doing tend to eliminate errors and improve the quality of the work.

It is essential that these procedures be written in clear and comprehensive language, that initial drafts be written with the help of those who are going to use them, and that after completion they be discussed in detail with the employees involved. Otherwise, the procedural manuals may raise more questions than they answer.

Control by a budget. A budget is a "blueprint" which is used to direct and control basic functions for one year. Three forms in use are:

- financial budget
- performance budget
- zero-based budget

The budget is very important for the quality function because it determines how much attention, if any, will be given to quality, whether resources will be allocated to the function or not, and whether a start will be made by hiring a consultant or preparing training and orientation materials, let alone whether a Quality Control or Quality Assurance Department will be established, continued, or expanded.

A financial budget sets the amount of money to be spent during one year on each department, division, or other organizational unit with regard to major items such as:

- personnel
- travel
- training
- equipment
- supplies
- rent

The budget involves expansions and contractions: new projects, new jobs, and new functions which are added to what now exists, or old jobs, tasks, operations, and functions which are reduced, phased out, or eliminated. The budget may call for a justification of every function, project, and job if there is any doubt about any of them.

A performance budget shows the major projects and functions planned for the year and usually includes: how personnel is allocated, time schedules, expected progress and completion dates, where applicable. It shows how total effective hours are allocated to projects and jobs over time. It is an attempt to direct and control performance, both supervisory and non-supervisory. A quarterly report may be required.

A zero-based budget, unlike the other budgets, calls for a full scale justification of the organizational unit under consideration. It means detailing functions, projects, and jobs, and justifying the expenditures proposed by what amounts to a clear cut cost benefit analysis.

It would appear wise for anyone connected with the quality function, regardless of what the particular budget process is, to be ready at all times to justify the position to higher level management, as well as any special group such as a Quality Control Department or a Quality Assurance Division.

Control by rules, standards, and specifications. Rules, standards, and specifications are used to exert control over specific service operations. Working rules may be incorporated into union labor contracts, or other contracts; they determine a wide variety of key factors such as number of hours worked per week, number of holidays, length of vacations, seniority rights to jobs and promotions, and how overtime is handled.

Standards involve such diverse things as bank checks using MICR (Magnetic Ink Character Recognition), air pollution, blood and urine tests, and sampling plans. Federal agencies like the Food and Drug Administration (FDA), the Environmental Protection Agency (EPA), and the Occupational Safety and Health Administration (OSHA) set numerous rules and standards for food, drugs, air, and water pollution, and conditions of safety for workers in service industries.

Specifications are required by a service company in connection with the purchasing of equipment, supplies, and goods in general; in connection with construction projects; and in connection with purchasing, installing, and operating various systems such as computer systems and electric power systems.

Time tables and time schedules. Time tables and schedules are common in a service industry like transportation where offering service at proper and convenient times and meeting a time schedule are of vital importance. This is true of all forms of transportation: local bus, inter-city bus, railroad, rapid transit, airline, and ship. Time is also a factor in pricing the service, such as ship travel: the longer the time spent on shipboard, the higher the fare.

In long-distance telephone service, time is an important factor because rates are lower for off-load hours. Other factors affecting cost of service include whether one has a private or party line, or uses the services of a telephone operator. Time is a very significant factor in the operation of an electric power plant since demand varies not only during the day but varies with months of the year. Demand is high during hot summer days due to air conditioning, and high during December and winter months due to the shortness of the days.

Time is related directly to billing customers. Many bills such as those for telephone service, gas, water, and electricity, as well as those for rent and mortgages, are on a monthly basis. Some insurance premiums are on a monthly basis, where others are on a semi-annual or annual basis. These time periods and dates determine when the customer has to pay his or her bills, determines when the billing agency or company engages in a sequence of operations, determines the nature and time of the peak workload, determines when delinquencies arise, and why a continuous set of quality controls has to be applied.

Time tables and schedules are goals or standards against which to measure quality of performance in terms of service time, delay time, excessive time, or meeting schedules on time.

Notes

[1]Walter A. Shewhart, *Economic Control of the Quality of Manufactured Products,* Van Nostrand, New York, 1931.

[2]Harold F. Dodge and Harry G. Romig, *Sampling Inspection Tables,* Wiley, New York, 1944.

[3]The source is the U.S. Bureau of Labor Statistics quoted in *The World Almanac,* 1976, p. 103. The ratio still held in 1979: 63.4 million versus 21.1 million.

[4]J. P. Werner, "Application of a QA program in a hospital facility," *34th Annual Technical Conference Transactions,* American Society for Quality Control, 1980, pp. 313-322.

[5]I am indebted to William Latzko for this statement.

[6]*The Denver Post* and *The Rocky Mountain News,* January 30, 1975. The abbreviation "pc" stands for picocuries, a unit of radioactivity; "kg" stands for kilograms.

2
Banking

Quality control in banking.[1] Banks have only recently taken an interest in quality control, but there are several reasons for this. Quality control has been associated with manufacturing and when applied to banking seemed nebulous and difficult, not only to describe but to explain how it could be applied to banking operations. In the past banking operations were manual and subject to continuous checking. It was hard to talk to managers about quality, and as far as customer relations were concerned they were taken for granted since they never seemed to pose any problems. There was an audit staff plus State or Federal bank auditors trying to find errors. There was a tradition of being accurate and careful. Moneys had to be accounted for daily and balance sheets had to be made public. Furthermore, there was no public hue and cry about banks in general and customer service in particular. Banks were rated high in public esteem in public opinion polls, but it was finally realized that public opinion can quickly erode. Quality control is just now beginning to receive attention, due to a number of reasons, including a shift from manual to computer operations and the realization that productivity needed to be improved.

Some important quality measures. *Time* is a very important quality characteristic since how fast a service is performed is not only important to the ordinary customer but can become very critical in many banking operations. Some services have to be performed very fast; a bank may have only two minutes to transfer funds. On the other hand, cashing a check on a foreign bank may require two to four weeks. The bank needs to explain to customers that time standards of performance may vary widely depending upon the specific operation.

Accuracy is a critical characteristic since mistakes are inevitable, time consuming, and costly. Since most processing is on line, an error cannot be corrected until the end to the line is reached. Once a transaction gets off the track, it is very difficult to get it back on. About 1,000 of the 12,000 employees or nearly 10 percent do nothing but correct errors. People are tolerant of mistakes if corrected quickly and courteously.

Courtesy is a characteristic rated highly by the public since tellers have direct contact with customers constantly. An example of undesirable contact with people is the case of a teller who never looked at any customer, did not smile, did not talk to a customer — just processed paper. Implied in this case is the need for human contact, for human reaction, for human communication, for human conversation — otherwise the teller could be a robot. An important element of quality in service industries, such as banking, which are people oriented is to limit, if not eliminate, the de-humanization that is characteristic of institutions both large and small. The need for this has become more and more important as applications of the computer have multiplied.

Starting a quality program. Quality assurance and customer relations were set up as functions to receive special attention. The vice president for these functions reports directly to the vice president for operations. It was recognized that quality control was a managerial responsibility. The plan was to have a small professional staff initiate the quality control program. This staff tried to get its bearings by exploring the following:

1. Determine what the present quality situation was.
2. How the managers perceived quality; what it meant to them.
3. How they measured quality, if they did.
4. What quality objectives they had.

These managers had no quality objectives, no standards, no concrete applications, and no error rates. Most of them had no knowledge of quality although they cared about it. There was no formal system or attack aimed at controlling quality.

After this initial survey it was necessary to develop a quality program in detail which was done along the following lines:

1. Arrive at specific measurements of quality.
2. Develop and refine standards to define quality desired.
3. Apply the quality concept across the entire company.
4. Motivate the employees, of which there were 12,000.
5. Develop the concept of quality assurance and have it endorsed.
6. Set up measurement of potential deviations from desired quality.
7. Use the participative approach to isolate deviations; meetings and discussions with employees.
8. Identify deviations that impair quality.
9. Appraise deviations that are most important, least important; establish priorities.
10. Introduce a quality cost program.
11. Develop reports.
12. Improve quality through training, etc.

Results and further developments. As the introduction of a quality control program advanced, specific results and developments emerged in four areas: reports, costs, long range quality planning, and customer relations.

A quality assurance report is issued every month showing managers "this is what exists" and here are the standards we are aiming at. These reports have been issued for 18 months and show quality characteristics such as percent defective or in error and a quality index.

Information is developed by a cost analysis staff on what the quality function costs. A profile of quality costs is prepared since quality costs are 25 percent of total operating costs. Failures or errors alone account for 50 percent of the quality cost and require about 1,000 employees working full time to correct them. Obviously this means a large savings if the failures can be reduced and an increasing number of the 1,000 can be shifted from corrective work to direct production work.

It soon became apparent that longer range quality control planning was necessary if improvement of quality was to be a realizable aim. Also, productivity as well as quality was in the air and desired by management. This was because cost and revenue trends and comparisons showed the need for more attention to productivity. There is a need to work smarter, but this requires knowledgeable employees with qualifications and training that fit in with personal satisfaction and decision making on the job. Actual quality and productivity are closely related, and it was found that quality control is meshing with productivity improvements.

Quality circles are being organized on a voluntary basis, but they call for drastic changes in management thinking and operating. Some manager training in quality circles is in-house, but some people are trained outside. They began with 11 circles in commercial banking, have trained about 1,000 employees in this concept, and plan to add 20 to 30 more circles.

Quality circles seem to fit into the new values many employees have about work: personal satisfaction, participating in decisions, and self determination. It is estimated that 70 percent of those working in the data processing group have these new values. Management has to learn how to cope with them, direct them, and use them to obtain the quality and quantity of work required in a successful business operation. A new approach to management is needed to counter the cynicism of employees on the line.

Careful attention needs to be paid to *customers*. Managers are asked, "What are you doing to correlate your operations and decisions with customer values?" There is a real need to stress the satisfied customer and to see that customer complaints are handled courteously and expeditiously. Customer complaints are feedback to management, and where they reflect a weakness in the quality of operations, the necessary corrective action should be taken.

Very rough standards of performance exist now based on past experiences, but these are expected to be refined as time goes on and the quality improvement program becomes effective. A continuous and progressive problem is to gather solid information that shows the quality of bank operations now.

Implementing a quality program. A report on the implementation of the preceding quality program in a large bank is now at hand.[2] Briefly, employees are assuming more responsibility for the quality of their work, quality levels are increasing, quality costs are being reduced, and quality improvements are being implemented. Features of the program are:

- Vital support by top officers right from the start.
- Establishment of a Quality Assurance Division.
- Group participation with four initial training and discussion sessions.
- Quality cost program.
- Quality improvement studies.
- Programs to improve employee performance.

Several factors gave impetus to the development of a strong formal program:

- The vital role of time as a characteristic in bank operations.
- Decentralizing the operating units.
- Expansion into diverse markets with competitive pressures.
- Rising customer expectations.
- Growth of electric transfer in bank operations.

Quality circles play a major role in this program. They are a management tool used to promote group participation and personal growth and improve performance. There are 135 quality circles averaging eight members each. They have been organized in such diverse areas as systems, check processing, operations, and accounting. About 65 percent of the employees volunteer to participate. The failure rate is very low — only one out of about 40 — and this has been found to be due more to individual causes than to any structural defects. These individual reasons are associated with one or more of the four groups of people involved: management, facilitators, leaders, and employees.

Specific problems handled by the quality circles include claims handling procedures, balancing with Federal Reserve, bank term glossary, reduction of paper costs, and improved handling of customer inquiries.

The Quality Assurance Division helped managers develop a quality cost program applicable to their operations. The usual four categories of quality costs are used: prevention, appraisal, internal failure, and external failure. Studies showed that about 50 percent of the total cost of quality is due to internal and external failures. These studies show how information about the division of quality costs can provide the basis for efforts to improve quality.

Two specific examples of quality improvement are cited:

- Over a period of five calendar quarters the error rate in service volume affecting customers in the bond department was reduced from about two percent to 1.4 percent with a goal of one percent.
- Over a period of five calendar quarters the average investigation turnaround time in the bond department was reduced from more than two days to less than one day.

Quality control in banking.[3] Two applications are described, one to the quality of checks using MICR (Magnetic Ink Character Recognition) and the other to clerical work. Using MICR printed checks enables banks to cope nationwide with the 26 billion checks written annually. In recent years banks realized how much money is being wasted because of poor quality printing of these checks. "When the devices sensing the magnetic characters fail to read one or more of the magnetic numbers, a check is rejected. When this occurs the cost of reconcilement, repairs, and re-entry of checks and documents back into the computer system is from 10 to 30 cents."

The selection of samples to measure quality is different for each of three types of documents: personal checks and debit tickets, non-continuous checks or forms, and continuous MICR checks or forms. Irving Trust Company has found that a sampling plan adapted from sampling plans in military standard MIL-STD-105-D covers most needs from an economic point of view.[4] Printers experience no difficulty in meeting a defective rate or average quality level (AQL) of 0.4 percent, while the long run protection to the bank is a maximum rejection rate of one percent or less defective fresh documents. During processing the reject rate will increase perhaps to 1.5 percent due to unavoidable mishandling of documents.

Sampling Table for MICR Documents
(Normal Inspection, Level I, AQL=0.4%)

Number in production run	*Sample size*	*Reject run if defective documents in sample equal or exceed this number*
Up to 3,200	32	1
3,201 - 35,000	125	2
35,001 - 150,000	200	3
150,001 - 500,000	315	4
500,001 and over	500	6

Personal checks and deposit tickets are usually ordered from a small number of printers skilled in MICR printing. By sampling a single document from every account number printed, a measure of quality is obtained with the average defect rate in the range from 0.25 percent to 0.50 percent. Deviations that are statistically significant can be noted and reviewed with the printer for his corrective action.

Non-continuous checks or forms are separate pieces of paper and can be sampled systematically by knowing the lot size and using the associated sample size from the above table, or by knowing the thickness of the stack in inches, one unit can be drawn at every inch, half inch, or whatevcr interval gives the sample size. In one instance, it is suggested that two units be selected from every inch.

Continuous forms are one continuous series of attached forms. Use is made of the printer's method of packaging to select enough off the top of each box to give the sample size. For example, if 40,000 forms are packaged 4,000 items to a box, 20 would be removed from the top of each of the 10 boxes to obtain a sample of 200. This is cluster sampling, not the random sampling of individual units assumed by MIL-STD-105-D. The cluster effect might be reduced by taking 10 off the top and 10 off the bottom, but even this is far from a random sample. These continuous forms need to be researched to determine what the clustering effect is and how it affects the sample.

Quality control improvement program in banking operations.[5] The purpose of this quality control program is to prevent low quality work from occurring, and taking prompt action when it does occur. Quality costs are divided into four categories, with the total quality cost divided as shown:

1. Appraisal cost (28 percent): checking, verifying, inspecting work.
2. Internal failure (41 percent): mistakes discovered in bank operations that require correction such as re-work.
3. External failure (29 percent): errors that get outside the bank and cause investigations, adjustments, penalties, and often lost accounts.
4. Prevention (2 percent): this is the cost of installing and operating the quality control program.

Since the results of poor quality work are often transmitted from stage to stage with an increasing cost, a continuing function of the quality control program is to prevent errors from occurring at the source, especially under the categories of appraisal and internal failure.

Stated another way, this program is to apply such procedures and techniques so that the quality of work produced meets quality standards or specifications, and to audit production constantly so it conforms to these specifications. If the work does not meet quality standards, then steps are taken to discover causes that give rise to these errors and correct them. An integral part of production control is the measurement of the capability of the system which is the best quality obtainable with existing personnel, procedures, and environment. This capability is used as a standard against which to measure performance. (Figure 2)

The quality improvement program operates at the supervisor or first-line manager level. Managers keep daily and weekly work records which are charted on a regular basis by computer to show process capability compared with actual group performance, as well as the performance and error rate of each operator. Figure 1 shows what happens when a control chart of the Shewhart type is applied to a typing section of the Cable and Telegraph department of the bank.[6] When first applied in mid-March, there was an immediate drop in errors as quality consciousness began. A sudden upturn in errors in early April indicated a problem which is temporarily corrected, resulting in a sharp drop from 11 to one percent. The pattern fluctuates in May and then in early June a rapid rise indicates another major problem, which, linked to the similar April rise, indicated an intermittent misunderstanding in the use of special forms. A permanent solution was devised. The error rate again dropped dramatically and remained continually low. The broken line indicates the highest acceptable tolerance range for error; the solid line the average rate. Following the introduction period, a new acceptable tolerance rate was applied in mid-July. As the error rate averaged around the one percent level, the section settled into a stable operation and problems are detected long before they reach a critical stage.

A weekly computer summary for the typing section which shows the process capability or quality target error rate and the actual error rate for the week is shown in Figure 2. Each typist is identified by name with the total number of documents sampled, the number of errors made, the error rate as a percentage, and the Q score — which is the average arbitrarily set equal to 100. The computer is programmed to show when the sample is too small to calculate the error rate, and when the error rate is so high that monitoring is required. The total volume sampled as well as the total number of errors are listed from which the actual error rate is calculated. Three operators averaged 1.35 percent, well within the quality target. The fourth operator had a high error rate so corrective action was indicated.

This program at Irving Trust Company, which has 5,000 employees, began with a professional and two clerks. Although application has increased ten-fold, only three analysts have been added. This shows how successful a small staff can be; other examples of the successful small staff can be cited. It is more a question of technical know-how, of capability of making successful applications, and of working with other professionals and management than size of staff. It is stated that the primary requirement is for management to want a quality control program and for quality control to emanate from the top down. In many instances, however, management will not want a quality control program until one or more lower level professionals demonstrate or convince management of the advantages and benefits of a quality control program. In many, if not most instances, it is middle or lower level professionals with technical know-how and experience in applying it who have been responsible for introducing programs of sampling, quality control, experimental design, and statistical analysis into an agency or company. This is because these professionals are familiar with the applications, including successful applications, in a wide variety of fields and management is not and is not expected to.

FIGURE 1

Work Record Chart for Communications Center
Cable and Telegram Processing Department
Typing Section

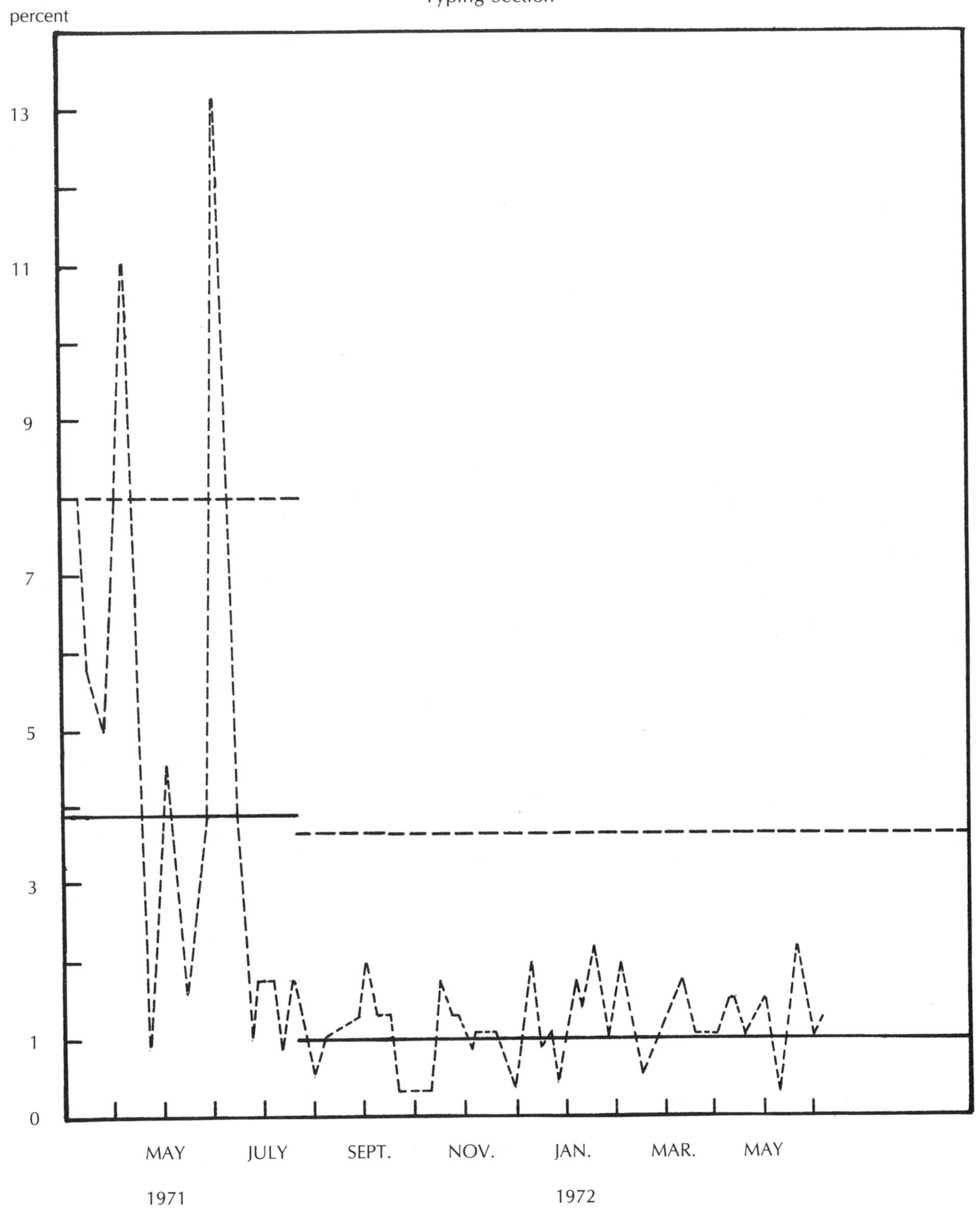

FIGURE 2

Quality Improvement Program Analysis

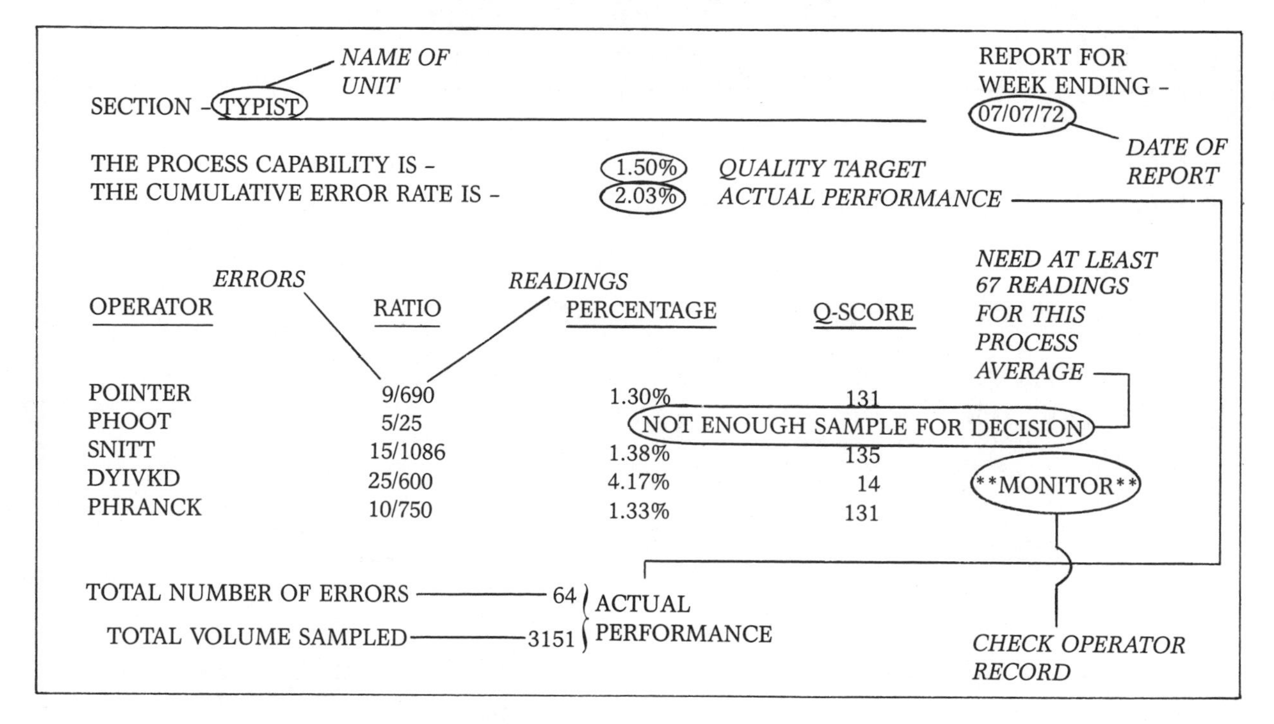

It is stated that there are three basic ways of setting up a quality control program in a bank: hire a quality control expert and teach him banking, take a banker and teach him quality control, or hire a consultant to explore banking operations and find the best places for introducing quality control. There is still another way, especially for a large company or agency. Hire a full time professional statistician qualified in probability sampling and quality control and have him study operations with banking personnel. He does not need to become an expert in banking, but like the lawyer he has to know how to ask the right questions, to work with other professionals, and be a practitioner interested in bringing statistical science to bear on the problems of management.

Notes

[1]This material is based on a talk by Lawrence Eldridge, Vice President, Continental Illinois National Bank, Chicago, given in Toledo, Ohio, March 13, 1981, at a conference sponsored by the Northwest Ohio Section and the Administrative Applications Division of the American Society for Quality Control.

[2]C. A. Aubrey, II and L. A. Eldridge, "Banking on High Quality," *Quality Progress,* December, 1981, pp. 14-19.

[3]W. J. Latzko, "The MICR Challenge for Bankers," Irving Trust Company, New York, 1976. Used by permission.

[4]For a description of lot sampling and MIL-STD-105-D, see Chapter 21.

[5]W. J. Latzko, "A Quality Control System for Banking," *The Magazine of Bank Administration,* November, 1972, p. 17. Used by permission of the Bank Administration Institute.

[6]For a description of Shewhartian charts, see Chapter 20.

3
Insurance

Nature and scope of quality control in insurance. There are three parties interested in the various quality aspects of the insurance business: the *seller* of insurance — the insurance company; the *individual buyer* of insurance such as an automobile owner or a home owner; and the *group buyer* of insurance such as a professional society, a fraternal organization, or an association of individuals.

Quality relates to six broad aspects of the business: quality of the raw or original data, quality of the derived data, quality of performance of employees at all levels, quality of performance of equipment and machinery, quality of decisions at all levels, and quality of the services relative to the financial aspects involved. If the seller builds quality into the data, the performance of employees and equipment, and the decisions of the employees, then these advantages will redound to the benefit of the buyers. The ultimate goals of the seller, however, are not the same as those of the buyers, so conflicts are bound to arise. These will be pointed out and discussed as we proceed.

Quality characteristics. A listing of the more detailed quality characteristics shows the specific questions to which quality control must be applied:

1. Relevance of basic data.
2. Accuracy of basic data.
3. Sufficiency of basic data.
4. Adequacy of data analysis and derived data.
5. Soundness and efficiency of technical performance.
6. Accuracy and speed of operating performance.
7. Adequacy of managerial performance.
8. Determination of fair and relevant rates and premiums.
9. Accuracy and speed of damage estimates.
10. Speed of damage settlements, of claim payments generally.
11. Risk characteristics of the buyer.
12. Buyer satisfaction.
13. Financial aspects: reasonable profits versus reasonable charges.

Goals in quality control. Ideal goals are set in quality control programs not because they can be realized but to call attention to some highly critical characteristics.

The concept of *Zero Defects* is critical because in many instances we must work toward Zero Defects in actual practice. Defects and errors are not only costly; they can be dangerous. A major problem in quality control in all of the service industries is the control, measurement, and elimination to the greatest extent possible of *human error.* The purpose is to stress the need for *accuracy.*

The concept of *Zero Delay* is critical because delay too can be costly and dangerous. Delay in taking action may simply mean errors and defects are accumulating. Delay in taking certain kinds of actions may endanger health, increase risks, increase costs, and promote ill-will. The purpose is to stress the need for *urgency.*

The concept of *perfect service* may sound far-fetched, but its purpose is to stress the need for careful, adequate, and timely performance, and to correct a complaint or deficiency without unreasonable delay and at a reasonable cost.

The concept of *minimum price* calls attention to the need for a *fair price* for services rendered. This is often one of the major complaints of buyers of services; the charges are exorbitant considering the type, kind, and amount of service involved.

The seller's interest in quality control. The seller's interest in quality control revolves around those characteristics and goals of paramount importance to the company: increasing revenues and increasing profits, reducing costs or at least controlling costs of various operations, reducing or eliminating errors in processing claims and in other operations, controlling costs by time studies, reducing risks by limitations and exclusions in policies, maintaining rates and premiums for profitable operations, and controlling damage appraisals by adjusters.

Examples of areas where the seller, the insurance company, has to be concerned about quality are the following:

1. **Technical methods.** The quality of the various technical methods used is vital because in the final analysis they determine the relevance, the accuracy, the efficiency, and the effectiveness of basic data and decisions. Technical methods include:
   a. Sample designs.
   b. Methods to control errors in sampling and data collection.
   c. Methods of estimation.
   d. Methods of allocation.
   e. Calculation of probabilities.
   f. Computer programming.
   g. Statistical analysis of the data.
   h. Criteria for estimating losses.
   i. Random time sampling using the minute model to measure utilization and costs.
   j. Design of computer editing methods.
   k. Design of controls over computer operations.

1. Determination of standards for quality of data, performance, decisions.

2. **Operations.** The best technical methods, plans, and designs are of little use if they are not properly implemented. This is the function of operations. Implementation requires that procedures be in detailed written form, that instructions are carefully followed, that deviations from instructions not be allowed except when authorized by a competent professional, that supervisors know how to explain procedures, and that supervisors are alert and capable of helping individuals and spotting and eliminating trouble spots before they grow. Operating problems involving quality tend to concentrate on avoiding errors, excessive service, and delay times.

a. Seeing that error standards are met.
b. Putting effective error controls into effect.
c. Putting controls on data processing.
d. Putting controls on data analysis and calculations.
e. Putting controls on the computer system.
f. Controlling computer edits.
g. Controlling time levels, time trends, and time standards.

3. **Actions of:**

a. Employees.
b. Managers.
c. Adjusters.
d. Agents.
e. Underwriters.
f. Customer service.

4. **Determination of premiums.** Of vital interest to the buyer of insurance is the way rates and premiums are determined, so that the cost to the buyer is commensurate with the risks covered. Has the company, for example, separated low risk groups from high risk groups and set rates and premiums to correspond?

Quality of data and data collection and processing. For quality control purposes it is necessary to know that the basic data are relevant, accurate, and sufficient, and that the sampling and non-sampling errors are kept under control so that they are reduced to acceptable levels. Specific questions to answer relating to losses, premiums, and revenues include the following:

1. How are losses determined? By past experience, sample studies, national figures, studies by others, or current records.

2. What are the probabilities of various kinds of losses and how are they determined? Are they based on sound and accurate data? To what extent are they derived from judgments and not on empirical data?

3. How are the loss classes determined or identified? Are frequency distributions of losses by classes constructed and analyzed? What are the classes that account for most of the losses, and what are these losses in absolute values and relative proportions?

4. How are the losses allocated or distributed to the various classes? How are they adjusted for time trends, if such time trends exist?

5. How are operating expenses allocated to these classes?

6. How are overhead expenses allocated to these classes?

7. What revenue is obtained from each of these classes, and what revenue should be obtained from each of these classes?

8. How is investment income from premiums allocated to these classes?

9. How are rates and premiums determined?

Quality of technical performance. Technical performance includes technical capability to perform the various specialized functions required by a total quality control program for an insurance company, and the extent to which these capabilities are used at an acceptable level. Aspects where technical performance is involved include: determination of what classes of losses are to be insured and what classes are to be excluded, determination of rates and premiums, determination of deductibles and trends, appraisal of claims, the quality control and quality assurance programs, the total sampling program including sample audits and random time sampling for costs, the statistical analysis (data analysis) program, the total computer program, and customer surveys and relations.

Technical insurance methods include those used by insurance analysts, underwriters, appraisers, and adjustors. These include the techniques used to determine rates and premiums, establish the level of deductibles and how and when they are changed, and determine what is to be included under "exclusions."

What criteria are used in appraising damages and losses and in examining and evaluating claims? What is the relation between repair and replacement? Although premiums are based on replacement, to what extent are losses based on repair? Is there any control exerted over appraisals? For example, how closely would two appraisers agree in their appraisals of the same situation done independently of each other? Is there an analysis of the actions taken by appraisers and adjustors? Is there a sample audit to insure that established criteria and procedures are being followed? To what extent are the estimates and adjustments made by appraisers and adjustors final? Or are they subject to review?

Quality control capability should include technical understanding of quality control and quality assurance methods, an ability to identify quality problems and quality areas, an ability to plan a total quality control program, and an ability to explain and implement this program. This understanding should include statistics, management, computer science, and human factors.

A separate statistical capability may be necessary since quality control is only one of many applications of probability and statistics. It is necessary to have the technical capability to design efficient probability sample studies, to write specifications for them, to see that they are properly managed and implemented, and to see that both sampling errors and non-sampling errors are controlled so that they are at acceptable levels. It is necessary to have someone knowledgeable in statistics to show the computer programmer how various mathematical and statistical equations and procedures are to be applied and programmed. These include various techniques of sampling using the computer, making estimates from samples, calculating various standard errors, making the calculations necessary for analysis of variance and regression analysis, and using methods appropriate for a statistical analysis of the data generally. For example, are frequency distributions being used to determine classes of losses, premiums, and deductibles?

If a computer is being used, and one or more usually are, then there has to be a capability to manage, program, operate, and use the computer effectively. This means quality control over programming, over operations, and over final products. Control has to be exerted in order to prevent errors and inefficiencies in programming and operations; furthermore, the computer itself should be used for error control by using computer edits and computer audits. A quality audit done on the computer is a post mortem and should not be considered a substitute for direct quality control over production directly at the work site or very near to it.

Quality of operating performance. Quality of technical performance deals with the quality of technical planning, technical capabilities, technical methods, technical procedures, and technical designs. It measures these in terms of efficiency, effectiveness, soundness, applicability, feasibility, and relevance.

Quality of operating performance deals with the quality of implementing plans, procedures, processes, techniques, designs, and methods. This is measured in terms of accuracy which is reflected in errors and error rates; in terms of operating times such as down time, delay time, service time, and processing time.

Quality of operating performance includes the quality of decisions made at every level: by clerks, supervisors, and managers; by underwriters, appraisers, adjustors, and agents; by quality control and quality assurance specialists, statisticians, and other specialists; and by top level managers and executives.

The buyer's interest in quality control. The buyer, no less than the seller of insurance, is interested in quality performance and quality control.

1. The buyer wants protection against common risks and against serious risks which may be rare.

2. The buyer desires premiums which are commensurate with the risk involved plus the cost to the insurance company of covering that risk.

3. The buyer wants to exclude all unnecessary and irrelevant protection that does not apply.

4. For insurance that is required by law, e.g., automobile insurance, the buyer wants to pay a premium commensurate with the risk associated with the class of drivers the buyer falls into.

5. The buyer wants excessive protection excluded, protection which is unreasonable considering the circumstances under which the buyer lives.

6. The buyer wants complaints handled seriously, politely, and in a timely fashion. He wants the proper corrective actions taken within a reasonable length of time.

7. The buyer wants a clear explanation of what the insurance policy covers and what it does not cover, without any gobbledegook or without any long legal obfuscations.

8. The buyer does not want the rules of the game changed without being informed in advance. The buyer does not want deductibles suddenly changed or his premiums suddenly increased without being notified in advance.

9. When a claim is submitted, the buyer wants an accurate settlement within a reasonable length of time, taking into consideration the nature of the claim.

10. The buyer does not want his insurance suddenly cancelled so that his livelihood is put in jeopardy, as it is when an insurance company suddenly cancels automobile insurance required by law. Actually, an insurance company should be required by law to give an automobile owner 60 days notice so that the owner can have a chance to obtain insurance elsewhere. Otherwise, as has occurred to many, cancelled insurance may stop the owner from going to work. If it is a family plan that is being cancelled, cancellation may stop children from going to school and older youth from going to work or to the college they are attending.

11. The buyer would like to have insurance companies give more attention to the satisfaction of the buyer and the way he perceives insurance. At the present time, of all the services the great mass of American people buy, insurance is the one they know the least about; the one about which they are informed the least, although insurance is supposed to be a regulated industry; and the one about which they have little freedom to choose or to make decisions although they are the ones who pay the bills.

The buyer and insurance practices. There are many insurance practices that raise questions about the quality of data, the quality of performance, the quality of decisions, and the quality of various practices.

Homeowners. With regard to homeowner's insurance, there are several practices which need attention.

1. Why is insurance issued in a "package" which may include all kinds of risks which the homeowner does not want? For example, a homeowner cannot buy fire insurance only; if he did he would be charged so much that it would be better to take the "package". Actual cases of this practice are known.

2. Why are premiums paid on the full value of a house when a house is rarely completely destroyed? Where this occurs, as in flood and hurricane areas, high insurance rates should be restricted to these areas. The homeowner for which the probability of these events is 0 (zero) should not have to pay for something that will never occur.

In this connection, it is imperative that the buyer know the distribution of losses or damages as a percent of the full value; is it 50 percent? 25 percent? or what? Why should the buyer pay premiums on full value when the insurance company is paying full value only five percent or 10 percent of the time?

3. Why is an insurance company allowed to increase the value of a house according to some national construction index? In this common practice the owner is dictated to; he is not allowed to make a decision about his own house. A national construction index has no validity in a local area; in fact, it is not an index of building single family homes in the local area in question.

4. Why is the value of the personal property within the house arbitrarily set at 50 percent (one half) of the full value of the house? There is considerable evidence that this value is much too high. Here again the owner of the house has no voice in determining the value of his personal possessions. Like the value of the house, this value also is arbitrarily determined by an outsider.

5. Why is "loss of income" included in a policy for a homeowner who is on an annuity or on a guaranteed income and therefore will never have a loss of income assumed in this insurance clause? This is another case where facts are brushed aside by the insurance company; it is another case of forced insurance.

6. Another example of this forced insurance is the provision for the insurance of "adjoining buildings" although there are none.

7. To what extent is repair substituted for replacement and how is this decision made? Is it a practice to maximize repair so as to reduce losses to the insurance company? Obviously, repair as a practice is quite contradictory to

paying premiums on the full replacement value whether of the house or of personal belongings.

Adjusting premiums to high and low risk groups. Insurance should be related to the probability of the occurrence of a loss event. To what extent this is done, and how accurately, the buyer of insurance does not know. The following are some illustrations of groups which have different probabilities or risks and whose premiums should reflect these differences:

1. Fire (homes): Smokers should pay higher premiums than non-smokers.
2. Auto accidents: Drinkers should pay higher premiums than non-drinkers. Those who drive only during the daylight should pay lower premiums than those who drive both day and night. Those with a no-accident history should pay lower premiums than those with an accident record, the same for careful versus careless drivers.
3. Flood: Those who live on a river bank or adjacent to a flood plain should pay higher premiums than those at well drained higher altitudes where the probability of a flood is zero.
4. Hail and wind damage: Those who live in a hail or wind belt should pay higher premiums than those who never have any damaging winds or hail storms.

Cases in Quality Control in the Insurance Industry

Case 1. Quality Control Program Stressing Individual Involvement, Quality Data, and Quality Audit

The United States Fidelity and Guaranty Company is a large multi-line insurance company with 61 branch offices and 135 service offices located throughout the country. Over 8,000 individuals are employed by the company of which 5,500 or about 70 percent are involved in clerical operations. The following is a description of their quality control program.[1]

An insurance company is a service organization. It operates in a labor intensive, high volume, production environment. Performance is based on productivity, cost and service levels. It is paramount that the customer, or user, be satisfied. The customers want to know that the service they receive reflects accuracy, not a number of errors and problems.

The biggest single problem the company faces today is the lack of a proper attitude toward *quality data.* Poor quality can cause damage to a corporation's reputation and . . . also provoke increased governmental regulation. By applying the principles of quality control, a properly managed quality operation can prevent this from happening as well as provide a valuable tool for management to meet its major objectives and future challenges.

The clerical operations include activities involving the issuance and renewal of policies as well as the collection and processing of insurance data, operations which are extremely complex. Accuracy and timeliness are essential to calculate proper rates and make proper underwriting decisions. These rates and decisions affect everyone of us.

When the company began the development of a quality control program for use in the branch offices and the home office operating departments, it had a number of goals:

1. To provide a quality control program which would enhance statistical accuracy.
2. To provide a program which would result in benefits to all levels of the company from the individual coder through the executive department.
3. To develop a program that would have as little negative impact on productivity as possible, and be accepted by the employees as a natural part of their processing procedure.

Very early in the project it was decided if any program of this type was to be successful, it would be necessary to involve the people who actually perform the work, not only in the development of the program but also in the day-to-day operation. The first step in initiating this employee involvement was to interview in depth some 200 employees involved in the day-to-day operation. We asked them what their problems were, what they felt were the major causes of the problems, and what could be done to aid them in solving their problems. When the quality control program was finally developed, a conscientious effort was made to address the major comments made by these people. As the program began to take shape, we consistently requested our company employees' comments and criticisms.

The different techniques developed were tested and proven. The high quality standards which were set must be built in as a natural part of the processing system. The quality control program is a prevention program, relying heavily upon employee commitment and involvement. Employees want to do their job right the first time. The employee interview confirmed this and also identified that to know whether the job was being done right or not, they must have immediate feedback on their performance.

The quality assurance program has three major components:

1. A branch office quality control program.
2. A quality control program audit.
3. A statistical audit.

There are six essential ingredients in the quality control program:

1. Involvement of employees.
2. Establishment of a realistic quality goal.
3. Establishment of an error identification system.
4. Establishment of analytical procedures.
5. Establishment of a corrective action program.
6. Establishment of a controlling or monitoring function.

Employee involvement. It is important that the employee be involved in the development of the day-to-day operation of the quality control program but this involvement must reach beyond the operation of the program; management must keep constantly in touch with employees to determine what they need to do their jobs in an efficient and effective manner.

Realistic quality goal. The quality goal is instilling in each employee the attitude that quality is in fact a requirement of the job. People are to process data that is both an accurate reflection of the transaction and arrives at its destination in usable format. Quality is not solely the responsibility of the person processing the data; it is the responsibility of the corporation.

Error identification system. This system must pinpoint to a supervisor what types of errors are being made and provide immediate feedback to the individuals on the kinds of errors being made. A three-part error memorandum is used: one copy for a log book, one for the audit file, and one for the individual. An individual daily performance sheet is a part of this system.

Analytical procedures. Trend analysis and Pareto analysis are made monthly. The trend analysis shows each employee how well they are doing. Pareto analysis is used in virtually every area in the quality control program.

Corrective action program. The actions taken by the supervisors and superintendents to address problem areas are actually a part of the corrective action. Personal error corrective action is expected. When people receive the error memorandums it is hoped that they will do some independent research and discover on their own the causes of their errors.

Controlling or monitoring function. Changes have to be monitored and brought to the attention of the employees. A monthly summary is compiled by the branch office and submitted to the Quality Assurance Department. If quality is to be improved and maintained, the quality control program must be ongoing.

A quality control audit is made by a team of quality assurance analysts who visit a location and conduct a two-day on site audit of the quality control program procedures. *The statistical audit* consists of a sample of files of each line of business, taken quarterly in each of the branch offices, and sent to the review unit for recoding. Any discrepancies are noted in a report and returned to the branch office for analysis and corrective action.

Case 2. Details of a Quality Control Program with Computer Operations[2]

Introduction. For over 20 years Blue Cross of Western Pennsylvania has promoted quality control to measure the performance of administrative and clerical functions. With the transition from manual operations to complexly integrated automated processes, management has expressed the need for timely and informative feedback. This has tripled the program's size during the past five years. In addition, increases in government regulations and a greater consumer awareness has mandated that quality control become an integral part of our highly competitive industry.

Quality control provides performance feedback, reduces errors through prevention, and identifies the cost associated with non-quality performance. In 1978, the company detected over $500,000 in errors which were subsequently corrected prior to being released. In addition to direct savings, there are intangible savings of far greater value achieved by:

1. Discovering and correcting error trends to reduce the percentage of error in the total output of work.
2. Reducing the amount of reprocessing which is highly dependent on costly computer time to correct files and generate new output.
3. Improving customer service to retain existing business and stimulate new business. Better service also means fewer inquiries and complaints, further reducing costs.

Organization. The following is a description of the organization for quality control, the objectives of the program, changes that have taken place in the program, how it is implemented, and various forms which are used to collect data to measure progress toward the goals of the program.

The organizational chart shows the quality control area as part of the audit division. The vice president and general auditor, who is the senior management member in the division, reports directly to the plan president and audit committee of the board. The general auditor is responsible for the internal audit and corporate performance evaluation departments. Performance measurement, which is headed by a manager and is part of the corporate performance evaluation department, has two units: the quality control section, and the process time measurement section. The quality control section is composed of one supervisor, one unit leader, eight inspectors, and two quality assurance reviewers. This discussion concentrates on quality control but process time measurement will be summarized.

Objectives of the program. There are two main objectives of the quality control program:

- To identify, correct, and report errors; basically this means inspecting bad quality out of the end product. It is informative, but probably not as instrumental to good quality as the second objective.
- To improve the quality by identifying causes of errors so they can be corrected.

A certain amount of improved quality results from just identifying and reporting errors. The number of carelessness-type errors are reduced because the processors are more aware and more careful. However, to reduce two additional types of errors, technique errors and training errors, requires a more formal organized approach.

Quality control changes from 1970 to 1980. The quality control program during the 1970s was characterized in four ways, each leading to, or resulting from another. The first characteristic is *centralization.* In the early years the unit leaders of the line areas did the actual quality control inspections. The quality control inspectors would subsample their work and then compare results. Having obvious flaws, this procedure led the quality control department to take over total responsibility for doing the actual inspections and all the reporting.

- As a result the quality program was recognized as being more objective because it was no longer under the influence of the line areas.
- It was also seen as more efficient because it offered continuity of reviews, standardized reporting, and statistically reliable results which, it was hoped, established its credibility.

After establishing this credibility, quality control was perceived not only as a function for reviewing documents for errors, but also as a way of establishing controls on certain functions. It could be used to check that authorizations were included on certain documents, sufficient documentation and investigation were a part of certain cases, and even a way of establishing controls on automated routines. The importance of having controls on automated routines cannot be overstated because one error can impact several claims or records, often involving large amounts of money. Even built-in computer program edits often do not function as intended. This new perception of the quality control function resulted in increased *utilization,* the second characteristic of the 1970s.

This utilization resulted in a tremendous *growth,* the third factor. The number of quality control studies increased from 25 to approximately 100 today. Initially, reviews were limited to the membership and claims areas. Later, quality studies were added in correspondence, the telephone area, liability, marketing, and even in word processing. Also, the scope of the reviews expanded to include controls on automated routines mentioned above, preliminary fraud screening for internal audit, and special testing as requested. This expansion caused the staff to grow from seven to twelve.

This tremendous growth required a more organized program with adequate internal controls. This *organization and control* is the fourth characteristic.

• The inspectors were cross-trained to assure continuity of the reviews. Absence or promotion of an inspector could not interrupt the inspection.

• A more effective method of monitoring the inspector's quality and quantity of work was necessary so the results could be used to improve staff efficiency.

• In order to maintain a given degree of reliability, a quarterly review of the total volume of work completed, the volume inspected, and the actual error rate was conducted so adjustments could be made to the sampling ratios.

• Acceptable quality levels were developed for all studies and used to determine if monthly reports would be issued. When the error rates exceed the acceptable quality levels, a monthly report was issued; and when the error rates did **not** exceed the acceptable quality levels, a monthly report was not issued. We labeled this concept exception reporting.

• With numerous changes being made, there was a need for procedures and formal documentation.

Quality control in 1981. The basis of the program are concepts for which definitions are needed.

The first definition is that of the primary error. A primary error can be broken down into two parts. The first part is those errors which in any way affect money: those which affect payment, those which cause financial loss to the corporation, or those which affect the disbursements of corporate funds. This goes beyond actual claim-payment errors to include errors affecting group premiums, utilization charged back to a group, or even refunds to subscribers. The second part of the primary error applies to service. It is any error which can result in inadequate service to subscribers, providers, government agencies, or other outside contacts.

An example of this type of error is a denial sent to a subscriber with an incorrect denial reason. Although denying the claim may have been correct, the subscriber may be misled by the information provided him and inquire further about it. This part of the definition is very judgmental and often causes a difference of opinion with the line areas. This type of error is indicated by "P" in Table 1.

Additional definitions:

. . . A secondary error is an error which violates an internal procedural rule, but does not affect money or service. This type of error is indicated by "S" in Table 1.

. . . An unacceptable document is any case or claim which contains one or more primary errors.

. . . Errors per 100 documents is an index used to measure the total number of errors per 100 units of work processed. It includes both primary and secondary errors.

. . . A process average is the percentage of work which is considered unacceptable. It takes into consideration only the primary errors.

. . . Acceptable quality level is the maximum level of error before a study is considered out-of-control.

When is a quality study necessary? Almost all quality activities result from a request by line management. These requests are sometimes generated because Internal Audit has discovered that sufficient controls are lacking. When a function is being considered for a quality control study, it is given an informal evaluation. The reason it is considered informal is because there is no statistical formula which is used, but certain factors are considered such as the amount of plan money involved, the amount of outside contacts, or any governmental or Blue Cross Association standards.

How is a study developed? The first requirement is a meeting between the line supervisor and the supervisor of the quality control area. They discuss how the particular function fits into total processing. This gives the quality control supervisor a good indication of the importance of the items being reviewed and how other areas may be using the information.

They develop a quality control checklist (Table 1) which is the list of specifications. Included on the checklist are the items that have to be checked on the document, the importance of each item (whether primary or secondary), and an explanation of how the document should be checked. The checklist is used by the quality control inspector as a guide for reviewing work and as a convenient way of categorizing errors for reporting. The quality control checklist is prepared by the word processing department. It is kept on file so future adjustments can be made to the checking procedures by programming them into the machine. A current checklist is then automatically generated.

Sampling is another item considered. An appropriate sampling time has to be established. It is usually scheduled to have minimum impact on the claims processing cycle. It is important that any work processed after the sample is taken be held until the following day's sample is pulled. It is decided where the activity will be located, the date sampling will begin, and whether sample selection is on an individual or departmental basis.

Table 1 Quality Control Checklist

No.	Name	Importance	Correct Entry
1	patient's name	P	verify patient's name on claim to name on status report
2	patient's birthdate	P	verify patient's birthdate on claim to birthdate on status; check to see that patient is not an over-age dependent or eligible for Medicare
3	admission date	P	compare admission date on claim to effective date on status; admission date must be after the effective and agreement must not be cancelled
4	total days stay	P	verify total days stay agrees with admission and discharge dates
5	approval	P	check that total days stay does not exceed available days
6	plan code	S	verify the plan code was coded correctly
7	payment	P	calculate the correct payment according to the total days stay and any special approvals

Later the quality control supervisor must prepare a formal checklist and issue it to members of management. A preliminary sample is pulled for review with the inspector to determine and work out the initial problems, and to use the results to set an initial sampling ratio. After a period of three months an acceptable quality level is recommended to the line areas. This period provides enough time for adjustments to be made in the quality checklist. Certain items may be deleted, other items may be added, and some specifications may have changed. When developing an acceptable quality level, the factors to consider are the number of items on the checklist, (and only primary items are considered), the degree of difficulty associated with each item (and that is weighted), and the grade level of the particular processor.

How is inspection performed? Presently all inspections are manual; however, one company is studying the Nolan AOC (Advanced Office Controls) as a way to simplify inspections and hopefully reduce inspection costs. Inspections are done daily on work in process instead of on the end product for a number of reasons:

- It provides immediate feedback to the line areas. This makes it easier for the supervisor to take action the same day as opposed to trying to review it with the processor sometimes two or three weeks later.
- It allows for easy adjustment. This is important for overpayments because money does not have to be recovered. It also eliminates various adjustments to manual and computer files.
- The average outgoing quality level (AOQL) is improved because mistakes are not released to providers or subscribers.
- Mistakes are not compounded. Incorrect information, resulting from a clerical error on a document, may cause the work of other employees using this information to be incorrect. In turn, when this incorrect work is used, additional work is generated in error. Thus, correcting errors before they are passed on prevents additional errors from being generated.

The samples are usually checked in the quality control area. One day turnaround time is customary so cases sampled one day are returned before the following day's sample is pulled. The study batch ticket is used for all samples that are taken. It shows when the work was received, the inspection starting time, when the inspection was completed, and also when the work was returned to the other area. It includes how many units of work were checked to gauge how long it is taking the inspector to do the reviews. It also shows the time that the sample is in the quality area so the delay to the processing cycle can be measured.

The quality control tally sheet (Table 2) provides daily feedback to the line areas. On the top, the last name of the subscriber is listed under document, the item number corresponds to the item number on the quality control checklist, the importance of the error is primary or secondary, the type of error is listed under entry, the explanation explains what is wrong; and on the far right is the individual processor who made the error. The bottom part of the tally sheet is summary information by processor. It shows the sample size taken for each processor that day, the number of unacceptable documents, and the total errors. The quality control tally sheet is a two part form. The original is kept in the quality area and the second part is returned to the line areas with the sample and the errors attached.

The information is taken from the tally sheet and posted on two forms. The first is the daily performance by clerk. The individual processors are listed across the top and the date down the side. There are three columns under the processor. The first is for the sample size pulled that day, the middle column shows the number of unacceptable documents, and the column on the right shows the total number of primary and secondary errors. This information is posted for each processor on a daily basis. Listed on the right-hand side is the total lot size available that day, the total sample size taken for all the processors, along with the total unacceptable documents and total incorrect entries. This information is added up at the end of the month by processor and total study and posted on the monthly reports.

The second form used is the quality control error chart. This form lists the errors by type of error. The numbers across the bottom of the form represent the numbers of the items on the quality control checklist for the study. Each time an error occurs, a block is completed along with the processors' initials and the date the error occurred. This information is also used for the reports. This form allows the inspector to balance the errors at the end of the month.

Table 2

Quality Control Daily Tally Sheet

Study Code 537-9 Date 3-1-198– Inspector PS Lot Size 1000 Sample Size 20

INCORRECT ENTRIES					
No.	Document	Item No.	Entry	Explanation	Clerk
1	SMITH	55	AGE	SHOULD BE 30	AB
2	JONES	24.8		ROOM RATE SHOULD BE $100 per DAY	CD
3					
4					
5					
6					

Clerk	Sample Size	Unacceptable Documents	Incorrect Entries	Clerk	Sample Size	Unacceptable Documents	Incorrect Entries
AB	10	0	1				
CD	10	1	1				

Table 3

QUALITY CONTROL/ASSURANCE Management Quarterly Summary

DEPT: INTER-PLAN BANK | Period Covered: 1ST QUARTER 198- | Prepared By: M. KING

STUDY NUMBER	STUDY TITLE	Available to Sample	Sampled	Number Unaccept Documents	Acceptable Quality Level	VARIANCE + over − under	Process Average	Prior Qtr Process Avrgs 4th Q 198-	3RD Q 198-	2ND Q 198-	Total Errors	Errors per 100 Document	Prior Quarter Errors 4th Q 198-	3RD Q 198-	2ND Q 198-
537-9	CLAIMS	5000	1000	36	3.00	+0.60	3.60	4.00	4.50	5.25	40	4.00	4.50	5.00	6.00

Presently, automated data retention is being considered to eliminate manual methods. The intent is to enter information into a minicomputer via CRTs on a daily basis so the minicomputer can store the information, calculate it, and automatically generate a report at the end of the month. Presently this is just in the planning stage.

What type of reports are generated? The first report is the Quality Control Management Quarterly Summary (Table 3). This form is distributed to director level and above. It lists the error rates by study and by department. The report shows the total number of documents that were available to sample that month, the number sampled, the number of unacceptable documents, the acceptable quality level, any variance under or over the error rate, the actual error rate or process average, along with three quarters prior data. The report also shows the errors per 100 documents, and three quarters prior data.

The Quality Control Detailed Report (Table 4) is issued to the managers and supervisors. It is issued on an exception basis (if the error rate exceeds the AQL) monthly; however, it is issued for all studies on a quarterly basis regardless of performance. At the very top is general study information such as the study number, the period covered, along with the AQL, process average, and any variance. The middle portion (indicated as study data) lists each processing area by clerk, sample size, number of unacceptable documents, total errors, process average, and errors per 100 documents. At the far right, the information is added together for the total study. The bottom part of the report shows the errors by type, broken down by individual. This information can be used by the supervisors to immediately identify where the errors are and who is responsible.

The Quality Control Graph (Figure 1) is issued on a yearly basis. It is plotted by a prime minicomputer over a two year period. Three years of monthly data along with a weighted yearly average is right below the graph. Also included is the slope of the trend line which can be used to predict future performance. The shaded area on the graph itself represents performance in the acceptable range so performance which rises above the shading is considered out-of-control. The broken line is the trend line indicating by its slope which way performance is heading.

What kind of impact has the quality control program had at Blue Cross? The dollar value of errors corrected prior to release from the corporation exceeded the department budget during the last two years. This is a very measurable impact, but there are also other intangibles or unmeasurables that have to be considered.

First, correcting errors improves the average outgoing quality level, so that a higher level of service is being provided. This includes errors affecting both money and service provided.

Second, since processors are aware that the work is being inspected, they tend to be more careful. Also, they benefit by seeing their mistakes, so fewer errors should be produced. This is especially important because not all of their output is inspected before it is released.

Another benefit is realized by minimizing the loss associated with overpayments which would not be detected or recovered. Savings are also achieved because of less re-work.

Finally, an overall improved level of service will hopefully minimize the number of unhappy customers and stimulate new business.

What are the major problems? Every quality control program has problems, but two are mentioned which the company finds significant. The first is the effect of management changes. Different people have different philosophies about all facets of life and quality control is no exception. Management changes bring with them a different view of what should be checked, how much should be checked, the way it should be checked, and the way it should be reported back. Often this requires significant changes in reviews and reporting. The company also has trouble insuring management understanding of statistical reporting. Presently a training program is being developed for them to help them understand the concept of statistical reporting as well as their role in quality control.

The second problem is the grade levels of the quality control inspectors. The majority of the inspectors is in grade 4 and grade 5. Blue Cross employees are grade 1 through grade 8, which puts the inspectors somewhere in the middle. As a result of the reviews they do, they receive a lot of exposure throughout the corporation and gain experience with many different corporate activities. As a result, they are prime candidates for openings in other areas. This presents problems with high turnover and the resulting training because every time a replacement is trained, knowledge is lost.

Table 4
Quality Control Detailed Performance Report

Study No. 537-9	Period Covered MARCH 198-	☐ Quarterly ☒ Interim
Study Title CLAIMS		
Prepared by M. KING	Date 4-6-198-	Page 1 of 1

SUMMARY DATA	
A.Q.L.	3.00
Process Avg.	3.60
Variance	+0.60

STUDY DATA	PROCESSING			WORD	
REFERENCE CODE	AB	CD	TOTAL	PROCESSING	TOTAL STUDY
SAMPLE SIZE	500	500	1000	1000	1000
UNACCEPTABLE DOCUMENTS	1	15	16	20	36
TOTAL ERRORS	5	15	20	20	46
PROCESS AVERAGE %	0.20	3.00	1.60	2.00	3.60
ERRORS PER 100 DOCUMENTS	1.00	3.00	2.00	2.00	4.00

INCORRECT ENTRY ERROR CHART

Item No.	Item-Description	Reference Code Errors			TOTAL INCORR
		AB	CD	WORD PROCESSING	
3P/S	I.D. NUMBER	1		13	14
55	AGE	4			4
8P	ADMISSION		1	2	3
18P	DAYS PAID		2	4	6
24P	BILLING		12	1	13
	TOTAL	5	15	20	40

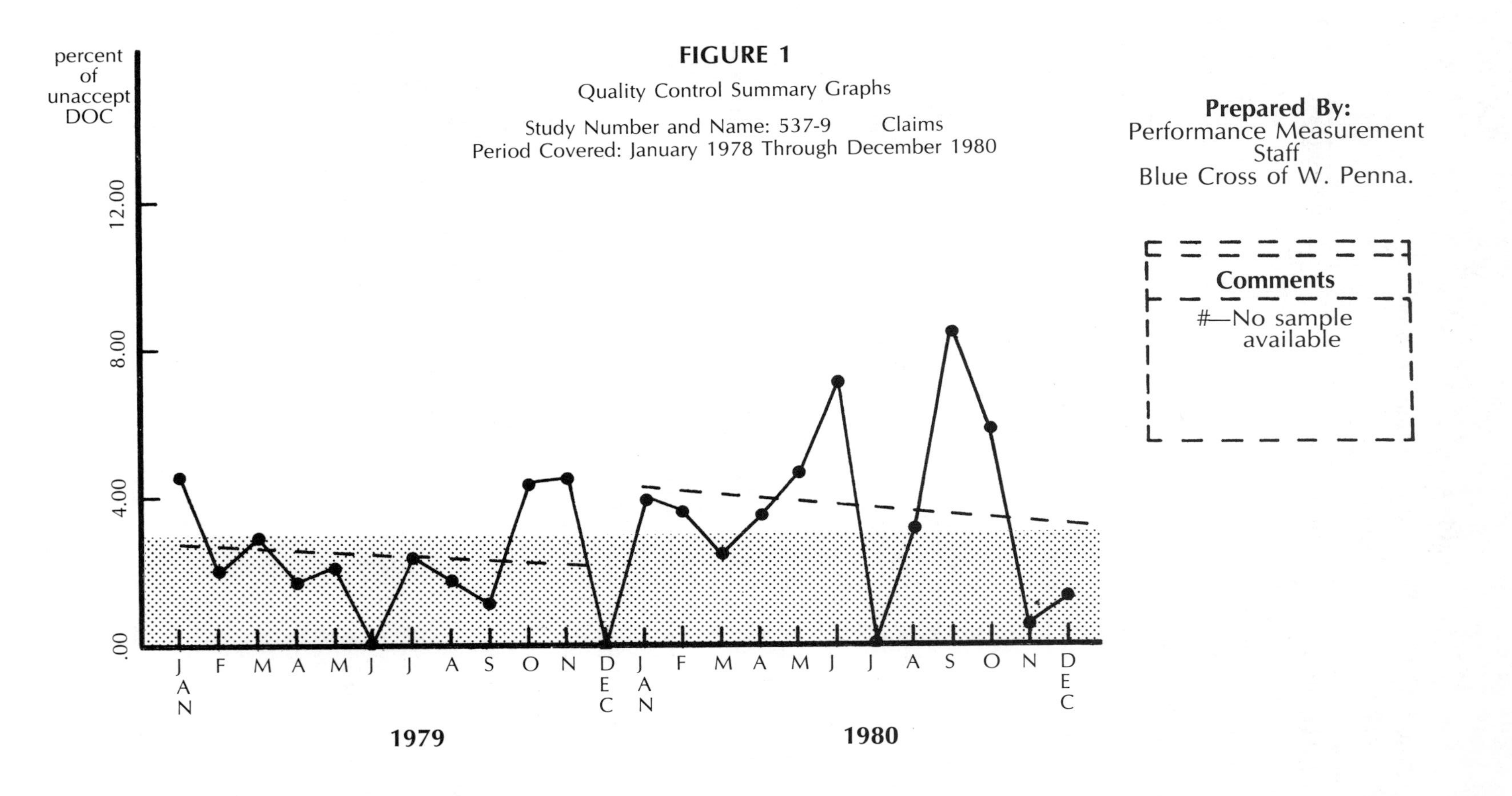

YEAR	JAN	FEB	MAR	APR	MAY	JUN	JUL	AUG	SEP	OCT	NOV	DEC	YEARLY PROC. AVERAGE (%)	AQL (%)	SLOPE OF TREND LINE
1980	3.95	3.58	2.46	3.51	4.62	7.06	.00	3.16	8.43	5.82	.57	1.39	3.97	3.00	−.065
1979	4.50	2.12	2.96	1.74	2.19	.00	2.37	1.74	1.22	4.31	4.46	.00	2.42	3.00	−.072
1978	3.64	#	3.57	4.17	4.48	#	6.34	.35	1.11	2.15	3.06	#	3.23	2.00	−.220

Quality Assurance

What is quality assurance? The company feels that the future lies with quality assurance. Quality assurance is a means of preventing errors (not only detecting them) by finding causes of errors and working with the line areas to make the necessary corrections.

Why is quality assurance necessary? There are several reasons why quality assurance is necessary. First, certain error rates which have remained high over a long period of time must be brought under control. It is uncertain exactly who or what is responsible. Operating areas have numerous other demands placed on their time and do not always have time to analyze complex quality problems. Quality control's involvement with inspection, reliability, and reporting has not provided time to analyze trends or determine causes of errors. Other problems such as turnover and changes in processing routines can also contribute to these consistently high error rates.

Another reason for quality assurance is the growing acceptance that quality is everyone's responsibility. This includes the quality control area, the line area, management, and the processor who is directly responsible for the quality of the product or service.

A third reason is the greater consumer awareness with service. Consumers are more quality-conscious, making good quality a necessity to remain competitive. Government is requiring it, groups are demanding it, and prospective groups often want to see examples of quality and production reports during negotiations.

A fourth reason is the realization that good quality is cost justified. Today's businesses are magnifying the search for cutting costs. Fewer phone calls from subscribers and less re-work of claims which is dependent on costly computer time are among those factors which can reduce costs.

What is Blue Cross doing regarding quality assurance? The first thing that must be done is to change the attitude of the people. Quality must be seen as something positive and the program must be responsive to peoples' needs. Changing the negative concept of acceptable quality levels to positive performance goals is one step which will be taken. There must also be incentives for the line areas and the quality staff to maintain a high level of quality.

Employees must see good quality as a necessity. The company has been promoting good quality by using posters which are being put up in the line areas, brochures, and other articles explaining why good quality is important. A training program will be developed for clerical staff so they can see the importance of good quality and the impact it can have. The quality staff must also be made aware that quality is their responsibility. They are being encouraged to take the initiative for identifying causes of errors and increasing their interaction with the line areas.

Action must also be taken to determine the customer's perception of what is important. The company needs this perspective to have something to measure its performance against. Presently a customer survey is being developed in conjunction with the Marketing Research Department to determine:

. . . Is the customer satisfied with the company's performance? — The customers' level of satisfaction should be consistent with what reports are telling the company. If they are satisfied, the reports should be indicating that performance is good and vice versa.

. . . What are the customer's concerns? — It is important that the major customer concerns are being measured. A great deal of the inspections being performed may be in areas of minimal customer interest.

. . . Is there a need to change? — The definition of what is an error may need to change to include what the customer sees as being defective. Other measuring or reporting techniques may have to change to accurately reflect Blue Cross performance as seen by the subscriber.

Assuming it is determined what is important, the next step has to be correlating the amount of involvement with the need for involvement. The study request form has recently been distributed to the line areas to help make this determination. It is to be completed for each new request for a quality control or process time study. The middle portion of this form is to include the importance of doing the study and any negative impact if it is not done. The dimensions of the study should provide some insight into whether the amount of plan money or the amount of outside contact warrants further consideration. If necessary, existing inspections will be reduced. A reduced inspection chart will be used when certain activities are being over-inspected.

Finally the amount of interaction with the line areas must increase. The results of the marketing research studies must be communicated to the line areas, so they know the comparative and relative importance of various activities. The line areas must be better informed of particular performance problems. They must let their needs be known regarding the type of feedback meaningful to them, so our reviews and reports can be designed to meet these needs.

Process time. Any good measurement of quality requires an equally accurate measurement of time. It is necessary to know how well something is being done **and** how long it will take to do it. The process time program experienced similar development through the 1970s. It was centralized, more utilized, and expanded. Process time studies are performed in the claim and membership areas, on inquiries, ID cards, and approvals. The number of process time studies increased from 28 to 85 during the 1970s.

The process time program uses basically the same guidelines to determine if a process time study should be done. Similar techniques are used to develop a study. However, what is more important with developing a process time study is what time should be measured. Is the overall processing time adequate or is it necessary to provide detailed breakdowns by area? Is it necessary to know how long the claim is in the mailroom, how long it is in liability or medical review for investigation, or how long it takes a check to be generated once a claim has been approved? It is necessary to know if all the dates are available on the document and how much is involved in extracting the dates.

Once sampling begins, dates are posted on the process time coding sheet. The number of breakdowns needed will determine how many dates are actually posted. This information is then keyed by the Keypunch Department into the computer. The computer at the end of the month does all the calculations and provides a listing used to generate the reports.

The Process Time Report (Table 5) is issued on a monthly basis to all levels of management. This particular study is Inpatient Claims. The top part of the form shows average times. The submission time shows how long it took for the claim to be submitted after the subscriber was discharged. This is either the provider's or subscriber's responsibility, but can significantly delay payment of the claim. The balance of the times shown by area(s) are Blue Cross Processing Times and include how long it took to process, to diagnose code, balance, etc. Monthly averages, "Year to Date" averages, and averages for the same month during the previous year are shown. The bottom part of the report is the percentile distribution, also broken down by function and total time. This distribution shows the percentage of activity which is completed in any given period of time. It is informative because it excludes isolated cases which take an exceptional amount of time to process or may have been keyed wrong. These cases will significantly impact the average, but have minimum impact on the percentile distribution.

Process time graphs are also issued on a yearly basis. They also cover a three year period, plotted for the most recent two. (Figure 2)

Table 5

PROCESS TIME REPORT — BLUE CROSS OF WESTERN PENNA

Report Number: A110
For: MARCH 198-
Prepared By: J. GUENTER
Date:

Study Title: REGULAR INPATIENT CLAIM PAYMENT PROCESSING

(Time is in days)

	March	1980 Year Through Feb.	1979 March
1. Submission Time — From Date of Discharge or Approval, Whichever is Later, to Plan Receipt	10.0	9.0	10.0
2. **Blue Cross Processing**			
A. Claim Processing Time — From Receipt to Completion of Processing	1.5	2.0	1.6
B. Diagnosis Coding Time — From Completion of Processing to Completion of Diagnosis Coding (including investigation lag time)	3.4	2.7	3.6
C. Claim Transfer Time — From Completion of Diagnosis Coding to Disbursements Receipt	0.8	1.0	0.8
D. Payment Balancing Time — From Receipt by Disbursements to Balancing of Payment	4.3	5.0	4.0
Plan Processing Time reported to BCA	10.0	10.7	10.0
E. Payment Time — From Balancing of Payment to Issuance of Check	4.0	3.3	3.0
3. **Total Blue Cross Processing Time**	14.0	14.0	13.0
TOTAL TIME	24.0	23.0	23.0

Percentile Distribution

Days Required To Complete	CLAIM PROCESSING TIME (From Blue Cross Receipt to Completion of Processing)		PAYMENT BALANCING TIME (From Disbursements Receipt to Balancing of Payment)		PLAN PROCESSING TIME AS REPORTED TO BCA (From Blue Cross Receipt to Balancing of Payment)		TOTAL BLUE CROSS PROCESSING (From Blue Cross Receipt to Issuance of Check)	
	Percentage of Activity Completed Within Specified Number of Days							
	Current Month	Prior Month	Current Month	Prior Month	Current Month	Prior Month	Current Month	Prior Month
0-1	0.0%	0.0%	0.0%	0.0%	0.0%	0.0%	0.0%	0.0%
0-7	54.0%	58.0%	75.0%	70.0%	31.0%	35.0%	15.0%	12.0%
0-14	92.0%	91.0%	99.0%	100%	65.0%	70.0%	52.0%	57.0%
0-21	99.0%	99.0%	100%	—	97.0%	98.0%	90.0%	92.0%
0-30	100%	100%	—		99.0%	100%	98.0%	99.0%

FIGURE 2

Process Time Measurement Summary Graph

Study Number and Name: A110 Regular Inpatient Claim Payment Processing
Period Covered: January 1978 Through December 1980

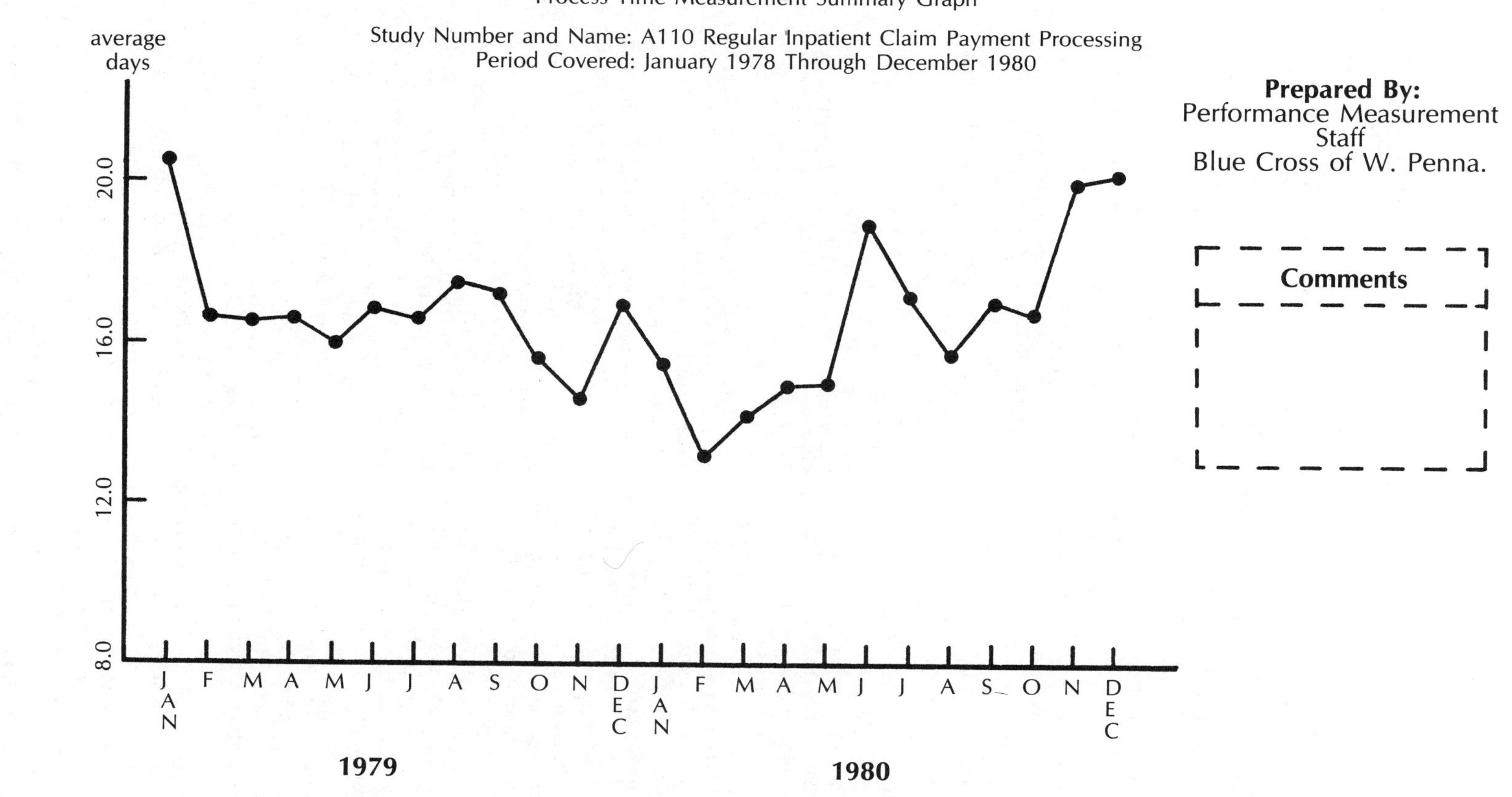

Prepared By:
Performance Measurement Staff
Blue Cross of W. Penna.

Comments

YEAR	JAN	FEB	MAR	APR	MAY	JUN	JUL	AUG	SEP	OCT	NOV	DEC	YEARLY AVERAGE
1980	15.5	13.3	14.2	15.0	15.1	19.0	17.3	15.8	17.1	16.8	20.1	20.3	16.6
1979	20.5	16.7	16.5	16.6	16.0	16.9	16.6	17.5	17.3	15.7	14.7	17.0	16.8
1978	17.5	15.3	16.5	15.9	15.4	16.5	16.7	16.7	18.5	17.7	17.4	19.3	16.9

Where to go with process time? The company is probably headed in the same direction as quality for many of the same reasons. The company has to determine where the delays or bottlenecks are in the system and work with the line areas to eliminate these delays.

Case 3. Large-scale Quality Management Program

For several years The Hartford Insurance Group has been developing a corporate quality program. The following is a detailed description of the major aspects of this program.[3]

The number of people in the working population involved in the business of providing service is estimated to range from 60 to 75 percent.[4] In addition to this large and growing number of workers, there is also a high percentage of personnel on the manufacturing side of industry that are involved totally in administrative tasks. This number exceeds 50 percent for many companies. Altogether, a large majority of working people in the United States are employed for the sole purpose of providing service to others. And most of them are not familiar with the meaning of productivity or quality management.

This is a message about quality management and it is for the managers of banks, insurance, financing, food service, lodging, real estate, publishing, broadcasting, retailing, leasing, repairs, utilities, hospitals, schools, transportation, communications, government, and other service operations, as well as for the managers of product manufacturing companies. In this era of strong foreign competition and growing customer demands for quality of service, the quality reputation of an organization is too precious an asset to leave to chance. Less than satisfactory service internal to an organization will result in higher operating costs. Less than satisfactory service as delivered to the customer will have a negative impact on revenues.

Consistent service quality will not happen accidentally. An organized, scientific approach to quality management is required to maintain or develop a quality reputation. That usually means an investment in the study and application of quality control principles to all areas of service work. That is, the work of the people who deliver service to the customer (or do the manufacturing work for products) and the people who provide support service to these front-line operators. Unfortunately, there are three obstacles to the straightforward fulfillment of quality management programs in the service business. The first is that the managers of service businesses are, generally speaking, almost totally unfamiliar with the substance and business value of quality control principles. The second is that all too often investments in control programs are viewed as unnecessary expenses rather than programs with a pay back. They are seen as having a negative rather than a positive effect on productivity. The third obstacle is that customers are not genuinely listened to. Their complaints are often seen as an irritant rather than as an opportunity.

The program. To understand the problem of unsatisfactory service quality and the solution of quality management, it is necessary to define the situation in terms that all parties can relate to. Unsatisfactory quality is described as, "undesirable results due to unwanted and unnecessary variations in performance." The degree of deficiency may be different but everyone has experienced it in hotels, restaurants, rental cars, appliance repairs, home maintenance and many others. It is usually not an out-and-out failure — just a completely unsatisfactory situation as measured by personal standards.

The cause of the problem is almost always attributed to standards of performance being weak or non-existent. It's very noticeable when service personnel seem to know exactly what they are doing — such as at Disney World or McDonald's. At the other extreme, customers are forced to complain. The unwanted difference occurs when individuals, managers and supervisors, as well as workers, are free to set their own standards of performance. ***Fortune Magazine,*** in its article, "Towards Service Without a Snarl" (March 23, 1981) stated: "For all the improvements made possible by technology, the quality of service still often depends on the individual who delivers it. All too often he is underpaid, untrained, unmotivated, and half-educated." This applies equally as well to the individuals who provide back-up support for the deliverers.

The solution to this predicament is to establish a quality management system. Very simply stated, this means establishing performance standards, measuring performance against the standard and then developing a quality improvement program. This solution is based on the premise that no matter where you are today in terms of performance, you can always get better. The benefits of this solution are that the unsatisfactory service prevented through performance improvement will not cost anything in extra service or in loss of future customers.

To bridge the gap between this simple solution to quality-of-service problems and the real world of organizational pressures and priorities, The Hartford's program will attempt to deal with all the realities of the business world as they are in fact. It will start with a sales pitch aimed at getting management interested. This will be followed by an explanation of the quality council and/or implementing team, establishment of performance standards and responsibilities, measuring performance, building "the improvement program," and implementing or managing the actions required for results. Then, the overview will examine the value of recognition, the use of quality costs, and finish with a summary of the results that can be expected.

Management commitment and the quality council. The job of the quality professional is to convince management that quality is what they need and show them how to build a quality system that can be measured, is guaranteed to produce improvement, and is cost-effective. In order to do this, the quality professional must be enthusiastic, project a "can do" attitude, and be specific with regard to the benefits for management. He must also build a very practical quality program.

The benefits to be used in selling management are:

- Improved image — quality improvement improves your reputation with employees and customers.
- Improved productivity — quality improvement means less re-work and reductions in your work process, thereby, improving productivity.
- Reduced expenses — less re-work requires less labor and material costs.
- Improved marketability of services which will increase market share/customer base — improved quality gives you the expense and service advantages over your competitors.
- Management of quality and quality costs — a quality improvement program will enable you to act rather than react to problems, to manage rather than be the victim of.
- Improved employee environment — quality improvement reduces employee frustrations leading to improvements in human relations, morale and self-worth of employees.

- Improved profitability — all of the above lead to this positive end result.

Although quality improvement is an economical means of achieving these benefits, quality is not free. The price the general manager must pay is:

—Visible ongoing quality leadership.
—A quality policy that is meaningful and practical.
—A budget to support up-front needs.
—Appointment of a quality council to carry out the quality policy.

Visible ongoing commitment is a non-transferable function and does not end with the appointment of a team to carry out the quality policy. It is important that employees recognize their chief executive officer as the leader in quality.

There is no one best quality policy. The words should be those which are most easily understood by the management and employees of an organization. A good example is The Hartford's quality policy, which is: "To perform each job or service in exact accordance with existing job requirements or standards." In setting a quality policy, it should be remembered that the policy itself and the definition of "quality" is applicable to all employees of an organization from the chief executive officer to the most recently hired individual. An area in which management often gets into trouble with a quality policy is to imply that the policy is only applicable to line-type individuals in the manufacturing, administrative, or service areas. It must be clearly understood that a quality policy is applicable to engineers, planners, all of management, marketing, and general managers themselves.

Appointment of a quality council by the general manager is a transfer of the quality program work task and not a transfer of the responsibility for visible leadership. The quality council should be made up of the same senior executives that the chief executive trusts in leading and managing major company functions — the individuals whose decisions and ideas make the company what it is today. The duties of the quality council are:

- Determine **what** should be done to implement the company quality policy.
- Determine **who** will be responsible for carrying out the actions decided by the council.
- Provide a communication link between the council, the general manager, employees, and the quality function.
- Assist **all** quality representatives (corporate office and profit centers) in identifying and resolving internal or interdepartmental implementation problems.
- Assign quality task teams to audit existing programs and to explore specific problems or opportunities.

The only limitations on the quality council's charter is that the program must:

—Adhere to the quality policy.
—Provide quality improvement.
—Develop meaningful measurements.

The role of the quality professional is to help create a system, attitude and environment that can assist all managers in carrying out the quality policy. This responsibility is carried out in conjunction with the quality council.

Setting quality standards. The first task of the quality council is to define the service quality standards. This means to clearly define exactly what is intended to be delivered to the customer. A service quality standard should be very similar to a product description or specification which, together with price, provides a specific choice to prospective customers. The exact requirements of the quality standard, the consistency with which it is met and the price for which it is delivered will have the greatest bearing on the growth or diminution of customer base. There are exceptions to every business rule such as, in this case, the initial attraction of something new or different, or the choice location of a particular hotel — but in the long haul, the quality of service and price will make the difference. It is well also to note that it is not possible to be a price leader in any industry without first having established a consistent and satisfactory quality reputation.

In the process of setting standards for quality the most important input is that which comes from the customer. It is always wrong to assume or guess about what the customer really wants or expects. The rolls of the bankrupt are replete with many examples of this mistake. Perhaps the best known example is the Edsel. On the other hand, understanding the customers' needs and expectations is no easy task. Customers are more sophisticated today. They are also better educated, more discriminating and more demanding than at any time in the past.

To truly understand the customer, an organized study of the marketplace must be undertaken from different points of view. Some of the most successful avenues of study include the following:

- Market research — done independently or as part of an industry association or agency.
- Public opinion polls — conducted by industry associations or agencies representing the customer or consumer.
- Review of competitive activities — an investment in comparing yourself with your competition.
- Analysis of complaints and compliments — learning from your customers' direct communication with you.

The objective of this study and analysis should be to allow you to establish quality standards or customer expectations for your product or service that are equal to, better than, or different in an appealing way from those of your competition. It can also be used to develop a special niche in the marketplace that can best be served by you. If you should discover that what is normally expected is considerably more than you are willing or able to deliver, your problem is beyond the solution being presented here.

The next step in the definition of quality standards is to convert the agreed-upon customer expectations into well-defined, specific requirements. This is needed for two reasons. The first is to serve as the finished criteria or target against which the entire service process must be designed. The other is that acceptance criteria or targets can be established for all operations leading up to the end product or service. As an example, Japan Air Lines Co. has established key quality requirements for flight operations (see Chapter 7).

Note that there is a target and control limit for each measurable performance standard established. The control limits are the acceptance standards for the operation and the targets are the (1980) goals for the improvement program.

To understand the service process as it affects the quality standard, consider the control limit of 3.00 baggage irregularities per 10,000 checked bags in the above example. In order for this standard to be met, every activity involved in the handling of baggage must be clearly defined. If it is not defined, merely left to the individuals involved to determine how to do the job, or to the whims of a supervisor on a given day, an infinite amount of variation will occur in the end results. This is in direct conflict with the need for consistency in operations in order to achieve predictable and consistent results. In manufacturing companies, detailed process sheets and operating procedures are used to assure the production of conforming end products. The same approach, detailed service process sheets, are needed for the service industry. These process specifications can then become the basis for employee position descriptions and overall training programs.

As detailed service process specifications are prepared back up the line from the delivered service to each activity required to achieve the desired end result, it will become clear that these specifications actually spell out the standards of performance for all personnel involved in the service process. Measurements of performance against these standards will indicate quite clearly whether or not the end result can be achieved. It is like examining all of the sub-assemblies that go into an end product. If any of them fail to meet their quality standards, it is not possible for the end product to meet its requirements. Even a succulent hamburger, to meet its quality standard, needs the right ingredients, cooked exactly to order and served in a timely manner in attractive surroundings.

In short, establishing service quality standards means first determining the quality image that you want your customers to experience. It should be an image that he will remember, return to and tell others about. Next, it means determining the exact standards of performance that each employee must fulfill in order that the desired image can be achieved and a quality reputation be established and maintained.

Measuring performance. Once quality standards are established, the next task of the quality council is to measure performance against these standards. Measurements are needed for three very important reasons. The first is to determine where a business stands against its standards in order to identify and justify specific improvement needs. The second is to establish a base-line for measurement of improvement progress and the third is to provide input for the identification of specific problems for improvement action.

It should be noted here that, generally speaking, people do not like to be measured. Foremost among the reasons is management's misuse, in a destructive rather than constructive sense, of measured results. Also, people never before measured may be well aware of the fact that they are under-producing. On the other hand, without measurements, management will not be able to identify performance problems and explore their causes. They will also miss opportunities to reward superior performance, a genuine quality improvement asset. Unfortunately, lack of measurement always penalizes the good performer and rewards the bad.

Quality measurement is an activity where a clear-cut level of investment must be determined. It is usually not practical or economical to measure detailed performance against each standard. One practical approach is to examine flow charts of service processes as they integrate to achieve the final result and to then carefully determine key points within the overall process. Key measures are those which provide the most appropriate and direct appraisal of the achievement of incremental performance objectives. For example, sampling inspection of hotel rooms ready for occupancy can serve as a measure of overall housekeeping and maintenance activities. Analysis of credit memos is a good measure in the distribution business. A sample survey of restaurant customers, after they have been served, can be a measure of order taking, food preparation, and the timeliness and quality of delivered service. Where a system is involved, such as a communications network, sample messages tracked through the network can measure how well each segment is performing.

One of the most important performance measures is the direct input of the customer. Some investment in this area is mandatory. Only about one in fifty customers will say what he really thinks, so it is very important to listen. Some of the methods employed are customer questionnaires, customer comment log books, telephone surveys, and analysis of complaint/compliment letters and phone calls. It is vital to progress that all customer complaints be objectively analyzed and resolved fairly and in a timely manner. If the concern for customer satisfaction is genuine, no customer need ever be lost.

Once it has been decided what performance measures are to be taken, the next step is to arrange for the data to be collected and organized for effective use. This generally means arrangements that show percentages of satisfactory performance at key points in the process and Pareto arrays of the frequency and distribution of all (measured) nonconformances to standard. Analysis of these arrays will lead directly to the identification of those specific elements of the service process in need of improvement, the perfect input for development of the improvement program.

Building a quality management system. The prerequisites for building a quality management system already discussed include general management commitment and personal leadership, establishment of a quality policy and a quality council to carry it out, setting quality standards and measuring performance against these standards. Additional prerequisites include understanding of the following quality elements by both management and employees:

A. Definition of quality.

Quality can never be achieved when everyone has his own idea of what it means. In terms of measured performance, quality can only mean conformance to a standard. That is, quality can only be achieved when the end-product or service, as delivered to the customer, looks and performs exactly like the instructions say it should. Quality is a noun, the result of clearly defined efforts. It is not an adjective — the degree of definition or effort.

B. First line manager's role in achieving quality.

Individual job specifications must be clear and understood by all employees that are part of the process. The end product must meet these specifications. If the specification is not correct, management must change it.

It is impossible to achieve quality if the individual hired cannot perform the functions necessary to meet the specification. It is the responsibility of first line management to hire the proper individual. Even with the most comprehensive testing available, management is not always sure that they have hired the right person. Therefore, it is a quality program failure when management fails to take remedial action when it becomes clear that they did not hire the right person.

If the individual hired to meet the job specification is not trained properly, then quality cannot happen. It is management's responsibility to train people properly.

Without communications, quality cannot happen. It is the responsibility of the first line manager to make certain communications not only take place, but are also effective.

C. Responsibilities for quality.

a. Quality is the responsibility of the individual who was hired to do the job. This is a non-transferrable responsibility. Management interviews people to fill a particular job and it is expected the person will do the job correctly. On the second day of work, people should not be told what their acceptable error factor is. Management cannot do the job for their employees, nor can they accept as a matter of course that errors are inevitable.

b. Managing quality is the responsibility of the first line supervisor. The management of quality includes:
   —Knowing the current quality level within the unit through measurement.
   —Knowing what the major error cause is through analysis.
   —Having in place a practical corrective action system which addresses the **elimination** of this error cause.

c. Monitoring quality is the responsibility of middle management. That is, they are responsible to know what the level of quality is and what corrective actions are in place to improve quality.

d. Questioning quality and quality measurements is the responsibility of the general manager. It is necessary for the general manager to show the visible leadership and involvement in the quality system by asking questions where slippage occurs, by challenging results when they are in conflict to what department heads and customers are saying and to recognize truly outstanding quality improvement.

e. Auditing is the responsibility of an independent functional and/or internal operation which specifically has the responsibility to take a snapshot profile of the quality level of a given function. This is an objective view of the quality performance of a given unit/function.

D. Quality goal.

The objective of a quality improvement program is to improve quality . . . period. The purpose of a quality management system is to identify for the first line manager what his major problem is in order that he can investigate the **cause** of the problem for the purpose of developing a preventative type of corrective action plan to **eliminate** the cause.

E. Quality control versus quality improvement.

Generally, the purpose of a quality control program is to develop a method to "insure" end products are going out correctly. A quality improvement program is a program based on historical fact, that is, analyzing what has been done for the purpose of identifying error causes to insure the same type of errors will not recur as frequently in the future, thereby, gaining consistent quality improvement. Quality improvement can be summarized as "do it right the first time, next time and every time."

With all of the prerequisites understood, you can now ***build the quality system.*** At best this is a "cookbook" approach to building a quality system. In applying it to your company, take into consideration the specialness of your company's product, service, process and personality and translate the recommendations into a language that can best be understood by your management and employees. An effective quality improvement program is something that must be unique to your company. Quality improvement is essentially achieved through a program of ongoing quality measurements, analysis and corrective action. The elements of the quality system are as follows:

1. Sampling (quality measurements).

Once you have identified the current level of quality performance and have made the management decision as to the target levels and variances that are acceptable (or we want to pay for), a standard sampling approach can be utilized for ongoing measurements. Sampling is done in lieu of 100 percent checking for the following reasons:
—Purpose of the quality improvement program is to identify error causes through trending. Sampling will do this.
—100 percent checking is only 85 percent effective.
—The cost of 100 percent checking in dollars and time is prohibitive.
—Studies have shown that if people know they are being 100 percent checked, they tend to make more errors rather than less.

2. Checking.

Very simply stated, the items which have been sampled out of the total work population are verified back to the written standard to determine if they meet the job specification or quality standard for that portion of the service process. If the item meets this standard — it is quality; if not — it is not quality.

3. Recording.

All errors identified regardless of source should be recorded as to where they occurred and if possible who made them. This includes errors which are detected in the processing (sampling checking method), errors from work rejected internally as a part of normal operations, and errors contained in service rejected externally by customers.

4. Analyze results and identify major problems.

Measurement data, to be of real value, must be analyzed and interpreted — differentiating between cause and effect. Examine and interpret each measure. Then, the results can be used for determining necessary corrective action.

Timeliness is of prime importance when problems exist. The results of data analysis, with proper identification of the major problems, must be provided to responsible personnel as quickly as possible. Quite frequently the real causes of a problem are unknown — and the measurements reflect only the effects. If the analysis and feedback are in "real-time" then there is a better probability that the causes of the problem can be identified and corrected.

Graphical displays of data are very effective. First, because the graphical displays are usually "self-analyzing" (i.e., trend charts/graphs) and second, management awareness, understanding and acceptance of the data are enhanced by the graphical display. In this way, the data is used as a measure of performance.

Periodically (weekly or monthly) a quality status report should be prepared and distributed to the management of the business. This report should include a narrative interpretation of results and actions currently required, in addition to the typical summaries (table, trend charts, etc.) of measurement data.

5. Corrective action.

In order to go from analysis to corrective action, to truly address the cause, it is necessary to integrate people as part of the quality cycle. The best person to identify why an error was made is the person who created it; therefore, a practical corrective action plan must include direct input from the error source. When the error source is not known, investigative problem-solving techniques must be brought into play. Every undesirable situation has a cause but it is sometimes hidden in a complexity of interactive functions. Until sorted out for corrective action, these hidden causes can plague a company for long periods of time. In order to achieve practical corrective action, the following are necessary:

—Organize investigative and corrective action teams — identify the real problems.
—Develop priority action plans — need to spend dollars wisely.
—Get action commitments and dates — assure that everyone understands the plan.
—Document and monitor actions taken — if expected results do not occur, continue the investigation.

Corrective action is the pay-off step. Without corrective change, the best you can hope for is the status quo. But with a viable corrective action system, constantly fed by employee recommendations, analyses of performance measurements and customer problems or complaints, performance improvement is assured. And performance with fewer defects will always be less costly and more on schedule.

6. **Verify action results.**
—Conduct follow-up audits. Avoid the tendency to assume because some action was taken, the result was the correction of a problem. Action does not always mean accomplishment.
—Verify effectiveness. Corrective action is not effective if it only fixes the symptoms. It must go to the real cause of error.
—Close out effective results. If the problem is really fixed, close it out and, thereby, close the loop of having, achieved performance improvement. You are now at an improved and more profitable level of performance.

Implementation and management of the improvement program. Once agreement has been reached on the exact make-up of the quality improvement program, the next step is to plan, schedule and promote the implementation phase. This phase should start with a communication to all employees of management's commitment to quality and their intent to initiate an "improvement program."

The kick-off event for the program can be done quietly or with a lot of fanfare. For example, management can communicate to the employees through a simple, written pronouncement circulated throughout the organization, or they can organize a series of promotional gatherings using audio-visual aids. It depends on the personality and style of the company. The important criteria is that every employee receive the message and that they become keenly aware that "how we collectively do our work" is going to be different.

When quality is viewed as conformance to a standard, it becomes readily apparent that the achievement of quality is dependent upon each and every employee, from manager to operator, performing up to the standards of their job. That is why quality improvement is a company-wide program. If the chain of quality is broken in any one area, the effect will be felt in the end results. In this respect, quality is no different than cost or schedule. That is, variations in the performance of individuals will have an effect on cost, schedule, and quality. The dilemma to be avoided here is that if cost or schedule objectives are achieved at the expense of quality, there will always be a penalty to be paid in future sales.

It is important at the outset of quality improvement to reaffirm basic responsibilities. Each person must be held accountable for the quality (conformance) of the work performed at his station or desk. It is also advisable to encourage personal commitments to quality to go along with management's commitment. It would be an ideal situation if each employee committed to 100 percent error-free work.

Following the kick-off or initiating event, the next step is the education and training of all involved personnel. Whenever quality improvement has not been an integral part of the company operations, its introduction will require a series of education and training programs for each element of organization as they are to become involved in the improvement program designed by the quality council. Usually, the key to this phase of the program will be the education of middle managers and first-line supervision. The workers normally respond with a positive attitude.

Training management personnel to respond differently to "quality" is not a task to be taken lightly. Basic attitudes about work and mind-sets that have developed over many years will be involved. There has to be a strong emphasis on the value and practicality of the program. And it must build a strong desire to identify and eliminate the causes of errors and nonstandard performance. It must also get into the mechanics of measurement, analysis, problem-solving techniques and corrective action. The best approaches involve the use of workshops where the managers and supervisors actually gain the experience of doing quality improvement.

The challenging aspects of management training for quality should provide a good justification for having a professional quality manager. As previously noted, a quality professional would be a valuable asset in developing the quality improvement program. He would be even more valuable in getting the program successfully implemented. In fact, it is not feasible to seriously consider a quality improvement program for a major enterprise without also considering the need for a key person to mold and guide the program through the intricate peculiarities of that specific enterprise.

In considering the role of the quality professional, the following responsibilities should be included:

1. Assure that quality is a built-in ingredient of all service processes and support systems by developing system quality procedures as part of the mechanism of company operations.
2. Become the internal "voice of the customer" through responsibility for the coordination and control of company responses to and resolution of all customer complaints and problems.
3. Assure that customer experience is summarized and used as important input to the establishment and improvement of service quality standards.
4. Provide education and training programs as needed for each element of organization to assure that all personnel from executives to operators understand the objectives of quality improvement and their individual responsibilities for the quality of the company's operations and product or service.
5. Establish and conduct a monitoring program to assure adequate quality controls of supplies and internal performance.
6. Keep abreast of quality control technology developments and assure maximum utilization of new ideas in the company quality improvement program.

While management of the quality improvement program is fundamentally the responsibility of the quality council, use of a quality professional will make the entire effort run more smoothly. The quality manager can act as a catalyst and help to crystalize the meaning of quality in each management area. He can also supply the additional leadership and day-to-day guidance necessary to keep the program on track. It is difficult to conceive of the implementation and management of a quality improvement program without a full time quality manager.

Communications. Quality cannot happen without communications. Communications is the life blood of a quality system. In order for communication to be effective, it must be reciprocal, that is, there must be a sender, a receiver and feedback to the sender to make certain the original communications and/or standards were fully understood. To make communications/quality even more effective, we must try to generate communications horizontally as well as vertically. Vertical communications usually means that there is a network in place to push communications downward, but in the spirit of quality improvement, there must also be a guarantee there will be an upward flow of communications. The horizontal communications needed are the communications that take place between individuals of the organization at a peer level, especially, when they are performing different functions. Increased horizontal communications will also help employees to understand their job better. Management consultants make many dollars in this area by recommending some of these activities under the title of job awareness, job enrichment, job enlargement, etc. Basically speaking, instant quality and productivity improvement could be achieved if each employee of the company knew the answers to the following questions:

What is my job? Where do I fit in general terms in the organizational structure? Who do I receive my work from? What do they do? Why do they do it? How do their errors impact me? After I finish my work task, who receives the work? What do they do? Why do they do it? How do my errors impact them?

Do not leave communications to chance. The quality professional can help identify communication gaps, but every manager and supervisor must participate in completing the communications cycle. Further, upward communications can be stimulated through short-term, highly visible programs, such as Buck-a-Day, Quality Improvement 15 and ZD/30 and then through a more permanent type of upward communications, such as quality circles and other participative management schemes.

Recognition. Recognition is the heart of a quality program. Recognition that is honest, deserving and sincere gives management the tool to truly integrate people as part of their quality program.

In an attempt to improve quality recognition at The Hartford, a questionnaire was sent to 10 percent of our employees countrywide for their input. The survey pointed out five areas in which our employees recommended improvement. They are as follows:

In order of priority —

1. To add monetary award.
2. To have unit awards as well as individual awards to stimulate the team approach.
3. More local control.
4. More publicity.

These four recommendations have been incorporated into Hartford's formal quality recognition program. Of more significance in this survey was the *second* most frequently mentioned item right behind monetary award — "All I want is to be recognized by my boss for doing the job right." Spontaneous recognition between first line managers and the employees is the most effective and inexpensive recognition available, and it should be done in front of as many peers as possible.

The cost of quality. Cost of quality techniques have not been universally applied to the service industry. This does not mean, however, that quality professionals have not been active in this area. In fact, the many efforts at application of quality costs to service and administrative activities could provide enough material for another chapter. There is a need for development of quality cost programs to support and justify the quality improvement programs for the service area. Here, however, we can merely share some of the conceptual thoughts that would probably underlay any serious development efforts.

One of the principal differences between a service business and a manufacturing business is the make-up of the customer base. Most manufacturing businesses operate from a large backlog of orders heavily dependent upon a few key customers, whereas service companies operate from a large customer base with many repeat customers. In the service business, more customers almost always mean more profit. For example, to a hotel more customers means a higher occupancy rate; to a telephone company, more calls per network; to a loan company, more contracts per employee. Because of this direct relationship, maintenance of the customer base is crucial to the survival and growth of a service business.

As amply supported in the main body of this case study, many service businesses now realize that "service quality" is a major factor in maintaining customer base and repeat customers. And they are coming to realize that a comprehensive quality improvement program is the best way to assure service quality achievement. To enhance the use of quality improvement programs, to lend credence to their business value and to provide cost justifications to the actions they propose, a supportive quality cost system should be developed.

The cost of quality was formulated many years ago in the manufacturing sector of American industry to assist the professional quality manager in getting the quality system understood and accepted. It was based on the premise that "there are no economics of quality" and it is intended to help sell the quality system on the basis of its impact on profit and loss. By identifying those costs directly affected, in a positive or negative way, by the quality program, the way was paved for better communications between the quality function and other functions, especially general management. The exact same concept can be applied to the service business.

In applying the concept of quality costs to the service industry, it is not nearly as important to the implementation of quality improvement as it is to the knowledge and support of top management. When the chief executive can visualize a program to improve customer satisfaction as being equally beneficial to profitability, it is easier to get his full support and participation. It can also lend support to the acquisition of a quality professional.

Service quality costs, in concept, are considered to exist in two distinct parts:

1. That portion of operating costs caused by "inadequate conformance to performance standards" — costs resulting from customer rejections or complaints, as well as costs incurred due to internal errors or substandard performance, requiring some re-doing of work.
2. Lost revenues due to unsatisfactory service quality — sales not achieved because of unhappy or lost customers.

Neither part of service quality costs, as defined above, can be measured directly. Therefore, the quality cost system will need to provide for an indirect means of identification. One way would be to estimate the values. Another would be to measure the difference between optimum, or perfect performance and actual performance in both areas. These differences or estimates could then be identified as "quality costs" and targeted for improvement efforts. The real value of this measurement effort is recognition of the opportunity for improvement that actually exists and the fact that actual results will be a matter of record.

In the internal operations area, quality cost measurements can be based on examining each element of operating budget as being made up of two parts. The first part of what the budget would be if the work could be performed perfectly. This is called the "ideal budget." The second part is the difference between the ideal budget and the actual budget (based on recent past history and current expectations). This difference can be called the "quality variance."

An alternative measure of quality costs in the operations area would be to estimate the costs related to quality. This estimate should include the known costs of errors and substandard work-failure costs, the cost of quality inspections or monitoring — appraisal costs, and the costs associated with the implementation of corrective action — prevention costs. The total of these three cost estimates are then designated as operations quality costs and targeted for improvement. This method is a close parallel to the concept of quality costs for manufacturing companies.

In the area of lost revenues, there is no precedent in the manufacturing sector quality cost system. Maximum achievable revenues can be estimated on the basis of market research, competitor evaluations, actual customer experiences, capacity limitations and other factors. Revenues are also affected by pricing and there is a direct relationship between quality leadership and price leadership which should also be taken into consideration.

For service quality costs, it is suggested that maximum possible revenues be carefully estimated and identified as "ideal revenues." Then a quality variance for revenues can be identified as the difference between budgeted (expected) revenues and ideal revenues. This variation, like the operations cost quality variance, can then be viewed as another opportunity for business improvement.

Establishing quality costs as a measure of the difference between an ideal budget and actual costs or revenues does not imply that the business can be run perfectly. It merely emphasizes the potential for improvement that actually exists. It can then be used to focus attention on and justify the quality improvement program.

Summary. The purpose of this study was to provide an overview from the inception of the idea through implementation of a practical quality improvement program that will guarantee improved productivity through quality management. It ran from the selling to management leading to the commitment and policy to the establishment of the quality organization, the setting of quality standards, the measuring of performance, the building of the quality program itself, the implementation and managing of the quality program through communications and recognition. It also introduced some concepts associated with service/administrative quality costs.

Some key points to remember are:

—Make quality improvement an ongoing part of your operations. An effective program is not something you turn on and off. Make it part of the woodwork.

—Customize it to your unique organizational needs. Service businesses differ greatly in structure and organization. There is no standard approach.

—Get all personnel to participate. The people deep inside a system are the ones that really know the weaknesses and soft spots. They are an untapped reservoir of knowledge and value. It is urgent they be involved.

—Keep people fully involved and informed. Communications is the life-blood of the quality improvement program.

—Recognize and applaud outstanding performance. Recognition truly is the heart of a quality program and gives management the opportunity of building quality for positive reasons.

—Concepts regarding quality costs have just been introduced in the service and administrative areas. Much more work is required in this area.

Once again, it warrants repeating that this study represents only the skeleton of a quality improvement program and it must be translated into the terminology which is most applicable to your operation and to the understanding of your management and employees. You must develop a unique quality improvement program for your company. You can achieve improved productivity through quality management.

Case 4. Quality Control Manual and Quality Audit Guide for a Large Organization

The United States Fidelity and Guaranty Company has 61 branch offices and 135 service offices located throughout the United States; its total employment is 8,000. The scope of its activities is illustrated by the fact that it has ten lines of business:

1. Casualty: commercial automobile
2. Casualty: general liability
3. Casualty: private passenger automobile
4. Casualty: worker's compensation
5. Other casualty
6. Commercial fire and property
7. Personal lines: fire, marine, and multi-line
8. Other fire, marine and multi-line
9. Fidelity coverage
10. Surety coverage

This firm now has a company-wide quality control program the scope of which is indicated by Document 1 which is the table of contents for the quality control manual and Document 2 which is the table of contents for the quality audit guide. These procedural manuals are characterized by a clarity and thoroughness far too often lacking in these documents; as proposed manuals and guides, they are an excellent start.

Document 1 explains the goals of the quality control program, what it is, and the implications for employees, rater, coder, underwriters, superintendents-supervisors, home office department heads, and branch office managers. The responsibilities are described in some detail. Included is a section on the purpose of inspection plans, why sampling is used, and three different types of sample inspection plans to give managers some flexibility in implementing the program. The several forms are described in detail showing the purpose of each and how it is to be filled out. Finally, evaluation of the sample data is described together with an emphasis on how and when corrective actions are to be taken.

Document 1. Table of Contents of the Quality Control Manual

Document 2. Proposed Quality Audit Guide

Document 2 explains why a quality audit is necessary and how each of the three kinds of audits are to be conducted: the program audit, the statistical data audit, and system audit.

In both the manual and the guide a glossary of terms is given at the beginning, an excellent idea in any procedural manual but especially necessary in manuals of this type which are being used to introduce a company-wide program to everyone in the company.

Quality is people. People produce quality. Without quality people running the computers, writing programs, and preparing manuals, we cannot achieve our goal of accurate and timely data for internal and external use.

Whether the function of improving the quality of statistical data is called "Data Quality", "Quality Control", "Quality Assurance", or some other name, basically the same question is being addressed — the development of sound, reliable, timely statistical data for use by the company. The industry embarked on a data quality improvement program in 1977. Since that time, much progress has been made in order to improve all aspects of statistical data so that we will not only be comfortable but confident that we are basing our management decisions, as well as rate-making processes, on reliable and meaningful data.

State regulators are becoming more and more aware of the need for accurate and reliable data. Examples are California Regulation No. 78.5 (described later in this chapter) that requires a quality control program which is subject to State audit, as well as the statistical data generated by the company, a North Carolina rate increase which was denied because the data submitted was not audited by external auditors, and indications from insurance commissioners that they are considering establishing statistical agencies of their own.

Quality control is the function where a control can be exerted over the processing or generation of data. The control must lie with the people involved in the function. Quality assurance is that management function which values the accuracy of statistical data as well as audits the quality control function.

The branch office quality control program introduces the quality control concept at a branch office-operational level. Emphasis is placed on the basic inspection process of quality control as well as the utilization of quality control techniques as they relate to personal performance evaluation, department evaluation, trend analysis, and Pareto analysis. Documentation by quality forms provides a standard to which a program may be audited, and provides data for the internal use of the department.

The program introduces the employee to the "attitude" we symbolically refer to as "Zero Errors" which is necessary to bring about a drastic reduction in the error percent for any employee. "Zero Errors" is an attitude and not an actual performance goal. If we are to produce data, we must get as close to Zero Errors as possible.

General comments on the cases. These cases reflect several commendable features:

1. Quality is an individual responsibility.
2. Quality is advantageous to the employee as well as to the company.
3. Documentation is thorough including written explanations, procedures, and instructions; manuals, forms, and reports are given.
4. The stress is on taking corrective action to eliminate the causes of error.
5. Quality responsibilities at various levels are described.
6. The need for quality in statistical data is emphasized.

These cases raise questions about several important aspects of a quality control program. Some of these may be explained by the fact that a quality control program is being introduced for the first time by the home office throughout a nationwide organization consisting of hundreds of offices and thousands of employees. This procedure is almost bound to mean imposing a simple common system which may be far from the best for any single office. An alternative procedure that is widely recommended and used is to test new methods and techniques in a pilot study in a limited area such as a few offices, before putting it into effect nationally. Some of these aspects are:

1. Use, design, and implementation of sampling.
2. Role and duties of supervisors are not clear; others seem to be doing tasks that properly belong to the supervisor.
3. There appears to be a lot of unnecessary paper work.
4. All applications seem to be limited to the control of errors in the processing of insurance claims. Control over other aspects such as rating, underwriting, and appraising for the purpose of determining premiums on the one hand and losses on the other, are not discussed.
5. Giving office managers "flexibility" may simply mean that in many cases they will choose to do nothing because they do not understand quality control techniques and how to implement them.
6. A question arises as to where and how a quality program is started. One position is that it should start at the top. Obviously top level approval is necessary if a company-wide program is to be planned and implemented. But if the proper technical and managerial foundations have not been laid throughout the company, this may easily lead to a superficial and ineffective "quality" program. This is because not only do top level officials not understand the full implications of what we mean by quality control, but neither do the lower level managers, professionals, and supervisors.

Many quality control programs in the past started where the technical and managerial know-how was, which was almost invariably middle level managers, specialists, and professionals. Furthermore, the initial applications were started in a small restricted but important area so that various techniques and procedures, such as sampling, could be designed and implemented so as to be most effective. With this sound initial application, which was modified until it was successful, it was easy to extend the application with adaptations to other parts of the organization.

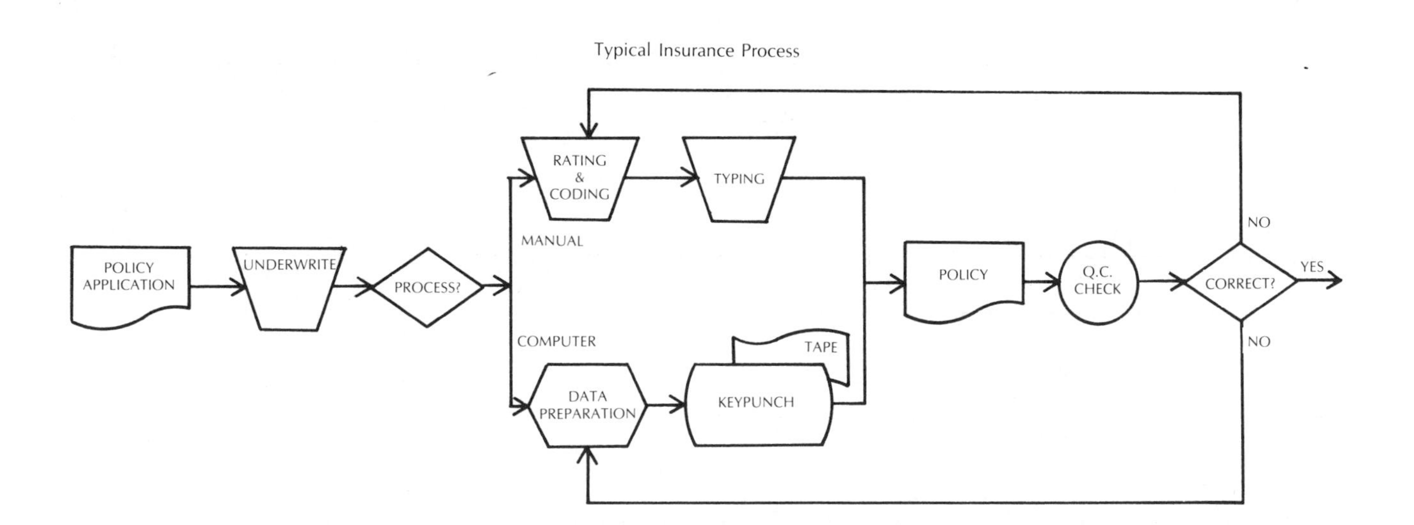

THE

USF&G COMPANIES

QUALITY ASSURANCE DEPARTMENT

CORRECTIVE ACTION REQUEST

B.O./H.O. DEPT.	AUDIT DATE
TYPE OF AUDIT ☐ PROGRAM AUDIT ☐ STATISTICAL DATA AUDIT	

SECTION I — DISCREPANCIES (TO BE COMPLETED BY H.O. Q.A. DEPT.)

This section contains a description of the area(s) noted in the above audit which require steps be taken to bring them into conformance with standards outlined in the Quality Control Manual.

DATE REPLY REQUESTED BY	SIGNATURE (REQUESTOR - RETAIN FOURTH COPY)	DATE REQUEST MADE

SECTION II — CORRECTIVE ACTION (TO BE COMPLETED BY FIELD)

This section is to be completed when corrective action has been implemented. After completing, return the original (white) and the canary copy to the Home Office Quality Assurance Department; retain the pink copy until return receipt of the canary copy.

CAUSE OF DISCREPANCIES

ACTION TAKEN TO CORRECT

ACTION TO PREVENT RECURRENCE

SIGNATURE	DATE

SECTION III — APPROVAL (TO BE COMPLETED BY H.O. Q.A. DEPT.)

This section, when completed and returned by the Home Office, is your permanent documentation.

☐ APPROVED BY H.O. Q.A. DEPT.	☐ NOT APPROVED BY H.O. Q.A. DEPT.; SEE ADDITIONAL QA 13, ATTACHED.		DATE

WHITE—H.O. Documentation Copy CANARY—Field Office Documentation Copy PINK—Field Office Control Copy GOLDENROD—H.O. Control Copy

QA 13 (HO) (7-79)

Detailed comments on the cases. In these cases quality control is applied to the identification, correction, and prevention of errors in processing claims. One case also emphasizes customer service measurements including time service studies, customer surveys, and customer complaints related to claims.

Application of quality control to other aspects of the insurance business such as underwriting, rating, and appraising are not discussed although California Bulletin No. 78-5, described later, calls specific attention to quality control applied to underwriting and rating. One company, United States Fidelity and Guaranty, does include raters and underwriters, but does not describe quality control applied to their work.

In most cases, quality control is a part of, or closely related to, internal audit. In one case, quality control is actually a part of internal audit. This is a drastic shift from the direct type of quality control originated and applied by Shewhart to a delayed indirect type of control represented by quality audit:

1. Quality audit is **not** quality control.
2. Quality audit assumes controls are already in place and working. The audit is for the purpose of verifying the existence of these controls and their effectiveness.
3. Quality audit is a post mortem.
4. Quality audit with its obvious delays is not compatible with the goal of "Zero Defects" or "Zero Errors".

Although time studies are used by one firm, the time element is an ignored dimension of quality. Two ZD's are necessary ideals: Zero Defects, a concept that stresses the need for *accuracy;* Zero Delays, a concept that stresses the need for *urgency.* Without a sense of urgency, errors accumulate due to delays in unnecessary paper work, error analysis, deciding on corrective actions, using a roundabout feedback route, and planning and organizing training classes. We show how to gain urgency by using the direct quality control cycle described below.

5. Sample plans should not only be used to discover errors, but to estimate the error rates. Sampling proportional to production in various categories of work is not the most efficient type of sampling; the effectiveness of a sample is determined by its absolute value, not its relative value.

The direct quality control cycle. The direct quality control cycle places a different emphasis on both operations and control activities by stressing urgent feedback and centering responsibility for eliminating sources of error on the supervisor. The steps in this cycle are given below:

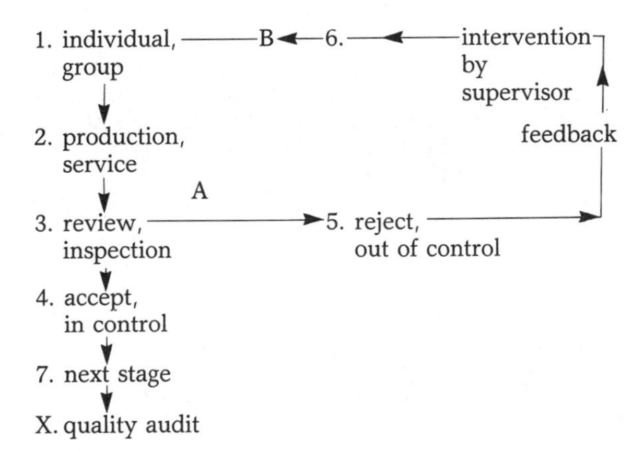

The basic requirements are:

1. Minimize the time that elapses between points A and B.
2. Immediate intervention by the supervisor to eliminate causes of error or subquality service by concentrating on individuals causing the errors.
3. Continuous daily control over the performance of individuals.
4. Maintaining detailed records of errors by type and by individual for accepted and rejected lots is the job of the reviewers at Step 3; they are used by the supervisor at Step 6 to eliminate error sources.
5. Use an effective sample plan together with charts or lot acceptance-rejection that indicates trouble, exerts control, or provides a means of estimating error rate in the outgoing acceptable quality work.

The quality control cycle by a direct head-on attack on errors has several features which distinguish it from an indirect roundabout approach which can be characteristic of quality assurance and quality audit:

1. There is a continuous control over the work of an individual.
2. The reviewers or inspectors, as well as the supervisor, must know what constitutes an error or subquality service.
3. The reviewers or inspectors accept or reject batches of work on the basis of a thorough examination of a sample of documents, and keep individual records on the number of errors found in every accepted and rejected lot or batch.
4. Rejected batches are fed back as quickly as possible to the individual through the supervisor for immediate correction and 100 percent re-work, including advice and instruction from the supervisor as to how to eliminate the errors.
5. The key person is the supervisor who knows:
    1. What is correct performance and what is not.
    2. How to explain to individuals various kinds of errors and how to avoid them.
    3. How to tutor individuals as needed.
    4. How to follow up on individuals making errors to see that they are actually eliminating them.
    5. How to use effectively the error and production data recorded by the reviewers or inspectors.

Sample plans and other technical aspects are furnished and explained by qualified professionals or specialists who monitor plans to see that they are implemented correctly. These technical decisions are not made by executives, managers, or supervisors. Although a technical consultant may initiate the professional aspects, in-house staff either developed or hired is usually more advantageous if the full benefit of the techniques of quality control is to be obtained.

Advantages of the quality control cycle. The quality control cycle described above has several advantages:

1. The supervisor concentrates on eliminating sources of error. No formal training courses are needed once the group is able to do acceptable quality work. A screening test determines when to shift a person from training to production.
2. Errors are found by the reviewers or inspectors and corrected immediately by the individual.
3. Records are kept at a minimum by the reviewers.
4. The job of the supervisor is not to correct the errors of others.

5. The supervisor uses the error records and instructs the individual in how to avoid specific errors.

6. The supervisor is faced with a teaching-learning situation in which individuals learn in different ways. Supervisors should be capable of handling this job since it is part of the supervisor's duties.

7. No formal training courses are needed; a formal error analysis is not needed either since the supervisor deals with errors and error sources directly and continuously and resolves them on the spot.

8. The goal is to turn out error-free products and services. It is not to produce records, analyses, and reports. Neither is it to turn out a mass of computer readouts. Excessive paper work and computer runs may easily defeat the purpose of obtaining urgent feedback and quick elimination of error sources.

If required, a quality audit can be made by taking an adequate random sample of work off the end of the line at the end of each month to estimate the average outgoing error rate to see if it meets acceptable quality standards. This information plus the record of the reviewers will furnish officials with accurate measures of the quality of performance.

It is also very easy to introduce quality costs into this plan. Adequately designed random time samples based on the minute model can be used to estimate costs of various quality and quality-related activities. It is the only known way of obtaining unbiased estimates of the various components of joint costs. This type of random time sampling is necessary to supplement the usual cost accounting records and methods.

What quality control is not.

1. Quality audit is not quality control.
2. Quality assurance is not quality control.
3. Correcting errors in samples is not quality control.
4. Screening errors and defects from a batch is not quality control.
5. Inspection or review is not quality control.
6. Spot checking is not quality control.
7. Inspecting something a dozen times is not quality control.
8. What is quality to the producer may not be quality to the buyer.
9. Fitness for use may be quality to the producer, but it may not be quality to the buyer.
10. Meeting specifications may be quality to the producer, but it may not be quality to the buyer.

Insurance Data from California and Colorado

California Bulletin No. 78-5.[6] This bulletin deals with underwriting and rating examination procedures. What concerns us here are the sections and parts of the bulletin that deal with quality control and the need to build quality into basic statistics.

The notice that is sent to the insurer or the insurance company will outline the types of information that the company is to make available to the rate analysts of the California Department of Insurance. These include:

1. Statistical data supporting pricing decisions.
2. Premium level review procedures.
3. Manuals of rates and rules.
4. Copies of policy forms and endorsements.
5. A description of the quality control program in use.

Items 1, 2, and 5 deal directly with quality control aspects or have quality control implications. Only the first and the fifth are discussed in the Bulletin.

"The Department's past enforcement of the McBride-Grunsky Insurance Regulatory Law of 1947 . . . has consisted of the application of tests to determine whether the rates adopted meet the statutory requirement that they not be excessive, inadequate or unfairly discriminatory."

"This was usually followed by an extensive policy file review to monitor the company's implementation of its plans and programs to determine how accurately they were applied to individual cases. Future examination will focus more intensely on information . . . for the purpose of producing a better understanding of the company's rate-making and rate-usage procedures."

Collection and use of quality statistical data. ". . . [G]reater emphasis will be placed on the company's uses of statistical data and the rate-making methodologies employed. . . . [A]lthough the Insurance Code recognizes the use of judgment factors in the rate-making process, the rate analyst will explore whether unfounded assumptions have been substituted for conclusions reached after a careful assessment of known or existing facts. . . . [W]hile judgment is a necessary element of the rate-making process when relevant statistical data is not available, or where the data available is so limited as to lack credibility, it is not an acceptable substitute for available data which has not been collected, or which has been collected but neither assembled nor analyzed for rate-making purposes. Failure to observe these elementary rate-making rules might result in adverse findings against the insurer."

"Premium and loss statistical data should be collected in a manner which will permit the insurer to continually monitor the validity of its rating classifications. . . . [I]t has repeatedly been observed in the past that the collection of private passenger automobile territorial experience by many insurers lacks the flexibility necessary to identify shifting loss trends within the rather large geographical area contained within a rating territory . . . confirming the view that this data should be collected on a subdivisional basis such as zip codes, in order to better assure equitable treatment in the rate-making process."

"Special emphasis will be placed on tests for unfair rate discrimination in areas where certain practices have been defined as unfair under regulations . . . such as discriminatory underwriting or the offering of different policy terms or conditions based solely upon a person's sex, marital status, or sexual orientation."

"Any adjustments to territorial or classification definitions should be initiated only when the rate-maker has gathered the evidence necessary to justify that change. The failure to implement an adjustment on a timely basis once the supporting data has been gathered could result in an excessive rate situation and the imposition of disciplinary measures by this Department."

"Our rate analysts will question company rate-making methodologies which perpetuate and aggravate such practices as the use of fixed percentage expense loadings. . . . Our staff will closely monitor these activities to ensure that practices which result in unfair discrimination will be eliminated. . . ."

"Data handling systems and procedures which impact the accuracy, validity, and timeliness of data used in rate-making will receive our increasing attention. Recent events indicate that the industry is making a major commitment to improve the quality of insurance data. We applaud this action. . . Reform in this area should help reduce the current credibility gap between (consumers and insurers)."

Quality control procedures. "Our second major area of emphasis will be to assess the insurer's quality control activities for effectiveness in assuring that its rating plans are being properly implemented and that an acceptable level of quality of service to policyholders is being maintained. This will also enable our rate analysts to uncover abusive or unfair practices while reducing the length of time otherwise necessary to perform an extensive policy file review."

" . . . [T]he term 'quality control' encompasses error elimination measures installed over the rating, coding, policywriting, and underwriting functions, including validity tests of the resulting statistical data as well as those steps taken to verify the rating of the policy and/or the condition of the object being insured. . . [V]erification procedures in private automobile insurance might include the use of new and renewal classification checks, verification of the garaging location, and physical inspection of the vehicle."

"The rate of insurer processing errors observed during past examinations indicates that the increasing complexity of the classification rating systems adopted have created a situation in which the plans cannot be administered with a high degree of accuracy. . . . If the complexity of a rating system is the cause of these errors, and the insurer proves unable to quickly neutralize their adverse effect, then it becomes necessary to simplify the rating system itself. Failure . . . to resolve this problem will leave the Department no choice but to conclude that the system is unfairly discriminatory."

". . . [I]n the future we will request copies of the insurer's quality control/audit procedures and will focus our attention on whether the programs adopted appear appropriate and whether they are diligently enforced and accurately used. Where the rate analysts are convinced that the insurer's practices are adequate to safeguard the consumer's interest and provide an acceptable quality of service, the extensive policy file review that would otherwise be made may be waived . . . [otherwise] an extensive policy file review will be initiated."

Comments on California Bulletin No. 78-5. This Bulletin reveals a number of significant points with regard to the insurance industry, certainly so far as California is concerned, and perhaps the entire country:

1. Quality of insurance data needs to be improved.
2. Discrimination in rates exists because:
   - judgment is substituted for data.
   - data may be available but they are inadequate.
   - adjustments for trends are inadequate.
   - certain rate methodologies are unsound.
   - data lack the necessary quality.
   - there are too many processing errors.

Evidence that these statements have a factual basis is seen by the statement that the industry is committed to improving the quality of data, that a number of companies have introduced revised expense loading methodologies in private automobile rate adjustments, and that a number of companies are also making private passenger automobile rate adjustments which reflect losses on a zip code basis, while others are conducting extensive studies of their existing territorial definitions.

Some Insurance Premium Data from Colorado[7]

Table 6 shows the insurance premiums charged for the same protection by 10 insurance companies doing business in Colorado in a large city (Denver) and a small city (Steamboat Springs) for the year 1980.

Part A of Table 6 gives the premiums charged for the following home protection: Special form HO 3 for a 12 year-old single family dwelling with an approved roof. The property is covered for $50,000 with basic liability coverage.

Part B of Table 6 gives the premiums charged for the following private automobile situation: 31 year-old married male, no accidents or convictions during the past 3 years, car is used solely for pleasure, 1978 Malibu cost $4,795, no youthful drivers, coverage includes $50,000/$100,000 bodily injury, $25,000 property damage, basic no-fault coverage, uninsured motorist protection $15,000/$50,000, comprehensive coverage with no deductible and collision with $100 deductible. (The highest published figure was $421 in Pueblo.)

Some obvious questions arise immediately. Why under homeowner's insurance in Denver do companies D and I charge over twice as much as company E — $314 vs $146? Similarly, in Steamboat Springs, why do companies D and I charge over twice as much as company A — $210 vs $100 — and nearly twice as much as company E?

Under automobile owner's insurance why does company C charge such low rates compared with all the other companies; in most cases less than half as much, and in Steamboat Springs only one third as much as company H? Furthermore, why should companies E, H, and J charge more for automobile insurance in Steamboat Springs than in Denver?

Table 6

Company	Denver	Steamboat Springs
A. Homeowner's insurance*		
A	$165	$100 low
B	168	133
C	203	164
D	314 high	210 high
E	146 low	106
F	295	198
G	149	102
H	180	115
I	314 high	210 high
J	189	126
B. Automobile insurance*		
A	275	260
B	296	282
C	144 low	131 low
D	317	294
E	312	319
F	283	278
G	348	318
H	388 high	403 high
I	237	231
J	299	304

*Details of the insurance policy are given in the text.

Needless to say the wide variations in the premiums charged in both cities for homeowner's insurance and private automobile insurance raise a whole host of questions that need to be answered, not with opinions, but with data and evidence. How these wide variations are justified remains an unanswered question.

Distribution of premiums for homes and automobiles.[8] Figure 3 shows the frequency distribution of premiums charged by 171 insurance companies for the same home policy in Denver in 1980. The range of the 171 premiums is $278 (the highest minus the lowest), the mean or average is $244, the standard deviation is $58, and the coefficient of variation is 24 percent.

Much more meaningful is the fact that the distribution has 3 peaks (modes) — at $170, at $230, and at $310 — showing the possible existence of three policy writing groups — a low group, a middle group, and a high group. Actually there is a fourth group of four companies that charged premiums above $340. How can this variation exist when writing the same policy?

FIGURE 3

Distribution of Homeowner's Premiums Charged by 171 Insurance Companies for the Same Policy—Denver 1980

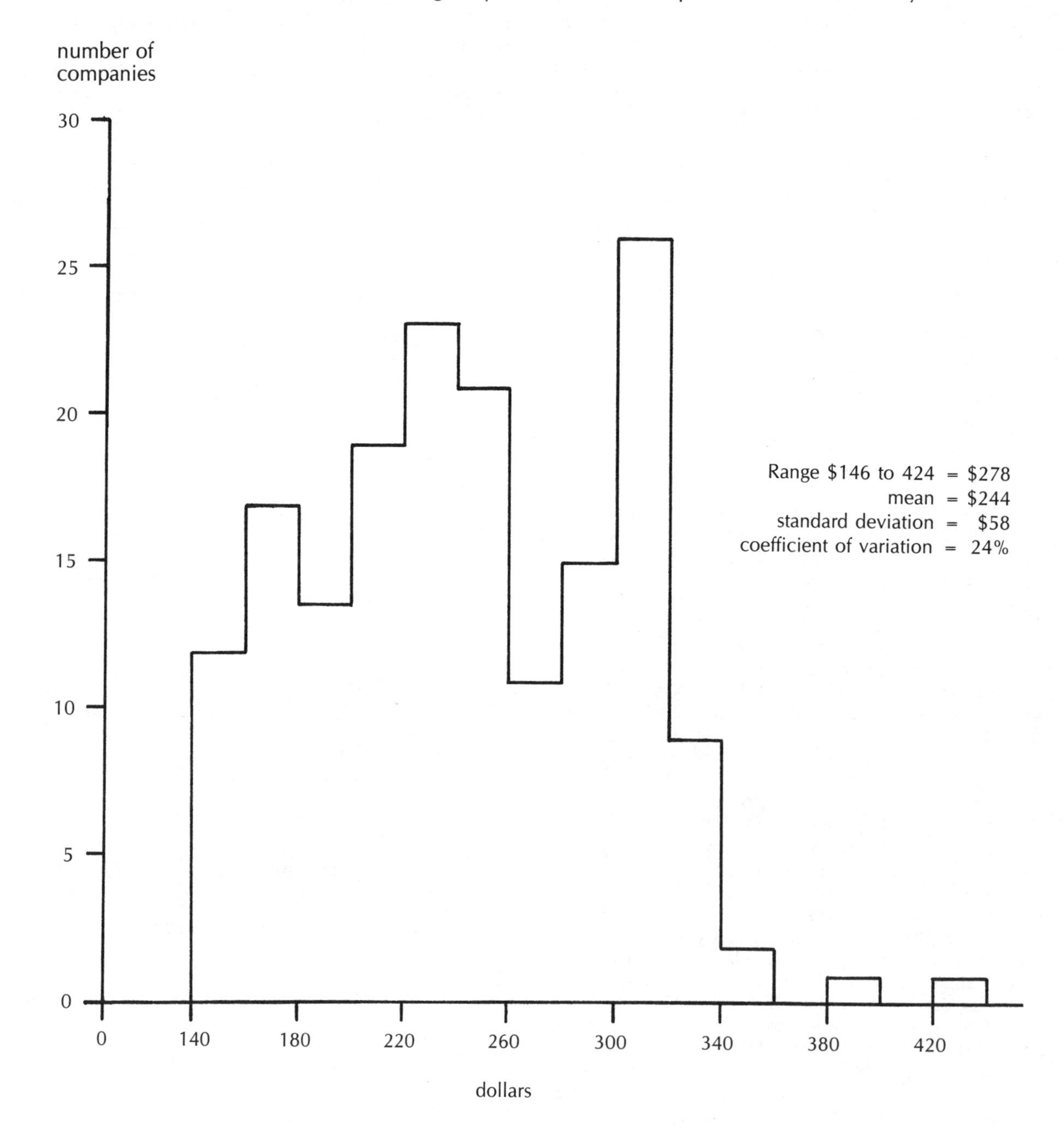

Figure 4 shows the frequency distribution of premiums charged by 180 insurance companies for the same automobile insurance in Denver in 1980. This distribution is strikingly different from that in Figure 3. If the eight companies above $500 are excluded, the distribution is single peaked with a mean of $302 and all values fall between 144 and 472; the mean falls very closely at the midpoint of the range.

Assuming a normal curve with a mean of 302, the three sigma limits include all companies but the eight outliers; one value of 492 is just outside the upper limit:

$$302 \pm 3\ (61) = 302 \pm 183 \text{ or } 119 \text{ and } 485.$$

With the exception of these outliers, this distribution seems to be evidence that a common standard of practice exists among these companies with regard to the writing of this policy. The distribution in Figure 4, with the extreme values excluded, would seem to indicate that insurance premium calculations are under control and varying according to expectation. On the other hand, this certainly cannot be said of the distribution shown in Figure 3.

FIGURE 4

Distribution of Automobile Premiums Charged by 180 Insurance Companies for the Same Policy—Denver 1980

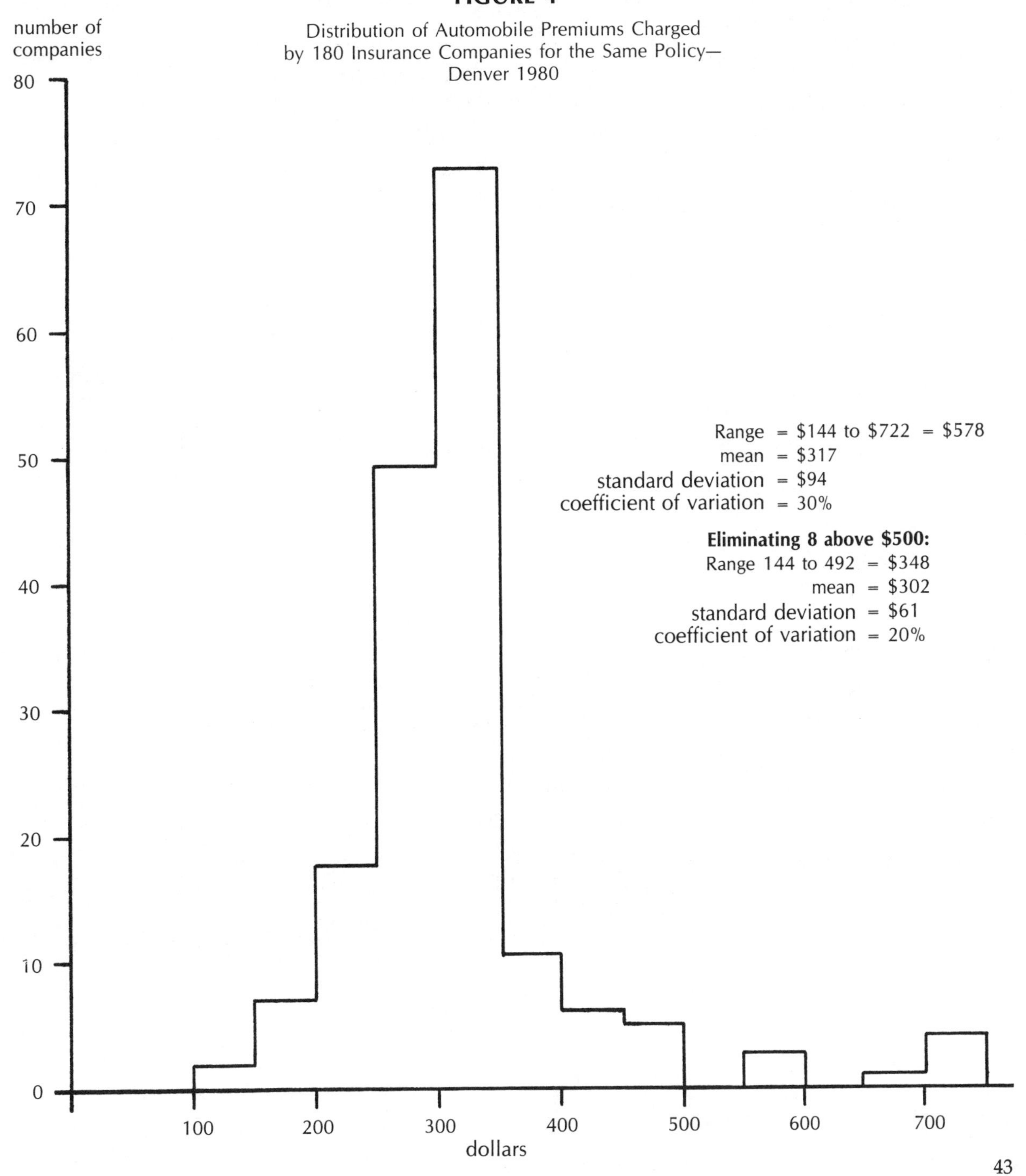

Notes

[1]This material is excerpted from a longer statement prepared by R. L. Rohrbaugh, United States Fidelity and Guaranty Company, Baltimore, Maryland, *AAD Newsletter,* Fall 1979, issued by the Administrative Applications Division of the American Society for Quality Control. There was some very minor editing.

[2]Material prepared by Peter S. Spynda, Blue Cross of Western Pennsylvania, including that which appeared in the *AAD Newsletter,* Fall 1979, issued by the Administrative Applications Division of the American Society for Quality Control. Used by permission.

[3]With only a few very minor editorial changes, this is a reproduction of a paper by Frank Scanlon and John T. Hagan, "Improved Productivity through Quality Management," *Annual Quality Congress Transactions,* 1982, American Society for Quality Control, pp. 491-505. Also in *Quality Progress,* (May/June 1983). Frank Scanlon is Secretary-Director Education/Quality, The Hartford Insurance Group; John T. Hagan is with ITT World Headquarters.

[4]The latest figure (1979) on an industry basis is 62 percent, labor force, 102.9 million, service industries including government, 63.4 million, manufacturing, 21.1 million. Source: *The World Almanac,* 1982, pp. 114-120.

[5]Manual and guide received from Charles Zimmerman III, Assistant Vice President — Quality Control Department, United States Fidelity and Guaranty, Baltimore, Maryland, 1980. He is the author of "Quality is People," and furnished other material from United States Fidelity and Guaranty.

[6]Bulletin No. 78-5, (February 17, 1978), *Underwriting and Rating Examination Procedures,* Department of Insurance, State of California, 100 Van Ness Ave., San Francisco, California 94102. Most of what follows is a direct quotation from the document.

[7]Data from two bulletins, *A Colorado Buyer's Guide to Homeowner's Insurance* and *A Colorado Buyer's Guide to Automobile Insurance,* both issued by the Division of Insurance, 106 State Office Building, 201 East Colfax Avenue, Denver, Colorado 80203.

[8]The source of these data is the Colorado Division of Insurance, Denver.

4

Federal Government

Quality control is nothing new to the Federal Government. Indeed, it was the Federal Government that really gave impetus to the application of quality control techniques on a large scale both in the civilian and military fields. Four events are evidence of these early developments:

1. Dr. W. Edwards Deming arranged to have Dr. Walter A. Shewhart of the Bell Telephone Laboratories give four lectures on statistical quality control in Washington during March 1938. The book, "Statistical Methods from the Viewpoint of Quality Control", was based on these lectures.[1] Hundreds of officials, middle level managers, and other employees attended these lectures which were open to the public.

2. The first large scale application of quality control techniques in the Federal Government was made by Dr. W. Edwards Deming in connection with the key punching of documents collected for the population census of 1940.

3. During World War II in the War Production Board, the Office of Production Research and Development (OPRD), headed by Vannevar Bush, sponsored a series of courses on quality control principles and techniques which were given across the country by professionals including Deming, Working, Knowles, and Olds. These courses exposed thousands of inspectors and managers to quality control. This group formed the basis for the American Society for Quality Control which was started in 1946.

4. The World War II experience showed that quality control techniques could not only be applied to control the quality of war material, but could be used to control the quality of goods in the civilian sector. There was also the need to continue the sampling inspection plans for military and defense contractors during peacetime. A series of separate sampling plans were finally incorporated into what was to become Military Standard 105; up to the present it has gone through four editions — A, B, C, and D.

In the examples described, quality is built into data and decisions by using improved data collection methods such as probability sampling, sample surveys, sample tests, regression analysis, laboratory tests, and sample audits. These resulted not only in new and improved estimates but in a better analysis of the data available.

Sometimes improvement in quality performance resulted because the information was used to reduce errors at the source, or to correct an unsafe condition or situation or to show that allotments were in balance, or to show that corrective actions were needed. Whether or not the necessary actions were taken in several cases is unknown, but at least the data and the analysis indicated that trouble existed and that steps should be taken to rectify it.

In the Federal Government, in most instances, quality control was initiated and promoted by professionals and middle management, mostly statisticians or those with training in statistics and mathematics, not by top level executives. This was expected due to the mathematical nature of the techniques. Managers who avoided mathematics in school and college suddenly found that it was needed in management. It was a real shock and required years of patient work to convince them that these new statistical techniques really worked. In many instances quality control was initiated by way of applied probability sampling which many managers could understand easier than they could understand and accept quality control since the latter dealt with errors which they did not like to talk about. Probability sampling on the other hand, saved paper work, saved tabulations, and enabled reports to be issued more promptly. These advantages they could understand.

Audit control program for tax returns. The first step to improved quality of performance is often to collect quantitative data to define the size, extent, and location of the problem. Management needs to be sure it has the right problem before it takes corrective action or proposes a solution.

The Internal Revenue Service did just that: it applied sampling to individual income tax returns for the purpose of defining the areas of non-compliance — its extent, location, and magnitude — so that audit resources could be allocated to correct the most tax error per man-hour of investigation. Also, the sample would feed back information as a basis for actions that would reduce, if not prevent, errors from occurring in the first place. There would be error prevention as well as error correction.

The first audit sample of individual income tax returns was a subsample of about 160,000 returns selected from the regular sample used for statistical reporting purposes; later audit samples would be smaller with more emphasis on business returns. To increase its effectiveness the sample was stratified by size of adjusted gross income, and by business and non-business types of returns.

The sample yielded nationwide estimates of the average tax error per tax return, and average tax error corrected per man hour for a wide variety of taxpayer characteristics hitherto unknown; such as type of tax return, type of business, kind of tax error, size of refund, standard deduction versus itemized deductions, and exemptions.

The major findings were released by the Internal Revenue Service and are as follows. These are nationwide estimates derived from the sample:[2]

1. About 25 percent of individual income tax returns had a tax error exceeding $2.
2. The total amount of tax error was $1.5 billion.
3. Of returns with tax error: 90 percent favored the taxpayer; 10 percent favored the government.
4. The $1.5 billion of tax error were made by taxpayers reporting $7.4 billion, so corrected tax was $8.9 billion, a 17 percent underpayment.
5. Average tax error increases in amount as income increases.

6. Frequency of tax error among business returns is about twice that of non-business returns.
7. The major source of tax error is unreported and underreported income. Personal deductions and exemptions are other major sources of error. Arithmetic errors made on the tax return are a minor source of error, despite the publicity and widespread feeling that Form 1040 involves very difficult arithmetic.
8. The frequency distribution of tax error on individual income tax returns, based on these data, is a single highly peaked, highly skewed distribution. The average of the entire distribution was -$26; the average error of those paying too much tax was $71, while the average of those paying too little was -$114.
9. 38.4 million tax returns had the correct tax as defined; 1.2 million paid too much, while 12.5 million paid too little.

The actions taken as a result of the feedback from this audit sample were:

1. Audit planning was revised in terms of the magnitude, extent, and location of non-compliance and of tax error.
2. A change was made in the allocation of audit resources.
3. A more detailed explanation of what constituted an exemption was included in the instructions.
4. A new schedule was added to the tax form.

This nationwide audit sample was of historic importance because for the first time the tax administrators had sound and accurate information defining the individual income tax problem. As a result, the quality of performance of both tax administration personnel and taxpayers improved.

Audit sample coverage.[3] A question arose in connection with planning the above nationwide audit sample as to whether 160,000 was an adequate sample to select from 52 million returns — an overall sampling rate of 1/325. Management was afraid many areas would not be represented. It was shown, however, by simple probability that this would not be the case: the probability of obtaining at least one tax return in the sample in a county of only 3,000 persons (1,000 tax returns) was slightly more than 0.95 or 95 percent.

A check of a few counties with about 20,000 people showed no less than three or four returns fell in the sample. Later, a map of the United States showing the county distribution of the sample was prepared which verified what the calculations predicted: the sample was distributed very well over the entire country. Texas, for example, was very well covered, as were other western states. Only 144 counties out of 3,100 had less than 3,000 inhabitants, or about 1,000 Federal individual income taxpayers or less, in the 1970 census according to the Bureau of the Census. No doubt many of these counties were represented but the figures are not available.

Control estimates — the cocoa bean survey.[4] A sample survey of imported cocoa beans was made by the Food and Drug Administration between January 1, 1959 and March 31, 1960 to measure the degree of infestation in cocoa beans, by country of origin. The purpose of the sample was to test the tolerance of 10 percent infestation (set in 1932 by the U.S. Department of Agriculture) and a tolerance of 5 percent moldy (set in 1933). During the period from 1950 to 1960, very few lots sampled by FDA contained over five percent moldy. Thus, there was a real need to review these quality standards or tolerances to determine what the present quality was.

The population of this study was all bags of cocoa beans imported into the United States. Sampling was limited to Boston, New York, Philadelphia, and San Francisco imports, since they accounted for about 98 percent of all cocoa beans imported. The frame sampled was about 67 percent of 1959 imports; judgment sampling of ships was used.

The main features of the sample plan are as follows:

The frame:

1. The time frame is 15 months.
2. The geographical frame was 22 countries with shipments to United States of more than 200,000 pounds in 1957.
3. The ship frame was not indicated.
4. The estimated bag population for 1959 was 3,452,000.
5. The 1959 bag frame was 2,325,819, averaging 140 pounds per bag.
6. The 1959 frame included 67 percent of the imported bags.

The sample:

1. The sample consisted of 627 ships for 1959 and 683 ships in all. How they were selected is not described other than they were selected on the basis of judgment.
2. The sample contained 18,009 bags of cocoa beans. How they were selected is not described.
3. The sample of bags was reduced to 6,003 subdivisions, by combining or compositing the sample from three bags. The reason for this is not explained.
4. The 18,009 bags represented 2,169 marks. The term "mark" is not explained, but apparently it identifies a shipper, grower, or exporter.

Sampling of beans from ships:

1. Sample quotas were established for each country exporting over 200,000 pounds of cocoa beans to the United States in 1957, roughly proportional to the volume of exports of each country.
2. For each selected ship, inspectors determined from the manifest the total number of bags of beans and the number of marks from each country.
3. A sub was approximately one pound of beans drawn at random from three bags of the same mark. How this selection was made is not explained. The beans from the three bags were combined to form a single sample or sub.
4. From one to 20 subs were drawn from every 250 bags or fraction thereof.
5. Subs were selected as equally as possible from all of the marks. Six countries accounted for 82 percent of the total imports of cocoa beans; about 73 percent of the sample was taken from lots from these countries.

Measuring quality of beans. Each sample pound (sub) was sifted on a three mesh screen to remove insects. One hundred beans were selected at random by using a special cracking board with 100 holes drilled in it in a symmetrical 10×10 pattern. The beans from each sub were poured on to this board and spread over it so that each hole contained one bean; excess beans were not used. Each bean was broken into small pieces and magnifiers were used as needed in order to classify the beans into six categories: slightly moldy, moldy, insect infested or insect damaged, both moldy and infested, rusty, and sound beans. Rejected beans included moldy, infested, and both moldy and infested.

The following gives estimates of rejects for all 19 countries, as well as selected countries, including the country with the minimum and the country with the maximum reject rate:

Country	No. of subs	Percent rejects	No. marks	Percent of marks with over 3% rejects
All 19 countries	6,003	1.08	2,169	10.0
Brazil	1,075	0.72	192	4.1
Ghana	1,024	0.80	252	1.2 min
Mexico	210	2.14	128	19.2
Guatemala	40	0.63 min	no data	no data
French Equatorial Africa	35	4.77 max	246*	11.7*
Trinidad and Tobago	no data	no data	92	28.7 max

*French West Africa

There is no estimate of the magnitude of the compliance problem. The number of shipments, or bags by marks or countries of acceptable and unacceptable quality, are not given. The quality of cocoa beans, however, varies widely and is much higher in some countries than in others. The Food and Drug Administration does have the power to reject an entire shipload if the quality is too low.

No sampling errors of estimates are given, nor explanations of the "mark" and why it is used, nor use of compositing, nor why a pound of beans was needed to obtain 100 beans for testing. Actually, sampling errors cannot be calculated because judgment sampling was applied to both ships and bags. Selecting samples by countries roughly proportional to past volume of annual imports was an inefficient method because it led to sampling most heavily those countries with the highest quality beans.

At least two steps can be taken to improve this situation. One is to research the entire sampling and testing procedure to find a more simple yet more efficient method of sampling and testing these cocoa beans. The other is to make arrangements with the country of origin to explain what constitutes acceptable quality so as to reduce if not prevent unacceptable quality from being shipped. This involves also the condition of boats.

Control of aflatoxin. The control of a very potent toxin called aflatoxin is very important in the food field. Two examples are described; one from shelled peanuts, the other from shelled corn. Peanuts are sampled to estimate aflatoxin concentration, or for accepting or rejecting lots of shelled peanuts from which the sample is drawn. Several papers have been written on this subject and various sampling plans and their operating characteristic (OC) curves have been studied, both by industry and the Federal Government.[5]

The sampling problem is complicated by the high variability between peanuts compared with the variability between lots. In one study it is stated that one contaminated peanut in 10,000 peanuts could result in an uncomfortably high level of aflatoxin of 50 parts per billion (ppb). Two major sampling problems are: how large a sample to select for estimation or for lot acceptance or rejection, and how best to prepare the sample for analytical purposes.

Sample sizes ranging from 10,000 to 36,000 peanuts are used. Operating characteristic curves are given for samples of 10, 30, 50, 70, and 100 pounds, the latter containing 50,000 peanuts (500 to a pound), or 90,000 peanuts (900 to a pound). In one study it was found that the standard deviation due to sample preparation and analysis was 18.5, but the standard deviation due to sampling was 53.2 or nearly three times as much. The result of this study was to show the improvement expected by increasing the sample size from 10 pounds to 50 pounds, especially in lots with aflatoxin of 50 ppb or more.

An interesting legal case arose in connection with aflatoxin in shelled corn. In 1966, four carloads or about 400,000 pounds of shelled corn were shipped to the Quaker Oats Company from a Kansas elevator. When the company examined the cars they found heat and odor due to molds forming in the corn. They rejected the four carloads and returned them to the elevator.

The Food and Drug Administration (FDA) received information about this shipment and selected samples from the bottom and top of the elevator when the 400,000 pounds were being unloaded. These samples were sent to the Kansas City and Dallas district laboratories for analysis. Tests showed the existence of 75 to 180 ppb of aflatoxin, a very high level. These laboratories sent samples to the headquarters laboratory at Washington, D.C. for confirmation. Dr. Verrett applied the chick embryo method to the material and verified, by this biological test, what the district laboratories had found.

Meanwhile, the elevator company had mixed 1,200,000 pounds of good corn with the 400,000 pounds to try to reduce aflatoxin to a safe level. FDA took additional samples as the mixture was being moved from one bin to another. The tests showed a range of aflatoxin in the samples from 0 to 80 ppb. The case went to court. Experts testified on both sides. Dr. Verrett and others testified for the FDA; the defense had various witnesses testify, including a veterinarian who submitted evidence to show that the corn in question when fed to chickens did not effect their growth.

The court ruled in favor of the FDA and ordered the entire mixed lot of 1,600,000 pounds of corn destroyed. No questions were raised about the soundness or accuracy of the sampling or analytical methods used by the FDA.

Control by means of the relation between two variables. A chart based on the relationship between two variables x and y, can be used for control. To be effective, the relationship between the two variables must be stable and pronounced. The discussion here is limited to the case where a straight line $y = a + bx$ is a good fit to the observed points (x,y); this is called linear regression.

An actual example is the relation between y, the output of whiskey of a distillery in gallons and the corn input in bushels. This relationship can be used not only by the distillery, but also by the Federal Government in monitoring distillery operations for diversion. Although this method has been suggested, there is no evidence that the government is using this simple and effective method.

During war production, close control is exerted by the Federal Government over critical materials by industries and companies in order to insure a fair allocation of these materials. The relationship between the consumption of two critical materials was used to test whether two quarterly allocations were under control or in balance.

The data in Table 1 show in net tons the carbon steel (x) and alloy steel (y) actually used to make a machine tool product. The relationship between these two characteristics was used to determine whether quarterly allocations made by the Requirements Committee for war production purposes were under control or were in balance based on past usage.[6]

The 16 pairs of values, the straight line fitted to the values by the method of least squares, the correlation coefficient, the vertical deviations of the points from the line (residuals), and the upper and lower control values for the two quarters are given below and in Table 1. The equation of the straight line is $Y = -53.95 + 0.2668x$; the correlation coefficient is 0.98.

	Allocations	Upper limit y	Lower limit y
	x y		
1st quarter	3286 932	960	686
2nd quarter	3095 772	908	636

Using the allocation value $x = 3286$, the limits for y are 686 and 960, so 932 falls within these limits; similarly, 772 falls within 636 and 908. Inventories could be adjusted up or down, and still be within these limits.

Control over product safety. The Consumer Product Safety Commission (CPSC) is charged by Federal law to keep certain dangerous products off the market. To implement the law means that data have to be collected that indicate the extent to which a product is dangerous to a buyer or user. This requires several things:

1. An explanation of what constitutes "danger" and what does not.
2. The identification of such products.
3. The extent of the danger.
4. Determining whether corrective action should be taken or not.
5. What the corrective action, if taken, should be.

The nationwide data collection system used to control product safety by furnishing information that throws light on the foregoing five steps is called the NEISS — National Electronic Injury Surveillance System. This is a continuous sample survey used to estimate the frequency of various types of product-related injuries by age and sex groups reported in the emergency wards of 119 hospitals in continental United States. The sample plan is as follows:

1. 1,900 primary sampling units are used from an area frame developed by the Bureau of the Census.

Table 1

Data for Straight Line to Control Allocations of Critical Materials

Observed values (net tons)			
Carbon steel (x)	Alloy steel (y)	Calculated value of y on line for value of x(Y)	Vertical deviations from line (residuals) (y–Y)
1050	172	226.17	-54.17
1678	373	393.72	-20.72
1798	512	425.73	86.27
1321	241	298.47	-57.47
3060	674	762.42	-88.42
3286	902	822.71	79.29
2238	577	543.12	33.88
3536	967	889.41	77.59
1488	327	343.03	-16.03
1774	359	419.33	-60.33
881	218	181.09	36.91
3298	836	825.91	10.09
833	241	168.28	72.72
3869	914	978.25	-64.25
1208	245	268.33	-23.33
3857	963	975.05	-12.05
Sum 35175	8521		-00.02*
Mean 2198.44	532.56		
Standard Deviation 1103.09	300.00		

*Theoretically, the sum of the residuals is zero.

Source of observed values: War Production Board.

2. These units are divided into 100 strata and reduced to 13 blocks with 3 subblocks per block.
3. The hospital frame is 4,906 hospitals with emergency rooms.
4. 119 hospitals were selected on a probability basis from the 39 subblocks.
5. The hospitals report on a continuous time basis.
6. A time sample is selected from sample hospital records for purposes of estimation.

Types of errors which arise are misclassification, miscounting, missing information, transcribing errors, and erroneous information. These are controlled by manual instructions, coding manuals, review of data at the source, review by field personnel, and computer audits. These procedures are prescribed to maximize the quality of the sample survey data.

This sample underestimates the total number of product-related injuries in the United States because the frame sampled excludes injuries never reported anywhere, those reported to doctors, those reported only to school nurses, those that are hospital treated but by-pass the emergency room, and hospitals that do not have emergency rooms.

Several questions arise in connection with this sample study, the data collected, and its interpretation. The soundness of the conclusions and the in-depth study are no better than the quality of the data on which they are based, and this is unknown. Controls were used, but there is no explanation of how effective they were. All large scale sample studies, but especially those transmitting data to a central computer, need to be carefully planned, designed, and tested to insure an effective sample plan, to identify major sources of bias, and to take the steps necessary to reduce if not eliminate bias.

• A major shortcoming is the limitation of the frame to emergency rooms.

• While controls were used, there is no indication how large the biases are in the estimates from the sample.

• Neither is there any indication how much bias is due to the restricted frame.

• Neither the tables nor the text has any information about the sampling variations in the various estimates derived from the sample.

• There is no information on how effective the method of stratification is, or whether it was worth the cost. Stratification by size of hospital is based on the assumption that the larger hospital has more product-related injuries reported in the emergency room. This may or may not be true.

• It may very well be, as it is in so many other sample studies, that sampling more hospitals and sampling less within hospitals would give better estimates without greatly changing the overall cost of the study.

• It is claimed that 100 percent time coverage, rather than sampling on a time basis within the hospital or sampling the records, is needed for in-depth studies. Whether this is really necessary depends upon the nature and the purpose of what is meant by an "in-depth" study.

• Another reason for very careful research into all aspects of a situation being studied and controlled on a sample data basis is that once a large nationwide sample is connected to a central computer, it becomes almost impossible to make any changes or make any improvements. The computer freezes everything: the sample plan, the biases, the sampling errors, the misleading data, and the poor quality data.

Effectiveness of social services.[7] The General Accounting Office (GAO), an agency of Congress, conducts audits and surveys of all aspects and agencies of the Federal Government. In some of these audits and surveys it uses random sampling of appropriate sampling units (individuals, families, or records) to obtain the data required.

Over the years it has conducted such sample audits of the AFDC program (Aid to Families with Dependent Children) which is under the Department of Health, Education, and Welfare. To be eligible for this program certain official rules have to be met: (1) income must be below a certain level, (2) children must be under 18 years of age, (3) no employable man lives in the house. The GAO has made sample studies to determine whether eligibility rules are being observed. An early example was a sample survey in Washington, D.C. which showed that about 55 percent of the families receiving aid were **not** eligible for aid under this program.

During 1972, the GAO conducted a sample study of both open and closed AFDC cases in five cities to evaluate the effect of services offered in connection with this program. The cities were Baltimore, Denver, Jefferson County Kentucky (Louisville), Orleans Parish Louisiana (New Orleans), and Oakland, California.

A table of random numbers was used to select random samples of 150 AFDC open cases and 150 AFDC closed cases at each of the five locations. The sample of closed cases was selected from all of the cases that were closed (AFDC money payments discontinued) during the period August 1, 1971 to July 31, 1972 and which remained closed on July 31, 1972. Open cases were selected from all those cases that received AFDC money payments August 1, 1971 and July 31, 1972, except in Denver where the sample cases were selected from those that received welfare payments January 1, 1971 and July 31, 1972. The population of cases N and the sample sizes are as follows:

Location	Number N		Sample size n	
	Open cases	Closed cases	Open cases	Closed cases
Baltimore	26,964	8,635	150	150
Denver	10,537	4,083	150	150
Louisville	10,092	2,037	150	150
New Orleans	14,612	2,833	150	150
Oakland	11,027	5,569	150	150
Total	73,232	23,157	750	750

Since the sample sizes are identical and the totals are all different, the rates of sampling n/N are all different; hence, the multipliers N/n required to "blow up" the sample to the total are all different. Due to different rates of sampling, the samples of 150 for each location cannot be added; they must be blown up to the location estimates before they can be added. (The GAO does not show these figures nor comment on this important point). The sampling rates and multipliers by location for open and closed cases are as follows:

	Sampling rates $f=n/N$		Multipliers $w=1/f=N/n$	
Location	Open cases	Closed cases	Open cases	Closed cases
Baltimore	0.00556	0.01737	179.76	57.57
Denver	0.01424	0.03674	70.25	27.22
Louisville	0.01486	0.07364	67.28	13.58
New Orleans	0.01027	0.05295	97.41	18.89
Oakland	0.01360	0.02693	73.51	37.13

For open cases, Baltimore is sampled at slightly more than ½ of 1 percent (0.56 percent); while Louisville is sampled at nearly 1.5 percent. For closed cases, Baltimore is sampled at 1.7 percent; while Louisville is sampled at more than four times that rate or 7.4 percent.

Information was obtained from case records and from interviews with case workers and recipients. In evaluating the effects of services, information was gathered for August 1970 through July 1972. The GAO recognized that certain factors such as high unemployment rates, limited job training openings, inadequate educational systems, and insufficient day care vacancies — some of which cannot be influenced by social services — play a major role in determining whether AFDC recipients obtain employment. The GAO did not determine the extent to which these factors directly affected the ability of recipients to obtain employment. Instead, rather general information, statistics, and opinions of welfare officials were obtained with regard to the extent to which these factors existed at each location. Although the positive effect of social services may not always be measured, the almost complete lack of data on the input of the program and the need to develop program accountability made it necessary to report on that portion of the program which is quantifiable — the direct impact of services.

In connection with **closed cases,** the primary goal was to determine whether services had directly assisted recipients to obtain employment. It was determined:

1. Why the recipients no longer needed AFDC, concentrating on cases closed because of employment.
2. Whether these recipients received services, and if so, whether the services were of the type that could help them obtain employment.
3. Whether the services helped the recipients to obtain employment.

The study shows the estimated number and percent of cases closed because of employment in each of the five locations.

Location	Number closed cases N	Estimated number and percent closed because of employment: number	percent
Baltimore	8,635	2,710	31
Denver	4,083	1,060	26
Louisville	2,037	270	13
New Orleans	2,833	510	18
Oakland	5,569	1,490	27
Total	23,157	6,040[b]	26[a]

[a]Chart showing 23 percent is in error since it was derived from pooled sample of 750, instead of weighted sample.
[b]Rounded.

From the foregoing percentages, the number in each sample identified as closed through employment, n_a, is as follows; also shown is $1\text{-}f_i$ and standard error of each proportion:

Location	Sample size n	Number in subclass n_a	Proportion p	$1\text{-}f_i$	Standard error of p
Baltimore	150	47	0.31	0.983	0.038
Denver	150	39	0.26	0.963	0.035
Louisville	150	20	0.13	0.926	0.027
New Orleans	150	27	0.18	0.947	0.031
Oakland	150	40	0.27	0.973	0.036

For the sampling variance of a proportion assuming a binomial distribution with $n=150$

$$s_p^2=\frac{pq(1\text{-}f)}{n\text{-}1}$$

For Baltimore, $p=0.31$ and

$$s_p=\left[\frac{(0.31)(0.69)(0.983)}{149}\right]^{1/2}=0.038 \text{ or } 3.8 \text{ percent.}$$

Other values for the standard error of p_i are calculated in the same way.

The standard error of the estimated total 6,040 is the square root of the sum of the variances of the five component frequencies, since these frequencies are independent and therefore the variances are additive. For each frequency the variance is:

$$s_i^2=w_i^2 n_i p_i q_i (1\text{-}f_i)$$

where:

$w_i=N_i/n_i$, $n_i=150$, $p_i=$proportion, $q_i=1-p_i$, $f_i=n_i/N_i=1/w_i$, and "i" indicates a location. The calculations are as follows:

Location	Calculation	s_i^2
Baltimore	$57.57^2(150)(0.31)(0.69)(1\text{-}0.017)$	104,532
Denver	$27.22^2(150)(0.26)(0.74)(1\text{-}0.037)$	20,592
Louisville	$13.58^2(150)(0.13)(0.87)(1\text{-}0.074)$	2,897
New Orleans	$18.89^2(150)(0.18)(0.82)(1\text{-}0.053)$	7,482
Oakland	$37.13^2(150)(0.27)(0.73)(1\text{-}0.027)$	39,659
Sum		175,162

The standard error is $(175{,}162)^{1/2}=419$. Hence, the estimate of 6,040 has a standard error of 419 and a relative sampling error of $419/6040=0.069$ or 6.9 percent. At the 95 percent level (2 standard errors), the confidence limits are 6040 ± 838 or 5,202 and 6,878. These give percentages of 22 and 30 which are the 95 percent limits of 26 percent. This estimate and these limits are appraised later.

These data show that in the five cities, 26 percent of the cases were closed because of employment and this varied from a low of 13 percent in Louisville to a high of 31 percent in Baltimore. The extent to which services directly helped recipients to obtain employment was 21 percent calculated as follows:

Location	Number	Percent of sample of 150
Baltimore	12	8.0
Denver	8	5.3
Louisville	4	2.7
New Orleans	4	2.7
Oakland	6	4.0

The estimated total receiving services which directly helped recipient is 1,262 using location weights for closed cases.

Location	Weight for closed cases	Size of sample	Estimated total
Baltimore	57.57	12	691
Denver	27.22	8	218
Louisville	13.58	4	54
New Orleans	18.89	4	76
Oakland	37.13	6	223
Total			1,262

Based on cases closed because of employment, services helped 21 percent (1,262/6,040). Based on the total number of closed cases, the percentage is 5.4 (1,262/23,157). The GAO gives 4.5 percent (34/750) based on the pooled same sizes and the pooled number reporting helpful services, but this is incorrect. The GAO, however, also gives the correct total of 1,260 and 5 percent.

Several reasons are given for closing the 750 cases; estimates of percentages are not proper because of the different weights for the five cities:

1. Employment or increased earnings	173
2. Moved	152
3. Absent parent returned, or remarried	83
4. Eligible children no longer at home	76
5. Recipients could not be located	72
6. Eligibility not established or maintained	59
7. Increase or receipt of other benefits	42
8. Voluntary withdrawal	24
9. Other: death, no longer incapacitated (18 categories)	69
Total	750

The GAO concluded that social services had little direct impact on helping recipients reduce their dependency, that most recipients received maintenance services which could not directly help reduce dependency, and that there was evidence that recipients did not request or receive services, did not understand how services could help them, and did not believe they needed services. Also, the impact of developmental services is limited. Recipients of these latter services did not reduce their dependency at a rate significantly greater than those not receiving these services.

In an answer to the GAO report, the Department of Health, Education, and Welfare (HEW) stated that the cities were not selected at random, the largest cities were not included, there was no assurance that samples from the five cities were representative of the country as a whole, that concluding that social services as narrowly defined by the GAO were having only a minor impact is not justified, and that HEW in appraising social programs is following the "state of the art" in this field.

It is quite true the cities were not selected at random, that the largest cities were not included, and that the five cities do not represent the entire country but just the five cities. Whether the GAO excluded social services which should have been included only a careful study can reveal; furthermore, there may be differences of opinion even among specialists about what "social services" should include. The "state of the art" in measuring progress toward a goal and in determining that laws and regulations associated with a social program are being observed, by using a sample audit such as the GAO used, may be much more advanced than the HEW officials state in their comments on the GAO report.

Two technical defects in the GAO study need to be noted. We have already pointed out that due to the different sampling rates in the five cities, each city has to be weighted in order to obtain a total; pooling the sample data is justified only if all the sampling rates are the same. There is an even more serious problem. Of the 750 closed cases in the sample, 173 were closed because of employment, 353 were closed for other reasons, and 224 were not contacted because they had moved or could not be located. How these are divided among the five cities is not given. In other words, no information is available for about one-third of the total sample of 750. This not only raises a question as to how the estimate of 26 percent would be affected if we knew what happened to the 224 closed cases who were not contacted, but it renders invalid the use of confidence limits. How many of these 224 are closed cases because of employment? These would increase the 26 percent. How many of these 224 are not closed cases at all, but open cases somewhere else? Do you ignore these cases?

Effectiveness of delinquent tax collection program.[8] This review of the effectiveness of the policies and procedures followed by Internal Revenue Service in handling and collecting taxpayers' delinquent accounts was made by the General Accounting Office at the request of the Joint Committee on Internal Revenue Taxation. The magnitude and importance of the problem is shown by data released by IRS for fiscal year 1972 relative to the disposition of delinquent accounts.

Type of disposition	Amount in $1000
collected	$2,232,953
uncollectible	518,787
abated	323,268
payment tracer or adjustment	168,309
Total	$3,243,317

As of June 30, 1972, there were 659,227 delinquent accounts representing assessments of about $1.9 billion. The $2.2 billion were collected from 2,163,749 delinquent accounts so the average amount collected per delinquent account was slightly more than $1,000.

Some delinquent taxpayers respond to the initial collection efforts by the district offices and pay the full amount owed at that time. Some cases, however, require collection through provision to pay in installments, offsets against funds due a taxpayer from the Federal Government, levies against taxpayers' assets, and sale of seized property, all methods allowed under the tax laws.

An account is classified as uncollectible when the likelihood of collection is so remote that IRS considers it unwise to devote further manpower to it or when the cost of collection does not justify the effort. Accounts are considered uncollectible if the balance owed is small, the taxpayer has no assets, the taxpayers cannot be located, or the collection would cause the taxpayer undue hardship.

The GAO reviewed randomly selected taxpayers' delinquent accounts actively pursued for collection and accounts classified as uncollectible. They did not contact or solicit the views of any taxpayers concerning IRS operations. Conclusions are based solely on information obtained by interviewing IRS personnel and reviewing

IRS records and internal operating policies and procedures. The report does not contain any taxpayers' views on how effectively or equitably the IRS is administering the delinquent accounts collection program.

The sampling of delinquent accounts. To ascertain the effectiveness and equitableness of IRS collection activities, a random sample of 437 taxpayers involving 670 delinquent accounts was selected from the automated records of individual and business delinquent accounts. Some manually maintained accounts, including accounts for wagering, estate, and gift taxes were also reviewed. At the end of the field work the status of the 670 accounts was as follows:

Status	Delinquent taxes, amount	percentage
Accounts closed:		
collected	$ 721,608	67.6
uncollectible	58,597	5.5
abated	23,757	2.2
Collection action pending	264,192	24.7
Total	$1,068,154	100.0

To collect the taxes due from the 437 taxpayers, the IRS filed liens against the property of 61 taxpayers, issued levies against the assets of 118 taxpayers, and seized cash or other property from six taxpayers. The 437 taxpayers were given adequate notice and reasonable opportunity to pay their tax delinquencies before liens were filed or levies were issued, and IRS procedures for collecting delinquent taxes were consistently applied.

The GAO concludes from their sample study that the IRS is collecting the major portion of the delinquent accounts, and that the 5.5 percent classified as being uncollectible is reasonable from a management standpoint. Since the taxpayers were given adequate notice and reasonable opportunity to pay their delinquent taxes, the GAO concludes that the taxpayers were treated equitably.

The sampling of uncollectible accounts. At four district offices, 426 delinquent accounts of 221 taxpayers were selected at random from the automated records of individual and business accounts classified as uncollectible. Apparently the taxpayers, and not the accounts, were selected at random, since it is stated that, "Our random sample of individual taxpayers whose accounts IRS had classified as uncollectible disclosed that about 13.6 percent of the individuals were self-employed and were liable for paying the self-employment social security tax."

The reasons for classifying the accounts as uncollectible and the amounts and percentages are as follows:

Reason classified as uncollectible	Number of accounts	Amount	Tax liability percentage
Defunct corporation	144	$352,673	75.0
Collection would cause undue hardship	123	83,890	17.8
Unable to locate taxpayer	103	26,213	5.6
Low dollar amounts—not worth collecting	46	404	0.1
Responsible persons deceased	10	7,102	1.5
Total	426	$470,282	100.0

The GAO concludes from this sample study that the procedures for administering uncollectible amounts provide for equitable treatment of the taxpayers and adequate safeguards to insure that the IRS does not prematurely suspend collection action on delinquent accounts. The review of the 426 cases also indicated that the prescribed procedures were generally being followed.

Delinquency in paying federal occupational tax by retail liquor dealers.[9] The General Accounting office used a sample based on zip code areas to measure the extent to which retail liquor dealers were delinquent for at least one year's federal occupational tax. The purpose was to determine the effectiveness of the law and its enforcement; the results led to a recommendation that the tax be repealed.

The details of the sample plan, with weights, estimated totals, and proportion delinquent added, are given in Table 2. Each of four states was divided into three population size groups giving 12 independent strata. A random sample of zip codes was selected in each stratum. Using State license data, the retail liquor dealers in each zip code sample were listed. A random sample of dealers to contact was selected from each list.

The number of delinquents in a stratum is $n_{ia} = n_i p_i$, where $p_i = n_{ia}/n_i$ and $q_i = 1 - p_i$. These values are estimated from the sample. The estimated number of delinquents in a stratum is $X_i = w_i n_{ia} = w_i n_i p_i$, where w_i is the weight derived from the inverse of the sampling rates. If this expression is summed over the 12 strata, the total estimated number of delinquents is obtained. This gives a total of 29,503 or 29.8 percent based on a population total of 98,934. This percentage is higher than the 27 percent derived from the unweighted sample figures (174/650) shown in Table 2.

In Table 2 the weight 341 $= \left(\frac{1024}{3}\right)\left(\frac{16}{16}\right)$, while the weight 443 $= \left(\frac{380}{3}\right)\left(\frac{175}{50}\right)$. It is assumed that the number of delinquents found in each sample is a binomial variate, so that the variance of the total estimate for a single stratum is $w_i^2 n_i p_i q_i$. Summed over 12 strata, this gives the variance of the estimated total number of delinquents. This total variance is 5,400,127, so the standard error is 2,324 while two standard errors is 4,648. The estimate 29,503±4,648 gives 24,860 and 34,150, compared with 24,000 and 33,500 as given in the report. The differences are of no practical importance.

The 174 delinquent taxpayers found in the entire sample gave the following reasons for not paying the tax:

Not aware of the tax	113
No renewal notice from IRS	22
Forgot to pay	19
No explanation	8
Other reasons	12
Total	174

On this unweighted basis which is only approximate, very nearly two-thirds of these retail dealers were not aware that they had to pay a federal occupational tax.

Table 2

Occupational Tax Study: Sample Plan and Estimates for Retail Liquor Dealers

State and population class[a]	No. retail dealers	No. zip codes in population	No. zip codes in sample	No. retail dealers in zip sample	No. retail dealers selected	No. retail dealers delinquent	Weights w_i	Delinquent estimates total	Delinquent estimates percent
California									
1		1,024	3	16	16	4	341		
2		380	3	175	50	16	443		
3		329	4	176	75	28	193		
total	48,725							13,856	28.4
Georgia									
1		476	3	22	22	8	159		
2		43	3	251	60	25	60		
3		55	4	119	57	17	29		
total	7,711							3,265	42.3
Illinois									
1		1,247	3	28	28	5	416		
2		170	3	168	66	11	144		
3		81	4	344	87	18	80		
total	19,484							5,104	26.2
Ohio									
1		1,042	4	47	40	16	306		
2		152	4	70	55	5	48		
3		167	5	286	94	21	102		
total	23,014							7,278	31.6
Total	98,934	5,166	43	1,702	650	174		29,503	29.8

[a]Under 10,000 population = 1; 10,000-100,000 = 2; over 100,000 = 3.

Auditing loan program of the Small Business Administration.[10] The General Accounting Office used a random sample to appraise the administration of the direct loan program of the Small Business Administration. The random sample was selected from the records of 24 district offices scattered throughout the nation, and was selected at different rates for each of the three groups or strata based on status of the loan to a small business. The basic data are as follows:

Loan status	Number of loans in population	Number selected at random
Group I: current loans	33,345	480
Group II: loans in liquidation	1,540	250
Group III: loans charged off	733	250
Total	35,618	980

The population and sample sizes for each of these three groups, as well as the amounts of the loans for the population and the sample, for each of the 24 district offices are given in Table 3. It will be noted that in nine cases the population was sampled 100 percent; in the others the rate of selection varied since the population sizes varied, but the sample sizes tend to be fixed for each of the three groups: 20 in Group I and about 10 in the other two groups. This means that each group in each district office has a different weight, and these weights have to be applied to obtain any population estimates requiring combining loan status groups or district offices, in whole or in part.

The sample data were the basis for recommendations to improve the operations of the agency in three major areas:

1. The problem of approving loans for questionable purposes.
2. The need to analyze adequately the financial condition of all prospective borrowers.
3. The need to help the borrower to increase chances of loan repayment.

The 24 district offices whose records were sampled, handled 44 percent of the loans approved and 47 percent of the total dollar value of loans approved by all SBA offices for the fiscal years studied, 1971-1974. Each district office has three basic functions — loan approval, loan servicing, and management assistance. The sample of 980 loans revealed SBA approval of practices such as transfer of risk of obtaining repayment from bank and creditors to SBA, refinancing of questionable loans, and making loans to wealthy borrowers not intended to receive SBA assistance.

Some of the important findings derived from the sample were:

1. Eighty-seven loans resulted in reduction in the bank's exposure, thereby transferring risk of loan default from banks to the SBA.
2. It was estimated from the sample that 3,865 loans were used to refinance prior SBA loans, of which 23 percent were questionable because the borrowers were delinquent in their loan payments or had little or no chance of recovery when the new loans were approved.

Table 3

Population and Sample Sizes by Loan Status, Amounts of Population, and Sample Loans, by District Offices

District office	Number of loans[a]						Loans all groups ($1,000,000)	
	population			sample				
	I	II	III	I	II	III	population	sample
Boston	1,755	163	59	20	11	12	$138.1	$2.0
Hartford	1,021	68	38	20	11	12	68.0	2.2
New York	2,010	61	19	20	10	10	238.1	3.6
Newark	1,038	63	9	20	11	9	111.7	4.1
Baltimore	579	14	12	20	14	12	46.7	3.5
Clarksburg	773	25	15	21	12	15	74.3	3.5
Birmingham	917	28	33	20	14	11	63.2	2.3
Charlotte	804	25	10	20	12	10	45.3	2.3
Detroit	1,269	105	24	20	10	12	121.9	3.5
Cleveland	840	70	38	20	10	13	81.8	3.0
Chicago	1,665	119	68	20	11	11	171.2	2.8
Indianapolis	2,291	91	18	20	10	10	180.2	3.0
Columbus	898	21	33	20	11	11	91.3	4.0
Houston	1,620	79	66	20	12	11	132.4	2.8
Kansas City	2,107	106	39	20	10	13	138.3	2.4
St. Louis	1,681	82	48	20	10	12	100.8	2.2
Denver	1,749	60	29	20	10	14	124.0	2.5
Salt Lake City	1,227	20	4	20	10	4	104.8	3.4
San Francisco	2,921	91	47	20	10	12	181.2	2.6
Los Angeles	2,606	136	53	19	10	10	299.1	4.3
Honolulu	600	3	2	21	3	2	45.8	1.7
Seattle	1,611	43	46	19	11	12	129.2	2.2
Philadelphia	1,138	59	22	20	11	11	102.3	3.9
Las Vegas	225	8	1	20	6	1	12.4	1.0
Total	33,345	1,540	733	480	250	250	$2,802.0[b]	$68.7[b]

[a]I stands for current loans, II stands for loans in liquidation, III stands for loans charged off.

[b]These are rounded from totals expressed in $1,000s.

3. The sample showed loan specialists of the SBA did not make thorough or adequate evaluations of the borrower's financial condition.
4. About 15 percent of the borrowers or about 5,500 did not submit the financial statements required.
5. The SBA verified about 15 percent of the loans to insure that the loan proceeds were used for authorized purposes.
6. Banks are to notify the SBA within 30 days of any delinquency or default by borrowers. From the sample, it was found that 1,845 out of 6,187 delinquency notices, or about 30 percent, were received within 30 days. The average time of those not reporting within 30 days was 113 days.
7. Field visits were made prior to loan approval in 20 percent of the loans.
8. Reasons were obtained from banks as to why the loan was denied the borrower in nine percent of the loans.
9. The number of loans not visited and which should have been was 36 percent.

Sample audit of small vouchers. For many years, federal agencies have been allowed to use sampling to post-audit vouchers with small money amounts. Now they are allowed to sample audit pre-payment vouchers of small money amounts. (See attached authorization dated November 20, 1980). The most recent authorization by the General Accounting Office has four provisos:

1. Vouchers above $750 have to be audited 100 percent.
2. A sample can be used to audit pre-payment vouchers of $750 or less.
3. The cut-off or upper limit for sampling is $750, or some smaller value.
4. A cost-benefit analysis must be submitted to justify the limit selected.

It is not clear whether the sample is to detect, estimate, control, or simply correct errors. If it is the latter, then the assumption is that the errors not found in the sample are not worth the resources required to correct them 100 percent.

There is no indication that the sample audit will be used as a control — that is, the errors will be traced back to their sources and an attempt will be made to correct the causes so that the errors do not occur again.

What is needed before a cut-off can be determined, and a sample plan designed, is a frequency distribution of the size of the vouchers and a frequency distribution of the money errors. Better yet, a joint frequency distribution needs to be constructed from past data to show the frequency and magnitude of errors as a function of the dollar magnitude of the vouchers. Any of these distributions will require a large number of vouchers so that the full range of the dollar value is covered.

Presumably, the reason sampling is proposed is that the dollar error on the small vouchers is so small that it does not pay to audit them 100 percent. The costs are as follows:

100 percent audit	**Sampling**
paper handling	paper handling
audit time	sample selection
	audit time

The high cost is audit time so all that is necessary to be cost-benefit is to sample so that tens of thousands of vouchers will not have to be audited. A frequency distribution will show any one of several limits that may be cost-benefit.

B-153509

November 20, 1980

HEADS OF DEPARTMENTS AND AGENCIES

Subject: Statistical Sampling Procedures in the Examination of Disbursement Vouchers

Based on the authority contained in Title I of Public Law 93-604 approved January 2, 1975, statistical sampling procedures authorized by Public Law 88-521 [31 U.S.C. 82-1 (a)] may be used in the prepayment examination of disbursement vouchers for amounts not in excess of $750.

The head of each department and agency will set his own limit within the $750 maximum. He will demonstrate by cost-benefit analysis that economies will result by use of the limit he selects. In our reviews of your accounting systems, we will evaluate the adequacy and effectiveness of your statistical sampling procedures.

Chapter 5 of Title 3 and Section 23.4 of Title 7 of the General Accounting Office Policy and Procedures Manual for Guidance of Federal Agencies will be amended accordingly.

Elmer B. Staats

Comptroller General
of the United States

Notes

[1]Published by the Graduate School, Department of Agriculture, Washington, 1939. Dr. Deming was the editor of the book.

[2]*The Audit Control Program — A Summary of Preliminary Results,* May 1951, U.S. Treasury Department, Bureau of Internal Revenue, Washington, D.C. The total number of individual income tax returns was about 52 million.

[3]The binomial, Poisson, and MF methods all give a value of 95 percent; see Chapter 14.

[4]*Cocoa Bean Import Survey, January 1, 1959 — March 31, 1960,* Food and Drug Administration, U.S. Department of Health, Education, and Welfare, Washington, D.C.

[5]L. Stoloff et al., "Sample Preparation for Aflatoxin Assay," (1969), *Journal of the American Oil Chemists' Society,* Vol. 46, pp. 678-684; T. B. Whittaker and E. H. Wiser, "Theoretical Investigations into the Accuracy of Sampling Shelled Peanuts for Aflatoxin," (1969), *Journal of the American Oil Chemists' Society,* Vol. 46, pp. 377-379; T. B. Whittaker, J. W. Dickens, and E. H. Wiser, "Design and Analysis of Sampling Plans to Estimate Aflatoxin Concentrations in Shelled Peanuts," (1970), *Journal of the American Oil Chemists' Society,* Vol. 47, pp. 501-504; P. Tiemstra, "Study of the Variability Associated With Sampling Peanuts for Aflatoxin," (1969), *Journal of the American Oil Chemists' Society,* Vol. 46, pp. 667-672.

[6]The data are from the official records of the War Production Board. These calculations are easily and quickly done with a hand-held calculator such as the HP 32 E which has programmed in it means, standard deviations, linear regression, the correlation coefficient, and calculation of Y value on the line for any value of x.

[7]*Report to the Congress,* "Social Services: Do they help welfare recipients achieve self support or reduced dependency?", *By the Comptroller General of the United States,* June 27, 1973. Publication number B-164031 (3).

[8]"Collection of Taxpayers' Delinquent Accounts by the Internal Revenue Service," *Report to the Joint Committee on Internal Revenue Taxation, Congress of the United States,* by the Comptroller General of the United States, August 9, 1973, p. 33.

[9]"Occupational taxes on the alcohol industry should be repealed," *Report to the Joint Committee on Internal Revenue Taxation, Congress of the United States,* by the Comptroller General of the United States, GGD-75-111, January 16, 1976, p. 6.

[10]"The SBA needs to improve its 7(a) loan program," *Report to the Congress,* by the Comptroller General of the United States, GGD-76-24, February 23, 1976, pp. 10, 71, 92, 98-100.

5

State and Local Governments

Auditing airport parking revenues.[1] Denver owns Stapleton International Airport, one of the 10 largest in the country. Among the concessions leased to contractors is the airport parking which has steadily expanded in recent years as air traffic has grown.

APCOA — the parking lot operator — collects parking fees as automobiles leave the airport using the ticket the driver picks out of a machine in order to open the gate when he enters the parking lot. Each ticket has stamped on it the exact time the parking lot was entered. From this, the exit cashier determines how long the lot was used and the charge.

According to the contract with the city, Denver receives 89 percent of the total receipts and APCOA receives 11 percent. During the last fiscal year, the total parking revenue was $5.28 million, of which Denver received $4.7 million (89% of $5.28 million). Average yearly revenues per parking ticket reported by APCOA are: $2.05 for 1978, $1.86 for 1979, and $1.66 for 1980. If the total number of tickets was known for each year, the number was not published. Serious doubt, however, arose as to the accuracy of these average amounts of revenue per ticket.

A former professor at the University of Denver, George Bardwell, sensed trouble when he paid a parking fee of $54 ($4.50 for 12 twenty-four hour periods) and saw the cashier ring up only $2. He reported this to city officials, but as a result of inaction, called it to the attention of the *Rocky Mountain News* which initiated an investigation of APCOA. It found by investigating the records that about 100,000 tickets during 1980 were reported as "missing" or "unaccounted for". They then undertook a sample audit at Stapleton, designed by Bardwell, with the following features.

1. Every fifteenth car leaving the parking lot was stopped.
2. The time frame was 24 hours every day for a week during February 1981.
3. The sample motorists were asked to show their parking tickets so the time in the parking lot could be calculated. They were stopped before they reached the cashier's check-out booth.
4. All parking lots were included: the long term, the short term, and the close-in lots.
5. Eight college students assisted in collecting the information.
6. More than 2,300 motorists were stopped so that the sample size exceeded this number.

The sample disclosed two things. The average revenue per ticket was $2.79, much higher than any average APCOA had ever reported, as the time line in Figure 1 shows. Furthermore, as soon as the inquiry started, which was prior to the sample study, the average per ticket suddenly increased to $2.39.

Two objections were raised by APCOA: all of the factors involved were not considered and the one week in February did not give accurate figures. Inquiries disclosed, however, that nothing unusual happened during the sample week such as snow storms, airport construction, competition changes, or flight changes.

Remarks. A monthly record of both the total number of parking tickets and the corresponding revenues would have helped to clarify the entire situation; furthermore, averages are not enough.

The hassle over the sample, in view of the large amounts of money involved, could have been silenced by repeating the sample during a week in March, April, and May.

With a mean of $2.79, and extreme values running above $50, the distribution of the sample of 2,300 was obviously highly skewed. No data are given showing the sampling error (standard error) in the estimate of $2.79 on the assumption that a random sample was drawn from a population of about 34,500 — 2,300 times 15.

The following figures are consistent with a sample size of 2,300: the standard deviation + $11, z = 1.96 for the 95 percent level, and d = 45 cents = $0.45:

$$n = \left(\frac{11 \times 1.96}{0.45}\right)^2 = 2{,}295$$

This gives $2.79 – $0.45 = $2.34 as the lower limit. All past averages were below this figure. This assumes that the $11 is about the right standard deviation.

Projecting the findings of the weekly sample audit gave a shortage of about $4 million during the past three years. This assumes that what was found during one week reflected the average of the past three years. This is a conjecture which may or may not be true. As stated above, with so much money involved, it would have been wise to have continued the sample until a broader generalization was technically justified.

The sample audit, however, did indicate trouble, but no corrective action was taken. Neither was the sample continued to substantiate the findings. The case is now in litigation. Clearly, a continuous quality control procedure is needed which will eliminate or greatly reduce the sources of error. Some feasible plans that might be put into effect are the following:

1. Continue the sample of one week each month for an indefinite period. With a continuous sample, the sample of cars could be greatly reduced, say to 1 in 50.
2. Use a balanced random day sample with a random car sample: sample cars each sample day at a rate of 1 in 25, and sample 7 days out of every 28 days for a rate of 1 in 4 so that every day of the week occurs once in the sample. In 364 days this gives 13 periods of 7 random days selected from each.
3. Use the same plan as item 2 above, but select four replicates — 4 cars out of every 100 — every sample day. In this way the count is always in blocks of 100 cars, from which 4 are chosen, say the 20th, 40th, 60th, and 80th car. For a total of 6,000 cars daily, the total sample is 4 percent or 240 cars, with 60 cars in each of the four replicates or

FIGURE 1

Average Revenue per Parking Ticket as Reported by Concessionaire APCOA 1978-1981 Compared with Average Revenue per Parking Ticket as Obtained from a Weekly Sample February 1981—Stapleton Airport Denver

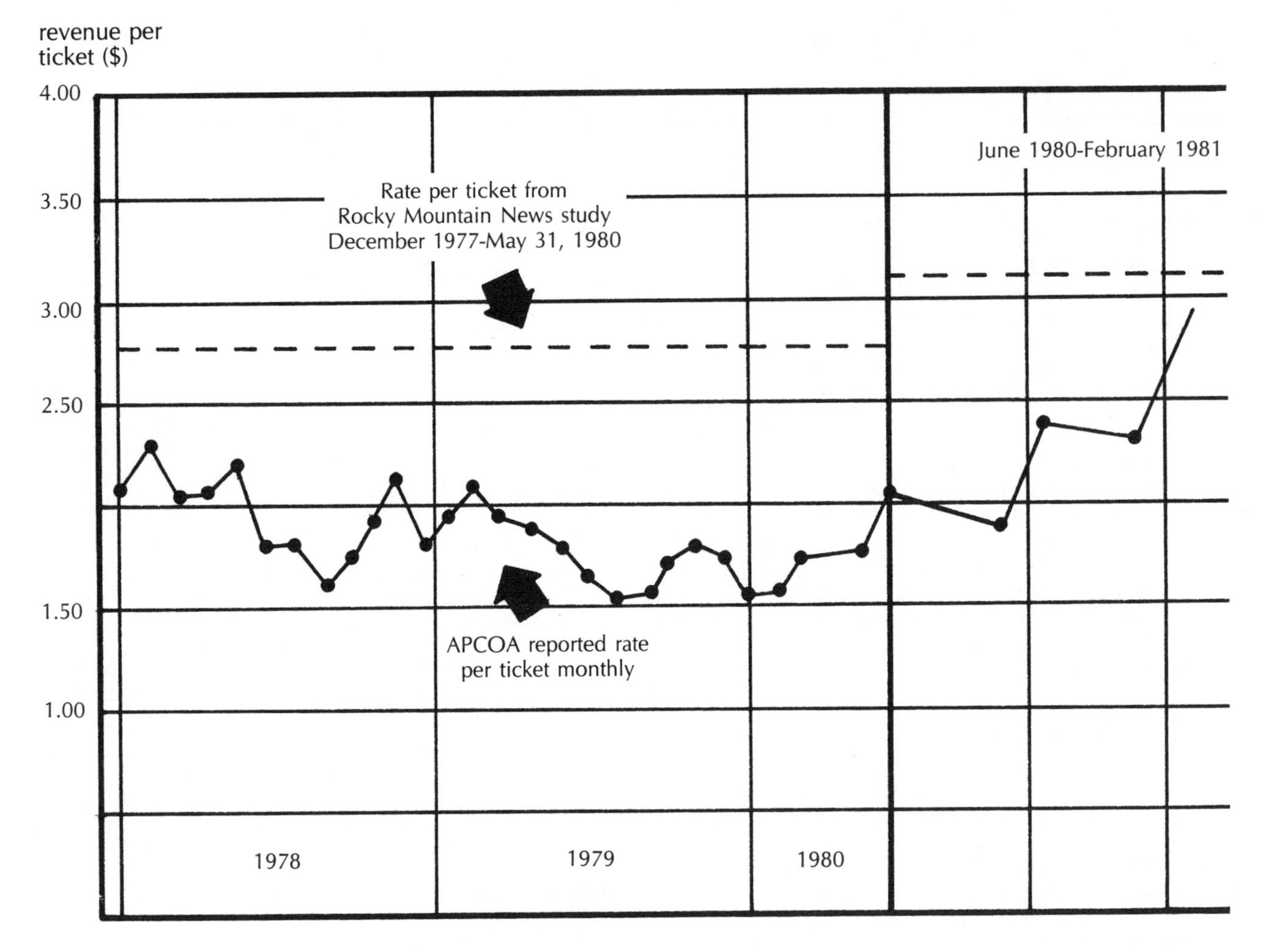

sub-samples. For each period of 28 days, each replicate covers 7 × 60 or 420 cars, for a total of 1,680 cars. Replication provides a measure of consistency and an estimate of the sampling variation using simple arithmetic because there are four independent random samples as well as the total sample.

Environmental quality control. The control of air, water, and radioactivity in a natural environment is quite different from the control of characteristics of products in manufacturing or services rendered by a service industry or a service operation. These differences are now described in more detail.

There is no single stable population; only a complex mixture of populations or sequence of populations. This is because air and water, especially streams, are in constant motion, they are infinite; they can be sampled at a very large number of points even within a specified geographical area because air and water exist in three dimensions plus a time dimension.

Air and water characteristics vary greatly within days, between days, and between months and seasons. Monthly and seasonal differences often arise from weather conditions which cannot be controlled, such as rain, wind, snow, temperature inversions, and temperature. Differences arise in three dimensions; air five feet above a street may be quite different from air 50 feet above the street; a point on the edge of a city may be quite different from a point in the downtown area.

Direct control is very limited since conditions are subject to meteorological factors, economics, politics, and legal actions. Control takes place, if at all, slowly over time, in such forms as control over factory or establishment discharges, automobile engine performance, and new housing, industrial, and business developments. It also includes exhortations of motorists to ride buses, form carpools, and drive less.

The present collection of air, water, and soil data is largely for the purpose of measuring levels of key or critical characteristics and for monitoring and warning the public, rather than control as such. Warnings are issued, for example, if the level of carbon monoxide in the air of a city rises to a critical value due to a temperature inversion. In Denver a "warning" is issued if the eight hour average is 30 parts per million; an "emergency" is declared if the average reaches 40 parts per million. The latter calls for shutting down all industry and banning all motor vehicle traffic. Neither a "warning" nor an "emergency" level has ever been reached for CO alone.

These conditions are quite different from those met in probability sampling, quality control, and design of experiments. The assumptions of a single population, independent observations, equal variances, normal distributions, and random sampling from a single population are not met. Time series analysis seems to be most applicable.

Sampling for environmental quality control. Sampling for environmental quality control, which means control over air, water, and soil for pollutants and other potentially harmful characteristics, is very much more difficult to plan and operate than control over products and performance in a factory or office, for obvious reasons.

1. There is no fixed or bounded population; air and water are in constant motion.
2. The geographical area involved is extensive and often difficult to define — such as a city or metropolitan area, or a river or creek or drainage basin.
3. The characteristic may vary greatly from point to point, from day to day, or even from hour to hour.
4. Control limits (standard levels not to be exceeded) require extensive research to determine and validate; there is a serious question about "national" standards being applicable nationwide. Sea-level standards for auto emissions have been found inapplicable to higher altitudes in the West.
5. Control usually involves governmental participation and action — local, State, and Federal — and hence can be accomplished only by a lengthy process if at all. Control is not in the hands of a scientist or engineer or manager as in an office or factory.
6. Automatic devices used to collect samples may produce a continuous stream of overlapping data to which standard methods of statistical analysis are not applicable.
7. Emphasis may be on collecting large masses of data for a remote computer data bank, rather than on collecting samples of data to which immediate analysis is applied and appropriate action taken. Analysis and use of data on the spot may be postponed, if not ignored.

In this chapter, we describe in detail some actual quality control operations; how the data are collected, analyzed, and used; and the problems which arise.

Air quality.[2] Two different methods are used in sampling air for purposes of pollution control — continuous and discrete. In the former, measurements are taken continuously from a flow of air; in the latter, a specified volume of air is captured in a container or flows through a filter in 24 hours. The continuous method is applied to the measurement of the level of carbon monoxide.

In one method of continuous sampling, a pump draws air from outside a building into a manifold to which is attached a carbon monoxide analyzer. The height of intake is 15 feet which is higher than the five to six foot level at which most persons on the street are breathing. It is claimed, however, that the difference found between these two levels are within measurement errors. Six of these fixed site sampling stations are scattered throughout the metropolitan area.

One type of analyzer has a test cell of about 500 cubic centimeters. A rotating arm breaks the infra-red rays passing through the air flow 10 times per second, so 10 measurements are taken per second. The air flows at the rate of 1,000 cubic centimeters (cc) per minute so 500 cc flow through in 30 seconds, or about 17 cc per second; hence, during an impulse of 1/10 second 1.7 cc flow out of the test cell and 1.7 cc flow into the test cell. This means that 300 readings are taken on a column of 500 cc of air or parts thereof before all of this air has moved out of the cell. Adjacent readings are practically identical (unless a change took place between two one-tenths of a second), a tremendous amount of redundancy exists, and observations are highly correlated. Different volumes of air will not be measured unless readings are separated by 30 seconds or more.

One minute values are averaged by computer for one hour, and these averages are printed out at the end of each hour from which hour to hours plots are made. The computer is programmed to reject a value that shows a 20 percent change from the preceding minute since it is claimed that experience shows that such a change is due to some stray extraneous effect.

The analyzer is tested daily to determine any malfunctioning. It is tested for accuracy of level indication using a test gas of known concentration. It is also checked for the zero level reading to prevent drift.

The one minute values which are averaged to obtain hourly averages represent 10×60 or 600 readings of the highly overlapping kind described above. Clearly, these "averages" are not the same as an arithmetic mean calculated from n independent values. They are not the average of the values obtained from 20 randomly selected minutes (instants) during the hour, nor the average of the 30 readings taken every minute so as to represent 30 separate half liter volumes of air.

Measuring particulates in the air. About 80 stations are located throughout Colorado to collect samples of air to measure the amount and nature of physical particles or particulates in the air. These include organic matter, mineral grains, and inorganic matter, such as soil, sand, and pollen. The reasons for making these continuous tests are to protect those subject to respiratory ailments such as asthma, to protect property from damage, and to protect esthetic values.

Pre-weighed filters are furnished each station and every fourth day a sample of air is drawn through the filter for 24 hours — midnight to midnight. The filter is then removed and weighed. Because the volume of air passing through the filter is known, the number of micrograms of matter per cubic meter of air is calculated. That wide differences exist in the amount of particulates at different places is very evident by the appearance of the withdrawn filter. A point in the mountains shows a very light grey, a point in the city is black, while a point on the plains where the wind is strong is a dirty sandy color.

Figure 2 shows a plot of the particulate monthly means for each of the months of 1974 for Station 5 which is located in downtown Denver. The points follow a concave upward trend, the lowest months are July, August, and September, while the highest month is January. A parabola fits the points very closely.

Several factors affect the quality of the data. Samples may be missed because of power failure, motor failure, and human factors such as illness; or they may be rejected because of a torn filter, loose gasket, or an excessive number of insects. The missing values reduce the effective size of the already small sample. It may be enough to test a 12 hour sample, or at most an 18 hour sample, rather than a 24 hour sample. The height at which air

FIGURE 2

Fitting Parabolic Model to Particulate Monthly Means
Station 5—1974

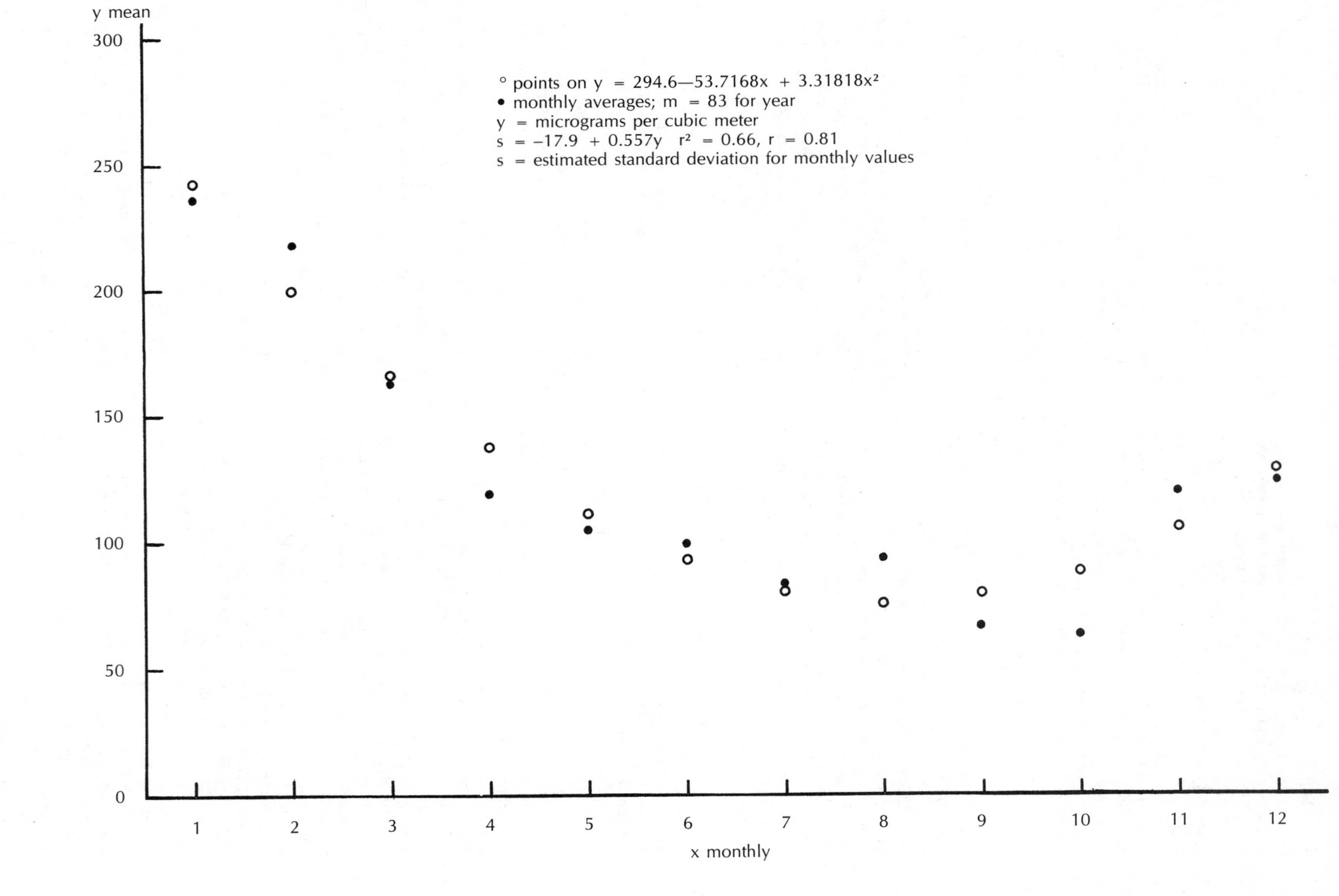

FIGURE 3

Residuals of Monthly Means from Parabolic Model
Station 5—1974

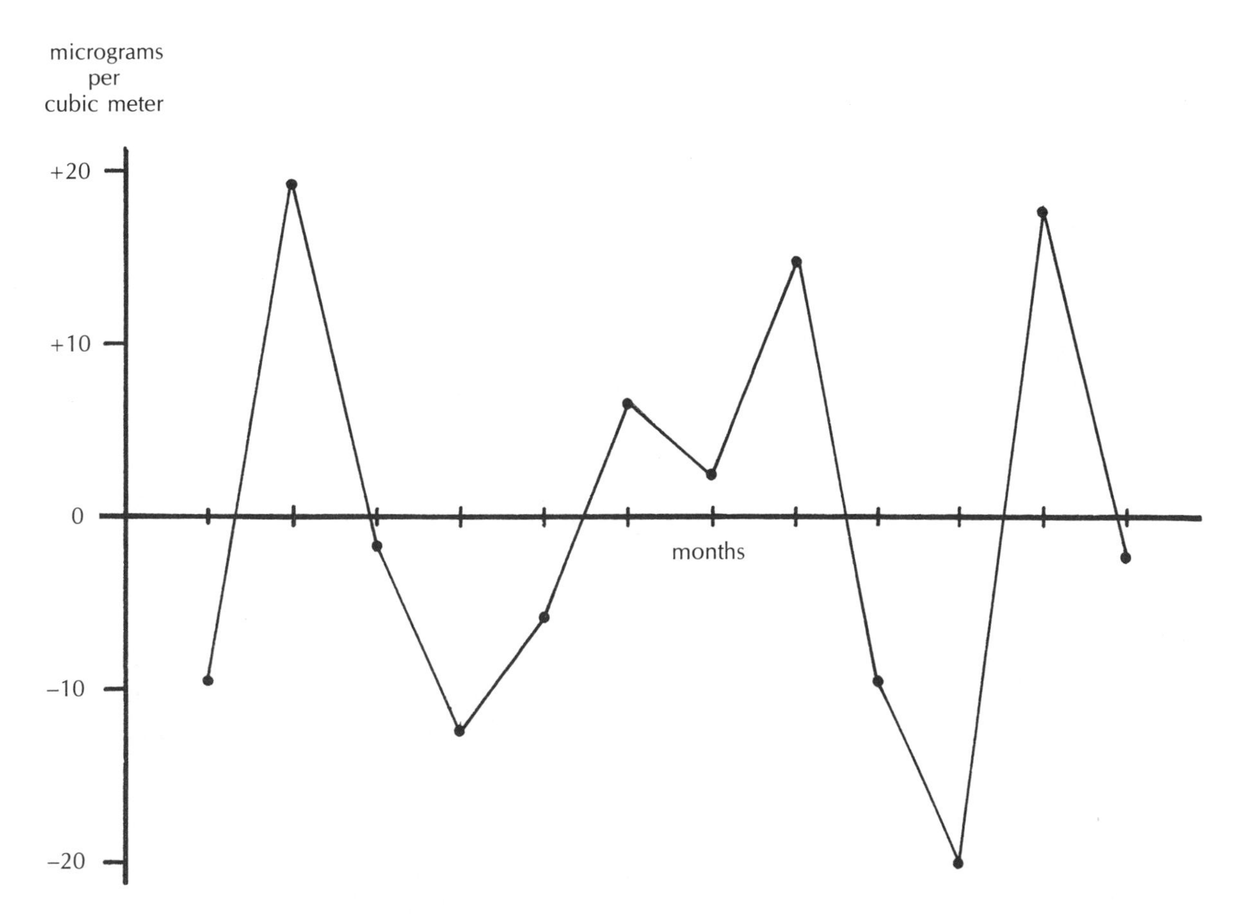

should be sampled is another problem that needs to be tested.

The air quality scale. The air quality scale developed by the Colorado Department of Health takes the following form, based on a pollution standard index:

Range of Index	Word Rating
0- 50	good
50-100	acceptable
100-200	poor
200-300	extremely poor
300-500	dangerous

When the index moves above 200, a **warning** alert to stay inside is issued for those who suffer from respiratory ailments or other related ills. A few such warnings have been issued in recent years.

When the dangerous level is reached, which has never occurred, the Governor of the State has the power to close factories, stop automobile and truck traffic, and take other necessary actions. The magnitude of the index as well as the word rating is broadcast daily over most TV stations with the weather reports.

The five pollutants which contribute to the pollution index are: carbon monoxide, nitrogen oxides, sulphur oxides, ozone, and particulates. The pollution level on the scale is based on the value found by dividing the actual level by the standard set for that pollutant. For example, the primary standard for carbon monoxide is 9 parts per million (ppm). If the current level lies between 0 and 4.5, so that the index ranges from 0 to 0.5 (4.5/9), then this level is assigned to the "good" category or rating. The other four pollutants are rated by following a similar method. The final rating assigned is the worst rating found in the five pollutants. If four are "acceptable" and one is "poor," the final rating is "poor".

Automobile air pollution test. Colorado requires annual inspection of passenger automobiles for control of air pollution, but the inspection is limited to counties along the Front Range; that is, along the eastern edge of the Rocky Mountains. The inspection is made by a licensed mechanic who tests the levels of carbon monoxide and hydrocarbons. Table 1 shows the form filled out by the mechanic for a car that had been in use for eight years. The car measured 2 percent carbon monoxide (CO), so it was well below the standard of 6 percent. The hydrocarbons measured 100, which was also well below the standard of 1300 parts per million.

There is no evidence that these are the standards that are required to clear out the "brown cloud" that hovers over the Denver area so frequently. Requiring an annual auto emissions test was a hot political issue, so that so-called "standards" were set that were not too strict. The assumption, which was no more than a guess, was that a level should be set so that no more than 20 percent would fail the test and require engine and other adjustments. It may very well be that these "standards" will fail to accomplish their objective, and require much more severe levels of compliance. It will be noted from Table 1 that an eight year old car was way below the carbon monoxide level, but even more removed from the hydrocarbon level, even though this car was using regular leaded gasoline.

Water quality. Water quality has been the concern of State and local health departments ever since it was discovered that water carried disease bacteria, such as the typhoid fever germ. These bacteria tests of water continue, but during the past decade or so other tests have been added to detect pollutants, chemicals, and radioactive substances that may be harmful to plants, animals, and fish as well as to human beings. The water quality data sheet used by the Colorado Department of Health, shown here, illustrates the wide variety of characteristics, solids, and inorganic chemicals that enter into a comprehensive analysis of water quality. (Table 2).

Table 1 Colorado "Air" Program Report

CERTIFICATE NO. ______ VEH. OWNERS NAME: ______

424226

(1) VEHICLE IDENTIFICATION NUMBER

ADDRESS: ______

CITY: ______

(2) AUTO MAKE: AMC, AUDI, AUHE, AUST, BMW, BUCK, CADI, CHEK, CHEV, CHRY, DATS, DODG, FIAT, FORD, HOND, INTE, JAGU, JEEP, LANC, LINC, MAZD, MEBZ, MERC, MG, OLDS, OPEL, PLYM, PONT, PORS, PUGT, RENA, SAAB, SUBA, TOY, TRIP, VOLK, VOLV, OTHR

(3) LICENSE PLATE

(4) DATE OF TEST — MO. DAY YR.

(5) STATION NUMBER

(6) MECHANIC'S NUMBER

(7) MOD. YEAR

(8) NO. OF CYL.

(16) CERTIFICATE ISSUED

COMPLIANCE ● (Passed stds.)
ADJUSTMENT ○ (Failed stds. after adj.)
DENIED ○

(14) FINAL TEST VISUAL

82 OR NEWER ○

PASSED		FAILED
(P)	CATALYTIC CONVERTER	(F)
(P)	FUEL RESTRICTOR	(F)
(P)	AIR SYSTEM	(F)

○ 81 OR OLDER

(9) INSPECTION COST — $ $ ¢ ¢

(15) FINAL TEST EMISSIONS LEVELS

2500 RPM*			IDLE		
% CO	*81 OR NEWER	ppm HC	% CO	1st / 2nd*	ppm HC
0–9	Compare these levels to the emissions standards for model year being tested.	0–9	0–9	Compare these levels to the emissions standards for model year being tested.	0–9
CO (P)/(F)	PASS / FAIL	(P)/(F) HC	CO (P)/(F)	PASS / FAIL	(P)/(F) HC

(10) FIRST TEST VISUAL

82 OR NEWER ○

PASSED		FAILED
(P)	CATALYTIC CONVERTER	(F)
(P)	FUEL RESTRICTOR	(F)
(P)	AIR SYSTEM	(F)

○ 81 OR OLDER

424226

(12) "AIR" PROGRAM ADJUSTMENTS

STATION NUMBER	MECHANIC'S NUMBER	ADJUSTMENT COST
		$ $ ¢ ¢

(13) EMISSIONS REPAIR COST (81 OR NEWER) — $ $ $ ¢ ¢

(11) FIRST TEST EMISSIONS LEVELS

2500 RPM*			IDLE		
% CO	*81 OR NEWER	ppm HC	% CO	1st / 2nd*	ppm HC
0–9	Compare these levels to the emissions standards for model year being tested.	0–9	0–9 (marked: 0, 1, 2)	Compare these levels to the emissions standards for model year being tested.	0–9 (marked: 0, 1, 0, 0)
CO (P)/(F)	PASS / FAIL	(P)/(F) HC	CO ● PASS	PASS / FAIL	● PASS HC

NOTE TO VEHICLE OWNER

If your vehilce did not pass the first test, you may make any adjustments or repairs yourself or have anyone else do them but, in order to receive a Certification of Emissions Control, your vehicle must meet at least one of the following conditions:

1) Pass both the visual and the emissons final tests, or
2) Pass the visual final test and be adjusted by a licensed emissions mechanic and, for 1981 and newer vehicles, have at least one hundred dollars ($100) in emissions-related repairs done.

FOR ANY ADDITIONAL INFORMATION, CONTACT ONE OF THE FOLLOWING EMISSION TECHNICAL CENTERS:

Denver Metro Area - 9640 E. Colfax Ave., Aurora 364-4135
Fort Collins Area - 429 N. College Ave., Ft. Collins 221-5324
Colo. Springs Area - 1403 S. Tejon St., Colo. Springs 633-2333

OR THE COLORADO DEPARTMENT OF REVENUE 866-5518

EMISSIONS STANDARDS FOR THIS VEHICLE

HC ______ CO ______

Table 2

COLORADO DEPARTMENT OF HEALTH
Water Quality Control Division

WATER QUALITY DATA

LAB NO. ____________

Station Designation:	Date of Sample:	REMARKS:
Town, County, etc.:	year | month | day	
Collector:	Time of Sample:	
Station Code Serial:		
Date sample received at Laboratory:	year | month | day	
Received by:	TIME:	

TEMPERATURE (°F)	TOTAL SOLIDS (mg/l)	CYANIDE (mg/l)
pH (Standard Units)	VOLATILE SOLIDS (mg/l)	SELENIUM (ug/l)
DISSOLVED OXYGEN (mg/l)	TOTAL COLIFORM (per 100ml)	SULFATE (mg/l)
CHLORINE RESIDUAL (mg/l)	FECAL COLIFORM (per 100ml)	TOTAL HARDNESS as $CaCO_3$ (mg/l)
TURBIDITY (ftu)	AMMONIA as N. (mg/l)	CALCIUM as $CACO_3$ (mg/l)
CONDUCTIVITY (micromhos)	NITRITE as N. (mg/l)	MAGNESIUM AS Mg (mg/l)
BOD (mg/l)	NITRATE as N. (mg/l)	SODIUM (mg/l)
COD (mg/l)	KJELDAHL NITROGEN (mg/l)	ARSENIC (ug/l)
COLOR (units)	TOTAL PHOSPHORUS as P (mg/l)	CADMIUM (ug/l)
OIL & GREASE (mg/l)	MBAS (mg/l)	COPPER (ug/l)
SETTLEABLE SOLIDS (ml/l	BORON (ug/l)	CHROMIUM, HEX. (ug/l)
DISSOLVED SOLIDS (mg/l)	CHLORIDES (mg/l)	IRON (ug/l)
SUSPENDED SOLIDS (mg/l)	FLUORIDE (mg/l)	LEAD (ug/l)
MANGANESE (ug/l)	TOTAL ALPHA RADIOACTIVITY (pc/l) ±	
SILVER (ug/l)	TOTAL BETA RADIOACTIVITY (pc/l) ±	
ZINC (ug/l)	RADIUM-226 (pc/l) ±	
SAR (ratio)		

Table 2 (continued)

MOLYBDENUM (ug/l)		
STREAM FLOW (cfs)		

SKETCH OF SAMPLING LOCATION:

STREAM OR PLANT CONDITIONS	SAMPLE APPEARANCE	WEATHER

Colorado has 127 permanent water sampling stations, 28 in a primary network and 99 in a secondary network. The sampling program is as follows:

Primary network: 52 samples per year for 23 characteristics.
6 samples per year for 22 characteristics.

Secondary network: 4 to 6 samples per year for 43 characteristics.

The major purposes of this sampling program are to test and control municipal wastewater treatment, to monitor water quality of streams and ground water, to conduct special investigations of water pollution and water-related situations, and to use this information system as a basis for inspections, planning and enforcement.

In 1974, two sections of land in Northwestern Colorado were leased to private corporations for oil shale development. Studies were conducted to obtain data needed to assess the potential impact of oil shale operations on the streams in these two areas. An aim of one of these studies was to estimate the quantity of plant growth by measuring daily levels and variations in dissolved oxygen, alkalinity, and chlorophyl at the developments. The data in Table 3 are measurements of dissolved oxygen in parts per million (ppm) obtained during a 24-hour period at four stations in the Piceance Creek system. Six samples were taken at each station about every four hours beginning at about 6 a.m. on one day and ending at about 2 a.m. of the second day so that each of the six measurements represent about the same time of day or night at each of the four stations.

Table 3

Dissolved Oxygen in Parts Per Million Measured about Every Four Hours for 24 Hours at Four Locations

Time order	Locations					
	1	2	3	4	Sum	Mean
1	8.3	7.4	8.3	8.4	32.4	8.10
2	9.4	7.5	8.7	8.8	34.4	8.60
3	8.7	7.8	8.5	8.4	33.4	8.35
4	7.7	7.0	8.0	7.8	30.5	7.62
5	7.6	6.4	8.1	7.7	29.8	7.45
6	7.9	6.4	8.0	7.6	29.9	7.47
Sum	49.6	42.5	49.6	48.7	190.4	
Mean	8.27	7.08	8.27	8.12		7.93
Range	1.8	1.4	0.7	1.2		

Analysis of variance gives the following results:

Component	df	ss	ms	F	P
between locations	3	5.870	1.957	31.1	p <1%
between hours	5	4.738	0.938	15.0	p <1%
interaction	15	0.945	0.063		
Sum	23	11.553			

Dissolved oxygen ranges from 6.4 to 9.4 ppm at these four locations. These are highly satisfactory levels since fish require about 3 to 4 ppm. At each location a peak is reached at the second or third hour due to plants beginning photosynthesis as the first light of the morning appears. Variation over the six time readings is therefore curvilinear and not linear, although linear regression does explain about 55 percent of the variation between the six means. Interaction is very low, giving very high and very significant values of F, because examination of the data shows that the differences between adjacent hours follow a similar pattern for each of the four locations: up, down, then level off. The report concludes that the immediate danger to the streams in this valley is increased siltation and sedimentation from construction projects related to oil shale development.[3]

Biomedical oxygen demand (BOD) is another important characteristic used in measuring water quality. It is the amount of oxygen required to break down the organic matter in the water. An initial reading is taken on each sample, the sample is allowed to stand in an incubator at 68 degrees Fahrenheit for five days, and then a final reading is made. The difference is used to measure the BOD in ppm (parts per million). A stream with good quality water measures less than 5 ppm and usually 1 or 2 ppm. Larger values mean that there is more organic matter to breakdown. This is illustrated by measurements of the BOD on the same day at two different points on the South Platte River — one above Denver and one below Denver. This reflects largely on the effects of sewage and irrigation ditches. Samples were selected on the same day at both points and the BOD is measured in ppm.

	South Platte River	
Day of Sample	Below Denver	Above Denver
1974		
April 1	19.0	4.0
11	27.0	2.0
16	20.0	1.0
24	19.0	1.0
May 2	20.0	9.0*
7	18.0	2.0
16	24.0	1.0
23	9.0	1.0
30	6.0	2.0
June 7	16.0	2.0
14	12.0	1.0
18	9.0	2.0
28	12.0	1.0
October 9	29.0	2.0
18	23.0	2.0
24	34.0	2.0
29	22.0	2.0

*Appears to be an error; 4 was maximum for the year.

The lower values in late May and in June are due to increased volume of river flow due to rains and melting snow.

A permit system is used to control discharges from point sources such as municipal treatment plants, factories and other industrial establishments, wells, ponds, irrigation ditches, and feed lots. The standards set to control pollutants vary with the source, the streams into which the effluent discharges, and whether protection is for persons, fish, plants, or all of them.

Measuring radioactivity. Measurements are made of radioactivity in samples of four common substances:

1. Air: alpha, beta, plutonium.
2. Water: plutonium, tritium, alpha, beta, uranium.
3. Milk: gamma, tritium, strontium 90.
4. Soil: plutonium.

Measuring and controlling radioactive substances in the Denver area is very important because of the Federal Rocky Flats plutonium plant 16 miles northwest of Denver, operated by a private contractor, and the Fort St. Vrain Nuclear Generating Station 35 miles north of Denver, a power station which is now in operation. During 1972, surveillance of the Rocky Flats area to determine the effects on the environment was based on 2,328 air samples, 305 water samples, and 700 soil samples.

While the Rocky Flats plant contaminates the adjacent area with plutonium, the 1972 report of the Colorado Department of Health found all levels below any existing standards for the general population, except soil in an area southeast of the plant site, an area known to be exposed to plutonium contamination. Total alpha and plutonium readings of water samples taken from Walnut Creek during the later months of the year exceeded previous averages. Plutonium from dust samples is reported to be much higher in plutonium than the usual soil samples, and this is another factor in maintaining a continuous sampling surveillance of the area. Furthermore, the question has arisen as to whether it is safe to live down-wind from the plant and whether it is safe to build a housing development in the vicinity.

Air. Air samples are taken from within the plant boundary, as well as in the Denver metropolitan area, and at stations in northern and southern Colorado. Most of these stations are included in the 80 stations mentioned earlier in connection with sampling for air quality characteristics. It has been found that beta concentrations follow a seasonal pattern reaching a maximum in late spring and a minimum during the winter months. Average plutonium concentrations in 1972 in picocuries per cubic meter were much higher for "on site" sample stations (one half mile within plant boundary) than for "off-site" stations (10 to 20 miles distant), as the following data show.

On site station	pCi per m^3 of air	Off site station	pCi per m^3 of air
1	0.00144	5	0.00014
2	0.00069	15	0.00005
56	0.00228	16	0.00005
3	0.00451	19	0.00003
4	0.00548	22	0.00003

The maximum permissible value of Pu 239 is 0.33 (insoluble form) and 0.02 (soluble) in pCi per m^3.[4]

Water. Three samples of one gallon each are taken weekly of water flowing from Walnut Creek into the Great Western Reservoir from which the city of Broomfield gets its water. These samples are taken on Monday, Wednesday, and Friday afternoons throughout the year.

Individual samples are tested for alpha, beta, and tritium, the latter being a radioactive form of hydrogen. The water samples are composited (combined) weekly for testing for plutonium and uranium. Tritium is in solution, but the others are in suspension attached to particles. Tests show that measurements of any radioactive substance tend to vary widely between samples obtained at the same place at different times of the year. Fallout is readily detected. In January 1972, the maximum value for beta activity was five times, and the average was twice, the value for other months due to fallout from the Chinese nuclear detonation. The February levels, however, were back to their usual values.

An example of how the measurements are used for quality control purposes is illustrated by an experience with Walnut Creek. Repeated samples showed the measurements of plutonium were steadily increasing in magnitude. Investigation showed that waste water in holding ponds was being released into Walnut Creek. Steps were taken to have this drainage stopped, and to have the company keep all such water out of Walnut Creek. The company is developing an internal re-cycling process which will solve this problem and eliminate all waste water flow into Walnut Creek.

The maximum permissible amount in picocuries per liter of water is 1,000,000 for tritium, and 1,600 each for plutonium 238 and 239.

Soil. Tests are run on sediment taken from streams and reservoirs all over the State. At Rocky Flats, 13 areas are identified and 25 samples of the top soil (about 1/8th of an inch) are taken from each area. The 25 samples are composited (combined), ground, and mixed. Tests run on each of the 13 areas. This is done once a year and three determinations are made on each test for each sample. The average uranium-238 concentration in soil in an arc five miles from the Rocky Flats plant was 0.77 pCi per gram of dry soil. This is stated to be about three times the level found in other parts of the world (0.25 pCi per gram). This is expected because of the high uranium content in the soil usually found on the eastern slope of the Colorado Rockies.

Cattle tests. Several samples of cattle grazing in various parts of Nevada as well as near the Rocky Flats plant have been tested for plutonium concentrations. The young cattle are less than one year old, while the mature

Table 4

Geometric Means in Picocuries Per Kilogram for Plutonium-239

Area	Young cattle			Mature cattle		
	no. of cattle	lung	TBLN[a]	no. of cattle	lung	TBLN[a]
Colorado						
1. Rocky Flats 1973	5	0.10	4.6	5	0.36	5.1
Nevada						
2. Roller Coaster 1972	4	0.19	0.51	6	0.73	4.47
3. NTS[b] 1972	3	0.69	0.94	7,6	0.75	1.34
4. NTS[b] 1973	2	0.28	3.55	9,8	0.99	12.0
5. Searchlight 1972	4	0.08	0.54	6	0.11	0.35
6. Reno 1971	—	—	—	6	0.027	1.32

[a]Tracheo-bronchial lymph nodes.

[b]National Test Site.

cattle may be 18 years old. Of the five areas, three are known to be contaminated with plutonium or other radioactive substances:

Rocky Flats: grazing area outside cattle fence surrounding plant; cattle drink water from streams draining Rocky Flats plant.

Roller Coaster: range contaminated with plutonium during 1963 tests.

NTS: National Test Site where several atmospheric nuclear tests were made early in 1966.

Searchlight: cattle exposed only to world-wide fallout and natural soil radioactivity.

Reno: cattle from Reno slaughterhouse presumably not near any nuclear tests of the past.

Table 4 shows geometric means in picocuries per kilogram for plutonium-239 found in lungs and tracheobronchial lymph nodes, of the number of young and mature cattle specified for each of the six areas, one in Colorado and five in Nevada.[5]

The high variability of plutonium-239 in fCi per gram of ash by area by tissue is shown by the following, where n = number of cattle, $\bar{x}$ is the *arithmetic mean,* s is the standard deviation derived from variance using n-1, and c is the coefficient of variation $s/\bar{x}$. Nevada Test Site figures for lymph nodes are expressed in picocuries per gram of ash and n = 5.[6]

		Lung			Lymph nodes		
	n	$\bar{x}$	s	c	$\bar{x}$	s	c
1. Rocky Flats	10	2.94	2.20	0.75	94.9	166.71	1.76
2. Reno	6	3.35	1.29	.39	41.8	13.4	.32
3. Roller Coaster	10	60.20	60.81	1.01	1540.0	3209.	2.08
4. Searchlight	10	63.0	18.3	.29	10.2	7.3	.71
5. NTS 1972	7	71.9	87.1	1.21	2.29	3.09	1.35

The press statement[7] that plutonium was four times as strong at Rocky Flats as at the Nevada Test site (1972), derived from a ratio of geometric means (see 5.1/1.34 in last column of Table 4) was erroneous due to high between-cattle variation within each herd. Furthermore, the Nevada Test Site geometric mean for 1973 was more than twice that at Rocky Flats — 12.0 compared with 5.1. In addition, in this test as well as in others, there is no mention of the amount of exposure being equated or even randomized in the tests. Using the coefficients of variation, Rocky Flats tends to fall in the middle of the values for the five areas given above.

Smith and Black report that the 90 percent confidence interval about the geometric mean for five organs is greater in human populations than in the cattle tested at Rocky Flats. They conclude that "the Rocky Flats cattle have tissue concentrations of plutonium-239 similar to those collected from the NTS and Roller Coaster herds. . . . The wide range of concentration among the various exposed cattle renders the common statistical comparison test impotent so that more definitive statements cannot be made. . . ." In other words, the variations about the means are so large that with the small samples used, the differences between the means can be explained by sampling variations, or by non-comparable herds of cattle, or by these factors plus others.

It should be noted that in comparing young cattle with mature cattle, in nine out of 10 comparisons of geometric means, the mature cattle have the large means. Length of exposure appears to be a real factor in determining levels in tissue. Some other comments about these tests follow:

The misreporting and misinterpretation of the data, and the scare-type headlines which were false, raise a serious question about reporting tests of this kind accurately in the press, especially in view of their immediate critical public importance.

The high variability of the individual measurements renders an average unsatisfactory, whether median, arithmetic mean, or geometric mean, unless it is used or printed with associated measures of variability, such as the range or standard deviation. In view of these large variations, two means may be widely different numerically without representing a real difference at all. This principle needs to be emphasized at all times in view of the wide publicity any data of this type receive in the press.

Serious questions arise in connection with the cattle tests. The data would be more useful and less ambiguous if standard experimental methods were used, such as larger samples, random assignment of cattle to sites, matching cattle or herds by length of exposure, long-time carefully controlled experiments, or the use of replication.

In this area, the measurements of substances such as plutonium are of critical importance, and making any transformations or manipulations of highly variable data distorts the real world. Public health is affected by 100 or 1,000 units of plutonium, not by 2 or 3, the logarithms of 100 and 1,000, respectively.

Notes

[1]*Rocky Mountain News,* March 22, 1981.

[2]Source of data is the Colorado Department of Health.

[3]"Investigations of the aquatic ecosystems of Piceance and Yellow Creeks, Northwestern Colorado, September and October, 1974," Water Quality Division, Colorado Department of Health, pp. 4, 8.

[4]"U.S. AEC Rocky Flats Plant-1972-Environmental Surveillance Summary Report," Colorado Department of Health, Division of Occupational and Radiological Health, Denver, Colorado, pp. 22, 24, 48.

[5]D. D. Smith and S. C. Black, "Actinide concentrations in tissues from cattle grazing near the Rocky Flats Plant," 1975, U.S. Environmental Protection Agency, Las Vegas, Nevada, NERC-LV-559-36, pp. 10, 13, 15, 16.

[6]D. D. Smith, S. C. Black, et al., "Tissue Burden of Selected Radionuclides in Beef Cattle on and around the Nevada Test Site," U.S. Environmental Protection Agency, Las Vegas, Nevada, NERC-LV-539-29, pp. 41, 42, 54, 55, 69, 70. Reno data are from NERC-LV-539-29, p. 28.

[7]*Rocky Mountain News,* January 30, 1975.

6

Health Services

Quality aspects of health services. Health services include a wide variety of quality aspects all of which are important; indeed, in spite of the long history of health services the concept of "quality" is still not carefully and completely defined. These aspects are outlined below and described in some detail in this chapter. A lack of detailed information about many, if not most, of these aspects simply means that we have not yet faced up to the need for quality control in numerous areas of vital interest to the buyer.

The vital aspects of health service may be described as follows:

1. A seller — doctor, hospital, nursing home, clinic, etc. — offers health services for sale at stipulated prices.

2. A buyer — client, patient, etc. — buys these health services at the stipulated prices, either directly, by means of insurance or subsidies.

3. The buyer wants acceptable quality services which are commensurate with what he or she is paying the seller.

4. Acceptable quality services not only include the quality of direct medical services such as diagnoses, medicines, surgery, and treatments, but indirect operations such as administration, purchasing, etc. whose costs are reflected in what the buyer pays. It also includes the quality of performance that is directly connected and closely related to health care such as food, housing, safety, security, attitude of employees, and other factors which arise in connection with hospitals and nursing homes.

5. A major factor of vital importance in much of health service is time — time to appointment, delay time, service time, timing with regard to medicines, treatments, and surgery.

The several major aspects and areas of quality control are outlined and described below:

1. Quality of administration and management.
   - 1.1 In a doctor's office
   - 1.2 In a hospital
   - 1.3 In a nursing home
   - 1.4 In a clinic
   - 1.5 In other health offices or institutions

The quality aspects include such operations as the following: staffing, purchasing, supervision, appointments, admissions, discharges, emergency room, physical arrangements, payments and insurance, record keeping, prescriptions, pharmacy and medicines, linens and laundry, housekeeping, sanitation, operating rooms, and laboratories.

2. Time factor in a doctor's office or clinic.
   - 2.1 Time required to obtain an appointment: t_0 to $t_1 = x$ days
   - 2.2 Waiting time in the office: t_1 to $t_2 = y$ minutes
   - 2.3 Actual service time: t_2 to $t_3 = z$ minutes
   - 2.4 Time required for one or more additional visits: t_3 to $t_4 = d$ days

The value of x might be 5 to 15 days, the value of y from 30 to 60 minutes, the value of z from 5 to 30 minutes, and the value of d may be anywhere from 1 hour to an indeterminate amount. A doctor or a group of doctors keep their schedules full by delaying the buyer, the would-be patient. There is deliberate overbooking because a doctor never knows how much time a caller requires, especially one he has never seen before. Hence, a certain amount of this delay is inherent in the business, although many times this delay in the office waiting room seems excessive. One report states that the average waiting time is 30 minutes which means many persons are waiting as long as 45 to 60 minutes.

On the other hand, an employed person wishing to see a doctor may suffer substantial loss if he or she has to wait a considerable length of time to receive medical care. This is especially true of those without paid sick leave. An even more critical situation arises if a person suddenly stricken ill has to wait to obtain medical care from a doctor or hospital. For example, it does not pay to get sick on weekends; the seven-day 24-hour a day doctor no longer exists — almost.

3. Quality of a doctor's services.
   - 3.1 Examination
   - 3.2 Laboratory tests: blood, urine, other
   - 3.3 Office tests: X rays, EKG, blood pressure, glaucoma, temperature, pulse, weight, other
   - 3.4 Diagnosis, accuracy of
   - 3.5 Effectiveness of medicines prescribed
   - 3.6 Effectiveness of surgery performed
   - 3.7 Prescriptions: fill and refill
   - 3.8 Anything unnecessary?
   - 3.9 Anything important omitted?

The quality of a doctor's service is measured by the accuracy of the diagnosis, the effectiveness of the medicines prescribed, the effectiveness of any surgery performed, the effectiveness of any measures or treatments prescribed, and the effectiveness of any regimen recommended: Is the trouble eliminated? Is the trouble brought under control? Does the medicine control or correct the situation? Does the surgery eliminate the source of trouble? Does the person or patient get well? Considering the situation and conditions, did the doctor do all that could be done?

4. Quality of hospital care.
   - 4.1 Emergency room
   - 4.2 Ambulance service
   - 4.3 Admissions
   - 4.4 Patient control
   - 4.5 Patient's room
   - 4.6 Patient's surroundings
   - 4.7 Patient's care: food
   - 4.8 Patient's care: medicines
   - 4.9 Patient's care: other: personal, bath, attention, cleanliness, security
   - 4.10 Patient's care: monitoring
   - 4.11 Central telephone control
   - 4.12 Visitor control
   - 4.13 Delivery control, mail control
   - 4.14 Discharge control, insurance and billing
   - 4.15 Operating room control
   - 4.16 Nurse performance
   - 4.17 Doctor's care
   - 4.18 Housekeeping

Quality enters into every aspect of a hospital directly from the time the patient is admitted until the patient is discharged. One control is simply knowing where the patient is: whether in a certain room in bed, whether preparing for an operation, or whether in the operating room or out of it. Another is keeping a central front office file of patient status, up to the minute literally, so as to handle mail, visitors, and deliveries promptly and accurately — a correct record of the patient's name, when admitted, what room, whether in the hospital, and whether discharged and when.

Consider an example. The key person is the patient, the key characteristic is the health of the patient, and a related characteristic is the cost to the patient. Five areas of quality are of vital interest to the patient: 1) the location and status of the patient which requires accurate and up-to-the-minute records; 2) the medical care which includes medicines, nursing care, and doctor's care; 3) supportive medical care which includes food, housing, personal care, treatment, and security; 4) financial controls which includes bills, claims, insurance, and Medicare; and 5) unnecessary medical care which includes unnecessary surgery, medicines, tests, and medical technology. Control over all of these reflects a concern for the patient's health, welfare, comfort, and pocketbook.

It is asserted by some that patients generally cannot appraise the health services they receive. Except for a very small percentage of patients and for highly specialized illnesses, this simply is not true. Patients are quite competent to pass sound judgments on the food, the medicines, the treatment received, the bills they pay, and on whether the medical services helped them, had no effect, or made them worse. Quality control needs to start with the patient and the quality characteristics associated with the patient.

Side effects of medicines and treatments. A very important aspect of the quality of medical service is the side effects of medicines and treatments. Due to strong individual differences to the dosages prescribed for drugs and treatments, what is prescribed as the dosage either by the label or by the doctor may be too strong. A patient may be aware of the wrong dosage almost immediately by the ill feeling or unusual reaction that occurs. The patient has to decide to reduce the dosage or even stop taking the medicine. Examples of such drugs are tagamet and hydrochlorothiazide.

Side effects may behave in the same or even in a more drastic way. Some doctors may even insist that there are no side effects when in fact they actually occur. An actual case was that of a woman taking estrogen. She broke out in a rash and developed small open red sores. One doctor laughed at the idea that this could be due to estrogen. Another doctor, a dermatologist, was just as insistent that this was due to estrogen and told her to stop taking the drug at once. Within a few days the rash and the sores disappeared.

This experience is leading to giving prominence to the rule that before one has any operation or any costly medical treatment, one confers with at least two independent physicians.

An example of a proposed quality control system.[1] Although the original quality assurance program was to include the development of several quality control reporting programs, it became evident early in the program that the hospital lacked the necessary resources for this expansion. Seven programs or systems however have been identified:

1. Patient and commodity transportation
2. Material supply and distribution
3. Patient dietary and food services
4. Medication ordering and distribution
5. Ancillary services, e.g., radiology and laboratory
6. Medical chart completion
7. Accounts receivable charge processing

As an example the following specifications were formulated for the first program or system; it should be noted that this specification deals with the very important characteristic of time and the significance of delay time as a quality characteristic:

Departmental Scope:

Central Escort service
Nursing units
Ancillary services

Proposed inspection characteristics:

1. Elapsed time from request assigned to arrival at pick-up area
2. Elapsed time from arrival at pick-up area to departure from pick-up area
3. Elapsed time from departure at pick-up area to arrival at destination
4. Number of assignment delays
   a. lack of personnel
   b. other ______________________
5. Number of in-route delays
   a. elevator delay
   b. other ______________________

6. Number of pick-up area delays
   a. lack of wheel chair
   b. lack of stretcher
   c. patient not ready
   d. patient receiving treatment
   e. patient not at pick-up area
   f. other ______________________

Proposed reports:

1. Variables control chart — request assignment to arrival at pick-up area
2. Variables control chart — arrival to departure from pick-up area
3. Variables control chart — departure from pick-up to arrival at destination
4. Assignment delays by day of week
5. In-route delays by day of week
6. Pick-up area delays by pick-up area

It is also recommended that the hospital consider implementing a receiving inspection program on a selected number of purchased products, since receiving inspection is unusual in the hospital industry. One reason for such a program would be to determine if receiving inspection of various purchased products including drugs and medicines was really necessary, or whether certification of quality could be obtained from the vendor.

The Front Desk in a Hospital[2]

The front desk is such a vital and busy operation in this hospital that it was staffed, depending upon the day of the week and the hour of the day, by anywhere from one to four highly competent volunteers in uniform.

The heart of the front desk is the ***patient control board*** on which is exhibited a card with name, home address, telephone number, and hospital room number for every patient. This board is updated continuously by picking up slips of paper from Admissions, each filled out for every new patient admitted to the hospital. The board is also updated continuously on the basis of telephone calls and patient movements from Discharges. Volunteers, in answer to telephone calls from Discharges, pick up patients and their belongings, put them in wheel chairs, and bring them down to the front desk. The date and time of discharge are entered on each name card which is pulled and put in a special file for a week or so, to be used to answer telephone or other inquiries. Patients are then moved out to the loading area, where they are picked up by auto or taxi. The board is used for a number of purposes:

1. Deliveries to patients. Deliveries to patients include mail, flowers, packages, clothing, and other articles. The name given is checked against the board and the room number identified to which the delivery is to be made, providing the patient has not been discharged. An exception is when a patient is in Surgery, Intensive Care, or Recovery. In these cases delivery is made to the central desk serving that unit.

2. Telephone calls. The board is used to answer telephone calls about patients — to determine if the patient is still in the hospital and in which room, to ascertain if and when the patient can be visited, the date and time of admission, and date and time of discharge if the patient has left the hospital.

3. Visitors. The board is used to answer questions about the patient including the room number and whether no visitors are allowed as would be the case if the patient was in Surgery, Intensive Care, or Recovery. Also, the visitor is told about the rules governing visitors and how to find the room in the hospital.

Other functions performed by those at the front desk include distributing mail for doctors, filing patient records alphabetically, moving quickly to Emergency if any emergency cases come through the front door, and directing persons to where lectures, seminars, and meetings are being held.

The mail for the doctors is separated from the patient mail at the incoming stage. The supervisor of volunteers has a special person take the mail to the doctor's mail room and sort the mail into the proper pigeon holes. This operation has to be performed quickly and accurately since a doctor may be waiting for a very important record, document, letter, or communication.

4. Quality aspects. Several quality factors are involved in these various operations:

1. Accurate alphabetical filing of any or all patient records.
2. No delays or errors in delivering mail to the doctors.
3. No errors in deliveries to patients.
4. Accurate and expeditious answers to telephone calls.
5. Relevant and expeditious explanations and guides to visitors.
6. Strict observance of hospital rules: e.g., with regard to visitors, discharges requiring use of a wheel chair, etc.
7. No delays in updating the patient control board with regard to admissions, deaths, and discharges.

Quality of Laboratory Tests and Other Tests

Examples are the following:

1. Blood and urine
2. Blood pressure
3. Temperature and pulse
4. Weight
5. X rays
6. EKG (electrocardiograph)
7. Glaucoma
8. Pap smear

Youden, scientists in the Food and Drug Administration, and others have found that different laboratories often give widely different measurements of the same characteristic of the same samples.[3] This experience raises a question about how much variation exists in the several measurements made by a laboratory testing blood and urine samples from the same patient. Other laboratories may obtain different results. Extensive tests need to be made to show the extent of this variation, and the extent to which a doctor's decision about a patient may be affected by these differences.

Laboratory tests. Some data on errors made in laboratory tests are available from the national Center for Disease Control (CDC) Bureau of Laboratories, an agency that checks interstate laboratories for the Health Care Finance Administration. Based on studies of about 3,000 clinical laboratories, it estimates that about 14 percent of the tests of blood and urine are unreliable. This error rate, it states, is due to several causes: incompetent laboratory personnel, bad management, sloppy supervision, no quality control, and bad reagents (test chemicals).

Until this error rate is close to one percent, CDC states that it will not be satisfied, even though one may raise the question that this is too high a rate for such important laboratory tests. Even so, CDC states that only about 15 years ago tests showed that the error rate was about 25 percent, an alarmingly high error rate.

About 12,000 to 13,000 major laboratories scattered across the country are subject to some kind of regulation and inspection by Federal health agencies, the States, or private medical organizations.[4]

This situation requires not more outside inspection, but a mandatory effective quality control operation in every laboratory whose goal is a substantial and continuous reduction in the errors made in these highly significant laboratory tests. Quality is built into these tests by careful and competent performance, not by inspection.

Example

On one medical data form which is used to record the various measurements for one patient, a range is shown for each measurement (Table 1). Presumably if a measurement falls outside the range given, it is considered an indication of some trouble that needs to be watched, if not corrected. We need to know how these ranges are set, e.g., for sugar, uric acid, cholesterol; whether they are derived from adequate experimental designs which yield ranges which apply to say 95 percent of the patients in the United States; and to what extent doctors' decisions based on these ranges are correct and to what extent they are wrong.

Table 1

Actual Record of Laboratory Tests on a Patient with Normal Limits*

Laboratory ______ Patient ______ Date ______

Test	Results	Normals (range)	Unit
1. glucose	139 xx	65-105	mg/dl
2. BUN	32 xx	5-26	mg/dl
3. uric acid	8.2 xx	2.5-8.0	mg/dl
4. tot prot	7.4	6.0-8.0	gm/dl
5. alb	4.5	3.4-5.0	gm/dl
6. glob	2.9	2.4-3.4	gm/dl
7. A/G ratio	1.55/1	1.07-2.00	
8. T bili	1.1	0.1-1.5	mg/dl
9. cholesterol	232	150-300	mg/dl
10. trig	141	10-170	mg/dl
11. calcium	9.5	9.0-11.0	mg/dl
12. phosphate	2.7	2.5-4.5	mg/dl
13. sodium	143	134-148	meg/l
14. potassium	4.7	3.5-5.0	meg/l
15. chloride	100	95-110	meg/l
16. carbonate	28	23-33	meg/l
17. T4(RIA)	7.6	5-13	u/dl
18. T3 uptake	25.5	23-37%	
19. FTI (I7)	1.9	1.1-4.8	
20. alk-phos	63	25-80	IU/L
21. SGOT	14	0-19	IU/L
22. SGPT	10	0-19	IU/L
23. WBC	6000	4.5-10+	1000
24. serum iron	74	60-135	mcg%

*Name of laboratory and name of patient deleted.
xx Value is above the upper limit on range or "normals".

It should be noted that these ranges are control limits and are used as such. Corrective actions are based on whether a measurement falls within the limits and no action is taken, or whether a measurement falls outside the limits and something is done, such as prescribing medicine, diet, exercise, or other health measures.

Quality of Nursing Home Care

There is an urgent need for an immediate in-depth survey of every nursing home in the 50 states. This should be done by independent groups or agencies, such as a Congressional committee that has no axes to grind and is willing to probe, examine, describe, and expose conditions as they actually are. A proposed quality service status sheet is given below which reflects a wide variety of news reports about nursing homes, as well as the experiences and observations of six nursing homes in four states. Several critical instances are cited to show the need for immediate change and improvement. A bill of rights for nursing home residents proposed by the Department of Health and Human Services is explained with implications for improving the quality of nursing home services. Finally, alternatives to the nursing home which have been suggested at various times are cited.

Several situations bearing directly on the quality of services need immediate attention because evidence shows they are far too common:

Stealing of resident's personal property. A man in one home has the following articles stolen: clothing, wallet, and shaving kit. He saw other patients wearing his clothing. He asked for a lock on his door, but was refused. A woman in another home had even more valuable personal property stolen: a Christmas gift sweater worth $25, a wig worth $30, jewelry worth $75, gowns worth $25, hose worth $20, and toilet articles worth $30. In White Plains, New York, the co-owner of a nursing home was sentenced to jail by a judge who said the defendant helped herself to the money of elderly patients. It was alleged that over $200,000 was improperly obtained from elderly persons.[5]

Inadequate food. There are many examples observed and reported of patients not receiving sufficient food. A woman in one home lost 6 pounds but quickly regained them when transferred to a home in another state. Even so, her son had to monitor the second home to see that his mother did not go hungry. A man in one home was always hungry, so the family finally took him out so that he could be properly fed. A woman in another home was supposed to be taken to the dining room by an aid, but sometimes was not. The daughter who was paying for her care discovered this practice on a visit. Reports indicate that this is often the case: a serious situation is discovered only when some relative or friend visits the nursing home and stays long enough to find out what is going on. A doctor was reported in the press as stating that a woman in an Eastern nursing home had actually died because of malnutrition.

False billing and misuse of funds. False billing for medical services and misuse of funds exists under both Medicare and Medicaid funds provided by federal taxpayers. A press report[6] states that a judge ordered the owner of five Jefferson County (Colorado) nursing homes to stand trial on charges of diverting $27,000 in Medicaid funds to his personal use. The owner and his accountant were ordered to be arraigned on 51 charges of theft and filing false reports.

The owner is charged with using $27,000 to obtain a bank loan on a 21-foot cabin cruiser, $12,000 to decorate his Denver home, monies for private school tuition for three of his children, and monies for electrical services for Aspen and Vail condominiums. False expense accounts were filed over a period of three years claiming expenditures on behalf of patients. The charges resulted from a year-long investigation by the State's Medicaid fraud unit.

The owner is also facing a $63 million lawsuit filed by the State of Colorado based on the claim he padded his expense accounts.

Patient abuse and neglect. Patient abuse and neglect include a wide variety of practices: removing trays before meals are eaten, threatening slow eaters, delays in answering the call light, refusing to give patients something they like to eat such as a slice of bread, isolating patients in a corridor and leaving them for hours, refusing to talk to patients, or employees taking long coffee breaks regardless of the patients' needs.

There is also the practice of forcing a person to stay in a nursing home much longer than is necessary. An injured rancher went to a nursing home to convalesce, but he was forced to stay two months longer than necessary, thus incurring an additional cost of $1,000. While waiting to get out, he folded towels for the nurses and did other odd jobs, since he was well and in good health.

This same person feared other residents who approached him in the corridors with abusive language. He was normal mentally and emotionally, they were not. They had no business being in a nursing home with normal people. In another nursing home, a woman who was also mentally alert and emotionally stable was placed in a room with an insane person whose behavior put her in constant fear. A forceful complaint by the daughter finally resulted in moving her to another more appropriate room.

Mixing normal and abnormal persons. The situation just described needs further attention and emphasis. A nursing home may contain everybody from a convalescent who is mentally and emotionally normal to the disorderly, aggressive, if not dangerously insane person. This indiscriminate mixing, including putting the two extremes in the same room, is a practice that should be stopped. The one group should not have to suffer because of the irrational and uncontrollable behavior of the other. It apparently grows out of a common but false set of assumptions that:

1. Everyone is equal.
2. "Abnormal" people are really not abnormal.
3. Abnormal people are simply mentally retarded.
4. Abnormal people should not be segregated, but should mix with normal persons.
5. All persons should be treated alike regardless of wide differences in mentality, emotional stability, and aggressive behavior.

Bill of rights eyed for nursing homes

WASHINGTON (UPI) — The government Wednesday proposed a bill of rights for 1.3 million Americans who live in nursing homes that receive Medicare and Medicaid funds.

"A person should not have to surrender the right to self-determination upon entering a nursing home," said Nathan Stark, undersecretary of the Department of Health and Human Services.

"We are proposing, for the first time, to make patients' rights a condition upon which nursing homes would receive Medicare and Medicaid funding."

Under the department's proposal, affected nursing home patients would be guaranteed, among other things, the right to:

— Access at all times to their families, nursing home ombudsmen, legal advocates and counsels.

— Full information about all decisions that affect them.

— Privacy and personal property; the right to be free from unnecessary chemical or physical restraints; to form resident councils, and to be involved in planning their care.

— "Free association" with other patients and visitors, and the right "to share a room with one's spouse." Under the regulations, nursing homes would have to allow at least 12 hours visiting time each day.

— "To know how much the government is being charged for the care provided and to receive itemized statements of charges" it does not pay.

Of 2.2 million people in nursing homes, 1.3 million receive Medicare or Medicaid. In fiscal 1979, state and federal expenditures for long-term care under the two programs were about $8 billion.

Stark said nursing home patients often are victimized by arbitrary visiting rules, a lack of control over such basic decisions as what to eat and where to hang pictures of their grandchildren.

He called the proposed rules — published subject to a 60-day public comment period — a step toward ending the "impersonality, indifference and isolation that characterize" their lives.

The government estimates the cost will be about 15 cents a patient per day, or $80 million a year.

But the National Council of Health Centers, which represents nursing homes, said the "new conditions . . . may cost more than $1 billion annually to implement and will impose significant burdens on nursing homes without significantly improving the quality of patient care."

It also said, "Neither the federal government, the states nor nursing homes have resources to make the regulations work."

Rocky Mountain News, July 10, 1980, p. 36

Protests and actions. To date, attempts to improve the situation in nursing homes have taken two forms: criticisms and legal actions. The latter include, as indicated above, legal investigations and indictments of individuals for theft and fraud in connection with Medicare and Medicaid. The protests include critical editorials in the press such as the one reproduced below, and the proposed bill of rights for residents initiated by the Department of Health and Human Services, the press announcement of which is reproduced here.

Nothing appears to have been done by either the federal government or by the state legislatures to place a *continuous control* on nursing homes to see that they: (1) provide *acceptable quality services* for residents, considering the $1,000 or more paid monthly for these services; and (2) keep accurate records and make proper charges to the state and federal governments. Nothing is done to compel the nursing home owners and operators to establish and maintain an *acceptable quality control system* along the lines outlined below, nor is anything done to establish quality standards for various types of services bought, nor is anything done to see that such *standards* are implemented and maintained in a satisfactory manner, nor is anything done to see that accurate and complete records are established, maintained, and audited periodically.

The obstacles to these two goals are the profit motive that underlies the establishment of many of these homes, the nursing home lobbyists in state legislatures and in Washington, the age bigotry that is all too common, the reluctance to really investigate in-depth the total operations of nursing homes, the false notion that a nursing home should be regulated solely by the State Department of Health as though quality factors in this institution deal only with health narrowly defined, and the mistaken belief that a "bill of rights" for residents will correct these deficiencies.

In all of this discussion and legislation, nothing is said about the *responsibilities of the seller* of the services — the nursing home — to deliver acceptable quality services to the buyers — the residents, patients, or their families who are paying the bills either directly or through insurance.

But even state and federal laws, bills or rights, quality control standards, quality control programs and plans, and quality audits are simply pieces of paper, simply fine-sounding plans unless they are implemented and vigorously enforced by federal, state, and local officials.

For example the proposed Department of Health and Human Services "bill of rights" *does not:*

- Protect a resident or patient from thieves, whether employees, other residents or patients; or protect residents from fire, smoke, and other hazards.
- Protect a resident or patient from insane or other abnormal residents.
- Say anything about *adequate meals,* say anything about *the right to be fed properly.*
- Protect a resident or patient from abuse or neglect.
- State the responsibilities of the nursing home.

The Department of Health and Human Services attacks the problem at the wrong end. It is not the rights of the buyer — resident or patient — that need to be stressed; it is the responsibility of the seller — the nursing home — to provide acceptable quality services to those or their families *who are paying for the services and who are keeping the nursing homes in business.*

Their bill of rights smacks of social and political institutionalism when it puts so much emphasis on full information, forming resident councils, self government, residents making their own decisions, access to legal advocates, and access to families. It is easy to divert one's attention to these aspects, however desirable they may be, and lose sight of the major issue — the responsibility of the nursing home to provide acceptable quality care.

Nursing Home Fraud[7]

"The nursing home industry, it sometimes seems, is rip-off city — a business where it's easy to cheat the government out of lots of money if you don't mind treating the elderly in your care like some kind of subhuman species."

"Oh, sure, we know there must be many nursing homes that are operated by honest, efficient, compassionate folks who are doing their best to provide first-rate care, but there's growing evidence that not a few of the facilities in Colorado need to be intensely investigated."

"Most recently, state grand jury indictments, charging Medicaid fraud, have been handed down against operators of a Steamboat Springs nursing home. A probe of the home was launched last summer after Health Department officials found unfit conditions there. Other allegations — some of them serious enough to make you shudder — have been made this past year about a number of other Colorado nursing homes, including some in the Denver area."

"The State's Medicaid Fraud Unit, it seems to us, has been doing a good job and deserves the support of the public and district attorneys. We urge all public officials with jurisdiction in the field to do their utmost to make sure our elderly are treated with dignity in these homes, and we urge the public to remain deeply concerned until it's incontestably clear that the industry is squeaky clean."

Comment. Despite the statements by this editorial writer that the government is being cheated, that the elderly are being treated like some kind of subhuman species, and that some conditions found are enough to make one shudder, the final appeal is to public officials and to the public to make the industry "squeaky clean". Not a word is written about those who are responsible for these conditions — the nursing home operators. Not a word is written about their duty and obligation to give acceptable quality services to those who are paying the bills. We present below a nursing home patient quality service status sheet, which if used to audit nursing homes, would go a long ways toward improving conditions, provided acceptable quality standards are enforced.

A Patient Quality Service Report

A quality service report made on an individual patient shows that many quality aspects need to be met before the service is considered of acceptable quality. The format shown and the items given cover major aspects but may need adaptation to apply most effectively to a particular situation or to a specified class of patients.

The sources of information for filling out this report are three: observation, interview with the patient, and interview with friends or relatives who have visited the patient and observed the daily routine, especially over a period of time.

At least three different groups of patients need to be studied because of the differences in their health and other needs:

1. Minimum nursing care with complete mobility.
2. Moderate nursing care with high mobility.
3. Full medical care; immobile, bed-ridden.

There are also mental gradations and abilities to communicate: the mentally normal, the retarded, and the abnormal, such as the insane. Consideration will have to be given to the ability to communicate in a meaningful manner; observations and interviews with friends and relatives are required where a meaningful interview with the patient is not possible.

The patient quality service report is quite distinct from the quality report required to measure the quality aspects of the administration, management, and organization of the nursing home. The latter requires a separate form with appropriate headings and quality characteristics on aspects such as:

1. Form of the organization
2. Administration set up
3. Staffing
4. Purchasing
5. Records
6. Financial
7. Supervision
8. Employee relations
9. Personnel
10. Budgeting
11. Patient controls
12. Other controls
13. Procedural manuals
14. Requisitions
15. Training

Quality of Services in a Doctor's Office: A Case Study

Gene Amole, journalist and radio station owner, describes in the following an ear ailment he has and what he had to do in order to receive medical relief.[8] It is significant because it reveals at least four important quality characteristics of medical services in a doctor's office that just about everyone sooner or later encounters: excessive paper work, long waiting time before service, service time, and the impersonal institutional atmosphere.

> "I have nerve deafness. Too many wars. Too many loud phonograph recordings in too many radio studios. It is one of those hearing loss problems for which there is no cure. A hearing aid doesn't help. It just makes the ringing louder. . . .
>
> "My left ear became completely blocked with wax last week. I couldn't hear anything with it, not ever gibberish. That's when I made an appointment with my ear doctor.
>
> "I had been going to this physician for maybe six or seven years. He is a nice man and a talented doctor. But when I arrived for my appointment, the receptionist acted as though I had never been there before.
>
> "She made me fill out new forms about insurance, whom to notify in case of an emergency, where I work, a list of my dependents — all that jazz. I tried to tell her she had the information, but she said there was no record of me. A triage nurse then filled out more forms about the wax in my ear. Then she put me in a little room. That's where my blood pressure went up. There was one stool, one examination bench and one stainless steel cart with little hoses coming out of it.
>
> "A total of 40 minutes had passed and no one had looked at the wax in my ear. I was left to read the instructions on the stainless steel cart and to look at the before-and-after photographs of nose jobs the doctor had done.
>
> "I got up and walked out. A nurse followed me down the hall, telling me it wouldn't be much longer. I quietly told her that 40 minutes was too long to wait. . . . . I had started to remember all the wasted hours I had spent in doctors' offices over the years and I really got steamed.
>
> "This is not a blanket indictment of all physicians. My regular doctor is very punctual. So is my dentist. But there are some doctors who take the attitude that there is nothing wrong with requiring a patient to spend a half-day just to get 10 minutes of treatment. I am not going to put up with that anymore.
>
> "The deadline at this newspaper is 7:15 p.m. If my column isn't ready by then, I get fired. The Federal Communications Commission will take away my radio station if I am repeatedly careless about timing. I have had to accommodate myself to those realities.
>
> "I understand that physicians, particularly surgeons and obstetricians, can't always maintain precise schedules. I am willing to give a little leeway. But I'll never again wait longer than 20 minutes for any medical service. Life is too short to spend reading old magazines.
>
> "The bright side to this is that I found a quick, easy and inexpensive solution to my ear wax problem. By the next day, Sunday, I was in real agony. I went to that little walk-in medical emergency clinic at South Wadsworth Boulevard and West Hampden Avenue.
>
> "There was a minimum of paper work. A medical doctor and three nurses took care of my problem in a matter of minutes. The whole thing cost 26 bucks. They even filled out my insurance forms for me. There was a little sign on the desk that said I was welcome. I appreciated that."

Millions have been through what Amole describes and many no doubt have felt the same way:

- Wasting hours in a doctor's office — spending half a day to get 10 minutes of treatment.
- Life is too short to waste time reading old magazines.
- Many of us cannot be careless about time, about deadlines and delays, and about time schedules without facing a severe penalty.

Amole, the buyer of services, from now on is setting the time specifications. He is setting a tolerance on waiting time: No more than 20 minutes or he leaves.

He perceives and appreciates the kind of acceptable quality service millions of others want:

- Minimum paper work.

PATIENT QUALITY SERVICE REPORT

Name ______________________ Room __________ Date ______________

Directions: Use words which show evidence of unacceptable quality service such as errors, delays, neglect, unsafe, theft, abuse, fear, complaints. Use words to describe acceptable quality service: these could be OK, yes, safe, no evidence, no complaint, cooperative.

1. Room
   1. furnishings ______
   2. bed ______
   3. bedding ______
   4. decorations ______
   5. attractiveness ______
   6. light ______
   7. ventilation ______
   8. heat ______
   9. privacy ______
   10. roommate(s) ______
   11. space ______
   12. closet, drawers ______
   13. other ______

2. Food
   1. as prescribed ______
   2. enough ______
   3. nutritious ______
   4. meals on time ______
   5. eating atmosphere ______
   6. effect on weight ______
   7. eating assistance ______
   8. other ______

3. Medical
   1. correct medicines ______
   2. correct dosages ______
   3. taken on time ______
   4. nurse's care ______
   5. doctor's care, surgery ______
   6. supervision ______
   7. other ______

4. Personal care
   1. bath ______
   2. dressing ______
   3. hair ______
   4. personal ______
   5. clean clothes ______
   6. sanitation ______
   7. assistance ______
   8. other ______

5. Security
   1. protection from stealing ______
   2. protection from people ______
   3. protection from abuse ______
   4. protection from fire ______
   5. other ______

6. Safety, convenience
   1. wheel chair ______
   2. attendant, aide ______
   3. assistance as needed ______
   4. ramps ______
   5. rods, special devices ______
   6. other ______

7. Personal attention
   1. response to calls ______
   2. response to requests ______
   3. treatment by employees ______
   4. conversation ______
   5. other ______

8. Financial
   1. money allowance ______
   2. SS checks ______ other checks ______
   3. medical bills ______
   4. nursing home bill ______
   5. Other ______

9. Employees' competence
   1. training ______
   2. attitude ______ ______
   3. reliability ______
   4. breaks ______
   5. other ______

10. Other patients
    1. attitude of ______
    2. treatment of ______
    3. conflicts with ______
    4. health problems of ______
    5. other problems of ______

11. Other
    1. treatment of visitors ______
    2. mail ______
    3. religious services ______
    4. entertainment ______
    5. recreation for mobile patients ______
    6. recreation for wheel chair cases ______
    7. changing services ______
    8. leaving, discharge ______
    9. other ______

- Short waiting and service times.
- Low cost.
- Helpful related services such as filling out insurance forms.
- A kind atmosphere: the patient is welcome, there is a sign to show it, and behavior to prove it.

Waiting Times and Costs of Medical Services

The National Health Center for Health Services Research of the Department of Health and Human Services in Washington made a health study of about 37,000 individuals during 1977 and 1978 which yielded the waiting times and costs shown in Table 2. Four important service times are missing:

1. Time the doctor actually spent with a patient (some data are given in Table 3).
2. Time required for emergency service.
3. Time required for outpatient service.
4. Time the dentist actually spent on a patient.

Also missing is a measure of the variations in service time. Averages are not enough; frequency distributions are required to give an accurate picture of service times.

Table 2

Average Waiting Times and Cost for Doctor and Dentist Services United States 1977-1978

Item	Average value		
	Total	White	All others
Waiting for doctor's appointment (days)	7	—	—
Waiting in doctor's office (minutes)	29	28	40
Waiting in emergency room (minutes)	38	—	—
Waiting as hospital outpatient (minutes)	45	41	61
Charges per doctor visit — 1977	$21.29		
U.S. region			
West	28.01		
North Central	18.08		
North East	21.02		
South	20.39		
Charges per dental visit — 1977	$31.71		

Table 3 shows the distribution of time actually spent with a physician; it is derived from a nationwide probability sample of physicians in United States for 1979. Excluded from the frame sampled are Alaska and Hawaii, as well as radiologists, pathologists, anesthesiologists, and federally employed doctors.

Part A shows the time interval distribution based on number of visits; the percentage distribution is also shown. A value of "0" means that the person did not see the doctor.

Part B shows how the average (mean) and standard deviation are calculated from the interval distribution. The value of "0" is omitted in these calculations. It is assumed in these calculations that time in minutes is read to the nearest minute — 10.6 minutes is recorded as 11 minutes; 10.4 minutes is recorded as 10 minutes. The interval of "31 or more" is assumed to have an average value of 40 minutes.

The estimates are as follows:

1. median = 11.6 minutes or rounded = 12 minutes.
2. mean $\bar{x} = 13.99$ minutes or rounded = 14 minutes.
3. variance $s^2 = 86.91$.
4. standard deviation s = 9.32 or rounded = 9 minutes.
5. coefficient of variation $cv = \frac{s}{\bar{x}} = 0.67$ or 67%.

The distribution shows that:

Thirteen percent were served in 5 minutes or less.

Forty-four percent were served in 10 minutes or less.

Seventy-two percent were served in 15 minutes or less.

Ninety-four percent were served in 30 minutes or less.

A "visit" is defined as a direct personal exchange between an ambulatory patient (one able to walk) and a physician, or between a patient and a staff member working under the physician's supervision. The time has three parts: waiting in cubicles; nurses or others taking weight, blood pressure, blood samples, etc.; and time actually spent with the doctor. These times need to be obtained to determine more exactly how much time the doctor actually spends with a patient.

Averages as standards are not enough. It would be dangerous to use these averages as standards because the reasons for variations are unknown. More detailed studies are needed to determine actual causes and factors affecting waiting times. An analysis of what causes of variation have to be accepted and what causes can be reduced or eliminated also has to be developed. As stated above, and as shown in Table 3, it is necessary to obtain frequency distributions of various times in order to obtain a more exact picture of performance; averages are not enough. Some major factors affecting times and charges are given below:

1. Time required to receive an appointment
   - 1.1 Doctor may be booked up
   - 1.2 Doctor may be on vacation
   - 1.3 Doctor may be going to a convention or at a convention
   - 1.4 Patient lacks timely transportation
2. Time waiting in a doctor's office
   - 2.1 Doctor over-booked
   - 2.2 Emergency
   - 2.3 Patients require more time than expected
   - 2.4 Doctor late in arriving at office
3. Charges
   - 3.1 Seriousness of the complaint
   - 3.2 Medical technology required
   - 3.3 Medical tests required
   - 3.4 Surgery or not
   - 3.5 Functional or structural trouble
   - 3.6 Difficulty of diagnosis
   - 3.7 Customary fee or not

Table 3
Time Person Actually Spent with Physician
—United States 1979

Part A

Duration in minutes	Number of visits	Percentage
0	18,997,000	3.41
1-5	67,610,000	12.15
6-10	169,217,000	30.42
11-15	149,291,000	26.84
16-30	118,171,000	21.24
31 or more	33,027,000	5.94
Total	556,313,000	100.00

Part B
(Excludes 0)

Interval (minutes)	Midpoint x	Frequency f (1,000)	fx	fx^2
Less than 5.50	2.75	67,610	185,927.5	511,300.63
5.5-10.49	8	169,217	1,353,736	10,829,888
10.5-15.49	13	149,291	1,940,783	25,230,179
15.5-30.49	23	118,171	2,717,933	62,512,459
30.5 or more	40	33,027	1,321,080	52,843,200
Total		537,316	7,519,459.5	151,927,026.63

Source (Part A): **1979 Summary National Ambulatory Medical Care Survey, from advance data, No. 66,** March 2, 1981, National Center for Health Statistics, U.S. Dept. of Health and Human Services.

Calculations

1. median $= 10.5 + 0.213(5) = 11.57$
2. mean $\bar{x} = 7{,}519{,}459/537{,}316 = 13.99$
3. variance $= 46{,}696{,}080/537{,}315 = 86.91$
4. standard deviation $s = 9.32$
5. coefficient of variation $cv = \frac{s}{\bar{x}} = 0.67$ or 67%

Medicaid Quality Control (MQC)

Medicaid quality control systems are required by Federal law and are in operation in all states except Arizona. Under this system, states are required to review a sample of certified Medicaid cases and claims to detect errors and incorrect payments. Then the federal Health Care Financing Administration (HCFA) selects a subsample of these cases to evaluate state findings.

The total number of cases in all of the state semiannual samples is about 75,000; they are selected from an estimated frame which totals about 9.2 million cases monthly. The federal subsample totals about 16,000 cases or about 20 percent of the state samples. The subsample is used to compute the total error rate, and the eligibility, claims processing, and third party error liability error rates for every state for each review period. Two types of error estimates are calculated for each state: the percentage of cases in error and the percentage of money payments in error.

A primary purpose of the MQC system is to identify case and payment errors by states so that corrective action can be taken to eliminate the source. Analysis covers characteristics of error cases, identifying trends, and comparing performance of states. Efforts are made to identify specific causes of both high and low error rates. The states are under pressure to keep their error rates down since target error rates are set for each state by the HCFA. Failure to meet these rates can result in a reduction of Medicaid payments to the state.

Target rates vary from 6.2 percent to 12.3 percent for eligibility payment error rates. No explanation is given as to why the rates are so high when money payments are involved, or why the target rates vary so widely for the several states. Why 6.2 percent is the lowest target error rate is not explained either.

The total sample size of 75,000 and the subsample size of 16,000 are of little value in evaluating the samples because error rates are estimated for each state. What is important is the sample plan used by each state, the size of the sample, how it is drawn, and how it is implemented and processed. No information is given on specific sample plans used by the states, or by the HCFA in selecting the subsamples from each state and estimating the error rates by states.

Table 4
Medicaid Case Error Rates by Type of Error for U.S. with High and Low Error Rates Identified by State

Type of case error	U.S. totals		Apr-Sept 1979		Oct 1979-Mar 1980	
	Apr-Sept 1979	Oct-Mar 1979-80	low	high	low	high
Ineligible	6.9%	5.5%	0.7 Nev.	22.2 Fla.	1.0 Nev.	24.1 N.D.
Liability overstated	0.3	0.4	0.0 many	1.5 Neb. Col.	0.0 many	7.8 Okla.
Claims processing	3.5	3.0	0.0 Tex. W.Va.	32.4 Neb.	0.1 Tex.	29.0 Neb.
Third party liability	0.7	1.0	0.0 many	3.3 Neb.	0.0 Nev.	5.1 Mich.
U.S. total	11.4%	9.9%	4.2 Nev.	39.3 Neb.	4.2 Tenn.	33.4 N.D.

Source: ***Medicaid Quality Control Report, October 1979-March 1980, April 1979-September 1979, July 1978-December 1978.*** Health Care Financing Administration, Department of Health and Human Services, Washington, D.C.

No sampling errors are given for the percent errors or for the bias that may be in the data and estimates due to variations in the interpretation of rules, regulations, and terms, and in reporting data at the source. The sample data are adjusted at the federal level to obtain the final state error rates. This method of adjustment is not described so we do not know how much bias this method may add to the final figures. The federal MQC may result in an undesirable situation, if it leads the states to expect the federal subsample to pick up the errors and thus become careless in their own operations. MQC should be a sample audit; the responsibility for doing an acceptable quality job lies strictly with the states. Quality control is the job of the state that does the original operation. There should be a continuous quality control program, not some twice-a-year sample test. This means corrective action is not immediate, as it should be, but delayed by six months, a year, or even longer.

It should be emphasized that these errors are limited to errors made by the state agency in processing claims and making payments. There is nothing given about auditing the medical bill at the source to determine whether the bill consisted of charges that were valid, accurate, and reasonable.

Table 3 summarizes national and state data by type of case error for two six-months periods in 1979 and 1980.

Stop Errors Because Errors Kill People

In some medical service operations where only one error (1 error) cannot only be dangerous but fatal, the goals of Zero Errors and Zero Delays should dominate every person every minute of every day. This means steps must be taken to prevent special errors due to individuals *and* system faults due to management. There are many situations, such as the three cases given below, where *both individuals and management are to blame.* We cite three cases reported in the press that occurred during May 1983.

Case 1. A man in a nursing home suffering from emphysema falls and breaks a bone. Four days later he is taken to a hospital. They see fit to put him on oxygen. An oxygen cylinder or tank is rolled out, and he is connected to it. Within 15 minutes he is dead. The cylinder contained carbon dioxide (CO_2), not oxygen (O_2).

Case 2. A girl suffering from pains enters the emergency room of a hospital. She is presumably given phenobarbital mixed with some other drug. Cocaine is kept in a locked cabinet in bottles very similar in appearance to those for phenobarbital. The cocaine bottle is grabbed instead of the correct drug. The girl goes into convulsions, and they give her another drug that only hastens her death.

Case 3. A military hospital has an oxygen supply system. It contracts to a private company to keep it supplied with tanks of oxygen. After two patients die and another becomes critically ill, an examination shows that a cylinder of argon was connected to the system, not a cylinder of oxygen.

Who is at fault? The system that stores carbon dioxide and oxygen together, does not label them prominently, and does not paint them sharply different colors? Or is it the individual that does not know the difference between carbon dioxide and oxygen, does not read labels, does not monitor carefully the behavior of the patient, and does not verify the action taken? Or are both of them to blame?

Who is at fault? The system with an emergency room in which cocaine and phenobarbital are kept in similar looking bottles in the same cabinet? Or is it the person who did not check the labels, did not realize these bottles should be widely separated, and was unaware of the dangerous situation that could arise? Or are both to blame?

Who is at fault when argon is substituted for oxygen? Is it the system that fails to sharply identify cylinders of argon from cylinders of oxygen? Or is it the individual who does not know how to distinguish one cylinder from another or who does not verify with the supervisor that the correct cylinder is being selected? Or is it the hospital that takes no steps to verify what is being delivered to the hospital's oxygen system? At all points, both the system and individuals are lax and are doing nothing to prevent a fatal error.

Error prevention. Below are some steps that management can take to insure that the probability of fatal errors occurring is very close to zero:

1. Storage facilities.
   Wide separation of containers with clear identification of what is in each area.
2. Labeling; identification.
   Use large easily read labels; use large painted names.
3. Containers.
   Avoid putting dangerous items in containers that look alike.
4. Eternal verification: The verifier.
   In any dangerous situation, application, and connection, be sure to verify 100 percent that which is being used, applied, and connected is what it is supposed to be. **Take nothing for granted.** This verification must be done by an independent person — foreman, supervisor, monitor, or verifier. It might pay to create the job of "verifier" to insure that no mistake is made.
5. Hiring personnel.
   Hire only those persons who are knowledgeable, careful, literate, and preferably technically trained in the chemicals and other materials they are handling.
6. Training.
   If personnel are not technically trained, then it is necessary to conduct safety courses periodically to insure that they are aware of the goal of Zero Errors and Zero Delays and the continuous program of verification that is necessary to insure that no errors are made.

What Deming's 14 Points mean in terms of health services received directly by the patient.

In concentrating on the initiation and improvement of quality in the services offered by hospitals and nursing homes and related agencies, it is easy to get so absorbed in vendors, purchasing, costs, supplies, inspection, testing, and other factory-like aspects that the Number One quality responsibility is forgotten: THE QUALITY OF THE MEDICAL AND NON-MEDICAL SERVICES RENDERED TO THE PATIENT. All other quality control is simply supportive. The following illustrate what the 14 points of Dr. Deming mean when translated into terms applicable to the quality of health services received by the patient.[9]

1. Constancy of purpose. The constant purpose should be rendering acceptable quality health service to patients at an affordable price. This means adequate attention and care for every patient despite the wide variations in physical and mental health. This purpose must include patient safety, security, and satisfaction.

2. New age. "We can no longer live with commonly accepted levels of . . . mistakes" such as:

- Using a cylinder of carbon dioxide instead of a cylinder of oxygen (1 dead).
- Connecting a source of argon to an oxygen line (2 dead).
- Mistaking a bottle of cocaine for a bottle of phenobarbital (1 dead).
- Oxygen valves that stick on an anesthesia machine (2 dead).
- Skimpy feeding in nursing homes resulting in malnutrition (x dead).
- Interchanging the operations of two women.
- Raping a patient in a nursing home.
- Stealing patients' belongings in nursing homes: money, clothing, jewelry, food, wigs.
- Fabricating services and bills for Medicare to pay.

3. Cease dependence on mass inspection.

- Insist on Zero Defects; eliminate inspection by proper quality control on vendors.
- Zero errors to be applied to all patients' services.
- Test all equipment daily to see that it operates properly and safely.
- Maintain a continuous error prevention program; keep daily time, error, and performance records of all personnel who are directly connected with quality of services rendered to the patient.

4. Select vendors based on quality performance as well as price. To the patient the health facility is the vendor.

- What the patient wants from the health facility is acceptable quality service at an affordable price — the same that the health facility wants from its vendors.
- What the patient wants is service within a reasonable time as well as the elimination of all unnecessary medical and non-medical services.
- The potential patient can select vendors of certain kinds, such as doctors and nursing homes that are available, but usually there is very little choice of a hospital, except in large cities. Usually the complete absence of knowledge makes it impossible to select a health facility on the basis of quality performance.

5. Constantly improve the system.

- By constantly searching for problems and trouble spots, and their elimination.
- By a continuous error prevention program and an adequate time-use program.
- By proper emphatic labeling, so cylinders of carbon dioxide are not confused with cylinders of oxygen.
- By proper shelving, labeling, and separation of drugs and medicines so mix-ups are prevented.

6. Training. To reduce the cost of unnecessary training, strict standards should be imposed when hiring all personnel with regard to technical knowledge, safety records, attitudes, work habits, and other relevant factors affecting patient service quality.

- Training in medical care as needed: error prevention, reducing delay times, and prompt responses to patients' needs.
- Training in non-medical care: importance of individual differences, attitudes, conversation, treatment, and adequate and prompt responses to patients' requests and needs.

7. Supervision. The supervisor must set examples, assist and coach, and do the training.

- Maintain and use daily error, time, and performance records on all individuals.
- Shift those who are not flexible enough to handle adequately the wide variety of individuals encountered in a health facility.
- Hold individual workers accountable for high quality work and let them know it.

8. Drive out fear. The fears to drive out are those which the patient has developed as a resident of the medical facility. These are actual cases:

- an insane roommate
- a thieving employee
- an aggressive patient

- an indifferent and callous aide
- delays in answering an urgent call

9. Break down barriers. A health facility is a human relations operation 100 percent. Products, equipment, and devices are vital, but only to the extent that they promote health, relieve pain, and treat patients as human beings.

- It is imperative to have the close cooperation of all units: front desk, admission, housekeeping, food service, nurses, doctors, surgery, radiology, laboratory, pharmacy, discharge, volunteers, and visitors. This cooperation must be aimed at rendering adequate service to the patient, not for the purpose of making it easier for others, unless justified.

10 and 11. Eliminate numerical goals, work standards, and slogans.

- Time, error, and performance records are used daily to eliminate errors and reduce delays.
- When one error can be fatal, you do not construct a time chart or control chart and wait for the error to occur.
- The 85-15 percent division does not apply since many, if not most, of these dangerous errors or fatal errors are made by individuals. Fifteen percent is too low and should not be used to prevent holding all employees strictly accountable for what they are doing. An error prevention program is imperative.
- Neither do you wait for a weekly or monthly report to find out what is wrong. Control must be continuous throughout the day.

12. Remove barriers that hinder the hourly worker.

- Remove barriers to job efficiency and job satisfaction; restore dignity to work.
- Hire those who do not bring these barriers with them.
- Be sure each person's job is described in writing, in detail, so each one knows the major tasks and what is expected, including attitudes as well as knowledge and skills.
- Do not forget that the patient, no less than the hourly worker, needs to be treated with dignity.

13. Education and training programs, including retraining.

- Two needs: polish present knowledge and skills; and learn new methods, techniques, skills, abilities, and arrangements as the need arises.
- Continuous training programs are needed in error prevention, delay time prevention, and optimizing service time.
- Training should be based on findings from daily time, error, and performance records.
- Training should be based also on patient's complaints, on relative's complaints, and on complaints from other sources.

14. Top management and all supervisors should push every day on the above to see that they are carried out, as well as all other aspects relating to quality of health care of the patient.

Notes

[1]John P. Werner, "Application of a QA program to a Hospital Facility," *34th Annual Technical Conference Transactions,* American Society for Quality Control, Atlanta, Georgia, May 20-22, 1980, pp. 313-322. Used by permission.

[2]Source of this material is a professional specialist, Margaret Rosander, who served as a volunteer at Porter Memorial Hospital, Denver, Colorado, for two years, 1971-1973.

[3]For a detailed description of Youden's test of the variability among 29 laboratories, see Chapter 25.

[4]*Rocky Mountains News,* March 24, 1980, p. 32.

[5]Associated Press Dispatch, *Rocky Mountain News,* April 5, 1980.

[6]*Rocky Mountain News,* April 10, 1980, p. 5.

[7]Editorial in the *Rocky Mountain News,* April 9, 1979, p. 48.

[8]*Rocky Mountain News,* Arnole's Corner, July 30, 1981, p. 4.

[9]W. Edwards Deming, *Quality, Productivity, and Competitive Position,* Massachusetts Institute of Technology, Cambridge, Massachusetts, 1982, pp. 16-17, 240-245.

7
Transportation

Airlines

Early applications. Lobsinger of United Airlines was one of the first, if not the first, to apply quality control techniques to airline operations. These applications included control over reservation errors, reducing the number of excess meals on flights, and comparing performance of one branch office with another. It was found, for example, that offices that were rated high by management did not rate so high when subject to objective quality control measurements of performance.[1]

Dalleck described how probability sampling of passenger ticket coupons was used to settle interline passenger accounts of an airline in cases where a passenger is carried by more than one airline. The coupons were stratified by four types of fares and sampled at different rates using serial number endings on the coupons. Since fares varied with the number of miles flown, a regression type estimate was used. The procedure was used successfully and at considerable savings by three airlines when the volume of interline passenger tickets increased greatly during the 1950's.[2]

Quality control areas. For the purpose of presentation, it is convenient to identify three distinct areas to which quality control can be applied and to list the several characteristics under each category: passenger service, airport facilities, and airline operations. Under **passenger service** it is convenient to list quality control aspects under three headings:

1. Before a flight:
   a. Reservations: methods of obtaining, methods of payment, errors, delay time, restrictions, service with various rates.
   b. Waiting time at check-in counter.
   c. Waiting time at ticket-buying counter.
   d. Security check: false checks; examining contents of bags, cameras and film; delay; inconvenience getting possessions after check; variations in detection in various airports; messing up bag contents.
   e. Waiting time in lounge.
   f. Getting seat assignments: smoking and non-smoking sections.
   g. Waiting time for flight.
   h. Waiting time for departure, take off.
   i. Bumping, cancellations, over-booking.
   j. Priorities in boarding.
2. During a flight:
   a. Seating.
   b. Storage of clothes, bags.
   c. Safety demonstrations.
   d. Flight information by flight attendant over public address system.
   e. Flight information by captain over public address system.
   f. Passing out reading materials, magazines.
   g. Serving alcoholic beverages: top priority, delays in food serving.
   h. Serving other beverages.
   i. Serving food, meals: delays, quality, service, nature, sufficient, meeting individual needs.
   j. Smoking and non-smoking sections: does this protect non-smoker.
   k. Giving connections with other flights and procedures to follow.
   l. Temperature and air: comfortable, too cold, too hot, drafty.
3. After a flight:
   a. Arrival time: amount of delay.
   b. Time required to walk from plane to carousel.
   c. Waiting time for baggage at carousel.
   d. Condition of baggage.
   e. Transportation connections: taxi, bus, limousine.
   f. Nearby accommodations.
   g. Waiting time for transportation.
   h. Courtesy buses.

Under ***airport facilities,*** the following need to be considered:

1. Parking: ease, convenience, accessibility.
2. Cost of parking.
3. Car rentals.
4. Airport restaurants.
5. Airport shops.
6. Transportation to and from.
7. Waiting rooms.
8. Rest rooms.
9. Courtesy buses to motels, hotels.
10. Solicitation.
11. Surface traffic control.
12. Paying for parking and its control.

Under ***airline operations,*** the following are included:

1. Flight scheduling: dropping, adding, shifting flights.
2. Reliability of service: breakdowns due to weather, strikes, mechanical failure, baggage system failure (belt breaks).
3. Overbooking practices.
4. Service relative to costs: convenience, safety, reliability, delays.
5. Percent of capacity utilized.
6. Quality control program, inspections.
7. Cancellation of flights.
8. Conditions for reduced fares.
9. Various service times.
10. Various delay times.

Under ***baggage handling,*** the following are important:

1. Time required from plane to carousel.
2. Time waiting for all pieces of baggage.
3. Extent of damage, if any.
4. Not delivered, not unloaded, misrouted.
5. Lost baggage.
6. Baggage on wrong carousel.

7. Damage and lost claims.
8. Time required to find and deliver baggage.

Complaints of consumers, presumably meaning passengers, to the Civil Aeronautics Board.[3] The Civil Aeronautics Board (CAB) which regulates interstate airlines in the United States issues monthly and annual reports on the complaints received from passengers on these airlines. The report contains three kinds of complaint data:

1. Number of complaints by 13 types, by airlines.
2. Number of complaints per 100,000 passengers, by airlines.
3. Number of complaints by 13 types, by industry group.

Data illustrating these three aspects of complaints are given in Tables 1, 2, and 3, respectively.

The frequency and percentage distributions of the 13 kinds of complaints for U.S. airlines for the calendar year 1980 are given in Table 1. The types of problems CAB places under each of the 13 categories are listed on the sheet entitled "Notes to Report".

Table 1 shows that the top three categories of complaints dealt with flight problems, baggage, and customer service. The record shows that these were the three most common complaints in the same order during calendar year 1979. Oversales also ranked high: fourth in 1980 and fifth in 1979. No data are given on the extent to which those who complained during 1979 also complained during 1980; in other words, there was no matching of complaintants for the two years. The six U.S. airlines with over 1,000 complaints during 1980 were, in order from the highest to lowest:

TWA	2,402	complaints
Pan Am/National	2,039	"
Eastern	1,783	"
American	1,477	"
United	1,202	"
Braniff	1,155	"

Definition of Complaint Terms Used in Civil Aeronautics Board Report

NOTES TO REPORT

This report is based on informal consumer complaints the Board has received by mail or telephone. We have not determined the validity of each complaint. The types of problems included in each category are:

Flight Problems: Cancellations, delays, or any other deviations from schedule, whether planned or unplanned.

Oversales: All bumping problems, whether or not the airline complied with CAB oversale regulations.

Reservations and Ticketing: Airline or agent mistakes in reservations and ticketing; problems in making reservations and obtaining tickets due to busy telephone lines or waiting in line, or delays in mailing tickets.

Fares: Incorrect or incomplete information about fares, discount fare conditions and availability, overcharges, fare increases and level of fares in general.

Refunds: Problems in obtaining refunds for unused or lost tickets.

Baggage: Lost, damaged or delayed baggage claims, charges for excess baggage, carry-on problems, and difficulties with airline claim procedure.

Customer Service: Rude or unhelpful employees, inadequate meals or cabin service, treatment of delayed passengers, and discriminatory treatment.

Special Passengers: Handicapped passengers, passengers on stretchers, children, elderly passengers, passengers requiring oxygen or other medical care.

Smoking: Inadequate segregation of smokers from non-smokers, failure of airline to enforce no-smoking rules, objections to the rules.

Advertising: Advertising that is unfair, misleading, or offensive to consumers.

Credit: Denial of credit, interest or late payment charges, incorrect billing, or incorrect credit reports.

Tours: Problems with scheduled or charter tour packages.

Other: Cargo problems, security, airport facilities, claims for bodily injury, lack of adequate service, and other problems not classified above.

Table 1
Consumer Complaints to Civil Aeronautics Board Relative to U.S. Airlines 1980

Rank	Category of complaint	Number	Percent
1	flight problems	4,169	22.2%
2	baggage	3,423	18.2
3	customer service	2,640	14.1
4	oversales	1,947	10.4
5	refunds	1,828	9.7
6	fares	1,278	6.8
7	reservations, ticketing	1,113	5.9
8	smoking	587	3.1
9	advertising	218	1.2
10	special passengers	206	1.1
11	credit	197	1.0
12	tours	137	0.7
13	other	1,033	5.5
	Total	18,776	100.0%

Source: Tabulated and summarized from Civil Aeronautics Board/Consumer Complaint Report for calendar year 1980.

Table 2 shows the number of complaints per 100,000 passengers for major U.S. airlines during 1979 and 1980. All airlines carried fewer passengers in 1980 than in 1979, except Northwest and Piedmont. Texas International is the only airline on the list for which the complaints per 100,000 passengers increased from 1979 to 1980.

Decline in these ratios from 1979 to 1980 does not mean that the quality of service necessarily improved; for one reason the three top classes of complaints were the same for the two years. Except for Texas International, the actual number of complaints decreased, in some cases very markedly, between 1979 and 1980. American, Western, and United, decreased more than 50 percent.

Since no study was made matching the 18,776 complaints in 1980 with the 28,848 complaints in 1979, one can only speculate on what happened. Plausible explanations are:

1. Complaints tended to come from a new group of passengers; biased samples.

2. Some complainants stopped complaining.
3. Some complainants stopped flying.
4. Some complainants shifted airlines.
5. Flights with higher complaint rates tended to be dropped.

Table 2
Number of Complaints per 100,000 Passengers on Major U.S. Airlines 1979 and 1980

Airlines (alphabetical order)	Complaints per 100,000 passengers	
	1980	1979
American	5.52	10.07
Braniff	9.26	14.41
Continental	5.09	8.59
Delta	1.41	2.20
Eastern	4.42	6.86
Frontier	3.45	5.09
Northwest	4.93	8.94
Pan American	12.77	15.68
Piedmont	2.95	5.49
Republic	5.33	6.36
Texas International	12.96	10.69
TWA	11.72	12.24
United	3.45	7.91
Western	4.83	9.10

Source: Tabulated from the Civil Aeronautics Board Consumer Complaint Report for calendar year 1980.

Table 3 shows the distribution of complaints by industry groups for the year 1980. Over 1,000 complaints were made against tour operators and travel agents combined, so clearly there is a need for more attention to the quality of services in these areas. With regard to foreign airlines, the first five major complaints against them were the same as those against U.S. airlines, except that the order is different with baggage being a strong Number 1:

	Number	Percent
Baggage	792	28%
Oversales	460	16
Flight problems	426	15
Customer service	347	12
Refunds	321	11
All others	506	18
Total	**2,852**	**100%**

Table 3
Number of Complaints to Civil Aeronautics Board by Industry, 1980

Industry group	Number	Percent
U.S. airlines	18,776	81%
Foreign airlines	2,852	12
tour operators	752	3
travel agents	358	2
cargo-freight fwd	198	1
other carriers	322	1
Total	**23,258**	**100%**

Source: Tabulated and arranged from data in Civil Aeronautics Board Consumer Complaint Report for calendar year 1980.

Table 4 gives the number of complaints reported to the Civil Aeronautics Board in Washington for January 1980 and for the year 1980 by an airline. The types are arranged in order beginning with the one most frequently mentioned, which was baggage. Tables of this kind have to be used with the caution in mind that those complaining to Washington may be quite different from those complaining directly to the company.

Table 4
Complaints Received by the Civil Aeronautics Board for December 1980 and the Year 1980 about a Major Airline

Type of complaint	Year 1980	December 1980
baggage	131	15
flight problems	127	16
customer service	92	7
oversales	89	10
fares	40	3
refunds	32	2
reservations, tickets	28	0
all other	38	4
Total	**577**	**57**

Source: Tabulated and arranged from data in the Civil Aeronautics Board Consumer Complaint Report for calendar year 1980.

A copy of an airline rating scale is reproduced below. There are 16 items to be rated poor, fair, good, or excellent. The categories are: telephone reservations, check-in service, flight information, cabin, cabin personnel, meal, inflight literature, and overall impression of the flight.

Since actual complaints are much more specific than the items on a rating scale, the effectiveness of a rating scale for quality control purposes is discussed in detail later in this chapter.

Sample Airline Rating Scale

	Excellent	Good	Fair	Poor	
					Telephone Reservations
1	○	○	○	○	courtesy of Reservation Personnel
2	○	○	○	○	efficiency of service
					Check-In Service
3	○	○	○	○	baggage handling
4	○	○	○	○	courtesy of Airport Personnel
5	○	○	○	○	efficiency of service
					Flight Information
6	○	○	○	○	en route
7	○	○	○	○	on the ground
					Cabin
8	○	○	○	○	cleanliness
9	○	○	○	○	comfort
					Cabin Personnel
10	○	○	○	○	appearance
11	○	○	○	○	courtesy
					Meal
12	○	○	○	○	appearance
13	○	○	○	○	quantity
14	○	○	○	○	quality
15	○	○	○	○	**Inflight Literature**
16	○	○	○	○	**Overall Impression of this flight**

Advantages and disadvantages of customer complaints. Customer complaints, especially if they are specific, have both advantages and disadvantages. Advantages are:

1. If the same complaint occurs with a high frequency, it may reflect a fault, defect, or weakness in the system or operation.

2. They may be serious enough so that, if corrective actions are not taken immediately, a loss of customers may result.

3. Some customers may be greatly upset and may consider a change in service company even though the management of the service company may think that the complaint is minor or even insignificant.

4. They give the company an opportunity to see the buyer's side of how good the service operations and transactions really are.

5. They may reveal trouble spots management is not aware of — areas that need careful and immediate attention.

Disadvantages are:

1. There is always the danger that the complaints come from a highly biased sample. This does not invalidate the complaints, but does mean that some complaints have to be used with caution.

2. Complaints to the Civil Aeronautics Board may differ sharply from those made directly to the airline. These differences need to be considered carefully.

3. Many passengers with complaints may never complain to the company. They may complain to others. If the complaint is serious, they may shift to another airline. This is why a probability sample study of *all* passengers is really necessary.

4. Those who complain may be a special group; they may be vocal and outspoken. This, if true, leads to the biased sample under item 1 above.

5. Complaints may be over minor or even trivial aspects or occasions as viewed by the airline. What the seller of a service thinks unimportant may not be unimportant to the buyer. This is why all complaints need to be answered.

Measuring airline performance. Airlines are faced with the problem of measuring and improving the performance of a wide variety of airline operations at all of their airports. The following partial list illustrates the nature and the scope of these measurements and activities. The measurements and observations would be taken at each airport and compared on a weekly and monthly basis with company standards of acceptable quality service. A distinction, however, needs to be made between situations and conditions which can be controlled by the personnel at each airport and those which are outside their control. The extensive use of time distributions and time deviation distributions is recommended because of the top significance these objective measures have in determining the quality of service. Examples are:

1. Passenger wait at the ticket counter: time distribution.
2. Number and percentage of flights overbooked, oversold.
3. Timely passenger boarding: time distribution.
4. Delay in scheduled departure time: time distribution.
5. Delay in scheduled arrival time: time distribution.
6. Summary of delay times (average and variability) by major sources.
7. Quality of cabin service:
   - announcements
   - beverage service
   - food service
   - passenger assistance
   - courtesy, attitudes
8. Baggage delivery time: time distribution of waiting time at carousel.
9. Number and rate of baggage claims: lost bags, damaged bags, bags with missing contents.
10. Baggage mishandling: cost to company, average amount of settlement, distribution of time required to settle claims.
11. Distribution of passenger complaints by service area and specific complaint.
12. Grounded flights: number, rate, delay time, how handled, cost.

Delay time and other measures of quality. Table 5 shows the distribution of times required to wait in line before receiving service at the ticket counter at Airport A during May 1981. Although the average waiting time was 2.5 minutes, about 22 percent did not have to wait at all. The proportion waiting 5.5 minutes or less was 0.88 or about 9 out of 10. Seventeen persons or about 12 percent waited more than 5.5 minutes because of long lines or because of the absence of sufficient agents.

Table 6 shows the distribution of waiting time at the carousel for passengers on 80 major flights in continental United States. The average time was 23 minutes, with a range from 4.5 minutes to 39.9 minutes. It shows that it is rare to obtain *all* of your baggage within 9.5 minutes; many times two pieces of baggage, although placed on the plane (apparently) at the same time, do not arrive at the carousel at the same time — most of the time they do not. The delay time shown is the time required to obtain *all* of your baggage. Distributions of this type are greatly needed because this delay time is crucial for those who have to meet the schedules of other flights, limousines, buses, and appointments.

Table 7 shows the distribution of baggage problems on 140 flights, including the number of passengers and the number of baggage problems per 1,000 passengers. The data are fictitious, but illustrate the type of distribution needed in order to show the nature and extent of the baggage problem.

Table 8 shows the distribution of delay times for 100 flights. This is a very important aspect of quality in airline operations. About 70 percent of the flights were delayed less than 14.5 minutes which would be considered satisfactory to most airlines. About 10 percent were delayed more than half an hour (29.5 minutes), which many would consider unsatisfactory, a view which needs to be modified to the extent that some delays were due to weather conditions or related airport conditions such as snow, which cannot be avoided.

Table 5
Waiting Time Required at Ticket Counter Airport A, May 1981 *

Minutes			Number of		Calculations	
Waiting	Interval	Midpoint x	passengers n	Percent	nx	nx^2
0	0	0	32	22.2	0	0
1-2	0.0-2.5	1.25	47	32.6	58.75	73.44
2-3	2.5-3.5	3	24	16.7	72	216
3-4	3.5-4.5	4	15	10.4	60	240
4-5	4.5-5.5	5	9	6.2	45	225
5-6	5.5-6.5	6	5	3.5	30	180
6-7	6.5-7.5	7	4	2.8	28	196
7-8	7.5-8.5	8	4	2.8	32	256
8-9	8.5-9.5	9	3	2.1	27	243
9-10	9.5-10.5	10	1	0.7	10	100
Total			144	100.0%	362.75	1,729.44

*It is assumed that time is read to the nearest minute so that 0 means no waiting time at all; four minutes means more than 3.5, but less than 4.5.

Calculations

1. mean $\overline{x} = 362.75/144 = 2.52$ minutes.
2. variance $s^2 = \left(\frac{1}{143}\right)(1729.44\text{-}362.75^2/144) = 5.70$.
3. standard deviation $s = 2.39$.
4. coefficient of variation $cv = 2.39/2.52 = 0.95$ or 95 percent.
5. proportion p less than 5.5 minutes $= 127/144 = 0.882$ or 88 percent.
6. proportion p 5.5 minutes or more $= 17/144 = 0.118$ or 12 percent.

Table 6
Distribution of Waiting Times at Carousel for 80 Flights

Minutes		Number of	Calculations	
Interval	Midpoint x	flights n	nx	nx^2
4.5-9.5	7	1	7	49
9.5-14.5	12	5	60	720
14.5-19.5	17	19	323	5,491
19.5-24.5	22	24	528	11,616
24.5-29.5	27	20	540	14,580
29.5-34.5	32	9	288	9,216
34.5-39.5	37	2	74	2,738
Total		80	1,820	44,410

Calculations

1. mean $\overline{x} = 1820/80 = 22.75 = 23$ minutes.
2. variance $s^2 = \left(\frac{1}{79}\right)(44{,}410\text{-}1820^2/80) = 38.04$.
3. standard deviation $s = 6.17$.
4. coefficient of variation $cv = 6.17/22.75 = 0.27$ or 27 percent.
5. proportion p less than $19.5 = 25/80 = 0.3125$ or 31 percent.

Table 7
Baggage Problems on 1400 Flights

Interval		Number	Estimated	Rate per
No. passengers	No. flights	baggage problems	no. passengers	1,000 passengers
under 40	50	5	1,500	3.3
40-59	200	60	10,000	6.0
60-79	300	105	21,000	5.0
80-99	400	216	36,000	6.0
100-149	200	200	25,000	8.0
150-199	150	308	28,000	11.0
200-249	80	180	18,000	10.0
250 and over	20	48	6,000	8.0
Total	1,400	1,122	145,500	7.7

Table 8
Distribution of Delay Times for the Departures of 1,100 Flights

Minutes			Number of		Calculations	
Minutes	Interval	Midpoint x	flights n	Percent	nx	nx^2
0	0	0	87	7.9	0	0
1-5	0.0-4.5	2.25	303	27.5	681.75	1,533.94
5-10	4.5-9.5	7	298	27.1	2,086	14,602
10-15	9.5-14.5	12	90	8.2	1,080	12,960
15-20	14.5-19.5	17	75	6.8	1,275	21,675
20-25	19.5-24.5	22	70	6.4	1,540	33,880
25-30	24.5-29.5	27	68	6.2	1,836	49,572
30-35	29.5-34.5	32	65	5.9	2,080	66,560
35-	over 34.5	42	44	4.0	1,848	77,616
	Total		1,100	100.0	12,426.75	278,398.94

Calculations

1. mean $\bar{x} = 12{,}426.75/1100 = 11.30$ minutes.
2. variance $s^2 = \left(\frac{1}{1099}\right)(278{,}398.94 - 12{,}426.75^2/1100) = 125.58$.
3. standard deviation $s = 11.21$ minutes.
4. coefficient of variation $cv = 11.21/11.30 = 0.99$ or 99 percent.
5. proportion p less than $14.5 = 778/1100 = 0.71 = 71$ percent.

Quality control in Japan Air Lines Company.[4] The objective of quality control is to consolidate a system to guarantee the quality of products offered to customers. In accordance with this underlying idea, each organization within Japan Air Lines is required to take responsibility in the building up of the quality of the 'soft' merchandise (service in a broad sense) called air transportation while it is still in the 'production' process, maintaining and improving its quality and guaranteeing it to customers just as a manufacturer of industrial products does with his merchandise. Quality control should be a company-wide activity, involving all managers and workers. In other words, the quality of the air transportation which we offer is achieved by a systematic coordination of all employees in every part of the organization. In this context the general quality of air transportation (as a service product) is analyzed below:

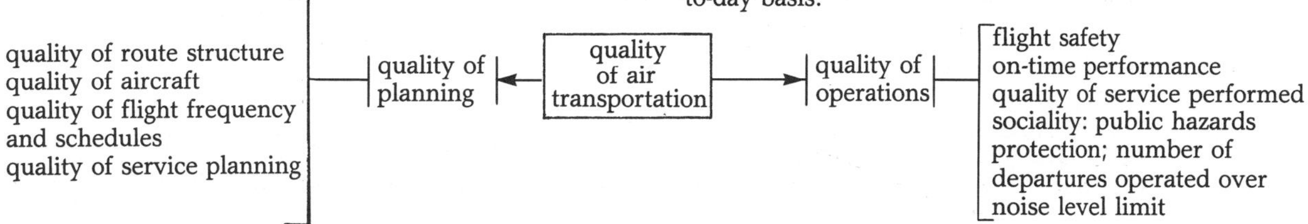

Ceaseless efforts by all management personnel, as well as every worker in the company, in consolidating and putting into effect various management systems have earned our company its present position. It is their efforts in achieving the targets for quality which are aimed at producing, and being able to offer to customers, a first class level of service that have determined the company's performance. As a tangible method of setting into motion quality control activities, the company as a whole and each division within it establishes policies of quality control and specific targets to be reached. These targets are divided and assigned to lower elements in the organization in a harmonious fashion, putting into practice the concept of "management by objectives" and hence creating a company-wide organization of quality targets. Table 9 shows some examples of these targets. Our Zero Defects program is where these ideas are put into effect on a day-to-day basis.

Table 9
Examples of Company Level Targets of Quality (1980)

QUALITY	QUALITY INDICES		CLASSIFICATION		TARGET	CONTROL LIMIT	DEFINITION
SAFETY	RATE OF ACCIDENTS		FATAL ACCIDENTS		0		NBR. OF FATAL ACCIDENTS
			NON FATAL ACCIDENTS		0		NBR. OF NONFATAL ACCIDENTS
					0		NBR. OF NONFATAL ACCIDENTS PER 100,000 HRS. FLOWN
	RATE OF INCIDENTS		A		0		NBR. OF INCIDENTS PER 1,000 HRS. FLOWN
			B		0		NBR. OF INCIDENTS PER 100 DEPARTURES
SERVICE (OPN.)	RATE OF FLT. OPN. IRREGULARITIES		A		0.74	0.95	NBR. OF FLT. OPN. IRREG. PER 1,000 HRS. FLOWN
			B		0.21	0.28	NBR. OF FLT. OPN. IRREG. PER 100 DEPARTURES
	RATE OF SCHEDULE PERFORMANCE		INTERNATIONAL		99.0	98.6	*ACTUAL NBR. OF DEPARTURES* / SCHEDULED NBR. OF DEPT. × 100%
			DOMESTIC		99.0	98.0	
	RATE OF ON-TIME DEPT.	OVER-ALL	INTERNATIONAL		82.0	79.5	*NBR. OF ON-TIME DEPARTURES* / NBR. OF ACTUAL DEPARTURES × 100%
			DOMESTIC		90.0	87.2	
		SPECI-FIED	STN. DELAY	INTERN'L	6.30	7.52	*NBR. OF DELAYED DEPT. (EXCLUDE CODE "D" DELAY)* / NBR. OF ACTUAL DEPARTURES × 100%
				DOMESTIC	5.00	6.75	
			MAINT. DELAY	INTERN'L	1.47	1.74	*NBR. OF DELAYED DEPT. DUE TO TECHNICAL & MAINT.* / NBR. OF ACTUAL DEPARTURES × 100%
				DOMESTIC	0.90	1.05	
	RATE OF ON-TIME ARRIVAL	OVER-ALL	INTERNATIONAL		77.0	74.0	*NBR. OF DELAYED ARRIVALS* / NBR. OF ACTUAL ARRIVALS × 100%
			DOMESTIC		79.0	73.5	
		RATE OF LONG DELAYED ARR.	INTERN'L		0.70	1.44	*TTL. NBR. OF LONG DELAYED ARRIVALS* / TTL. NBR. OF ACTUAL ARR. × 100%
		AVR. OF DE-LAYED DURATN.			1.32	1.64	*TTL. DURATION OF LONG DELAYED ARRIVALS* / TTL. NBR. OF LONG DELAYED ARRIVALS
SERVICE (PAX.)	COMPLAINT PASSENGER RATE		INTERN'L	TO	0.40	0.54	NBR. OF COMPLAINTS PER 10,000 PAX.
				APO	0.70	0.84	
				FLT	0.70	0.86	
			DOMESTIC	TO	PENDING		*TTL. NBR. OF PAX. NEGATIVE REPLY* / TTL. NBR. OF PAX. REPLIED
				APO			
				FLT			
	BAGGAGE IRREGULARITY RATE		INTERNATIONAL		15.7	18.7	NBR. OF BAGGAGE IRREG. PER 10,000 PASSENGER
			DOMESTIC		2.50	3.00	NBR. OF BAGGAGE IRREG. PER 10,000 CHECKED BAGGAGES
SERVICE (CARGO)	CARGO IRREGULARITY RATE		INTERNATIONAL		6.81	8.20	NBR. OF BAGGAGE IRREG. PER 10,000 AIR WAY BILLS
			DOMESTIC		0.74	0.69	NBR. OF CARGO IRREG. PER 300 TONS OF CARGO
	CARGO COMPENSATION RATE		INTERNATIONAL		REFERENCE		TTL. AMOUNT OF COMPENSATION PER Y 100 MILLION OF CARGO SALES
			DOMESTIC				

Explanatory notes for Table 9

1. Abbreviations

FLT	flight	ARR	arrival
OPN	operation	AVR	average
STN	station	DURATN	duration
MAINT	maintenance	PAX	passenger
DEPT	departure		

2. Control limit
   The control limit is calculated from past data just as in the case of a control chart, and this value means 1 (one) standard deviation from the target. The present rule is as follows:
   > If the value of some quality falls outside the control limit more than twice in three successive months, we should investigate and analyze the status precisely and take necessary action to maintain the quality.

Characteristics of the Quality of Air Transportation

Generally, the same concept of quality control applied to industrial products may also be applied to the quality control of service products. However, at the same time, there are certain distinctive features in the quality control of service products which should be noted.

(1) Service products cannot be produced in advance and stored before being offered to customers.
(2) Service products do not lend themselves to a system of inspection and discarding of defective elements.
(3) The quality of service is affected by many human factors which are very difficult to standardize and evaluate. Some elements of this quality are determined by the personalities of those workers actually involved in offering the service.

On a more concrete level, quality control in the field of air transportation must have as its core the necessity to ensure the absolute safety of flight operations, which means a complete elimination of fatal and other accidents, because loss of human life can never be compensated for. This has been a fundamental part of our concept of air transportation since our company's foundation, and the basic tenet, *"Flight Safety is Our Supreme Responsibility,"* will never change.

To offer good service to passengers, each job must be performed carefully by following the standards set and avoiding mistakes. Therefore, in the quality control of our service, it is most important for each worker to be actively concerned with improving the quality of his own job area. For this purpose, the motivation, participation and teamwork of every worker are required.

In the case of quality control of industrial products, the following concept has been already generally recognized in Japan; *"Promote Quality During the Production Process, Quality is Not Attainable by Inspection."*

Introduction of the ZD Program to Japan

In the 1960's the Zero Defects program was developed and already widely applied by many manufacturers as a method of quality and reliability assurance, assisted by the Department of Defense in the U.S.A. In 1965, the ZD program was introduced into Japan, maintaining the original concept as it was, but, at the same time, also emphasizing several other items.

Original Concepts:

(1) Do the job right the first time.
(2) Motivation towards error-free performance.
(3) Improvement by objective oriented activities.
(4) Recognition for achievement.

Added and Modified Concepts:

(1) Respect for humanity.
(2) Voluntary activities of groups.
(3) Participation in job and management.
(4) Self and mutual development.
(5) Application of many analysis and improvement techniques.
(6) Challenge improvement not only in terms of quality but also in areas such as cost, production, and other points.

In Japan, the ZD program was introduced based upon a solid system of quality control just as was the case in the U.S. Therefore, the fundamental and most important concept of a ZD program, *"Do the job right the first time,"* was kept and the ZD program's activities were very effective in improving quality, especially the elimination of defects in products. Moreover, the basic ideas together with the modified concepts have brought us valuable by-products such as good communication, good human relations and a greater vitality in all our employees.

Introduction of the ZD Program to JAL

In our quality control system, the setting of targets for quality in many areas is very important but also very difficult. Of course the quality of flight safety should be absolutely perfect. However, although 100% perfection (defect rate is zero) may be accomplished by some special method such as 100% precise inspection, it is impossible for us to inspect the performance of all the facets of air transportation before we offer the services to customers. When some defects have no relation to human safety, we may approve such defects within certain designated limits as a common practice in quality control. However, we recognize that defects originating from human error should be eliminated as much as possible, and this concept is very important, especially for our service. In 1967, JAL introduced its ZD program, the concept of which was very suitable for the requirements of our quality control from the following viewpoints.

(1) For JAL, what was first required was absolute safety; the elimination of fatal accidents was a paramount target. The phrase *"Zero Defects"* in its meaning and feeling was very suitable to JAL.
(2) An error by only one employee could take the lives of a number of people and fatally damage the destiny of the company. Therefore, it was quite natural to apply the concept of Zero Defects *(Do the job right for the first time)* in preventing errors in order to improve the quality of the air transportation offered, and in every activity in which the company is engaged.
(3) Management systems for quality, cost, production and other areas would all be strengthened by objective oriented activities which permeated to the first line stratum in the form of voluntary groups.

Accomplishments of the ZD Program

It is extremely difficult to quantitatively evaluate the achievements of a ZD program independently. However, the following items can be singled out as some of the results of our ZD program, more than 10 years after its inception.

(1) The spirit of Zero Defects (the elimination of errors and the challenge to perform jobs perfectly) has infiltrated into all departments and sections in the company, and this spirit has been reflected in all the activities of the company at every opportunity.

(2) Through ZD group activities, each individual employee has developed an awareness of the importance of being vitally and actively involved in all aspects of his work. Through such activities, targets of higher and higher levels are being achieved throughout the company.

(3) ZD group activities have both improved the communication among the workers themselves as well as having enabled the smooth transmission of policies from managers and supervisors.

(4) Our Error Cause Removal suggestion system, which is an integral part of the ZD program, has been merged with our general suggestion system which had existed earlier in order to promote suggestions. We consider that objective oriented activities and suggestions are the most important ZD program activities. Suggestions from ZD groups have remarkably contributed to the improvement of quality and productivity.

Some Examples of Results in Quality Control Assisted by the ZD Program

(1) Flight Safety

We have established a flight safety record of about 800,000 flight hours (Revenue) as of December, 1980, since the last fatal accident at Kuala Lumpur, Malaysia in 1977. Although this is directly the result of cooperation from employees participating in operations, maintenance, engineering, service and traffic, the existence of a ZD spirit (in other words "Challenge Perfection") in all the staff of the company should not be forgotten. Although such a concept may be very difficult to understand for those from other nations, there is no doubt that it exists. The rate of fatal accidents (a target of flight safety) is, of course, zero, and no tolerance is allowed. We would like to extend this record to 1,000,000 hours as a first step and then extend it further to promote our survival in the 1980's.

(2) Reliability of Aircraft

Reliability of aircraft is one of the most important factors determining the quality of air transportation. It is closely related to safety, schedule performance, on-time departure and other areas. We usually use "Dispatch Reliability" as a measurement of aircraft reliability.

Dispatch reliability is calculated as follows:

$$\left[1-\frac{\text{Number of Delays Due to Technical and Maintenance Factors}}{\text{Number of Actual Departures}}\right]\times 100$$

This quality is mainly related to design and maintenance in main bases and airports, but the cooperation of flight crews is also required. The dispatch reliability of our B-747's has improved remarkably since their first flight in 1970, as shown in Figure 1.

JAL's reliability is superior to the average of other airlines. This must be a result of company-wide activities in quality control, and ZD groups in maintenance shops

FIGURE 1

Reliability of Aircraft

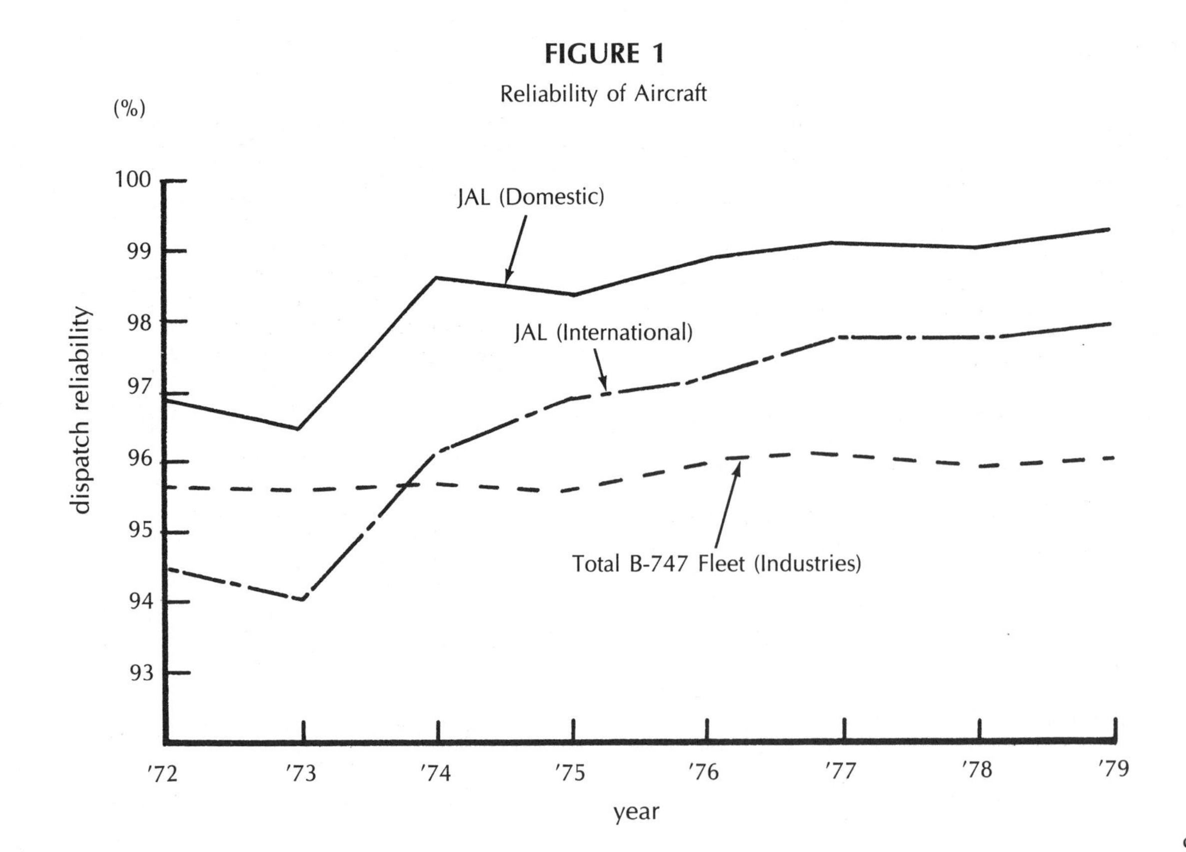

and maintenance sections of other stations who are constantly challenging their own detailed targets which have been broken down through the company, shop, and section level targets.

(3) Baggage Irregularities

Baggage irregularity is one of the factors most detrimental to good service and often creates problems for passengers in their travels. Therefore, from a passenger's point of view, its occurrence should be zero. However, baggage irregularities cannot be eliminated completely due to congestion of airport facilities and several kinds of errors committed not only by airline personnel but also by contracted workers. In reality, almost all causes of such irregularities could be eliminated by careful attention to detail and precise confirmation of procedures by all employees in ideal circumstances. However, we have in fact to operate under many difficulties such as insufficient time, large numbers of passengers and inadequate facilities as well as other problems. To solve such problems is precisely the object of a Zero Defects program.

We have improved procedures and facilities for baggage handling, and are developing a new computerized baggage tracing system to reduce the cost involved in cases of mishandled baggage and make possible quicker and easier retrieval of missing bags and improved customer relations. But the most important aspect of the effort to attack this problem is the development of a high morale among employees which will help to prevent the occurrence of such cases. Again, in this case, the motto, "Do the Job Right the First Time," holds the key to the solution of the problem.

The "Baggage Irregularity Rate," a measurement of baggage irregularities, is calculated as follows:

$$\frac{\text{Number of Baggage Irregularities}}{\text{Number of Items of Baggage Handled}} \times 10{,}000 \text{ (Domestic)}$$

The Baggage Irregularity Rate (Domestic) has improved in the past ten years as shown in Figure 2, and ZD groups' contribution to this improvement has been remarkable at many airports.

Railroads

Quality control areas. Freight operations to which quality control is, or can be, applied includes:

1. Errors on waybills, cargo statements, bills of lading.
2. Revenue accounting.
3. Delay time in filling freight car orders.
4. Shipper satisfaction with cars placed: defective cars, dirty cars, wrong cars, substitutes.
5. Freight car utilization.
6. Employee utilization.
7. Equipment utilization.
8. Track maintenance.
9. Freight car life and cost.
10. Per diem costs versus freight car purchases.
11. Freight car maintenance, repairs, renovations, conversions.
12. Derailments: number, type of cars, products, loss, location, causes.
13. Lost shipments.

FIGURE 2

Baggage Irregularity Rate (Domestic)

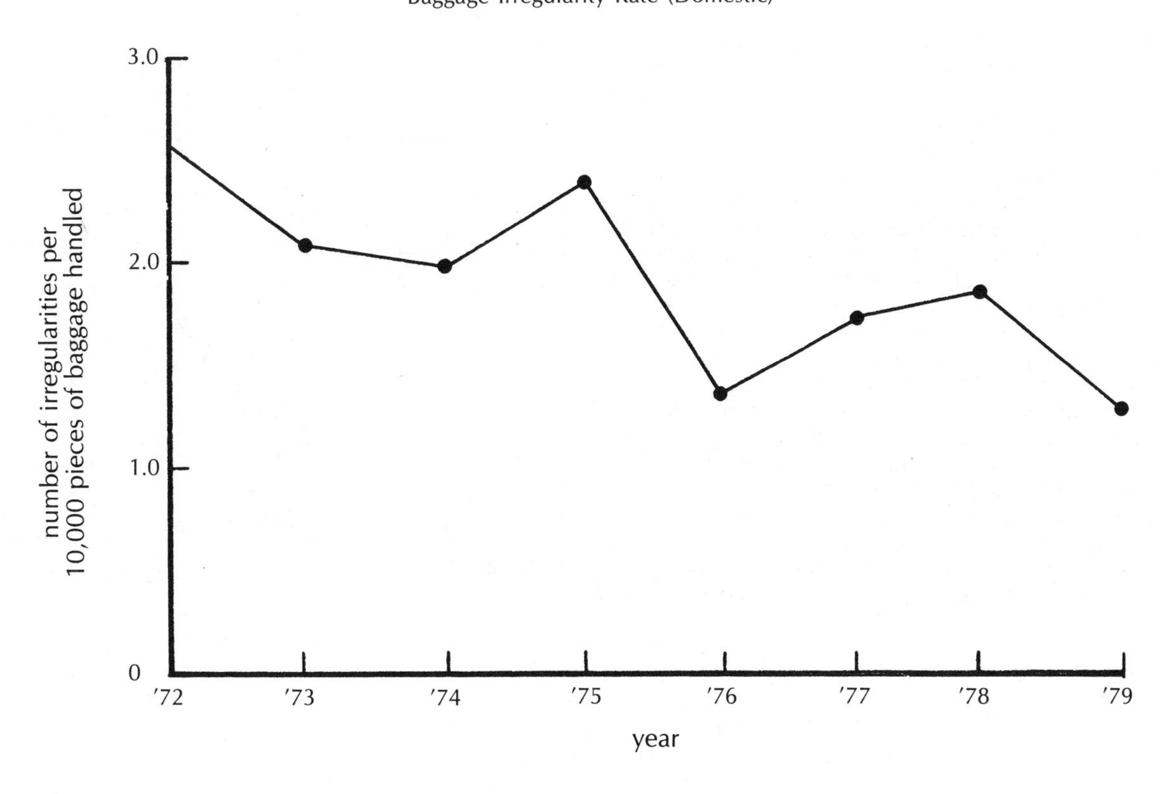

14. Car location.
15. Shipment location.
16. Interline settlements.
17. Stealing, vandalism.
18. Crossings: controls, accidents, traffic problems.
19. Delay in shipment deliveries.
20. Freight car shortages.
21. Assigned versus non-assigned cars.
22. Terminal loading and unloading.

Freight car service. Freight car service can be measured by the delay that occurs between the time the shipper wants the freight cars and the time they are placed where required. The delay can be shown objectively by comparing the frequency distribution of car orders by days of 24 hours with the frequency distribution of car placements, on this same day basis. Figure 3 shows two such distributions, one representing the service requested and the other the service received. The data come from the distributions in Table 10.

Both Figure 3 and Table 10 show that the 280 freight cars were requested over a period of 10 days, but it actually took 17 days to place the cars. The difference between the two curves is a measure of the service quality rendered. It is a measure of the delay time elapsing between when the shipper wanted cars and when he received them. The farther apart these two distributions are, the slower and poorer the service obtained. For these distributions, the average number of days requested to fill the order is 4.9 days or 117 hours, while the average number of days required to fill the order is 6.9 days or 165 hours. The average delay is two days of 24 hours or 48 hours.

What constitutes satisfactory quality service depends upon many factors:

1. How quickly the shipper needs cars.
2. What type or types of cars are needed.
3. Availability of these cars to the railroad servicing the shipper.
4. Whether the demand for various types of cars is steady, seasonal, or emergency.
5. Whether the shipper is big enough or influential enough to get assigned cars — that is preferential treatment for the type of cars needed.

The type of car required has considerable bearing on the service: a special type such as a covered hopper may be harder to obtain than a box car, gondola, open hopper, or flat car. A 50-foot-wide door boxcar usually is harder to obtain than an ordinary 40-foot boxcar. Heavy seasonal demand for boxcars, as in the Midwest grain belt, poses a problem practically every year. A large delay in service — weeks or months — leads shippers to claim that a freight car shortage exists. Grain elevator operators and farmers from the Midwest and plywood manufacturers from the Pacific Northwest expressed such a view some years ago before a Congressional Committee. One complained of waiting 30 to 60 days for cars.

Table 11 shows the delay distribution for general service boxcars for the Middle Atlantic and Great Lakes regions for a period of about six months.[5] The number of cars delayed 11 or more days was about 33,000. Consider-

Table 10
Frequency Distribution of Freight Cars as Requested and as Placed *

Day (x)	Number of cars (y) Requested	Placed
1	5	2
2	32	12
3	46	26
4	54	32
5	45	38
6	34	34
7	28	30
8	16	25
9	12	20
10	8	16
11		13
12		10
13		8
14		6
15		4
16		2
17		2
Total	280	280
Mean days	4.9	6.9
Mean hours	117	165

*"x" stands for days of 24 hours. Relative to time, "requested" means the number of cars, "y", is to be delivered or placed on the designated day. For example, 46 cars are to be delivered on the third day, eight on the tenth day.

Table 11
Delay Distribution for Unequipped General Service Boxcars — Great Lakes and Middle Atlantic Zones — 1968*

Delay time days	Number of cars	Percentage distribution
Less than 1 day	668,112	80.1
1-2	92,496	11.1
3-4	21,336	2.6
5-6	5,304	0.6
7-8	8,016	1.0
9-10	5,400	0.6
11-12	960	0.1
13 or more	32,240	3.9
Total	833,864	100.0
Mean	2 days	

*Actual period is July 15, 1968 to January 23, 1969, inclusive.

FIGURE 3

Distribution of 280 Freight as Ordered and as Placed or Received, by Days of 24 Hours

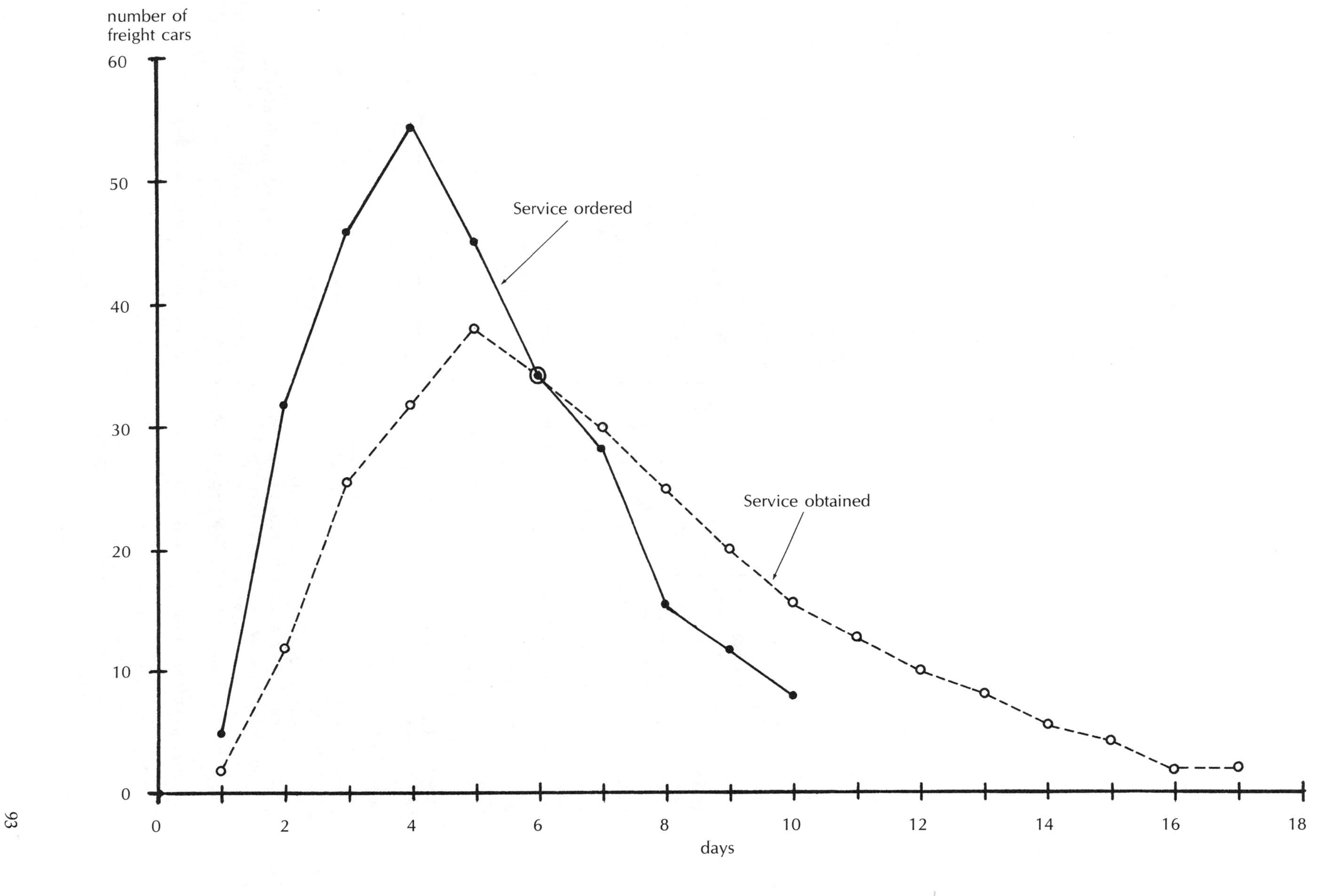

ing the fact that boxcars average 60 tons of wheat, and a bushel of wheat weighs about 60 pounds, this means that this number of cars could move close to 70 million bushels of wheat. Since the total for the United States for 11 or more days was 39,000, this region alone accounted for 85 percent of the cars delayed this length of time. This shows why delay is equated to shortage.

In this distribution and in other delay distributions, it is not the average that is important. It is the absolute number which is in the upper ranges of the frequency distribution because these are the long delay times which indicate a problem or serious trouble.

Buses

Local and intercity buses. The quality areas involved in local and intercity bus travel include the following:

1. Frequency of service.
2. Waiting times.
3. Meeting time schedules.
4. Delays.
5. Transfers, connections.
6. Convenience: coverage, interchange, network, walking distances.
7. Fare.
8. Mechanical condition of bus.
9. Interior condition of bus.
10. Heating and cooling.
11. Ventilation.
12. Driver: competence, attitude, helpful.
13. Tickets: fare, buying, delays, waiting, errors.

Quality of service can be measured in a variety of ways, of which the following are illustrations:

1. Recording and plotting arrival times at designated points, compared with planned times. An actual time plot covering a period of two months for a specified express bus route is given in Chapter 19.

2. Time performance versus actual performance of a fleet of buses.

3. Actual departure and arrival times of intercity buses, compared with scheduled times.

4. Loading factors for local buses on various routes for different times of the day.

5. Loading factors for intercity buses on various routes for different times of the day.

6. Providing enough buses during rush hours or on special occasions.

Sight-Seeing Tours. Quality control applies to a wide variety of operations and activities which arise in connection with a sight-seeing tour. Some of the major aspects are the following:

1. Reservations: arrangements, errors, delays.
2. Transportation: nature, comfortable, convenient.
3. Hotels: accommodations, convenience, service.
4. Rooms: clean, comfortable, service.
5. Food: served on time, variety, choice, service, timing.
6. Meals: lunches, dinners, breakfasts: time schedule.
7. Baggage handling, baggage limitations.
8. Time scheduling: arrival and departure times.
9. Tour guide: knowledge, presentation, courtesy, communication.
10. Driver: competent, safe, communication, courtesy.
11. Stops for: rest, scenery, points of interest, picture taking, meals.
12. Keeping passengers properly informed.
13. Cost relative to quality of service and value of the tour.
14. Physical arrangements on bus: sufficient room, conveniences, cleanliness.

Rating scale for quality of an intercity tour. A rating scale to be filled out by passengers at the end of an intercity tour is given below. One is asked to rate as substandard, fair, good, or excellent the following items classified under "quality control":

1. Rooms and food under Hotels.
2. Personal service and narrative under Tour Director.
3. Driving and coach cleanliness under Driver.

Space is also provided for additional comments.

Example of a Rating Scale Used by a Touring Company

2 QUALITY CONTROL. We welcome your frank evaluation of our tours. Your remarks aid us in making our tours more enjoyable in the future and are CONFIDENTIAL. Check (✓) your evaluation below.

	Excellent	Good	Fair	Substandard
A. HOTELS				
Rooms	____	____	____	____
Food	____	____	____	____
B. TOUR DIRECTOR				
Personal Service	____	____	____	____
Narrative	____	____	____	____
C. DRIVER				
Driving	____	____	____	____
Coach Cleanliness	____	____	____	____

Space below and back of top flap for additional comments.

____

____

____

____

____

Source: Johansen Royal Tours, Seattle, Washington.

Household Goods Moving

Quality control enters into each of the two usual ways of moving household goods: renting a truck such as U-Haul and hiring a household goods mover. The quality aspects may be itemized as follows:

1. Rent a truck:
   a. Condition.
   b. Time required to obtain size required.
   c. Cost.
   d. Reliability: breakdown.

e. Provision for return.
f. Insurance.
2. Hire a household goods mover:
a. Initial estimates of weight, time cost.
b. Time schedule.
c. Packing by owner.
d. Packing by mover.
e. Loading.
f. Travel time, delay time.
g. Unloading: "early" means storage; "late" means motel, hotel bills.
h. Unpacking.
i. Deviation from estimates: time, weight, cost.
j. Cost of insurance, insurance coverage.
k. Damage, claims, settlement, repair.
l. Customer satisfaction.

ICC study of household goods moving 1968.[6] The Interstate Commerce Commission (ICC) is required under the law to regulate household goods moving in interstate commerce, that is between states. It has established rules and regulations to carry out this responsibility; it also receives complaints from householders relative to the quality of service rendered by various carriers. Most of these complaints are about initial estimates of time and cost, which were so much in error as to create a very substantial additional cost to the householder, and about the slowness and inadequacy of settling damage claims. If the initial cost estimate is too low, the bill will turn out to be much higher than expected. If the timing is greatly in error, additional bills are incurred at hotels, motels, and restaurants. Because of these complaints, the ICC designed and implemented a nationwide sample study of interstate household goods movements, using moving company bills or invoices as the population of documents to be sampled. A stratified replicated sample of about 8,000 was selected from a total population of about 1,200,000 for 1968.

The estimated totals for United States are given below for delivery date and cost.

	percent
1. Delivery date	
promised delivery date met	59%
delivered early — put in storage	4
delivered late	27
no data on delivery date	10
2. Cost estimates	
too low by 10% or less	22%
too low by 11 to 19%	12
too low by 20% or more	12
same as bill, or over by 10% or less	22
over by more than 10%	28
deviation not shown	4
3. Average claim	
number of days for settlement	63

Four percent of the goods were delivered early and put in storage; this adds more cost in the form of storage and moving. Sometimes the householder may want this done. The cost estimate figures show that 44 percent of the estimates fell within 10 percent of the actual; while another 24 percent were too low by more than 10 percent. This latter group could add an extra payment up to $500.

Appraisal of rating scales as quality measures. Rating scales are used to rate individual work performance, financial credit, performance in diving and figure skating, competitive products, and competitive vendors. The purpose of a rating scale is to rank individuals, companies, or products in order of their superiority beginning with the best. Quality of performance is reflected in the rating, but in most cases all kinds of corrective actions have already been taken to improve performance.

The rating scales used by airlines, sightseeing tours, and other service companies are used to rate several classes of services and activities, but they are primarily opinion polls with some gradations thrown in. In many cases, the ratings are made hastily by persons in a hurry, or on short notice, so there is a tendency to base judgments on last impressions, some unusual event, or what is remembered.

Rating scales are inefficient as quality control devices. Actually, rating scales are not used as quality control devices except in the case of rating vendors. In this case the ratings are based on past experience with special reference to how the vendor's product affected the quality of the final product. Weaknesses of the rating scale are:

1. The classes or categories or items are too broad or too general.
2. Ratings give no clue as to the nature of "quality", whether rated "excellent" at one end or "poor" at the other. This is because individuals are rating a class or category on the basis of different experiences, activities, or services and not necessarily on the same service. It all depends upon what kind of "sample" is being observed. This raises a question of accuracy, soundness, of validity.
3. Ratings provide no basis for action. If something is rated poor, what is going to be done about it? Nothing, because management does not know what service activities gave rise to a rating of "poor".
4. Ratings conceal the specific activities on which the ratings are based. It is these specific activities which are needed, not how the individual rates them.
5. It is difficult, if not misleading, to attempt to interpret a count or tabulation of these ratings.

Activities dealing directly with quality should be reported, not rated. This means that unsatisfactory service should be pinpointed:

1. Describe in detail the nature of the poor service: delay, errors, damage, indifference, neglect, overcharge, rudeness, irresponsibility, carelessness, penalties suffered because of delay and errors and neglect, additional costs, and additional expenses.

2. Describe not only the nature of the poor service, but also when, where, who, and the conditions.

It is also necessary to pinpoint various examples of satisfactory service in order to have some objective basis for distinguishing between what is acceptable and what is unacceptable.

Note instances where the service time was acceptable, where the delay time was acceptable, where no additional expenses were incurred, and where service was pleasant and timely at an affordable price.

Customer complaints may be used for quality control purposes, but caution will be needed. Is the complaint accurate? Is the complaint rare and unimportant? Is the complaint and complaintant important enough to call for an investigation and corrective action? Is the complaint one which falls in an area of service where many complaints have been received? If so, an investigation is needed and corrective action taken.

If customer complaints are used only to issue a monthly report on various types of complaints and comments on the trend of these complaints, then they are of little use in improving the quality of service. There has to be feedback; something has to be done to meet and eliminate complaints if quality of services is to be improved.

One should not overlook what happened when customer complaints to the CAB were compared with customer complaints to a large U.S. airline. The complaint pattern was quite different in the two cases suggesting that one group was sending complaints to the CAB, while another group was sending them to the company. Location and distance may have been a factor influencing which one of the two was complained to. This raises a serious question about customer complaintants being biased samples. The complaints of a random sample of passengers on various flights should be made and compared with those received by telephone or by mail over the same length of time, say one year.

Testing the soundness of a sample used in a motor carrier rate case. In a motor carrier rate case that came before the Interstate Commerce Commission (ICC), a very critical question was raised by a party opposing the rate increase. The party argued that the sample of freight bills used in the traffic study was not random, but biased. Hence, the results were suspect and should be rejected. The entire outcome of the case depended upon whether the sample was sound or not.[7]

A consultant was called in to study this question. He went to the officials involved, talked to the operating people, selected a sample of terminals (both large and small), and had the key characteristic (weight of shipment) recorded in the order in which they were recorded and received. The companies and terminals from which samples were selected, included both the larger and the smaller terminals.

The median test was then run on 221 sets of shipments. The number of values in each shipment ranged from as low as 50 to well over 900. These calculations gave 221 values for the normal deviate. A comparison of the actual values with the expected values is given in Table 12; the differences are shown as well as the value of chi-square. If the sample is random, then the distribution of normal deviates derived from the actual sample should follow a normal curve. The chi-square value is not significant; hence, the hypothesis of randomness was a reasonable one to assume and the administrative judge so held.

Table 12
Test for Normal Distribution of Test Values

Intervals on Normal Distribution	Actual Frequency	Expected Frequency
2.5 and above	0	1.3
2.0-2.49	4	3.8
1.5-1.99	7	9.7
1.0-1.49	21	20.3
0-.99	80	75.4
−.99 to 0	76	75.4
−1.0 to −1.49	12	20.3
−1.5 to −1.99	12	9.7
−2.0 to −2.49	7	3.8
−2.5 and below	2	1.3
Total	**221**	**221.0**

$\chi^2 = 9.37$ for 9 degrees of freedom.
$P = .40$.

Notes

[1]Dale L. Lobsinger, "Air Transportation Finds New and Lucrative Use for SQC," *Industrial Quality Control,* Vol. 6, May 1950, pp. 76-78.

[2]Winston C. Dalleck, "Inductive Accounting — Settling Interline Accounts by Sampling Methods," *Industrial Quality Control,* December 1956, pp. 12-16.

[3]Source is *CAB News,* dated March 4, 1981, issued by the Civil Aeronautics Board, Washington, D.C.

[4]This is a paper, with some slight editing, by Hideo Takakuwa, "Service Quality Improvement by Circle Activities," *35th Annual Quality Congress Transactions,* American Society for Quality Control 1981, pp. 709-715. Mr. Takakuwa is General Manager, Overall Safety, ZD Program and Quality Administration Department, Japan Air Lines Company, Tokyo. Used by permission of author.

[5]A. C. Rosander, *Case Studies in Sample Design,* Marcel Dekker Inc., New York, 1977, pp. 341, 345, 348, reprinted by courtesy of Marcel Dekker Inc.

[6]*Ibid.,* pp. 221-224. Original data are from an Interstate Commerce Commission published study, Washington, D.C.

[7]*I and S Docket M 22930,* Interstate Commerce Commission, Washington, D.C. Testimony of A. C. Rosander, June 25, 1969 is designated Exhibit H 38.

8
Retail Trade

Some actual quality problems encountered by the buyer (customer). Some idea of the quality problems existing in a retail store can be obtained from the experiences of the buyer (customer). We describe some actual experiences in supermarkets and department stores.

Supermarkets.

1. The checker rings up the old price, not the sale price or the sticker price. New prices need to be posted daily so the checker (cashier) knows them. Otherwise, the unwary buyer thinks the price rung up is the lower sales price, when it is not. The wary buyer watches the checker and corrects any error the latter makes. It is a management responsibility to see that charges made are those advertised. When price scanning is used, it is necessary to see that the computer has been re-programmed to reflect the lower price.

2. A paper bag is not properly packaged and breaks because of too much weight, e.g, canned goods and meat, sharp corners of boxes, or frozen foods moisten and weaken the paper. Many checkers and baggers do not know how to package a wide variety of products. It also is a management responsibility to see that employees are properly trained to package groceries. Otherwise, many of the employees simply pile the articles into bags helter-skelter. Sometimes double bagging is necessary, and sometimes it is necessary to place frozen foods and ice cream in special and separate bags.

3. Waiting time at checkout is too long due to long lines. There is a delay in opening new checkout lanes. This again requires watchful management.

4. The so-called "express" lane is far from a fast checkout lane because checks, credit cards, and food stamps are accepted; also, the number of items accepted often exceeds the posted maximum of eight or 10. There is a need to monitor waiting lines, the length of lines, and open new checkout lines as needed. This includes making the express lane a fast lane in fact as well as in theory.

5. Some generic foods, such as canned fruits and vegetables, may be of poor quality. Examples are canned pears that can be cut only with a knife and canned tomatoes that are partly green and tough. This seems to indicate that some generic foods are rejects; they are not the quality food in ordinary cans, including company brands. This does not mean some generic products, such as paper towels, may not be of acceptable quality. What it does mean is that the broad-brush advertising that claims generic food products have the same quality as regular products, with only different labels, is false.

6. A check cashing permit is refused because the check number has only five digits in it; computer requires six or seven. The request was tossed back at the buyer (customer) because the check number had only five digits in it! Wasted time and effort on the part of everyone concerned occur because the clerks could not see that all one needed to do was to put one or more zeros (0) up front! There was nothing wrong about the check number, although this was the implication.

7. Erratic and large increases in the price of an item, increases of 15 to 20 percent — 45 to 80 cents — even though inflation rate is about 13 percent. Then, in a few days, the price is reduced for a big discount of 10 cents! It is stated that these price changes are received daily from some central office which may be 100 miles away; the prices are not set locally.

8. Food is spoiled: milk is tainted, butter is rancid, or corn meal is wormy.

Department stores. A department store advertised a sale on drapes; as a result a buyer ordered a number of sets of drapes which were to be ready in about three weeks. The reason for the delay was that the drapes were ordered from a quite distant company. After the allotted time, the buyer checked a number of times with the store, but was always told they had not arrived. Finally, after about six weeks, the buyer went to the drapery department and told her story. A thorough check was made and the drapes were found in some remote corner where they had been since they were received two weeks before. Not only had the store not informed the buyer as promised, but the store could not find their sales slips so they had to use the buyer's! This large store obviously did not have a good control system covering outside orders, nor were they able to control their own sales slips. Furthermore, it was the buyer and not the store that had to take the initiative and responsibility for locating and finding the order. If there was a control system, there was no evidence of it, nor was there any evidence of proper supervision of employees involved in the transaction and in the receiving and delivery system.

A large department store sold a kit for using needle work to sew on colored beads and sequins for a calendar. The kit included thread as well as a picture of what the finished work looked like. Three serious defects were found in the kit:

1. Small sequins and beads were all mixed together; this created a very tedious job of separating them by colors. Why were not the separate colors placed in different packets in the first place?

2. Some sequins did not have any holes in them; this created an extra job before any sewing could be done.

3. Sometimes the thread breaks, showing that it is not strong enough.

Some stores serve their customers by agreeing to mail out gift packages from their mail room. One customer had three experiences in which a check at the destination showed that the package had never been received. Obviously the store had no control over this operation, nor did they attempt to introduce any controls after these complaints.

General factors in quality service. The previous examples show how local managements and local employees enter into the problem of quality of service rendered by retail stores. There are, however, more general factors that have to be taken into consideration, not only by management of large scale retail operations but also by local outlets of chain stores.

1. **Variety of merchandise.** The variety of merchandise calls for a staff of buyers who have knowledge of suppliers, on the one hand, and what the various publics buy, on the other.

2. **Consumer preferences.** Market studies and market trends have to be studied and analyzed carefully. Carefully designed market studies are the beginning of defining the quality product problem, and fine-tuning production and distribution so as to meet these needs. Trends in buyer (consumer) requirements, including trends in taste, style, types, models, and shifts in demand, are crucial. The increased demand for small and more efficient automobiles, jeans, and more nutritious food are examples. Price is a major factor in consumer buying in times of inflation, high taxes, and high cost of living. Affordable price plus acceptable quality is a combination to work toward.

3. **Population trends.** Population trends need to be watched and studied carefully to anticipate changing needs, changing buying habits, and shifting demand. The baby boom of the late 1940's and 1950's called for drastic changes in the selling of food, clothing, and housing. So does the increasing number and proportion of buyers (customers) 65 years of age and older.

4. **Economic changes.** Economic conditions such as inflation, increased cost of living, and increased cost of such necessities as automobiles and gasoline caused millions of Americans over the years to buy smaller low cost more reliable automobiles, take shorter automobile trips, and turn to other and cheaper modes of transportation. The auto manufacturers were victimized by the cost-benefit formula — they maximized short-run profits by making larger cars — ($1500 profit versus $400 profit), but this decision proved fatal in the long run. Another business assumption also failed: that the sales department could sell anything the manufacturer made regardless of what the buyer wanted. Market trends since 1965, issued by the auto makers themselves, were ignored.

Quality aspects of a supermarket's operations. Quality enters into all of the operations of a supermarket, from the ordering of stock to the checking out of buyers to the handling of customer's complaints. It is convenient to show the aspects which directly affect management as well as the service operations as the customer (buyer) sees them:

Management Aspects

1. Ordering stock
2. Receiving
3. Unloading: damaged goods
4. Unpacking: damaged goods
5. Delivery schedules: delays in store receiving needed stock
6. Financial
7. Personnel hiring
8. Personnel training
9. Personnel supervision
10. Personnel work scheduling
11. Assignment of checkers; opening and closing lanes
12. Personnel promotions
13. Express checkout arrangements
14. Maintaining and checking inventory
15. Housekeeping functions
16. Checks and credit cards: policy, procedures, control
17. Store hours
18. Customer conveniences: rest rooms, duplicating
19. Customer service provisions
20. Provision for, and handling of, customer complaints
21. Outside arrangements: parking, snow removal, cart assistance
22. Employee facilities: rest rooms, lounges, etc.
23. Provision for special orders by customers

Service operations. In setting up a quality control program in a retail store, it is necessary to consider the service operations which the customer-buyer is concerned about. The following is a list of services which are important to the customer-buyer.

1. Availability of sales persons: Is a sales person handy or does one have to hunt for one? Are there convenient cashier locations? Is there a considerable delay because no cashier is around? Is it difficult to get waited on? Does a customer have to take a number to insure that service is in order of arrival?

2. Availability of checkout cashiers: Are enough lines open so that long delays are eliminated? Is there an express line for those with only a few items?

3. How does management handle the waiting line problem?

4. What practices are followed in filling orders that have to be filled outside? How much delay is there? Are these orders carefully controlled and monitored; and is the customer-buyer informed when they arrive? Are realistic dates of delivery given to the customer?

5. What about the adequacy of the stock on hand? What is the policy when a customer wants to buy an item that is "already sold out"? Is it re-ordered? Is the situation ignored, as in some bookstores? Is a record made to determine customer-buyer demand?

6. What are the practices with regard to financial arrangements? Check handling? Credit cards? Credit practices? Food stamps? Buying by personal check? Are credit card customers preferred to cash customers?

7. What are the practices with regard to returning items? Making refunds? Exchanging items? Replacing items?

8. What are the practices with regard to wrapping and packaging? Are bags overloaded? Do the sacks rip and break unless carried by the bottom? Is the wrapping appropriate for the article purchased?

9. What about pricing policies? Are price changes erratic and increases large? What about sales? Are prices reduced on damaged goods? On rejects?

10. Is there a customer service office that handles customer problems and inquiries?

11. Is there a customer complaint procedure? How are customer complaints handled?

Quality and price. In retail trade, millions of customers buy what they can afford because their incomes are modest and relatively fixed, whereas necessities are many. They cannot give attention to better quality that may come, but not always, with higher prices. They make do with the quality they can buy. Furthermore, millions of customers are not educated in buying, and have to learn, if they learn at all, by many years of buying. Buying can be rational, but most of the time is more irrational than rational due to advertising, television, habit-buying, loyalty buying, social pressure, buying what is in style, and buying what others are buying.

Attempts to make and offer products that entice customers to buy include a large number of selling practices and gimmicks:

1. Discount stores
2. Discount prices
3. Generic foods and other merchandise
4. Generic drugs
5. Special sales
6. Premiums
7. Coupons
8. Trading stamps
9. Weekly specials
10. Reduced bakery goods prices
11. Reduced prices on damaged goods
12. Free samples

UPC scanning system and quality. UPC stands for Universal Product Code which is used in a supermarket computerized system to itemize and total sales tickets at the checkout point. The UPC consists of a pattern of numbers and vertical black bars of different widths, separated by white spaces also of different widths, which are printed on labels or some part of the package or container. This identifies the name of the product as well as its weight or volume. This information is read when the cashier runs the UPC of the item to be purchased across the scanning device and is sent to a central computer where the price is stored. The item description and price are then displayed on a screen as well as printed on the sales ticket, so the customer has an itemized bill such as in Figure 1.

Items which do not carry the UPC, such as fresh fruits, vegetables, and baked goods, have to be handled in the usual way.

An example of a conversion is a store with 16 checkout stands with regular cash registers. The company estimated it would cost about $9,000 per stand or a total of $144,000 to convert to the UPC system. The major reasons for changing to the new system are lower cost, higher efficiency, and the fact that cash registers are no longer manufactured like they once were, making it difficult to obtain repair parts.

The savings to the company comes from the fact that individual items will no longer be price tagged; the price will appear only on a tag below the item on the shelf. It is claimed that this will reduce much of the human error that arose in pricing each item manually, since the price occurs only in the memory of the central computer.[1] It is expected that each store will save about $2,000 a week by using the new system.

Advantages of this system for the customer are:

1. Faster service.
2. More accuracy and a reduction of errors.
3. Detailed product and price sales slips for checking and comparison buying.

Advantages of this system for the company are: reduces costs because fewer employees are needed to stamp the price or fasten a price label on every individual item; faster service; and better control over inventory.

The UPC system has a number of actual, if not potential, disadvantages for the customer:

1. Due to a centrally located computer where all prices are programmed, all stores will be charged the same prices whether in a low cost town or in a high priced city.

2. Local discount sales are impossible unless they are handled as they are now, outside the UPC system.

3. The computer program may not contain reduced sales prices because someone forgot to program the computer for these changes.

4. If the computer makes an error, how long is it going to take to correct it?

5. Is it not possible for the express line to be just as slow as it is now because, as the speed of the scanning increases, the manager may allow the lines to get longer? Will the maximum number of items be raised to 12 or 15, instead of eight or 10 as now?

6. What will be done about items on the shelf that have not been sold when the price is raised in the computer? For example, an item is $1.79 on the shelf; but the computer price is $1.89. What about the reverse situation?

7. Will fewer lanes be used now that the service is supposed to be faster? If so, the time of service may not be improved substantially, if at all, because faster service leads to longer lines.

FIGURE 1

Example of a UPC Customer Sales Slip

KING SOOPER #44
LOVELAND, COLORADO
07/06 9:46 2 117 38
1981

BUGLITE YELLOW	1.53 TX
SAFEGUARD SOAP	1.68 TX
NIAGARA STARCH	1.13 TX
UTILITY BRUSH	1.05 TX
RENUZIT FRESHENR	.79 TX
RENUZIT FRESHENR	.79 TX
DOW BATH CLEANER	1.77 TX
KELLOGG ALL BRAN	1.34*
DOW BATH CLEANER	1.77 TX
QUAK YEL CORN ML	.74*
PLAYTEX GLOVES	1.99 TX
COLGAT .15 OFF	.94 TX
PLAYTEX GLOVES	1.49 TX
CUP-A-SP CHIX VG	.79*
CUP-A-SP CHIX VG	.79*
CS CUP-SOUP VEGT	.57*
LIP CHIX NOODLE	.88*
LIP CHIX NOODLE	.88*
MAX HSE IN COFFE	3.08*
TOP SCOUR SPONGE	.47 TX
FELS NAPTHA SOAP	.47 TX
PRODUCE	.39*
2.20 LB @ 1/.59	
YELLOW PEACHES	1.30*
3.13 LB @ 1/.39	
GOLD DEL APPLES	1.22*
GENER PAPR TOWEL	.48 TX
BAKERY	.45*
PRODUCE	2.49*
PRODUCE	2.49*
1.25 LB @ 1/.69	
NECTARINES	.86*
1.54 LB @ 1/.49	
PINEAPPLES	.75*
TOTAL	36.73
CASH TEND	40.00
SUBTOTAL	35.37
TAX PAID	1.36
3.27 CHANGE	

TX means taxable. Products without a UPC, such as bakery products and produce, are simply added on.

What a customer wants are two things: acceptable quality items at an affordable price. Will the UPC system help the customer in either one or both of these requirements?

This raises a basic question: Is this technological device introduced primarily, if not solely, to help the supermarket? Will the millions spent installing this system result in higher prices to the customer, since the costs of doing business are passed on to the customer? How will the company recover the costs? Two ways are by using fewer employees or increasing prices.

There is no indication that the UPC system will have any effect on the quality of the products purchased. With regard to price, the customer will be worse off if the prices are increased to cover the costs of the system. Neither will they be any better off if service is faster but the lines are longer. Nor will they gain anything of real significance if the actual waiting time is reduced by a minute or two. It should not be overlooked that a considerable amount of the time spent now at the checkout point is due to checks, credit cards, food stamps, and making change. UPC will not change this part of the service time.

An example of the analysis and use of customer complaints. Table 1 shows the distribution of all customer complaints received by a large supermarket chain over a period of three months. The classification, interpretation, and allocation of complaints to categories were made by the company. The distribution given was constructed from the original data.

Table 1
Distribution of Customer Complaints in All Stores for Three Months

	Customer complaint	Number	Percent
1.	Stock condition	176	20.7
2.	Product request	105	12.3
3.	Product quality	77	9.1
4.	Checker	69	8.1
5.	General	60	7.1
6.	General service level	58	6.8
7.	Prices and price marking	38	4.5
8.	Policy and procedures	33	3.9
9.	Queueing (waiting line)	33	3.9
10.	Human error	27	3.2
11.	Sanitation	27	3.2
12.	Courtesy clerk (bagger)	22	2.6
13.	Parking	18	2.1
14.	Customer service promotions	17	2.0
15.	Check cashing	15	1.8
16.	Generic	10	1.2
17.	All other	64	7.5
	Total	849	100.0%

The major complaints dealt with products — their condition, quality, absence from the shelves, and generic brands. This group accounted for about 43 percent of the total. About 21 percent dealt with employees: checkers, general service, human error, and baggers. The remainder dealt with other problems related to management such as prices, policy, waiting lines, sanitation, and product promotions.

A manager reported that 90 percent of the complaints are justified. Action is taken on every complaint by calling or otherwise contacting the customer and discussing the complaint. If at all possible, action is taken to satisfy the customer. In this system customer complaints very definitely lead to a maximum amount of corrective action, and therefore, hopefully, a maintenance, if not improvement of the quality of service. Personal contact with customer certainly maximizes the probability that the customer will be held and will not be lost because of some complaint.

Some complaints are about conditions that cannot be changed or which are outside the jurisdiction of the manager, such as shifting the front entrance of the store, rain or water standing in the parking lot, or customers having to walk too far in the parking lot.

Complaints about employees include talking, not being courteous, conflicts over a check or credit card, charging the regular rather than a reduced sale price, and bagging or packaging. The complaints about price do not involve stores where the UPC scanning system is being used.

Classification of complaints for retail trade. The following is a check list for identifying major areas of complaints for retail stores, especially department stores and supermarkets:

1. Products:
   a. Stock shortages
   b. Quality of stock
   c. Stock condition
   d. Generic brands
   e. Unit prices
   f. Diet brands
   g. Use of UPC
   h. Damaged products, defective products
   i. Product directory
2. Employees performance, attitude, competence:
   a. Cashier
   b. Checker
   c. Packaging, wrapping
   d. Dress, appearance
   e. Human error
   f. Availability of sales persons; waiting for service
   g. Helping with packages, with carts
   h. Courtesy
   i. Response to questions, inquiries
   j. General attitude
   k. Ability to answer telephone inquiries
   l. Filling out sales slips
3. Prices:
   a. Price tags
   b. Sales prices
   c. Price reductions for rejects, damaged goods
   d. Price changes
4. Service:
   a. Fast lane; waiting
   b. Length of lines; waiting
   c. Opening up new lines
   d. Check cashing
   e. Refund policy and practice
   f. Availability of carts
   g. Availability of cashiers
   h. Availability of sales persons
   i. Customer service office
   j. Pharmacy practice: hours, delays, delays of refills

k. Information, answering inquiries
5. Facilities, arrangements:
   a. Parking lot
   b. Rest rooms
   c. Hours
   d. Store access
   e. Layout
   f. Duplicating machine service
   g. Gift wrap
6. Internal environment:
   a. Sanitation
   b. Smoking
   c. Identification of major classes of merchandise, groceries, etc.
   d. Product directory
   e. Lighting
   f. Ventilation, heating, cooling

Quality control statement to Sears' suppliers.[2] The following statement, which describes the quality control standards to be followed by suppliers of merchandise to Sears, is reproduced verbatim. Some of the highlights of this statement are the following:

1. The supplier agrees to furnish a product which meets the standards of quality and is free from defects; if the standards are not met, the supplier agrees to correct the condition in a manner mutually satisfactory to the supplier and Sears.

2. A knowledge of customer requirements, which comes as a result of experience, should be built into factory capability. Sensitivity to these requirements distinguishes the outstanding supplier from the average one.

3. Because a product bears a Sears brand does not reduce the responsibility of the supplier to meet the quality standards acceptable to a Sears customer.

4. While Sears has a large product testing laboratory, this facility supplements the product test facilities of the source which should be able to test many more products than Sears does.

5. Sources should have an established quality system, including receiving inspection and in-process controls.

6. Sears recommends the use of a quality audit based on a formal continuous inspection of a random sample of a finished product.

7. Sources have available the assistance of the staff of Sears quality control engineers.

8. Quality standards to be met must include safety, proper labeling, color, flammability, and perma-prest.

9. Sources of customer use data which are available to suppliers include catalog returns, service calls, and market surveys.

10. Source requests for services of laboratory personnel and test findings are to be made through the buyer assigned to the supplier.

Product Quality

1. Responsibilities. It is the responsibility of each source to assure that products which it manufactures for Sears meet quality standards as defined by the Sears buyer at the time of contract negotiations. While some product lines have definitive quality standards that can become a part of the purchase contract, it is customary in many industries and product lines to "buy by sample" or other means in which details of product quality requirements are not reduced to writing. It should be the mutual objective of Sears and its sources to provide to Sears customers product that will be free of defects. Where product received by Sears does not meet standards of quality as agreed to with the buyer, or where shipments contain manufacturing defects clearly below standards that customers should expect, (even though this standard was not specifically referred to previously by the buyer) the source shall be expected to take the responsibility of correcting the condition, in a manner mutually acceptable to him and Sears.

   Sources must have a clear understanding of product requirements. Such knowledge is not developed solely by Sears, but should be a part of the factory capability, qualifying it as a reliable source. Knowledge of customer requirements is gained through experience in an industry, and is one of the intangibles that distinguishes an outstanding source from an average one.
2. Sears private brand policy. Most merchandise that Sears sells is identified by a Sears brand, as compared to manufacturer brands or unbranded goods. The fact that merchandise bears Sears' name does not reduce the responsibilities of the source in assuring that product meets quality standards acceptable to the Sears customer.
3. Product testing. Sears maintains a large product testing Laboratory, with facilities in Chicago, and Fort Myers Beach, Florida, that offers buyers and sources a wide range of technical services, including product evaluation. The activities of Sears Laboratories generally supplement, rather than take the place of source product test facilities. Test facilities at sources, while specialized in their own requirements, should generally be capable of performing tests on a greater number of units of products than Sears test facilities will permit.
4. Required or recommended factory facilities for conducting tests of manufactured product.
   a. Sources must have the capability to continuously judge the quality of products that they manufacture. This should entail evaluation of completed product, through an effective 100% inspection, or by use of recognized statistical sampling procedures, where they apply.
   b. Sources must have sufficient additional elements of quality control, including the inspection of purchased parts and materials, in-process controls appropriate to the product, and a quality organization that will assure the continued proper administration of the established quality system. (This varies today).
   c. Sears recommends that factories maintain formal, continuous inspection of a random sample of finished product. This activity, called a finished product Quality Audit, is for the purpose of aiding the factory in effectively evaluating product that it manufactures against agreed-to standards.
   d. Sears maintains a staff of Quality Control engineers to aid sources in establishing adequate quality control systems and techniques.
5. In certain specialized areas, Sears has established additional product test requirements:
   a. Safety. Where electrical, thermal or mechanical characteristics are unusually important for user safety, requirements for continuous testing of manufactured product have been established.
   b. Fabric labeling requirements. It is the responsibility of the source to adhere to requirements of the Fiber Products Identification Act and the Wool Labeling

Act. Control procedures sufficient to validate labeling statements are essential.

c. Perma-Prest apparel and other soft goods. In addition to meeting Sears quality standards for appearance and performance, items bearing the Perma-Prest label must be continuously tested in all manufacturing facilities. Requirements are detailed in "Sears Perma-Prest Quality Manual."

d. Color control. Effective color control is essential in many product categories, such as major appliances and soft goods, and in other lines where color compatibility of products made by various sources is expected by the customer. Sears provides definitive color standards and limited technical assistance to sources in these product categories.

e. Flammability requirements as in manual.

6. Usage data available to suppliers. Sears is increasing the amount of data available that reflect the quality of products as observed by customers. Types of data include:
   a. Catalog returns. Rate to sales, for quality reasons.
   b. Service calls. Ratio to sales, reasons that are manufacturers' responsibility.
   c. Market surveys.

   Where such data is available in a form that can be relayed to a specific supplier, it will normally be done so at the direction of the Sears buyer.
7. Procedures for sources to work with Sears technical staff activities. Just as all product requirements are relayed to sources via the buyer, source contacts to Sears concerning product quality are to be channeled through the buyer. This includes requests for services of Sears Laboratory personnel, as well as requests for reports of test findings.

The use of customer returns to improve quality.[3] A paper by R. W. Peach describes in detail how customer returns to a retail store can be used as feedback to improve the quality of a manufactured product. In this paper, he discusses customer expectations, the nature and significance of complaints, quality characteristics, customer usage data, returns to the factory, and the technique of measuring quality by testing garments on selected users.

Some of the highlights of the paper are given before presenting the paper in full. Peach's perceptive analysis of the role and nature of customer complaints not only applies to the specific problem in retail trade which he is discussing, but to numerous other service industries which sell goods and services in a mass market.

Customer expectations are dynamic, not static, e.g., as prices rise expectations rise. Therefore, the product standards should reflect these changes. This means that garments should be tested on the kinds of users that are going to wear them.

With regard to customer complaints, most dissatisfied customers do not complain at all. They accept the article, but the store loses a customer. The merchandise does not return and neither does the customer. "Certainly there are several dissatisfied customers for each return that reaches the factory." Even though the rates of complaints may be low, the information obtained is nevertheless important. For example, a complaint may reveal a quality characteristic that has been overlooked, neglected in laboratory tests or design, or in quality control programs.

The basic product guide is: Produce an acceptable quality product at an affordable price. Characteristics that must meet customer expectations include fit, wear, appearance, care, fasteners, and color. A critical question is: Are the characteristics important in the factory or laboratory important to the customer? (Many are just now in 1981 becoming aware of what Peach called attention to in 1969.) Conformance to specifications does not necessarily mean that the customer is going to buy the product or the service. Standards of acceptance and standards of requirements are set by the customer.

It is important that customer usage data be timely and accurate. Even though usage data are not a fair sample, but usually are biased they can be helpful. There is a need to compile such data for use in new garment design. In addition, there is a need to combine test garment analysis with actual field data.

Returns to the factory, as well as complaints, should be traced back to their origin. This requires an identification of some kind on the garment which shows when and where it was produced (month or quarter of manufacture and source). The factory needs to keep a detailed record of each return. Peach gives a list of such items. These records can be a useful feedback as a returns history is accumulated.

Another source of data is the test garment program. In this program, samples of various garments are provided to a selected group of users who put the garments through a customary wear history. In this way use data can be obtained on a more controlled basis, to be used in conjunction with data from customer complaints and customer returns.

Customer returns

One of the more important tasks of the apparel manufacturer is to stay in touch with the ultimate user of his product. All manufacturers are aware that they must offer the ultimate customer a product that she wants, and at a price that she is willing to pay. Great effort is expended in determining what the customer will buy, for it is clearly recognized that these decisions affect the growth of a company and its ultimate profitability.

But far less attention is given to the problem for establishing whether the user of a product is satisfied with the product that she purchased; whether the product, as she receives it, meets her expectations. The characteristics of the product that may be of interest to the customer are familiar:

- Is it worth the price?
- Does it present well as the customer receives it, in its package or on its hanger?
- Is it adequately identified to the customer, as to size and features?
- Are adequate care instructions provided? Are they clear?
- Does it fit — not just adequately, but exceptionally well, so that it is comfortable and seems to be made for the wearer?
- Is the workmanship right, for the product and the price?
- Do collars, cuffs and flaps lie neatly in place, as the garment is worn, or do these and other workmanship details detract from the appearance of the garment?
- Do the buttons, zippers and other fasteners work easily and positively?
- Is it warm enough, or cool enough, for its intended purpose?

These characteristics are observable to the wearer as the garment is first tried on. They may affect the actual purchase decision, and thus become important to the manufacturer and retailer. But of even more importance to the customer is the manner in which the garment performs in service, after a number of wearings and cleanings. The characteristics that most customers place at the top of the list are fit and wear. Is the amount of shrinkage that occurs when the garment is cared for in the recommended manner such that it will still fit the wearer comfortably? And will the garment wear adequately over a time period that satisfactorily meets the customer's expectations? Appearance after laundering is also important today, particularly with easy care fabrics that promise the customer excellent appearance without ironing. But certainly, all of the characteristics that are considered by the customer at the time of purchases are, to some degree, factors that affect customer satisfaction throughout the life of the garment.

Product Requirements

The task of establishing actual customer experience is a continuous one. Admittedly, as much information as possible should be obtained about a garment before it is offered to the customer. And most quality characteristics are such that tests of one or two sample garments will tell a great deal about the way production garments will perform in use. But the problem remains: are the characteristics that are being observed in a factory or laboratory the ones that are important to the customer? The two mistakes that can be made here are to overemphasize characteristics unimportant to the customer and to play down, or overlook entirely, characteristics of considerable concern to the user. Even when these characteristics are properly identified, what are the standards of acceptance that are being used by the customer? What are the actual product requirements, as they are set by the ultimate judge, the customer?

These are complex questions. Standards of acceptance should reflect customer expectations. As a result an effective advertising campaign stressing specific product characteristics can raise product requirements for a specific garment type. There is no question that as prices rise, expectations of customers rise. We must recognize that standards of acceptance are not static, that it is not possible to "write a book" of garment acceptance standards and have it remain current indefinitely. If standards used in judging customer acceptability are to mean anything, there must be a flow of information from actual users to influence these standards and to keep them up to date.

Whatever efforts are expended in determining actual customer experience with garments, we must recognize that user opinion is going to influence the results. Ideally, those wearing the garments that are to be evaluated should represent the actual or potential market for the product, economically as well as physically. This should be kept in mind by manufacturers when children in orphanages are used for garment try-ons.

Role Of Complaints

While most customer complaints about products that come to the retailer's attention probably do not get relayed to the factory, more important is the fact that most dissatisfied customers don't complain at all. For example, consider the purchaser of an item that costs one dollar and who finds that it is not what he expects. Very likely he will not return it, but may avoid repeat purchases in the future. One way to express this is to say that the merchandise doesn't return and neither does the customer. This means that the retailer loses a future sale. If the customer talks to anyone about his disappointment, it will most likely be to his friends and neighbors.

The fact that customer complaints and returns do not often reach the apparel manufacturer should not result in the relatively few that reach the factory being ignored. The *rate* of complaint will probably be low, but the information that is contained in those few that are returned may be quite valuable. We no longer freely quote standard multipliers to be used to determine actual number of dissatisfied customers based on the number of complaints reaching the factory. Certainly there are *several* dissatisfied or disappointed customers for each return that reaches the factory.

The subject of garment fit deserves special comment. This is unquestionably one of the most important areas of quality to the customer, because a garment must fit in order for the sale to be made, or for the garment to stay sold. The technology of mass producing a garment that will fit the customer is a complex and demanding one, well beyond the scope of our current subject. But good fit involves much more than just manufacturing to a specification. A poor fitting range of garments may never leave the sales floor, or if sold may cause considerable dissatisfaction and returns to the retailer. As a rule, high returns to the retailer caused by poor fit come to the attention of the manufacturer only in extreme cases. So the absence of size and fit returns to the factory probably should not in itself be taken to mean that customers are finding garment fit to be satisfactory.

Usefulness Of Data

While there is seemingly no limit to the amount of customer usage data that can be generated and made good use of, the actual development and use of such data must depend upon the cost of obtaining such information as compared to its potential usefulness. The major factors to consider when developing and responding to customer usage data are time and accuracy of the data. The earlier in the design and production sequence the usage data becomes available, the more useful it is in affecting production garments. But usage data developed in advance of production requires the construction of test garments (which are more expensive and may not be completely like production garments) and the finding of test users, who may not be completely typical of actual users.

A practical answer to the problem of test data not being representative, while actual field data arrives too late for action to be taken, is to make effective use of both techniques in an organized, co-ordinated program, and to recognize the limitations of each. A planned program of test garment analysis, including laboratory testing and controlled usage, is a key to providing customer satisfaction with the product. But these tests do not take the place of an orderly analysis of returns and complaints. For, even though there are dangers in reaching conclusions from customer return analyses, because their representativeness may be badly distorted by the time garments or data are returned to the factory, it is essential that a history of garment returns be kept, for customer returns remain an important source of information about the acceptability of garments to the user. Furthermore, histories of returns on similar garments provide a useful background as new garments are designed.

The impact that returns information will have on the analysis of laboratory tests results falls into two areas.

First, properly analyzed, it will point out the areas of customer concern that are important enough to customers to make them complain. This may help put laboratory and experimental usage test programs into focus by calling attention to quality characteristics that would otherwise be largely overlooked in the test program. Second, customer returns can help establish the degree of acceptability to the customer of certain characteristics which have already been identified for analysis, where insufficient information has been available about the standards that the customer expects to be maintained.

Returns Can Help

Most factories receive few returned garments. Many act as though they want to receive even fewer. Since returns and allowance are an expense, essentially a subtraction from net profit, there is a natural tendency for factory management to discourage the acceptance of returns from retailers whom it serves. A more effective way for a manufacturer to reduce returns is to avoid the conditions that cause them. But manufacturers should actively attempt to receive all, or a representative cross section of garments returned for quality reasons, while limiting allowances or credits to retailers where the reason for return is clearly not manufacturer responsibility, or where the return privilege of the retailer has been abused. Accepting returned garments should usually include receiving them physically, not just processing paper-work documentation of a return.

Defective garments manufactured by a specific production facility should be returned to that location. Where manufacturing is done in several locations, returns may be accepted and recorded at one location, but the garments should be quickly relayed to the manufacturing facility. Complaints that are not accompanied with a garment should receive the same handling and documentation as the physical return.

Basic to effective returns analysis is retaining the identity of the garment, on the garment, so that it is adequately legible even after a period of use by the customer. The RN registry number or other code numbers or symbols may be used. What is important is that the production facility producing the garment be identifiable, including the production line or department, if that detail of identification is meaningful. Style or range number may help garment identification. A permanent size identification in the garment will also help in establishing its full identity, to say nothing of the value such identification may be to the customer.

Record Systems

If the factory receives a number of garments each week, it should consider maintaining a record system that charges the return to the production period in which the garment was made, rather than just recording the date the return was received. This may be a new concept to some, so I will describe it in some detail, after contrasting it to the conventional (and not incorrect) method of computing returns rate. Typically, factories keep track of the dollar value of returns received in a given calendar period, such as a month, and compute a rate of return to the dollar value of shipments for that same month. By charging returns to the production period in which they were actually made, a more accurate measure of the volume of returns can be obtained.

To establish the production period in which a garment was made, it is necessary to use a marking system suitable to the product. Generally, identification of the quarter of the year in which the garment was produced will be satisfactory for returns analysis, although some may wish to identify the month of production. High style garments may be in production in only one quarter, and it may be possible to readily identify the period of production by establishing style identity. For the majority of relatively stable garment styles that may have repeat production runs over an extended time, the production date should be clearly identified. Changing the color or style of an identification tag or label, or the printing used on it, may be all that is necessary. While it is a good idea to have a similar system of production date identification on outer packing as well, to identify the product while in inventories, marking on packaging will not be sufficient in itself, since customers returns may not be in their original pack. Code letters or numbers may be used to identify dates, if they are not confused with style identification numbers. Elaborate woven labels which are ordered in advance should not have a date code included on them, for this would require perfect prediction of label usage or result in surplus unused labels. But it may be possible to include code symbols representing production dates on labels serving other functions.

In addition to keeping a master record of the date coding system in use, a record of the number of garments produced should be preserved for each production period, including a breakdown of the quantity produced of each style, if that is necessary for later analysis.

Each time a return is received, descriptive detail should be recorded on a suitable form. Information that should be included on the form includes:

- Product description, style number, fabric, size
- Date the item was received by factory
- Retailer returning item
- Value of item, when new
- Allowance to retailer by factory
- Date produced, to nearest quarter
- Reason for return, as stated by retailer
- Condition of garment as received
    - In original pack
    - Out of pack, but apparently not worn
    - Apparently worn, but not laundered
    - Laundered, but no extensive wear
    - Worn out
- Observed reason for return, as judged by factory inspector
    - Fabric flaw
    - Workmanship
    - Shrinkage
    - Appearance after washing
    - Mismarked
    - Other reason
    - Reason not established
- Additional comments of factory inspector
    - Presence of manufacturing defects, deterioration in washing by customer, evidence of abuse
- Recommended follow-up action
    - Examine addition stock
    - Launder garment from stock
    - Other
    - None
- Results of follow-up action by factory

Item return records created during each month should be sorted by production period and product type, and a monthly returns summary prepared that provides the detail for returns and allowance expense appearing on the monthly accounting statement. Thus, not only is the total value of returns and allowances reported, but management is told which production periods and product types were responsible for current returns. Rate of return, as a percent of current shipments, may also be computed.

At the end of each quarter, a more detailed analysis of returns should be prepared, actually charging returns to their production period, and computing rate of return against production. This is done by sorting all item returns records received during the quarter by period of production. A permanent quarterly returns record card is created for each past production period. All returns are then posted to this permanent card as they occur, making the record a history card for the production of one complete quarter.

Only returns pertaining to the production period identified at the top of the quarterly returns record are posted to the card. Returns are posted by reason code. Total production for this period is posted to the card as a one-time entry. Each quarter, as new entries of returns are made, the increasing cumulative returns may be observed as well as the cumulative figures for returns by reason.

Meaning Of Returns

Whatever system of returns recording and analysis is used, the *ratio* of returns to production is essential in order to take into account changing levels of factory volume. But the very calculation of this ratio of return may take away from the effectiveness of return analysis, since the *ratio* of returns will be such a small fraction of factory volume that it will seem quite insignificant, possibly even a source of pride to factory personnel. We have already pointed out that *level* of returns is a poor measure of customer dissatisfaction. But the *content* of returns data can indicate the more serious defects coming to the attention of customers, while less serious defects that affect customer attitude toward a product may not be returned. As a returns history is accumulated, sufficient understanding of its significance will be gained so that a *change* in rate of returns can be a valuable indicator of customer acceptance, even though the percentage of production is low.

While we have described a returns analysis system in some detail, there are other approaches to customer returns analysis that may be considered. Since only those returns that "fight" their way back to the factory are included in the report described, they cannot indicate the complete customer attitude toward the quality of garments, nor does this actually convey all that retailers know about the product, as they face the ultimate purchaser. Special arrangements with selected retailers can be established so that they become test locations for reporting all complaints and returns. The key to such special arrangements is the retailer's understanding that complete information concerning customer reaction to product is needed. If the retailer repairs garments, he should report it. Careful records should be kept of products sold to the test retail locations. Data from the test locations should be totaled at intervals, and the retailer should see evidence of how the data he provides is being used. Retailers are aware of a significantly higher level of returns and customers dissatisfaction than has ever come to your attention.

Controlled Usage

Customer usage data determined under carefully selected conditions, where the user of each test garment is identified, and the number of wearings and launderings is recorded, provides excellent control of test results. On the other extreme, usage data developed from analysis of customer returns provides little or no control over usage conditions. There are various intermediate alternatives which provide relative degrees of control of garment usage. These should be considered when additional data is desired. Providing garments to test users, perhaps employees, to use until worn out or for a specific period of time, will provide garments that can be evaluated after a period of service. Lack of control of number of washings and wearings limits the ability to compare results with laboratory data, but may provide more normal conditions of usage. By using visual seconds for such a program, the costs need not be unduly high, and administrative costs will be low. This is no substitute for a controlled test program, but it can be a useful supplement to one.

Many of the procedures discussed may not seem practical to use on a continuous basis, because of the ratio of cost to the information to be gained. Certainly any test program must be geared to the potential benefits to be received. The objective of garment test programs and returns analysis is to provide the maximum information to factory design and production personnel about products they have produced or plan to produce. There is a multiplying effect gained when test data is properly developed and analyzed, for it adds to the fund of experience upon which key personnel can draw to make correct decisions in the future. Even though it may not be possible to conduct extensive field tests on all product categories at the same time, information gained on one product can contribute to the knowledge about other items.

The analysis of customer returns is a valuable source of product usage information when included as a part of a broader plan of product test and analysis. Returns data can mislead, low returns levels can give a false report of customer satisfaction. They can distort, by reporting only serious defects, when there may be many minor areas that are causing customer dissatisfaction. But with all of these potential pitfalls, the factory management that is determined to listen to the customer cannot ignore those garments that are returned.

Quality control in a mail order house. The first application of quality control techniques in the retail field appears to have been made by James Ballowe at Alden's mail order house in Chicago about 1945.[4]

Error control was applied to a large group of employees, not to individuals, who were filling mail orders for shipment. A random sample of 100 orders was inspected daily for errors relative to size, color, style, price, etc., and the number of orders in error recorded. The error rate was plotted daily on a very large wall chart that everyone could see. The method was very effective in reducing the error rate, which changed from about six percent to about one percent in a few months. It turned out to be so successful in this one department that it was later applied to other departments.

Other quality problems facing management in retail stores. Other problems involving quality aspects in retail trade include the following:

1. Purchasing and deliveries:
    a. Vendors
    b. Lead time in ordering
    c. Purchase order control
    d. Receiving and shipping
    e. Defective products purchased:
        - disposal method
        - stopping them from being put on sale
        - selling defective products at reduced prices

f. Delivery system:
   scheduling
   delivery times
   delay time
   delivery damage
   reducing costs of delivery

2. Inventory:
   a. Balancing inventory by customer demand
   b. Perpetual inventory
   c. Computer control
   d. Eliminating manual inventory
   e. Profitable versus non-profitable items
   f. Minimizing carrying charges on inventory
   g. Inventory control for re-ordering:
      keep record of items customers request which are not in stock
   h. Ordering at customer request

3. Sales and costs:
   a. Losses: damage, breakage, pilferage, shoplifting, stealing by employees
   b. Pricing system and practice
   c. Advertising and sales
   d. Accounting system
   e. Sales inducements: discounts, coupons, stamps, price cuts, bonuses, refunds
   f. Employee wages and salaries
   g. Security
   h. Warranties and their costs
   i. Servicing own sales
   j. Cost reduction program
   k. Maintenance contracts

Repair Services

Auto repairs. Complaints about the unsatisfactory nature of automobile repairs have become so serious that several state legislatures and large cities have taken legal steps to reduce auto-repair fraud, overcharging, and poor quality service.[5] The states include Florida, Michigan, Ohio, Georgia, and New York; the cities include Chicago, New York City, Wichita, and Jacksonville, Florida. One of the major requirements is that the repair shop must furnish the customer with an itemized written statement of estimates before any repair work is started. This does not prevent overcharging, but it does help eliminate unnecessary work. It is a step in the right direction.

Obtaining an acceptable quality repair job on an automobile will be difficult as long as automobiles are so complicated; so few persons know how an automobile operates, let alone what constitutes a satisfactory repair job; and the growth of self-service gasoline stations where a check-up on oil, battery water, radiator, and tires are eliminated.

The cause of the auto repair, or an excessive charge for auto repair, may be due to several causes: 1) a defect or defects in manufacturing; 2) a wrong diagnosis of the trouble by the garage or repair shop; 3) the practice of replacing rather than repairing; 4) including unnecessary repairs in the repair bill; 5) faulty repairing by the garage or repair shop; 6) the failure of an automatic indicator; and 7) neglect by an auto owner or driver to take proper care of the automobile, including a proper amount of oil, a proper amount of water in the battery, a proper amount of anti-freeze in the radiator, a proper amount of air in the tires, and proper brake adjustments.

Examples of various kinds of auto repair. Listed below are various examples of repair situations classified according to the major causes given above.

Manufacturer as cause

1. An air pollution system does not work. An attempt to accelerate may kill the engine, making driving dangerous in heavy and fast traffic. The trouble was not fixed in three trips to the regular dealer. The car was traded for a foreign car.

2. The paint peels off in large strips from the hood and the top of the body after only a few years. A body expert says that the steel was never treated properly in the factory, and that any new paint job will last no more than one year.

3. A car has a handle-type gear shift fastened to the steering column. This handle shears off completely when the driver shifts gears during rush-hour traffic. Luckily, the gear is in neutral and the car can be pushed onto a side street.

4. The linkage in the gear shift locks in second gear. A black attendant at a nearby gasoline station shows the driver how to unlock the linkage and get it off dead center. He says he has seen a number of such cases.

5. A new car leaks while being driven in a rainstorm. It takes the dealer about three days to seal the car, something that should have been done in the factory.

Wrong diagnosis

1. When a car does not seem to have any power, it is driven into a regular dealer's garage. A mechanic looks at it, tests the car, and says it needs a new automatic transmission. This is a Friday afternoon and the driver cannot leave the car. He takes it to a small two-man auto repair shop, and they find that all the automobile needs is 75 cents worth of transmission fluid.

2. The garage mechanic says a car has a leaky radiator, but a delay in having it fixed showed only that the so-called leak was no more than an over-flow of the radiator.

Faulty repair or faulty installation

1. When the ammeter read zero, the driver drove the car into a small garage to determine the cause. The mechanic replaced the generator, which had burned out, and showed the driver that the voltage regulator had been set at 50 amperes when it should have been set at 30 amperes. The higher amperage caused the solder to melt and the armature to burn out.

2. The air conditioning system in a car blows out because when it was installed one of the hoses was placed over one end of the battery. When the acid ate through the hose, the gas was released and the air conditioning system failed.

3. Before a long trip, a car is taken to the dealer's for a complete check-up. After a hundred miles, the front wheels begin to shimmy, necessitating a slowing down. A stop at a small-town garage discloses that the wheels are unbalanced and that the brakes are almost worn out. The wheels are balanced and new brakes are installed. Where a large city dealer failed, a small-town garage performed a high quality job.

4. When a tire blew out in the vicinity of Bismarck, North Dakota, it was discovered that what was alleged by the dealer to be a regular tire was nothing but a retread.

Unnecessary charges

1. In a collision on an icy street, the front end of a car was damaged to the extent of $80 according to the esti-

mate made at a garage. However, the estimator examined the entire car and included dents on the rear fenders and other places; he estimated the bill at $180. This was bill bloating. The driver told them to fix the front end and forget about the rest.

2. A driver found that the steering mechanism was not working right. He drove to a garage where a mechanic found a cracked part and said the entire steering column must be replaced. The driver suggested welding the part instead. The mechanic said he would not guarantee it. The driver said, weld it and I will take the risk. After some grumbling by the mechanic, this was done. It lasted until the car was traded in for another.

Lemon laws for new automobiles.[6] Pressure on state legislatures has been growing because of customer discontent with new automobiles. This discontent has given rise to "lemon" laws and also customer appeals programs, although it is reported that the customer wins rather infrequently in the latter. A "lemon" apparently is defined as one which meets either of the following conditions:

1. Has been in repair four times and still is not working properly.
2. Has been in repairs 30 days or more during the first year.

The status of the actual or proposed lemon laws is as follows:

1. State laws passed (2 in 1982; 7 in 1983)	9
2. Under consideration	13
3. Awaiting governor's action	4
4. Not approved, recent session	11
5. Not introduced	14
Total	51

(including District of Columbia)

A lemon is a candidate for replacement or refund, but the customer may have to hire a lawyer and threaten to go to court to get satisfaction.

The fact that an automobile contains 14,000 parts, plus or minus, is no justification for producing an unacceptable quality automobile, or for not giving adequate service the first time it is brought to the dealer. This whole situation is very unfortunate, but reveals another serious consequence of not making an acceptable quality automobile in the first place.

Television repair. This actual case reveals how the individual buyer may be treated by a large company and how the buyer can avoid paying for an error the company makes. The picture tube on this 17 inch television set went blank. The manufacturer's service repair station about 10 miles away was called and the trouble explained. They came to the house and fixed it. That night, when the set was turned on, the picture was there, but the sound had disappeared. The next day the service station was called and the new trouble was explained. They came and fixed the audio portion, but the customer was billed for two repair calls.

The customer refused to pay for the second trip since it had to be made to correct some mistake the repair man made on the first trip. The service office kept insisting on the double billing. The customer then tried a method which a friend had used successfully to obtain an automobile from a local dealer — he telegraphed the Detroit headquarters. This customer, however, wrote a letter to the headquarters addressed to the Vice President in charge of Customer Relations and explained in detail the situation. In about a week, he received an apologetic letter saying that the double billing was a mistake and that the customer was right in insisting on paying for only one trip. This the customer did, and that ended the matter.

In another case, the on-off switch of a television set did not operate properly; it was loose and apparently worn out. The set was taken to a large nearby television repair shop, where it was agreed that a new switch was needed which would be ordered immediately. They claimed this would require two weeks. The customer returned in two weeks, but the set was not fixed. Finally, after two more weeks, the set was fixed, but not with a new switch. Three different persons were involved and the claim was that a new switch could not be obtained. Cost $48. The set was taken to another repair shop, where a new switch was ordered and promptly installed.

Typewriter repair. A regular model office typewriter made by a well-known company was purchased new with a 90-day guarantee. Some minor adjustments had to be made on the typewriter before it would perform properly. The most surprising defect that appeared was the failure of the on-off switch, after only about six months! Of all parts on a typewriter, or anything else for that matter, the most reliable part one would expect would be a simple electric switch.

Corrective actions that can be taken by buyer or customer. There are a wide variety of actions which the buyer or customer can take to correct an error, to replace a defective product, or to obtain favorable action. This list is illustrative, not comprehensive:

1. Return defective goods to store or to manufacturer and obtain either credit or a refund.
2. If a food is inedible, such as tainted milk or rancid butter, return it to the supermarket to obtain a replacement or credit.
3. If a generic product is inedible, such as canned pears that have to be cut with a knife, return it to the supermarket for a replacement and a reminder that, contrary to a commonly expressed view, "generic" products are not always the same quality as other brands or even another can of the same generic product.
4. If double billed for something purchased, or charged with not paying, such as an insurance premium or a book purchase or a magazine subscription, mail them a *copy* of your cancelled check or other evidence of payment.
5. If a magazine is persistently mailed to you which you did not order and for which you are billed, write REFUSED across it, leave it in the mail box, and refuse to pay. It usually takes months, but they finally discover their error.
6. If a firm refuses to return your money for an advertised article not in stock, threaten to report them to the U.S. postal inspectors for misuse of the mails. This brought action from a book company that wanted to credit an account rather than give a refund, even though the customer seldom ordered books from this company.
7. If a company double bills you or refuses to fulfill a promise that is important, write or telegraph the President or a Vice President at the headquarters office and explain in detail the situation. Be sure to attach *copies* of any relevant documents. This worked wonders when a customer who was promised an automobile, but was being put off, telegraphed the President of a large automobile company in Detroit. He received his automobile within 24 hours. It also worked in the case of a television

double billing situation when the customer wrote a letter to the Vice President in charge of Consumer Affairs at the headquarters office. The customer was not only assured he was right, but received an apology in addition.

Notes

[1]From a description of a conversion in a King Soopers store, Loveland, Colorado, *Reporter Herald,* June 6-7, 1981, p. 6. Data and material from King Soopers used with the permission of Jim Baldwin, President of the Company.

[2]This material was prepared by R. W. Peach, and used by his permission and that of Sears Roebuck and Company. Mr. Peach, formerly quality manager for Sears for many years now has his own firm, Robert Peach and Associates, Inc., 541 North Brainard Avenue, La Grange Park, Illinois 60525.

[3]This is a paper by R. W. Peach appearing in *Textile Chemist and Colorist,* Vol. 1, No. 23, (November 5,1969), pp. 23-26, reprinted by permission of the author and the American Association of Textile Chemists and Colorists, Research Triangle, North Carolina.

[4]This application was described by James Ballowe in a paper issued by the Office of Production Research and Development of the War Production Board, No. 5 of a series, dated September 1945.

[5]*U.S. News and World Report,* "Ripoffs in Auto Repairs — Any Safeguards in Sight?", December 1, 1980, pp. 64-65. The complaints listed later are from other sources.

[6]*USA Today,* May 31, 1983, pp. 4A, 1B, 2B.

9

Business Services

Business services denote those services which are rendered by one business to another, usually a specialty or a technical or scientific service that a firm does not have the capacity to perform or needs special assistance to perform. This service may be rendered by an individual consultant or by a company of some size. Examples of this type of service can often be found within a company or agency — a group with special know-how such as accounting, laboratory testing, or statistical expertise assists another part of the company or agency. The services described in this chapter include:

1. Data processing
2. Computer edit
3. Retail store inventory

Data Processing

The operation. This operation featured the following:

1. A very large sample.
2. Starting and implementing the project: organizing, hiring, training, planning and designing and implementing controls, etc.
3. Continuous quality control by lot sampling; keeping individual error records.
4. Continuous quantitative controls: keeping individual production records.
5. Technical assistance by a consultant.
6. Complex data documents.
7. Training inexperienced transcribers; supervisor has a major role.
8. Designing an effective sample plan.
9. Immediate feedback to individuals; analysis of error records.
10. Supervisory tutoring to eliminate sources of errors.
11. Charting daily production per person per day, and daily average transcriber cost per acceptable quality document.

A nationwide sample of over 100,000 shipping documents received during a year from the 70 largest railroads in the United States had to be transcribed to a standard format so that the freight traffic information could be key punched for card input to a computer. These documents were carload waybills or equivalent statements which varied widely in composition and arrangement although all of them contained the necessary information somewhere. Since the documents were not standardized, every one of them had to be studied carefully before transcription could take place.

The data transcribed included both numbers and words, to the extent of two punch cards, and required careful examination and reading of every part of the document which consisted of a single page. The task was not a simple copying job from a standardized document; it required careful study, a substantial amount of special knowledge, the use of reference books, and accurate transcribing. Items to identify and transcribe included origin and destination stations, weight — which sometimes had to be converted from cords or gallons or some other unit, revenue, type of rate, and origin and terminating railroad. In addition, it was necessary to code from the standard commodity code book the commodity being shipped.

A 95-page procedural manual had to be mastered and used as a reference continuously, so that the transcribed data were comparable and additive. Numerous abbreviations were used on the document and these had to be interpreted correctly, adjustments and calculations had to be made in connection with different units of measurement and with transit movements, and the type of rate had to be identified properly. Obviously, this was no simple clerical job.

Because of the complexity of the data and its presentation on the document and the need for a fairly low error rate (below five percent), a training period of about three or four weeks was necessary. Training continued until each individual could pass a test given by the supervisor. Even so daily assistance and tutoring by the supervisor were necessary.

Procedure

The consultant designed and prepared procedures and plans for implementing the subsampling plan and the quality assurance plan, designed and implemented a sample receipts control system, designed forms for production and error records for every employee, analyzed production and error records by individuals weekly and advised supervisor with regard to actions to be taken, calculated average quality level weekly, analyzed individual learning (production) curves weekly, and advised management and supervisor on various aspects and implications of sampling, quality control, and the error rates and production curves. The consultant spent considerable time demonstrating the cost effectiveness of using sample review as a basis for controlling the quality of the work of each transcriber, so that the overall error rate would be in the neighborhood of five percent.

The production goal was to transcribe 100,000 sample documents with acceptable quality within 24 weeks. Due to the time required to hire, organize, and train the clerical staff, and bring them to a level of acceptable production, most of the actual production took place during a much shorter period of time — about 16 weeks.

The overall acceptable quality level was set between four and five percent in view of the following considerations: (1) the documents were samples and therefore any data from them were already subject to sampling errors, (2) a major data item was a seven digit commodity code which experience showed a five percent variation was not unusual even among experts in this field, (3) the computer edit to be applied later was expected to catch one to two percent of these errors, and (4) the final data were to be estimates of population parameters which meant this was to be primarily a statistical job from beginning to end. It was not a bank check or payroll job where error free data and error free performance were required.

The Staff

As indicated above, this was not a simple clerical job like stuffing envelopes or running a duplicating machine. It was a difficult job made even more difficult by the use of temporary employees. At maximum there were 23 persons — the supervisor, three reviewers, and 19 others. These represented a wide range of talent — a few housewives, a college graduate, high school graduates, and college students interested in working during the summer. With few exceptions, all were inexperienced in difficult clerical work and for most of them this was their first job.

All were given about three weeks of intensive training by the supervisor and had to pass a written test before being put on production. This test, together with the difficult nature of the work, eliminated several persons early in the project. (These are not included in the figure of 23 given above.)

Every employee did the complete transcription job and worked independently of the others. This was an individual production job, not a sequential line production job where each employee does only a fraction of the entire job.

Motivation was a real problem. While most of the employees were able to master the essentials, some were not motivated by money, some became bored with the routine of production, and with few exceptions, all of them had difficulty adjusting to the discipline of the work situation in an office. Evidence that motivation was a problem included absenteeism, sickness, unnecessary talking, and unnecessary travel.

A major advantage of having inexperienced persons was the absence of fixed ideas about the methods of work. Hence, gains could be obtained from the learning process (learning curve) which might not have been possible otherwise.

The supervisor was the key to the success of this operation. The supervisor had the rank of manager and reported directly to the vice president. She had five major tasks:

1. To train the clerical force to the point where acceptable quality output could be produced.
2. To maintain quality levels and still meet production goals.
3. To resolve all technical questions raised by the source documents.
4. To maintain, for this large group, a satisfactory working atmosphere with regard to discipline, motivation, and attitude. This included attention to physical arrangements, supplies, and personnel problems.
5. To analyze and use production records.

The supervisor had over 20 years of experience in training and supervising persons engaged in an identical type of production work. The supervisor was a transportation specialist with a certificate in traffic management and had a successful career in this work. On this job, she not only had to be a regular supervisor, but also had to be a combination psychiatrist, counselor, traffic expert, and teacher.

Training the clerical force required the use of a 135-page procedural manual which had already been prepared for past work on this type of work. For this job the manual was revised and reduced to about 95 pages. This manual alone was evidence that the type of work involved was not a simple clerical job.

The training included all-day classes for two weeks for some and three weeks for others, plus continuous tutoring and assistance during the entire production operation. There were group presentations as needed as well as lengthy individual conferences on problems and difficulties. The supervisor processed the more difficult documents such as transits and rebills.

Persons had to pass a written test before they were put on production. Furthermore if their initial error rate was high — above 10 percent — they were tutored. If no improvement occurred, they were released or shifted to other work.

Even some who performed satisfactory were disillusioned about the work — it was too difficult or too monotonous — and so they quit very shortly. In all about 15 persons were trained who for one reason or another never produced any satisfactory data sheets or worked only a few days. Included in this number were about five persons who had to be released because they were really not interested in working on such a demanding job.

Training included explanations of the shipping document, what the various parts meant, what the abbreviations stood for, how to identify various types of documents requiring special treatment, identification and verification of the commodity and its coding, and many other aspects including type of freight car, weight, rates, revenue, and stations.

The Quality Control Plan

On this job, several standard methods were available to measure and control the quality, that is the error rate, made by the individual transcribers:

1. The p chart for every individual.
2. The p chart for the entire group.
3. Acceptance sampling for every individual.
4. Acceptance sampling for the entire group.
5. Continuous sampling.

Exploration of the work situation, the resources available, and other factors indicated that a sample review of the individual lots or batches (item 3 above) would appear to be the most feasible. Actually, it was the only method that would be likely to prove successful under the circumstances.

The final quality control plan consisted of an acceptance sampling plan at a normal level of inspection — Plan H II in Military Standard 105-D. It was used as a process control as is shown below. This plan has the following specifications:

1. Batch or lot size = 300 data sheets.
2. Sample size = 50 data sheets.
3. Sampling rate = 1/6 with random start.
4. AQL = 4 percent.
5. Acceptance rule: five or less data sheets in error in sample.
6. Rejection rule: six or more data sheets in error in sample.
7. AOQL: about 6.3 percent.

This plan was used because the supervisor decided that a batch or lot size of 300 was convenient to handle, process, and control. Furthermore, this batch size had advantages when it came to rejected lots that had to be completely redone. Lots or batches two or three times as large would not only have been unwieldly to handle and to sample, but would have created serious psychological effects.

The four criteria used to determine the error rate to be tolerated on the average in the transcribed data form have already been described. The important aspect to stress in this job is that we are dealing with a large nationwide sample aimed at producing a large number of estimates of population parameters. The problem is to keep total uncertainty in the final estimates at a low level, not to try to eliminate it.

The fact that the input data are derived from a probability sample survey puts the entire process in a different light and makes it basically statistical, not only in the collection of the data but in all of the succeeding stages as well. Computer practices and procedures designed for payroll, bank checks, accounts receivable, ticket reservations, inventory control, personnel records and similar non-sampling situations are not applicable, although in these operations quality control is just as necessary as in collecting, processing, and analyzing sample data.

On recommendation by the consultant, a subsampling plan was used to reduce the size of the sample to be processed by about 90,000 documents. This subsampling reduced the size of the total job by about 40 percent without any significant loss in basic information. This was possible because of gross inefficiencies in the original sample plan under which the documents were collected. The consultant designed the subsample plan, prepared the detailed instructions necessary to implement it, instructed the supervisor in its use, and set up the necessary control counts and records for both the frame and the subsample. The consultant monitored the operation from time to time to see that instructions were being carried out according to plan.

The job of sample review posed a number of problems. It meant that two to three persons had to be instructed and assigned to this function. These persons had to be carefully selected because they not only had to understand the work, but they had to do an accurate, fair, and meticulous job of review, since a reviewer always knew whose batch of work was being reviewed. Control was exerted over individuals, not over the group as a whole.

Both production and error records were maintained daily for every person. Total production was posted daily and a graph was maintained showing production as a function of time. A graph of daily production was maintained for each person and was used by the supervisor, when necessary, as the basis of a personal interview with the individual. Individual error records were used for a similar purpose.

Records were kept on every batch of 300 data sheets transcribed by every person. The number of errors found in a sample of 50 selected from that batch was also recorded. A rejected batch was returned to the individual immediately for complete review and correction of all of the errors in the batch of 300. Accepted batches were filed for the next operation.

Production was studied daily and weekly, and the error rate for the group computed weekly on a cumulative basis. Production levels were met by making use of the learning curve, special instructions on difficult points, personal conferences, and by eliminating persons who had serious learning or emotional problems.

Before production really started, management raised the question of the need for an incentive system to reach the production goal. The consultant recommended that any incentive pay system should carry penalties for rejected lots, and, furthermore, that any such plan should be postponed until after it was seen how production progressed. It was discovered very soon that an incentive pay system was unnecessary for two reasons: additional money was not a motivation for some persons, and, furthermore, the learning curves indicated that the production goal would be met without any incentive other than close personal supervision. The amount of absenteeism showed that money was not a strong motivating force for several, although it was a strong motive for others.

The Learning Curve

The production curve for the group is the sum of the individual learning curves. These curves played a very important role in this project because they showed how much production could be obtained once the learning process stabilized.

Furthermore, they showed what effect the rejection of a lot had upon the individual learning curve. In this project, a specified quality control plan was imposed on the learning process, but as will be shown later both sets of specifications were met simultaneously. The learning curve applies to every individual. The composite for the group constitutes the production curve wherein production y is a function of time t. Production means acceptable quality production.

The learning curve arises when the following conditions exist:

1. The individual lacks a specified ability, skill, or understanding of a substantial set of facts or knowledge. The task is short and repetitive.
2. It takes considerable time to gain the required ability, skill, or knowledge before a start can be made on the learning curve.
3. As time progresses under proper guidance and supervision, the sources of error are eliminated or minimized, understanding is clarified and expanded, and common difficulties are resolved.
4. In other words, as time progresses, useless and erroneous knowledge and acts are eliminated. Correct repetition plus memory effects help both in reducing errors and increasing production.

This means that under certain conditions the speed of transcribing can be increased with a decrease in the number of errors made. The learning curve is a measure of the capability of an individual or group expressed in terms of actual sustained performance.

A study of individual curves (learning curves) showed that it required from 30 to 40 days for most individuals to reach a stable maximum. Some never reached their peak before they left. The output per person after 40 weeks was close to 200 units per day. Everyone reached or exceeded 200 per day, but not everyone was able to maintain this level of production day after day.

Some persons varied widely from day to day, but still continued to move upward. Others were more stable and showed a more narrow range of variation. Some slowed down their production much more than others once they received a rejected lot.

Analysis of the group production curve revealed the following:

1. The production curve was maintained at a practically constant acceptable quality level — between four and five percent.
2. Nearly half of the total growth took place during the first five days.
3. The maximum level stabilized after about 40 days.
4. Production was actually higher with 11 persons after 40 weeks than it was with 16 persons after 25 weeks.

5. The production goal was met without any special incentive pay plans.

The phenomenon of getting more production with fewer employees is not unusual in work performed under these conditions. It was found to exist in many aircraft and other plants during World War II. In the present case it was due to four causes: (1) the learning effects of those remaining more than compensated for the loss of those leaving, (2) the removal of low producers, (3) a sharp reduction in the amount of inter-person interference resulting in an actual increase in the numbers of hours worked per day; and (4) a removal of persons with emotional problems and antagonistic attitudes or the resolving of these difficulties in potential high level producers.

Results of the Quality Control Plan

The sample review quality control plan called for selecting every sixth data sheet using a random start, from each batch or lot of 300, for a total sample of 50. Lots were accepted if there were five or fewer data sheets in error; they were rejected otherwise. The characteristics derived from 188 lots or 56,400 data sheets are as follows:

1. Fourteen percent or about one in seven of the lots were rejected. All errors were to be corrected for these lots according to plan.
2. The average quality level for the first 188 lots or 56,400 data sheets was 4.8 percent.

The use of sample review on every lot of transcribed data sheets was an effective control since it put the brakes on what could otherwise have been a runaway error rate. The effectiveness of a rejected lot in reducing the error rate is illustrated by the following individual examples; lots are shown in time sequence:

Transcriber No. 1		Transcriber No. 2		Transcriber No. 3	
Lot no.	No. of errors	Lot no.	No. of errors	Lot no.	No. of errors
1	2	1	10 reject	1	4
2	5	2	13 reject	2	11 reject
3	9 reject	3	2	3	0
4	0	4	1	4	1
5	4	5	1	5	2
		6	0	6	4
				7	1
				8	0
				9	3
				10	1

Table 1 shows for each of 45 days the number of transcribers, the total number of acceptable quality records transcribed, the average number per transcriber, and the cost per record per transcriber. Figures 1 and 2 were derived from this table; the first shows average production per day, the second the cost in cents per record per day. Table 1 shows:

1. A total of 88,554 acceptable quality records were transcribed in 45 days at a total transcriber cost of $10,290; this is an average of 11.6 cents per record.
2. The average daily production per transcriber increased from 64 on the first day to 227 on the 40th day and to 213 on the 45th day.
3. The average transcriber cost per document decreased from 26.3 cents on the first day to 7.4 cents on the 40th day and 7.9 cents the 45th day.
4. Three broad levels of performance are discernable with about 200 records per day being the actual capability of these trained and experienced workers for this type of document.
5. Most of the daily gain was made in the first six days. A gain of 57 records on the daily average was made during the first six days compared with 106 for the next 34 days.

Figures 1 and 2 show the following:

1. Transcriber cost per record per day decreased from 26.3 cents to 13.9 cents during the first six days. The total reduction in cost was more than one half.
2. From the seventh to the 16th day, production tends to level off below 120 records per day at a cost of about 14.5 cents per record.
3. From the 17th to the 36th day, a new level varies about 150 records per day at a cost of about 11 cents per record.
4. From the 37th through the 45th day, the level averages slightly above 200 records per day at a cost of slightly more than eight cents per record.

Factors operating to create variations, plateaus, and higher levels included transcribers being required to correct rejected batches of work completely which stopped direct production whenever such rejects occurred, less capable transcribers leaving the project, continuous personal assistance by the supervisor thereby progressively reducing the probability of errors, reduction of mutual interference between employees as their number decreased, and the effectiveness of rejection of batches in reducing errors.

It should be noted that Figure 1 is really a process capability curve for this group of employees processing a specific type of document under a given set of conditions, but at the same time having to meet specified quality goals. The vice-president wanted to set 2,000 records per day as the performance goal right from the start, but the group did not reach this level as a total until the 18th day. Clearly, 2,000 was an arbitrary and unrealistic goal to set as the learning curve-production curve shows.

Controlling computer edit readouts. A large nationwide sample had to be prepared for computer processing, including a computer edit. Records were kept of the total number of sample documents received each month, the number read out by the computer edit, and the percent. The edit included apparent deficiencies such as impossible values and codes, inconsistent values or relationships, extreme values, and other violations of the editing rules.[1]

A high degree of stability existed in the proportions read out without any control chart. A p chart was constructed using an average value of five percent since this is what past records showed. Hence, p = 0.05, 1-p = 0.95, and n = 20,000 per month. The value of 3 sigma is:

$$3 \text{ sigma} = 3\sqrt{\frac{p(1-p)}{n}} = 3\sqrt{\frac{0.05 \times 0.95}{20{,}000}} = 0.005.$$

Hence,

1. The center line is 0.05 or 5 percent.
2. The upper control limit UCL = 0.05 + 0.005 = 0.055 or 5.5 percent.
3. The lower control limit LCL = 0.05 − 0.005 = 0.045 or 4.5 percent.

FIGURE 1

Production per Day per Person from Start Through 45th Day

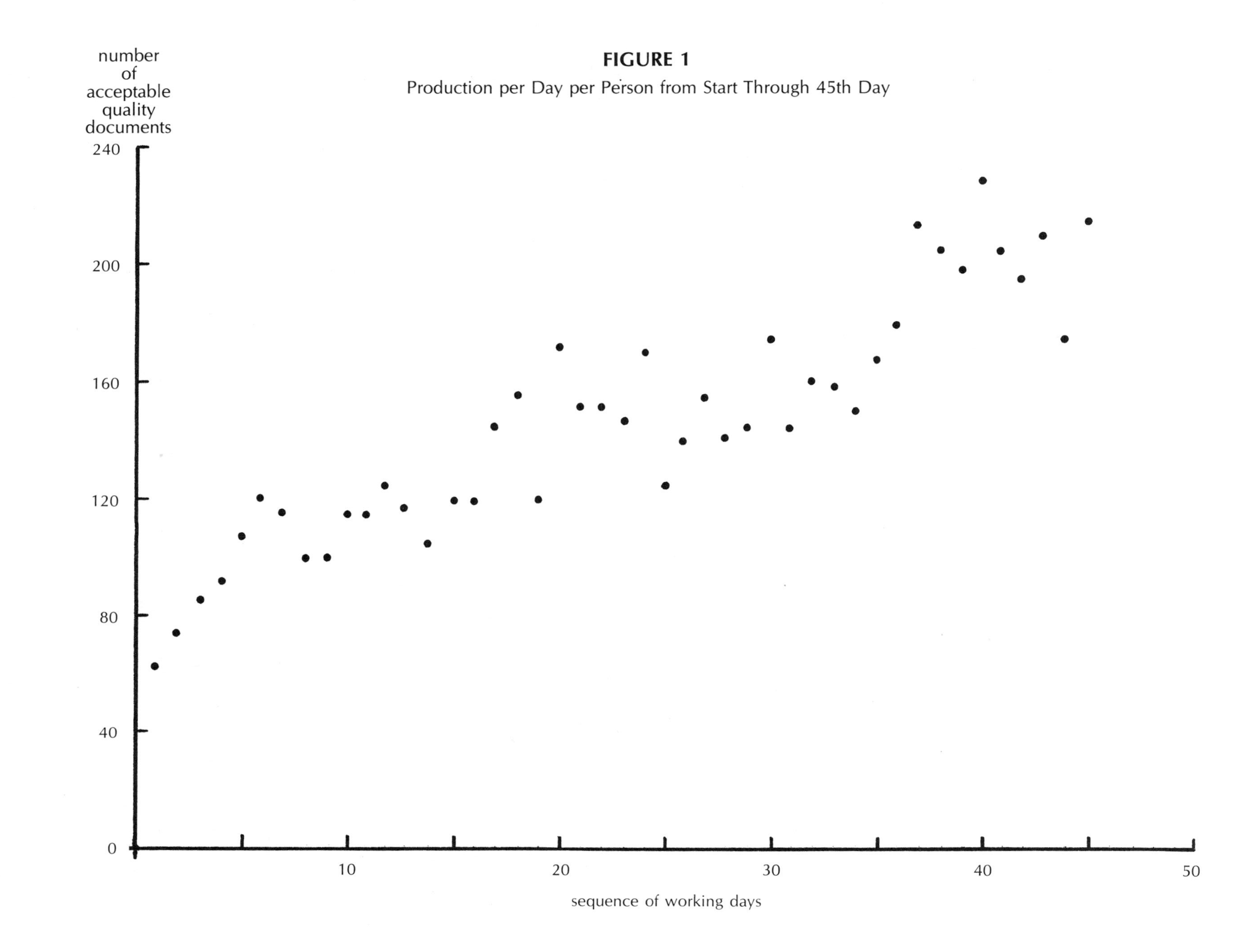

Table 1
Results of Quality Control Plan

Day x	Number of transcribers[a]	Number records transcribed	Average number per person per day y	Cents per record per person[b]
1	8	515	64	26.3
2	12	929	77	21.8
3	12	1,028	86	19.5
4	14	1,322	94	17.9
5	15.5	1,696	109	15.4
6	15	1,814	121	13.9
7	17	1,977	116	14.5
8	16	1,575	98	17.1
9	16.5	1,651	100	16.8
10	15	1,742	116	14.5
11	13	1,507	116	14.5
12	15	1,864	124	13.5
13	15	1,767	118	14.2
14	16	1,782	111	15.1
15	16	1,932	121	13.9
16	12	1,460	122	13.8
17	13	1,870	144	11.7
18	14	2,214	158	10.6
19	16	1,976	124	13.5
20	13.75	2,383	173	9.7
21	13.75	2,104	153	11.0
22	13.75	2,127	155	10.8
23	14.75	2,204	149	11.3
24	12	2,061	172	9.8
25	16	2,022	126	13.3
26	14	1,938	138	12.2
27	14	2,188	156	10.8
28	15	2,092	139	12.1
29	15	2,170	145	11.6
30	13	2,260	174	9.7
31	14	2,006	143	11.7
32	15	2,426	162	10.4
33	14	2,227	159	10.6
34	14	2,148	153	11.0
35	12.5	2,117	169	9.9
36	12.5	2,240	179	9.4
37	11	2,328	212	7.9
38	13.5	2,764	205	8.2
39	14	2,774	198	8.5
40	11	2,495	227	7.4
41	12	2,444	204	8.2
42	12	2,351	196	8.6
43	11	2,293	208	8.1
44	12	2,071	173	9.7
45	8	1,700	213	7.9
Total or average	612.50	88,554	144.6	11.6

[a]Number of actual transcriber days; variations due to hiring, absenteeism, and leaving; 67 transcriber days absent or about 1.5 per day for about an 11% rate.
[b]Based on a $16.80 daily rate for each transcriber working.
Source: Compiled from original records in the possession of the author.

The 25 monthly values were:

1. 4.7	4.7	5.1	5.5	6.3 too high (error)
2. 5.8 too high	4.4 too low	4.8	5.4	4.8
3. 5.0	5.1	5.0	5.2	5.0
4. 4.7	5.6 too high	4.8	5.2	5.5
5. 4.9	4.9	5.1	5.6 too high	4.8

The stability of these data is surprising in view of the absence of any formal Shewhartian charts or other controls at the source. In any event, actions at the source were quite difficult since thousands of persons throughout the United States were involved in preparing, selecting, and shipping these sample documents to be processed by the computer. Direct control had to be limited to the computer and processing operations. (Some actions were taken to correct errors at the source.)

The error made for the 21st month — 6.3 percent — was discovered by an expert reviewer who showed that 176 documents had been read out by the computer by mistake. When this correction was made, the percentage was reduced to 5.3 percent which was within limits. If the reviewer had not discovered the error, both the time chart and the above p chart would have indicated trouble at this point. Reasons for other out-of-limit points were not discovered. The p chart would indicate trouble regardless of whether review was by an expert or not.

Accounting services: sample audit of a store inventory. Accounting is a field that has so many aspects and applications, and is so important in any business or agency, that a company of any size usually has its own accounting staff or department. Special situations arise, however, which often call for special outside expertise in such fields as cost accounting, tax accounting, inventory control, and application of computers to accounting.

The following example is based on an actual situation in which an inventory of merchandise had to be taken of several retail stores in order to settle an estate. Two accounting firms were involved: the firm which was to make the inventory and the firm which was to audit the inventory. Under court order, this latter firm was hired to see that the inventory count and amount were accurate and trustworthy. Since the stores were not to be closed, the inventory was taken at night and the sample audit had to be made right behind and simultaneously with the actual inventory count and extension.

FIGURE 2

Daily Average Transcriber Cost per Acceptable Quality Document

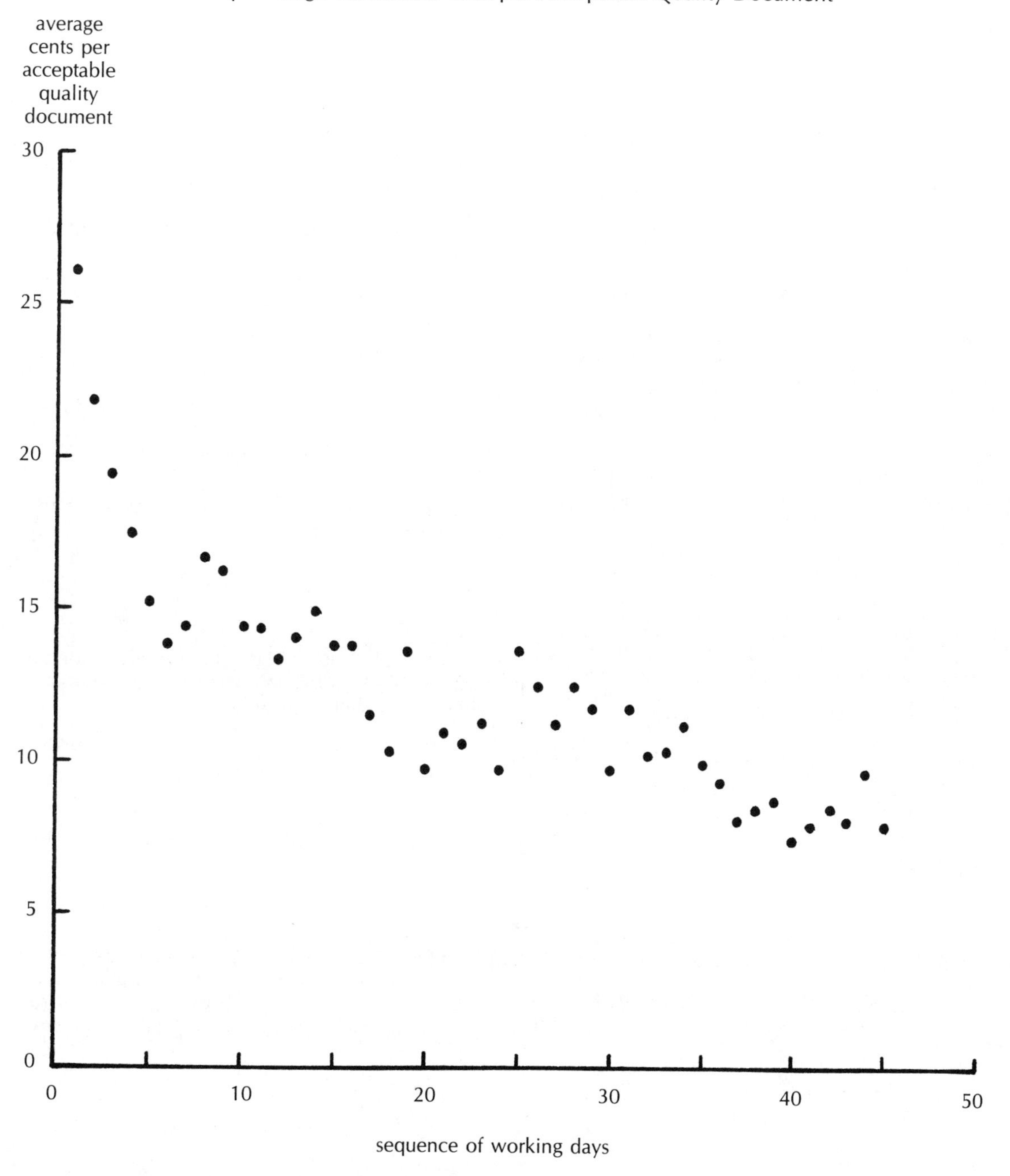

An audit sample based on stratified sampling was designed using three strata derived from the dollar value of the items or segments. The estimated number in the population, the average dollar value, and the total dollar value were as follows:[2]

Stratum	Average dollar value; range	Total number	Estimated dollar value
small	$ 40 (1-100)	5,000	$ 200,000
medium	$ 300 (100-999)	2,000	600,000
large	$1,500 (1,000 and over)	800	1,200,000
		7,800	$2,000,000

The total number of items and the average dollar value of each stratum must be known, estimated, or guessed at using persons familiar with the merchandise. Rough information is sufficient for stratification purposes; highly accurate figures are unnecessary.

The 7,800 items or segments must be identified by size — small, medium, and large — and numbered or counted carefully so effective sampling rates such as the following can be applied:

	Rate	Multiplier	Sample size
small	1 in 100	100	$50 = n_1$
medium	1 in 20	20	$100 = n_2$
large	1 in 3	3	$270 = n_3$
			$420 = n$

The sample is used for three purposes: to audit behind a 100 percent inventory count and valuation, to determine the error rate as well as the amount of error made, and to estimate the total dollar value of the inventory and its sampling variance to determine if the 100 percent inventory valuation falls within the sampling variations about the estimated total from the sample (falls within the 95 percent confidence limits).

This design results in a feasible sample size which can be carried out behind a 100 percent inventory if properly designed, carefully planned, and carefully and competently managed.

Each stratum has to be weighted by its multiplier before any total can be calculated due to the varying sampling rates in the strata:

$$\text{Estimated aggregate } X = 100\sum_{1}^{n_1} x_i + 20\sum_{1}^{n_2} x_i + 3\sum_{1}^{n_3} x_i.$$

$$\text{Estimated number } N_{est} = 100n_1 + 20n_2 + 3n_3.$$

$$\text{Estimated grand average} = \frac{X}{N_{est}}.$$

The standard error of the estimated aggregate X for k strata is the square root of s_x^2:

$$s_x^2 = \sum_{i}^{k} \frac{N_j^2 s_j^2}{n_j}\left(1 - \frac{n_j}{N_j}\right)$$

where $j = 1,2,...,k$ strata, and s_j^2 is the variance within the jth stratum. $1 - n_j/N_j$ is the correction in each stratum for sampling from a finite population.

Notes

[1]The data are given in Chapter 19 and plotted for each of the 25 months.

[2]See Chapters 15 and 23 (case 4) for detailed descriptions of stratified random sampling and its applications.

10
Personal Services

Personal services. Companies or individuals engaged in rendering personal services deal directly with individuals every day. Therefore they have to be highly sensitive to personal wishes, personal desires, personal attitudes, and personal tastes. The individual buyer, on the other hand, is sensitive not only to the quality of the service being rendered but, even more important, to the employees who are rendering the service. First impressions, personal likes and dislikes, the tone of the voice, and the expression of cooperation and courtesy are very important.

The company has to cater to these individual differences for several reasons. Repeat business is extremely important and may be the difference between a successful business and a mediocre one. Also, the satisfied guest or customer is the best kind of advertising, while the disgruntled one can seriously hurt business. Flexibility is needed, not only in the amount, but in the kind of service because service that will satisfy one person may not satisfy another. Customers have different standards depending upon their occupation, standard of living, income, expectations, and a host of other factors. Some customers know exactly what they want, others are demanding, and still others accept whatever is offered without any complaint or criticism.

The major types of personal services described in this chapter include some of the following:

Hotels and motels
Restaurants and cafeterias
Cleaning and laundry
Traveling and tours
Beauty and barber shops

Other kinds of direct personal services, such as health services and transportation services, are of such major importance that they are described in separate chapters.

Services which deal directly with *products* owned or rented by the customer, such as the repair of automobiles, TV sets, radio sets, dishwashers, stoves, disposal units, refrigerators, air conditioning units, furnaces, and the like are discussed in the chapter on retail trade.

Hotels and Motels

The quality problem of hotels and motels is developed by listing customer complaints, by presenting in detail quality characteristics that need to be considered, by presenting a short quality questionnaire for a customer to answer, and by presenting five actual case studies of customer questionnaires, including one distributed by a company that conducts intercity sightseeing tours.

Customer complaints. The complaints listed below are those which have been gathered from customers, including tourists, conventioneers, and business men:

1. Cancellations of reservations at 6 p.m. because the airplane was late or because some other factor over which the customer has no control delayed arrival. In one case, this caused several parties to have to hunt for accommodations that were very difficult to find; one party had to travel 10 miles to find a place to stay.
2. Tough and rough treatment received by cash customers who are forced to pay the entire amount due in advance. The assumption is that the credit card holder is honest, the cash customer is dishonest.
3. Long wait for a room even though room was reserved and arrival was before 6 p.m. or paid for. In one case, several parties had to wait about an hour before a room was available.
4. A serious error was made on the reservation dates which were arranged by long distance telephone. The person taking the telephone call was not the person filling out and mailing notice of the reservation. A letter and another telephone call were necessary to straighten out the matter.
5. A high price was charged, but the room was small, plain, and cramped; the low quality did not justify the price charged.
6. In one room, there was only one chair for two persons; in another case, the lights were very dim, making reading impossible.
7. Did not give room at the price advertised, but at a much higher one despite ample advance notice.
8. Room was very cold and drafty with no control over air conditioning.
9. Thermostat does not work so there is no control over heat.
10. In an expensive hotel, heat unit does not work so unit was replaced with a new unit.
11. Trouble and delay in obtaining extra blankets for cold nights.
12. No notice that there would be no Sunday breakfast served in the hotel, even though the nearest open restaurant was several miles away.
13. Worn out black and white television set.
14. Restricted variety of food available with little or no choice, even though prices are high. Food was largely gourmet type. No variety for those who prefer something else.
15. For a high priced hotel, the choice at the salad bar was a disaster. Some small town restaurants have better salad bars.
16. Room was dirty; carpet needed cleaning since it was covered with food and drink spots.
17. Room was simply too low quality for the price charged: a complaint made by two business men who traveled widely and stayed in the better hotels.

A quality control inventory. The following is a detailed list of items which are to be considered when measuring the quality of service. All of them are involved when the buyer (customer) comes to appraise the service he or she is receiving:

1. Reservations:
   a. Ease of obtaining

b . Ability to get what is wanted
c . Delay time
d . Error free
e . Attitude
f . Written confirmation
g . 6 p.m. rule
h . Check in, honoring reservations
i . Baggage handling

2. Room:
   a . Size
   b . Furnishings: table, chairs, beds, carpet, dresser, drapes, curtains
   c . Cleanliness
   d . Heating, cooling, ventilation
   e . Lighting, lamps, windows
   f . Radio, TV, clock
   g . Linens, blankets
   h . Noise, disturbances
   i . Bathroom: shower, tub, plumbing, towels, hot water, soap
   j . Linen changed daily
   k . Room service
3. Food:
   a . Availability of restaurants
   b . Hours
   c . Prices
   d . Variety, choice
   e . Table service
   f . Waiting time for service
   g . Waiting line to enter
   h . Salad bar
   i . Quality of food
   j . Are menu substitutions allowed
   k . Atmosphere, environment
   l . Attitude of employees
   m. Courtesy
4. Bar:
   a . Cocktail lounge, bar
   b . Availability
   c . Hours
   d . Variety, choice
   e . Quality
   f . Cost
   g . Delay in service
5. Other services:
   a . Parking
   b . Transportation information
   c . Sightseeing information
   d . City information
   e . Newspapers and magazines
   f . Common medicines
   g . Gift shop
   h . Mail box
   i . Telephone service
   j . Laundry
   k . Baggage service, baggage storage
   l . Souvenirs
   m. Postage stamps
   n . Snacks
   o . Elevators: location, adequacy, waiting time
   p . Taxi service
   q . Other transportation service
   r . Airport service
6. Check out:
   a . Baggage service
   b . Transportation arrangements
   c . Billing, cashier
   d . Attitude
   e . Waiting line

A short questionnaire of quality questions. The following contains some of the major questions which bear on the quality of the service rendered which a customer needs to answer:

1. Room: Did you receive your money's worth?
   Was anything lacking?
   Were you satisfied?
2. Food: Did you receive your money's worth?
   Was anything lacking?
   Were you satisfied?
   Did you have the choice you wanted?
   Was the quality acceptable?
3. Drinks: Did you receive your money's worth?
   Were you satisfied?
   Was anything lacking?
   Did you have the choice you wanted?
4. Were you treated courteously: at reservation desk, by room attendants, by bell hops, by waiters, by waitresses, by the hostess, at the check out, by sales persons, by others?
5. Reservation and check out: Were these prompt? error free? did you get what you wanted? what you requested?
6. Would you come back? Would you patronize this chain elsewhere?

Case 1. Marriott Hotels

The Guest Comment Form is for guests to appraise the services received at a Marriott hotel. There are 10 questions, including space, to describe any unresolved problem. Questions 1, 3, 5, and 6 call for one of five ratings for each of the services listed: excellent, good, average, fair, or poor, and relate to the following:

Question 1: overall rating of the hotel
3: check in, check out, room, room price
5: rating of eight staffs or individuals
6A: rating of restaurant
6B: rating of room service
6C: rating of cocktail lounge
6D: rating of banquet/convention event

Ratings as quality measurements are fine so long as everybody and every characteristic is rated excellent or good. But difficulties arise when a group of employees or a service is rated poor or fair. This is because the latter ratings give management no basis for taking corrective actions because the specific behavior, specific service, or specific situation that led to the rating is unknown.

Questions 2 and 3 deal with reservations, checking in, and checking out. If errors are made, delays occur, or the guest has a complaint, they will show up here or under question 7.

Questions 3 and 4 deal with the condition of the room, the price paid, and the working condition of the equipment and facilities in the room. Question 7 shows whether or not the "hot line" was used to report complaints or dissatisfactions.

Under question 5 there is space to record the name of a person who was especially helpful, and what was done to be helpful. Provision should be made for entering the name of more than one person. This provision is commendable because it stresses the positive approach to quality performance by identifying those employees who were unusually good or helpful.

Bethesda Marriott Hotel

OCA 728

1. How would you rate our hotel on an *overall* basis?

☐ Excellent ☐ Good ☐ Average ☐ Fair ☐ Poor

2. Was your room reservation in order at check in?

☐ Yes ☐ No

3. How would you rate the following?

	Excellent	Good	Average	Fair	Poor
Check in, speed/efficiency	☐	☐	☐	☐	☐
Cleanliness of room on first entering	☐	☐	☐	☐	☐
Cleanliness and servicing of your room during stay	☐	☐	☐	☐	☐
Decor of your room	☐	☐	☐	☐	☐
Check out, speed/efficiency	☐	☐	☐	☐	☐
Value of room for price paid	☐	☐	☐	☐	☐

4. Was everything in working order in your room?

☐ Yes ☐ No

If you checked NO, would you please tell us what was *not* in working order?

☐ Room air conditioning
☐ Room heating
☐ Bathroom plumbing
☐ Television
☐ Light bulbs
☐ Other ______________________

5. How would you rate the following in terms of their friendly and efficient services?

	Excellent	Good	Average	Fair	Poor
Reservation staff	☐	☐	☐	☐	☐
Front desk clerk	☐	☐	☐	☐	☐
Bellstaff	☐	☐	☐	☐	☐
Housekeeping staff	☐	☐	☐	☐	☐
Telephone operators	☐	☐	☐	☐	☐
Gift shop staff	☐	☐	☐	☐	☐
Engineering staff	☐	☐	☐	☐	☐
Front desk cashier	☐	☐	☐	☐	☐

If any members of our staff were especially helpful, please let us know who they are and how they were helpful so that we can show them our appreciation.

Name ______________________

Position/Comments ______________________

____

____

6. Please rate the following which you have used on this visit:

A. Restaurant

Please indicate name of restaurant.

____

☐ Breakfast ☐ Lunch ☐ Dinner

	Yes	No
Were you seated promptly?	☐	☐
Was your order taken promptly?	☐	☐
Was your food served promptly?	☐	☐

	Excellent	Good	Average	Fair	Poor
Friendly service	☐	☐	☐	☐	☐
Quality of food	☐	☐	☐	☐	☐
Menu variety	☐	☐	☐	☐	☐
Value for price paid	☐	☐	☐	☐	☐

B. Room Service

Prompt service	☐	☐	☐	☐	☐
Friendly service	☐	☐	☐	☐	☐
Quality of food	☐	☐	☐	☐	☐
Menu variety	☐	☐	☐	☐	☐
Value for price paid	☐	☐	☐	☐	☐

C. Cocktail Lounge

Prompt service	☐	☐	☐	☐	☐
Friendly service	☐	☐	☐	☐	☐
Quality of drinks	☐	☐	☐	☐	☐
Value for price paid	☐	☐	☐	☐	☐

D. Banquet/Convention Event

Prompt service	☐	☐	☐	☐	☐
Friendly service	☐	☐	☐	☐	☐
Quality of food	☐	☐	☐	☐	☐

7. Did you use "The Marriott Hot Line" to register any dissatisfaction with our hotel?

☐ No
☐ Yes . . . problem was resolved.
☐ Yes . . . but problem was *not* resolved.

Please explain any problem which remains unresolved.

____

____

____

____

____

____

____

____

____

____

____

8. What was the primary purpose of your visit?

☐ Pleasure
☐ Convention/group meeting/banquet
☐ Business (other than above)

9. Have you stayed at this hotel previously?

☐ Yes ☐ No

10. If in the area again, would you return to this Marriott?

☐ Yes ☐ No

PLEASE PRINT THE FOLLOWING INFORMATION

Departure date: ______________________

Length of stay: __________ days. Room number __________

☐ Mr. ☐ Mrs. ☐ Miss ☐ Ms.

Name ______________________

Home address ______________________

______________________ Zip __________

Company or organization ______________________

____

Business address ______________________

______________________ Zip __________

THANK YOU VERY MUCH FOR YOUR RESPONSE.
YOUR EVALUATION *WILL* MAKE A DIFFERENCE.

Welcome to our hotel. We are pleased to have you stay with us.

It's very important to us that we do things right for you. That means . . .

- Reservations should be easy to make
- Quick, hassle-free check-ins
- Rooms that are pleasant, immaculate
- Restaurants so good that many times they're local favorites
- A staff brimming with vitality — and smiles
- Expertise in handling meetings
- Fast, painless check-outs.

You're the one who can tell us whether we have been successful or not. We value your comments and want to hear about the things we do right and the things we need to improve.

Will you let me know?

I have to make *sure* we do things right. After all, it's my name over the door.

Bill Marriott

President, Marriott Corp.

Case 2. Hyatt Regency

This questionnaire has several items relating to the front desk, receiving assistance, housekeeping, restaurants, cocktail lounges, room service, other hotel services, observations and suggestions for improvements, and how the guest learned about this Hyatt Hotel. The form of the questions is such that the answers are "yes", "no", or some modification thereof. Observations and suggestions for improvements are to be given at the end of the questionnaire, not under each item. One reason for brevity, if not the major reason, is that the card is to be mailed to headquarters; it is not deposited at the hotel. The form therefore could be filled out and mailed either before leaving the hotel or afterwards.

Some questions seem general, but these are satisfactory if the particular service is acceptable. Specific instances, incidents, and cases are most helpful to management so that proper corrective actions can be taken to improve service. It would also be very helpful to management, in all of these service companies, if guests were asked to list the three or five things about the hotel and its services that they liked best.

Examples of specific items that guests are concerned about are the following:

- Was the room large enough? attractive? comfortable? properly heated? properly cooled? properly ventilated? properly and sufficiently lighted? reasonably priced?
- Did the following work adequately: TV? radio? clock? telephone? plumbing?
- Were the linens and towels satisfactory? changed daily? were extra blankets available?
- With regard to restaurants, did the menu give you the choices you desired? Was the food satisfactory? Was the service satisfactory? Were the prices reasonable?
- With regard to room service, did the menu give you desirable choices? Was the food satisfactory? Was the service satisfactory? Were the prices reasonable?
- Were the cocktail services satisfactory? prices reasonable?
- Were the services, prices, and atmosphere such that you would return?

Front Desk Was check-in and check-out satisfactory?

Bellstand and Doorman Were they efficient and friendly?

Concierge Department Did you use this service?

Was the staff courteous?

Were they helpful?

Housekeeping Was your room spic and span?

Restaurants Was the food good? Did you like the decor?

Hugo's

Clock of 5's

Polaris

Kafe Kobenhavn

Cocktail Lounges Did you enjoy yourself?

Le Parasol

Club Atlantis

Ampersand

Room Service Was the service efficient? How was the food?

Other Hotel Facilities Did you find everything you wanted?

Your observations and suggestions for improvement are appreciated.

How did you find out about this Hyatt Hotel? COMPANY ☐ ADVERTISING ☐

FRIENDS ☐ AIRLINE ☐ OTHER HYATT HOTEL ☐ Other ☐

Thank you for your comments. We hope to have the pleasure of a return visit from you in the near future. Please take advantage of our free advance reservations to all Hyatt hotels. Call 800/228-9000.

Thank you for staying at the HYATT REGENCY ATLANTA
In Peachtree Center

Room Number Check Out Date

Name

Company

Address

City State Zip

Telephone No. ( )

Reproduced by permission of Hyatt Hotels, Rosemont, Illinois.

Case 3. Holiday Inn

The "report card" for guests of Holiday Inns to fill out appraising services and rates is a school-type grading scale with each item rated from excellent A through B, C, and D to F which is designated bad. The items differ from those on other rating scales in two very important respects:

1. It includes items on prices of the room and prices of food in the restaurant.
2. It includes a question as to whether the guest is returning to the area, would come back, or look elsewhere.

In addition to 14 specific items to grade and three questions to answer, there is space for comments.

The most important statement, however, occurs in the letter to the guest from the president of the chain, especially the following sentence:

> "Also, your opinions are especially helpful in making sure everything in this Holiday Inn measures up to *your standards*". (emphasis added)

Note that it is the buyer's standards, the customer's standards, that are to be met. *It is not our standards that are to be met.* The buyer, not the seller, sets the specifications. This is a radical departure from current and past practices in many other companies and industries. This is a real break-through for the customer-buyer.

The report card grading method, however, suffers from the same limitations as a rating scale. If an item receives a rating of D or F, there is nothing management can do about it since there is no detailed information to be used as a basis for corrective action. It would greatly improve this report card as a control, if the guest were asked to describe in detail under "comments" any item rated D or F.

Case 4. Canadian National[1]

This is a "yes" or "no" set of questions dealing largely with prompt and courteous service with regard to reservations, reception and service personnel, and quality and service in the restaurants and bars. "Prompt" and "courteous" service are combined, although one may exist without the other. Likewise "courteous" and "efficient" service are combined, although in a restaurant or bar one may receive one without the other.

DEAR GUEST:

Thank you for staying at a Holiday Inn hotel. We want you to feel welcome and comfortable.

Should you encounter a problem during your stay, or if the hotel staff can be of service to you, please contact the manager who is on duty.

Also, your opinions are especially helpful in making sure everything in this Holiday Inn measures up to your standards.

So, we hope you will take the time to complete and detach The Holiday Inn Report Card. Just drop it in the mail and tell us how we're pleasing you.

Sincerely,

JAMES L. SCHORR
President, Hotel Group
Holiday Inns, Inc.

Holiday Inn

Reproduced by permission of Holiday Inns, Memphis, Tennessee.

BSA 2392 **HOLIDAY INN REPORT CARD** 4473

	EXCELLENT				BAD	
1. Overall how would you grade *this* Holiday Inn?	A	B	C	D	F	1
Now, please grade:						
2. YOUR ROOM:						
Appearance	A	B	C	D	F	2
Cleanliness	A	B	C	D	F	3
Comfort	A	B	C	D	F	4
Furnishings	A	B	C	D	F	5
Bathroom	A	B	C	D	F	6
3. RESTAURANT:						
Food Quality	A	B	C	D	F	7
Service Quality	A	B	C	D	F	8
4. PRICE/VALUE:	EXCELLENT VALUE				BAD VALUE	
Of your Room	A	B	C	D	F	9
In the Restaurant	A	B	C	D	F	10
5. FRONT DESK PEOPLE:	EXCELLENT				BAD	
Friendliness	A	B	C	D	F	11
Efficiency	A	B	C	D	F	12
6. SERVICES: (Messages, Wake-Up Calls, Bellman, etc.)	A	B	C	D	F	13
7. OTHER FACILITIES: (Pool, Lounge, Lobby Parking, etc.)	A	B	C	D	F	14

8. If you were to return to this area, would you stay at *this* Holiday Inn or look elsewhere? **Stay here** ☐ **Look elsewhere** ☐ 15

9. Number of people staying in room: **One** ☐ **Two** ☐ **More than two** ☐ 16

10. Have you stayed at *this* Holiday Inn before? **Yes** ☐ **No** ☐ 17

Comments ______________________________

Today's Date ______________ Room # ______

Your Name/Address (Please) ______________________________

______________________________ Zip ______

Location of Inn: **BELOIT, WI**

4473

This questionnaire is also too short and brief to be fair to the service company (hotel) and to the customer who clearly will encounter numerous aspects of quality service which are not listed on this short form. We have already described these specific questions and items which relate to the measurement of quality of services rendered by a hotel, so they will not be described here.

To our Guests

Service is what a good hotel is all about. We'd really like to know how you rate the service you've received on this visit. Would you please take a moment to complete this questionnaire and return it to me?

Gordon J. Trainor, Acting General Manager, CN hotels

Reservation

Was your reservation handled —		Yes	No
	Promptly	☐	☐
	Courteously	☐	☐
	Efficiently	☐	☐

Reservation made through ______________

Reception and Services

Did you receive prompt and courteous attention from

	Yes	No		Yes	No
Parking Attendant	☐	☐	Elevator Service	☐	☐
Doorman	☐	☐	Telephone Service	☐	☐
Room Clerk	☐	☐	Room Service	☐	☐
Bellman	☐	☐	Laundry Service	☐	☐
Maid Service	☐	☐	Valet Service	☐	☐
Cashier	☐	☐	Other (specify)	☐	☐

Restaurants and Bars

	Yes	No
Was service courteous and efficient?	☐	☐
Was food attractively served?	☐	☐
Was quality satisfactory?	☐	☐

Name of Restaurant or Bar ______________

Case 5. Tour Quality Control[2]

This is a rating form for six service items performed in connection with a bus tour through Canada. There is space to enter comments or remarks, and it is here where helpful specific complaints are to be found. The ratings are Excellent, Good, Fair, and Substandard.

This rating scale has the usual weaknesses of rating scales for quality measurement: it is ambiguous, general, and vague. If anyone checks "Substandard", what is management going to do about it without specific instances, detailed situations, or actual behavior? There is very little or nothing that can be done. Furthermore, what one person considers substandard, another person may consider fair or even good, and vice versa.

Only if everyone rates an item excellent or good can management feel safe that the customers are satisfied and pleased with what they are receiving. To be a sound and valid basis for corrective action, performance, and behavior, specific services must be described in detail indicating the individual or individuals, the time, the place, and the situation. Only with such specificity has management any valid basis for taking action or not. Just because a customer makes a complaint and describes it in detail, does this mean that corrective action is needed or even necessary? It all depends upon the nature of the complaint.

Quality Control is Part 2 of a longer questionnaire which is designed to obtain an appraisal by every one on a tour about what was satisfactory and what was not. Included are three very important aspects of a tour, aspects which make or break the enjoyment of a tour: the hotel accommodations, including food; the tour director, whether knowledgeable, interesting, and helpful; and the driver, whether careful and communicative. This part is fine so far as it goes, but it could be improved by using more specific aspects and expanding its coverage. Items such as rotation of seats on a bus, the hour at which meals are to be served, and other specific situations could be added.

2 QUALITY CONTROL. We welcome your frank evaluation of our tours. Your remarks aid us in making our tours more enjoyable in the future and are CONFIDENTIAL. Check (✓) your evaluation below.

	Excellent	Good	Fair	Substandard
A. HOTELS				
Rooms	____	____	____	____
Food	____	____	____	____
B. TOUR DIRECTOR				
Personal Service	____	____	____	____
Narrative	____	____	____	____
C. DRIVER				
Driving	____	____	____	____
Coach Cleanliness	____	____	____	____

Space below and back of top flap for additional comments.

Restaurants and Cafeterias

Quality control areas. In a restaurant or cafeteria, there are many areas in which quality is involved. There is the quality of purchased products, materials, and ingredients. These include all kinds of food products — everything from raw fruits, vegetables and meats to products of a bakery such as pies, rolls, bread, and cakes. There are also cleaning and maintenance supplies. There is the problem of damage caused to products during shipment, unloading, and unpacking. These are returned for credit.

Quality of employee performance also has top priority. Without high quality products — meat, vegetables, fruits, bakery goods, canned goods, etc. — not even the best person can produce high quality food. Quality performance applies to everyone: chefs, cooks, assistants, dishwashers, janitors, handymen, waiters, waitresses, servers, hostesses, counterpersons, cashiers, bartenders, busboys, busgirls, and anyone else who works anywhere in the restaurant or cafeteria.

The goal of quality performance is to give the customer good quality food at an affordable or reasonable price. To reach this goal means:

1. Giving the customer what he or she wants.
2. Serving the customer with a minimum amount of delay: waiting to be seated, waiting to have order taken, waiting to receive food, waiting for a bill.
3. Serving the customer with good quality food and beverage.
4. Serving the customer with consideration and courtesy.
5. Listening politely to any customer complaint and resolving it immediately.
6. Listening to customer's questions carefully and answering courteously.

Quality control check sheet. A quality control check sheet that includes the major characteristics involved in quality performance, quality management, and quality audit is outlined below. This check list includes, but goes far beyond, what a local or state health inspector is authorized to examine and approve or disapprove. This quality check list is aimed at helping management to improve the quality aspects of the entire operation or establishment, not just one part of it.

1. Location:
   - a. Accessibility
   - b. Location
   - c. Parking and parking lights
   - d. Surroundings
   - e. Condition of parking lot
2. Receiving:
   - a. Facilities
   - b. Unloading
   - c. Controls
   - d. Distribution and storage
3. Kitchen:
   - a. Layout, arrangement
   - b. Equipment
   - c. Sanitation
   - d. Cleanliness:
     - personal including clothing, hair
     - dishes
     - equipment
     - utensils
     - floors
     - cabinets
   - e. Division of work
   - f. Work places and areas
   - g. Safety measures:
     - fire extinguishers
     - exits
     - exhaust systems, ventilation
     - handling hot substances
     - clothing
     - swinging doors, traffic routing
   - h. System for handling orders: verbal
     - written slips of paper
     - checked ✓ slips
     - other
   - i. Delay time:
     - time to fill various orders
     - time waiting for pickup
   - j. Preparation of orders:
     - time sequence
     - proper preparation
     - cooked as ordered
     - served as ordered
     - food quality
4. Serving line (cafeteria):
   - Arrangement
   - Identification of dishes with prices
   - Service time
   - Speed at which line moves
   - Variety of dishes to select from
   - Posting of entries with prices near entrance to speed up selection
   - Are there waiters or waitresses to take trays to tables?
   - Arrangements to speed up movement of line
   - Do servers understand requests and serve promptly?
   - Does cashier itemize dishes correctly and promptly?
5. Dining room service:
   - Cleanliness of tables, chairs, booths
   - Variety of menu: is there a choice
   - Waiting time to obtain a table
   - Waiting time to have order taken
   - Waiting time to have order prepared
   - Waiting time to have waitress or waiter serve
   - Truth in menu: are dishes served for what is on the menu
   - Is food hot or cold; satisfying quality
   - Did customer receive special dinner that was advertised?
   - Attitude of hostess
   - Attitude of waitress, waiter
   - Are exits clearly indicated
   - Was bill correct
   - Atmosphere: noise, quiet, relaxed
   - Ventilation, drafts, comfortable
   - Was salad bar adequate
   - Waiting time to pay bill
6. Counter service:
   - Cleanliness of counter
   - Variety of menu, choice
   - Waiting time to have order taken
   - Waiting time to have order filled
   - Waiting time to have order served
   - Attitude of counter person
   - Courtesy of counter person
   - Is check correct
   - Quality of food
   - Cleanliness of dishes
   - Waiting time to pay bill

Quality control in hotels and restaurants. Quality control in large restaurants and hotels is interpreted to mean product oriented controls such as a test kitchen for food and operation audits for other operations.[3]

The operations audit is an inspection by the regional manager or other company official using a detailed check list. Premises are examined. Operating records are checked for accuracy and completeness, ongoing activities are observed, and key employees are interviewed. This inspection group may include a test kitchen standards person or an auditor who reviews internal controls, security, and cash handling practices. A report is prepared; points are assigned. This is the basis for manager and supervisors to take action on items that do not pass. Hotel chains call this audit "quality control".

This is inspection, not quality control. Quality is not a continuous program involving every employee with the aim of improved service. No attempt is made to eliminate the causes of errors, deficiencies, and troubles that affect the customer-buyer. It is management and product oriented, not customer and service oriented.

Four methods are used to measure the quality of service to the customer or guest:

- Unsolicited guest complaints and compliments: These may or may not be used to identify problems and improve service.
- A customer comment card: This is subject to response bias, even though it may be mailed to headquarters.
- A shopping service may be hired to test the performance of the operation.
- A customer survey may be made: This can be effective if properly designed and implemented, and if it obtains customer preferences, sources of dissatisfaction and complaints, and what the customer considers acceptable quality service.

These four methods leave much to be desired. There is no quality policy covering the entire organization and every employee. There is no quality program aimed at eliminating or even reducing sources of trouble. Furthermore, new techniques are required. Two are suggested: *the service report card* and *the room-to-room survey.* Both of these make use of the customers that are in the hotel or restaurant. Getting a continuous record of current customers' appraisals of service on a day-by-day basis is the best and cheapest way of detecting deficiencies, errors, shortcomings, and troubles. It also gives management fresh evidence on which immediate actions can be taken.

The *service report card* can be filled out when a person checks out of the hotel. This could be done at a special service window preceding payment to the cashier. A discount or other incentives could be used to foster taking time to fill out this brief report.

Another possibility is the five minute *room-to-room survey.* One or more full-time persons contact every room for a five minute interview to discover specific deficiencies, shortcomings, and complaints. This should be done by trained personable interviewers. The data are compiled, analyzed, and acted upon daily. Here again, some incentive can be used to increase response rate.

An example of a filled out *Service Report Card* is given below:

SERVICE REPORT CARD — HOTEL X　　Room 999

Name JOHN BROWN　　DATE 8-1-8-

Number in party ______

List specific items such as defects, deficiencies, errors, or shortcomings.

1. leaky faucet
2. no heat, too cold
3. very noisy Thursday night
4. waiter makes $1.40 error; sales pad needs re-designing
5. cashier makes error
6. too long a line for lunch
7. soupy oatmeal
8. skimpy serving of cold cereal
9. housekeeping fine
10. restaurant service fine
11.
12.

The *restaurant business* is not yet fully aware of the advantages of the concept and techniques of quality control.[4] They tend to copy the methods used in the factory which are primarily product oriented rather than customer-buyer oriented. They confuse inspection with quality control. It is not enough to pass Board of Health inspections or any other inspections.

The quality control system must be based on two fundamental factors: customer satisfaction with both the product sold (meal) and the service rendered, and preparing and pricing these meals so that the business can make a profit. The latter cannot be attained unless the restaurant serves what people want at a price they can afford in such a manner that customers return. Repeat business is what very often makes or breaks a company. Quality performance means gaining customer approval by preventing any deficiencies from occurring, and by correcting them immediately when they do occur.

Four aspects of a quality program need to be planned and implemented. The *environmental* aspect means provision for proper housekeeping, safety, decor, lighting, uniforms, menu printing, and furnishings which appeal to the eye and promote pleasure, relaxation, and comfort. Esthetics are important, but good food and service are even better.

The *behavioral* aspects include the behavior of employees and their interaction with patrons: conversing, taking orders pleasantly, offering to meet any special needs of an individual (catering to likes, dislikes, and the need for diet food) offering to make substitutions, and taking other actions that please the customer. Some customers may not like the rigid "package" menu; they want flexibility.

The *mechanics of delivery* means providing the right products (food) in the right sequence at the right time. This includes recipes, service details, job description, how orders are taken by waiters or waitresses, clearing tables, seating patrons, and much more.

Internal standards involve standards which are set for performance, productivity, and costs. They involve size of the staff, assignments, hours of work, and other activities involving employees and management directly and the customer indirectly.

Example of a simple change that will improve quality performance. An example of where a simple change in a form will reduce, if not prevent, errors and delay time in recording an order for a meal is the form used by a waiter or waitress to write down the order and price it out. What is lacking on this form is a systematic arrangement of columns for entering the price of each item — the arrangement used in all accounting and sales slips. A simple format is given below showing the necessary columns on the right to facilitate a correct listing of prices and correct addition to obtain the total in dollars and cents:

Order and Sales Slip No. 1-3-4

Location X
Name or initials AB
Date 8-1-8-

Item	Price			
A		1	4	0
B			6	0
C		3	2	5
D		2	4	0
E		1	2	0
F		3	6	0
G			6	0
Sum or total	1	3	0	5

Because the waiter did not have columns on the form to enter the price of each item in a vertical arrangement, he made an error of $1.40 in the total. This is the fault of the system; the waiter can do nothing about the format of the sales slips he has to use.

Notes

[1]This guest questionnaire of Canadian National is reproduced here with the permission of Gordon Wheatley, Assistant General Manager.

[2]The quality control rating scale and questionnaire are those of the Johansen Royal Tours, Seattle, Washington, and is used by permission of Jon R. Bates, Director of Operations.

[3]Carol A. King, "Quality Controls in the Hospitality Service Operations," *37th Annual Quality Congress Transactions,* American Society for Quality Control, 1983, pp. 412-417. The service report card and the room-to-room survey are not in the reference; these are methods proposed by the author.

[4]Carol A. King, "Quality Control Imperative," *Restaurant,* September 15, 1982, pp. 110, 114, 118.

11
Public Utilities

Public utilities. Two major public utilities are considered: the telephone company, and the electric light and power company. They have a number of common characteristics:

1. They are monopolies.
2. They are regulated by public service commissions.
3. They have millions of guaranteed customers or buyers.
4. Rates are set according to the amount of service rendered (energy consumed) or the amount and kind of service purchased, as in the case of telephone service.

There is some competition in these industries, but to date it is either not serious or it is in a field where it is not profitable for private industry. The telephone industry has some competition on long distance rates, but the most significant competition seems to be in the future as data transmission by telephone lines becomes more and more important.

In the electric power and light field, there are the federally subsidized agencies like Rural Electrification and the Tennessee Valley Authority (TVA) which provide service in competition with private companies. The former provides service to farmers and rural areas, where service is not profitable for private utilities without charging very high rates. TVA provides electrical service to both homes and private business. In both cases, rates charged are below cost and the difference is paid by federal taxpayers. TVA was set up originally for flood control with hydroelectric power as a by-product. Now it generates most of its power in coal-burning plants and is the chief user of waterways to move in the coal needed.

Despite the monopolistic nature of public utilities, there is a real incentive to perform a continuous and reliable service if for no other reasons than to reduce maintenance, eliminate complaints, reduce criticism generally, and convince the public utility commissions to allow them to earn a good return on their investment.

Strange as it may seem, it is in these monopolistic industries where customer complaints about the quality of the service are the least. In the more competitive industries, such as retail trade, repair service, and personal service, the complaints are more common. This statement refers to the adequacy and reliability of the service, not the price. The latter, which is greatly influenced by both state and federal legislation, rules, and regulations, is subject to constant protest, especially in gas and electricity. One reason for the protests is the wide variations in different parts of the country in the price of a kilowatt hour of electricity and a cubic foot of gas. Some bills are very high, especially in all-electric houses, because space heating by electricity currently costs three or four times as much as gas.

Public utilities are subject to a demand which is not placed on the great mass of services which the individual or company buys. One expects an electric light to go on whenever the switch is snapped, at any time during 24 hours of the day, seven days of the week. The same is true of a gas stove or gas furnace or gas burner: it is required to function on demand, regardless of the time of day or night. The same is also true of telephone service; it must be available continuously and constantly. The same is true of water; it is expected to flow whenever the faucet is turned on. This means uninterrupted service, not intermittent service at certain hours at the supermarket, the garage, the drug store, the doctor's office, the bank, the insurance office, and the restaurant. The company must be organized and administered, and the machinery and equipment installed and operated, so that continuous uninterrupted service with extremely high reliability is available to every customer.

The electric power industry.[1] The unique characteristics of the electric power industry are described by Nagel. They are of paramount importance in considering the question of quality of service and how the company has to be organized, planned, and directed if this goal is to be attained.

"In any consideration of an electric utility's ability to provide a continuous and uninterrupted supply of electric power, it is necessary to keep in mind the unique nature of electricity, a uniqueness frequently lost in public discussions even within the scientific community and most often ignored in political debate concerning energy. This uniqueness is marked by the twin facts that electricity by and large cannot be stored and yet must be instantaneously available in the quantities demanded by the consumer whenever and wherever he needs it.

Every other industry in our society can schedule its service and exert substantial control over the use of its product. This is true even for service industries such as transportation and communication. The transportation industry can limit the number of passengers on its conveyances, and the communications industry can prevent overloads by a busy signal. For electric power, however, the consumer controls the consumption of the product by a simple flip of the switch.

Hence the capability to serve — to make certain the switch turns on the light — must be planned years in advance. Overloading facilities has dire consequences for the quality of service to the consumer and threatens the very integrity of the power system itself. These stringent requirements are made increasingly difficult by the uncertainties now confronting the industry."

Quality service means that customers expect a continuous and uninterrupted flow of electricity whenever they snap a wall switch or throw a power line switch. To provide quality service means that the realities must be faced that electricity cannot be stored and that it must always be available, regardless of the load on the system and the time of day or night.

The electric power industry, while characterized by complex machinery and equipment, by the pooling of power by interconnections and regional networks, by an emphasis on reliability of electrical service, and by planning and implementing the necessary expansion of capacity, still must rely in the last analysis on human judgment and performance. In other words, this means the quality of the service, the reliability of equipment, and the prevention of interruptions are dependent upon careful if not error-free performance by people. In the last analysis, quality is determined by human factors. Nagel describes it in these terms:

"The planning and operation of a major electric utility, while basically technological in nature, combines both engineering and art. Engineering aspects include the complex technology of its components, that is, its generating plants, transformers, transmission lines, circuit interrupters, relays, and controls. Mathematical techniques, supplemented by modern computational aids, provide the tools to design and analyze the behavior of such equipment, both separately and in combination.

The art in power system planning arises largely in the synthesis of equipment components into power systems. Although much theoretical work has been done toward the digital programming of system synthesis, *the human mind* — with its capabilities of imagination and creativity, bolstered by knowledge and judgment — *remains the critical element in overall power system design.*

This is also true of system operation, wherein final decisions, especially in times of emergency, *depend on human intervention* despite all informational aids and computational assistance" (emphasis added).

The national network. The map on the following page shows the territory controlled by the nine electric reliability councils which encompass essentially all of the electric power systems of the United States and Canada. The purpose of these councils is to ensure the reliability of the interconnected transmission network by eliminating or otherwise preventing the causes of widespread electrical power outages.

In all electric power systems there are three basic types of facilities: generation, transmission, and distribution. The public is largely aware of distribution outages. Occasionally, a transmission failure may cause a "blackout". These latter stoppages are usually arrested rather quickly despite the inconveniences and disruptions created. Electric service failure due to a lack of generating facilities is relatively rare and has never been significant although currently this type of facility poses a serious problem for the future. The major reasons are: the problems of using coal instead of oil, the problems in connection with nuclear power plants, and the problems of financing.

Quality aspects of an electric power company. The quality of electric service, whether to industry or household, depends solely on the reliability of every point in the system if service is to be continuous. This means that the goal has to be error-free performance of machinery, equipment, and people. Major quality and reliability aspects involve the following:

1. Source of power now existing:
   a. Generated: hydro, nuclear, oil, coal, gas for steam driven generators
   b. Purchased: interconnections, network
2. Source of power for future demand:
   a. Present plant utilization
   b. Expanded plant
   c. Purchased power
3. Expanded power capability
   a. Forecasting growth, increased load
   b. Lead time to plan expansion, determine added capacity, and added demand
   c. building new facilities:
      1) design
         generating system, type of plant
         transmission lines
         transformers
         substations
         control centers, circuit breakers
         distribution lines
      2) construction and installation
      3) operation
4. Causes of failure, of unreliability:
   a. Plant failure
   b. Equipment failure
   c. Transmission line failure
   d. Distribution failure
   e. Operating failure:
      human error, accident, overload
      human factors, carelessness, lack of careful monitoring
   f. Network failure
   g. Failure due to extraneous factors:
      weather: lightning, floods, winds
      damage by construction and other machinery, sabotage, strikes
5. Reducing causes of failure:
   a . Use of auxiliary or stand-by equipment
   b . Preventive maintenance
   c . Continuous monitoring, checking, and inspecting
   d . Replacements before failure
   e . Careful and improved supervision of employees
   f . Training classes and seminars
   g . Emphasis on error-free performance
   h . Adhering to operating manuals and standards
   i . Training, practice, and rehearsal of what employees should do in an emergency
6. Measuring and billing demand and usage:
   a . Controlling new installations in houses and buildings
   b . Reading meters
   c . Keeping accurate customer records
   d . Billing: metered billing, average billing, special billing
   e . Revenue accounting
   f . Customer complaints
   g . Customer service
   h . Customer surveys
   i . Variations in demand, load
   j . Rates:
      for industry
      for business
      for private dwellings
   k . Protection: fuses, circuit breakers, enclosures, warning notices
   l . Efficiency:
      for cooking stove
      for space heating
      for appliances
   m. Customer information about electric service, saving energy, rates, peak loads, and demand

NATIONAL ELECTRIC RELIABILITY COUNCIL

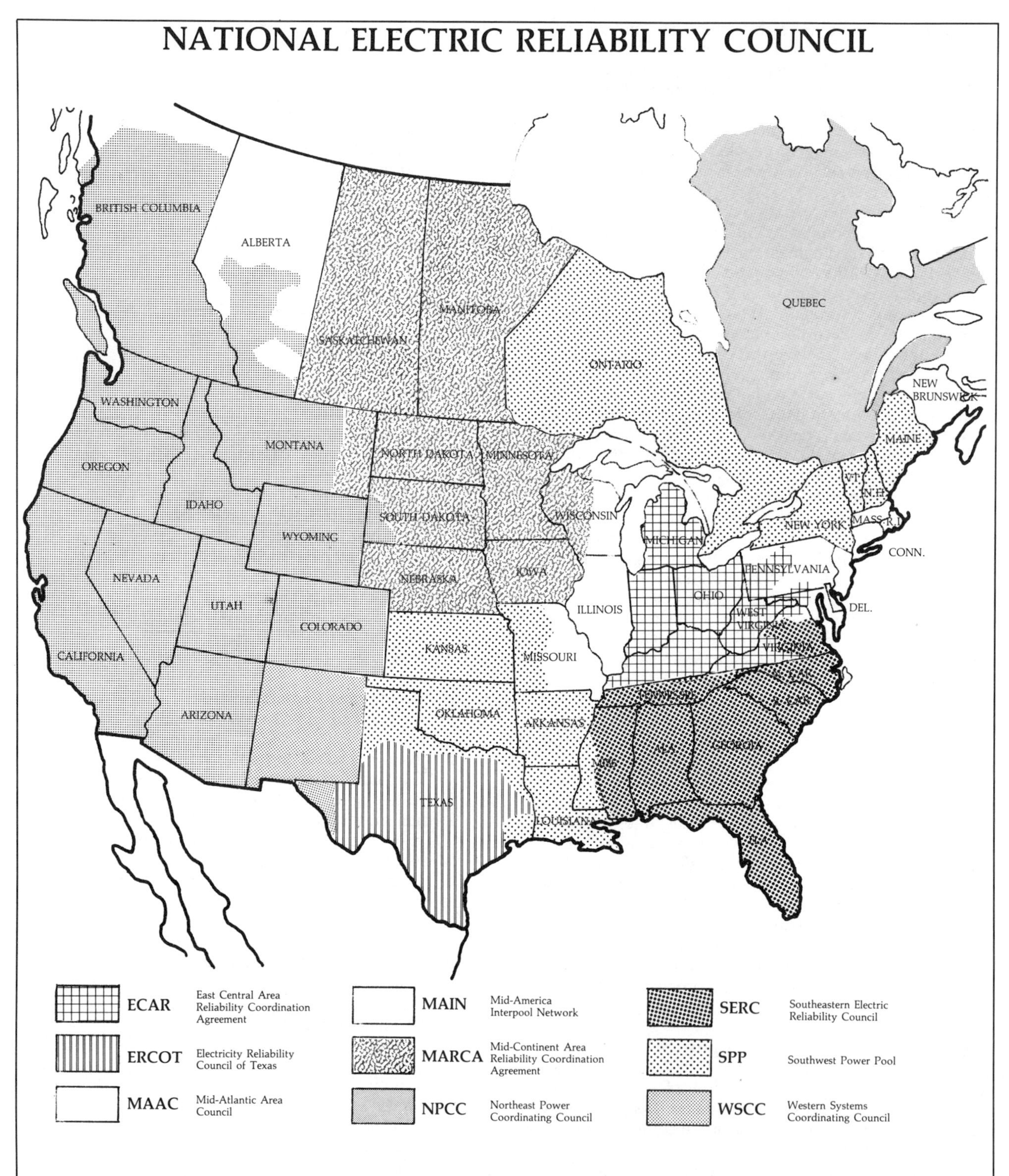

The National Electric Reliability Council (NERC) was formed in 1968 with the stated purpose: ". . . further to augment the RELIABILITY and ADEQUACY of bulk power supply in the electric utility systems of North America." It consists of nine Regional Reliability Councils and encompasses essentially all of the power systems of the United States and Canadian systems in Ontario, British Columbia, Manitoba, New Brunswick and Alberta.

RELIABILITY and ADEQUACY are two separate but interdependent aspects relating to the bulk power supply system of the electric utility industry in North America. RELIABILITY involves the security of the interconnected transmission network and the avoidance of uncontrolled cascading tripouts which may result in widespread power outages. ADEQUACY refers to having sufficient generating capability to be able at all times to meet the aggregate electric peak loads of all customers and supply all their electric energy requirements.

Rates for electric service. In receiving electrical service, as with other services, the buyer wants acceptable service at an affordable price. Money is so limited and so scarce to the great mass of people, that getting your money's worth and receiving satisfying service for what you can afford to pay is the supreme concern of most buyers and customers.

This is often hard to do because modern living is dominated by the gasoline automobile on the one hand, and electrical devices on the other. Without them the clock is turned back to 1900 and before. We live in an electrical age, an age in which lifestyles are dominated by electricity: electricity for light, electricity for heat, electricity for a wide variety of appliances such as electric irons, electric refrigerators, electric toasters, electric air conditioning and electric driven furnaces; electricity for entertainment devices such as the radio, the television set and the hi-fi set; electricity for washers and dryers; electricity for typewriters, elevators and subways. When electricity fails, living is literally blacked out. This makes it hard if not impossible, for those on low incomes to afford to buy all of these devices in the first place and to afford the electricity needed to run them in the second place. This is why the rates paid for electricity by the ordinary householder are so important.

Electricity is sold by the kilowatt hour — one kilowatt (1000 watts) of electricity consumed for a period of one hour. This is the amount of electricity consumed by ten 100 watt electric light bulbs in one hour. The kilowatt hour is a unit of electricity — the amount of energy used or consumed. It is identical everywhere regardless of what the electricity is being used for. What the buyer pays for a kilowatt hour of electricity varies widely, not only in different parts of the country, but also within a state as Table 1 shows.

Table 1
Highest and Lowest Electricity Rates in United States*
(Rates in cents per kilowatt hour)

States with highest rates			States with lowest rates		
State	low	high	State	low	high
New York	1.34	10.36	Washington	1.15	3.17
Hawaii	7.01	9.31	Oregon	1.36	3.68
Vermont	4.14	9.00	Washington, D.C.	4.01	4.01
Missouri	3.16	8.78	Alabama	2.68	4.64
New Jersey	4.92	8.68	Tennessee	2.68	4.80
Alaska	4.14	8.50	Idaho	2.00	5.12
Delaware	5.37	8.20			
Massachusetts	5.05	8.18			

*Derived from data given in the *Statistical Abstract of the United States, 1979,* p. 493, U.S. Department of Commerce, Bureau of the Census, Washington, D.C.

The rates by states fall into two well defined groups:

1. The low rate group:
   a. Washington and Oregon are low because of federally built hydroelectric plants in the Pacific Northwest.
   b. Idaho buys considerable power from Washington and Oregon.
   c. Alabama and Tennessee are in the area served by the Tennessee Valley Authority which uses coal as fuel for steam driven generators much more than it uses hydroelectric power. This source also is a federal agency not operating on the same financial basis as a private utility. It sells to private companies and government agencies as well as to private householders.
2. The high rate group:
   a. States with high rates have to rely on high-priced oil to produce steam. This applies to Hawaii, Alaska, the New England States, New York, and New Jersey.

A third group that receives low rates includes those living in farm and rural areas, and companies, served by the Rural Electrification Administration (REA), a federal agency. In 1978, the REA received 3.54 cents per kilowatt hour of electricity sold. The charge is calculated as follows:[2]

$$\frac{\text{Revenue}}{\text{kwh sold}} = \frac{\$\ 5{,}017{,}000{,}000}{141{,}900{,}000{,}000} = 3.54 \text{ cents per kwh}$$

The buyer of electricity will just have to face the economic reality that the price of electricity to consumers depends upon where they live. In some areas, the price is three or four times as much as it is in other areas. This is due to federally financed electric plants, on the one hand, and the cost of oil, gas, and coal, on the other.

An individual buyer's monthly use of electricity. Figure 1 shows the number of kilowatts of electricity used monthly by a homeowner for 32 months from January 1979 through August 1981. The data were derived from measurements obtained by reading the electric meter on the first day of every month. The record serves several purposes:

1. It shows very clearly what the usual bill is going to be based on past usage. The householder will know in advance just about what to expect on the bill, making adjustments for any unusual conditions, including an increase in rates. This will prevent one from being surprised at the bill received.
2. It detects any major error made by the company employee reading the meter or major error made in preparing the bill.
3. It shows the variations to be expected due to monthly or seasonal variations — variations due to holidays, absences from home, very hot weather requiring more use of air conditioning, or very cold weather if the house is heated by electricity. The range on the chart is from a low of 250 kwh to a high of 470 kwh.
4. It gives the average level of use per month. This is found by adding all the monthly totals and dividing by 32 months; in this case, the average was 368 kwh per month.
5. It indicates the magnitude of changes in the monthly billing, including the effect of changing the basic rates.

Telephone service customer complaints. All of the various complaints described below are actual instances and show that a wide variety of factors can affect the quality of telephone service. Some directly affect telephone service; others involve interference with the subscriber's time and privacy. Four involved errors in billing long distance calls; others involve telephone conversation practices, while others involve action that requires calling the police.

In no case did the complaint result in any corrective action being taken by the company. In two instances, it was the subscriber who requested the changes that eliminated the trouble. Admittedly, some of these complaints

FIGURE 1

Monthly Use of Electricity by Householder 1979, 1980, 1981

Family of 2 adults

Note: Electric stove, central heat by gas, air conditioned only occasionally

Low points are due to away from home

$\bar{x}$ = 368 per month

air conditioner

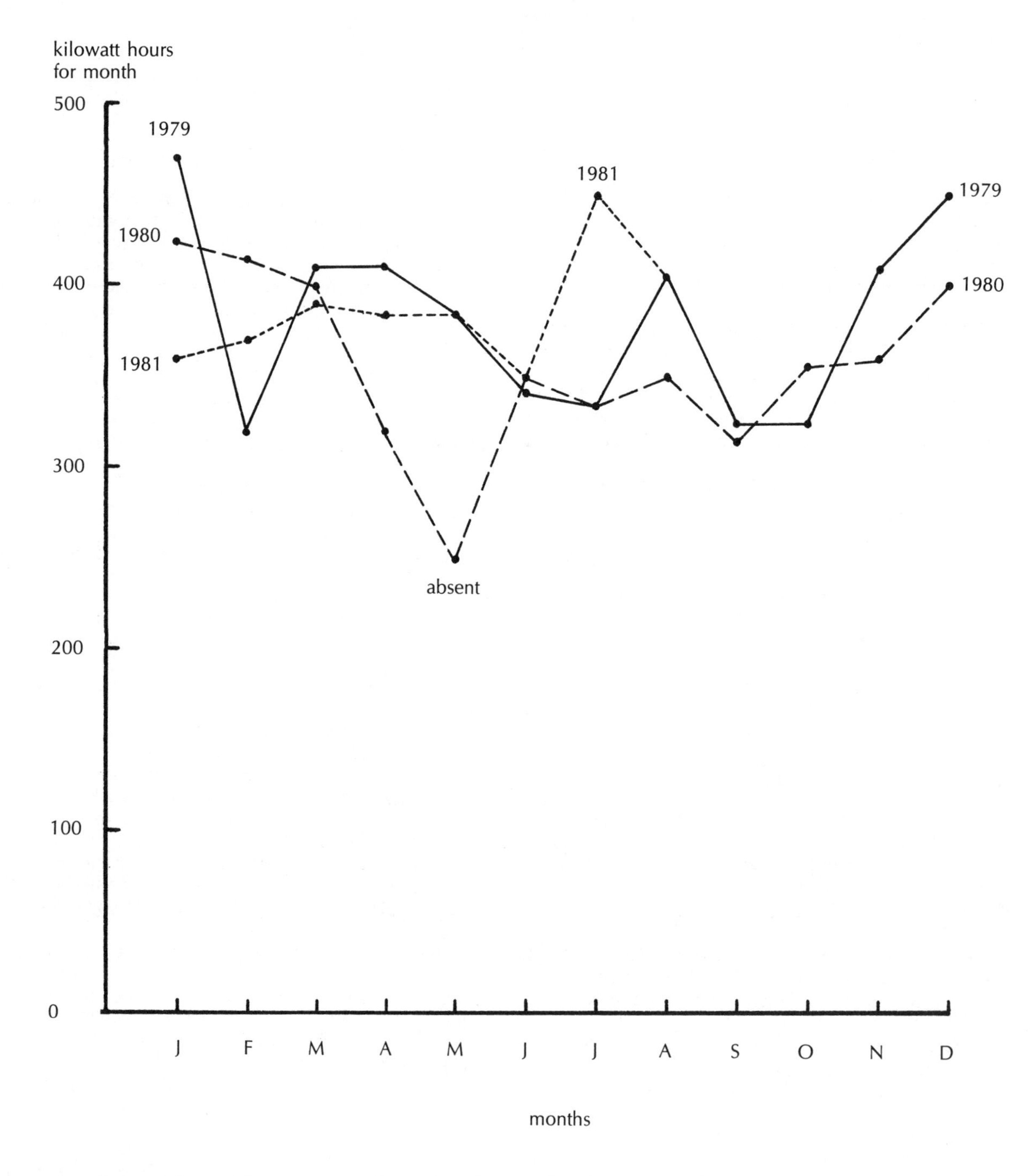

Source: Derived from meter readings taken the first day of every month.

involve freedom of speech under the First Amendment, complaints which the company can do nothing about. In connection with improving telephone conversation practices, the company could issue aids and suggestions and recommendations, but there is no evidence that this company or any company is doing this yet. Some of these could be called nuisances because they do not affect telephone service per se. But why should the telephone subscriber paying for a service have to put up with nuisances? To be sure, some are the penalties subscribers have to pay at the present time for having a telephone and their names in a telephone directory. Few persons are so rich or well known that they can afford an unlisted number; for the common person, this limits the very service the telephone is being installed to provide.

1. Delay in installing a telephone. A small business moved and called the telephone company to install a telephone as quickly as possible because the business depended to a large extent on the telephone. The telephone company gave a number of reasons why the phone could not be installed for at least one week — holiday, vacations, and shortage of installers. As a result, the business operated out of a telephone booth nearly a mile away for over a week.

This illustrates a common situation today; numerous services are a five day week excluding holidays. So it pays not to get sick, or have your automobile go dead or run into trouble, or want a lawyer, or have an emergency repair trouble arise during the weekend or on a holiday.

2. Failure of a two-party line. A two-party line is risky at best, but often a subscriber is forced to take it because a one-party line is not available, nor will it be. The risk lies in the conflicting times the two parties want to use the line. This often occurs during the early evening and at night, although it can arise at other times, especially if the parties indulge in long conversations.

Two severe cases arose that involved deliberate interruptions of telephone conversations by the second party — telling the party using the line to get off and similar obnoxious behavior. This was reported to the telephone company with a request to shift line to a new two-party line. Eventually this was done but the second party was not much better than the first.

This type of service is especially unsatisfactory for a professional person operating out of his home and having not only professional society work to do but also business or semi-business work that requires reliable telephone communication. This not only applies to outgoing calls, but to incoming calls as well as to where delays are built in by a busy signal due to the second party.

3. Impossible to obtain a one-party line. As indicated above, there are times when the company will not give the subscriber a one-party line because none is available and never will be. This means a party will have to settle for less than the quality of service demanded.

4. Four telephone numbers in two weeks. A consulting firm received four different telephone numbers in two weeks which nearly ruined their business. First, the telephone company changed the telephone numbers in the entire locality. Then, the consulting firm had to move and obtain a new telephone number. The telephone man came and installed the phone and assigned a new number at about noon. Immediately, the consulting firm called friends and clients in the area to give them the new telephone number. But at 3 p.m. the same day, the telephone man came back and said the number assigned could not be used; he had to give them a new number! So all the calls made earlier in the day had to be repeated to give the parties the latest telephone number. A professional man who had been trying to contact the firm for days finally contacted them and told them to have the telephone company correct the matter or they would lose all their business. All he was getting was an old recording!

5. Four long distance bills in error. Examination of a telephone bill by an individual householder showed a charge for a long distance call that had never been made. This was reported to the telephone company who traced the call and found it had been made by someone else, but charged to another subscriber. Shifting the charge for long distance calls to this one subscriber continued for three more bills; long distance calls to New York and Florida were included. A check by the telephone company in each case showed that someone was shifting the charge for long distance calls to this one subscriber. By this time, the subscriber was incensed about the whole matter. He called up the telephone company and requested an immediate change in his telephone number. After this was done the trouble disappeared. This was not an error made by the telephone company but a fraudulent use of the system then in use to shift long distance telephone charges to someone else.

6. Duplicate telephone numbers. In a suburban city of about 35,000 people, the telephone directory, unbeknownst to the telephone company, carried the same telephone number for two parties. This was discovered by party one when calls were received for party two who was listed, but whose telephone had been disconnected. This duplication, discovered by party one, was reported to the telephone company, but nothing could be done about it. Party one kept receiving the calls intended for party two for about two years.

7. Reassignment of telephone numbers previously used. Under the last item, calls were received intended for another party due to the printing of duplicate telephone numbers in the telephone directory. A similar situation arises when a subscriber is assigned a number which some years before had been used by another party. In this case, the previous party was a Volkswagen auto repair shop; the number was now assigned to a private home. Over a period of nearly three years, this private home received scores of telephone calls intended for the auto repair shop. Customers of the repair shop had written down the telephone number but neglected to check the yellow pages which would have shown that the repair shop was now out of business. There is one thing the telephone company could do in this connection: it could carry a warning in both the directory and yellow pages relative to the re-assignment of telephone numbers and the need to verify old numbers to see that they are still being used by the same party.

8. Number in telephone directory has been changed. Another disturbance to smooth service is to find that the number given in the directory has been changed by the telephone company. This can be a real inconvenience, especially if a large number of telephone numbers have been changed, as is sometimes the case. This may mean that a score of telephone numbers have to be changed.

9. Receiving anonymous calls with threats, warnings, or obscenities. The telephone company is vigorously opposed to this type of call and advises those who receive them to hang up and report the call to the police. The telephone company should be informed also.

10. Annoying or irritating telephone conversation practices. There are several kinds of annoying telephone calls a subscriber receives:

1. A subscriber receives a telephone call but it is the wrong number. Receiving wrong number calls is now a common practice of which both adults and children are guilty. It is not uncommon to receive one or two of these calls every week. The telephone company could carry a suggestion to those dialing that they write down the telephone number before they dial it. This might help eliminate this annoyance in telephone service.

2. In the "silent call", the subscriber picks up the telephone and says "hello", but is given the silent treatment: no one answers.

3. In the "who's this" call the caller does not ask "I would like to talk to So and So", or I am _____ and would like to talk to _____". All they ask is "Who's this?" This shows the need for instructing people on how to use the telephone, especially what to say when a person has answered. This training should also include how to answer a call one receives as well as how the caller should respond. The telephone company could be a big help in presenting such instructions in the directories as well as holding classes in schools or at their offices.

4. Another annoyance is comparable to receiving junk mail — it is receiving calls which are selling talks, solicitations for charities or for other purposes, market surveys, and the like. These calls are increasing in number.

Quality control aspects for a telephone company. Outlined below are various aspects of the management and operation of a telephone company which have quality aspects, that involve the quality of plant and equipment, the quality of plant operations, the quality of financial operations, the quality of employee performance, the quality of customer service, and the quality of data and decisions.

1. Plant and equipment:
   a. Construction: design, building, contracting
   b. Expansion: market trends, lead time, estimated capacity needed
   c. Line system: underground, overhead, microwave
   d. Reliability of the components
   e. Reliability of the system
   f. Failure analysis feedback
2. Operations:
   a. Type of system: direct dialing, assisted calls, long distance, local, PBX
   b. System failure: floods, wind, other; construction accidents, other
   c. Equipment failure: failure analysis
   d. Feedback of failures
   e. Prevention of blackouts
   f. Customer service: directories, yellow pages, installations
   g. Customer complaints about service, rates, telephone practices, installations
   h. Assigning telephone numbers
   i. Preventative maintenance
   j. Waiting time distributions: for installations, for telephone repair, for local calls, for long distance calls; for two or more party lines; for different kinds of service in the rate structure
3. Financial:
   a. Rate structure: special rates for nightly, weekend, holidays calls; local call area, measured rates versus fixed rates; direct dialing versus operator assistance; person to person; station to station
   b. Billing, delinquent accounts, revenue accounting
   c. Auditing: financial, property, inventory
   d. Appraisal of public utility property for rate purposes
4. System audit and follow-up:
   a. Reliability of components and system
   b. Reliability of operations
   c. Time distributions of all major operations
   d. Life distributions of components
   e. Analysis and verification of waiting time distributions
   f. Customer complaint analysis and follow-up
   g. Audit of billing and revenue operations

Quality control for the buyer of telephone service. The following are some of the major items a buyer of telephone service is greatly concerned about since they relate directly to the quality of the service:

1. Waiting times: this is a major quality characteristic.
   to install a phone
   to complete a long distance call
   to fix a defective or non-operating telephone
   to correct an error on a bill
   to install a one-party line instead of a two-party line
   to restore service after a failure
2. Rates: the price is just as important as the waiting times.
   local area; who establishes this area
   long distance calls
   rates for different kinds of service: person to person, station to station, operator assisted, one-party versus two-party, conference hook-ups
3. Billing:
   type of billing: staggered, average, measured, fixed cost
   errors in billing
   total monthly cost
4. Restoration of service after breakdowns:
   failure due to weather
   failure due to construction accidents, other accidents which break lines
   failure of individual telephones
   failure of switchboard
5. Public telephone service booths:
   location; convenience, where they are most needed, noise
   number
   accessibility
   availability of telephone directories and yellow pages at these booths
6. Instructions to the public:
   information in the directory: does it answer common questions, does it explain clearly the different types of services and the rates associated with them, does it give examples of how to make various types of calls, does it explain how to report trouble and how to terminate service.

Applications of quality techniques. Several applications of techniques which improved the quality of operations, provided some significant information not otherwise available, or reduced costs are cited.

McMurdo described how a comparative test run for several months on the telephone billing operation showed that the use of work sampling (random time sampling) applied to employees and machines gave much better results at a reduced cost than the use of time sheets reporting how 100 percent of the time was spent each working day. By changing to random time sampling, 40 percent of the time formerly spent on these time reports was saved.[3]

A wide variety of applications have been described by Magruder.[4] The experience with quality control applied to large volume clerical work and operations showed that an error rate of 1 to 1.5 items per 1000 items processed was about the best that could be expected with employees who are carefully trained and closely supervised. This error level is much lower than the one many companies and agencies obtain and live with. These low error rates, however, are quite consistent with the present stress on Zero Defects and Zero Errors. It is also consistent with another similar goal — Zero Delay Time.

The following are areas in the record-keeping or report field where sampling was applied to detect errors, make estimates, monitor operations, and improve performance:

1. Verification of property records
2. Accounting work
3. Inventories
4. Appraising public utility property for rate purposes.

In the clerical field, several specific characteristics were discovered that need to be controlled:

1. Clerical production and accuracy
2. Inventory turnover rate
3. Audit of accounts and records
4. Property appraisal and inventories
5. Wage structure
6. Absence, sickness, and accident rates
7. Operating reports

Example. A book record was kept of all telephone poles and their locations. A sample audit showed that hundreds of telephone poles no longer existed. The book record had not been kept up-to-date; the sample data were much more accurate and led to corrective action relative to the book record.

Example: a survey of residence customers. A stratified systematic sample of residence customers was selected monthly to measure, by means of a mail questionnaire, customer opinion with respect to satisfaction with the service and with the company (see table below).

Stratification was based on the number of customers in an area that contained 39 exchanges; this led to four size groups or strata. Sample selection was proportional to the number of customers in each group or stratum. The sample plan on a monthly basis was as follows: the sampling rates and the number in the sample are on a monthly basis but the study was continued for 12 months.

"Typing and clerical work for initial mailing, check-off returns, follow-up of non-returns, editing and coding questionnaires, and processing the survey data are functions that each require proper planning to assure accurate and prompt attention. These matters (are) covered by appropriate departmental instructions."[5]

It should be pointed out that stratification will be no better than simple random sampling unless the proportions in the various strata are widely different, from say 20 to 80 percent. Proportions that vary between 40 and 60 percent give so little gain that simple random sample should be used.[6] It is for this reason that most opinion polls use simple random sampling.

One may want to stratify because estimates are needed for each of the strata or groups. In this case, the sample sizes for each of the groups or strata must be large enough to obtain estimates with acceptable standard errors.

Other examples. Operating areas to which sampling techniques are used to control and improve the quality of operations include the following:

drop and block inventory
analysis of a billing system
control of billing errors
station apparatus survey
survey of residence customers
physical condition of the plant
typing toll bills
measurement of customer opinion
PBX extensions
appraising public utility property
verifying billing
verification (audit) of property records
analysis of telephone inquiries at the business office

Example. In this example, the billing system of a large corporation is subject to a detailed analysis. The basic data are given below:

1.	Number of items	17,659
2.	Dollar value of billing	$3,204,457
3.	Number of errors found	52
4.	Number of errors per 1,000 items	2.944
5.	Dollar error value per $1,000	0.633
6.	Number of items under $100	15,477
7.	Dollar value of billing for these items	$256,460
8.	Number of items $100 or more	2,182
9.	Dollar value of these billings	$2,947,997
10.	Dollar value of the errors	$2,029

Those items $100 and over accounted for 12 percent of the number, but 92 percent of the gross billing. There were 44 items in error under $100, yielding $329 of error. For items over $100, the figures were eight and $1,700, respectively.

Overall, the errors amounted to 63 cents per $1,000 of billing. It is very likely that it would cost more to correct these errors than to accept them. But this statement might

Size group	Approx. no. res. customers	Sample rate	Number	Expected number	Percent sample	Distribution frame
3,000 and over	29,000	1/300	100	70	35.0	33.7
1,000-3,000	11,000	1/180	60	45	22.5	23.0
400-1,000	7,000	1/120	60	45	22.5	23.6
under 400	3,000	1/60	50	40	20.0	19.7
Total	50,000		270	200	100.0	100.0

have to be qualified after a detailed analysis of the source of each error. A simple audit, however, should be applied periodically to maintain this level of "acceptable" quality, if closer study and analysis reveals this to be the case.

Notes

[1]This material is based on an article by Theodore J. Nagel, "Operating a Major Electric Utility Today," *Science,* Vol. 201, September 15, 1978, pp. 985-993. It is quoted by permission.

[2]*Statistical Abstract of the United States, 1979.* U.S. Department of Commerce, Bureau of the Census, Washington, D.C. This calculation does not appear in the source.

[3]C. E. McMurdo, "Work Sampling for Distributing Machine and Clerical Time," *Proceedings of the Middle Atlantic Conference,* American Society for Quality Control, March 1962, pp. 305-310.

[4]E. T. Magruder, *Some Sampling Applications in the Chesapeake and Potomac Telephone Companies.* The Chesapeake and Potomac Telephone Companies, Washington, D.C., 1955.

[5]Magruder, *ibid.,* p. 99.

[6]William G. Cochran, *Sampling Techniques,* 2nd edition, Wiley, New York, pp. 107-109.

Part 2
Statistics

12

Basic Concepts in Statistics

Statistics in quality control. In 1924, the Western Electric plant, which manufactured telephone parts for the Bell System, sent a problem to the Bell Telephone Laboratories. From this problem, Walter Shewhart developed the control chart and a complete theory of quality control in manufacturing based on statistics and probability. It was not based on any principles of management or engineering. Statistics provided an objective method of separating variations in a characteristic which were due to chance causes from those variations due to assignable causes which could be eliminated.

Applications soon showed that this statistical theory was sound and that it could be successfully applied to a wide variety of measurements and counts in manufacturing operations. This break-through came from a brilliant and innovative application of the science of statistics. Indeed Shewhart's classic, *Economic Control of Quality of Manufactured Product,* is so overwhelmingly statistical that it might as well be described as a statistical approach to the control of quality in manufacturing.

It was not long before it was discovered that some of these statistical techniques used to control quality in the factory could be applied, with some adaptations, to the operations of a service company, including government. Some early applications include the work of Deming in applying acceptance sampling to key punch operators in the 1940 census; Rosander in applying regression analysis to a pair of critical war materials in 1943 to control allocations to industries; Ballowe in applying the group error chart in 1945 to control the errors made by clerks filling orders in a mail order house; Halbert of American Telephone and Telegraph Company in applying acceptance sampling to revenue accounting in 1947; Jones of Illinois Bell Telephone Company in applying quality control to clerical work; and Lobsinger of United Airlines in applying these techniques to control reservations and plan meals.

Statistics as a science of control. Statistics provides a wide variety of techniques which can be applied to build quality into data, performance, and decisions, and to control key characteristics. Sampling, in its various forms, can be used to detect, estimate, control, compare, and measure effectiveness — all of which are related directly to some aspect of quality. Examples are:

- Estimating error levels, key characteristics
- Controlling error rates
- Detecting an undesirable situation
- Comparing different methods, machines, sample designs
- Measuring the effectiveness of a program, e.g., training

Control is implied in the definition of statistics as the science of sampling, estimation, and inference. This means there is a theory of statistics and a practice of statistics. Quality control is a field of statistical practice in which a practitioner bridges the gap between theory and practice. Shewhartian charts are excellent examples of how this gap is bridged very successfully. Other examples of how this gap is bridged in connection with real-world problems are described in detail in the literature.[1]

Cautions from practice. Several cautions from practice are described before basic concepts are discussed.

- Bias in the original data is a major problem, often the number one problem, in statistical practice. It behooves those in charge to take every precaution and use every control to collect low-bias, if not bias-free, data. This is crucial since the quality of the original data determines the accuracy of the estimates and the soundness of the inferences and decisions.

Bias in the original data can be so great that it invalidates the use of any measure of the random sampling variations, such as the standard error, as well as confidence limits which are derived from the standard error. Confidence limits assume that bias in the data is negligible. Hence, bias can invalidate a control chart or other techniques used in quality control.

- In quality control work, one starts with the problems facing management, not with data. A careful and detailed analysis of a problem leads to the kinds and amounts of data required. Data are a means, not an end. Problem analysis means that data are collected in terms of needs, requirements, conditions that need to be corrected, and important problems.
- Whenever possible, a pre-designed sample study or experiment should be used to collect data based on a detailed analysis of the problem. This tends to build quality into data and produces correct results right from the start. Collecting masses of data is not necessarily a virtue; instead, it may be inefficient and wasteful. Immediate problems, purpose, and need should govern.
- Sampling is greatly misunderstood and misused. There is probably more malpractice associated with sampling than with any other single concept in quality control, statistics, or science. This is because sampling is a science in itself, a very serious and complex subject matter that requires thorough study and experience in order to apply it effectively.[2] Sampling involves at least four concepts: population frame, judgment sampling, accessibility sampling, and probability sampling.
- The expanding use of the computer has given rise to the use of new and often ambiguous and superfluous terms such as the following:

computer term	term in statistics
data set	data, sample data, subset, population, frame
data element	measurement, count, attribute, characteristic, class, variate, variable
data bank	(no single term), could be a frame
data management	(no term)

A computer can multiply waste, as well as multiply efficiency. No amount of manipulation by statistics, mathematics, or by a high speed computer can make ambiguous data clear, irrelevant data relevant, invalid data valid, or erroneous data accurate.

Statistics. Statistics is the science of sampling, estimation, and inference based on probability. Probability gives it the power to deal with variability and uncertainty, basic characteristics of the real world. This definition is unique; it is not applicable to other sciences. Stated another way, statistics is a science that deals with quantitative logic; it derives from logic as well as from mathematics.

To define statistics, as some do, as decision making under uncertainty is to leave the term hazy, ambiguous, and too narrow. It is presumptious because decisions are made in every field, and vary greatly in their nature and importance. Furthermore, statistics involves much more than making decisions.

Attesting to the power and versatility of this science is the fact that it has developed as a solid and serious science only since the 1920s, due largely to the work of R. A. Fisher and his followers. In the 1930s there were only three universities offering degrees in statistics, whereas now there are more than 100, the field has expanded so much both in theory and in practice.

Probability. The empirical basis for probability is the frequency distribution of countless characteristics everywhere, even though it (probability) is now developed on axiomatic grounds. It is used in practice in the objective sense to mean the proportion of events in a subset or class relative to the total number of such events. The objective approach has proven to be highly successful in solving problems in the real world. We shall not discuss subjective probability, not only because of differences of opinion by experts in this field, but because its effective application to problems in the real world has not been demonstrated.

Applications of probability to quality control are far reaching and include:

- The probability of detecting at least one event in sample detection.
- The use of theoretical probability distributions such as the normal, the Poisson, the hypergeometric, the binomial, and the exponential.
- The use of distributions of test statistics to test hypotheses such as t, chi-square, and F.

It should be noted that, except when dealing with reliability, a probability figure is seldom the answer to a problem. Rather, probability is used to measure or establish the risk or degree of uncertainty in using a sample of a certain size, of making a certain estimate, or of testing a certain hypothesis.

These distributions form a solid foundation for numerous applications of statistics in just about every subject matter field and to a wide variety of problems encountered in research and operations, in private industry, and in government:

- quality control
- market research
- opinion polls
- sample studies
- sample surveys
- sample audits
- random time sampling
- random space sampling
- experiments
- field tests
- statistical analysis
- operations research
- input-output analysis
- regression analysis
- multi-variate analysis
- reliability

Three concepts closely connected with probability are *permutations,* P; *combinations,* C; and n *factorial,* n! Consider n objects, events, or elements taken x at a time. Then, the number of possible permutations:

$$_nP_x = n(n-1)(n-2)(n-3)\ .\ .\ .\ (n-x+1) \quad (1)$$

The first object can be selected in n different ways. When this is done the second object can be selected from the remaining n-1 objects; when this is done the third object can be selected from the remaining n-2 objects, and so on. Since these events are independent, the number of ways are multiplied as indicated. Permutations include all possible *orders.*

Combinations consist of all possible groups or sets of a specified size *without regard to order.* If $n=4$(a,b,c,d) and $x=3$, then

$$P=4\times3\times\ .\ .\ .\ (4-3+1)=4\times3\times2=24.$$

But C = abc, abd, acd, and bcd = 4. The x! or $3\times2\times1=6$ orders are ignored since abc, acb, bca, bac, cab, and cba are considered 1 group or set, abc. Therefore:

$$C=\frac{P}{x!}$$

where $\quad x!=x(x-1)(x-2)(x-3)\ .\ .\ .\ 1. \quad (2)$

But

$$_nP_x=\frac{n!}{(n-x)!};$$

then,

$$_nC_x=\frac{n!}{x!(n-x)!}=\binom{n}{x}$$

The symbol $\binom{n}{x}$ is used in mathematical equations instead of the longer expression. It always means the number of combinations of n things taken x at a time. C and x! occur in the equations for the binomial and hypergeometric probability distributions, while x! occurs in the Poisson probability distribution. These distributions are described in Chapter 14.

At this point we describe a simple problem in probability that has wide application in quality control. It concerns the size of a random sample required to detect at least one object, element, or event with a specified probability. The reasoning is as follows:

1. Let the probability of occurrence of a single event A be p.
2. Let the probability of occurrence of non-A be q, where $p+q=1$.
3. Then the probability of occurrence of non-A in n independent events is $q^n=p_0$.
4. Then the probability of obtaining *at least 1 event A* is $1-q^n$.
5. To solve for n set $1-q^n$ to 1-a, where 1-a is a high probability such as 0.95 or 0.99, so $a=0.05$ or 0.01, respectively:

$$1-q^n=1-a \text{ or } q^n=a.$$

Therefore,

$$n=\frac{\log a}{\log q},$$

where the base of the logarithms can be either 10 or e (2.71828...).

Example. In an actual work sample, a random minute sample of eight per day for five days ($n=40$) revealed that a Graphic Arts unit was idle, due to poor scheduling of work, 10 percent of the time. Was this a fluke or was this according to expectation? How large a sample was needed to detect this situation with a probability of 0.99?

$$a = 0.01 \text{ and } q = 0.90$$

$$n = \frac{\log 0.01}{\log 0.90} = 44.$$

We would expect to discover this situation with a very high probability with a sample of 40 random minutes; we could bet on discovering this situation with $n = 40$.

Example. In testing for the effect of a rare event, as in testing a chemical additive or testing a carcinogen (a cancer-causing material), it is not uncommon to have the event occur only once in 100 times ($p = 0.01$), or even once in 1,000 times or even less. If an event occurs once in 100 times, how large a sample is required to detect this event with a probability of 0.99?

$$a = 0.01 \quad q = 0.99$$

$$n = \frac{\log 0.01}{\log 0.99} = 458$$

This shows that a relatively large sample of 458 elements, e.g., test animals, is required to detect at least one instance of the event (cancerous cells). If the event occurs only once in 1,000 times, then the sample will have to be about 10 times as large or about 4,600. (See Chapter 22 for details.) This illustrates a very serious problem facing those doing research or testing in the cancer field, testing for air and water pollution, or testing in other fields where the objective is to discover highly rare but highly dangerous events.

The size of the sample becomes so large that experimenters or field test people are faced with a sample problem they are unfamiliar with — they have been accustomed to using 10 or 20 test animals in many experiments. This has given rise to the practice of using relatively small samples of experimental animals, but resorting to *massive dosages,* an experimental procedure some scientists reject. For example, it was this procedure which put cyclamates off the market; in one test, the dosage given test animals was 50 times the normal dosage. Suffice it to say, however, that the massive dosage technique, which is closely analogous to accelerated life testing, has its supporters as well as its critics.

Our purpose is simply to show that by some simple probability reasoning an extremely important equation is obtained for the sample size which applies directly to a very important aspect of health and environmental control. It shows that there is no alternative but to use large sample sizes if rare events are to be detected with a high probability. We should beware of small samples leading to false conclusions.

Population or universe. The population or universe specifies the particular objects under study to which the estimates and inferences apply. It establishes limits and boundaries, whether the study is based on sampling or is derived from a 100 percent coverage as in a census, complete count, or 100 percent tabulation.

The population is very important because this is what estimates, inferences, and generalizations apply to. Without a well-defined and bounded population, no one knows to what the findings of the study or research apply. Generalizations are restricted in value, may be misleading, or may be false because of bias. Only when a large number of similar studies produce similar and consistent results can any credence be put in the findings.

In practice, a population may be easy to define on paper but extremely difficult to isolate because it is dynamic not static, changing not constant, inaccessible not accessible, and dispersed not concentrated. Furthermore, populations may be infinite as well as finite; some finite populations may be so large in numbers that for all practical purposes they are infinite. As a result of this nature of finite populations, recourse is made to a *frame* which is selected to approximate the population as nearly as possible. It is a set of symbols which is available or is constructed in such a way as to reflect the population under study.

Analysis of the problem or problems under consideration leads to the identification of the population, on the one hand, and possible frames, on the other. An explanation of purpose and need in connection with problem analysis will bring out the scope of the study or research which identifies the population.

A random sample of time is an exception because the population is known exactly both now and in the future. In the service industries, there are numerous frames subject to study which may be files, storage, computer memory, or microfilm. Examples are:

checks	leases	charge accounts
claims	invoices	credit cards
bills	waybills	certificates of deposit
premiums	freight bills	savings accounts
tax returns	bills of lading	payroll
orders	inventory	personnel records
policies	sales slips	warranties
customers	vendors	purchase orders
contracts	equipment lists	licenses, registrations

A population or frame must include a number of specifications, but not all of those are listed below:

1. Object identification: required.
2. Time interval: required if not automatically taken care of by studying a frame covering a specific time period, e.g., all bills for one month, all tax returns for one year, or all insurance claims for one year.
3. Geographical area: required in those studies which cover areas such as drivers' licenses for a state, individual tax returns for the United States, or automobile accident records by state.
4. Individual locations: required in some studies where the study applies to individual units such as retail stores, offices which take freight car orders, nursing homes, or hospitals.
5. Object location: this can include a listing, files, computer tapes or drums or disks, punch cards, directories, records, or books of record.

It should be noted that in some cases the frame identifies only the object to be studied; the characteristics of interest have to be obtained in a separate operation. A directory may give the name of an individual, family, hospital, nursing home, retail store, etc. which has to be contacted in order to obtain the information needed.

In other cases, the frame not only identifies the object but also contains the information required. Examples are individual tax returns, data on computer tapes or disks or drums, punch cards, waybills, freight bills, invoices, or personnel records.

Time may be involved in a number of ways. The frame may cover a time period, e.g., tax returns for one year. In random time sampling, time is both the population and the frame; a daily time frame may be 480 working minutes, a working week may be 2,400 minutes, and a working year may be 120,000 minutes. A frame for studying trips out of a truck terminal might include a listing of all the trips on each of 24 randomly selected days, each of six days of the week occurring four times during the year. If the number of daily trips is large, the trips may also be sampled.[3]

In a process, the population is the work submitted to the employee or the jobs put on the computer; it is the input whatever its nature. In this case, methods are used to control the process, generally sampling, in order to keep error rates within specified limits and rejected work at a minimum. Examples are:

- typing letters
- reviewing health insurance claims
- transcribing waybills
- editing source data
- key punching bills
- auditing car rental bills
- updating personnel records

A sharp distinction needs to be made between the population or frame object and the population characteristic under study for which the object is the carrier. Examples are:

population or frame object	characteristic
nursing home	monthly cost, quality of care, medical errors, adequacy of the food
tax return	amount and kinds of errors, business and non-business, income, taxes
health insurance claim	items disallowed, errors, amounts, items allowed, percentage of total medical expenses paid
customer	income, renter or owner, credit rating, credit cards, amount of business, kinds of merchandise, etc. purchased, satisfaction with goods bought
waybills	commodities, weight, freight charges, rate, origin, destination, miles
freight car orders	kinds of cars ordered, waiting time, shipper satisfaction
airline passenger	trips, miles, cost, baggage handling, baggage damaged or lost, delay times, meals, satisfaction

Randomness. One of the most important concepts in probability and statistics is randomness or randomization, which includes random selection and random assignment. Random selection is basic because so many concepts and distributions in probability and statistics derive from random variation. It can be demonstrated by actual experiments that the actual results obtained can be predicted very closely by mathematical expressions derived on the assumption of random variation.

Random selection is applied to the units or elements of a population or frame in order to eliminate bias due to selection and give every unit or element an equal, or known, probability of being selected. Random assignment is used for a similar purpose in experimentation: to give every experimental factor an equal chance of being assigned to each unit of experimental material and to provide a valid experimental error.

This process of random selection and random assignment is facilitated by using tables of random digits or random numbers. Tippett prepared the first table of random sampling numbers in 1927. The tables by Fisher and Yates[4] and by Kendall and Smith[5] have been subjected to extensive tests to insure random order and random arrangement. The former has 15,000 digits, the latter 100,000. Rand's table of a million digits is also available in the form of 20,000 punch cards.[6] Arkin also has prepared a table of random digits.[7] The author has found Kendall and Smith's little published table the most convenient to use. However, many computers either have a built-in random number generator or a piece of software containing such a program. A table of 4,000 random minutes and a table of 4,000 random days are also in print.[8] In addition, random days are also a subset of the table of random minutes. Random minutes run from 1:00 to 12:59, while random days run from 1-01 to 12-31 for a 366 day year.

Sampling. In sampling, four aspects are discussed: the purpose of the sample, the sampling unit, the sample selection method, and some special sampling techniques. The design of a sample depends upon what the sample is going to be used for since a design which is efficient for one purpose may be very inefficient for another. Furthermore, the statistical theory and formulas applicable to one sample design are quite different for another design. These *purposes* are:

- Sampling for detection
- Sampling for control (in a narrow traditional sense)
- Sampling for estimation
- Sampling for comparison
- Sampling for measuring effectiveness
- Sampling for developing and testing mathematical models

The *sampling unit* or element refers to what is selected from the frame or population, or from a process. It takes a wide variety of forms which can greatly influence the design of the sample and the nature of the study. Examples are:

Discrete objects	Non-discrete objects	Time
inventory	air	minutes
tax returns	water	days
drivers' licenses	soil	hours
nursing homes	chemicals	
bank checks	drugs	
patients	medicines	
offices	powders	
private homes		
customers		
areas such as farms, city blocks, docks		

A sample design is greatly influenced by whether the sampling unit is an intact unit or element, or whether it is a cluster of such units or elements — the "peapod" model. Examples are:

Intact units	Cluster units
individuals	families
one bank check	a checkbook, a shipment of checkbooks
student	classroom of students, class, grade, school
resident patient	nursing home, hospital
one telephone subscriber	names on one page of directory

A critical point in sampling is the method of selection from the frame. Four methods of selection are:

- Accessibility sampling
- Judgment sampling
- Systematic sampling
- Probability sampling using random selection

Accessibility sampling means selecting a sample from the most accessible sampling units or elements. It includes selecting the most cooperative persons, using only those who respond, and only those who are readily available. Therefore, it can be subject to unknown amounts of bias. In *judgment sampling,* personal judgment, including the judgment of experts, is used to select the sampling units. This leads to all kinds of sampling such as selecting "typical" units, selecting "average" units, selecting "bell wether" units, selecting 100 percent of the frame for two days, or selecting all units for "typical" days in each of several months. This too can be very biased.

In *systematic sampling,* a sample unit is selected in a systematic way from the beginning to the end of the frame, such as every 10th element or every 20th element. This method may give adequate results providing there is no cyclical arrangement in the frame. If the frame is well mixed, it may be very nearly equivalent to a random sampling with some possible gains if the frame is partially stratified.

Probability sampling using random selection is the best method from both a theoretical and practical point of view. In probability sampling every unit in the frame is given a known probability of being selected. The use of random selection guarantees that this occurs with any bias in sampling eliminated. Probability sampling has several advantages the other methods of sampling do not have. The most important are: it eliminates bias, it provides a sound method of estimation, it provides an estimate of the sampling error (standard error), it allows the applications of the powerful methods of analysis of statistics, and it minimizes the sample size for the amount and quality of information obtained.

In selecting from a process for control purposes, a sample is selected periodically from the last elements processed. Then, a mean, range, or proportion is calculated and plotted on a Shewhartian control chart. When an out-of-control point occurs steps are taken to determine if an assignable cause exists; if it does, try to eliminate it. This keeps variation within an acceptable range. No random sampling is involved here. We want to get into the sample as soon as possible elements that are out of control; objects are selected in the order produced.

When process control is attained by lot acceptance or rejection, a random sample must be selected from the lot. In practice, a systematic sample may be valid and a lot more convenient. A random sample is prescribed because these sampling plans are based on probability assuming random selection from a lot. For example, lots of work are assembled in lots of 300; a systematic sample of 50 is selected (every 6th unit) and the lot is rejected if four or more units are found in error. The lot is immediately referred to the individual who did it and the lot processed 100 percent.

Special sampling techniques that are powerful and effective in quality control include replication, stratification, and random time sampling. In *replication,* the total sample is selected as several independent subsamples, usually four to 10. This should not be confused with replication in experimental design. It has several merits: it can be applied to complex sampling situations, it provides an easy way to estimate a frame characteristic, it provides an easy way to calculate the standard error, and it measures the internal consistency of the data. The basic equation is:

$$Z = \frac{kN}{n}$$

where Z = zone width in number of frame units; k = number of replicates (subsamples); N = frame or population size; and n = sample size.

Example: Select a random sample of 2,000 from a computerized frame of 200,000 records using four replicates.

$$Z = \frac{4 \times 200{,}000}{2{,}000} = 400$$

Select four random numbers between 001 and 400 and record them in the order found as replicate values. Then add 400 successively to these numbers to identify sample elements. This requires the frame units be numbered or at least counted. A frame on a computer is ideal since it can be programmed to count records, identify replicate units, and punch a card for each replicate sample record properly coded. It can also give sample counts, total counts, and count and reject records not in the frame.[9] The first two zone intervals showing the four random starts are as follows. 500 lines are needed for n = 2000:

	Replicates			
Z	1	2	3	4
001-400	289	345	135	76
401-800	689	745	535	476
etc.		etc.		
(500 lines)				

To estimate the grand *mean,* compute the mean of the four replicate means:[10]

$$\bar{\bar{x}} = 1/4(\bar{x}_1 + \bar{x}_2 + \bar{x}_3 + \bar{x}_4).$$

Its standard error is $\sqrt{s^2}$: $s^2 = \frac{1}{3 \times 4} \sum_{1}^{4} (\bar{x}_j - \bar{\bar{x}})^2$.

A proportion is handled in the same way by substituting p_j for $\bar{x}_j$.

The standard error is also estimated by the range method:

$$1/4(\bar{x}_{max} - \bar{x}_{min}).$$

Stratified random sampling means dividing the frame into mutually exclusive groups by some trait that correlates highly with the characteristics being estimated. If effective, it reduces the sample size but yields estimates with acceptable sampling errors (standard errors). Very often the extremely large values in the frame are all included in order to reduce the sampling error. Stratification does not help in estimating proportions unless they vary greatly; if they fall between 0.20 and 0.80 nothing will be gained.

Characteristics that are effective strata include both the quantitative and the qualitative:

Frame	Possible stratifying characteristics
tax return	size of income, size of business receipts, taxable and non-taxable, business and non-business, industry or type of business
waybill, freight bill	weight of shipment, size of company, commodity, distance of shipment (origin-destination area)
health insurance claim	size of claim, hospital service or not, age of patient, sex of patient
retail store inventory	type of business, classes of items, volume sold, unit price

Given k strata of size N_j, each either known or approximated, different sampling rates f_j in each of the j strata, and sample size n_j. The proportion of the frame in each stratum is $p_j = N_j/N$. The weight applied to obtain the aggregate in each stratum is the inverse of the sampling rate of $w_j = 1/f_j$.

Then, the aggregate X is:

$$X = \sum_{1}^{k} \left[w_j \sum_{1}^{n_j} x_j \right] = \sum_{1}^{k} X_j .$$

The variance of X is:

$$s_X^2 = \sum_{1}^{k} \frac{N_j^2 s_j^2 (1 - f_j)}{n_j}$$

Since the strata are independent, the total variance is the sum of the strata variances of the means:

$$s_{\bar{x}}^2 = s^2/n.$$

The variance of the mean $\bar{\bar{x}}$ is the variance of X divided by $N^2 = s_X^2/N^2$.

$$N_1^2 s_{\bar{x}_1}^2 + N_2^2 s_{\bar{x}_2}^2 + \ldots = \sum N_j^2 s_{\bar{x}_j}^2 = \sum N_j^2 s_j^2/n_j,$$

Random time sampling uses a population of minutes. The minute model not only estimates proportions, as the orthodox tour method does, but estimates means, aggregates, and ratios as well. The hypergeometric model or distribution is applied to calculate probabilities and to obtain sample sizes while tables of random minutes are used to select sample minutes from the specified time population.[11]

The advantages of random time sampling over the orthodox tour method are several: it is based on sound sampling theory; its estimates are not limited to proportions; the sample design provides a sound method of estimation; a valid standard error can be calculated for each estimate; salary and wage and rental costs can be estimated directly from the sample; it is the only method that can decompose joint costs in an unbiased manner; and it provides a maximum amount of quality information at a minimum cost.

It can be used to detect, to estimate, and to control. It relates costs to performance of personnel, machinery, and equipment. It is the only technique available that adequately solves certain problems facing management, problems for which the accounting office and other departments have no data.

It can be used, and has been used, to study costs relative to performance in inspection, auditing, verification, review, and many other operations for the purpose of discovering where quality control either needs to be improved or introduced. It can be used effectively in connection with cost benefit studies and surveys.

Example: A graphic unit is idle 60 minutes a day due to poor work scheduling. If 16 random minutes are selected in a day of 480 minutes, two per hour, what is the probability of detecting this situation? Use the following hypergeometric format:

Class	Population	Sample	Non-sample
A	60	0	60
not A	420	16	404
Total	480	16	464

Required is $1 - p_o$, where p_o is the probability of getting no A's in a sample of 16.

$$p_o = \frac{420!464!}{404!480!} = \frac{420 \times 419 \times \ldots \times 405}{480 \times 479 \times \ldots \times 465} = 0.11.$$

By the median fraction MF method:[12]

$$p_o = \frac{(420 - 7.5)^{16}}{(480 - 75)^{16}} = \frac{(412.5)^{16}}{(472.5)^{16}} = 0.11 = (.8730)^{16}.$$

where $7.5 = (n - 1)/2$.

Hence, $1 - p_o = 0.89$. The probability of detecting this event with a random minute sample of 16 in one day is 0.89. The chances are high, therefore, that it will be discovered in one day. If a smaller sample, say eight random minutes per day, is used, the probability of detection is 0.66 in one day, 0.88 in two days, and 0.96 in three days.

Variability. Variability is a universal attribute of nature. What is not variable is very rare indeed. Very few constants exist outside the speed of light, planetary motions, and certain constants in physics and chemistry. Even some of these are not easy to measure. The speed of light is an example since measurements of it over the past century varied greatly. Statistics is the science of variability.

Variability is the basis of probability, of sampling, and of control. Characteristics vary; objects in the same population, universe, or frame differ. Individual differences seem obvious, but there are many who ignore these differences or want to wish them away. These differences are extremely important in connection with quality control in the service industries and operations.

A *variate* is any characteristic that has a probability distribution. A random variable means the same thing. The concept of variate leads directly to four methods of recording, studying, and presenting data:

- A table of actual counts or measurements in time order.

• A table of intervals of counts and measurements, with the frequency or count for each interval. Intervals are usually equal and convenient in number.

• A graph of the data points in order of their collection.

• A graph of the frequencies by intervals.

The second is a frequency distribution table; the last one is a frequency distribution chart or graph. Examples of variates include: amount of error on individual income tax returns, amount of income on these same tax returns, amount requested on a health insurance claim, amount allowed on such a claim, difference between estimated cost of moving and the final bill, number of days between a freight car order and the placement of the cars.

Frequency distribution. A frequency distribution, whether in a table or in a graph, shows 1) the extreme values, 2) the range of values, and 3) the concentration of values. A graph shows these easier than a table. This is a simple and convenient way of understanding the specific nature of the variability of the data and what it implies: What values are above or below a standard or tolerance limits? What questions are raised by the extreme values and what action if any should be taken? Do very large values need special attention, such as large errors on a tax return or a large value on an insurance claim? If the distribution represents a frame, extremely large values may call for a special stratum to be included 100 percent instead of being sampled.

A graph of a frequency distribution may take four forms: it may be a *histogram,* where frequencies are rectangles erected over each interval; it may be a *polygon,* in which midpoints of intervals are connected by straight lines; it may be a *curve,* where a curve is drawn through these midpoints; or it may be a *point* distribution, where a frequency is a point erected above each value of the variate which is a count (a discrete number rather than a measurement) such as size of family.

Figure 1 is an example of a histogram showing the distribution of heights of about 100,000 male white draft registrants in 1940-1941 for World War II. One inch intervals are used beginning with 55.5 inches and expressed from half inches to half inches, e.g., 67.5-68.4 because heights were recorded by doctors to the nearest inch. This shows that intervals for a frequency distribution have to be selected so as to reflect the way the measurements were made and recorded. The values fall in the middle of an interval rather than at the boundary. Unless this is done, a bias is introduced into the data from estimates made from grouped data such as the arithmetic mean.

The mean is 68.23 inches, the median is 68.25 inches, and the standard deviation is 2.792 inches.[13] For a normal probability distribution, 99.73 percent of the values should fall in the interval $\bar{x} \pm 3s$. In this distribution $68.23 \pm 3(2.792)$ gives 59.85 and 76.61 inches.

Outside of these limits are 293 below and 206 above for a total of 499 or a proportion of 0.005 or 0.5%. Included in the limits are 99.5% of the measurements instead of 99.7% according to the normal distribution. This goodness-of-fit of the height distribution to the normal distribution is discussed at length in a later chapter on Theoretical Distributions.

Since the median is practically equal to the mean, 50% of the distribution is below the median and 50% is above the median. This is true of the mean as well as the median and indicates a symmetry of the distribution about the mean.

Independent trials and events. Independent events and independent trials are assumed in many aspects of probability. Whatever the model is in probability theory, whether coin tossing, throwing dice, selecting cards, or drawing balls from an urn, it is assumed that the tossing, the throwing, the selecting, and the drawing results in independent events. This is expressed in probability theory without explaining how it is done.

Independent means that the probability of a combined event is the product of the probabilities of the separate events:

$$p = p_1 p_2 p_3 \ldots p_k.$$

$$p_{ab} = p_a p_b.$$

It also means that events a and b are independent if the conditional probability of a, given b, is the probability of a

$$p_{a/b} = p_a.$$

Independent events, trials, observations, or samples are assumed in the development of the probability distributions for the *binomial,* the *Poisson,* and the *hypergeometric.* Because these distributions are so widely used in sampling and statistical practice, including quality control, it is imperative that the concept of independence be understood and practiced. A departure from independent events and trials arises in cases of *conditional probability.* In problems of this type, the principles of conditional probability must be applied.

Two examples are cited to illustrate an absence of independent trials or events. The tour method in work sampling may not yield independent observations because the presence of the observer may change the behavior of two or more persons in a group. That is, a cluster effect is created. Throwing 10 pennies at once may violate independence because of mutual interference which creates a clustering effect. This is avoided by throwing each coin separately. For the Poisson distribution to apply, events must be independently dispersed, such as errors on printed pages of a book, automobile accidents, telephone calls at a switchboard, and viruses in a batch of vaccine.

Estimation. Estimation is one of the most important concepts in statistics. Its present development has been due largely to the work of R. A. Fisher, Egon Pearson, and J. Neyman. A big step forward is the sharp distinction between an estimate and the population or frame parameter it approximates. In the real world, only estimates exist, population or frame values and "true" values rarely exist. In practice the goal is to obtain the best estimates needed for the purpose. This means designing probability samples, tests, and experiments, and planning sample studies and analyses so that both *consistent estimates and efficient estimates* are obtained.

Consistent estimates are those which, with indefinitely increasing size samples, approach the population or frame value. Efficient estimates are those with the minimum variance or standard error. Most of the common estimates meet these criteria: the arithmetic mean, the binomial proportion p, and the Poisson mean m.

Estimates are of two kinds: *point* estimates and *interval* estimates. A point estimate is the best single estimate of the population or frame parameter derived from a sample. It is the estimate management most often requires for operations, policy making, and planning such as total tax error, error rate, amount and percent of down-time, shortage in number of boxcars, amount of unemployment, cost of correcting errors, airline load factor, and quality control costs.

FIGURE 1

Heights of 99,712 Male White Draft Registrants—U.S. 1940-41

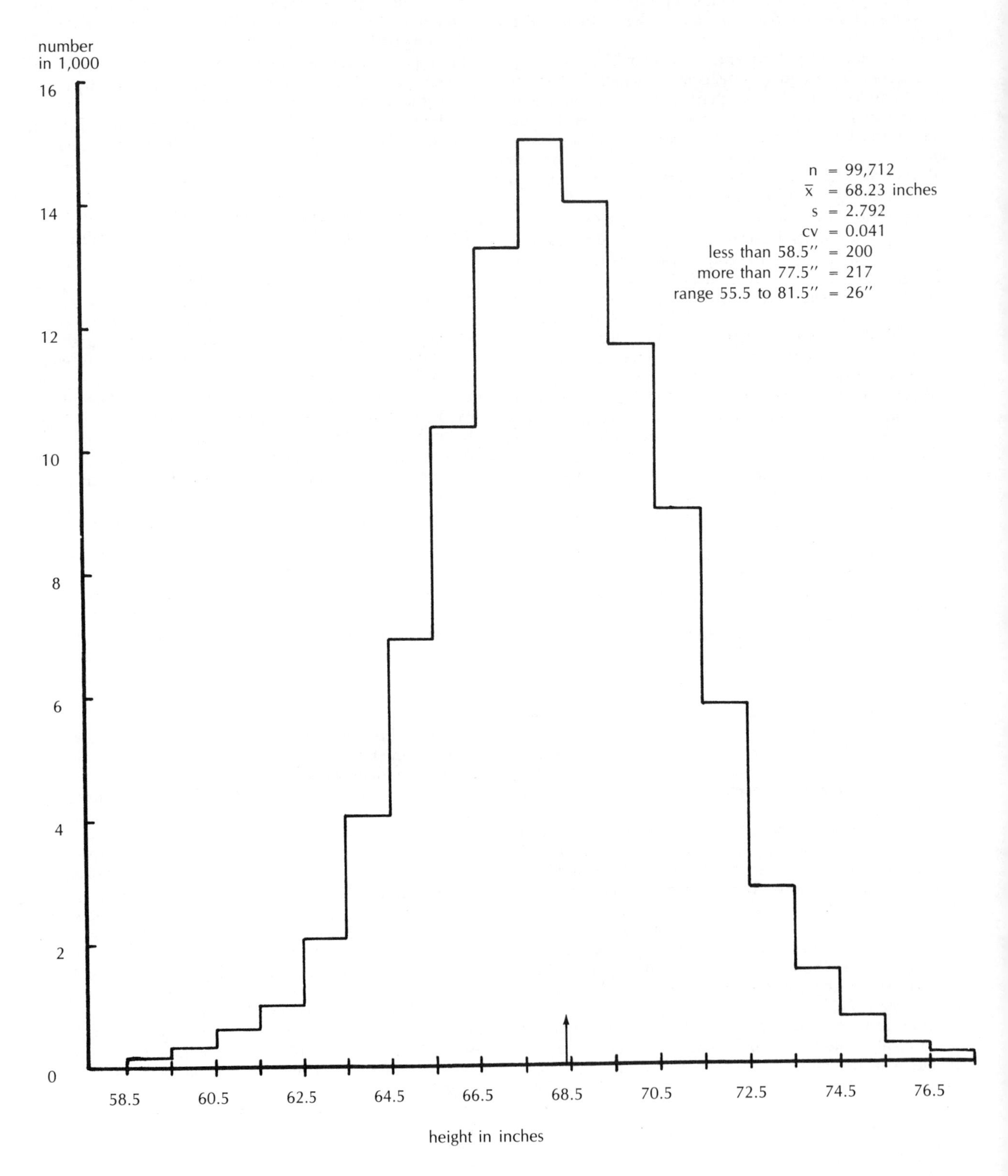

An interval estimate is the interval within which the population value falls providing little or no bias exists in the data. A common interval is one in which, for repeated samples of the same size, the interval includes the population value 19 out of 20 times — the 95 percent confidence interval. This interval estimate is favored more by academicians than by practitioners because the latter are well aware of the fact that, outside a few rather rare instances, management wants a single best estimate.

The concept of *estimator* is important. An estimator of a characteristic may take more than one form. The problem in practice is to use the best estimator:

Average: estimators are arithmetic mean, median, geometric mean.

Variability: estimators are range, standard deviation, mean deviation.

There are also *biased* and *unbiased* estimators. An unbiased estimator is one in which the average value of all possible samples is the population value; in other words, it is a consistent estimate. An arithmetic mean is an unbiased estimator. A biased estimator is one for which the average value of all possible sample values is not the population value. A ratio estimate is an example. However, the bias may be so small compared with the random sampling variation that this estimator may be preferred to an unbiased estimator.

An estimate has several important properties:

- It is a variate with a frequency or probability distribution; hence, it has a mean and a variance or standard deviation (standard error).
- It tends to be normally distributed in a large number of random samples of the same size, regardless of how the original variate is distributed (central limit theorem).
- Probability samples can be designed to reduce the random variation in an estimate to whatever limits are specified.
- An estimate is subject to random variation due to random sampling (the standard error), but this can be controlled by the sample design. This is unbiased since in a large number of samples this variation goes to zero (0).
- An estimate is subject to non-random variation due to *bias* from three major sources: the basic data, the method of estimation, and the processing errors.
- The bias in an estimate may greatly exceed the random variation in an estimate and invalidate the use of the standard error and confidence limits. This means that a sample study has to be properly designed and managed in order to control and reduce bias in the data, so that the use of the standard error and statistical analysis generally is valid. This means tight control over all possible sources of error in the original data, as well as similar control over all processing operations, including calculations.

Some examples of estimators:

variance: $s^2 = \frac{\Sigma(x_i - \bar{x})^2}{n-1}$ (unbiased)

binomial proportion: $p = x/n$ (unbiased)

standard deviation: $s = \sqrt{s^2}$ (biased)

mean: $\bar{x} = \Sigma x_i/n)$ (unbiased)

ratio: $f = \frac{\Sigma x_i}{\Sigma y_i}$ (biased)

Error. Error includes random or unbiased error and non-random or biased error. Random error means random variation in an estimate due to a random process, such as random selection. The expected value of such random deviations from an estimate is zero since in a sample of any substantial size the deviations on the plus side tend to equal and neutralize the deviations on the negative side. Hence, it is an unbiased error or variation. "Error" is hardly appropriate because this variation is due to a random process not to human error. It is kept because of common usage.

This *random variation,* due to random sampling, is measured by the standard deviation of the estimate obtained from a large number of random samples of the same size. It is called the *standard error.* It is sometimes referred to as the sampling error; the standard error squared is called the *variance* of the estimate. This variation, due to random sampling, is expressed by the mathematical equation which is different for different estimates. The standard error of several of these is given below. The expression $1 - n/N$ is the correction for sampling from a finite population, where n is the total sample size and N is the population or frame size from which the sample is selected at random. It can be ignored if n/N is 2% or less. Sometimes n/N is written "f":

1. Arithmetic mean $\bar{x}$:

$$s_{\bar{x}} = \frac{s_x}{\sqrt{n}} \sqrt{1 - n/N},$$

where

$$s_x^2 = \frac{1}{n-1} \sum_{1}^{n} (x_i - \bar{x})^2.$$

2. Binomial proportion p:

$$s_p = \sqrt{\frac{p(1-p)}{n}(1 - n/N)},$$

where $1 - p = q$.

3. Binomial frequency n_a:

$$s = \sqrt{np(1-p)(1 - n/N)}.$$

4. Poisson count n_a:

$$s = \sqrt{np} = \sqrt{m},$$

where $m = np =$ mean.

5. Ratio

$$f = \frac{\sum x}{\sum y} = \frac{\bar{x}}{\bar{y}},$$

including the ratio proportion, where $f = p$.

$$s^2 = \frac{(1 - n/N)}{n(n-1)\bar{y}^2}\left(\sum x^2 + f^2 \sum y^2 - 2f \sum xy\right).$$

Non-random or biased error. This refers to a wide variety of human errors, machine errors, and mistakes made in counting, measuring, remembering, recording, collecting, posting, and calculating. It includes errors or mistakes in programming a computer, in selecting equations for use in statistical analysis, in using the wrong sample design and estimating equations, and in designing a faulty test or experiment. In practice, this is the number one problem whether it is quality control, sample studies, design of experiments, or other quantitative work. This is because bias has so many sources, it involves human error, it requires a wide variety of controls, it requires careful planning and management, it requires close supervision, and it requires alertness and close attention to details.

This contrasts with random variation which can be closely controlled by the proper sample or experimental design, and by careful management to implement it. In this connection, it should be emphasized that implementing *a sample study is not simply another clerical study.* This is because biases — human mistakes and errors — multiply the same as accurate data. If a sample multiplier 1/f or N/n is 200 and if 1,000 is recorded as 100; then the error is not 900, it is 180,000.

Regression and correlation. Statistics is concerned with the relationship between two or more variates. The simplest and most widely used form is the straight line involving two quantities x and y:

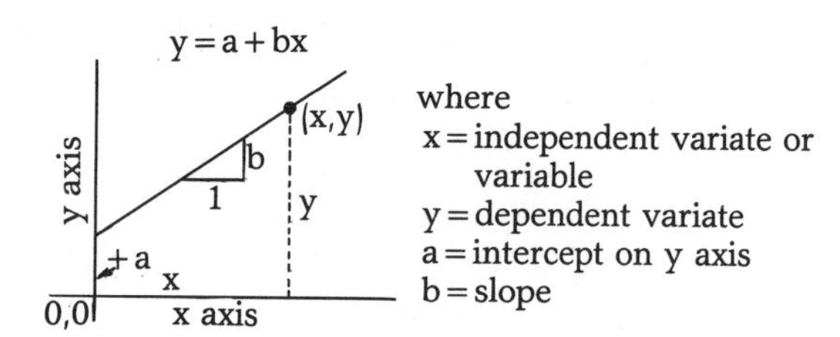

where
x = independent variate or variable
y = dependent variate
a = intercept on y axis
b = slope

and both a and b can be either positive or negative.

Long method of calculation. Rarely do all observed points (x,y) fall on a straight line — even when the relationship is strictly linear and not curvilinear. So a straight line is fitted to the data, using the method of least squares, to calculate values for a and b, the expressions for which, in terms of the observed pairs of values x and y, are as follows:

$$a = \frac{\sum y \sum x^2 - \sum x \sum xy}{D};$$

$$b = \frac{n\sum xy - \sum x \sum y}{D};$$

$$D = n\sum x^2 - \left(\sum x\right)^2.$$

The values needed are n, Σx, Σx^2, Σy, and Σxy.

It is necessary to know how closely a straight line fits the data. Two measures are used: the residual variance and the correlation coefficient. The smaller the former and the larger the latter, the better the fit.

If y_i is an observed value and y_c is the calculated value for each value of x_i from the least square line, then $\Sigma(y_i - y_c)^2$ is the sum of the squares of deviations from the line. Then the residual variance is:

$$v = s_v^2 = \frac{\sum(y_i - y_c)^2}{n-2} = \frac{\sum y^2 - a\sum y - b\sum xy}{n-2}$$

and $s_v = \sqrt{s_v^2}$ is called the *standard error of estimate.* The better the fit, the smaller the value of s_v.

The variance of the observed values of y_i is:

$$s_{y_i}^2 = \frac{\sum(y_i - \bar{y})^2}{n-1} = \frac{1}{n-1}\left[\sum y_i^2 - \frac{(\sum y_i)^2}{n}\right].$$

The correlation with x has reduced this variance, depending upon the extent that the observed points lie near the line. Set:

$$r^2 = \frac{s_{y_i}^2 - s_v^2}{s_{y_i}^2} = \frac{\text{variation taken out by line}}{\text{total variance}}$$

so that

$$r^2 = 1 - \frac{s_v^2}{s_{y_i}^2} = 1 - \text{proportion variance left.}$$

Hence, r^2 is the proportion of the variance in the observed y's accounted for or taken out by the relationship or linear correlation with x. The *correlation coefficient* is:

$$r = \sqrt{r^2}$$

and is commonly used, although r^2 is the more meaningful of the two.

Short method of calculation. The *computer, desk computer,* and *hand calculator* all make these calculations very simple. Examples are the desk calculator HP 9810 A with a stat rom and the hand calculator HP 32 E. Both of these give the following calculations very quickly once the n pairs of x and y are entered which is also a simple operation:

estimate	x	y	
mean	$\bar{x}$	$\bar{y}$	
variance	s_x^2	s_y^2	
standard deviation	s_x	s_y	or by using $\sqrt{x}$ key on s^2
coefficient of variation	c_x	c_y	divide standard deviation by mean
linear regression	a		intercept on y axis
	b		slope $(y_2 - y_1)/(x_2 - x_1)$
correlation coefficient squared	r^2		by using x^2 key on r if needed
correlation coefficient	r		by using $\sqrt{x}$ on r^2 if needed

Also in storage are n, Σx, Σy, Σx^2, Σy^2, Σxy for use in calculating s_v^2. It is easy to plot the points, plot the graph, calculate the residuals, and plot the residuals $y_i - y$ against x for residual analysis. If a plotter is available, it may do all the plotting including plotting of the residuals against x.

Linear transformations. Many non-linear functions can be converted to linear form by some transformation, thereby simplifying the analysis and presentation of the data. A common use of the linear plot is to show whether the data can be expressed by the function. This is measured by the extent to which the transformed data fit a straight line.

Hypotheses and inference. An inference is a conclusion based on experience such as observation, reading, study, conversations, participation, and experimentation. One infers that automobile A is better than automobile B, that prices for gasoline are too high, or that busing school children improves their education.

Hypothesis is a formal statement of an inference in the form of a proposition. In science, it is an assertion to test, not a prejudice to defend. It grows out of an analysis of a problem or from some other source. Usually hypotheses are in a general form and need to be analyzed into *statistical hypotheses* which can be tested by means of statistics. In some cases, this may be difficult if not impossible to do. When analysis reveals one or more statistical hypothesis, usually there is one or more test statistics which can be derived from the sample data and whose distributions are known when the hypothesis is true.

Hypothesis may be tested for several reasons: differences in levels, differences in variability, differences in effects, the existence of a relationship or independence, the goodness of fit of observed to expected frequencies, differences between and among groups and classes, and comparison of observed experimental results with expected theoretical values.

In many real-world problems and situations, the differences observed between levels, variability, effects, and the like are so great that it is obvious a real difference exists and thus statistical tests are unnecessary. This is the case, for example, where two frequency distributions do not overlap or overlap very little. The value of statistical tests lies in those numerous cases where it is not apparent that a difference exists or not.

The following steps are required to test a statistical hypothesis:

1. An equation is used to calculate the test statistic in terms of observed counts or measurements or estimates derived therefrom.
2. Then, the test statistic is referred, using the necessary number of degrees of freedom, if required, to its table.
3. Determine the probability of occurrence of a larger or smaller value.
4. Reject hypothesis if the test statistic falls in the region of rejection. This means that for the 95 percent level, that test statistic falls in 2.5 percent area at both ends for a two-tail test or in the 5 percent area for one-tail test.

Figure 2 illustrates a number of basic concepts in testing statistical hypotheses. The *critical region* or *region of rejection* of the hypothesis is the part of the probability distribution specified in advance which calls for the rejection of the hypothesis if the obtained value of the test statistic falls in this area. This probability is usually five percent or one percent to correspond to the 95 percent and 99 percent levels, respectively, or 0.3 percent to correspond to the 3 sigma level on quality control charts. This is the *level of significance* and it can be either a one-tailed test as in Figure 2A or a two-tailed test as in Figure 2B.

Figure 2C shows the two errors that arise in connection with the testing of statistical hypotheses: *Type 1 error*, which is the probability of rejecting a true hypothesis and is designated α, the level of significance and *Type 2 error*, which is the probability of accepting a false hypothesis and is designated β.

In Figure 2C, the hypothesis $H_1{:}u = u_1$ is tested against the alternative hypothesis $H_2{:}u = u_2$. The test of significance or Type 1 error is set as α, meaning this is the

Table 1

Examples of Function Converted to Linear Form

Function	Linear form
1. exponential: $y = ab^x$	$\log y = \log a + x \log b$ $Y = A + Bx$
2. exponential: $y = ce^{bx}$ (take logarithms to base e)	$\ln y = \ln c + bx$ $Y = A + bx$
3. power: $y = ax^b$	$\log y = \log a + b \log x$ $Y = A + bX$
4. hyperbola: $y = \dfrac{x}{a + bx}$	$\dfrac{x}{y} = Y = a + bx$
5. normal: $y = \dfrac{1}{\sigma\sqrt{2\pi}} e^{-\frac{1}{2}\frac{(x-u)^2}{\sigma^2}}$ let $z = (x - u)^2$	$\ln y = -\ln \sigma - \ln \sqrt{2\pi} - z/2\sigma^2$ $Y = a + Bz$
6. Weibull (take logs to base e twice) $F(x) = 1 - \exp - \dfrac{(x - \gamma)\beta}{\alpha}$ $\ln \ln \dfrac{1}{1 - F(x)} = - \ln\alpha + \beta \ln(x - \gamma)$ $Y = A + \beta X$	

FIGURE 2

Critical Regions for One- and Two-Tailed Tests with Type 1 and Type 2 Errors

A. One-tailed test with probability of rejection ∝

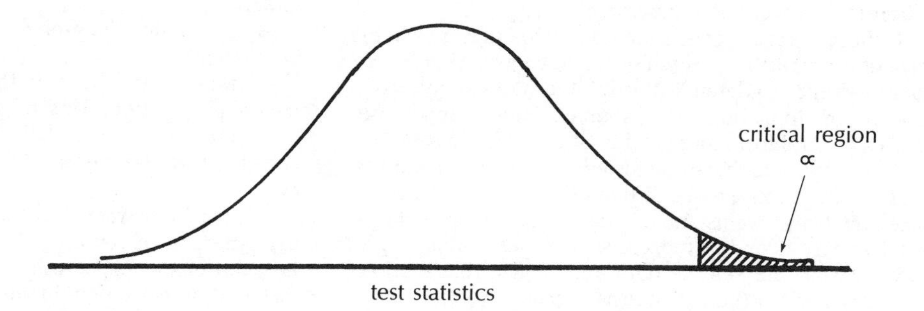

B. Two-tailed test with probability of rejection ∝/2 at each end

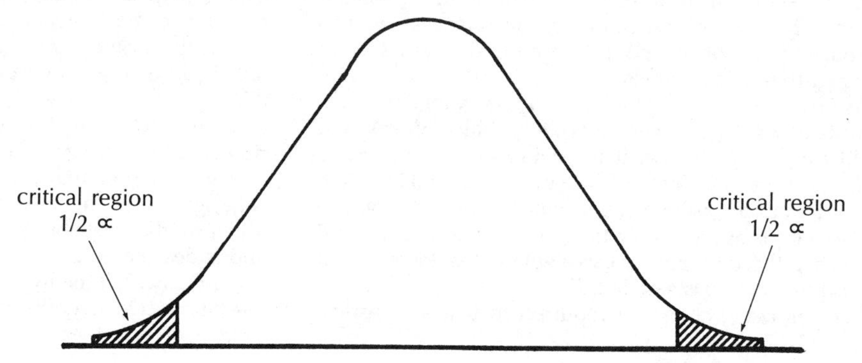

C. Testing hypothesis H_1 against hypothesis H_2 showing Type 1 and Type 2 errors

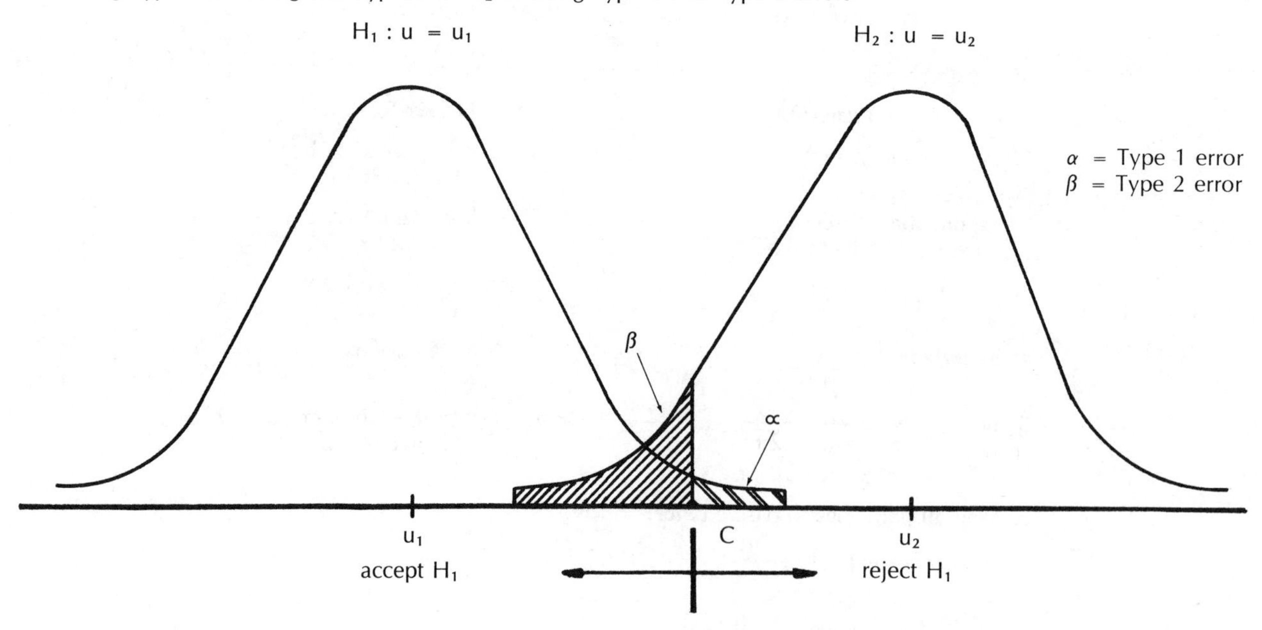

probability H_1 will be rejected when it is true. This is at point C on the u scale. But the distribution of H_2 extends to the left of this point with a probability β. This is the probability of accepting H_2 when in reality H_1 is the true hypothesis; hence, this is the Type 2 error.

The easiest way to control Type 2 error is to increase the random sample size so that the amount of overlapping of the distributions H_1 and H_2 will be at an acceptable level.

Several test statistics are available to test various hypotheses. Tables of their probability distributions are available to determine the probability of getting a larger, or smaller, value than the one obtained, of determining whether the value falls in the critical region or not. Six common test statistics are the following:

1. The binomial distribution: for counts $x=np$; for proportions $p=x/n$.
2. The normal distribution (z test): large sample means, differences.
3. The t distribution (t test): small sample means, regression b's, correlation coefficient.
4. The chi-square distribution (χ^2 test): frequencies, variances, goodness of fit, independence.
5. The F distribution (F test): variances, means, regression, analysis of variance.
6. The Poisson distribution (Poisson variate): counts, mean counts.

Warnings about tests of significance. Significant differences may be of no importance in practice or action. An example is the test of the two ends of a three minute timer which showed a significant difference of about two seconds. This small difference is of no significance in timing telephone calls or in boiling eggs.

A significant difference may have very limited application because the population is so uncertain that no broad generalizations are justified. An example is the superiority of wheat variety A at Rothamsted Agricultural Station, England. This does not mean that it is also superior in Kansas and Saskatchewan. Consistency of response in many places is needed to justify generalization to a broad population.

This warning applies to all inferences from agricultural, biological, and medicinal tests and experiments because the frame or population to generalize to is unknown, guessed at, assumed, proclaimed, or derived from personal opinions and bias.

Inferences may be false because the sample or experiment was unsound or biased and the analysis faulty. An example was the plutonium test at Rocky Flats, Colorado where 10 head of cattle were compared with 10 head of cattle in Nevada. An average of 5.2 picocuries per kilogram was found at the first location compared with 1.3 picocuries per kilogram at the latter site; this led to the false conclusion that the former was four times as strong as the latter. Not only did this apple arithmetic not apply, but the variability of measurements about the averages were so great as to erase any meaning to the difference 5.2-1.3. A later test in Nevada yielded a value of 12 picocuries per kilogram.

Significance depends upon the size of the standard error which is greatly influenced by the sample size and sample design. Two sample designs give a mean of 102; but the first has a standard error of 8, the second only 2. Test that these came from a population mean of 106. The first case (102-106)/8 does not reject the hypothesis, but the second one (102-106)/2 does at the 95 percent level.

There are an infinite number of alternative hypotheses. In practice, what management wants is the best point estimate. The job of the statistician or other professional is to see that this estimate has acceptable Type 1 and Type 2 errors by designing a sample plan with an ample size, or efficient design to accomplish this.

Table 2

General and Statistical Hypotheses and Their Appropriate Test Statistics

General hypothesis	Statistical hypothesis	Test statistic
Drug 1 is better than drug 2.	$u_1 - u_2 = 0$ $\bar{D} = 0$ (differences)	t
Auto 1 gets better gasoline mileage than auto 2.	$u_1 - u_2 = 0$	t or z
These two groups of students have the same distribution of IQ's.	$u_1 - u_2 = 0$ (means) $\sigma_1^2 - \sigma_2^2 = 0$ (variances)	z F
Deaths due to automobile accidents have declined during past year.	$m_1 - m_2 = 0$	Poisson

Notes

[1]A. C. Rosander, *Case Studies in Sample Design,* Marcel Dekker, Inc., New York, 1977, especially Chapter 2, Chapter 6 and Chapter 10, reprinted by courtesy of Marcel Dekker, Inc.

[2]A. C. Rosander, "Sampling Misunderstood," *Quality Progress,* June 1979, p. 4.

[3]A. C. Rosander, *Case Studies in Sample Design,* Marcel Dekker, Inc., New York, 1977, Chapter 13, reprinted by courtesy of Marcel Dekker, Inc.

[4]R. A. Fisher and F. Yates, *Statistical Tables for Biological, Agricultural, and Medical Research,* Oliver and Boyd, Edinburgh, 6th edition, 1963.

[5]M. G. Kendall and B. Babington Smith, *Tables of Random Sampling Numbers,* Cambridge University Press, Tracts for Computers, No. XXIV.

[6]Rand Corporation, *A Million Random Designs,* Free Press, Glencoe, New York, 1955.

[7]Herbert Arkin, *Handbook of Sampling for Auditing and Accounting,* McGraw Hill, New York, 1974, 2nd edition.

[8]A. C. Rosander, *Case Studies in Sample Design,* Marcel Dekker, Inc., New York, 1977, appendix contains tables of 4,000 random days and 4,000 random minutes. Reprinted by courtesy of Marcel Dekker, Inc.

[9]*Ibid.,* pp. 160-162, 362-364.

[10]*Ibid.,* p. 368.

[11]*Ibid.* The appendix contains tables of 4,000 random minutes and random days. Chapters 14, 15, and 16 describe in detail successful applications of random time sampling.

[12]The median fraction (MF) method for approximately P_0 originated with the author. It gives excellent approximations in most of these problems and is much simpler than using direct multiplication or logarithms of factorials. See Chapter 14 for a detailed description of the method. The average binomial is also excellent.

[13]The National Selective Service, which is the source for these figures, gives a mean of 68.47 but no explanation was ever found for the difference.

13

Empirical Distributions

Empirical distributions. Given a set of real-world numerical data, an empirical distribution shows the frequency with which the values are distributed. For measurements, it shows the frequency for each of several intervals of the measurements. For whole numbers or counts, it shows the frequency of occurrence of each number or count, such as size of family or number of errors.

An empirical distribution is the only real-world distribution we have. Probability distributions such as the binomial, normal, and Poisson are theoretical distributions based on certain assumptions that may or may not be met by real-world data. For this reason they are often called mathematical models. They are powerful and useful in interpreting real-world data because experience has shown they do approximate many situations in the real-world.

An empirical distribution is constructed with intervals of measurements, or whole numbers, measured horizontally on the x axis and the frequency of occurrence or count measured vertically on the y axis. Since there always is a total number, either a sample or a population or other aggregation, the numbers or frequencies can be converted into proportions to give a proportion distribution or into percentages to give a percentage distribution.

The following are suggested steps in interpreting or analyzing a set of empirical values, assuming we have a listing or record of the individual values x_i's:

1. If the values are a time sequence, plot a time chart, time order horizontal and the x_i value vertical.
2. Determine the range R: largest value minus smallest value in the set of n.
3. Determine intervals for x_i and construct a frequency distribution *table;* that is, the number of x_i's falling in each interval or for each whole number.
4. Plot a frequency distribution *graph* from the table: intervals horizontal and count or frequency vertical, or whole numbers horizontal and frequency vertical, above each whole number. Construct rectangles above intervals and vertical lines above whole numbers.
5. Make the basic calculations:

- The mean:

$$\bar{x} = \frac{1}{n}\sum_{1}^{n} x_i.$$

- The variance:

$$s^2 = \frac{1}{n-1}\sum_{1}^{n}(x_i - \bar{x})^2 = \frac{1}{n-1}\left[\sum_{1}^{n} x_i^2 - \frac{\left(\sum x_i\right)^2}{n}\right].$$

- The standard deviation:

$$s = \sqrt{s^2}.$$

- The coefficient of variation: $c_v = \frac{s}{\bar{x}}$.

This may have no meaning where $\bar{x}$ is negative as in an error frequency distribution.

- The proportion or percentage that is more, or less, than a specified value of x_i.
- The proportion of percentage that is more, or less, than a specified value of x_i.

These distributions reveal at a glance variability, range of the measurements or counts, extreme values, bunching, degree of skewness or lack of symmetry, or more than one peak. They show many things:

- Human variability doing the same, identical job.
- Variability around an average: mean or median.
- Extreme values suggesting "out of line", "out of control", different criteria or standards.
- Extent to which tolerance limits or critical values are being exceeded, such as error rates.
- Departure from acceptable quality performance, such as error rates.
- Biased data and biased methods.
- Human capability under a specified set of conditions.
- A means of measuring stability of performance or of an operation.

The plotting and analysis of frequency distributions are strongly recommended regardless of the small number of values. It should not be limited to cases where the number of values is in the hundreds, thousands, or millions. It is a powerful method of exploratory analysis and can easily be an initial stage in making quality control improvements or additions.

Characteristics for which empirical distributions may be helpful in obtaining important and useful information include the following:

1. Number of complaints by type, by months, by years.
2. Delay time.
3. Waiting time.
4. Service time.
5. Health, home insurance claimed (dollars).
6. Health, home insurance allowed (dollars).
7. Percent dollar claims allowed.
8. Premiums charged for same policy.
9. Amount of tax error.
10. Amount of tax error, by type.
11. Percent tax error.
12. Dollar amount, bank checks.
13. Assessed valuations for tax purposes.
14. Property taxes.
15. Delinquent property taxes.
16. Percent taxes delinquent, by years.
17. Monthly usage of electricity, monthly bill.
18. Monthly use of gas, monthly bill.
19. Monthly use of water, monthly bill.

20. Size of repair bills.
21. Error rates in clerical work.
22. Daily production by employees: individual and group.
23. Errors made by interviewers (amount).
24. Errors made in predictions, forecasts.

In this chapter several examples of empirical distributions are presented, not just to illustrate a frequency distribution but to show how such an empirical distribution can be helpful in understanding and improving operations. They are classified under the following headings:

1. Distributions of human variation in judgments and mechanical performance:
   1.1 Distribution of tax assessments made on same property.
   1.2 Distribution of error rates made by key punch operators.
2. Comparative distributions:
   2.1 Distribution of punch card production for two different jobs.
   2.2 Comparison of miles per gallon for four, six, and eight cylinder automobiles.
   2.3 Differences between check interviewers and original interviewers.
3. Appraisal of predictions, forecasts, and regression estimates:
   3.1 Distribution of 67 regression estimates (predictions).
   3.2 Distribution of 48 state errors made in a presidential opinion poll forecast.
4. Testing a procedure:
   4.1 Distribution of numbers drawn in the World War II lottery, in order of selection.
   4.2 Testing the distribution of 221 values of a test statistic with that of a normal distribution, to test random order of selection of a set of samples. See Table 12, Chapter 7.
5. Throwing light on a common practice:
   5.1 Distribution of bank checks, by dollar size.

Distribution of human variation. Variations in human performance arise in a wide variety of situations:

1. Differences in error rates, e.g., key punching, coding.
2. Differences in making measurements of the same dimension.
3. Differences in making a measurement or test on identical samples, e.g., laboratory tests.
4. Differences in auditing, e.g., auditing a tax return.
5. Differences in number of defects found in a lot or sample by different inspectors.
6. Differences in the number of errors found in a file or sample by different reviewers.
7. Differences in making ratings or judgments on the same thing:
   7.1 Rating figure ice skaters.
   7.2 Rating divers.
   7.3 Judging extent of the diversion of carload traffic in a merger case (Chapter 25).
   7.4 Appraising the value of a piece of property for tax purposes or for selling.

We illustrate this latter example of variation in human performance in the histogram shown in Figure 1; the individual values are given in Chapter 25. It is the frequency distribution of 148 assessed valuations made independently by 148 tax assessors on the same residential property. The distribution shows at a glance the range, the extreme values, and the bunching of the assessments. It shows that most of the values lie between $10,000 and $13,000. From the 148 individual values given in Chapter 25 the following estimates are obtained:

1. Range = $20,000 − $4,000 = $16,000.
2. Mean $\bar{x}$ = $11,462.
3. Standard deviation s = $1,847.
4. Coefficient of variation $cv = s/\bar{x} = 0.16$ or 16 percent.

If the seven extreme values are excluded, three at the lower level and four at the upper level, the mean is $11,375 and the standard deviation is $1,187 for a coefficient of variation of 0.10 or 10 percent. Assuming $\bar{x} = 11{,}375$ and $s = 1{,}187$ for a normal distribution, then 3s or 3 sigma gives 3,561, so that $11{,}375 \pm 3{,}561$ gives an upper limit of 14,936 and a lower limit of 7,814.

Six out of the seven extreme values fall outside these limits and hence can be considered "out of control". The seventh value is 7,900 and is just inside the lower limit of 7,814. If this value is included, then all seven values may be considered as being out of line with the other values and due to "assignable causes". It appears on the face of the distribution and without any statistical tests that at least the seven assessors were not using the standards and criteria to make judgments that the others were using.

What this frequency distribution shows is that the 148 tax assessors are not using the same criteria for assessing the value of this piece of residential property. It may be that those who are valuing the property between $10,000 and $13,000 are using about the same standards and are about as accurate as can be expected; again, they may all be subject to some bias. The only way to determine this is: 1) develop an objective set of criteria and standard with rules and examples; 2) train the assessors in learning and applying these rules and criteria; and 3) apply these rules and criteria in the field to actual properties. Explain to those who still have extreme values where they are deviating from the rules and standards, if they are. In this way, there is no doubt that the variability in the assessments made on the same property by these 148 assessors could be greatly reduced.

Figure 2 shows the variation in the error rates of 50 key punch operators. The range is from a low between one and two percent to a high between 10 and 11 percent. The proportion in error is the number of punch cards with one or more errors divided by the total number of cards punched. For example, over a period of 20 days the total number of cards punched by an operator was 12,594; the number in error was 327. This gives a percentage of $327 \times 100/12{,}594$ or 2.6 percent, which was highly satisfactory in a statistical division where acceptable quality was set at five percent or less. In this group 24, or 48 percent, had error rates above this level.

Actually, this group of key punch operators had been subject to a serious turnover because of the transfer of about 900 employees to various field offices. Older tenured employees "bumped" younger more efficient operators with less tenure.

What this graph shows is that at least 24 of these employees needed special intensive training in key punching, covering a period of at least one month, to improve their skills and to become more familiar with the documents and data punched to bring all of the error rates below five percent. The last available information was that this was not done, greatly delaying the publication and release of the annual reports and special studies produced from these punch cards.

FIGURE 1

Valuations of the Same Property by 148 Tax Assessors

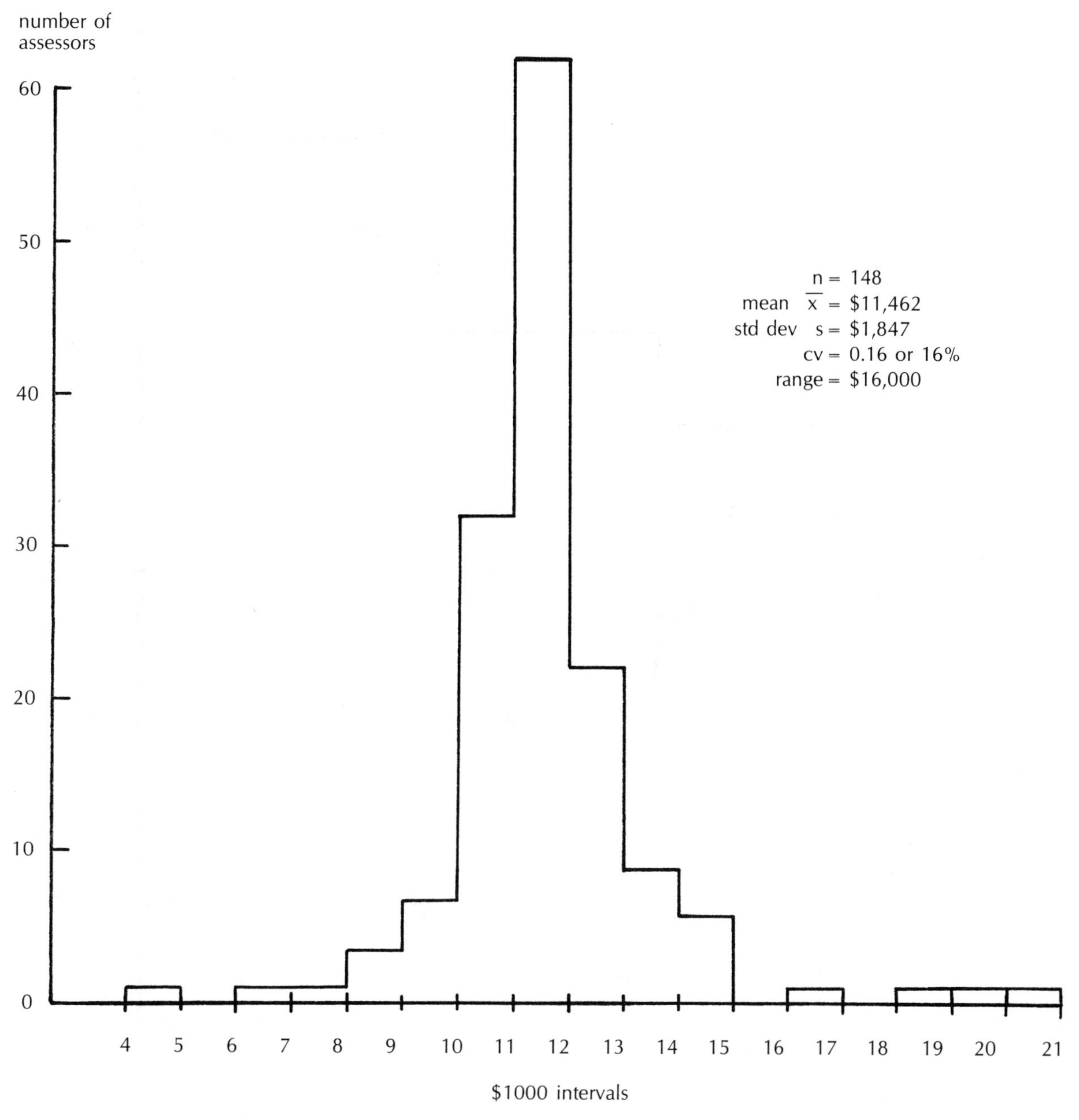

Source: Figures furnished to author by W.G. Murray and G.E. Bivens, Iowa State College. See their paper in *National Tax Journal,* V, December 1952, pp. 370-375.

Before the change, 16 operators had an error rate of less than four percent. The best one, who was released but later brought back, had a fantastic performance record. Her error rate, even with the most difficult punching, was consistently less than one percent; furthermore, her production was the greatest! In a period of 11 weeks, she punched 49,535 cards from the hardest material with an average error rate of 0.76 percent. For eight of the 11 weeks, her error rate was less than one percent and for two weeks she made no errors at all. There was no point in doing a 100 percent inspection of this employee's work nor that of about 15 other persons. A sampling inspection system based on lots was introduced that was very effective in keeping error rates below two percent, but due to the "bumping" mentioned above the use of lot sampling for control had to be abandoned. The error rate went up so high that the sample plan rejected all lots. More training was the answer before any control methods based on sampling could be used.

FIGURE 2

Error Rates of 50 Card Punch Operators

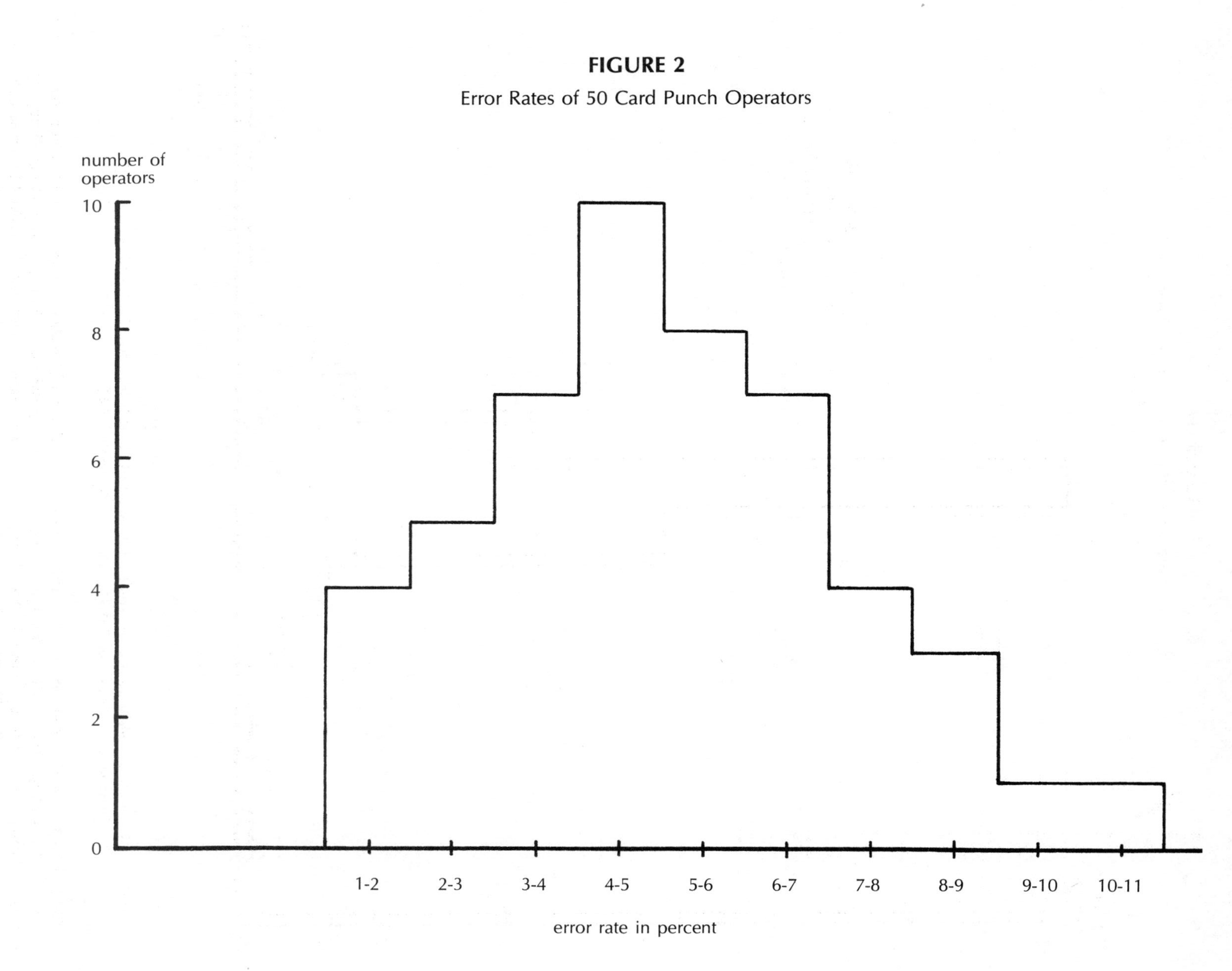

Comparative distributions. Frequency distributions of the same characteristics for two or more groups can be compared in order to study differences, relationships, and similarities. As with the distribution for a single group, the frequency distribution can be used to show what the actual capabilities of a group are, how much difference exists between the poorest and the best performer, and what steps might be taken to move performance of the group toward that of the best performer. Obviously, this latter step is a very delicate one, especially where employees, including unions, may be opposed to any such move. On the other hand, it needs to be pointed out that any improvement in the performance of employees, which increases productivity, almost always rebounds to the benefit of the employee both as a worker and as a consumer.

Comparative distributions can be used to give a "before" and "after" picture after some program or procedure of intervention, such as a training course, a "refresher" course, substituting a quality control system for 100 percent inspection, individual instruction or tutoring by a supervisor or other person, group discussions of the types of errors being made and how to avoid them, preparation of procedural manuals for employee use with examples as how to do a job correctly, or development of standards and criteria, a lack of which is a cause of variability as was pointed out in the case of the tax assessors.

Comparative distributions are not only an analytical tool to be used to measure improvements, to measure effectiveness of interventions, and to indicate human capability of a group of employees as of the present. It can be used as a presentation device to higher level officials to show on a factual and objective basis that certain procedures, interventions, methods, and the like are effective. Discussions can then take place, not on the basis of opinions and conjectures, but on the basis of what the actual distributions show. One could use averages, but averages are seldom enough; it is necessary to impress professionals, as well as high level officials, with the basic idea that it is variability that is at the heart of the real-world.

Figure 3 shows the frequency distributions of the daily production rates of two groups of key punch operators — one punching individual income tax returns and the other punching corporation income tax returns. The characteristic is cards punched per day and is derived from the total production of each operator over a period of three months divided by the total number of working days which was about 65.

The 16 operators punching corporation income tax returns punched from 200 to 600 per day, but the best operator punched 700 to 800 per day or about 4 times as many. The range was even greater for the 24 operators; six of them punched less than 300 daily. In this case, the best operator punched about eight times as many as the poorest, despite the fact that individual income tax returns were much easier to punch than corporation income tax returns. This fact raised serious questions about the skill level of the latter group. On the face of the evidence shown by this distribution, several of these operators need some intensive training in punching individual income tax returns to bring their production to an acceptable level.

Indeed, distributions of this kind, showing wide variations in human performance, raise a critical question: Is this variation fixed and unchangeable? Or is it at an easy level based on habit, which can be reduced and moved to a higher level in this case, or to a lower level in the case of an error distribution?

Figure 4 shows three distributions of miles per gallon for four, six, and eight cylinder model year 1976 automobiles tested by the Environmental Protection Agency (EPA). The measurements are obtained from engineering laboratory tests, *not* from a random selection of drivers using a random sample of dealer's models and driving under actual traffic conditions. The figures plotted are for highway driving at an average speed of 49 miles per hour as simulated in the laboratory.

These measurements, therefore, are not valid nor reliable measurements of actual gasoline consumption per mile of actual driving. Even for advertising purposes, for which they are widely used, they are admittedly false.

They can be used, as plotted, only if one assumes that all models tested were tested by the same methods, by the same operators, and that the same adjustments were made to obtain "highway" miles per gallon. Then the figures for different models can be compared on the assumption that all figures are about equally biased. Assuming this, an examination of the three distributions shows:

1. The miles per gallon increases as the number of cylinders decreases.
2. There is very little overlapping of the three distributions; only seven values for the four cylinder overlap the six cylinder distribution and only three values for the six cylinder distribution overlap the eight cylinder distribution.
3. The variability of a cylinder type or class increases as the average miles per gallon increases.

The numerical results derived from raw values are as follows:

Number cylinders	No.	Overlap	M/gal.	Diff.	Standard deviation	Coefficient of variation
8 cylinders	63		18.3		1.98	0.11
		3		8.1		
6 cylinders	34		26.4		3.02	0.11
		7		7.5		
4 cylinders	32		33.9		3.73	0.11

The values for both six and eight cylinder models are suspect: for the former, 25 is recorded 14 times and 30 is recorded nine times showing a very strong tendency to record either 25 or 30. This is why six cylinder models are plotted on a different interval. For eight cylinder models, 32 or half of the values are either 18 or 19. The values for the four cylinder models do not show this concentration.

This example of empirical distribution analysis shows the value of this type of analysis, even to the extent of throwing doubt on the quality of the measurements themselves.

FIGURE 3

Distribution of Cards Punched per Day from Federal Income Tax Returns
(based on 3 months)

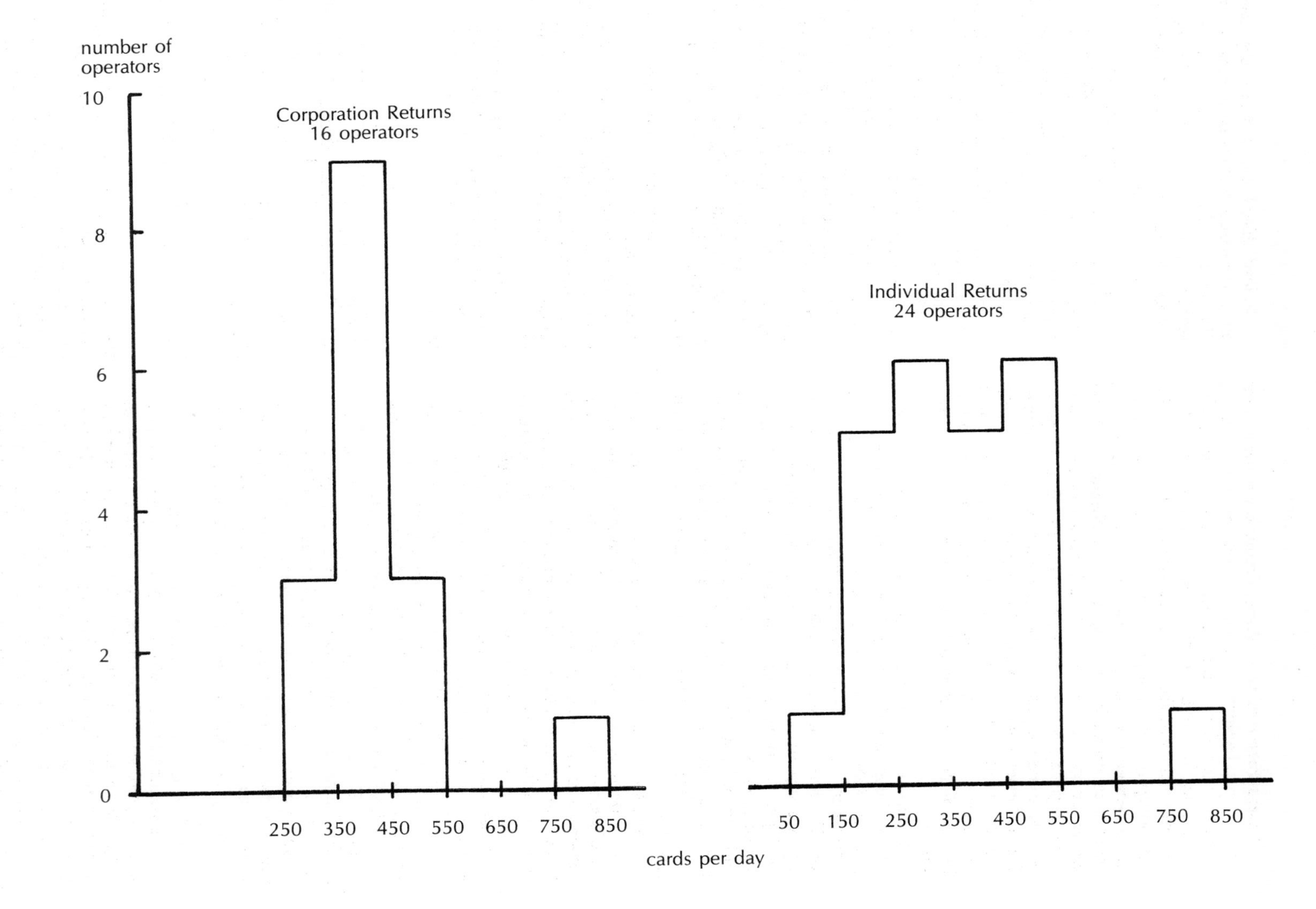

FIGURE 4

Distribution of Miles per Gallon of Automobile Models by Number of Cylinders

	no.	mean	standard deviation	CV
8 cylinder	63	18.3	1.98	0.11
6 cylinder	34	26.4	3.02	0.11
4 cylinder	32	33.9	3.73	0.11

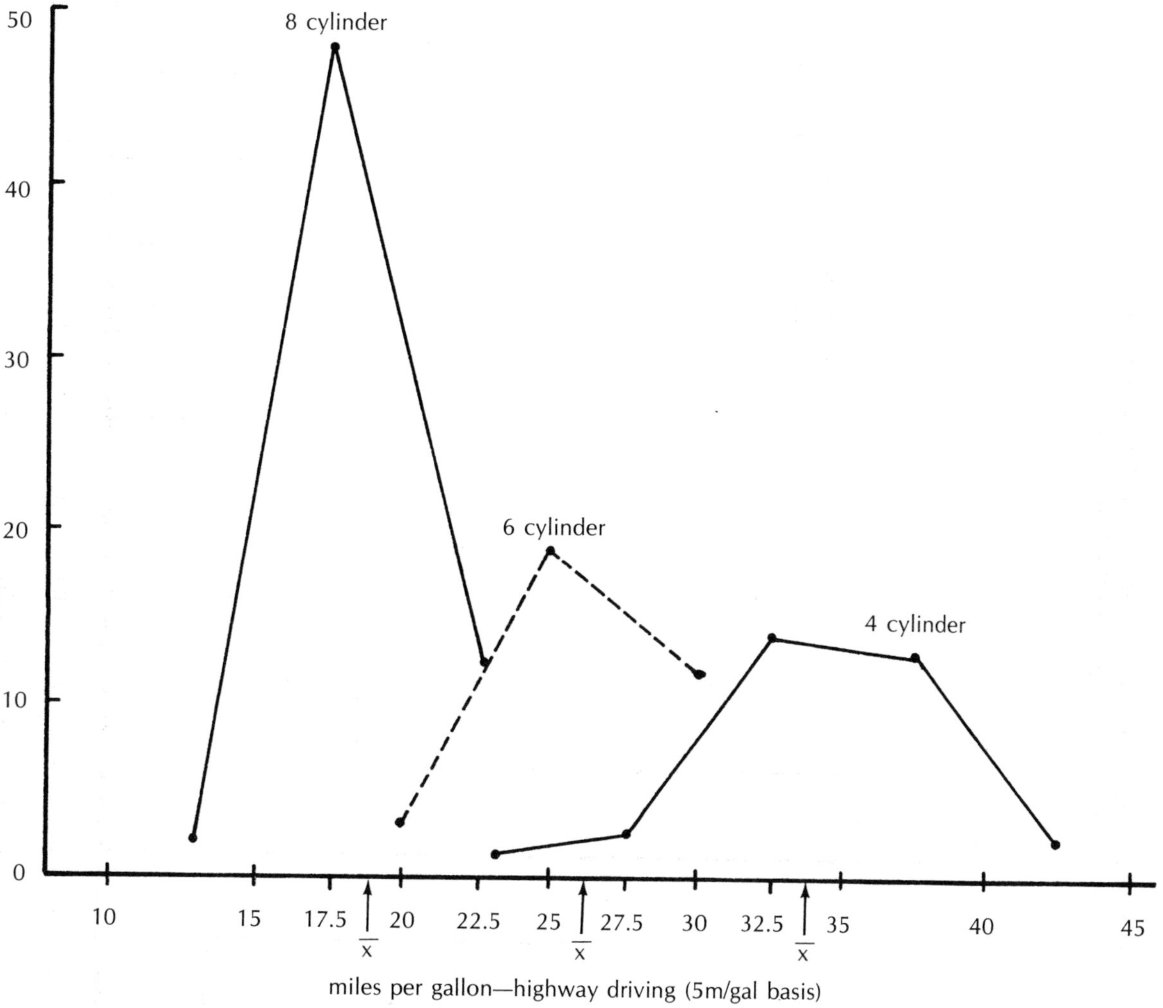

number of automobile models

miles per gallon—highway driving (5m/gal basis)

Source: EPA, *The World Almanac,* 1976, p 97.

Figure 5 shows the distribution of the differences between two distributions, rather than the two distributions themselves. In a nationwide study of families, the money values obtained by the original interviewers were audited by check interviewers who recorded the individual items where the amounts of money were changed. The check interviewers were supervisors or the best interviewers so that the differences found by the check interviewers were real errors in collecting the original data. Figure 5 shows the differences between the values obtained by the check interviewers and the original interviewers for 54 families in a city in New England. Minus values mean that the original figures were too high; positive values mean that the original values were too low. There were 16 of the former and 38 of the latter.

The mean of the original interviewers was $1,460, that of the check interviewers was $1,627, or a difference of $167. The standard deviation of the former was 628, that of the latter 684. If we take $1,627 as the nearest value to the "true" value, then the mean difference of 167 represents a 10 percent bias in an estimate of the mean.

Suppose the 1,460 was taken at face value, as it actually was, and confidence limits were applied to it. Then, at the 95 percent level:

$$1460 \pm 1.96 \frac{(628)}{\sqrt{54}} = 1292 \text{ and } 1627.$$

FIGURE 5

Errors in Dollars Found by Check Interviewers of 54 Families in City A
Note: A value of + $1427 is not shown.

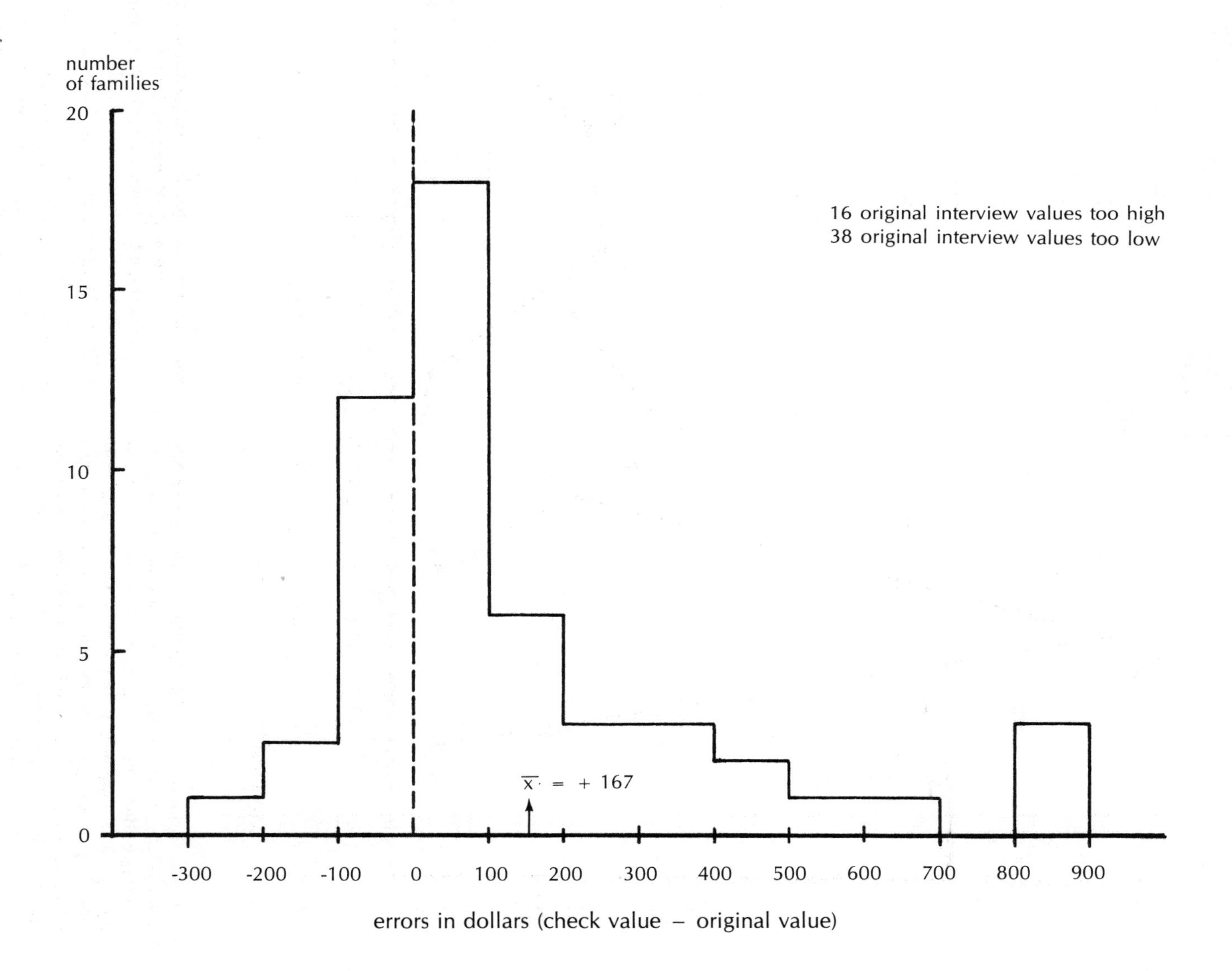

The upper limit of the 95 percent confidence limit is 1,627, the same as the best "true" value. Hence, the statement that repeated samples of 54 with confidence limits calculated as above will include the population or "true" value 95 percent of the time is false, since the actual probability is zero. Confidence limits assume the bias is 0; in this actual case, it is 10 percent.

Actually, it was this very case, as well as other findings that came out of the use of the check interviewers, that led to some very important conclusions:

1. There is no point in getting all excited about sampling errors of four or five or six percent, when the bias in the original data is 10 percent.
2. The original data and the estimates in this case should be corrected because they were underreported to the extent of 10 percent on the average; in one city, the bias was 20 percent.
3. Confidence limits are invalid and false where bias exists to any substantial degree, as it did in this case.
4. The time to apply these tests checks was while the study was being made, not after it was completed when it was too late to take any corrective action.

Appraisal of predictions, forecasts, and regression estimates. In the preceding example, a comparison was made between measurements obtained by a set of original interviewers and a group of check or audit interviewers. They were compared by taking the differences found for each of 54 families and analyzing the frequency distribution. In using a frequency distribution to appraise predictions, forecasts, or regression estimates, the performance of two groups is not being compared. The differences in paired values, one of which is obtained by some estimation technique and the other which is the actual or population value, are being compared. In cases where the units may differ greatly, the differences between the obtained value and the population value may be expressed in percentages; the distribution variate is the latter.

Figure 6 is a distribution of the percentage error in regression estimates of civilian requirements for 67 specific commodities, projected one year in advance. In all cases, the actual magnitude of the commodity was eventually available for comparison with the estimate from the regression line. The regression line took several different forms. Some were simple straight lines extrapolated to the projected value of x. All regressions were derived from the relationships between two or more variates or variables; no estimate was a simple projection of a time trend. Forms were $y=a+bx$, $y=a+bx-ct$, where $t=$ time and $y=a+b_1x_1+b_2x_2$. Both graphic and algebraic methods were used; some non-linear relationships were transformed into a linear form.

The percentage errors were calculated as follows:

$$\text{Percentage error}=\frac{(\text{predicted value}-\text{actual value})}{\text{actual value}}\times 100.$$

Hence, negative errors were those in which the predicted value was too low; positive errors were those in which the predicted value was too high. The results are reflected in the frequency distribution: the mean $\bar{x}=-0.10$ percent, the standard deviation $s=6.24$ percent, and the range was from -16 percent to $+16$ percent. The distribution is single-peaked and fairly symmetrical about the center. There were 21 values between 2 and -2, 23 values below this range, and 22 above this range. With a practically zero mean, a 6 percent standard deviation, and all values within three standard deviations of the mean, this distribution approximated a normal curve, thus indicating absence of bias in the methods of estimation.

Figure 7 is a distribution of the differences in percentages in a presidential opinion poll by states, between the predicted percentage that would vote for the Democratic candidate, and the actual percentage as shown by election returns. For each of the 48 states the difference was calculated (estimated percentage minus actual percentage) and the proper sign attached. The examination of the 48 pairs and calculations from the differences revealed the following:[1]

1. The mean of the differences was -1.90 percent.
2. The standard deviation was 2.23 percent.
3. The coefficient of variation is -1.17 or -117 percent; this has no meaning here.
4. The range is from -6 percent to $+2$ percent in percentage points.
5. The variation of the estimates or predictions is as follows:
   - underestimates 35
   - zeros 5
   - overestimates 8
6. Modes appeared at -3 (12 states) and at $+2$ (7 states).

If the estimates were unbiased, the expected mean would tend to be zero as in the preceding example, with an equal division of the 48 states about this point. In this case it is -1.9 which shows a strong tendency to underestimate the Democratic vote. Other tests indicate a highly significant statistical test including the chi-squared test and the binomial frequency test. An unbiased method of estimation would tend to give 24 above zero and 24 below zero; the five zeros are divided two and three:

Chi-square test

Actual frequency	Expected frequency	Difference	$\frac{\text{Difference}^2}{\text{expected}}$
37	24	13	169/24
11	24	-13	169/24
48	48		$\chi^2=$ 14.1

Chi-square of 14.1 for one degree of freedom gives a P of 0.00018.

Binomial frequency test

1. The mean $np=48\times\frac{1}{2}=24=u$; $\sigma=\sqrt{npq}=\sqrt{48\times\frac{1}{2}\times\frac{1}{2}}=3.464$.

2. Using the normal approximation, omitting continuity correction,

$$z=\frac{x-u}{\sigma}=\frac{37-24}{3.464}=3.75.$$

3. For a two-sided test, $P=0.00018$, the same as the chi-square test.

Figure 8 shows two frequency distributions, the frequency of occurrence of numbers in the first 1,000 numbers and the last 1,000 numbers drawn in the national draft lottery for World War II. Since no draft board included more than 9,000 males of draft age, the order in which the males were to be drafted was determined by drawing capsules from a bowl containing 9,000 capsules each containing a number from 1 to 9,000.[2] To construct

FIGURE 6

Distribution of Percent Error in Regression Estimates of Civilian Requirements for 67 Commodities One Year in Advance—U.S.

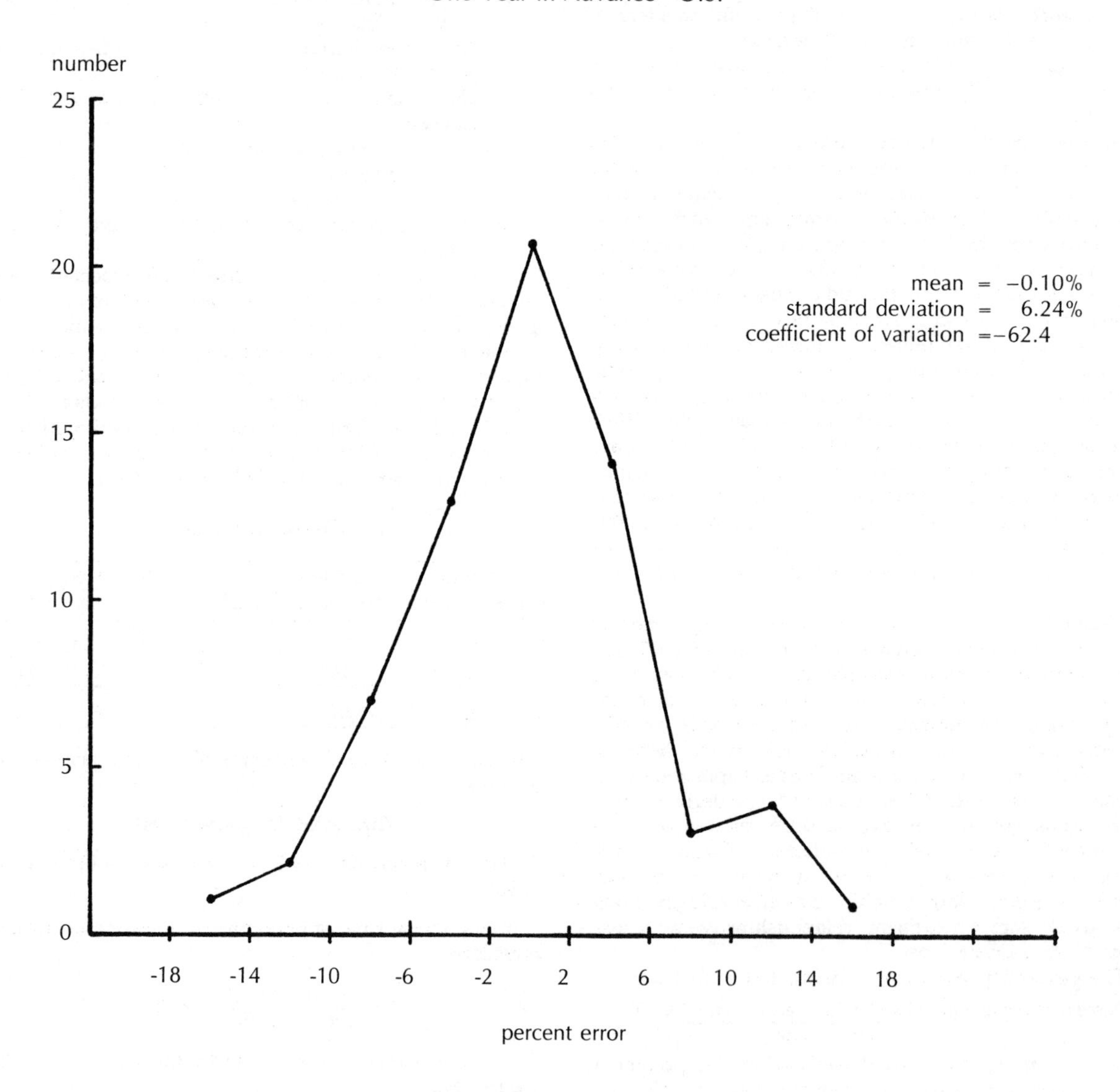

FIGURE 7

Distribution of Errors in Predicting 1940 Presidential Election by States, Based on Democratic Vote—U.S.

mean = −1.90
standard deviation = 2.23
cv = −1.17 = −117%

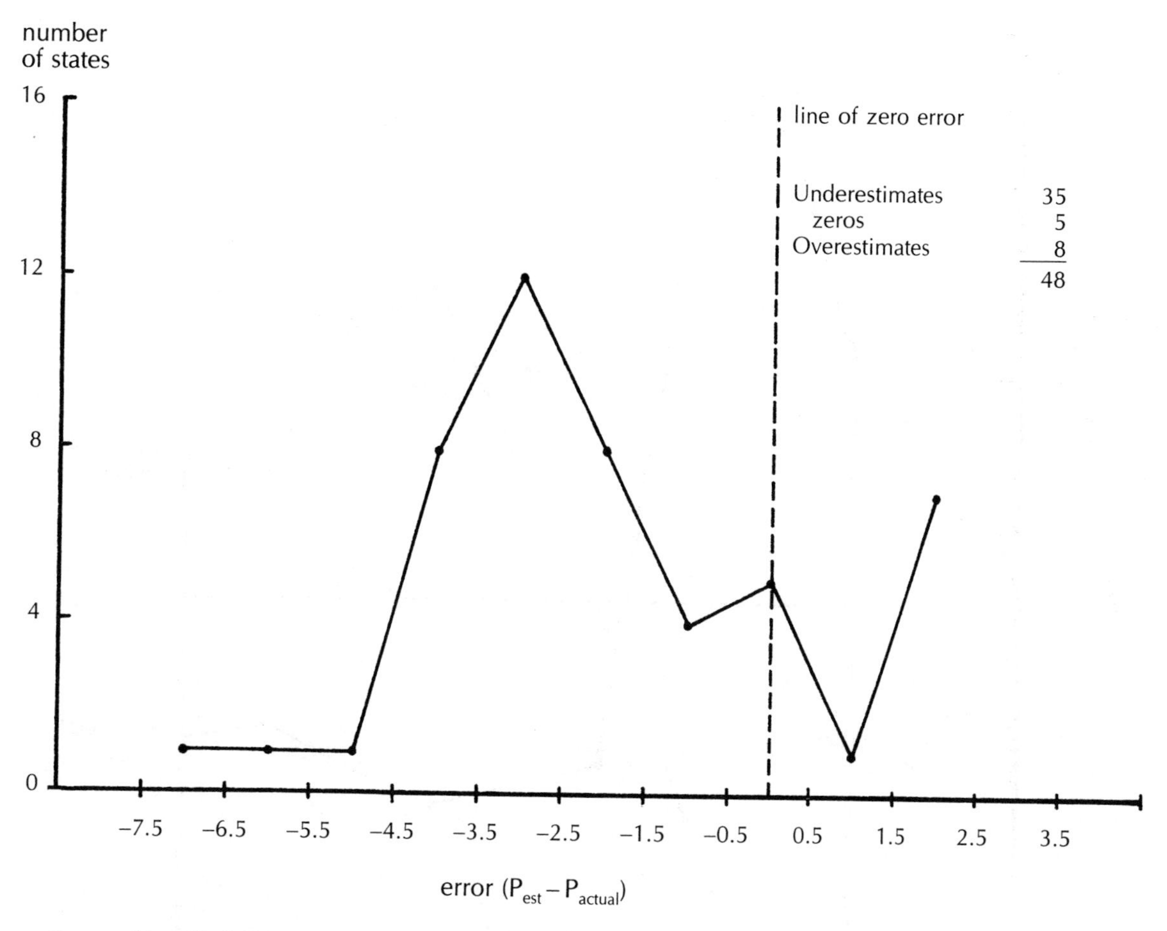

Source: New York Times November 9, 1940

the frequency distribution, nine intervals of 1,000 numbers each was used: 1-1,000; 1,001-2,000; 2,001-3000; ....8,001-9,000. The frequency was determined by how many numbers fell in each interval for each 1,000 numbers drawn in the order in which they were drawn. If the numbers were equally distributed over the 9,000 within each draw of 1,000, then the expected number or frequency in each interval of 1,000 is 1,000/9 or 111. This expected number based on the hypothesis of random order is shown in Figure 8 as a horizontal line at a frequency of 111; in other words, the expected frequency distribution of the serial numbers by number intervals is a rectangular distribution.

The capsules were prepared by clerks, placed in boxes, and then emptied into the bowl just prior to the drawing which was a dramatic public affair. The capsules were mixed with a paddle after the capsules were placed in the bowl which was filled to the top and tall relative to the diameter. This meant that the paddle-stirring had very little effect on mixing the capsules.

In the first 1,000 numbers drawn, only about 30 were numbers from 1-1,000, but about 150 were between 8,001 and 9,000. The last 1,000 numbers in the bowl reflected the reverse: 160 were between 1 and 1,000; about 75 were between 8,001 and 9,000. This was bound to happen since all numbers were selected. Under-representation at one

FIGURE 8

Plot Showing Nonrandom Order of Draft Numbers

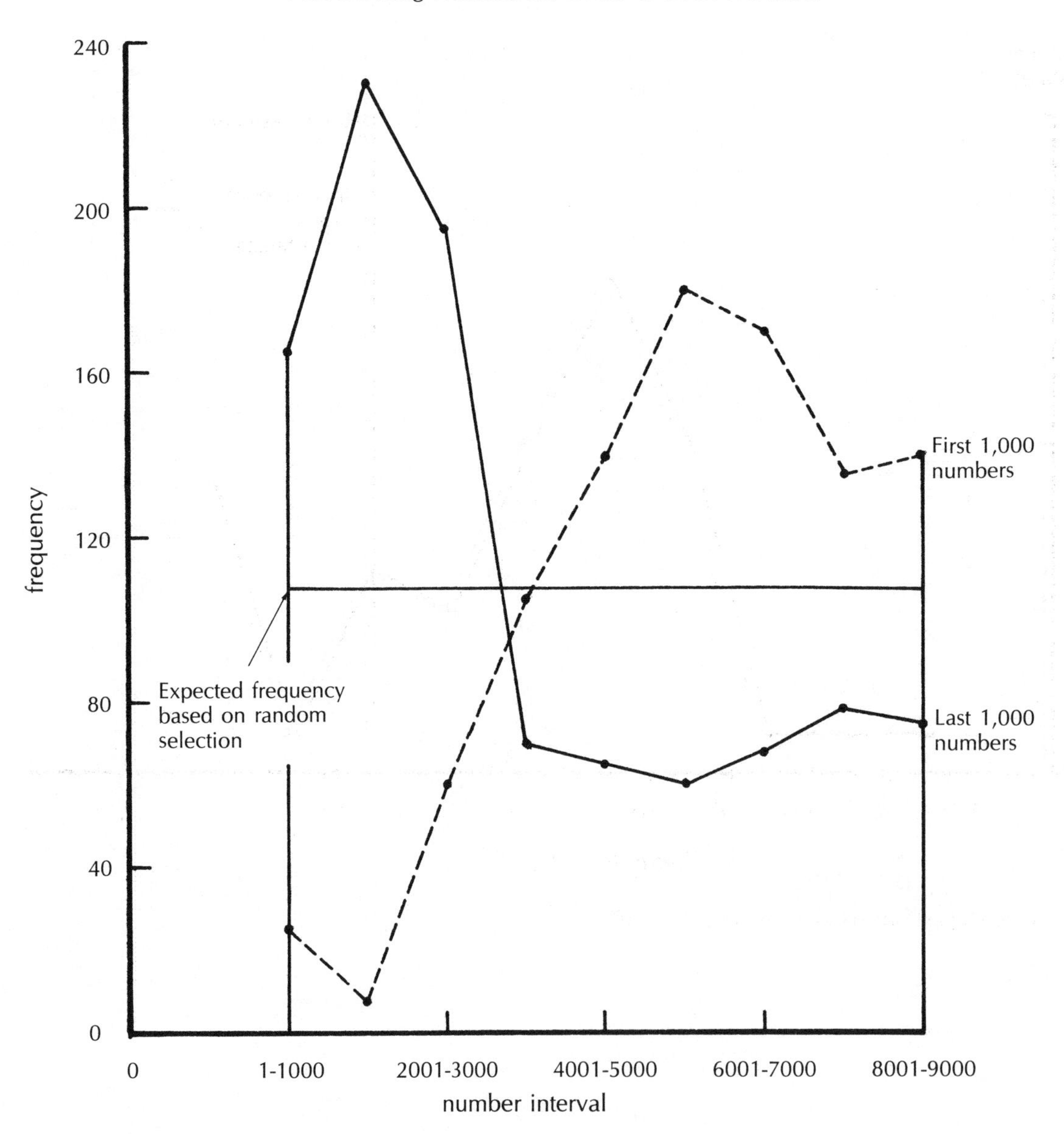

Frequency distributions of the first and last 1000 numbers drawn in the National Draft Lottery, October 29-30, 1940, United States

part of the draw meant over-representation at another point and Figure 8 clearly shows this.

Non-random order was due to the built-in stratification when the capsules were placed in the bowl. Neither stirring with a paddle nor dramatic selection by blind-folded officials corrected for this basic defect. The frequency distribution indicates that low numbers tended to be placed in the bottom of the bowl and high numbers tended to be placed at the top. There was some mixing, but clearly it was very inadequate. A much better, but less dramatic method, would have been to use a table of random numbers — four digits 0001 through 9000 rejecting duplicates. Two such tables were available at the time: Tippett's, and Fisher and Yates'.

Figure 9 shows the frequency distribution of 105 bank checks, by amount of the check, drawn over a period of seven months. It shows that while most of the checks, 50 percent, were for less than $50, the other half had a very wide range — actually from $50 to $3,433. The number over specified amounts was:

over $100	40
over 200	18
over 500	6

FIGURE 9

Distribution of 105 Bank Checks Written During 7 Months—1980

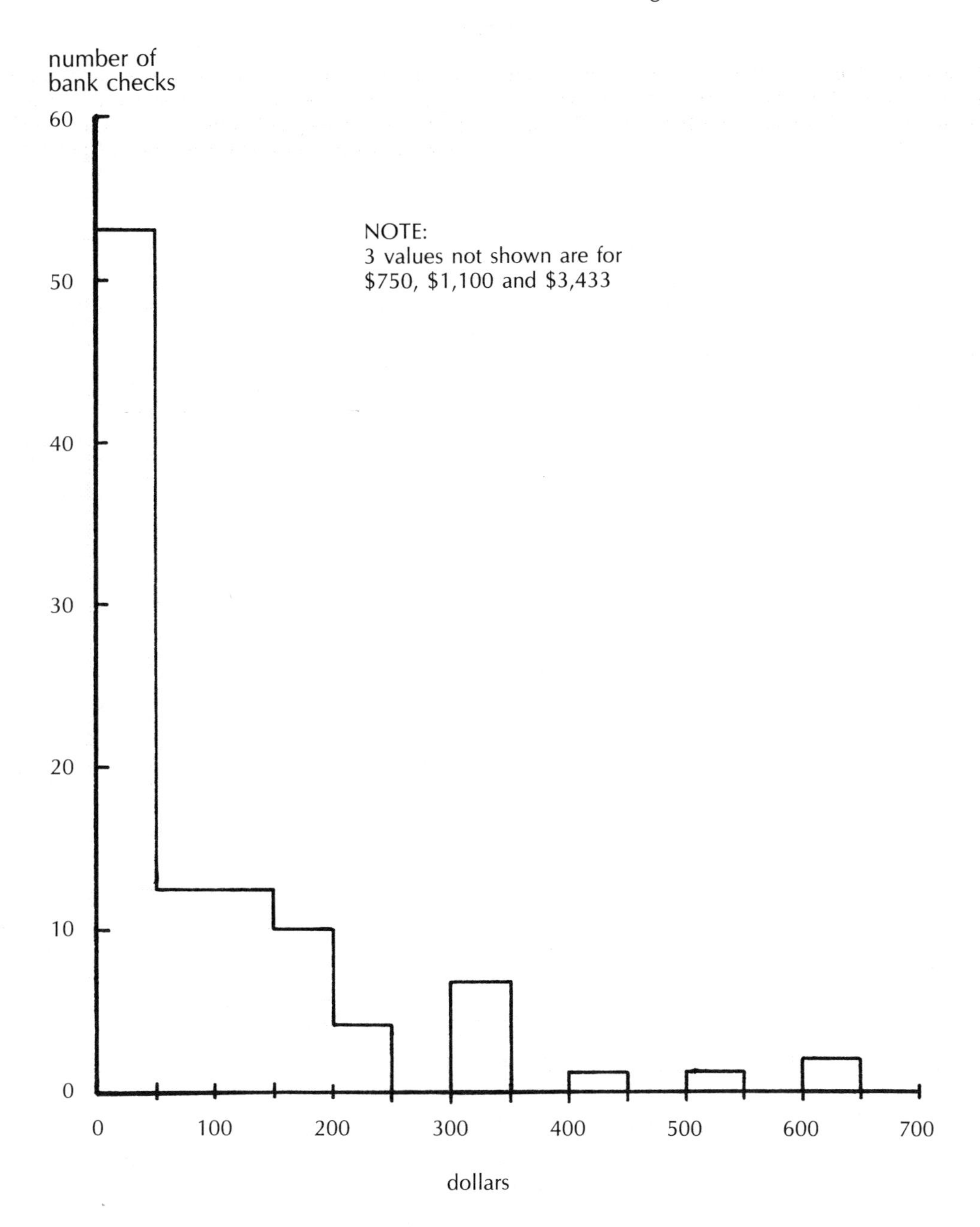

This shows that the bank check was used not only for relatively small amounts, but also for substantial amounts. The total amount of money involved in these 105 transactions was close to $15,000.

The bank check is the only real positive protection the customer or consumer has against error, because it provides a means of error control. Some examples of where cancelled checks have protected the customer from double billing, as well as downright error by omission or neglect, include the following:

1. Protection against an insurance company that insisted certain premiums had not been paid.

2. Protection against a book jobber that insisted that $33 worth of books had not been paid for.

3. Protection against a magazine that insisted that a gift subscription had not been paid.

4. Protection against the Internal Revenue Service and the Social Security Administration when the latter insisted, and their computer records showed, that social security taxes for the self-employed amounting to about $500 for 1972 had never been paid.

5. Protection against a book publisher that insisted that a book had not been paid for.

Notes

[1]A. C. Rosander, *Elementary Principles of Statistics,* Van Nostrand, New York, 1951. Copyright by Wadsworth, Inc. Used by permission of Brooks Cole Publishing Co., pp. 423-424.

[2]It is interesting and amusing to note that there were 8,994 capsules in the bowl; six numbers were missing at the end so they were assigned to the last six orders. It shows once more how difficult it is for persons to count accurately.

14

Theoretical Distributions

Theoretical distributions. Theoretical probability distributions are not only basic to statistical theory but form the scientific foundations on which the practice of statistics is built. These distributions are not just theory, but theory that a half century of practice has shown to be powerful and versatile, as well as successful in solving a wide variety of problems in the real-world.

The value of these theoretical distributions lies in their application to real-world situations, such as the following:

1. Use in interpreting and analyzing empirical distributions.
2. Use in designing a wide variety of probability samples.
3. Use in testing statistical hypotheses.
4. Use in making and appraising estimates from sample data.
5. Use in testing and interpreting differences and relationships.
6. Measuring uncertainty in estimates and inferences.

Five distributions are described. Three of these are based on counts or discrete variates: the *binomial, hypergeometric and the Poisson.* The other two are based on measurements which can take on, in theory at least, an infinite number of values: the *normal* and the *exponential.*

The binomial distribution. The binomial distribution is derived from counts or frequencies, not from measurements, and is characterized by the "heads or tails" situation or model. In this model the population under study is divided into two classes: A and non-A. Examples of this model or situation are: in error or not in error, defective or not defective, idle or not idle, working or not working, approve or not approve, operating or not operating, accepted or not accepted, fits or does not fit, cured or not cured, helped or not helped, like or dislike, favorable or not favorable, smokers or non-smokers, and voters or non-voters.

If an independent event can be classified into one of two mutually exclusive classes, so that it occurs either in class A or in class non-A, and it occurs with a constant probability P and does not occur with probability $Q = 1 - P$, then the probability of getting exactly x occurrences of the event in a random sample of n drawn from an infinite population, or with replacements from a finite population, is given by the *binomial distribution:*

$$Pr(x) = \binom{n}{x} P^x Q^{n-x}$$

where x = 0,1,2,3, . . . , n, and

$$\binom{n}{x} = \frac{n!}{x!(n-x)!}.$$

It can be shown that the expansion of the binomial in Q and P gives the probability of the individual terms *in the order* x = 0 through x = n; (note Q must be first):

$$(Q + P)^n = \sum_{0}^{n} \binom{n}{x} P^x Q^{n-x}$$

The following gives the value of x and the corresponding probability; note that the exponent of P equals x:

Value of x	Probability of occurrence
0	$Q^n \; P^0$
1	$nQ^{n-1} \; P^1$
2	$\frac{n(n-1)Q^{n-2} \; P^2}{2}$
3	$\frac{n(n-1)(n-2)Q^{n-3} \; P^3}{6}$
.	
.	
.	
x	$\frac{n!}{x!(n-x)!} Q^{n-x} \; P^x$
.	
.	
.	
n	P^n
Sum	1
1 or more	$1 - Q^n$

Example. Given P = 0.5 and n = 6, find the probability of occurrence for each value 0 through 6. The calculations are as follows; the distribution is plotted in Figure 1:

Value of x	Probability of occurrence	In 64ths	Proportion
0	$1 \times .5^6 \; .5^0$	1	.016
1	$6 \times .5^5 \; .5^1$	6	.094
2	$15 \times .5^4 \; .5^2$	15	.234
3	$20 \times .5^3 \; .5^3$	20	.312
4	$15 \times .5^2 \; .5^4$	15	.234
5	$6 \times .5^1 \; .5^5$	6	.094
6	$1 \times .5^0 \; .5^6$	1	.016
Sum		64	1.000

This distribution is symmetrical about the mean $nP=3$. Note that a^0 is always 1.

Example. Given $P=0.25$ and $n=6$, calculate the individual probabilities. (Figure 2.)

Value of x	Probability of occurrence	Proportion
0	$1\times.75^6\times.25^0$	.178
1	$6\times.75^5\times.25^1$	.356
2	$15\times.75^4\times.25^2$	.297
3	$20\times.75^3\times.25^3$	.132
4	$15\times.75^2\times.25^4$	.033
5	$6\times.75^1\times.25^5$	.004
6	$1\times.75^0\times.25^6$	.000
Sum		1.000

This is a skewed distribution with a mean $nP=1.5$. Algebraic expressions are given below:

Population basis	Mean	Variance	Standard deviation	Coefficient of variation
frequency	nP	nPQ	$(nPQ)^{1/2}$	$(Q/nP)^{1/2}$
proportion	P	PQ/n	$(PQ/n)^{1/2}$	$(Q/nP)^{1/2}$
Sample basis				
frequency	np	npq	$(npq)^{1/2}$	$(q/np)^{1/2}$
proportion	$p=x/n$	$pq/n-1$	$(pq/n-1)^{1/2}$	$(q/np-p)^{1/2}$

where $P+Q=1$ and $p+q=1$.

In practice P and Q, the population or frame values, are unknown so the estimates p and q derived from a sample are substituted for them. P and Q are used, however, as hypothetical or postulated values in testing estimates or hypotheses, e.g., P might be equated to ½ for testing. In sample designs, P might be equated to ½ because PQ is a maximum when $P=Q=1/2$.

If the sample size n is a substantial proportion of the frame or population N, then the variance of every estimate is reduced by multiplying by the factor $N-n/N = 1-f=1-n/N$, where f is the sampling rate n/N applied to the frame or population or some other rate set independently of N which may not be known.

In large samples, it is common practice in calculating the variance of a binomial proportion to divide by n instead of the more correct $n-1$, as given above.

Characteristics of the binomial frequency x:

1. Since $\sigma^2=npq$, the variance increases as n increases, p constant.
2. $(p+q)^n$, $x=0,1,2...n$, is the reverse of $(q+p)^n$ for the same n.
3. $(1/2+1/2)^n$ is symmetrical about the mean $np=1/2n$.
4. As p increases from ½ to near 1, the distribution becomes more and more highly skewed.
5. $(p+q)^n$ approaches the normal probability distribution for very large samples even though p is not ½, but 0.10 or 0.20.

Characteristics of the binomial proportion p:

1. Since $\sigma^2=pq/n$, the variance decreases as n increases, p constant.
2. σ^2 is the maximum when $p=q=1/2$ for constant n.
3. pq/n applies to a binomial proportion $p=x/n$; it does **not** apply to a ratio proportion, such as the following, where both x and y are random variables or variates:

$$p = R = \frac{\bar{x}}{\bar{y}} = \frac{\sum x}{\sum y}.$$

Assume that a random sample of size n is selected at rate f from an unknown frame or population and that n_a units fall in class A. Then an estimate of the total count or frequency in Class A is:

$$X_a=wn_a, \text{ where } w=1/f.$$

The variance of this estimate is:

$$s^2=w^2npq\ (1-f),$$

where $w=$ multiplier $1/f$, $1-f$ is the correction for sampling from a finite population, $p=n_a/n=x/n$, and $q=1-p$.

The hypergeometric distribution. If n sampling units are selected at random, *without replacement,* from a lot, frame, or population of N units of which N_a fall in a specified Class A, then the probability of getting exactly x units from Class A in the sample of n is given by the hypergeometric distribution.

Selecting a sample without replacement of sample units after each selection contrasts with binomial sampling where the lot, frame, or population size is held constant. If the sample size is large compared with the lot or frame or population, the characteristics may change appreciably due to the sampling process.

The hypergeometric distribution takes into consideration this depletion. Hence, it is applicable to sampling from lots of manufactured product, lots of clerical work, blocks of paper records, a large interval of time such as 480 minutes in a working day, or other similar situations where replacement of a sampling unit once drawn is impractical, difficult, costly, or impossible as is the case with a unit of time.

The format for determining hypergeometric probabilities is as follows where interest lies in Class A and all symbols refer to counts or frequencies:

Class	Lot, frame, population	Sample	Non-sample
A	N_a	x	N_a-x
non-A	$N-N_a$	$n-x$	$N-N_a-n+x$
Sum	N	n	$N-n$

The probability of getting exactly x units in a random sample of n, assuming Class A and Class non-A are independent in the probability sense, is:

$$Pr(x) = \frac{\binom{N_a}{x}\binom{N-N_a}{n-x}}{\binom{N}{n}}$$

Expanding this expression in terms of factorials gives a three-term product:

$$Pr(x) = \frac{N_a!}{x!(N_a-x)!}\ \frac{(N-N_a)!}{(n-x)!(N-N_a-n+x)!}\ \frac{n!(N-n)!}{N!}$$

All nine values in the format appear in the above equation. This is why the format should include the non-sample counts as well as the sample counts.

Interest lies in Pr(0), the probability of getting no units in Class A in the sample; in 1 − Pr(0), the probability of getting *at least one Class A unit* in the sample; or in using Pr(0) in a recursion formula to obtain the probabilities for x = 1,2,3, . . . n. Set x = 0 in the foregoing expression; then:

$$Pr(0) = \frac{(N - N_a!)(N - n)!}{(N - N_a - n)!N!}$$

If N_a is greater than n, start with $(N - N_a)!$. If the reverse is true, start with $(N - n)!$. [Pr(0) means the same as p_0 and we will use the latter hereafter.]

A number of methods can be used to approximate p_0: the median fraction method, the average binomial method, the regular binomial, and the Poisson. These methods are compared in an example below. The *median fraction method* is based on the fact that when x = 0, p_0 reduces to k diminishing terms in both the numerator and denominator. The value of p_0 is the quotient of the middle-most term raised to the kth power.

$$\text{exact } p_0 = \frac{v(v-1)(v-2)\ldots(v-k+1)}{y(y-1)(y-2)\ldots(y-k+1)}$$

$$\text{median fraction } p_{omf} = \frac{(v-a)^k}{(y-a)^k} = A^k,$$

where $a = \frac{k-1}{2}$, k can be n or N_a.

$$\text{average binomial } p_0 = \frac{q_{max} + q_{min}}{2},$$

where $q_{max} = \frac{v}{y}$ and $q_{min} = \frac{v-k+1}{y-k+1}$.

$$\text{regular binomial } p_0 = q^n = (1-p)^n.$$

Poisson $p_0 = e^m = e^{np}$, where m = np.

The average binomial method averages two probabilities, one that exists at the start of sampling and one that exists at the end of sampling. Like the median fraction method, this method adjusts for the changes that take place as a result of sampling.

Example. A lot of 1000 items has eight percent in error. What is the probability that a random sample selected from this lot contains no item in error (x = 0) if n = 20? n = 50? The format, the exact probability, and the four approximations are given below:

Class	Lot size	Sample of 20		Sample of 50	
		sample	non-sample	sample	non-sample
A (in error)	$80 = N_a$	0	80	0	80
non-A	920	20	900	50	870
	1000 = N	20	980	50	950

For a sample of 20: p = 0.08, q = 0.92

$$p_o = \frac{80!}{0!80!}\,\frac{920!}{20!900!}\,\frac{20!980!}{1000!} = \frac{920!980!}{900!1000!}$$

For a sample of 50:

$$p_o = \frac{80!}{0!80!}\,\frac{920!}{50!870!}\,\frac{50!950!}{1000!} = \frac{920!950!}{870!1000!}$$

1. Exact:

$$p_o = \frac{920x919x\ldots x901}{1000x999x\ldots x981} = 0.185559 \qquad p_o = \frac{920x919x\ldots x871}{1000x999x\ldots x951} = 0.013851$$

2. MF method:

$$p_o = \frac{(920\text{-}9.5)^{20}}{(1000\text{-}9.5)^{20}} = 0.185571 \qquad p_o = \frac{(920\text{-}24.5)^{50}}{(1000\text{-}24.5)^{50}} = 0.013865$$

3. Average binomial:

$$p_o = \frac{.92^{20} + .918451^{20}}{2} = 0.185567 \qquad p_o = \frac{.92^{50} + .915878^{50}}{2} = 0.013911$$

4. Regular binomial:

$$p_o = .92^{20} = 0.188693 \qquad p_o = 0.92^{50} = 0.015466$$

5. Poisson:

m = np = 20 x 0.08 = 1.6 $\qquad$ m = np = 50 x 0.08 = 4.0

$$p_o = e^{-1.6} = 0.201897 \qquad p_o = e^{-4.0} = 0.018316$$

The MF method and the average binomial give excellent approximations to the exact value. The regular binomial and the Poisson give too large a value, especially the Poisson. Once p_o is calculated, the other probabilities for x = 1,2,3, . . . , n can be easily found by using the recursion formula. This problem shows that even with an eight percent error rate, a random sample of 50 is necessary to detect at least one error with a probability of about 0.98.

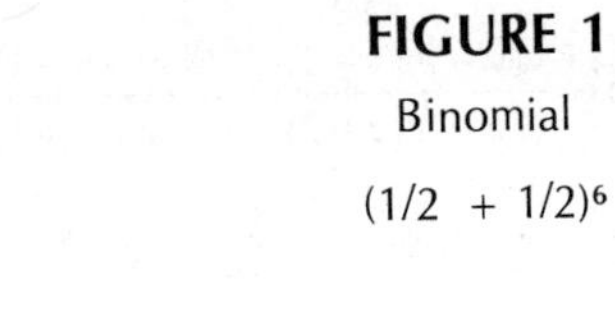

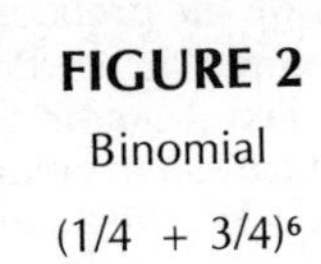

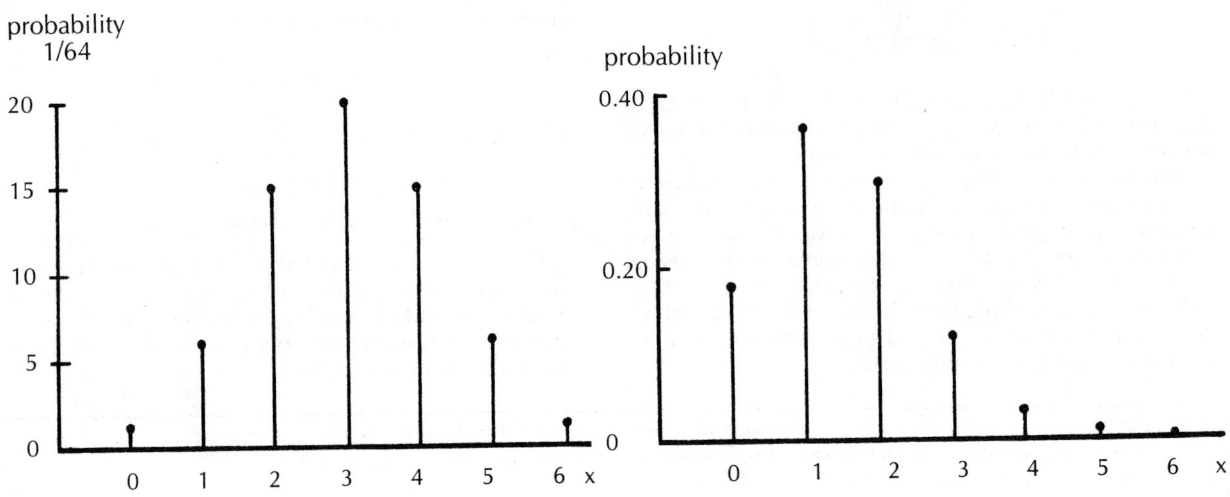

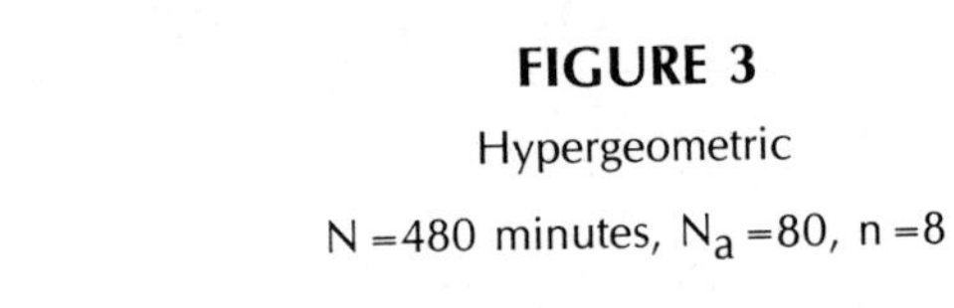

FIGURE 4

Poisson

• m = 0.8
o m = 3

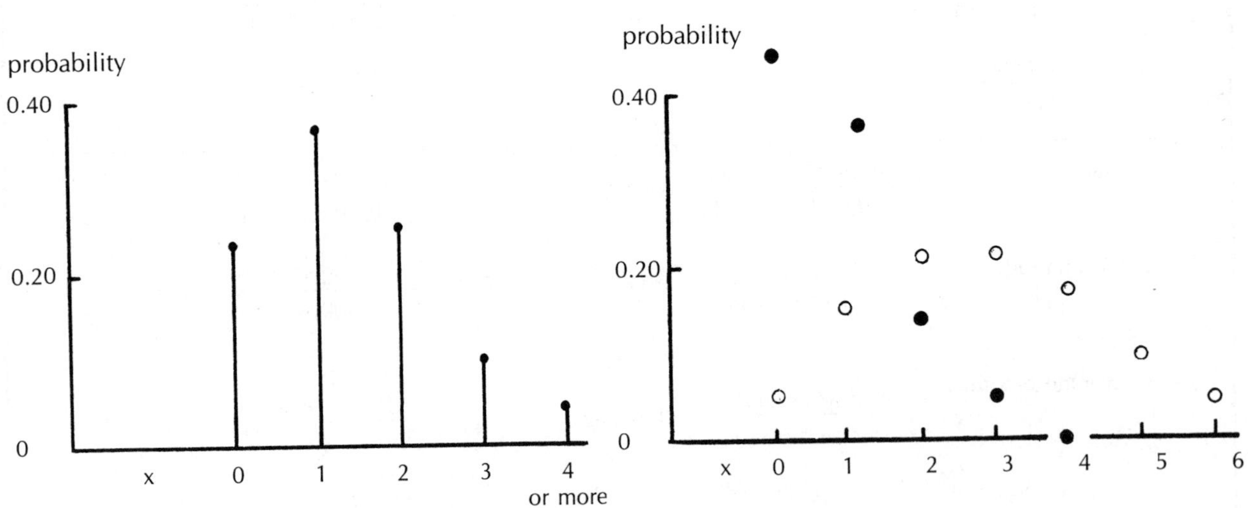

Once p_o is known, p_1 can be calculated; once p_1 is known, p_2 can be calculated; and so on until all probabilities are calculated. In terms of p_o, the value of p_1 is as follows:

$$p_1 = p_0 \frac{nN_a}{1(N - N_a - n + 1)}$$

The two rules to apply successively to calculate the next value of p are:

1. Subtract 1 from n and subtract 1 from N_a.
2. Add x to $N - N_a - n$ and multiply this value by x.

Hence, for x=2 and x=3, the values for p_2 and p_3 are respectively:

$$p_2 = p_1 \frac{(n - 1)(N_a - 1)}{2(N - N_a - n + 2)},$$

$$p_3 = p_2 \frac{(n - 2)(N_a - 2)}{3(N - N_a - n + 3)},$$

Example: Due to poor work scheduling, workers in a graphic arts unit are idle on the average of 10 minutes every hour. Find the probability distribution of $x = 0,1,2...,8$ if $n = 8$ random minutes during a day of 480 minutes.

The hypergeometric distribution format is:

Class	Population	Sample	Non-sample
(idle) A	80	x	80 − x
non-A	400	8 − x	392 + x
Total	480	8	472

$$\text{Exact } p_0 = \frac{400 \times 399 \times \ldots \times 393}{480 \times 479 \times \ldots \times 473} = 0.22984$$

Average binomial gives the same value:

$$p_o = \left(\frac{400^8}{480} + \frac{393^8}{473}\right) / 2 = 0.22984$$

Therefore,

$$p_1 = .22984 \frac{8 \times 80}{393} = .37429$$

$$p_2 = .37429 \frac{7 \times 79}{2 \times 394} = 0.26267.$$

The other values of p are calculated in the same way.

The distribution is:

x	Probability
0	0.230
1	0.374
2	0.263
3	0.104
4 or more	0.029
	1.000

Comments
The probability of not detecting this situation in eight random minutes is 0.23. The probability of detection $1 - 0.23 = 0.77$. In two days, the probability is $1 - .23^2 = 0.95$. In three days, the probability is $1 - .23^3 = 0.99$.

This distribution is graphed in Figure 3.

The basic characteristics of the hypergeometric distribution are:

Population frame:

mean: $nP = n\frac{N_a}{N}$

variance: $nPQ\frac{(N-n)}{N-1}$

standard deviation: $\sqrt{\frac{nPQ(N-n)}{N-1}}$

coefficient of variation: $\sqrt{\frac{Q(N-n)}{nP(N-1)}}$

where P + Q = 1 and n/N is the sampling rate.

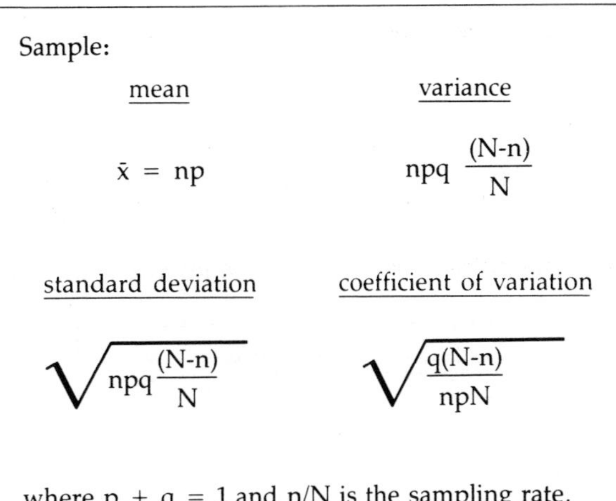

Sample:

mean: $\bar{x} = np$

variance: $npq\frac{(N-n)}{N}$

standard deviation: $\sqrt{npq\frac{(N-n)}{N}}$

coefficient of variation: $\sqrt{\frac{q(N-n)}{npN}}$

where p + q = 1 and n/N is the sampling rate.

Since $\frac{N-n}{N-1} = \frac{1-n/N}{1-1/N}$, this expression approaches one when the value of N increases indefinitely, so that the population variance approaches nPQ (the same as the variance of the binomial distribution).

Example. In the foregoing problem, the population values are:

$$N = 480, N_a = 80, n = 8, P = 1/6, Q = 5/6.$$

Therefore:

1. Mean: $\frac{8 \times 80}{480} = 1.33.$

2. Variance $= 8 \times 1/6 \times 5/6 \left(\frac{480-8}{480-1}\right) = 0.0913.$

3. Standard deviation = 0.30.

4. Coefficient of variation $= 0.23 = \frac{0.30}{1.33}.$

The Poisson distribution. The Poisson distribution can be derived in two different ways: one as a limiting form of the binomial distribution and the other as the distribution of counts in unit elements where the probability of occurrence is constant. In both cases, it is assumed that the events are independent and distributed with constant probability which is relatively small.

The probability of getting exactly x occurrences when the Poisson conditions are met is:

$$Pr(x) = e^{-m} \frac{m^x}{x!} \quad \text{where } x = 0,1,2,3,\ldots \text{infinity and } m = \text{the mean} = np.$$

Theoretically, the distribution has to be summed to infinity in order for the total probability to be one, but in practice a relatively small number of values of x usually accounts for practically all of the probability.

$$\sum_0^{\infty} Pr(x) = e^{-m} \sum_0^{\infty} \frac{m^x}{x!} = e^{-m} e^m = 1.$$

The individual probabilities for values of x are calculated as follows:

$$\Sigma Pr(x) = e^{-m} \left(\frac{m^0}{0!} + \frac{m^1}{1!} + \frac{m^2}{2!} + \ldots + \frac{m^x}{x!} + \ldots \right)$$

The probabilities for the various values of x are:

x	Probability
0	e^{-m}
1	me^{-m}
2	$\frac{m^2}{2} e^{-m}$
3	$\frac{m^3}{6} e^{-m}$
.	.
.	.
.	.
k	$\frac{m^k}{k!} e^{-m}$

The probability of getting a value of x *less than k* is the sum of all terms up to and including $x = k - 1$:

$$Pr(x<k) = \sum_0^{k-1} e^{-m} \frac{m^x}{x!}.$$

Figure 4 shows two Poisson distributions; one with a mean of 0.8, the other with a mean of 3. The two equations are:

$$y = e^{-0.8} \frac{(0.8)^x}{x!} \text{ and } y = e^{-3} \frac{3^x}{x!}.$$

The Poisson distribution is determined by one constant, the mean m, since the variance equals the mean and the standard deviation equals $m^{1/2}$.

The *population values* are:

$$\text{mean} = u = nP$$

$$\text{variance } \sigma^2 = u$$

$$\text{standard deviation } \sigma = \sqrt{u}$$

$$\text{coefficient of variation CV} = \sigma/u = \frac{1}{\sqrt{u}}$$

The *sample values* are:

$$\text{mean} = m = np$$

$$\text{variance} = m$$

$$\text{standard deviation} = \sqrt{m}$$

$$\text{coefficient of variation cv} = s/m = \frac{1}{\sqrt{m}}$$

For a binomial frequency, $\sigma^2 = npq$, but this means that as $q \rightarrow 1$, $\sigma^2 \rightarrow np$ which is the mean for the Poisson. For small p's the Poisson distribution is used to approximate the binomial. One of the major reasons for this is that the Poisson is much simpler and easier to calculate.

If the mean $m = kt$, where t measures time, length, area, volume, or other appropriate characteristics and k is the number of occurrences of the event per unit of time, length, area, volume, etc., then the Poisson distribution becomes:

$$Pr(x) = e^{-kt} \frac{(kt)^x}{x!}$$

where, as before, $x = 0,1,2,3, \ldots,$ infinity.

The probability of $x = 0$ is $Pr(0) = e^{-m}$. Hence, the probability of getting at least one event is:

$$Pr(x > 0) = 1 - e^{-m}.$$

These correspond to the binomial expressions:

$$Pr(0) = q^n,$$
$$Pr(x > 0) = 1 - q^n.$$

Testing two Poisson counts using normal approximation. Under certain conditions the square root of a Poisson variate tends to be normally distributed with a variance of 0.25. We consider the difference between two such independent variates x_1 and x_2.[1]

Sample count	$\sqrt{x}$	Variance
x_1	$\sqrt{x_1}$	1/4
x_2	$\sqrt{x_2}$	1/4
Difference $x_1 - x_2$	$\sqrt{x_1} - \sqrt{x_2}$	1/2

The normal deviate for testing the difference, $d = x_1 - x_2$, is:

$$z = \left(\sqrt{x_1} - \sqrt{x_2}\right)\sqrt{2}$$

For any assigned value z, x_2 can be expressed in terms of x_1, and vice versa. The equations show 1) how much x_1 must be reduced to obtain a significant test at the two and three sigma levels and 2) how much x_2 and x_3 have to be increased to obtain a significant test at the same levels.

Two sigma level (95 percent level; z = 1.96)

$x_2 = x_1 - 2\sqrt{2x_1} + 2$ where x_1 is given and x_2 is the reduced frequency.

$d_2 = \sqrt{2x_1} + \sqrt{2x_2}$ minimum difference between x_1 and x_2 for significance at two sigma level.

The expression for d_2 is derived by setting the equation for z at 95 percent equal to two (actually z = 1.96). This gives:

$$z = \sqrt{2x_1} - \sqrt{2x_2} = 2.$$

Multiply both sides of the equation by $\sqrt{2x_1} + \sqrt{2x_2}$ and divide out the 2. This gives:

$$d_2 = x_1 - x_2 = \sqrt{2x_1} + \sqrt{2x_2}.$$

This equation can be used to test two values x_1 and x_2 which are not given in Table 1.

Three sigma level (99.73 percent level; z = 3)

$x_3 = x_1 - 3\sqrt{2x_1} + 4.5$ where x_1 is given and x_3 is the reduced frequency.

Also $x_3 = x_2 - \sqrt{2x_1} + 2.5$ three sigma level in terms of the two sigma level.

$d_3 = \frac{3}{2}(\sqrt{2x_1} + \sqrt{2x_3})$ minimum difference between x_1 and x_3 for significance at the three sigma level

The expression for d_3 is derived by setting the equation for z equal to three corresponding to three sigma or 99.73 percent.

$$z = \sqrt{2x_1} - \sqrt{2x_3} = 3.$$

Multiply both sides of the equation by $\sqrt{2x_1} + \sqrt{2x_3}$ and divide out the 3. This gives the value for $d_3 = x_1 - x_3$ as shown above.

In the absence of a table or chart, the equations for d_2 and d_3 can be used to make a quick calculation to determine if two given frequencies are significantly different at the two or three sigma levels. By using these same equations, Table 1 can easily be expanded to include a finer scale of values of x_1 and the corresponding values of x_2 and x_3 if practice warrants such a step.

Using a chart to test the differences. Figure 5 shows three lines for testing the significance of the difference between two Poisson counts. The calculations on which these charts are based are given in Table 1. On the chart the x axis shows the x_1 count, the y axis shows the x_2 and x_3 counts; x_2 is the count at the two sigma level and x_3 is the count at the three sigma level. The upper line is $x_3 = x_2 = x_1$. The middle line shows the *reduction* that must take place in the count to be significant at the two sigma level. The lower line shows the *reduction* that must take place to be significant at the three sigma level.

These lines can also be used in reverse to test for a significant *increase* in the lesser count *reading the initial*
The first difference is not significant at the 95 percent level; the second one is.
value on the two sigma or three sigma lines and testing the increased valued on the upper line $x_1 = x_2 = x_3$.

Example: If $x_2 = 80$ and increases to 100, is this a significant change at the 95 percent level? No, since the chart shows that 80 on the two sigma line (read 80 horizontally) must increase to 108 to be significant at this level. Actual calculations check this.

$100^{1/2}$	= 10.00		$108^{1/2}$	= 10.39
$80^{1/2}$	= 8.94		$80^{1/2}$	= 8.94
Difference	1.06			1.45
$z = 1.06(2)^{1/2}$	= 1.50		$z = 1.45(2)^{1/2}$	= 2.05.

The first difference is not significant at the 95 percent level; the second one is.

FIGURE 5

Testing Difference of Two Poisson Counts at 2 and 3 Sigma Levels

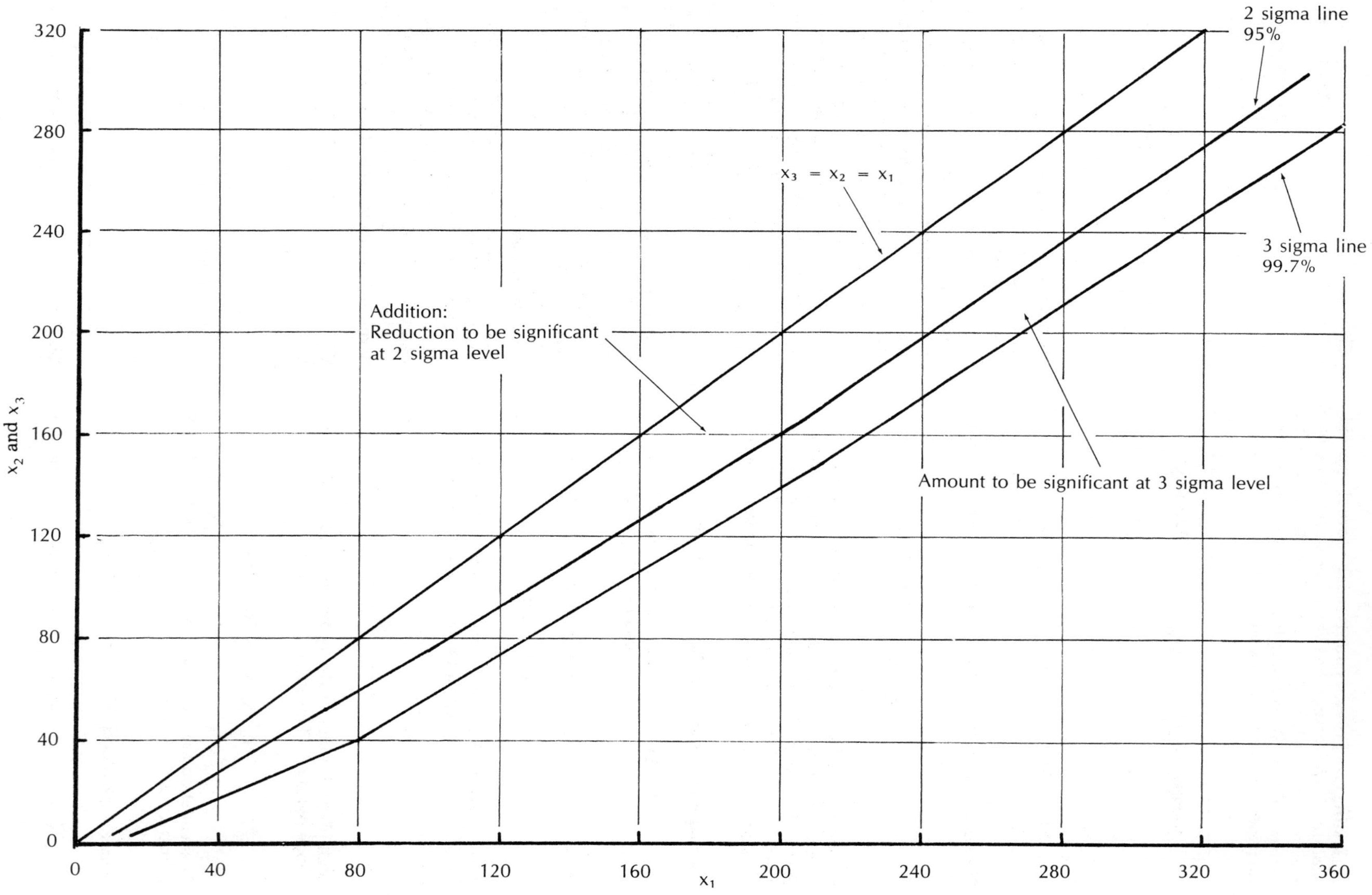

Table 1

Poisson Counts at Two Sigma Level (x_2) and Three Sigma Level (x_3) to Test Deviation from x_1

x_1	x_2	x_3	Two sigma difference $x_1 - x_2$	Three sigma difference $x_1 - x_3$
8	2	0.5	6	7.5
18	8	4.5	10	13.5
32	18	12.5	14	19.5
50	32	24.5	18	25.5
72	50	40.5	22	31.5
98	72	60.5	26	37.5
128	98	84.5	30	43.5
162	128	112.5	34	49.5
200	162	144.5	38	55.5
242	200	180.5	42	61.5
288	242	220.5	46	67.5
338	288	264.5	50	73.5
392	338	312.5	54	79.5
450	392	364.5	58	85.5
.	.	.	.	.
.	.	.	.	.
.	.	.	.	.

Note: To extend this table $x_{1i} = x_{2(i+1)}$; $(x_1 - x_2)$ increases by four and $(x_1 - x_3)$ increases by six.

The normal probability distribution or normal curve is a very important distribution for several reasons:

- It is the basis of control charts such as the $\bar{x}$ and p charts.
- It is the basis for sample designs for estimation of means, aggregates, proportions, differences, ratios.
- Test statistics such as t and F assume x is normally distributed.
- It approximates the binomial, hypergeometric, and Poisson distributions.
- A transformation is used to normalize an empirical distribution so the properties of the normal distribution can be applied to it.

It takes three forms where u = mean, σ^2 = variance, n = sample size, N = population or frame size, s^2 = variance from sample, w = data interval width, and $z = \frac{(x-u)}{\sigma}$.

1. Standard form where $u = 0$ and $\sigma = 1$ in $z = (x - u)/\sigma$:

$$y = \frac{1}{\sqrt{2\pi}} e^{-\frac{1}{2}z^2}.$$

2. Mean = u and variance = σ^2; if $u = 0$ and $\sigma = 1$ this reduces to the above:

$$y = \frac{1}{\sigma\sqrt{2\pi}} e^{-\frac{1}{2}\frac{(x-u)^2}{\sigma^2}}.$$

3. Fitting a normal distribution to an empirical distribution:

$$y = \frac{nw}{s\sqrt{2\pi}} e^{-\frac{1}{2}\frac{(x-\bar{x})^2}{s^2}}.$$

Figure 6 shows the normal distribution or normal curve in standard form; the characteristics of this curve are those given in the table for the normal distribution. It has the following very important properties:

1. The mean $u = 0$.
2. The variance σ^2 and the standard deviation $\sigma = 1$.
3. The curve is symmetrical about a vertical line through the mean.
4. This symmetry means that only half of the distribution needs to be tabulated.
5. The area under the curve is the probability; the total area is 1.
6. Very little of the distribution is outside the limits 0 ± 3; only 0.0027.

The normal probability distribution table shows:

1. The normal variate or deviate z,
2. The ordinate y for each value of z,
3. The area (probability) between mean 0 and $z = z_1$ (area A), or
4. The cumulative probability to $z = z_1$, which is $\frac{1}{2} + A$.

The probability that z is greater than some value z_1 is $\frac{1}{2} - A$. The probability that z is less than some value z_1 is $\frac{1}{2} + A$. To determine the probability of a value of z exceeding, or being less than, some value z_1 calculate:

$$z_1 = \frac{x - \bar{x}}{s}$$

and refer z_1 to the normal table, determining probability A. Then use $\frac{1}{2} - A$ or $\frac{1}{2} + A$. These characteristics are shown in the following graph.

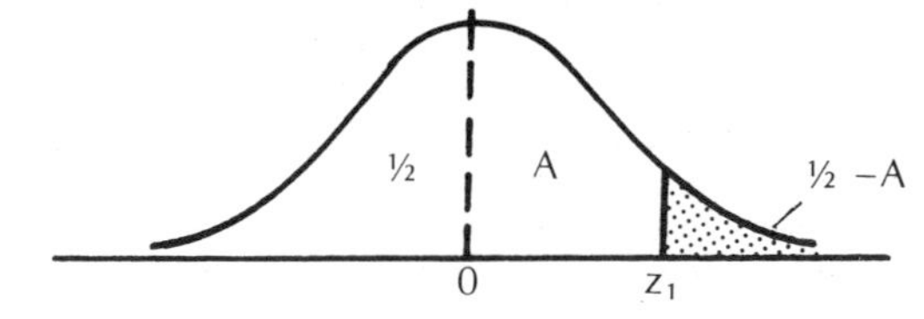

The probability that $z > z_1 = \frac{1}{2} - A$; the probability that $z < z_1 = \frac{1}{2} + A$. If z_1 is negative these relationships are reversed. The most common values of the normal deviate z and the probabilities connected with them are as follows:

Normal deviate z	Probability exceeding z	Area outside $0 \pm z$	Area inside $0 \pm z$
1.28	0.100	0.200	0.80
1.645	0.050	0.100	0.90
1.96	0.025	0.050	0.95
2.33	0.010	0.020	0.98
2.58	0.005	0.010	0.99
3.00	0.00135	0.0027	0.9973
3.09	0.001	0.002	0.998

FIGURE 6

Standardized Normal Distribution $y = \frac{1}{\sqrt{2\pi}} e^{-\frac{1}{2}z^2}$

mean u = 0 variance σ^2 = 1
standard deviation σ = 1
area = 1

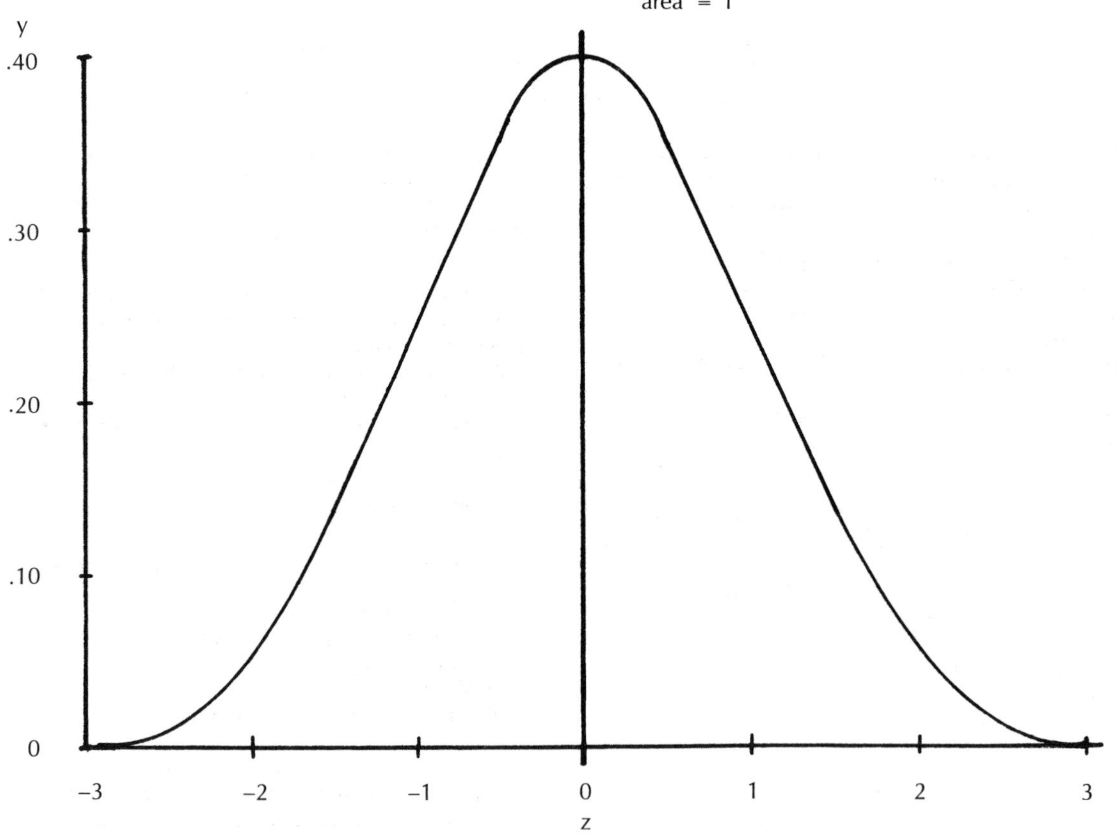

Normal probability paper. A graph paper has been constructed which can be used to test the normality of a frequency distribution very quickly by eye, or by algebraic equation. If the frequency distribution is normal, the data will plot on a straight line on the normal probability graph paper.[3]

Figure 7 is such a plot of the heights of 99,712 male white World War II draft registrants.[4] An equal division y scale is constructed for heights, while cumulative proportions are measured along the x axis. The line is quite linear from about 61.5 inches to 76.5 inches. It departs from normality at both tails of the distribution, but especially at the shorter heights where the frequencies are too high. This excess at the tails was observed in the chi-square test described above. This excess includes about 1 percent of the distribution (1056 measurements) at the lower end and about 0.2 percent of the distribution (206 measurements) at the upper end. The departure from normality is created by about 1.2 percent of the distribution (1262 measurements).

The value of chi-square for this distribution, divided into 13 equal intervals of x giving 12 degrees of freedom, showed a significant departure from the normal curve. This test shows what the graph shows: that the fit is poor at both tails of the distribution, but that the fit is very good otherwise.

FIGURE 7

Testing Male Height Distribution for Normality.

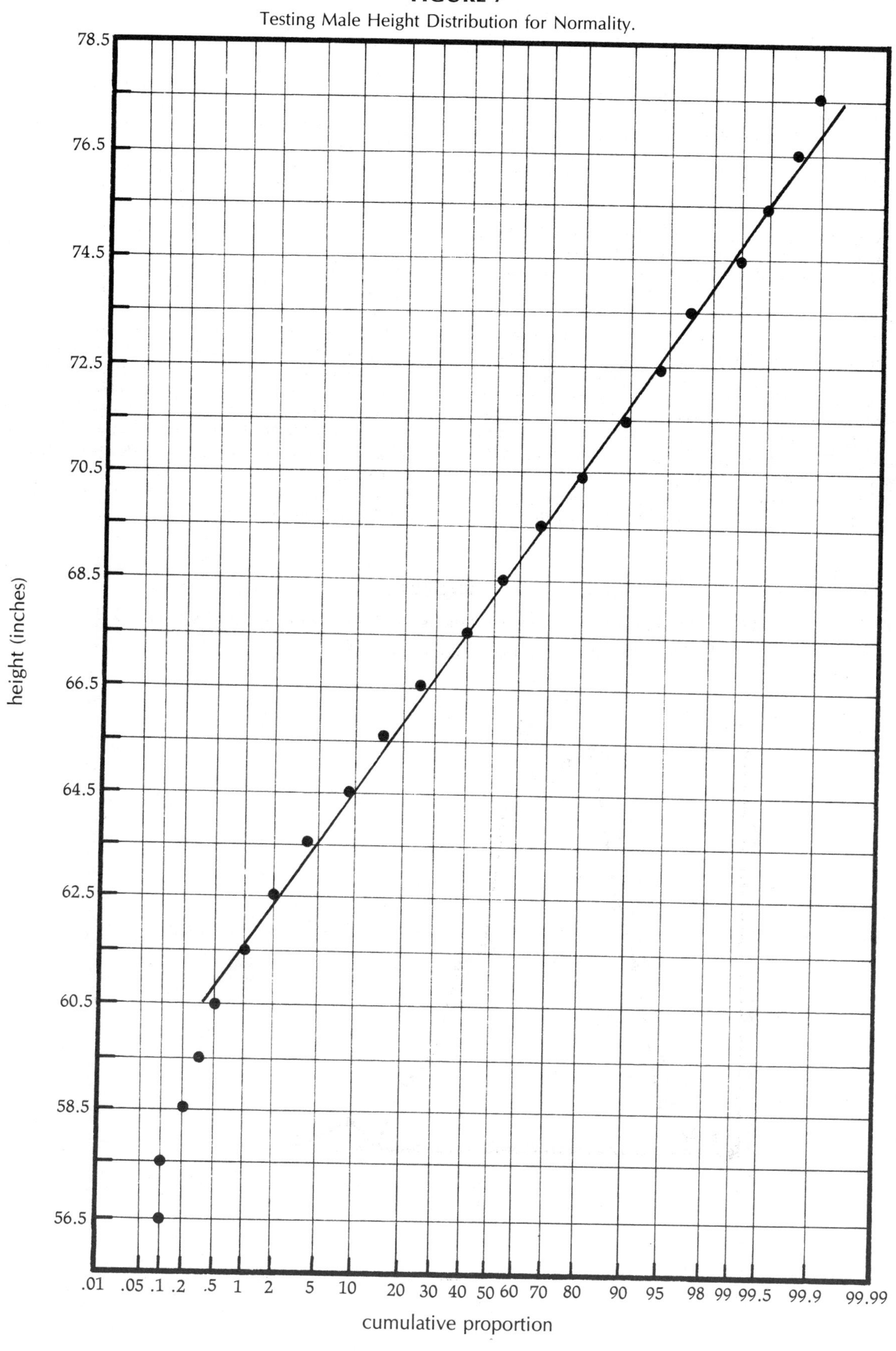

This situation raises a serious question: What does one conclude from a test of this kind? Do a few extreme values invalidate the assumption of a normal distribution? In any problem where these extreme values are of no importance in practice, one would be justified in assuming that the distribution is normal. If interest centers on the extreme values of the distribution, then this fact must be taken into consideration. In some problems, as in the distribution of delay times in delivering freight cars to shippers, *the extreme upper end of the distribution is the problem.*

Sampling distributions and the central limit theorem. One of the most important theorems for use in statistical and quality control practice is the central limit theorem. It states, in essence, that as the size of the random sample of x increases indefinitely, the distribution of the *estimates* derived therefrom tends to be distributed as the normal probability distribution regardless of how the variate x is distributed. Distributions in Figures 8, 9, and 10 are empirical evidence that this theorem actually applies in practice. The frequency distributions of estimates for the mean, standard deviation, and coefficient of variation are single peaked even though they are derived from a rectangular distribution. The mean is distributed likewise, even though x is a discrete triangular distribution. This theorem is the reason why Shewhart devised an $\bar{x}$ chart and not an x chart. With an $\bar{x}$ chart one can assume a normal distribution and set control limits at $\bar{x} \pm s_{\bar{x}}$.

This theorem means that in designing a sample to estimate a population or frame parameter, the characteristics of the normal curve can be used to set errors and risks. These errors and risks, together with the variance and coefficient of variation of x, can be used to determine the size of the sample.

If the distribution of x is highly skewed and covers a wide range of values, then the size of the sample would have to be extremely large to obtain normality. In this case, the extreme values are isolated and included 100 percent. The rest of the population or frame is divided into enough strata so that the variance is reduced to an acceptable value. Stratified sampling is described in the next chapter.

Even without stratification, the standard deviation and the skewness of the distribution of an estimate, say $\bar{x}$, varies inversely as the square root of the sample size. If the skewness of a variate x is 1, then a sample of 100 will reduce the skewness of the distribution of $\bar{x}$ to 0.1, a skewness that is hardly noticeable. To the extent that this skewness is due to assignable causes, effective quality control will eliminate it so that the distribution of the mean $\bar{x}$, for example, will tend to be distributed symmetrically as the normal curve.

FIGURE 8

Sampling Distribution of a Binomial Proportion P = 0.15 and Average Number = 7.5

200 samples of 50 each

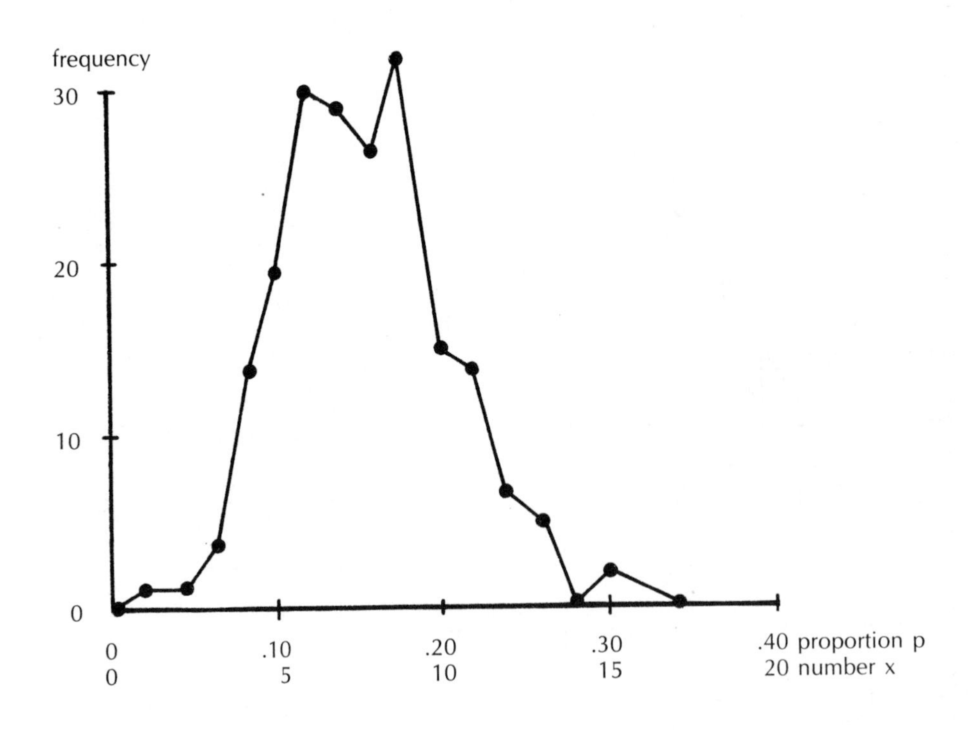

FIGURE 9

Sampling Distribution from 60 Samples of 10 each Selected from a Rectangular Distribution

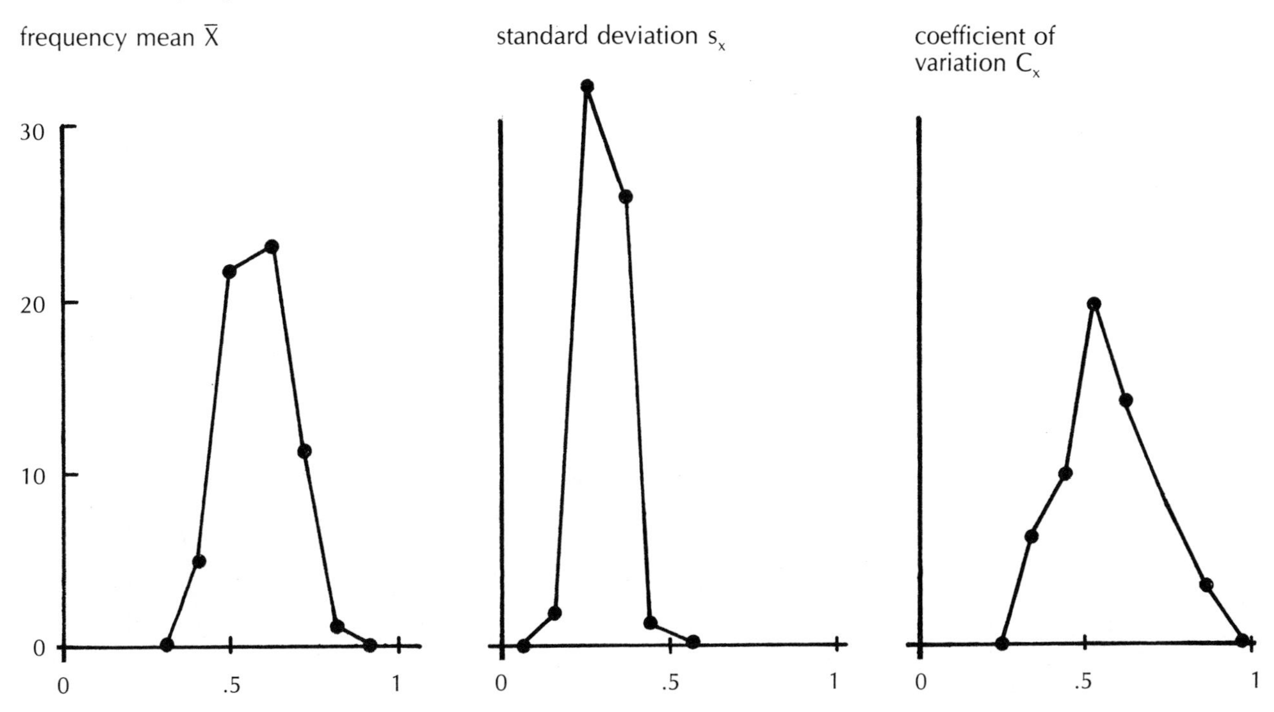

FIGURE 10

Sampling Distribution of 100 Means $\overline{x}$ from Samples of 10 from Triangular Distribution of x

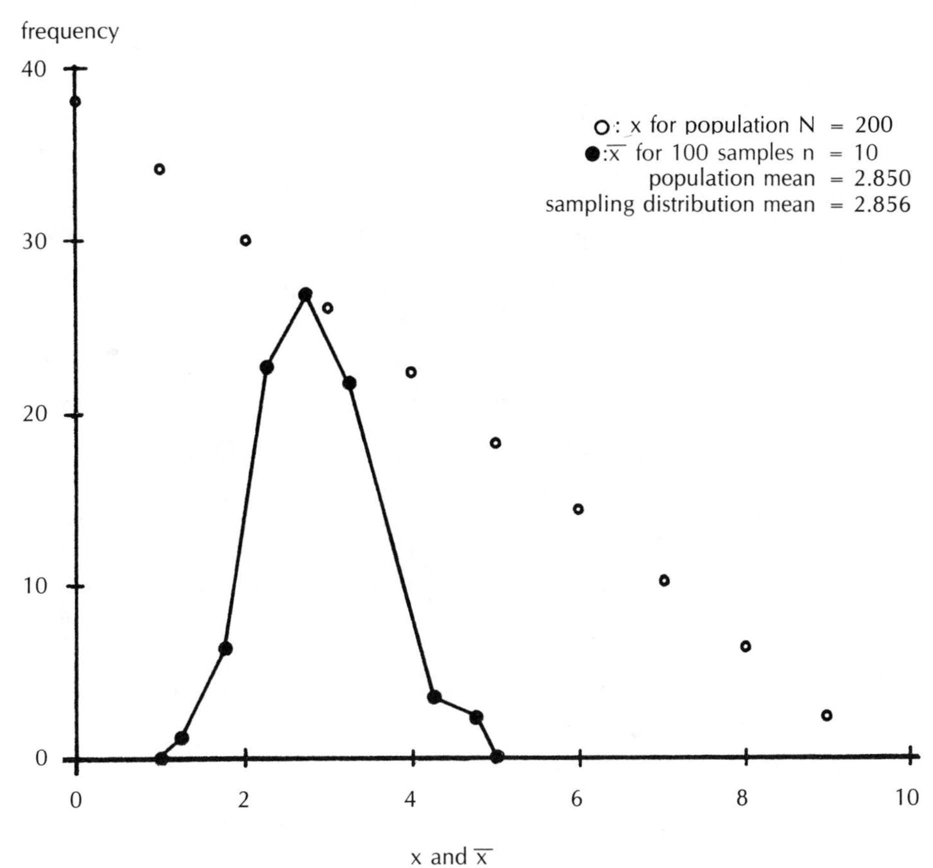

FIGURE 11
Reduction in Errors in Batches of 500 Documents over 102 Days

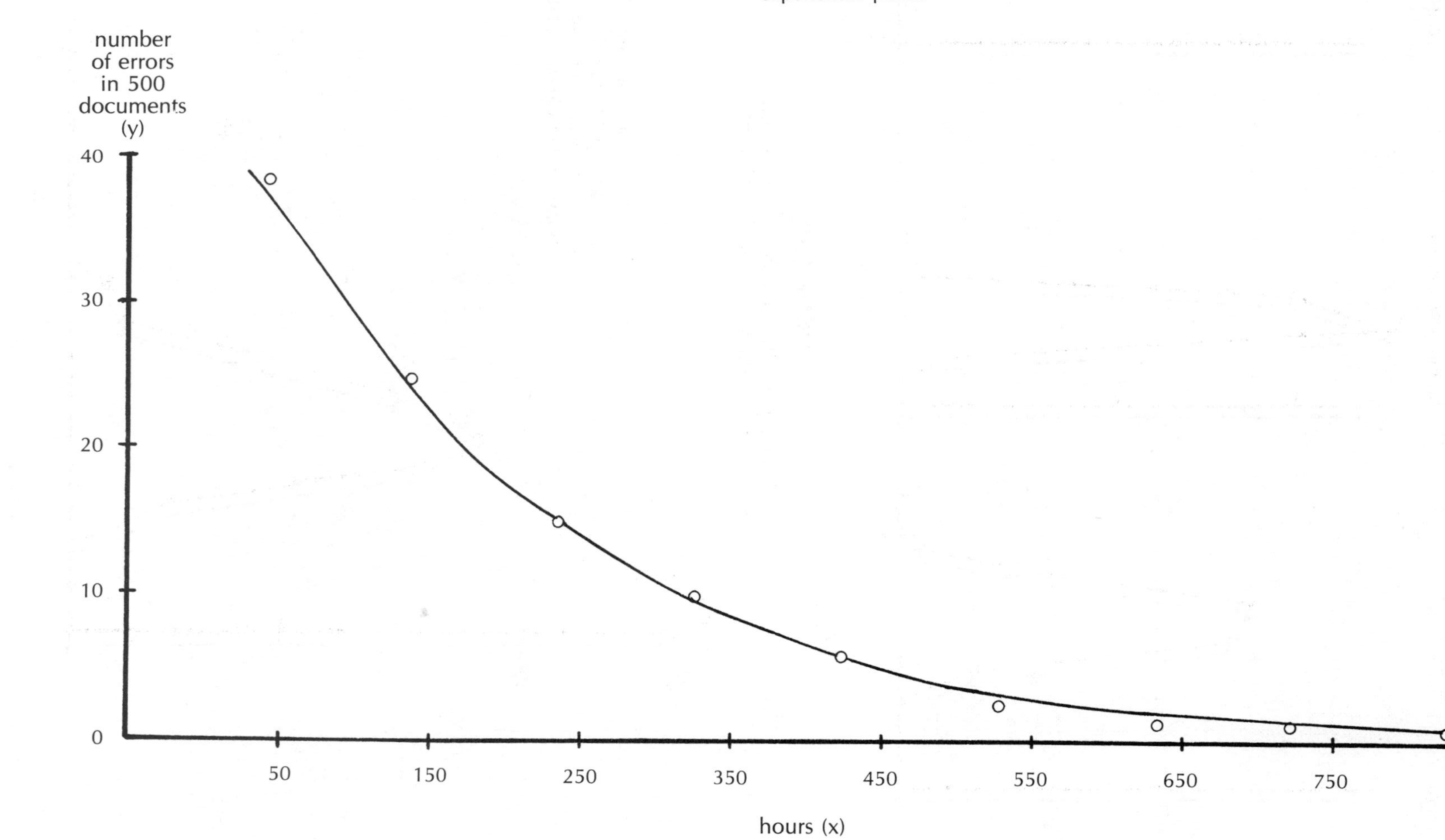

The exponential probability distribution has the form:

$$y = p(x) = ae^{-ax}$$

where a is a constant. The mean of the distribution is 1/a and the variance is $1/a^2$; the standard deviation is 1/a, and the coefficient of variation is 1. It applies to failure rates in reliability and life testing problems, to negative growth or decay situations, and to reduction in errors and error rates over time.

In a reliability problem, "a" is the failure rate and equals $1/\bar{x}$; $1/a = \bar{x}$ is the mean time to failure (MTTF) and is estimated from a frequency distribution of x, the number of hours required for an object or device to fail a life test. The frequency is the number of such objects or devices failing in each interval of time where x ranges from a small duration of time to a maximum; x cannot be zero in a real problem because some finite amount of time is necessary for the first failure.

The reliability of an object or device characterized by the exponential model is:

$$R = e^{-T/u}$$

where $u = \bar{x} = 1/a$ = mean life or mean time to failure, and T = time device is desired to operate.

Figure 11 is a plot of the reduction in the numbers of errors made in batches of 500 documents during time intervals of 96 hours, beginning with an interval centered on 48 hours. The mean number of hours based on 100 batches is 217. The equation is:

$$y = p(x) = \frac{1}{217}e^{-x/217}.$$

x hours	Error(f) frequency	p(x)	100 p(x)	Difference
48	34	0.369	37	−3
144	23	0.237	24	−1
240	15	0.152	15	0
336	10	0.098	10	0
432	7	0.063	6	1
528	5	0.040	4	1
624	3	0.026	3	0
720	2	0.017	2	0
816	1	0.011	1	0
	100	1.013		

$$\bar{x} = \frac{\sum fx}{100} = 216.96 = 217.$$

If the rate of change of y with time t is proportional to y and k = a is the rate of change, either growth or decay or decline, then:

$$y = y_o e^{kt}$$

where -k is for decay and +k is for growth; $y = y_o$ when t = 0.

The time required to reduce y by ½ or 50 percent is therefore:

$$\frac{y}{y_o} = ½ = e^{-kt}.$$

Hence:

$$t = \ln 2/k,$$

where $k = a = 1/\bar{x}$.

Example: In Figure 11 how many hours are necessary to reduce the error rate 50 percent?

$$t = \frac{\ln 2}{1/217} = 150 \text{ hours}.$$

For x = 0, y = 46, and ½ of 46 = 23. A t of 150 has an ordinate very close to 23.

Example: In Figure 11 how many hours are required to reduce the error rate one third?

$$t = \frac{\ln 3}{1/217} = 238 \text{ hours}.$$

$1/3\ (y_o) = 15$, and 15 corresponds to about 240 on the chart.

Example: If Figure 11 is a life curve, what is the reliability of a device if t = 50 hours?

$$R = e^{-50/217} = 0.79.$$

Notes

[1]Mean of the Poisson distribution should exceed 10. See Deming, *Some Theory of Sampling,* Wiley, 1950, pp. 420-421. The mean can be as low as three or four if Hald's correction is used. See Deming, *Sample Design in Business Research,* Wiley, 1960, pp. 461-462.

[2]W. Edwards Deming, *Sample Design in Business Research,* Wiley, New York, 1960, p. 45.

[3]This paper as well as other useful special graph paper can be obtained from TEAM, Tamworth, New Hampshire 03886.

[4]The source of these data and the graph are given in Chapter 12.

15

Sampling, Sample Designs, and Sample Sizes

Why sampling? There are many reasons for using sampling, but a very sharp distinction needs to be made between the science of probability sampling and various kinds of intuitive sampling. We are concerned here only with random and probability sampling, although the limited, but useful applications of non-probability sampling are described briefly.

1. Due to the redundancy in counts and measurements, the sample contains most of the information in the population, thereby making 100 percent collection unnecessary, e.g., adult height, individual income, and individual taxes.

2. 100 percent testing is destructive or fatal: life tests, human blood tests, strength of materials tests, drug tests, food tests, and taste tests.

3. 100 percent testing is impossible: pollution tests of air, water such as lakes and rivers and oceans, and soil; testing chemical additives; tax audits; and experiments.

4. 100 percent testing is not feasible: judging diversion of freight traffic under a proposed rail merger, e.g., involving three million carload movements when a replicated sample of 5,000 is ample; nationwide measurement of unemployment; a nationwide opinion poll.

5. 100 percent coverage is too costly: estimating the traffic characteristics of rail shipments, truck movements, airline travel, interstate household goods movements; estimates from mass records such as tax returns, bills, checks, invoices, sales slips, purchase orders, deposits and withdrawals, premiums, etc., whether on computer tapes or in files.

6. 100 percent coverage is unnecessary: this applies to most of the characteristics given above, except in those cases such as personnel, payroll, tax accounting and billing, where 100 percent coverage is a "must". This also includes estimating error rates, auditing financial operations, auditing procedures, inspecting and verifying; revenue exchange; quality control; quality audit; and work and activity analysis and control.

7. 100 percent coverage does not work: random time sampling is required to decompose joint costs.

Critical nature of sampling. No subject is more misunderstood nor misused and misapplied than sampling. This seems to be due to several causes. One cause is the widespread use of opinion polls based on so-called "scientific" sampling which is not scientific at all. High level officials and professionals received many false notions about sampling from these opinion polls — which sampling specialists have had to combat for decades. This was because opinion polls were based on judgment sampling which was shown time and again to be highly biased nationwide as well as by states.[1]

Opinion polls, plus other factors, have led many to believe that sampling is a matter of common sense, intuition, or judgment — something that can be picked up on the run or in a "quickie" course. Overlooked is the fact that it is a science; that it is complex subject matter that must be mastered, the best evidence of which is the large number of books on sampling theory and practice which have been published since World War II. Some of these are listed below:

1. Frank Yates, *Sampling Methods for Censuses and Surveys,* London, 4th edition, Charles Griffin, 1981.
2. W. Edwards Deming, *Some Theory of Sampling,* Wiley, New York, 1950.
3. W. G. Cochran, *Sampling Techniques,* Wiley, New York, 1953.
4. M H. Hansen, W. N. Hurwitz, and W. G. Madow, *Sample Survey Methods and Theory,* Wiley, New York, 1953.
5. P. V. Sukhatme, *Sampling Theory of Surveys with Applications,* Iowa State College Press, Ames, 1954.
6. W. E. Deming, *Sample Design in Business Research,* Wiley, New York, 1960.
7. Leslie Kish, *Survey Sampling,* Wiley, New York, 1965.
8. Des Raj, *Sampling Theory,* McGraw Hill, New York, 1968.
9. A. C. Rosander, *Case Studies in Sample Design,* Dekker, New York, 1977.

A sample is the key factor determining whether the results have any meaning or not, regardless of whether it is being used for a customer survey, for an opinion poll, for testing a food additive, for comparing two methods or machines, for auditing a set of transactions or procedures, for collecting evidence for a regulatory proceeding, for measuring the quality of performance of an employee or supplier, for estimating the level of air or water or radioactive pollution, or for measuring the efficacy of a drug.

The sample is critical for several reasons:[2]

1. It is the source of the data on which estimates, inferences, conclusions, decisions, and actions are based.

2. The quality of estimates, inferences, conclusions, decisions, and actions can be no better than the quality of the data from which they are derived.

3. If the sample does not represent a bounded frame or population, generalizations can be made but their area of application is really unknown.

4. If a sample allegedly represents some assumed population or frame, or one which is proclaimed or asserted, the findings are suspect until others in different locations obtain independently similar results.

5. Different methods of sampling applied to the same conditions, areas, products, customers, services, or transactions may lead to different conclusions and estimates.

6. If the data lack relevance, accuracy, validity, and clarity, no amount of "exploratory data analysis" or high speed computer manipulation or calculating can correct these deficiencies.

7. Quality of results from a sample involve the quality of the data at the source, the quality of performance in collecting the data, the quality of the sample design, and the quality of performance in processing, analyzing, and interpreting the sample data.

Quality control and sampling. Sampling is applied to quality control in the service operations and service industries in a wide variety of ways because it is the universal way to collect data on which to base estimates, inferences, decisions, and actions. On certain occasions, however, information is collected and analyzed on 100 percent coverage. The variety of applications is reflected in the following list.

1. Samples are the source of data used to construct control charts such as the $\bar{x}$, R, p, and c charts. For process control the sample units are *selected in the time order in which they are produced.* This is **not** a random sample. As Shewhart showed, destroying the time order may eliminate the very fluctuations we are trying to detect.

2. Random samples are, or should be, selected where sample plans are aimed to accept or reject a lot or batch. The same is true where a sample from a lot or batch is used for process control where rejection immediately calls for a 100 percent processing of the lot by the person involved. In the latter case, practice has shown that a systematic sample with a random start is much easier and just as effective.

3. Sampling customers to obtain their appraisal of the services rendered. This may take the form of a sample survey by mail or interview or both, or it may be an attempt to contact all customers on a probability sample of days. The goal is to give every customer an equal chance of being heard. Customer complaints are not enough because they tend to be highly biased. This is a sample survey of actual customers.

4. A market survey using sampling goes way beyond actual customers to the larger potential market. This sample survey is aimed at obtaining preferences for competing products, how potential customers would like products or services improved, what specific characteristics are preferred and at what price, and other important items.

5. Random time sampling is a powerful technique which can be used to measure down time, idle time, working time, project time, quality versus non-quality working time, decomposing joint times, and determining the wage and salary cost at the same time. This is because the method of estimation in the sample plan becomes the method of estimating costs; intuitive and common-sense methods of estimation are not needed.

6. Examples using sampling to construct an accurate *distribution of waiting or delay time* are:

- A check-out point in a supermarket.
- A check-in point for ticket and baggage at an airline counter.
- A doctor's waiting room.
- An emergency room in a hospital.

This type of sample requires a very careful study of the physical and human situation before a sound sample can be designed. This is a much more difficult technical problem than appears on the surface. For example, such questions as the following have to be considered:

- How is waiting time defined operationally? It is extremely important to obtain an accurate measure of time.
- What are the different sampling schemes which are both feasible and technically sound?
  - •• Do you cover lines 100 percent on random days or times?
  - •• Do you select lines at random?
  - •• Do you stratify time into rush and non-rush hours? days?
- How long should sampling continue? Here quality control charts are used until measurements are stabilized.

Clearly, in order to obtain the total waiting or delay time, it is necessary to count those who are in line from beginning to end, not those already in line.

7. Examples using sampling to construct an accurate *distribution of service times* are:

- Departure time and arrival time of airlines, trains, and other modes of transportation.
- Time actually devoted to medical service while in a doctor's office or in an emergency room; waiting periods which are not a part of the service should be excluded and be considered more delay time.
- Time devoted to repair an automobile, a radio, or a TV set.

8. Using sampling to *audit* an operation; this includes all kinds of sample audits:

- Sampling to detect an error or several errors.
- Sampling to detect deviations from or violations of instructions, specifications, and procedures.
- Sampling for a quality audit to determine if the quality program is being implemented as planned.

9. Designing and using a sample for a *field test* such as an environmental test of air, water, and soil for gases, particulates, and chemicals in solution. A sound field test is imperative because far-reaching decisions and actions of a public nature may be taken as a result of the findings. Numerous examples of inadequate sample studies of this nature already exist where both the sample design and the analysis were deficient. There is room for tremendous improvements in this area.

10. Carrying out *laboratory tests* of blood and urine for health purposes and various chemicals for environmental effects. The high error rate in the blood and urine tests reported by the Center for Disease Control in Atlanta shows that a tremendous amount of work has to be done to improve the accuracy of these very vital laboratory tests. There is also a need to validate the ranges of specific measurements, e.g., uric acid, cholesterol, glucose, potassium, and iron to determine what percentage of the population they apply to, to what extent they apply to all ages, and to what extent measurements outside these ranges are significant. Apparently, these ranges are "control limits" and are used by doctors as such, but we need to know how they were arrived at and how sound they really are for individuals.

11. The problem in designing an *experiment* is not only to insure that the experimental techniques are sound, but if A is found to be better than B or if A has some causal relation to B, any inference is limited to the scope of the experiment and not to a vast population. One of the serious objections to so many social, economic, health, and biological studies is that the findings are generalized far beyond what the study justifies. These broad brush generalizations may be fine journalism, but they are poor science.

12. Design and implement a test to determine *human variability* of persons independently doing the same thing, or random samples from the same population:

- Tax assessors appraising the same piece of property.
- Tax auditors auditing the same tax returns.
- Interviewers interviewing persons or families for the same information.
- Reviewers or inspectors reviewing or inspecting the same lots for errors or defects.
- Scientists in laboratories testing identical samples.

Use of the same test material instead of random samples from the same population eliminates the variation that is bound to exist between the random samples. The use of the same material shows at once the inherent variations existing in the group variations due to:

- Use of different criteria as the basis of judgments.
- Different interpretations of the same instructions.
- Differences in applying the same techniques.
- Differences in carrying out instructions.
- Different interpretations of the same situation.

13. Measuring learning progress and capability by means of *learning curves.* There may be no sampling of either persons or time in this situation. Records are kept daily of errors, error rates, and production during the learning cycle. It is assumed that the job is new to everyone concerned and that they all have to go through a training period, possibly as long as three or four weeks, before they are able to be assigned to production. These situations include coding, transcription to a standard form, other types of clerical work, and any job where a period of learning is necessary. The learning curve shows the capability of the individual; the composite shows the capability of the group under existing supervisory and other conditions. Over time it is assumed that due to training and coaching both the number and rate of errors will decline, that production will increase, and that quality will improve as sources of errors and delay are eliminated. The learning curve shows the actual level of acceptable quality performance management can expect from the group.

Some requirements for effective sampling. We describe here some of the major requirements for an effective sample study which have been developed from actual sampling practice.

The most effective sample, in practice, to aid management is one which is predesigned for a specific purpose. The data are determined by means of a detailed and progressive analysis of the problem or problems to be resolved, or questions to be answered by the sample. This means a sample yielding acceptable quality data of an optimum amount at a minimum cost in a timely fashion.

This approach eliminates the waste inherent in collecting masses of data as ends in themselves. It builds sense into the data; one does not have to waste time and resources trying to "make sense out of the data". It also eliminates futile analysis of large masses of dubious data collected with only a vague problem or purpose in view, or possibly with nothing in view except to "analyze" the data.

A sample has to be designed in terms of its major purpose which can be in one or more of six forms. An effective sample design for one purpose is not an effective sample design for some other purpose:

1. To *detect* the presence of some condition, such as an error in a set of books, the presence of a carcinogenic, or a group of employees without work.

2. To *estimate* the value of a characteristic, such as the error rate, the amount of money error on an insurance claim or a tax return, or the average amount of time required to perform a specified service.

3. To *control* the value of some measurement or count, or the level of employee performance, such as the errors made in coding or otherwise processing some document.

4. To make *comparisons* to determine which of two or more products, methods, machines, or procedures is better, such as comparing two computers, two pieces of software, or comparing two desk type electronic calculators.

5. To test *effectiveness,* such as the use of a new drug or vaccine, or to test whether some social or economic program produces the results it is supposed to, such as Medicare, public housing, aid to families with dependent children, and the 55 mile per hour speed limit.

6. To *audit* for the purpose of determining whether certain prescribed procedures are being followed, as in a bank, or whether certain levels of quality are being attained, as in the processing of insurance claims.

In any sample study or test, two sources of error need to be controlled: sources of random variation which are usually due to random sampling or random assignment and sources of non-random variation or bias. The number one problem in most, if not in all, sample studies is to *control sources of bias in the data,* which include bias at the data source, bias due to sampling, bias due to collection, bias due to tabulation, bias due to calculation, and bias in the analysis.

Sampling error can easily be controlled by the sample design; non-sampling bias cannot. There is no point in stressing sampling error in an estimate when bias can easily be two to 10 times as large. If the latter is true, and it can be, then the use of confidence limits and intervals is not justified, since the latter assume that bias is zero. Also, if bias is large, it may easily overwhelm the standard error and the random sampling variation, making the latter measure very deceptive.

There are two types of errors in conjunction with testing statistical hypotheses: Type 1 and Type 2 errors. Type 1 error is the probability of rejecting a true hypothesis, this is the probability of a test of significance such as that at the 95 percent level, which means 5 percent of the time we reject a true hypothesis. Type 2 error is the acceptance of a false hypothesis or concluding that an estimate is some value when the real value is quite different. The easiest solution to the Type 2 error is to make the sample large enough to reduce this error to a value that is acceptable. An example from a real problem illustrates this.

Example: A nationwide grocery chain handled items subject to federal excise taxes. The chain wanted the Internal Revenue Service (IRS) to allow 10 percent for breakage and pilferage; the IRS allowed three percent. A sample of 225 stores was designed to distinguish between these two values; if the real value was 10 percent the sample would not give three percent and vice versa. The overlap of the two distributions was only about one percent at z=2.25 sigma. Testing the difference, 10−3=7 percent, gave:

$$s_d^2 = \frac{0.03 \times 0.97}{225} + \frac{0.10 \times 0.90}{225} = 0.000529,$$

hence, $s_d = 0.023$. Using the normal curve

$$z = \frac{0.10 - 0.03}{0.023} = 3.04; P(z > 3.04) = 0.0012.$$

Hence, a sample of 225 would give an adequate point estimate by sharply distinguishing three percent from 10 percent. The lawyers finally decided to leave the figure at three percent!

An effective sample study requires the following: a careful analysis of the problem; the identification of the statistical aspects; a properly designed probability sample; properly designed non-sampling aspects such as questionnaire or data sheet and various procedural manuals for collection, tabulation, and calculation; and careful management and implementation of the sample plan so that sources of bias are kept under control. Indeed, it is much easier to design the technical aspects of a sample study than it is to put it into effect and see that it is carried out according to plan. Many a sample plan may fail not because of technical deficiencies but because it is not carefully managed and implemented.

Probability sampling. Probability sampling is a sample in which every sampling unit in the population or frame is given a known non-zero probability of being selected. Probability sampling takes many forms:

1. Simple random selection using a table of random numbers and a sampling rate of n/N, where N is the known or estimated frame size and n is the sample size.
2. Simple random selection using a table of random numbers at rate f, where N is unknown and f takes the form of 1 in 20, 1 in 100, or 1 in 2000.
3. Systematic selection with the starting number being selected from a table of random numbers. If the rate is 1 in 20 and the random start is 14, then the sample consists of units numbered 14, 34, 54, 74, 94, 114, etc.
4. Systematic selection with some convenient or easy start such as 50 selecting 1 in 100; then the sampling units are 50, 150, 250, 350, 450, 550, etc.
5. Replicated sampling with random starts in the first zone and a Z interval for each replicate thereafter. For $k=4$ and $Z=40$:

	replicate			
Z	1	2	3	4
1-40	23	5	14	38
41-80	63	45	54	78
81-120	103	85	94	118
etc.				

$$Z = \frac{kN}{n} \quad \text{where} \quad \begin{aligned} k &= \text{number of replicates,} \\ n &= \text{sample size, and} \\ N &= \text{frame or population size.} \end{aligned}$$

6. Stratified random sampling with rates n_j/N_j, or f_j, used in different strata.
7. Probability of selection is proportional to the size of the sampling units.

An initial step in sampling is the determination of the population or frame of *sampling units,* since the *sample size is the number of sampling units selected at random.* Sampling units include objects, areas, time, and operations. Objects include persons, families, households, tax returns, invoices, insurance claims, bank checks, passenger tickets, and sales slips. Areas include city blocks, farms, loading docks, offices, and construction sites. Time includes half minutes, minutes, days, hours, weeks, and months. Operations include freight movement, mail movements, travel trips, employees at work, and machines in operation.

It is very important to distinguish two major kinds of sampling units: the individual intact unit and the cluster or pea-pod unit. Intact units include all kinds of individual records whether in a file, in the form of punch cards, or in the memory of a computer (on tapes, disks, or drums). These include employees, tax returns, bank checks, automobile registrations, automobile driver's license, bills, claims, invoices, and licenses. A cluster unit includes two or more individual elements such as households relative to individual members, an office relative to its occupants, a school relative to the students, a school grade relative to the individual students, or a page of a telephone directory.

A different theory of sampling has to be applied to cluster sampling units than to individual type sampling units so the distinction is very important. Examples of this difference are given later in this chapter.

Non-probability sampling. It is not always necessary to use a probability sample as the following situations show, but the distinction needs to be kept clearly in mind.

1. Drawing a sample of work from a production line for an $\bar{x}$, R, or p control chart. The last units processed or produced are selected *in the order in which they were processed* to discover quickly when a process is out of control or headed that way. Mixing the objects or sampling units destroys the variation we are trying to detect. Eliminating assignable causes tends to leave random variations distributed as a normal curve, the assumption underlying the use of the three sigma limits.
2. In testing a questionnaire or a data sheet, it is highly desirable to use a wide variety of persons similar to those who are going to answer it. These should include the not highly literate, those with foreign language backgrounds, those unfamiliar with the language used. The purpose is to discover ambiguities, misunderstandings, and misinterpretations of words, phrases, and sentences. The goal is to debug the questionnaire or data sheet, not to make population or frame estimates.
3. In making a pilot test of a study, whether based on sampling or not, the purpose is to test procedures, obtain time estimates, discover difficulties and deficiencies, measure employee and supervisory and other person's reactions, to note criticisms and objections, and uncover various factors affecting the quality of the final data. The purpose is to test feasibility and acceptability and discover where changes and improvements should be made. It is not an exaggeration to state that pilot studies should be a "must" wherever possible, because experience and practice show that they pay off in a greatly improved study.
4. In certain tests and experiments, one may have to test what is available in the form of machines, equipment, and materials. This should not be taken to mean that an inefficient or inadequate test or experiment is justified. Quite the contrary, the need is to design and implement tests and experiments making full and better use of the principles and practices of experimental design. Too many experiments and tests in the fields of medicine, biology, radioactivity, air and water pollution, and social and economic intervention, which have highly significant public implications and effects, need to be planned, designed, and implemented using the best technical knowledge available. Too many are based on a test or experiment which is too small, biased in terms of what is most available, and exaggerated results which are highly publicized, all leading to unwarranted generalizations.

Different sampling methods and designs. A wide variety of sampling methods and designs are available to obtain information for quality control and quality assurance. These take the form of sample studies, sample surveys, sample audits, and sample tests depending upon the nature of the problem. Several of these are described in some detail below.

1. Simple random sampling
2. Systematic sampling
3. Systematic sampling with random start
4. Stratified random sampling
5. Cluster sampling
6. Replicated sampling
7. Sampling proportional to size (pps)
8. Two stage sampling
9. Sequential sampling

Simple random sampling. Simple random sampling is selection of a random sample without any restrictions on the population and frame or on the method of selection. This is in contrast to stratified random sampling where the frame of population is stratified or divided into independent parts with different rates of sampling for each stratum or division.

The frame used to illustrate simple random sampling consists of 40 actual incomes on federal individual income tax returns with total incomes less than $5000. These are listed in Table 1 in the order in which they occurred in the file. Each income was given a serial number one through 40 so a table of random numbers could be used to select a sample.

A random sample of five was selected using the first five random numbers in the Kendall-Smith tables of random numbers which fell in the range 01-40. This sample is used to estimate the population mean, the population aggregate or total income, and the standard errors (sampling variations) of these estimates. These five samples are selected in different ways in order to show the effect of sample design on estimates using the same sample size. The random numbers and the corresponding incomes are given below:

Order	Random number	Sample income x from Table 1
1	23	$ 2,210
2	05	1,880
3	14	1,102
4	38	3,581
5	11	1,710
	Sum	$10,483

The calculations are as follows:

1. The arithmetic mean

$$\bar{x} = \frac{\sum x}{n} = \frac{10,483}{5} = \$2097.$$

2. The aggregate income

$$X = w\sum x = 8(10,483) = \$83,864,$$

where $w=8=\frac{1}{f}$, since f = rate of sampling = 1/8 = 5/40.

3. The variance

$$s^2 = \frac{1}{n-1}\left[\sum x^2 - \frac{(\Sigma x)^2}{n}\right] = \frac{3,401,907}{4} = 850,477.$$

4. The standard deviation

$$s = (s^2)^{1/2} = 922.21.$$

5. The coefficient of variation

$$cv = s/\bar{x} = 922.21/2097 = 0.44 \text{ or } 44 \text{ percent.}$$

The estimate of $2,097 compares with the true mean of $1,955; the estimate of $83,864 compares with a true value of $78,205; and the estimate of $922 compares with a true value of $955.

The standard error of the mean

$$s_{\bar{x}} = \frac{s}{(n)^{1/2}}(1-f)^{1/2},$$

where $(1-f)^{1/2}$ is the correction for sampling from a finite population and is equal to

$$\left(1 - \frac{1}{8}\right)^{1/2} = 0.875^{1/2} = 0.9354.$$

Hence

$$s_{\bar{x}} = \frac{922.2}{5^{1/2}}(0.9354) = 385.78.$$

The relative sampling error is $\frac{385.78}{2096.6} = 0.1840$ or 18.4%. The estimated aggregate, $83,864, has the same relative error so its standard error is 83,864 (0.1840) = $15,431, which is the same as $Ns_{\bar{x}} = 40(385.78) = 15,431$.

The table of frame values (Table 1) shows the frame divided or stratified into four groups by income size. This will be used later to illustrate stratified random sampling and compare estimates obtained with those obtained from simple or unrestricted random sampling.

Rarely is the sample size n an exact multiple of the frame or population size N, but in practice this does not introduce any appreciable error because of the large sizes of both N and n. Furthermore, other problems in sample design and execution are much more serious.

Systematic sampling. In systematic sampling, the rate of selection and the starting point or unit are determined. It is assumed that the frame units are discrete and if not numbered they can be counted. The starting point can be selected at random — a random start — or it can be some convenient number. It must fall, however, within the range from which one sample unit is to be selected. If the rate is one in 200, then the random start must be a random number between 001 and 200. If the rate is one in 80, then the random start must be a random number between 01 and 80. If a random start is not used, some convenient number such as 100 may be selected in the first case and 50 in the second case. The sample consists of the sampling unit selected at the start plus sampling units at successive skip intervals, which is 200 in the first case and 80 in the second. In the first case, if the random start is 137, then the successive sample units selected are those numbered 337, 537, 737, etc. If the starting unit is 100, then the successive sample units are those numbered 300, 500, 700, etc.

A systematic sample is relatively simple and easy to administer. It can be selected with a minimum amount of error by clerks and others not familiar with sampling. If the frame units are thoroughly mixed or in no particular order, as a sample of tax returns received in the mail by the IRS or freight bills prepared by a motor carrier, then a systematic sample is equivalent to a simple random sample.[3]

On the other hand, if the frame values are correlated, then a systematic sample may be better or even worse than a simple random sample depending upon the nature of the correlation. Even more serious, if the frame values follow some periodic pattern, the results are not only biased but there is no way of calculating a valid standard error.

Table 1

Table of Frame Values—Total Income on 40 Federal Individual Income Tax Returns with Incomes Under $5,000

Serial number	Income	Serial number	Income
1	$1187	21	$2443
2	3319	22	4293
3	2390	23	2210
4	1046	24	1089
5	1880	25	1005
6	1383	26	1284
7	2344	27	1012
8	1199	28	2377
9	1332	29	3196
10	2473	30	2036
11	1710	31	2174
12	1439	32	1000
13	1906	33	1152
14	1102	34	1273
15	2113	35	1098
16	1628	36	1302
17	1641	37	1023
18	1039	38	3581
19	4532	39	3386
20	1858	40	3750

	Frame values
N	40
aggregate	$78,205
mean	1,955.13
standard deviation	954.64
coefficient of variation	.4883

A. Stratified Frame

	$1000-1999		$2000-2999	$3000-3999	$4000 and over
	1880	1039	2390	3319	4532
	1383	1858	2344	3196	4293
	1199	1089	2473	3581	
	1187	1005	2113	3386	
	1046	1284	2443	3750	
	1332	1012	2210		
	1710	1000	2377		
	1439	1152	2036		
	1906	1273	2174		
	1102	1098			
	1628	1302			
	1641	1023			
Number N_j		24	9	5	2
Sum Σx_i		31,588	20,560	17,232	8,825
Mean $\bar{x}_j$		1,316.17	2,284.44	3,446.40	4,412.50
Variance s_j^2		86,152.22	21,422.91	38,573.84	14,280.25
Weight w_j		12	9	5	2
Sample n_j		2	1	1	1

An extreme example of bias due to periodicity is a random time study in which the random start is 10:23 am and the skip interval is 480 minutes, a working day. This makes 10:23 am the sampling unit every day, obviously a very biased sample. The solution is to use a replicated sample with a random start, a procedure described later in this chapter.

Systematic sampling does not require that we know N, the frame size, although we ought to have some idea of its magnitude to determine the sample size. Using the sampling rate f which gives a multiplier $w = 1/f$, an estimate of $N = wn$ and an estimate of the aggregate $X = w\Sigma x_i$, where $i = 1,2,...,n$.

Systematic sample with a random start. From Table 1, a systematic sample of five is selected with a random start between one and eight, since $f = 1/8$. This means a skip interval of eight. From a table of random numbers, 3 is selected. Then the sample units by number and value are:

Serial number of sample elements	Sample value x
random start 3	$ 2,390
11	1,710
19	4,532
27	1,012
35	1,098
Sum x	$10,742

Calculations:

1. Mean: $\bar{x} = \frac{10,742}{5} = 2148.4$; true value 1,955.

2. Aggregate:

$$X = 8(10,742) = 85,936; \text{ true value } 79,205.$$
$$X = 40(2148.4) = 85,936;$$

3. Variance:

$$s^2 = \frac{1}{4}\left(\sum x^2 - \frac{10,742^2}{5}\right) = 2,081,715.$$

4. Standard deviation: $s = 1442.81$; true value 954.64.

5. Coefficient of variation: $cv_x = 1442.8/2148.4 = 0.67$ or 67 percent.

6. Standard error of the mean:

$$s_{\bar{x}} = \frac{1442.8}{5^{1/2}}(0.875)^{1/2} = 603.57.$$

7. Coefficient of variation of the mean:

$$cv_{\bar{x}} = 603.57/2148.4 = 0.2809 \text{ or } 28 \text{ percent.}$$

8. Standard error of the aggregate 85,936:

$$s_X = 0.2809(85,936) = 24,143.$$
$$s_X = 40(603.57) = 24,143.$$

The estimates are high because the sample included one very large value, 4532; this shows the danger of using a small sample selected from a highly variable unstratified frame or population.

Stratified random sampling. This type of sampling is used to obtain acceptable estimates from a highly variable or heterogenous frame or population. This is accomplished by dividing the frame or population into two or more divisions or strata so that each division or stratum will be less variable than the whole. Since each division or stratum is mutually exclusive of the others, and also independent, each division or stratum is treated as a subframe or sub-population. Although the determination of whether to stratify or not, the number of strata, and the boundaries of each stratum are questions of professional judgment, certain guilding principles apply. To be effective, the stratifying variable or attribute has to correlate substantially with the characteristics which are to be estimated from the sample. If the distribution of the stratifying variable is highly skewed, containing a substantial number of extreme and large values, then it is necessary to combine these into a single stratum and include them 100 percent. Stratification is effective in at least four situations.

1. The population or frame includes one or more classes or categories that contain a substantial proportion of the sampling units. Many classes or categories that are important contain only very small proportions of the sampling units. A simple random sample is a proportional sample so it yields a large number of sample units from classes of high proportions and a small number of sample units from classes of low proportions. Hence, the large classes may be oversampled and the small classes may be undersampled. The nationwide one percent sample of rail carload waybills is an excellent example since coal accounts for nearly 20 percent of the total carloads and iron ore accounts for about another five percent. These waybills are oversampled; those for more than a hundred other commodities are undersampled. Stratification by commodity groups would correct and improve this situation; sampling rates would be reduced for sampling coal and iron ore and increased for more than a hundred other commodities. This would result in more and better quality information without any increase in the total sample size. This assumes that the various commodity groups are of equal importance, or approximately so, and this is the actual situation.

2. The means of a basic or important characteristic vary greatly between subdivisions of the population although the variance is about the same in each.

3. The means and variances of a basic or important characteristic both vary greatly between subdivisions of the population or frame.

4. Stratification may be necessary to obtain acceptable quality estimates for small classes and rare groups which are important.

Examples of characteristics with widely varying means and variances, where stratification is effective, include: 1) dollar volume of business in an industry, 2) weight or shipments by truck or rail, 3) income of families by city, state, and nation, 4) state and federal individual income taxes, 5) number of families or households by cities, 6) size of business by assets, 7) number of students in a high school or college, and 8) number of emergency cases in hospitals.

Examples of where stratification is necessary in order to obtain acceptable quality data for small as well as large divisions include: 1) stratification by states in order to obtain good estimates for the small states as well as for the large ones, as in a study of automobile passenger car and truck registrations, a study of unemployment, or a study of family income and expenditures; 2) stratification by size of company in an industry, as in a nationwide study of movers of household goods between states where estimates were required for small, medium, and large companies.

Stratification is used in quality control and laboratory testing. Samples are drawn not from a mixture of products from several machines or sources, but from sources that may be directly associated with assignable causes of "out of control". Examples are individual workers, machines, shifts, departments, lots, suppliers, and production lines. Similarly, in making laboratory tests of chemical and physical properties, samples are selected on a stratified basis from, for example, each carload, each source, each supplier, each lot, or each batch. This procedure tends to pinpoint sources of trouble, whereas mixing the product from many sources — workers, machines, lots, etc. — may conceal the very trouble the quality control system and the laboratory tests are designed to detect. Shewhart called these specific sources, which may be related to assignable causes, "rational subgroups".

Other examples of stratification include: stratifying by country of origin a sample of imported cocoa bean shipments made by the Food and Drug Administration, stratifying by male and female adults a sample study designed to obtain body measurements for the purpose of setting size standards in clothing manufacturing, and stratifying children by sex and age in a sample study for the same purpose. In these instances, stratification is required by the very nature of the problem.

Examples of characteristics where variability is usually so limited that stratification is unnecessary, or the gain is so slight as to preclude its use, include:

1. Male adult height: About 95 percent of male adults are between 60 and 72 inches in height, a range of only 12 inches. The unit "inches" is immaterial; it is the variability that counts. (A random sample of only about 150 will give a range of uncertainty of 1/2 inch on either side of the mean that will be exceeded in only one out of 20 times.)

2. Proportions: Proportions can vary between 0 and 1, which explains why stratification seldom is effective in improving an estimate of a proportion or the gain is so slight as not to warrant the cost. Stratification is effective only if some strata have proportions below 0.20 and others have proportions above 0.80.

3. Measurements under very close control: Examples are weights and volumes of containers filled with liquids or solids, such as weights of barrels of whiskey and carloads of a certain grade of coal which are subject to close manufacturing and packaging control.

Four examples: Four examples given below show concretely the conditions under which stratification is effective and the extent to which it is superior to simple random sampling. They illustrate the sampling situations mentioned above: the case of high-proportion classes that are oversampled, the case of different means but same variances, the case where both means and variances differ, and a comparison of the results obtained from a random sample of five and a stratified random sample of five applied to the frame data of 40 individual incomes given in Table 1.

The following, which compares closely with an actual frame, shows 10 strata, the number of sampling units in each stratum, the expected distribution of a random sample of 2,000, the sampling rates to be applied to each stratum to give 200 sample units in each stratum, and a second plan which gives more weight to the large strata and less weight to the smaller strata.

Class or stratum	Frame size N_j	5% random sample	**Stratified Plan 1** Rate	**Stratified Plan 1** Sample size	**Stratified Plan 2** Sample size
A	15,000	750	1/75	200	300
B	7,000	350	1/35	200	300
C	5,000	·250	1/25	200	200
D	4,000	200	1/20	200	200
E	3,000	150	1/15	200	200
F	2,000	100	1/10	200	200
G	1,200	60	1/6	200	100
H	500	25	2/5	200	100
I	1,500	75	2/15	200	100
J	800	40	1/4	200	100
Sum	40,000	2,000		2,000	1,800

Assume that 200 sampling units per stratum give satisfactory results. Then groups A, B, and C are oversampled, and groups E, F, G, H, I, and J are undersampled, based on a simple random sample of 2,000 selected from a frame of 40,000. The sample sizes given for a five percent random sample are the expected sizes for a simple unstratified random sample. Stratified plan one applied a sampling rate to each stratum so that the expected sample size in each stratum is 200. This means that satisfactory data are obtained from all 10 strata and not just from four. This assumes that information for each of the 10 classes or strata is equal in importance to the user, to management. If not, the sample size can be increased in the more important classes, the larger ones, and decreased in the less important (smaller) classes. This is done in Plan two. The sample is set at 300 for the first two classes, at 200 for the next four classes, and at 100 for the last four classes. This plan would still give better data than those from a simple random sample of 2,000 and the sample would be 10 percent smaller at $n = 1800$.

An example of the effectiveness of stratification, where the means are different in the groups or strata but the variances are the same, is the following. It looks like an unreal situation, but really is not. Many actual problems are like this — the entries are values of x, the stratifying variate:

Group 1	Group 2	Group 3	Group 4
1	11	51	101
2	12	52	102
3	13	53	103
Sum 6	36	156	306

The frame aggregate or total is 504; the mean is $\frac{504}{12} = 42.00$. Two sample plans are compared: draw a random sample of four from the entire 12 values and then draw a random sample of four, one from each of the four strata. We calculate the maximum and minimum values of the mean of 42 estimated from these two samples of four:

	maximum value	minimum value
Stratified random sample of four		
aggregate	172	164
mean	43.0	41.0
Simple random sample of four		
aggregate	359	17
mean	89.75	4.25

The stratified random sample of four will give an estimate very close to 42, but the simple random sample will not in most instances. The stratified random sample gives a highly superior estimate of the frame mean because the estimate is subject only to the variation *within* groups or strata; the variation *between* groups or strata, which is very large and is reflected in the estimate derived from the simple random sample, *is eliminated by stratification.*

The effectiveness of stratified random sampling is evident even if both the means and variances are widely different in the groups or strata. In the following example, the total or aggregate of x is 636 and the frame mean is 53. In a random sample of four, selected in the two ways described in the previous example, the mean estimated from a simple random sample varies from 2.75 to 131.25, while the mean estimated from a stratified random sample of four ranges from 32.75 to 73.25 which indicates much less departure from the frame mean of 53.

Group 1	Group 2	Group 3	Group 4
1	5	25	100
2	10	50	150
3	15	75	200
Sum 6	30	150	450

Earlier in this chapter a simple random sample of five was selected from a frame of 40 values (Table 1) giving a mean of 2,097 and an aggregate of 83,864. We now want to compare these results with those obtained from a random sample of five selected from the same frame of 40 values stratified by size of incomes into four strata: $1000-1999, 2000-2999, 3000-3999, and 4000 and over. To obtain a sample of five, two values are selected at random from the first stratum and one value was selected at random from the other three. Estimates of the mean, the aggregate X, the standard error of X, and the relative sampling error of X are calculated from this sample. Population variances are calculated for each stratum and corrections are made in each stratum for sampling from a finite population. The five sample values and other pertinent data are:

Strata j	Sample size n_j	Sample values x_{ij}	Population N_j	N_j^2	Population variance s_j^2	$1-f_j$	w_j
1	2	1710;1039	24	576	86,152	11/12	12
2	1	2377	9	81	21,423	8/9	9
3	1	3319	5	25	38,574	4/5	5
4	1	4532	2	4	14,280	1/2	2
Sum	5		40				

The estimates are:

1. Aggregate $X = \sum_i \sum_j w_j x_{ij} = 12(2749) + 9(2377) + 5(3319) + 2(4532) = 80{,}040$. The population value is 78,205, a difference of +1805 or 2.3 percent too high.

2. Mean $\bar{x} = \frac{80{,}040}{40} = 2{,}001$. The population mean is 1,955, a difference of 46 or 2.3 percent too high.

3. Standard error of X: $s_x^2 = \sum \frac{N_j^2 s_j^2 (1-f_j)}{n_j} = 25{,}086{,}624$ (summing over four strata); $s_x = 5009$

4. Coefficient of variation (relative sampling error):

$$c_x = \frac{5009}{80{,}040} = 0.063 \text{ or } 6.3 \text{ percent.}$$

The simple random sample of five gave a standard error of x of 15,431 or a relative error of 18.4 percent: Stratification gave both a mean and an aggregate closer to the population values and reduced the standard error of X by two-thirds for the same sample size (n=5).

To obtain the standard error of the mean, divide s_x^2 by N^2 and obtain

$$s_{\bar{x}}^2 = \frac{25{,}086{,}624}{40^2} = 15{,}679.14;$$

hence, $s_{\bar{x}} = 125.22$ and $c_{\bar{x}} = \frac{125.22}{2001} = 0.063$ or 6.3 percent, the same as that obtained for X.

Cluster sampling. In cluster sampling, a cluster is not an intact indivisible unit but a collection, group or bunch of elements. The number of elements in a cluster may or may not be equal. One physical analogy of cluster sampling is the *pea pod model,* where interest may lie in sampling the pods or the peas within each pod or both. Examples of cluster sample situations are:

cluster	**elements**
city blocks	households
farms	cattle
bundles or blocks of tax returns	individual tax returns
rail freight train	freight cars
factory	machine operators
store	sales persons
lots of a product	contents of a lot
day's production	hourly production

Some basic questions to be answered about clusters are:

1. When should a cluster be used as a sampling unit and when as a stratum?
2. How large should a cluster be?
3. What if the clusters contain an unequal number of elements?
4. When is a sample cluster sub-sampled for elements?

These questions are very important as examples from sampling practice show. Bundles or blocks of 100 tax returns were used as sampling units since this method appeared to management to be easy and cheap. Bundles, however, were segregated according to four sizes of income and also by business and non-business within these income groups. Other bundles were sorted according to size of assets and still others were sorted in terms of size of gross receipts or total income. The variation between bundles tended to be much larger than the variation within bundles; therefore, these bundles were much better strata than they were sampling units. The sample plan was changed so that bundles were strata and a sample selected from every bundle. This considerably reduced the number of sample returns needed to obtain the same precision in the basic characteristics.

Sampling a mixture of products produced in one or two days — that is, setting up a large cluster or composite to sample from — is not as effective as sampling the product every hour or half hour, since the use of a large cluster or mixture may conceal the differences and out-of-control situations we are using the sample to detect.

Cluster sampling units can be of various sizes depending upon the nature of the characteristics being estimated. It is a difficult task to determine the optimum size of cluster to use where size is subject to the sampler's control. Jessen found, for example, that the relative sampling error of several farm characteristics using simple random sampling was less for sampling units of quarter sections (160 acres) and half sections than for one section or two sections (1,280 acres). One agricultural sample used in a midwest agricultural state was based on cluster units of three to five adjacent farms. The Bureau of the Census, in its nationwide unemployment sample, forms about 450 primary sampling units consisting of counties or groups of counties, enumeration districts within these primary units, blocks within these districts, and dwellings within blocks. The first three are stratified and the aim is to obtain about six dwellings from every segment, for a total sample of about 60,000 households.

Sometimes clusters may be equal or nearly equal in numbers of elements as in the case of the 100 tax returns per bundle, the number of freight cars in a unit train, and the number of names on each page in a directory. Very often, however, they are not equal; steps are taken where situations are under our control to try to equalize the occurrence of important characteristics in various clusters which we can create. In general, this tends to reduce the sampling errors of estimates of certain characteristics as examples given later show.

Use of one or more clusters with successive sub-sampling is necessary in many problems, especially in those where sampling is applied to dwellings or households in areas as small as a county — let alone a state or the entire country. The estimation of employment and unemployment for the United States is an excellent example of the use of successive sub-clusters or stages of sampling: large areas (one or more counties), enumeration districts within these areas, blocks within districts, and then households within blocks. The areas, districts, and blocks are clusters.

Since the values of the variate x in each cluster may be correlated, a cluster sample is affected by what is called *intraclass correlation.* This is the correlation that exists between all permuted pairs of values in each cluster; the correlation coefficient is calculated for all such values for all clusters. If s_b^2 is the mean square between clusters, and s_w^2 is the mean square within clusters, then the intraclass correlation coefficient is estimated by:

$$r_a = \frac{s_b^2 - s_w^2}{s_b^2 + (M-1)\, s_w^2},$$

where M is the number of elements in a cluster and the mean squares are obtained by applying analysis of variance. Calculations are illustrated below. If $s_b^2 = 0$, then

$$r_a = -\frac{1}{M-1}. \text{ If } s_w^2 = 0, \text{ then } r_a = 1.$$

The sampling variance of the mean $\bar{x}$ can be expressed in terms of r_a:

$$s_{\bar{x}}^2 = \frac{s_s^2}{nM}\left[1 + (M-1) r_a\right],$$

where n is the number of clusters in the sample and nM is the total number of elements in the cluster sample. If $r_a = 0$, then the term in brackets is one and the sampling variance reduces to that of a simple random sample of nM elements.

The sampling variance is reduced only if the value of r_a is negative. If r_a is positive, the sampling variance will exceed that of simple random sampling.

Four examples illustrate these and other characteristics of cluster sampling:

Example: The population consists of four unequal clusters, the rate of sampling $f = 1/2$, the multiplier $w = 1/f = 2$, and the estimate required is that of the population or frame aggregate X. The entries are x values; the population or frame total = 187.

Cluster 1	**Cluster 2**	**Cluster 3**	**Cluster 4**
8 12 20	6 14 8 9 10 47	10 16 4 8 12 6 56	12 6 14 8 4 20 64

All possible samples of two clusters give the following estimates since $f = 1/2$ and $w = 2$:

Clusters 1+2 (20+47)2 = 134
1+3 (20+56)2 = 152
1+4 (20+64)2 = 168
2+3 (47+56)2 = 206
2+4 (47+64)2 = 222
3+4 (56+64)2 = 240
Sum of all samples = 1,122
Mean of 6 samples = 1,122/6
= 187
= population aggregate

Hence, a random sample of two clusters gives an unbiased estimate of the population aggregate, 187, since the expected value or average of all possible samples of two is 187. This is simple random sampling applied to clusters since individual elements are not sampled. The variate dealt with is the aggregate associated with each cluster. The sample would be the same if the cluster totals were associated with some intact indivisible object.

Example: The following example shows why clusters may be useless sampling units but excellent strata:

Cluster 1	Cluster 2	Cluster 3	Cluster 4	Sum
1	11	51	101	164
2	12	52	102	168
3	13	53	103	172
4	14	54	104	176
Sum 10	50	210	410	680

Draw a random sample of four in two ways: one cluster, and one from each of the four clusters (strata) to estimate a total of 680:

Cluster: Four possible estimates: 40; 200; 840; and 1,640: mean = 680: range = 1,640 − 40 = 1,600.
One from each cluster:
minimum estimate = 164 × 4 = 656: mean = 680.
maximum estimate = 176 × 4 = 704: range = 48.

The difference arises from the fact that stratification by clusters *eliminates the variation between clusters* which in this case is very large. Using clusters as sampling units introduces this large between-cluster variation into the sampling error. Stated another way, the intraclass correlation coefficient is positive and very high, practically one, since the within cluster variance is so small.

Example: This is an example of where a cluster is better than a simple random sample of the same size. We assume four clusters of four elements each:

Cluster 1	Cluster 2	Cluster 3	Cluster 4
1	2	2	1
3	4	3	4
5	8	6	5
7	6	7	8
Sum 16	20	18	18
Mean 4	5	4.5	4.5

Select a random sample in two ways: select one cluster, and select four elements at random ignoring clusters. One randomly selected cluster gives four possible estimates of the aggregate of 72: 64; 80; 72; and 72, with a range of 16 and an average of 72 showing estimates are unbiased. The lowest estimate from a simple random sample of four elements is 1 + 2 + 2 + 1 = 6 so 6 × 4 = 24; the highest estimate is 7 + 8 + 7 + 8 = 30 so 30 × 4 = 120, a range of 96 compared with a range of 16.

The gain comes from the fact that the clusters are very much alike with about the same variances and means. This gives a negative intraclass correlation coefficient which reduces the sample variance below that of simple random sampling. Analysis of variance gives:

	Degrees of freedom	Sum of squares	Mean square
Between clusters	3	2	0.667
within clusters	12	82	6.833
Sum	15	84	5.600

Since M, the number of elements per cluster, is four:

$$r_a = \frac{0.667 - 6.833}{0.667 + 3(6.833)} = -0.291.$$

The standard error of the mean for four randomly selected elements is:

$$s_{\bar{x}}^2 = \frac{5.60}{4}(1 - 0.25) = 1.050;$$

$$s_{\bar{x}} = 1.02;$$

$$c_{\bar{x}} = \frac{1.02}{4.5} = 0.23.$$

The standard error of the mean for 1 randomly selected cluster where $M = 4$, $n = 1$ is:

$$s_{\bar{x}c}^2 = 1.05[1 + (4\text{-}1)(\text{-}0.291)] = 0.133;$$
$$s_{\bar{x}c} = 0.365;$$
$$c_{\bar{x}} = 0.08.$$

The standard error is reduced almost two-thirds by using the cluster sampling unit. This latter can be verified by calculating directly from the four cluster means 4, 5, 4.5, and 4.5. Their mean is 4.5; their variance is 0.125; the standard deviation is 0.354; and the coefficient of variation $c = 0.354/4.5 = 0.08$ or 8 percent — the same end results obtained from using the intraclass correlation coefficient.

We will now summarize some basic ideas about cluster sampling. If every cluster in the frame or population reflects exactly the characteristics of the population or frame, then any sample of one cluster will yield all of the information. This situation does not exist, but in some sample designs we may attempt artificially to develop heterogeneous clusters resembling the frame or population. More often, the opposite is the case; clusters, groups, classes or areas tend to be more alike within and different between units. Positive intraclass correlation dominates with both negative intraclass correlation and zero intraclass correlation being rare in actual practice.

If the positive intraclass correlation is high, the cluster should be a stratum in a stratified random sample, not a sampling unit. We may have to live with positive intraclass correlation in many situations because we have no alternative. If the intraclass correlation is close to zero, there may be little or no difference between using a cluster sample and a simple random sample of the same number of elements. A very low intraclass correlation, whether positive or negative, may be misleading because its effect is determined by the number of elements in a cluster. Where negative intraclass correlation exists, the use of clusters is better than the use of a simple random sample. In the example where r_a is −0.29 and the number of elements in a cluster was four, the gain was very substantial.

Elements in a cluster are selected at the same rate as the cluster itself. Random selection of cluster units gives unbiased estimates of aggregates and means providing the entire cluster is included and the proper weights are applied to the sample of clusters.

Replicated sampling. In replicated sampling two or more independent samples of the same size are selected at random from the total frame or population. Usually, systematic sampling is used with random starts for each replicate, but if there is any danger of periodicity in the values of the frame or population, then an independent random sample has to be selected from each zone.

Replicated sampling is simple and powerful allowing estimates, including their standard errors, to be made directly from replicate values. This eliminates the complexities of sample design and calculation in designs involving clusters and stages. Replicated sampling requires a frame of elemental sampling units in the form of records, files, listings, computer tapes, etc. It requires careful planning and supervision and the coding of each replicate sampling unit since each replicate has to be tabulated separately, as well as all of them combined. If stratification is used, a replicated sample is designed for each stratum as though it was a separate frame, but the number of replicates is kept the same.

Example: Five replicates of five values each were selected independently with random starts from the frame of 40 values given in the section on stratified random sampling. Systematic sampling was used with random starts and a skip interval of eight since the sampling rate is five in 40 or one in eight. The results are as follows using random starts of 3,4,5,7, and 8:

Selection number	x	Selection number	x	Selection number	x	Selection number	x	Selection number	x
3	2390	4	1040	5	1880	7	2344	8	1199
11	1710	12	1439	13	1906	15	2113	16	1628
19	4532	20	1858	21	2443	23	2210	24	1089
27	1012	28	2377	29	3196	31	2174	32	1000
35	1098	36	1302	37	1023	39	3386	40	3750
Mean	2148		1604		2090		2445		1733

The frame mean is 1925 so two estimates are low and three are high. With such a small sample, we do not expect to obtain estimates of either the mean or variance without considerable sampling variation as has been pointed out in the section on stratified random sampling. In that section, it was shown that stratified sampling, using four income classes as strata, gave very much better estimates than simple random or systematic sampling.

The basic equations in replicated sampling are:

1. Width of zone Z:

$$Z = \frac{kN}{n} = \frac{k}{f} = kw,$$

where Z is the interval of selection numbers and is constant; k = number of replicates; N = number of units in frame; n = sample size; f = sampling rate = n/N; and w = 1/f = weight or multiplier.

2. Number of zones m:

$$m = \frac{N}{Z} = \frac{n}{k}.$$

3. Weight for entire sample w:

$$w = \frac{N}{n} = \frac{Z}{k}.$$

4. Weight for one replicate w_i:

$$w_i = kw = \frac{kN}{n} = Z.$$

5. Estimate of aggregate X for entire sample:

$$X = wx = \frac{N}{n}x = \frac{Z}{k}x,$$

where $x = \sum\sum x_{ij}$, and summation is over entire sample: i replicates, j zones.

6. Estimate of aggregate X_i for one replicate:

$$X_i = w_i x_i = Zx_i,$$

where $x_i = \sum x_j$, where summation is over j zones in each replicate.

7. Standard error of estimate Θ:

$$\text{range method: } s_\Theta = \frac{\Theta\ \text{max} - \Theta\ \text{min}}{k}$$

$$\text{sum of squares method: } s_\Theta^2 = \frac{1}{k(k-1)}\sum(\Theta_i - \bar{\Theta})^2,$$

where i is summed over k replicates.

$$s_\Theta = (s_\Theta^2)^{1/2}.$$

8. Estimate of aggregate X for stratified sample for one replicate:

Assume h strata, then $Z_1, Z_2, \ldots, Z_h$ are strata multipliers for one replicate. Let $x_1, x_2, x_3, \ldots, x_h$ be the sample sums for each of the h strata for one replicate. This assumes Z's are different for different strata.

Estimate of aggregate X for replicate one for all strata is:

$$X^{(1)} = Z_1x_1 + Z_2x_2 + \ldots + Z_hx_h.$$

Similarly, calculate $X^{(2)},X^{(3)},X^{(4)},\ldots,X^{(k)}$ for each of the k replicates.

9. Estimate of X for the frame:

In terms of replicates:

$$X = \frac{X^{(1)} + X^{(2)} + \ldots + X^{(k)}}{k}.$$

In terms of the entire sample:

$$X = w_1x_1' + w_2x_2' + \ldots + w_hx_h';$$

where $w_1,w_2, \ldots, w_h$ = stratum weights, $x'_1,x'_2, \ldots, x'_h$ = stratum sums — all replicate sums combined in each stratum.

In terms of the entire sample:

$$X = \frac{N_1}{n_1}x_1 + \frac{N_2}{n_2}x_2 + \ldots + \frac{N_h}{n_h}x_h.$$

All of these give the same answer if $w_v = \frac{N_v}{n_v}$.

10. For the case where k = 2:

In general, $s_{\bar{x}}^2 = \frac{1}{k(k-1)}\sum (x - \bar{x})^2$.

For k = 2, this reduces to $\frac{1}{2}\sum (x - \bar{x})^2$.

If $d = x_1 - x_2$, then for two replicates $\sum (x - \bar{x})^2 = \frac{d^2}{2}$.

For the total sample:

$$X = \frac{Z}{k}x = Z\bar{x},$$

$$s_X^2 = Z^2 s_{\bar{x}}^2 = \frac{Z^2d^2}{4},$$

$$s_X = \frac{Zd}{2}.$$

For g independent groups with a common Z:

$$s_X^2 = \frac{Z^2}{4}\sum d_i^2,\ i = 1,2,3,\ldots,g;$$

$$s_X = \frac{Z}{2}(\sum d_i^2)^{1/2}.$$

Example: Replicated sampling with random selection in every zone is illustrated by a program which was written for a programmable calculator to generate and to select a random number in every zone for every replicate, for as many replicates as desired. The basic equation for the zone width is $Z = \frac{kN}{n}$, where k is the number of replicates, N is the number of sampling units in the frame or population, and n is the total sample size. In each replicate there are n/k units if n is an exact multiple of k; otherwise, one replicate may have one more value than another. Even if N and n are known, Z may not be integral so some convenient but close number to the calculated Z is used. If Z is 39, 41, or 42, it is likely 40 would be used for ease of application; with a computer this would be unnecessary. The limits of the zones are 1 to Z, Z+1 to 2Z, 2Z+1 to 3Z, 3Z+1 to 4Z, etc.

A random number generator is in the machine and the generating procedure is used in the program. This establishes each zone interval and reads out the first random digit which falls in the interval. Since many random numbers fall outside the interval, some random numbers in an interval appear quickly while others require several seconds. In the program Z is calculated and divided by 10,000. Limits are set on each zone. Random numbers are generated in decimal form and then converted to whole numbers by using the "integer x" program step. The program handles N up to 10,000 but the program can be easily changed to include N's up to 100,000. Each replicate requires insertion of a new 12 digit decimal number in the program to generate a new set of random numbers. The decimal digits used in the random number generator for each of the four replicates in the example are given in Table 2 at the top of the column for each replicate.

The program was applied to a problem where N = 2000, n = 200, k = 4; hence Z = 40: $Z = \frac{4(2,000)}{200}$, and f = 1/10. Random numbers were generated and posted for four replicates of 50 values each since there are 50 zones, for a total sample of 200 as required. These random numbers by replicate by zone are given in Table 2. Five random numbers occurred at an interval break; these were raised by one. Random substitutions were also made for duplicates in eight zones, although some may prefer to leave them alone.

The table can be used to select and to code manually four replicates of 50 units each from a frame of 2,000 units, providing the 2,000 numbers are identified with serial numbers 1 through 2,000, or some other series of 2,000 consecutive numbers. It can be applied to a frame divided into sets of 2,000 serial numbers.

Estimating with a sample with probabilities proportional to size (pps). This method of sampling is illustrated by selecting a random sample of five colleges with probabilities proportional to the number of teachers x in order to estimate the total number of students Y and the average number of students per college.[4] The frame values are: N = 52 colleges, X = 5218 the total number of teachers, Y = 100,600 students, and mean u = 1936 students per college. The purpose is to estimate Y with a random sample of n = 5 selected with probabilities proportional to the number of teachers. The number of teachers was cumulated by college from 1 to 5218; the first and last two intervals are:

1-56	5085-5164
57-123	5165-5218

Table 2

Random Numbers for Four Replicates of 50 Each

Z interval		.623950481725	.312795480638	.312795480681	.623481950752
1-	40	17	3	12	29
41-	80	71	70	65	64
81-	120	100	99	94	106
121-	160	154	160	150	132
161-	200	188	179	195	168
201-	240	206	223	231	215
241-	280	258	278	280	263
281-	320	298	313	282	297
321-	360	344	355	323	321
361-	400	392	393	363	397
401-	440	409	426	408	418
441-	480	479	480	447	470
481-	520	481	485	499	519
521-	560	537	551	535	556
561-	600	575	586	582	588
601-	640	606	631	624	630
641-	680	641	668	670	654
681-	720	710	691	682	698
721-	760	732	742	758	735
761-	800	785	773	796	775
801-	840	830	813	824	805
841-	880	878	872	865	858
881-	920	910	914	894	893
921-	960	941	947	949	944
961-	1000	996	986	972	961
1001-	1040	1038	1028	1020	1015
1041-	1080	1076	1080	1077	1074
1081-	1120	1113	1090	1081	1111
1121-	1160	1122	1134	1127	1139
1161-	1200	1183	1179	1174	1181
1201-	1240	1219	1215	1236	1201
1241-	1280	1249	1268	1279	1267
1281-	1320	1312	1309	1291	1282
1321-	1360	1330	1350	1357	1356
1361-	1400	1375	1386	1370	1382
1401-	1440	1414	1419	1413	1401
1441-	1480	1459	1448	1442	1445
1481-	1520	1514	1497	1508	1494
1521-	1560	1528	1537	1546	1547
1561-	1600	1576	1560	1595	1562
1601-	1640	1629	1603	1612	1601
1641-	1680	1657	1646	1651	1668
1681-	1720	1713	1706	1702	1686
1721-	1760	1753	1756	1742	1729
1761-	1800	1766	1784	1778	1786
1801-	1840	1836	1803	1838	1807
1841-	1880	1841	1862	1853	1845
1881-	1920	1900	1911	1904	1881
1921-	1960	1933	1932	1926	1951
1961-	2000	1980	1998	1996	1987

The five random numbers falling in the sample are 554, 1174, 1487, 2315, and 3897. The number of teachers x_i and students y_i associated with these colleges are listed in the table below.

The probability of selection of each sample college is $P_i = x_i/X = x_i/5218$. The estimates are:

$$\text{Aggregate } Y_{est} = N\bar{z} = \frac{1}{n}\sum \frac{y_i}{P_i} = \frac{X}{n}\sum \frac{y_i}{x_i}.$$

$$\text{Mean } \bar{z} = \frac{1}{n}\sum z_i = \frac{1}{n}\left(\frac{X}{N}\sum \frac{y_i}{x_i}\right),\ i = 1,2,\ldots,n.$$

The basic data are: $n=5$; $X=5218$; $N=52$; $X/N=100.346$:

no. teachers x_i	$P_i = \frac{x_i}{X}$	no. students y_i	$\frac{y_i}{x_i}$	$z_i = \frac{Xy_i}{Nx_i}$
342	0.065542	5689	16.635	1669.256
539	0.103296	9753	18.095	1815.761
188	0.036029	5512	29.319	2942.044
134	0.025680	2236	16.687	1674.474
385	0.073783	5518	14.332	1438.159
			sum 95.068	9539.694

Calculations: $Y_{est}=N\bar{z}=52\left(\frac{9539.708}{5}\right)=52(1907.937)$
$=99,213.$

Mean $\bar{z}=\frac{1}{5}(9539.694)=1907.94=1908,$
$1\text{-}f=1\text{-}5/52=0.9038.$

$$s_{\bar{z}}^2=\frac{1\text{-}f}{n(n\text{-}1)}\Sigma(z_i\text{-}\bar{z})^2=63,719.65;\ s_{\bar{z}}=252.43;\ c_{\bar{z}}=0.13230.$$

$$s_{Y_{est}}=99,213(0.1323)=13,126.$$

The mean is 1,908 compared with 1,936, the number of students is 99,213 compared with 100,690—about a 1.5 percent difference with a sample of only 5.

Sampling with probabilities proportional to size has real advantages if the frame is relatively small, but accumulating the sampling units of the frame and identifying the sample becomes prohibitive for very large frames — those with tens or hundreds of thousands of units or tens of millions of units. In these latter situations, stratification is not only easier but more feasible, cheaper, and just as effective.

Two stage sampling. In a two stage sample, the first sample is selected from a frame of first stage units; then a second sample of entirely different units is selected from each of the first stage sample units. Examples are:

First stage	Second stage
city blocks	houses
city blocks	households
city blocks	dwelling units
sections (640 acres)	farms
counties	farms
farms	fields
schools	classes

This method is illustrated by selecting at random 15 blocks in a city residential community consisting of about 270 blocks; then selecting houses at random within each of the 15 sample blocks. The purpose is to estimate the total number of persons. The number of houses in the 15 blocks is an actual count, but the number of persons living in a sample house is assumed for illustrative purposes. (See Table 3.) The basic characteristics are:

First stage: $N=270$ blocks; $n=15$ blocks; rate $= f=1/18$

Second stage: total number of houses in 15 blocks $M_j=314$; rate of selecting houses within a sample block $=1/5$; $m_j=$ sample size within each block and varied from 2 to 8.

The number of persons in each sample house in each sample block, with intermediate calculations, are given in the table. The calculations are as follows, where $f=n/N$; $j=1,2,3...,15$; $\bar{x}_j=$ average number of persons in m_j houses; $m_j=$ number of sample houses in jth block; $f_j=m_j/M_j$;

$$f_j = m_j/M_j;\ s_j^2 = \frac{1}{m_j\text{-}1}\sum(x\text{-}\bar{x})_j^2,$$

where summation is over the m_j houses and x is the number of persons in a sample house.

1. $X'=\sum_j M_j\bar{x}_j=1,150$ persons in 314 houses in 15 blocks.

2. Mean number of persons per house

$$\bar{x} = \frac{1149.6}{314} = 3.667 = \frac{231}{63}.$$

3. Estimated total number of persons

$$X = \frac{270}{15}(1149.6) = 20,693.$$

4. Estimated total number of houses

$$Y = \frac{270}{15}(314) = 5,652.$$

5. The variance of $X=s_b^2+s_w^2$, where the first is the variance between blocks and the second is the variance within blocks.

$$s_b^2 = \frac{N^2(1\text{-}f)}{n(n\text{-}1)}\sum_j(X_j - \bar{x})^2 = 5,593,561,$$

where $\sum_j(X_j - \bar{X})^2 = 17,061.$

$$s_w^2 = \frac{N}{n}\sum_j\frac{M_j^2s_j^2}{m_j}(1 - f_j) = \frac{270}{15}(3,431.43) = 61,766.$$

6. $s_X^2 = 5,655,327;\ s_X = 2378;$

$$c_X = \frac{2378}{20,693} = 0.115 \text{ or } 11.5 \text{ percent.}$$

Practically all of the variation is between blocks due to their variation in size; the within variance is small due to the highly limited range of persons in a house, usually from one to six. Clearly, selecting more houses within blocks will not improve the estimate. What is needed is to divide the larger blocks to obtain a more uniform distribution of houses by blocks or parts of blocks and to select a larger sample of blocks.

Other approximations to X, the total number of persons, are:

1. $X = 231(5)(18)$
$= 20{,}790$

Number of persons in the sample (231) multiplied by the inverses of the two sampling rates, 1/18 and 1/5; unweighted by number of houses in a block.

2. $X = \frac{231}{63}(314)(18)$
$= 20{,}724$

Average number of persons per house in entire sample times 314 sample households times block weight. This is very close to 20,693 derived from weighted block size in the sample blocks.

Table 3

Data for Two Stage Sample

Sample block number n = 15	Total houses in block = M_j	Number sample houses = m_j	x = number of persons living in house								Sum
1	12	2	3	2							5
2	39	8	5	3	4	1	6	3	4	4	30
3	24	5	2	4	3	5	5				19
4	10	2	2	4							6
5	11	2	6	2							8
6	18	4	3	6	4	2					15
7	15	3	2	3	5						10
8	14	3	2	4	6						12
9	27	5	6	2	5	1	3				17
10	27	5	3	5	5	2	4				19
11	30	6	5	2	4	6	3	4			24
12	25	5	4	3	5	2	6				20
13	17	4	2	2	5	4					13
14	15	3	6	2	3						11
15	30	6	4	6	2	5	3	2			22
Sum	314	63									231

Sample block number n = 15	$X_j = M_j\bar{x}_j$	$\bar{x}_j$	s_j^2
1	30.00	2.50	0.50
2	146.25	3.75	2.21
3	91.20	3.80	1.70
4	30.00	3.00	2.00
5	44.00	4.00	8.00
6	67.50	3.75	2.92
7	50.00	3.33	2.33
8	56.00	4.00	4.00
9	91.80	3.40	4.30
10	102.60	3.80	1.70
11	120.00	4.00	2.00
12	100.00	4.00	2.50
13	55.25	3.25	2.25
14	55.00	3.67	4.33
15	110.00	3.67	2.67
Sum	1149.60 $\underline{\ }1150$ $\bar{X} = 76.64$		

Sequential sampling. Sample size is usually fixed but in sequential sampling it is a variable — the random sample either in the form of individual elements or in groups is continued until a decision is reached. Its primary advantage is that it requires a smaller sample, sometimes only one half as large a sample, but this merit is offset by difficulties which usually arise in administering or implementing this type of sample.

This technique provides for taking a series of random observations, assuming they come from a population of values normally distributed, and making a test after each observation or group of observations until a decision is reached to accept or reject some hypothesis. This hypothesis may be that a lot of manufactured or purchased products are of acceptable quality or that an experimental or new method or product is better than a standard method or a present product.

Sequential sampling for lot acceptance or rejection assumes a binomial model and four points on the operating characteristic curve, values which are prescribed to give the necessary protection against rejecting good lots or accepting poor lots:

A = producer's risk of rejecting a good lot, corresponding to proportion defective p_1.

B = consumer's risk of accepting a poor lot, corresponding to proportion defective p_2.

The points are: $(0,1)$ $(p_1, 1-A)$ (p_2, B) $(1,0)$.
A and B are expressed as probabilities of acceptance; clearly, we want both of these to be low. In practice, they are both less than 10 percent or 0.10 and often less than five percent or 0.050.

The control lines are two parallel straight lines which establish a zone of rejection (above upper line), a zone of acceptance (below lower line), and a zone between these two lines which calls for continuation of sampling as long as a point falls in this zone.

Equations. The plan requires that values of A, B, p_1, and p_2 be specified. These values are based on experience, on established practice, on the risks to be taken, on the quality desired, on the costs and other effects of rejecting acceptable quality product, and on accepting an inferior quality product.

1. $c = \log \frac{1 - B}{A}$.

2. $d = \log \frac{1 - A}{B}$.

3. $g_1 = \log (p_2/p_1)$.

4. $g_2 = \log \left(\frac{1 - p_1}{1 - p_2}\right)$.

The two parallel lines with slope b and y intercepts a_2 and $-a_1$ are:

1. upper line:

$$y_2 = \frac{c}{g_1 + g_2} + \frac{g_2}{g_1 + g_2} n = a_2 + bn.$$

2. lower line:

$$y_1 = \frac{-d}{g_1 + g_2} + \frac{g_2}{g_1 + g_2} n = -a_1 + bn.$$

In these equations, n = sample size and is cumulative either by single units or by groups and y = cumulative number of errors, defects, or failures. Each point n_i, y_i represents a point on the chart. When a point falls below the lower line, the lot is accepted; when it falls between the two lines, sampling is continued; and when it falls above the upper line, the lot is rejected.

The average sample sizes required to reach a decision relative to p_1 and p_2 are:

$$\bar{n}_{p_1} = \frac{(1 - A)d - cA}{(1 - p_1)g_2 - p_1 g_1}.$$

$$\bar{n}_{p_2} = \frac{(1 - B)c - dB}{p_2 g_1 - (1 - p_2)g_2}.$$

Values are assigned to n and the upper and lower line values of y calculated from the equations. Then y_i values are assumed and the appropriate action noted. *Note that rejection values can be computed for various values of n* by assuming whatever values are desired for A, B, p_1, and p_2.

Example: Let A = 0.02, B = 0.04, $p_1 = 0.02$, and $p_2 = 0.06$. Then:

$$c = \log \frac{0.96}{0.02} = 1.68124,$$

$$d = \log \frac{0.98}{0.04} = 1.38917,$$

$$g_1 = \log \frac{0.06}{0.02} = 0.477121,$$

$$g_2 = \log \frac{0.98}{0.94} = 0.018098,$$

$$g_1 + g_2 = 0.495219.$$

The upper line equation is:

$$y_2 = 3.395 + 0.0365n.$$

The lower line equation is:

$$y_1 = -2.805 + 0.0365n.$$

Points n_i y_i	Lower line y value	Upper line y value	Action
20 4	-2.08	4.13	continue sampling, y_i within lines.
30 5	-1.71	4.49	reject, y_i above upper line.
40 6	-1.35	4.86	reject, y_i above upper line.
50 4	-0.98	5.22	continue sampling.
60 5	-0.62	5.59	continue sampling.
70 6	-0.25	5.95	reject, y_i above upper line.
100 1	0.85	7.05	continue sampling.
150 2	2.67	8.87	accept, y_i below lower line.
200 4	4.50	10.70	accept, y_i below lower line.

Note that with a sample of 100, a value of seven will reject the lot showing seven percent in error in the sample; with a sample of 150, a value of nine will reject the lot showing six percent in error in the sample; and with a sample of 200, 11 will reject the lot showing 5.5 percent in error in the sample. All of this can be done by simply using the equations derived above.

Sample Sizes

1. To detect at least one (1) event:

$$1.\ n = \frac{\ln a}{\ln q},$$

where n = sample size: 1-a = probability of detection and is set or assigned: q = 1 - p; p = probability of the event.

$$2.\ n = \frac{\ln b}{\ln x - \ln (x-1)},$$

where p = 1/x (x and b are integers); a = 1/b; 1 - 1/b = probability of detection.

2. To estimate mean and aggregate:

$$1.\ n = \frac{s_x^2 z^2}{d^2},$$

where s_x = standard deviation of x; z = normal deviate, e.g., 1.96 for 95 percent level; d = $\bar{x}$-u; *absolute deviation* from mean; use Nd for aggregate.

$$2.\ n' = \frac{s_x^2 z^2}{d^2 + s_x^2 z^2/N},$$

where N = population or frame size; this corrects for sampling from a finite frame or population.

$$3.\ n = \frac{c^2 z^2}{p^2},$$

c = coefficient of variation of x; z = normal deviate; p = **relative error** to be tolerated in the mean.

$$4.\ n' = \frac{c^2 z^2}{p^2 + c^2 z^2/N},$$

n' = sample size taking into consideration sampling from a finite frame or population N.

$$5.\ n' = \frac{nN}{n + N} = \frac{n}{1 + n/N},$$

n = sample size ignoring finite population correction.

Table 4 shows the coefficient of variation of six different distributions which can be used in designing samples. This is a very important use of the coefficient of variation of a variate, a use that apparently was overlooked by its originator Karl Pearson.

Table 4

Means, Standard Deviations, and Coefficients of Variation for Six Distributions*

Shape of distribution	mean u	standard deviation σ	coefficient of variation Cv
1. Exponential $y = a e^{-ax}$	$\frac{1}{a}$	$\frac{1}{a}$	1
2. Right triangle	$\frac{h}{3}$	$\frac{h}{3(2)^{1/2}}$	0.71
3. Rectangle	$\frac{h}{2}$	$\frac{h}{2(3)^{1/2}}$	0.58
4. Ellipse	$\frac{h}{2}$	$\frac{h}{4}$	0.50
5. Isoceles triangle	$\frac{h}{2}$	$\frac{h}{2(6)^{1/2}}$	0.41
6. Normal (6 sigma)	$\frac{h}{2}$	$\frac{h}{6}$	0.33

h = range of distribution along x axis from x = 0 to x = h

*The original table in W. Edward Deming, *Some Theory of Sampling*, Wiley, New York, 1950, p. 62, has been expanded and arranged in descending order of the coefficients of variation which were not given in the original table. The exponential distribution extends from x = 0 to x = + infinity; this distribution was added.

3. To estimate a binomial proportion:

1. $$n = \frac{pqz^2}{d^2},$$

where p = proportion known, assumed, estimated; q = 1-p; d = *absolute deviation* allowed in estimation proportion.

2. $$n' = \frac{pqz^2}{d^2 + pqz^2/N},$$

where n′ = sample size, taking into consideration sampling from a finite population of size N.

3. $$n = \frac{(pz^2 - p^2z^2)}{d^2},$$

where variates are the same as in 1. above, but changed to facilitate calculating and computer programming.

$$n = \frac{pz^2 - pz^2(p)}{d^2},$$

$$n = \frac{z^2(p - p^2)}{d^2},$$

where this form is the easiest and fastest to use on a desk or hand-held calculator: square z, multiply by p, move up, multiply by p, subtract, square d, divide.

4. To control both Type 1 and Type 2 errors on the mean and proportion:

1. $$n = \frac{(z_1s_1 + z_2s_2)^2}{d^2},$$

where $d = u_1 - u_2$.

$$\alpha = Pr(z > z_1),$$

where s_1^2 = variance of x_1.

$$\beta = Pr(z < z_2),$$

where s_2^2 = variance of x_2.

2. $n=\frac{s^2}{d^2}(z_1+z_2)^2$, if $s_1=s_2=s$.

3. $n=\frac{4s^2z^2}{d^2}$, if $z_1=z_2=z$.

This shows that the sample size is increased four times if $z_1=z_2$ and the variances are the same. Also, any combination, $z_1+z_2=$constant, gives the same sample size n, if s and d remain the same. In this case, increasing the risk of one type of error decreases the risk of the other. For the binomial proportion set, $s_1=(p_1q_1)^{1/2}$ and $s_2=(pq)^{1/2}$, and $d=p_1 - p_2$.

5. To estimate the ratio $R=\frac{X}{Y}=\frac{\bar{x}}{\bar{y}}$, or an aggregate or total $X=\frac{x}{y}Y$:

1. $n=\frac{1}{c_R^2}(c_x^2+c_y^2 - 2r_{xy}c_xc_y)$,

   where X=estimated total; Y=known total; x=sample total of x_i; y=sample total of y_i; c_R=coefficient of variation of R and is desired, set; c_x=coefficient of variation of x and is known or estimated; c_y=coefficient of variation of y and is known or estimated; and r_{xy}= correlation coefficient of x and y and is known or estimated; positive correlation increases precision and reduces sample size.

2. If necessary, recalculate sample size to take into consideration sampling from a finite population of size N. This reduces sample size:

$$n'=\frac{n}{1+n/N}.$$

6. To determine sample size for the mean using the t distribution for small samples:

1. $n=\frac{t^2s^2}{(x - u)^2}=\frac{t^2s^2}{d^2}$,

   where d=assumed deviation from the mean u; s^2= variance of x.

   Solve for $c=s^2=\frac{1}{n-1}\left[\Sigma x^2 - \frac{(\Sigma x)^2}{n}\right]$.

   In the table, find $\frac{n}{t^2}=c$. At the desired level in the t table, find $\frac{n}{t^2}$ to the nearest c.

   Example: d=2, s=3.4, one percent level and two tailed test:

$$\frac{n}{t^2}=\left(\frac{3.4}{2}\right)^2=2.89.$$

   Therefore, n must equal $2.89t^2$. For 22 degrees of freedom (n-1) at P=0.005, t=2.819, and

$$\frac{23}{2.819^2}=2.89.$$

   Therefore, n=23.

Sample sizes and related data for lot sampling by *attributes* are given in MIL-STD-105-D, a reproduction of the major parts are given by Burr[(5)]. The tables for single sampling are given in the appendix of this book. Sample sizes and related data for lot sampling by *variables* are given in MIL-STD-414. An excellent book on this latter subject is by Bowker and Goode;[6] Burr also has a chapter on it.

Notes

[1]A. C. Rosander, *Elementary Principles of Statistics,* Van Nostrand, New York, 1951. Copyright by Wadsworth, Inc. Used by permission of Brooks Cole Publishing Co., pp. 412-415, 423-424.

[2]For detailed examples of where sampling can go wrong in practice see A. C. Rosander, *Case Studies in Sample Design,* Marcel Dekker, Inc., New York, 1977, Chapter 6, reprinted by courtesy of Marcel Dekker, Inc. Also see "Sampling Misunderstood," *Quality Progress,* June 1979, p. 4.

[3]See the test for random order of a large systematic sample by A. C. Rosander, ICC, *I and S Docket M 22930,* Exhibit H 38. The data are given and discussed in Chapter 7.

[4]*The World Almanac,* 1971, p. 159.

[5]I. W. Burr, *Statistical Quality Control Methods,* Marcel Dekker, Inc., New York, 1976, Chapter 10.

[6]A. W. Bowker and H. P. Goode, *Sampling Inspection by Variables,* McGraw Hill, New York, 1952.

16

Relationship Analysis

Relationship analysis is for the purpose of determining whether or not a relationship exists between two or more factors, attributes, or variables, and if it does exist, to what degree. It provides a means of:

1. Measuring progress.
2. Measuring trends.
3. Setting standards.
4. Measuring and appraising performance.
5. Isolating potential causes, e.g., a major cause of error.
6. Measuring and controlling a variable y, e.g., output, knowing variable x, e.g., input.
7. Measuring the effectiveness of corrective actions, e.g., training.
8. Controlling costs.
9. Measuring the effectiveness of different kinds and degrees of interventions, e.g., procedures, processes, methods, and dosages.

Methods of measuring relationships. Several methods are available ranging from simple plots to regression analysis to model building. The simplest methods are often the most effective, are easy to apply, and a first step toward a more effective statistical treatment if this is justified.

1. The *time plot* or time trend is a simple graph with time plotted horizontally (x axis) and some count, measurement, or estimate plotted vertically (y axis). *Time* can be measured in minutes, hours, days, weeks, months, quarters, or years. *Counts* can include number of errors, numbers of defects, number of customers, number of patients, number of flights, number of policies, number of loans, number of checks, number of tax returns, number of sales, number of orders, number of claims, or number of employees. The possibilities are numerous. *Measurements* include dollar volume of sales, bills, claims, checks, taxes, premiums, receipts, expenditures, costs, salaries, or wages. *Estimates* include averages, percentages, ratios, totals, or aggregates.

Examples are error rates, average error rates, number of errors, number of units produced, number of units produced of acceptable quality, processing time, average processing time, production time per unit of output, delay time, average delay time, percent of produced units meeting standard, percent of specific operations meeting time standards, air pollution as a function of the hour of the day (CO, nitrogen oxides, and ozone), or number and severity of automobile accidents as a function of each hour of the day and night.

2. Two important time plots are the *growth curve* and the *learning curve.* The growth curve shows growth, improvement, change, progress, or an increase in capability as a function of time. The learning curve shows the reduction in errors and error rates as a function of time during which learning is taking place and causes of error are being eliminated as a result of training, tutoring, special instruction, pay incentives, and the like. The learning curve is also reflected in the production curve, which shows the increase in acceptable quality production per person as a function of learning time. This is to be expected as the errors are eliminated, unnecessary steps are eliminated, and the time required to produce an acceptable quality unit is reduced.

3. The *two variable plot* is a plot of empirical data consisting of pairs of values (x_i, y_i) which represent one point on the chart (rectangular graph paper). The purpose is to determine if variables x and y are related, and if so to what degree and whether the relationship (plot of points) falls in a linear rectangular zone or on a curve or in a curved zone.

Linear plots are easier to work with because of their relative uniqueness; the difficulty with non-linear or curvilinear plots is the countless paths they take. Linear plots, however, should be used with caution outside the range of the data available, because outside the range the relationship may be curvilinear. Indeed, over a limited range, many curves can be closely approximated by a straight line.

Examples of two variable plots are: volume of production versus number of employees, volume of production versus number of man hours, output versus input, delay time versus length of line, processing time versus volume of backlog, miles per gallon versus speed of automobile, processing time versus difficulty of document, and quality costs versus volume of errors made.

4. The 2×2 or r×c frequency table. In a two way frequency table in which the entries are counts, the relationship between the row factor and the column factor is measured by testing for the independence of the two factors using the probability test for independence.

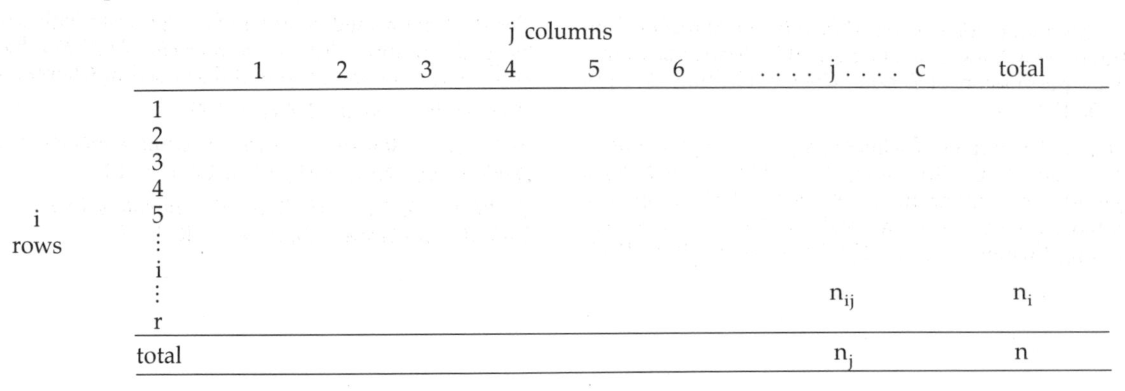

	j columns							
	1	2	3	4	5	6	. . . . j c	total
1								
2								
3								
4								
i rows 5								
⋮								
i								
⋮							n_{ij}	n_i
r								
total							n_j	n

With fixed marginal totals n_i and n_j, if the frequency at the intersection of a column and a row is independent, then its probability is:

$$p_{ij} = p_i p_j = \frac{n_i}{n}\frac{n_j}{n}.$$

If independence exists, then the expected frequency in the ijth cell is:

$$n_{ije} = n_{ij}\text{ (expected)} = np_{ij} = n\frac{n_i}{n}\frac{n_j}{n} = \frac{n_i n_j}{n}.$$

This expected value is used to compute the chi square with (r-1)(c-1) degrees of freedom. Without correction for continuity:

$$\chi^2 = \sum_i\sum_j\frac{(n_{ij} - n_{ije})^2}{n_{ije}}.$$

The 2×2 table is a special case with one degree of freedom. In this table only one expected value need be calculated; all the others are found by subtraction since the marginal totals are fixed. Examples are fatal and non-fatal auto accidents versus seat belt used or not used, fatal and non-fatal auto accidents versus driver training or not, age and sex versus driving under influence or not in auto accidents, and type of error versus training or no training.

5. Linear regression. In linear regression of two variables a straight line,

$$y = a + bx,$$

is fitted to the data using the method of least squares; it yields values for a, the intercept on the y axis, and for b the slope of the line. The correlation coefficient squared (r^2) measures the extent to which the variable x accounts for the variation in y; it is a proportion varying from 0 to 1.

For three variables the linear relation is a plane $z = a + bx + cy$ which is usually written in the form:

$$y = a + b_1x_1 + b_2x_2,$$

where y is the dependent variable and x_1 and x_2 are the independent variables which presumably correlate with the characteristic y. This equation can be extended to any number of variables and is known as multiple regression.

6. Non-linear regression. Many relationships are curvilinear or non-linear, which cannot be adequately expressed by the equation of a straight line or plane. Those most often used in practice include the following:

1. The parabola	$y = a + bx + cx^2$
2. The square root parabola	$y = a + bx + c(x)^{1/2}$
3. The hyperbola	$y = \frac{x}{a + bx}$ or $xy = c$
4. The power function	$y = ax^b$
5. The exponential function	$y = ae^{-ax}$
6. The exponential function	$y = ab^x$

A test for each of the last four functions is whether each forms a straight line when expressed in terms of logarithms. These are described later in this chapter. When linearized, the values of the constants a, b, and c can be determined from the straight line fitted to the data. For the hyperbola $z = \frac{x}{y} = a + bx$; a and b can be determined by a least square fit to the paired data (z_i, x_i). The constants a, b, and c for the parabolas can be determined by the method of least squares; this leads to three equations in three unknowns so unique values are obtained for the three constants.

Some examples. Examples of relationship analysis in the service industries are the following:

1. Learning curves for individuals or groups: error rate y as a function of time t.

2. Learning curves for individuals or groups: number of units of acceptable production y as a function of time t.

3. Learning curves for groups: average number of units of acceptable quality per day per person y as a function of the number of days x.

4. Average error rate y as a function of time t.

5. Production time per unit y as a function of time t.

6. Individual time to perform operation y as a function of time t.

7. Average time to perform operation y as a function of time t.

8. Number of days to pay claim y as a function of the size of the claim x and the type of claim.

9. Average cost per unit of acceptable quality product y as a function of time t.

The straight line. The straight line shows how a variate y measured vertically on ordinary graph paper is a function of a second variate x measured horizontally. A point on the line requires a pair of values (x,y). Many of the types of relationships described in the preceding section may be represented by a straight line, that is, have a linear as contrasted with a curvilinear relationship.

A common form of the straight line is $y = a + bx$, where a, the y intercept, is the distance to the origin $x = 0$, $y = 0$ from the point where the line intersects the y or vertical axis. If a is above the origin, it is positive in sign; if it is below the origin, it is negative in sign. The letter b is the slope of the line; it shows how much y increases or decreases for a unit change in x. A negative b means y decreases as x increases; a positive b means y increases as x increases.

An excellent example is a distillery where y is the output in 1,000 gallons of alcohol and x is the input in 10,000 pounds of corn which were recorded for each of 12 months in a year. These 12 points all practically fell on the straight line $y = 0.209 + 1.1939x$. (See Figure 1.) If we set $x = 500$, then $y = 0.209 + 1.1939(500) = 597.157$. Therefore, when $x = 500(10,000) = 5,000,000$ pounds, $y = 597.157(1000) =$ 597,157 gallons. The error in this latter figure is practically negligible since 99.7 percent of the variation in y is accounted for by x; it is subject only to measurement and recording errors.

$$\text{slope } b = \frac{\text{gallons of alcohol}}{\text{pounds of corn}} = \frac{1.1939\ (1{,}000)}{1\ (10{,}000)} = \frac{11.94}{100}$$

Hence, 11.94 gallons are produced per 100 pounds of corn or 8.38 pounds of corn are required to make one gallon of alcohol. These factors and this line can be used to predict and to control production with a very high degree of accuracy, providing, of course, that this relationship continues to hold. Sample tests can be run from time to time to determine if these factors are holding or whether a definite shift has taken place.

Data from five other distilleries showed that the straight line was a valid and excellent expression of the relation between x and y. Even though no two lines were identical or even close to one another, the x values explained 96 to 99 percent of the variation in y. In the author's experience, it is very unusual to find that the straight line expresses the relationship between factory production characteristics, or any economic characteristics so well. It shows that the straight line is not always a rough approximation to a complex relationship between x and y, although it often is.

Since distilleries are subject to Federal Government regulation, this method was proposed as an effective but simple means of controlling one important production process. There was nothing to stop a distillery from using it also. It shows that the input-output relationship can be a powerful means of control.

The correlation coefficient. The correlation coefficient, r, is a number between -1 and $+1$ that measures the direction and extent of the relationship between two variables x and y. Three simple examples show how x and y are related when $r = +1$, $r = -1$, and $r = 0$.

	$r = +1$		$r = -1$		$r = 0$
x	y	x	y	x	y
1	7	1	15	1	9
2	9	2	13	2	10
3	11	3	11	3	12
4	13	4	9	4	11
5	15	5	7	5	8.5
Sum 15	55	15	55	15	50.5

FIGURE 1

Alcohol Output Versus Corn Input in an Actual Distillery

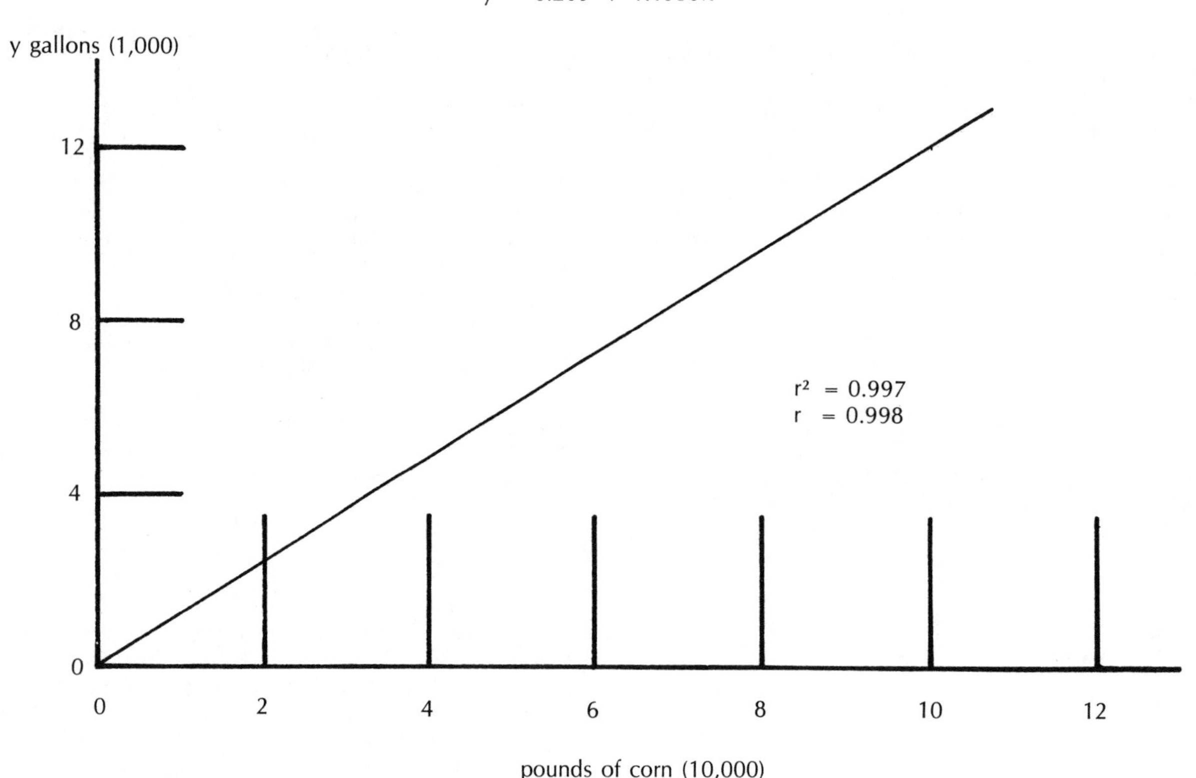

For r to be 1, every point has to fall on the straight line. This means that every pair of points has to give the same slope, or value of b. Hence, if (x_1,y_1) is one point and (x_2,y_2) is a second point, then $(y_2 - y_1)/(x_2 - x_1)$ = a constant. In the first example, $\frac{9-7}{2-1} = 2$ and so does $\frac{15-13}{5-4}$ and $\frac{15-7}{5-1}$. In the second example, $\frac{13-15}{2-1} = -2$ and so does $\frac{11-13}{3-2}$ and $\frac{7-15}{5-1}$. The first example illustrates positive correlation, where y increases as x increases; the second example illustrates negative or inverse correlation, where y decreases as x increases. The equation of the first line is $y = 5 + 2x$; the equation of the second line is $y = 17 - 2x$.

In the third example, the slope $b=0$; this is always true when $b=0$, $r=0$ and vice versa. Note that the slope between two points is positive as well as negative so they balance out to a zero slope. Obviously, all these points do not lie on a horizontal line of slope zero since $b=0$ and $y=a$, but what value does a take? A reasonable value for a is the arithmetic mean $\bar{y}$ or 50.5/5 or 10.1; hence, we can think of the line $\bar{y}=10.1$ as a "fit" to the data. Another reason for using $\bar{y}$ is that the deviations of the five points from this value equal zero, so the deviations above this line equal the deviations below the line, as the following calculations show:

y	y − 10.1
9	−1.1
10	−0.1
12	1.9
11	0.9
8.5	−1.6
Sum	0

This correlation coefficient is not the same as the intraclass correlation described in previous pages. It is an *interclass* correlation coefficient because each pair of x and y values is used only once. The correlation coefficient r can be calculated directly from the data using six values:

$$\sum x \quad \sum x^2 \quad \sum xy$$

$$\sum y \quad \sum y^2 \quad n$$

since

$$r = \frac{A - B}{(C)^{1/2}(D)^{1/2}},$$

where

$$A = n\sum xy$$

$$B = \sum x \sum y$$

$$C = n\sum x^2 - \left(\sum x\right)^2$$

$$D = n\sum y^2 - \left(\sum y\right)^2$$

Example: A random sample of 10 businesses is used to determine the correlation between the net income in millions of dollars for 1974, x, and similar values for 1973, y. The basic values are:

$n = 10$ $\qquad \Sigma x = 71.8$

$(\Sigma x)^2 = 5155.24$ $\qquad \Sigma y = 50.7$

$\Sigma x^2 = 1545.82$ $\qquad \Sigma xy = 863.47$

$\Sigma y^2 = 591.85$ $\qquad (\Sigma y)^2 = 2570.49$

Therefore:

$$A = 10(863.47) = 8634.7$$

$$B = (71.8)(50.7) = 3640.26$$

$$C = 10(1545.82) - 5155.24 = 10{,}302.96$$

$$C^{1/2} = 101.50$$

$$D = 10(591.85) - 2570.49 = 3348.01$$

$$D^{1/2} = 57.86$$

$$r = \frac{8634.7 - 3640.26}{(101.50)(57.86)} = 0.85$$

$$r^2 = 0.72.$$

This high correlation points to the superiority of the ratio estimate and why stratification of a frame based on the data for one year tends to remain effective for the next year.

Correlation coefficients should not be used without being familiar with the nature and relationship between x and y. This is why a plot of x and y on graph paper is highly recommended. Spurious or highly misleading correlation coefficients can arise from several sources:[1]

- by correlating ratios
- by having x and y correlate through a third variable such as time
- by having one point with very large values of x and y
- by mixing groups which have correlated means

For example, mental age and height of children between the ages of five and 12 correlate very highly as a usual rule because both of these characteristics increase with age. If time is held constant, no such relationship exists.

Least square equations for a straight line. It is convenient when fitting a straight line by the method of least squares to a set of pairs of points (x,y) to solve for both a and b in $y = a + bx$ directly from the data. The quantities needed are: n the number of parts of values, the sum of x, the sum of y, the sum of x^2, and the sum of the products xy.

$$a = \frac{1}{D}\left[\sum x^2 \sum y - \sum x \sum xy\right]$$

$$b = \frac{1}{D}\left[n\sum xy - \sum x \sum y\right]$$

$$D = n\sum x^2 - \left(\sum x\right)^2$$

Two independent calculations of a and b make it possible to use the equation $\bar{y} = a + b\bar{x}$ as a check on the calculations, since the two means $\bar{x}$ and $\bar{y}$ are a point on the line. This is preferred to using the latter equation to solve for either a or b, as some authors do.

Correlation and the straight line. The straight line $y = a + bx$ fitted to a series of points (x_i, y_i) does not fit the points exactly. There is a residual variation around the fitted line derived from the difference $(y_i - y_c)$ for any given x_i. Values of y_c are calculated from the equation:

$$y_c = a + bx_i,$$

where a and b are obtained by the method of least squares. Then the

1. Residual variance:

$$s_v^2 = \frac{1}{n-2}\sum_i (y_i - y_c)^2, \quad \text{where } i = 1, 2, 3, \ldots, n.$$

2. Residual variance:

$$s_v^2 = \frac{1}{n-2}\left(\sum y_i^2 - a\sum y_i - b\sum x_i y_i\right),$$

where a and b are from the fitted straight line. Several decimal places are needed in these two constants to avoid large rounding errors. The sum of y_i^2, sum of y_i, and the sum of $x_i y_i$ are from the observed data; n is the sample size, the number of paired values.

3. Standard error of estimate:

$$s_v = (s_v^2)^{1/2}$$

4. Correlation coefficient squared r^2:

$$r^2 = 1 - \frac{s_v^2}{s_{y_i}^2}, \quad \text{where } s_{y_i}^2 = \frac{1}{n-1}\sum_i^n (y_i - \bar{y})^2.$$

The quantity r^2 measures the proportion of variation taken out of y_i by the relationship between x and y. If $s_v^2 = 0$, then $r^2 = 1$ and all of the variation in y is accounted for by x. If $s_v^2 = s_{yi}^2$, then $r^2 = 0$ and no variation in y is taken out by the relationship with x. We should not become so engrossed with the correlation coefficient r that we overlook the extreme importance of r^2.

Calculations by means of a computer. Some desk and hand-held computers now have linear regression and the correlation coefficient programmed into them. If it does not, the necessary programs can be purchased or one can write the program required using the equations we have already presented. In any event, to obtain the most efficient and accurate use of any computer, it is necessary to understand the basic principles of mathematics and statistics which underlie the operations involved.

Once the data are entered into the computer according to instructions, the following may be obtained by pressing the proper key:

1. The values a and b in $y = a + bx$ based on method of least squares.
2. The correlation coefficient r or r^2.
3. The value y_c, the calculated value on the line for a given x; this symbol may be $\hat{Y}$ on the computer keyboard.
4. The two means $\bar{x}$ and $\bar{y}$.
5. The two variances s_x^2 and s_y^2 or their square roots which are the two standard deviations s_x and s_y.

From these easily obtained values, it is possible to make quickly three more useful calculations: the coefficients of variation, the individual residuals which show for each observed y value its deviation from the fitted straight line, and the residual variance which can be obtained by two different methods as shown above.

1. The coefficient of variation of x is $s_x/\bar{x}$; the coefficient of variation of y is $s_y/\bar{y}$. All these require is a division of values already in memory.
2. The residuals are deviations of actual values y_i from calculated values y_c on the line: residual $= y_i - y_c$, which is positive when the observed value is above the line and negative if it falls below the line. The algebraic sum of these residuals is zero or close to it depending upon rounding errors. These residuals should be plotted on graph paper against observed values of x.
3. The residual variance is the sum of the squared residuals divided by $n - 2$:

$$\Sigma(y_i - y_c)^2/(n-2).$$

Linearization of curvilinear equations. The following equations of curves can be reduced to linear form, as shown, by using some transformation; all but one is a logarithmic transformation:

		linear form
1. Hyperbola	$xy = c$	$\ln y = \ln c - \ln x$ $Y = A - X$
2. Hyperbola	$y = \dfrac{x}{a + bx}$	$Y = \dfrac{x}{y} = a + bx$
3. Power function	$y = ax^b$	$\ln y = \ln a + b \ln x$ $Y = A + bX$
4. Exponential	$y = ab^x$	$\ln y = \ln a + x \ln b$ $Y = A + Bx$
5. Exponential	$y = e^{-ax}$	$\ln y = -ax$ $Y = -ax$
6. Exponential	$y = ae^{-ax}$	$\ln y = \ln a - ax$ $Y = A - ax$

Example: An actual example of a learning production curve that followed the hyperbola $y = \dfrac{x}{a + bx}$ is the relationship between number of employees y and the number of units produced per month x for 10 months:

x:	1.0	1.2	10	18	24
y in 100:	3	4	24	31	36
x:	32	38	40	46	50
y in 100:	39	47	46	48	47

Production started from zero with all new employees requiring training. The curve measured the capability of the group under a certain set of conditions as well as the increase in productivity. It is a composite learning curve showing how production increased as individual proficiency increased; it also shows that during the last four months it increased without any increase in the number of employees.

The data were converted to ratios $Y = \dfrac{x}{y}$, and as Figure 2 shows, a straight line was a good fit, thus indicating that the data were following a hyperbolic equation whose equation was found to be

$$y = \frac{x}{0.002986 + 0.0001473x}$$

Linear regression: the complete analysis. The complete analysis of linear regression applied to a sample of n pairs of values x and y can readily be made on either a desk or hand-held electronic calculator. Such an analysis includes the following:

1. The mean of x and the mean of y.
2. The variance of x and the variance of y.
3. The standard deviation of x and the standard deviation of y.
4. The coefficient of variation of x and the coefficient of variation of y.
5. The correlation coefficient squared r^2 and the correlation coefficient r.
6. The coefficients a and b.
7. The standard error of b.
8. The residual variance about the line; the variance of Y as an estimate of mean and as an estimate of an individual value.
9. The standard error of estimate.
10. The confidence limits of B and the confidence limits of Y_{ave} and Y_{ind}.
11. The t test of b: Is $B=0$? Is $B=B_1$?
12. The t test of a: Is $A=0$?

The equations for solving the first six quantities are given on preceding pages; those for the last six are given below. A complete example is given in Table 1 and Table 2.

Equations: x_i and y_i represent actual values, not deviations from means $\bar{x}$ and $\bar{y}$.

1. Estimate:

$$Y = \bar{y} + b(x_i - \bar{x}) = a + bx, \text{ where } i = 1,2,\ldots, n.$$

2. Residual variance:

$$s_v^2 = \frac{1}{n-2}\left[\sum y_i^2 - a\sum y_i - b\sum x_i y_i\right].$$

3. Variance of b:

$$s_b^2 = \frac{s_v^2}{\sum x_i^2 - (\Sigma x_i)^2/n} = \frac{s_v^2}{D}.$$

4. t test of b; df = n - 2:

$$t_b = \frac{b - B}{s_b}.$$

FIGURE 2
Relation Between Employment and Production in a Learning Curve Situation

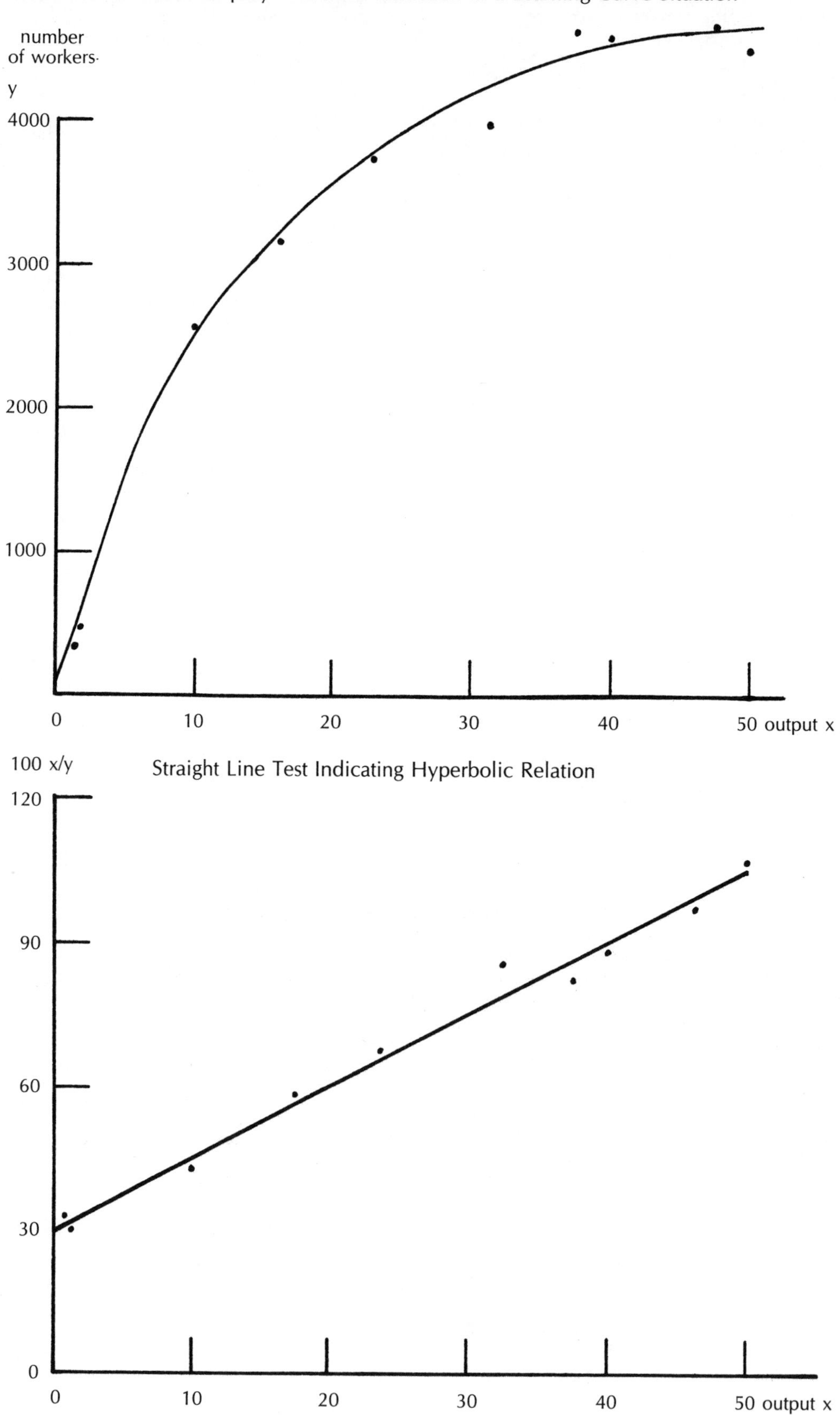

5. Interval estimates of B:

upper value: $b + ts_b$,

lower value: $b - ts_b$,

where t taken at some level such as 0.95 or 95 percent, with df = n - 2. Read t at 0.050 in the t table for the 95 percent level to give a two-sided test.

6. t test of the intercept a in y = a + bx; df = n - 2. Test the hypothesis that the straight line goes through origin x = 0, y = 0.

$$s_a^2 = s_v^2 \left[\frac{1}{n} + \frac{\bar{x}^2}{\sum x_i^2 - (\Sigma x_i)^2/n} \right],$$

$$t_a = a/s_a.$$

7. Variance of Y as an estimate of the *mean* at x_i:

$$s_{Y_{ave}}^2 = s_v^2 \left[\frac{1}{n} + \frac{(x_i - \bar{x})^2}{D} \right]; \quad D = \sum x_i^2 - (\sum x_i)^2/n \cdot$$

8. Variance of Y as an estimate of an *individual value*, at x_i:

$$s_{Y_{ind}}^2 = s_v^2 \left[1 + \frac{1}{n} + \frac{(x_i - \bar{x})^2}{D} \right].$$

9. Confidence limits of Y_{ave}:

upper limit: $Y_{ave} + ts_{Y_{ave}}$

lower limit: $Y_{ave} - ts_{Y_{ave}}$

10. Confidence limits of Y_{ind}:

upper limit: $Y_{ave} + ts_{Y_{ind}}$

lower limit: $Y_{ave} - ts_{Y_{ind}}$

Linear function in three variables. We have described and illustrated y as a linear function of x, where a straight line is an appropriate expression of the relationship. We now turn to the case in which z is a function of two other variables x and y. We write this function in the following form, which is the equation of a plane in three dimensions:

$$z = a + bx + cy.$$

In the two dimensional case, two lines intersect at right angles forming an xy plane on which points are plotted. The point of intersection, x = 0, y = 0, is called the origin. In three dimensions, three planes intersect at right angles, as at the corner of a box or room so three planes are formed, three lines or axes intersect at right angles, and the origin is where x = 0, y = 0, and z = 0. The three planes are the xy plane, the xz plane, and the yz plane. From the above equation:

1. If y = 0, z = a + bx which is a straight line in the xz plane.
2. If x = 0, z = a + cy which is a straight line in the yz plane.
3. If z = 0, $y = -\frac{a}{c} - \frac{b}{c}x$ which is a straight line in the xy plane.

In an actual sample of size n, each sample unit will yield a value for x, y, and z so there will be n sets of these three characteristics. Each set of x, y, and z will determine a point in three dimensional space. The question is: Is this n set of actual points in space closely approximated by a plane? This is strictly analogous to the previous question as to whether n set of points in a plane (two dimensional) can be closely approximated by a straight line, y = a + bx. At this point we present an actual sample of five sets of values for three characteristics of rail carload movements of wheat: revenue in dollars, weight in tons, and distance of the movement in miles. The data for the five carloads of wheat are as follows:

Carload	Revenue z	Tons x	Miles y
1	$911	56.95	498
2	555	50.00	217
3	761	58.05	381
4	898	60.70	476
5	269	44.15	96

The equation obtained by using the least square methods is:

$$z = -178.852 + 8.2082x + 1.2430y.$$

An idea of what this plane looks like is shown in Figure 3. The relationships are correct even though it is not drawn to scale. Two points are within 10 percent of the plane and three points are within two percent of the plane — this shows that the plane is not a bad fit to the five points. This example is for illustrative purposes only; in actually estimating the revenue from carloads of wheat a much larger sample would be used. A more complete analysis of these data, including the method of arriving at the equation, are described in detail later in this chapter. That this equation makes sense is illustrated by the value of a which is − 179 the value of z when both x and y equal 0. This may be interpreted as the overhead loss ($179) assignable to a carload of wheat when no carloads of wheat are being moved.

In the foregoing example, the variables were quantified and expressed in units which can be accurately measured; furthermore, we know that the cost of a rail car movement is related to both weight and distance. These requirements become increasingly difficult to meet as the relationship becomes more complex: it may be difficult to obtain variables which relate to the major characteristic we are trying to estimate or predict. It may be difficult to quantify and to measure them accurately. The problems encountered in measuring air and water pollution are good examples:

Air pollution (oxides, ozone, particulates) is a function of location, time of day and year, auto traffic, type of industry, weather, and wind. Furthermore, different pol-

Table 1
Linear Regression with t Tests and Confidence Limits of B*

Shipments (million dollars) x	Production 1000 units y		Basic values and tests		
82	12.0	1. a		−0.989819	
100	15.3				
92	12.2	2. b		0.1542	
102	16.9				
114	16.7	3. r^2		.9531	
128	18.6				
		4. r		.976	
136	20.3				
106	16.4	5. s_v^2		1.21186	
77	9.5				
130	17.8	6. s_v		1.1008	
131	18.3				
27	3.0	7. s_b		0.0108	
Mean 102.08	14.75	8. upper B		.1783	95% limits
Variance 942.8106	23.5155	lower B		.1301	df = 10 t = 2.228
Std. dev. 30.71	4.85	9. t tests of B			
cv 0.30	0.33	B = 0	t = 14.265		reject H
		B = 0.10	t = 5.013		reject H
		B = 0.14	t = 1.312		do not reject H
		10. t test of intercept			
		A = 0	$t_a = -0.862$		do not reject H

*The data show for a given dollar volume of output in a production schedule how much production is necessary. The data are excerpted from a larger sample of data.

Table 2
Linear Regression with Confidence Limits of Estimates*

Basic values of problem		Values of x_i: $19 = \bar{x}$	24	30	35
1. $n = 12$					
2. $s_v = 5.2334$					
3. $D = 924$	$s_{Y_{ave}}$	1.51	1.74	2.42	3.14
4. $a = 64.25$	$s_{Y_{ind}}$	5.45	5.51	5.77	6.10
5. $b = -1.013$	Y	$45 = \bar{y}$	39.94	33.86	28.79
6. $s_b = 0.172$	95% confidence intervals				
7. $t_b = -5.89$ (reject H: B = 0)	Y individual				
8. $t_{.05} = 2.228$ df = 10	upper	57.14	52.22	46.70	42.39
	lower	32.86	27.65	21.01	15.19
9. $Y = 64.25 - 1.013x_i$	Y average				
	upper	48.37	43.81	39.25	35.79
10. $\bar{x} = 19$; $\bar{y} = 45$	lower	41.63	36.06	28.46	21.79

*Adapted and enlarged from G. W. Snedecor and W. G. Cochran, *Statistical Methods,* Iowa University Press, Ames, 1967, 6th edition, pp. 150-156.

lutants may be and are correlated with different variables.

Water pollution (oxygen level, nitrates, toxic metals, organic matter) is a function of location, time of day and year, type of industry, weather, precipitation, soluble chemicals in the soil, sewage disposal, and tributaries. Here again, different pollutants may arise from different causes and relate to different variables.

Multiple regression. $Y = a + b_1X_1 + b_2X_2$. In statistics reference is made to multiple regression which is illustrated here by a plane in which Y is a function of two variables X_1 and X_2. The constants are a, b_1, and b_2 which are determined by the method of least squares. This is the statistical equivalent, with a different set of symbols, of the mathematical equation given in the preceding section. Equations are given below for solving the values of a, b_1, and b_2, the standard errors of the regression coefficients b_1 and b_2, the deviation mean square or the residual variance created by fitting a plane to the data, the correlation coefficient squared which shows what proportion of the variance in Y is taken out by X_1 and X_2, and the correlation coefficient.

The notation is changed: capital letters Y, X_1, and X_2 are observations or sample values; the lower case letters stand for deviations from their respective means:

Equations:

1. Basic equation:

$$Y = a + b_1X_1 + b_2X_2.$$

2. Equation, origin at means:

$$Y - \bar{Y} = b_1(X_1 - \bar{X}_1) + b_2(X_2 - \bar{X}_2).$$

3. Equation in terms of deviations from means:

$$y = b_1x_1 + b_2x_2.$$

4. Normal equations for solving for b_1 and b_2:

$$b_1 \sum x_1^2 + b_2 \sum x_1x_2 = \sum x_1y.$$

$$b_1 \sum x_1x_2 + b_2 \sum x_2^2 = \sum x_2y.$$

5. Determinant D:

$$D = \sum x_1^2 \sum x_2^2 - \left(\sum x_1x_2\right)^2.$$

6. Equation for

$$b_1 = \frac{1}{D}\left(\sum x_2^2 \sum x_1y - \sum x_1x_2 \sum x_2y\right) = \frac{A}{D}.$$

7. Equation for b_2:

$$b_2 = \frac{1}{D}(\Sigma x_1^2 \Sigma x_2y - \Sigma x_1x_2 \Sigma x_1y) = \frac{B}{D}.$$

8. From equations 1. and 2. above, solve for a:

$$a = \bar{Y} - b_1\bar{X}_1 - b_2\bar{X}_2,$$

where $\bar{Y} = \Sigma Y/n$, $\bar{X}_1 = \Sigma X_1/n$, and $\bar{X}_2 = \Sigma X_2/n$.

9. Sum of deviations squared about the plane (residual variation):

$$\Sigma d^2 = \Sigma y^2 - b_1\Sigma x_1y - b_2\Sigma x_2y.$$

10. Sum of squares due to regression:

$$b_1\Sigma x_1y + b_2\Sigma x_2y.$$

11. c values:

$$c_{11} = (\Sigma x_2^2)/D,$$
$$c_{12} = (-\Sigma x_1x_2)/D,$$
$$c_{22} = (\Sigma x_1^2)/D.$$

12. Standard errors of the regression coefficients:

$$s_{b_1} = s(c_{11})^{1/2}$$
$$s_{b_2} = s(c_{22})^{1/2},$$

where

$$s^2 = \text{deviation mean square} = \text{residual variance} = \frac{\Sigma d^2}{n-3}.$$

Example: In the example, the results are recorded as indicated in the answer table at each of the computer program steps two through nine, as well as at the end of the program. The equation for Y can be written directly since a, b_1, and b_2 are displayed by the calculator as programmed. The equation is plotted in Figure 3.

The data are from a sample of five railroad waybills for the shipment of wheat. Y is the revenue in dollars, X_2 is the miles traveled by each shipment, and X_1 is the weight of each shipment in tons. What is required is the equation of revenue as a function of tons and miles. This small sample is used to illustrate the method; ordinarily a much larger sample would be used. Summarized below are the raw data, the regression equation, the revenue calculated from the equation, the residuals or deviations of the points from the plane, the sum of the squares of the residuals, and the results.

	Tons X_1	Miles X_2	Actual revenue Y	Calculated revenue Y_c	Difference d (residuals)
	56.95	498	$911	$907.6	-3.4
	50.00	217	555	501.3	-53.7
	58.05	381	761	771.2	10.2
	60.70	476	898	911.0	13.0
	44.15	96	269	302.9	33.9
Sum	269.85	1668	3394	3394.0	00.0
Mean	53.97	333.6	678.80		$\Sigma d^2 = 4318.1$

Results

$b_1 = 8.2082$; $s_{b_1} = 10.8424$; $r^2 = 0.9704$.

$b_2 = 1.2430$; $s_{b_2} = 0.4243$; $r = 0.9851$.

$a = -178.8520$; $s = 46.4658$; $\Sigma \frac{d^2}{2} = s^2 = 2159.0666$.

$\Sigma d_2 = 4318.1331$; $s_y = 270.1781$; $s_y^2 = 72{,}996.2$.

The final regression equation is $Y = -178.8520 + 8.2082X_1 + 1.2430X_2$. This equation is plotted in Figure 3.

FIGURE 3

Three Dimensional Drawing Showing Plane $z = -178.9 + 8.208x + 1.243y$ and Point P on Plane
(not drawn according to scale)

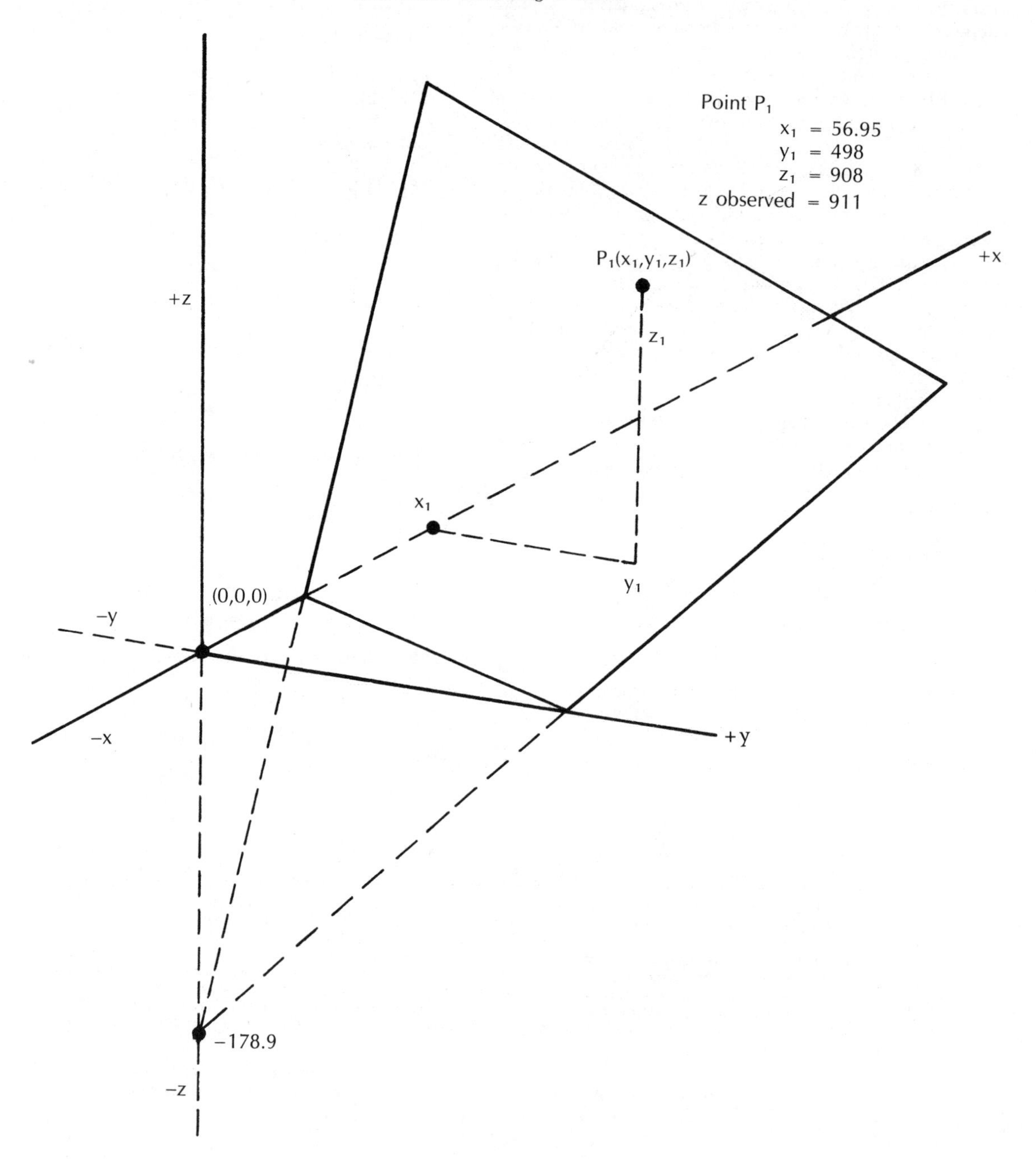

Note

[1]A. C. Rosander, *Elementary Principles of Statistics,* Van Nostrand, New York, 1951. Copyright by Wadsworth, Inc. Used by permission of Brooks Cole Publishing Co., pp. 351-353.

17

Tests and Experiments

Tests and experiments are a special kind of sample study used for the purpose of determining whether a significant *difference* exists which can be used to make a preference, decision or take other action; estimating a *level* which is important to compare with one or more other levels; determining whether or not a *relationship* exists between or among factors under study; and measuring whether or not an intervention really produces a significant desired or claimed *change*.

Examples of where tests or experiments are made to measure whether a significant *difference* exists which justifies a preference decision are the following: different products, materials, machines, and instruments; different methods, procedures, processes, techniques, formulas, and recipes; the new versus the old; and different experimental factors or treatments.

Examples of where interest centers in the determination of *levels* includes: obtained values to be compared with a standard or critical or safe level in terms of health, safety, cost, efficacy, reliability, or strength; a comparison of values in different areas, locations, and places; a comparison of values at different times or for different time periods; and a comparison of values for different classes, classifications, or groups.

Examples of where interest lies in determining the variation in *performance* includes: different laboratories testing samples of the same materials; different analysts and experts making comparative analyses on the same materials; different kinds of equipment, materials, and machines; and different groups of people.

Examples of where interest lies in determining whether a significant *relationship* exists between two or more categories or variables includes: age and sex of automobile drivers and fatal automobile accidents; horsepower of automobile and consumption of gasoline per mile; busing of students and their school grades; effect of vitamin C on prevention of colds; and the effect of price or rates on the consumption of water, electricity, and natural gas.

Comparison criteria. In making comparisons three questions arise. Does a real difference exist between what is being compared or could the difference arise from variations due to sampling, measurements, human error, or extraneous factors? If a significant difference is found, is the difference of any importance in practice, in the real world? This has to be answered by determining if the difference is significant in terms of some standard or critical value involving health, safety, performance, cost, price, and production. Finally, if a difference is found in a test or experiment, to what domain or universe or population can it be generalized?

Levels criteria. Here interest lies in determining whether a level is such as to call for some action. Is a level higher in one geographical area than in another? This is important for levels of unemployment, freight car shortages, errors on tax returns, consumer preferences, and air and water pollution. Is a level higher at one time than at another? This is important for time related levels such as air and water pollution, seasonal buying, prices, employment in certain industries, travel, and insurance claims. Is a level higher than a standard? This is important where standards are set for various levels of gases, radioactive substances, and other pollutants of air, water, and soil with the purpose of protecting human health, plants, animals, and fish. Are the variations in several levels so great as to call for action? This is illustrated by variations in levels obtained from the same samples by different laboratories and scientists.

Effectiveness criteria. The question arises as to how effective is a law, a social, economic, or political program, or some other form of intervention. Effectiveness is determined by how closely the program, law, or intervention accomplishes its stated goal or objective. This calls for two kinds of studies or audits: one calls for measuring whether the intervention made a real *change* or created a significant difference; the other calls for an audit of the effectiveness of the operations used to put the intervention into effect. These call for sample audits to determine deviations from any numerical goals set, eligibility rules, legal money payments and receipts, accurate billing, accurate allocation and spending of funds, sound and efficient accounting and management, and the extent to which laws, rules, and prescribed procedures are ignored, misunderstood, or violated.

Examples of Actual Tests

Test material	Characteristic tested	Location
1. automobiles	miles per gallon of gasoline	EPA
2. air above street	carbon monoxide level	Denver
3. soil	plutonium intensity	Rocky Flats, Colorado
4. food additive	carcinogenic effects	FDA
5. water	impurities	any city health department
6. computer programs	errors	any good computer set-up
7. MICR bank checks	errors	banks
8. individual	glaucoma	ophthalmologist's office
9. automobile driver	percent alcohol in blood	police station, doctor's office or hospital

Examples of Experiments. In experiments the goal is to determine differences in experimental factors: to isolate what is superior, to determine what is better, and to determine cause and effect:

Experimental material	Experimental factor
1. patients	drugs (kinds); dosages: side effects
2. patients	treatments
3. adults	diets
4. students	teaching methods
5. cattle	plutonium levels in two different areas
6. laboratory animals	chemical additives, drugs
7. desk computers	statistical problems and operators
8. typewriters	prepared material and typists
9. revenue operations	two different methods of measuring time and costs
10. customers	standard brands versus generic brands
11. customers	different kinds of packaging
12. customers	effect of prices on sales of product A.

Two factors. "Is A better than B?" is a common question that occurs in business, industry, and government. "Better" may be defined in a variety of ways: more productive, safer, cheaper, superior, more reliable, more effective, easier to use, more efficient, longer life, or less repair and maintenance. The two factors include a wide variety of objects: two machines, materials, methods, drugs, treatments, instruments, procedures, techniques, systems, products, offices, chemicals, or physical or environmental conditions. The two factors may be two different levels or dosages of drugs, medicines, chemicals; two different strengths, densities, colors, speeds, temperatures, or altitudes. Actual cases of two factor comparisons are the following:

1. Do chain grocery stores short-weigh more than non-chain grocery stores? (Federal Trade Commission).
2. Is an automobile tire using rayon cord better than an automobile tire using cotton cord? (World War II Congressional investigation).
3. Is a new type of shipping container more protective than the present container? (Described in Chapter 25).
4. Does one type of packaging for retail sales sell better than another type? (Case involving two ways of packaging candy).
5. Is a dry method of printing dollar bills better than a wet method? (Question faced by U.S. Bureau of Engraving).
6. Is one typewriter better than another? (Test run in a number of offices).
7. Is the use of probability sampling and regression analysis better than using bills of materials to estimate certain industry material requirements? (Raised in War Production Board during World War II).
8. Is selecting draftees using numbered capsules in a bowl better than using random numbers? (Question arose during draft in 1960s and 1970s).
9. Is the quota method of sampling as good as the method of probability sampling in opinion polls? (Question raised with regard to opinion polls since 1930s).
10. Is it better to test for plutonium contamination by using a dust sample or a soil sample taken at the same place? (Unsolved question raised at Rocky Flats, Colorado).
11. Is a platinum converter superior to other methods in reducing auto emission pollutants? (Raised in connection with auto pollution control).
12. Is random time sampling (replicated) better than a 100 percent time recording system to measure work performance of employees and machines? (Test run by Chesapeake and Potomac Telephone Companies and reported by McMurdo).

Judgment sampling in experiments. It is pointed out from time to time that all the great discoveries of chemistry and physics were made without the use of modern experimental methods. The nature of these discoveries show marked differences between phenomena under study then and phenomena under study today. These differences show why modern experimental methods cannot be abandoned in favor of the judgment sample approach. Examples which illustrate this conclusion are given later.

To discover the *existence* of a principle in physics, for example, requires a demonstration, not an experiment. To discover this uniformity requires one demonstration since replication produces the same results. Examples are the lever, pulleys, falling bodies, atmospheric pressure, buoyancy, reflection of light, magnetic attraction and repulsion, static electricity, electromagnetic induction, the voltaic cell, and electrolysis. In addition, homogeneous materials make it possible to isolate phenomena with a very small sample, even as small as one. Examples are the reaction of zinc in diluted sulphuric acid and copper electroplating using a copper sulphate solution. *Relationships* can also be established with a sample of one or, at most, a very few samples. Examples include determining how the volume of an enclosed gas changes with the pressure exerted upon it, the principle of the parallelogram of forces, and the principle of electromagnetic induction.

In measurements, scientific research has not been so fortunate. Shewhart[1] showed that the measurements of three of the five fundamental constants — velocity of light, gravitational constant, and Planck's constant — did not come from a chance system normally distributed; they came from a system of assignable causes (biases). Youden pointed out that this was also true of the measurements of the mean distance of the earth from the sun. Not only were the measurements of the velocity of light out of control, but about 92 percent of Michelson's 2885 measurements were less than the present standard of 299,792.5 kilometers per second.[2]

Problems today tend to involve the living world, with all of its variations and individual differences, in contrast to the physical world where objects and characteristics are not subject to this high variability. Present day problems involve many variables and many possible factors which may influence the outcome or interaction between factors which introduces complexities and rare but significant effects that can be measured only over long time periods, such as the effects of air and water pollutants, food additives, agricultural pesticides, and radioactive contamination and fallout. It is no accident that early progress was more prevalent in many aspects of physics and chemistry than in biology, genetics, medicine, and biochemistry.

Four examples of the use of judgment sampling in experiments or field tests are cited from press accounts to show how incomplete and inconclusive, if not misleading, these experiments and tests can be. The brevity in reporting, the basic weakness of the experiment or test itself, and faulty analysis and presentation are all involved.

1. A school busing study made at one Illinois school is asserted to prove that bused black students outperformed students who were not bused. No details or data were given.[3]

2. A study of 12,076 women ages 18-54 in California showed that those who used birth control pills had a 2.4 percent infection rate of the urinary tract while those who did not use birth control pills had a 1.6 percent infection rate. No details or data were given.[4]

3. A four-year study of 400 elderly patients, 200 of whom received annual health check-ups and 200 of whom went to the doctor only when they felt it necessary, refutes the idea of the need for an annual check-up. No details or data were given.[5]

4. A test of 461 women using birth control pills and 1,300 women not using birth control pills showed that the danger of blood clots were six times as great in the former as in the latter. No data were given other than that the dosage levels had been reduced because of the danger of blood clotting.[6]

These data provide a basis for an exploratory analysis. Assuming that the number of women susceptible to blood clots is a Poisson variate, the number in each group required to give a real difference at the 95 percent level with a six to one rate is as follows:

1. Let $\frac{x}{461}$ be the blood clot rate in the pill group.

2. Let $\frac{y}{1300}$ be the blood clot rate in the non-pill group.

Then $6y/1300 = x/461$ to meet a six to one rate. This gives $x = 2.13y$. To obtain a normal deviate significant at the 95 percent level, $x^{1/2} - y^{1/2} = 1.96\ (0.5)^{1/2}$. Solving these two equations in x and y gives rounded $x = 20$ and $y = 9$. A minimum of 20 out of 461 and at least nine out of 1300 should be blood clot cases to give a six to one rate. There was no indication that these minimum counts should be obtained.

Some basic problems. The problem in making comparisons is to make a true comparison of A and B with other effects being eliminated. *Other factors may be concealed in A and B.* Factor C may be associated with factor A and factor D may be associated with factor B; C and D may be different, but A and B are not. Hence, the difference assigned to A and B really arises from C and D. The problem is to eliminate or neutralize the effects of C and D by seeing that both C and D affect A as well as B equally.

A or B may receive preferential treatment due to such factors as time, space, position, arrangement, testing materials, and the like so that one factor is favored over the other. *Random assignment* of experimental factors A and B to test materials and other possible influences insures conditions of fair comparison.

It is also necessary to have the same amount of exposure of the factors and to have equal numbers of such exposures. If this is not done, the observed difference may simply be due to the fact that one factor was exposed more or was applied to more testing material. This calls for *balancing* of the exposure of the experimental factors.

Replication, or repetition, of the test is also necessary in order to obtain an estimate of the experimental error, since any difference obtained between A and B must exceed the experimental error to an appropriate extent to have any significance. Replication also makes it possible to determine the extent of the consistency of the behavior of the factors being tested. If there is a *consistent* difference between the two factors, inference is on a much sounder basis than if the factors behaved one way on some replicates and the opposite on others.

Factors at different levels. In the two factor situations above, the factors are unique — each factor does not exist at quantitative levels, although there may be some subclassifications within each factor. The question here is: At what level or levels is factor A most effective in connection with other factors such as B and C? These factors exist at different quantitative levels such as density, drug dosage, chemical composition, speed, temperature, percent concentration, and other similar characteristics. Included are different exposures to air, water, and soil pollutants including gases, chemical liquids, and radioactive substances. Comparisons of factors and levels are made by sample surveys, field tests, and experiments including factorial designs.

Two situations arise which require different approaches to sample design, experimental design, collection of data, and analysis of data — situations involving measurements and those involving counts or frequencies.

Paired comparisons. In this method, each testing unit is a pair — one value for each factor A and B — and the test uses differences between each pair of values. Various extraneous characteristics affecting the differences between each pair have to be equalized or eliminated to obtain a valid comparison between A and B. A and B may be experimental factors in the form of two drugs; the experimental material is patients. The administering of the second drug must be such that its effects are independent of the effects of the first drug. This is a case when the treatments are applied to the same experimental or test units.

The method is effective only if the correlation between the two measures x and y is enough to offset the loss in degrees of freedom (df): for paired, $df = n - 1$; for unpaired, $df = 2(n - 1)$. The layout is as follows:

experimental material	experimental factors		difference d	d^2
	A	B	$d = x - y$	
1	x_1	y_1	d_1	d_1^2
2	x_2	y_2	d_2	d_2^2
⋮	⋮	⋮	⋮	⋮
n	x_n	y_n	d_n	d_n^2
Sum			Σd_i	Σd_i^2
Mean			$\bar{d}$	

The sum of d_i is the algebraic sum of the d_i's, with signs of d_i taken into consideration; the mean $\bar{d}$ is the sum of d_i divided by n the number of pairs.

If no real difference exists, the individual differences will vary around 0 in a random manner. The t test, with n-1 degrees of freedom, tests whether there is a real departure from the hypothesis that $\bar{D} = 0$:

$$t = \frac{\bar{d} - \bar{D}}{s_{\bar{d}}} = (\bar{d} - 0) \frac{n(n-1)}{\Sigma d_i^2 - (\Sigma d_i)^2/n}, \quad i = 1,2, \ldots, n.$$

If this hypothesis is rejected at the five percent or one percent level, this is evidence that a real difference exists between the two experimental factors with regard to the testing material used. How far one can generalize depends upon the scope of the experiment.

Unpaired comparisons. In this method, one sample is used to measure A and another is used to measure B; the samples can be equal or unequal in size. Both cases are described.

sample elements	A (sample 1)	B (sample 2)
1	x_1	y_1
2	x_2	y_2
⋮	⋮	⋮
n	x_n	y_n
Sum	$\Sigma\ x_i$	$\Sigma\ y_i$
Mean	$\bar{x}$	$\bar{y}$

Equal sample sizes. The t test is used to test the hypothesis that $D = u_x - u_y = 0$. From the sample data, the difference between the two means is $\bar{d} = \bar{x} - \bar{y}$. The variance for the test is found by pooling the two variances since the hypothesis assumes that both x and y come from the same population. Therefore, the variance is:

$$s^2 = \frac{1}{2(n-1)}\left[\Sigma(x - \bar{x})^2 + \Sigma(y - \bar{y})^2\right].$$

The variance of the *difference between the two means*, since the n's are equal, is:

$$s_{\bar{x}-\bar{y}}^2 = s^2\left(\frac{1}{n} + \frac{1}{n}\right) = \frac{2s^2}{n}.$$

Hence,

$$s_{\bar{x}-\bar{y}} = s\sqrt{\frac{2}{n}}.$$

The significance of the difference between the two means is found by entering the t table with $t = \frac{\bar{d}}{s_{\bar{x}-\bar{y}}}$ with 2(n-1) degrees of freedom.

	Identification number	Determination 1	Determination 2	Difference d_i	d_i^2
	1	x_1	y_1	$x_1 - y_1$	d_1^2
batches	2	x_2	y_2	$x_2 - y_2$	d_2^2
samples	3	x_3	y_3	$x_3 - y_3$	d_3^2
analysts					
laboratories	⋮	⋮	⋮	⋮	⋮
etc.					
	k	x_k	y_k	$x_k - y_k$	d_k^2
	Sum	Σx_i	Σy_i	$\lvert\Sigma d_i\rvert$	Σd_i^2
	Mean	$\bar{x}$	$\bar{y}$		

Unequal sample sizes. The t test and pooled variances are used the same as above, but the number of degrees of freedom is $n_1 + n_2 - 2$, where n_1 is the size of one sample and n_2 is the size of the other.

$$s^2 = \frac{1}{n_1 + n_2 - 2}\left[\Sigma(x - \bar{x})^2 + \Sigma(y - \bar{y})^2\right],$$

where x is summed over n_1 and y is summed over n_2. The variance and standard error of the difference is:

$$s_{\bar{x}-\bar{y}}^2 = s^2\left(\frac{1}{n_1} + \frac{1}{n_2}\right) = s^2\frac{(n_1 + n_2)}{n_1 n_2}$$

$$s_{\bar{x}-\bar{y}} = s\sqrt{\frac{n_1 + n_2}{n_1 n_2}} \quad \text{and } \bar{d} = \bar{x} - \bar{y}.$$

Therefore:

$$t = \frac{\bar{d}}{s_{\bar{x}-\bar{y}}},$$

with degrees of freedom $n_1 + n_2 - 2$.

Paired or duplicate determinations. This method can be used to show differences between batches of a product: where the analyst takes two samples from each batch and makes one determination on each; where the same sample is divided into several parts and several analysts make two determinations from each sample for the purpose of comparing analysts; or where two reference materials about the same chemical composition are submitted to several laboratories for testing and one determination is made on each material for the purpose of comparing laboratories.

These differences measure analytical or measurement variations, not variations due to sampling. If determinations are made on independent samples, then both sources of variation can be measured and controlled. The estimate of the analytical variance for k pairs of determinations is:

$$s^2 = \frac{\Sigma d_i^2}{2k}.$$

Another method of estimating s is by using the absolute values of the differences to obtain the average range $\bar{R}$.

$$s = \frac{\bar{R}}{d_2} = \frac{\lvert\Sigma d_i\rvert}{kd_2},$$

where $\bar{R} = \lvert\Sigma d_i\rvert/k$ and $d_2 = 1.128$ for two determinations.

Control limits are set up as follows:

$$d_i = x_i - y_i,$$

$$s_d^2 = s_x^2 + s_y^2 = 2s^2,$$

$$s_d = s(2)^{1/2}.$$

The control limits, assuming a normal distribution of d, are $0 \pm 3\, s(2)^{1/2}$, using s estimated from the differences squared or from the mean range $\bar{R}$.

Testing for variations among laboratories and analysts using identical samples. Assume that the pairs of values are independent and form a bivariate normal probability distribution with means u_x and u_y and with correlation coefficient $\rho = 0$, circular control limits can be set to measure the extent to which bias or constant errors have been eliminated and the extent to which variations are random and normally distributed. For example, this method can be used to measure the differences between analysts or laboratories. The equation of the circle is:

$$(x - u_x)^2 + (y - u_y)^2 = r^2 = \lambda^2\sigma^2$$

where r = the radius of the circle about the center (u_x, u_y) and is estimated by λs and s is calculated from the data as described above. If deviation from a standard is desired, then $u_x = u_y - u$, where u is the standard mean; otherwise, the means $\bar{x}$ and $\bar{y}$ are used. Circles can be drawn which, if the conditions are met, will contain 95 percent or 99 percent or 99.73 percent of all the points — the same as the limits set in the usual normal curve. The constants necessary are:

Percent within circle	Center of circle: Standard	Center of circle: No standard	λ	Radius of circle
95	u,u	$\bar{x},\bar{y}$	2.448	2.448s
99	u,u	$\bar{x},\bar{y}$	3.035	3.035s
99.73 (3σ)	u,u	$\bar{x},\bar{y}$	3.441	3.441s

The constant $\lambda = \sqrt{\chi^2_{1-\alpha}}$, where $\alpha = 0.05$ for 95 percent, 0.01 for 99 percent, and 0.0027 for three sigma limits. The value of the chi-square is taken at two degrees of freedom. In an application made of circular probability controls by Youden and described in Chapter 25, the circle is centered at the median values of x and y.

2×2 table. This is a table of counts or frequencies with two classifications under each of two factors, A and B. The two classifications may or may not be the same and include such dichotomies as absent or present, error or no error, defective or not defective, fail or not fail, or treated or not treated. They can also be more general and include treatment one and treatment two, male and female, product one and product two, or vendor one and vendor two.

To test whether factors A and B are independent, the marginal totals are fixed (that is, the row totals and the column totals) and used to calculate the expected frequency or count in each cell. Then the usual chi square test is applied with one degree of freedom, since with fixed marginal totals only one cell frequency needs to be known in order to fill in the others.

		Actual frequencies factor A			Expected frequencies factor A	
		1	2	Sum	1	2
B	1	n_{11}	n_{12}	n_1	n_1m_1/n	n_1m_2/n
	2	n_{21}	n_{22}	n_2	n_2m_1/n	n_2m_2/n
	Sum	m_1	m_2	n		

The sum of the frequencies in the ith row is n_i, the sum of the frequencies in the jth column is m_j, and the total size of the sample is n. If independence exists, then the probability of an event occurring in the ith row is n_i/n; in the jth column it is m_j/n. The probability p_{ij} of an event occurring at the intersection of the ith row and the jth column which is the ijth cell is the product of these two probabilities or n_im_j/n^2. The expected frequency is:

$$n_{ije} = np_{ij} = n_im_j/n.$$

For a 2×2 table the four differences between actual and expected frequencies are identical except for sign; thus, the calculation of chi square is simplified.

$$\chi^2 = \left(\left|n_{ij} - n_{ije}\right| - 0.5\right)^2\left(\sum_i\sum_j \frac{1}{n_{ije}}\right), \text{ with df } = 1.$$

The value 0.5 is subtracted from the absolute value $n_{ij} - n_{ije}$, disregarding sign, to correct for continuity.

If the chi-square is significant, this is evidence that factors A and B are not independent but are showing some degree of relationship between the two classes of each.

r×c table. This extends the 2×2 table to any number of rows r and columns c. As before, n is the total sample size, n_i is the total frequency for the ith row, and m_j is the total frequency for the jth column. The actual frequency in a cell is n_{ij}, where $i = 1,2,\ldots,r$ and $j = 1,2,\ldots,c$. The expected frequency in this cell, assuming independence of rows and columns, is:

$$n_{ije} = n_im_j/n,$$

and the deviation of an actual cell frequency from an expected cell frequency is $n_{ij} - n_{ije}$. Hence:

$$\chi^2 = \sum_i\sum_j \frac{(|n_{ij} - n_{ije}| - 0.5)^2}{n_{ije}},$$

with $(r-1)(c-1)$ degrees of freedom.

Experimental and control groups. A method used to test the effect of some factor A is to compare the differences between a control group and an experimental group of test units or elements. The factor is applied to the experimental group but not to the control group; hence, the control group measures the absence of the factor and the experimental group measures the presence of the factor.

Test unit	Experimental	Control	Difference E-C
1	x_1	y_1	$x_1 - y_1$
2	x_2	y_2	$x_2 - y_2$
3	x_3	y_3	$x_3 - y_3$
⋮	⋮	⋮	⋮
k	x_k	y_k	$x_k - y_k$
Mean	$\bar{x}$	$\bar{y}$	$\bar{d}$

If the experimental and control units are paired, then the data are treated as described under paired comparisons above. If they are not paired, but are two independent samples, then the procedure described under unpaired comparisons should be used.

Extreme care has to be taken in determining which elements or units are assigned to the control group and which are assigned to the experimental group. Inherent differences in response to the experimental factor can be reduced, if not eliminated, by insuring that both groups are selected at random from the same population if two independent samples are used or by careful matching if the paired method is used. If a fixed number of test units are available, such as test animals, each one can be assigned at random to one of the two groups with equal numbers in each group. Characteristics that may influence the outcome have to be equalized and balanced; otherwise observed differences between experimental and control units may not be due to the factor under consideration but to differences in characteristics in the two groups before the experiment started. The use of cattle at two sites to test plutonium concentration in tissues, as described in Chapter 5, illustrates the problems encountered in using the experimental-control method.

Analysis of variance for one-way classifications.[7] Analysis of variance is a powerful method of testing differences between measurement data from samples collected from experiments or tests. Analysis of variance provides the following:

1. An estimate of the residual variance or experimental error.
2. The F test of the null hypothesis that all the class means come from the same population mean. This is a test of means, not variances.
3. It shows how much of the total variance is due to each component (variance component analysis).
4. A method of estimating the intraclass correlation coefficient.
5. An F test of the variances of two independent samples assumed to come from normal populations with the same variance.

Analysis of variance is based on four assumptions:

1. Normal distributions: transformations are used to correct for non-normality that is deemed too great.
2. Independent classes and observations.

3. Equal class variances: Bartlett's test of the homogeneity of k variances can be applied if there is any doubt.

4. Additivity of effects, which is assumed in the mathematical models most often used.

A distinction has to be made between whether the classes constitute the entire population, or are a random sample from a population, or are a combination of both. The first is called the fixed model, the second the random model, and the third the mixed model.

Analysis of variance is discussed in connection with measurements arranged into one set of classifications, data arranged into classes of two categories, and data obtained from successive subsamples called a nested design.

The calculations and layout of the analysis of variance table are the same for the three models, but the expected values of the mean square between classes are different.

Fixed model.

$$x_{ij} = u_j + z_{ij} = u + d_j + z_{ij},$$
$$j = 1,2,\ldots,k \text{ classes}$$
$$i = 1,2,\ldots,n_j \text{ values}$$

where u = population mean, u_j = means of normal population with variance σ^2 (constants), z_{ij} = independent and random normal deviate with mean 0 and variance σ^2, $d_j = u_j - u$ and $\Sigma d_{ij} = 0$. The mean squares are as follows:

	Mean square (ms)	Expected ms
between k classes	s_2^2	$\sigma^2 + \frac{1}{k-1}\left[\Sigma n_j(u_j - u)^2\right]$
within k classes	s_1^2	σ^2

Random model.

$$x_{ij} = u + y_j + z_{ij},$$

where j = population mean, y_j is distributed $N(0,w^2)$, and z_{ij} is distributed $N(0,\sigma^2)$. The mean squares are as follows:

	Mean square (ms)	Expected ms
between k classes	s_2^2	$\sigma^2 + n_j w^2$
within k classes	s_1^2	σ^2

where n_j = same number of observations in each of the k classes, w^2 = component of variance between classes, $w^2_{est} = (s_2^2 - s_1^2)/n_i$, and s_1^2 = unbiased estimate of σ^2. The components of variance are:

between k classes	w^2_{est}
within k classes	s_1^2
Total variance (Sum)	s^2

Equations for sum of squares (ss).

1. Within classes: $$ss_w = \sum_i^{n_j}\sum_j^{k} (x_{ij} - \bar{x}_j)^2$$
$$= (n_j - 1)\sum_1^k v_j \quad (n_j \text{ is the same for all k's}).$$

Also, i stands for rows, $i = 1,2,\ldots,n_j$ values; j stands for columns, $j = 1,2,\ldots,k$ classes; and v_j = variance within jth class.

2. Between classes:

Let $$C = \frac{\left(\Sigma\Sigma x_{ij}\right)^2}{n}$$

and $$n = kn_j = \Sigma n_j$$

then $$ss_b = \sum_{j=1}^{k} \frac{\left(\sum_i^{n_j} x_{ij}\right)^2}{n_j} - C.$$

3. Total: $$ss_t = ss_w + ss_b = \sum_i\sum_j x_{ij}^2 - C.$$

Equations for mean square (ms);

1. Within: $$ms_w = \frac{ss_w}{k(n_j - 1)}.$$

2. Between: $$ms_b = \frac{ss_b}{k - 1}.$$

3. Total: $$ms_t = \frac{ss_t}{n - 1}.$$

Example: Analysis of variance is applied to the following data, where an entry is grams of fat per batch less 100 grams, to determine if there is any difference between fats. Batches have the role of replicates or separate determinations. [8]

		Kind of fat				Sum
		1	2	3	4	(all kinds)
batch	1	64	78	75	55	272
	2	72	91	93	66	322
	3	68	97	78	49	292
	4	77	82	71	64	294
	5	56	85	63	70	274
	6	95	77	76	68	316
Sum		432	510	456	372	1770

Table totals.

$n = 24;$ $C = 1770^2/24 = 130{,}537.5;$

$\sum x^2 = 134{,}192;$ $ss = 134{,}192 - 130{,}537.5 = 3654.5;$

$\sum x = 1770.$ $\bar{X} = 73.75$ $S_{\bar{x}} = 4.77$

Analysis of Variance

Source of variation	ss	df	ms	F
Between four fats $\frac{432^2 + 510^2 + 456^2 + 372^2}{6} - C$	1636.5	3	545.5	5.41
Within fats	2018.0	20	100.9	
Total	3654.5	23		

For the first fat, the "within" sum of squares is:

$$64^2 + 72^2 + 68^2 + 77^2 + 56^2 + 95^2 - \frac{432^2}{6} = 890.$$

The three other within fats sums of squares are calculated similarly and added to give 2018.0. The F test for fats is 545.5/100.9 = 5.41 which for three and 20 degrees of freedom is significant at the one percent level ($F_{.01}$ = 4.94).

The standard deviation of a single determination is $(100.9)^{1/2}$ or 10.0. The standard error of a mean derived from six batches is $10.0/(6)^{1/2}$ = 4.10. Pooling the variances for the four fats assumes that there are no differences between the variances of batches within fats.

A component analysis of the variances gives the following:

fats:	$s_b^2 = \frac{545.5 - 100.9}{6} =$	74.1	42%
within fats between batches	$s_w^2 =$	100.9	58%
Total variance		175.0	100%

This shows that 42 percent of the variance is accounted for by fats; 58 percent is accounted for by the variation between batches within fats.

Analysis of variance for two classifications. The general model assumes an r×c table of r rows and c columns forming rc cells, with two or more observations per cell (n_{ij}). Algebraically:

$$x_{ijk} = u + v_j + y_j + w_{ij} + z_{ijk},$$

where u = population mean, v_i = row effect, y_j = column effect. w_{ij} = interaction between rows and columns = interaction constants, and z_{ijk} = random variable $N(0,\sigma^2)$. This model is additive in the main effects only if the interaction $w_{ij} = 0$ or approximately so. Three specific models have to be distinguished:

Model I: Both rows and columns are populations; this is the fixed model.

Model II: Both rows and columns are random samples from infinite populations. This is the random model.

Mixed model: Row is a random sample and column is a population, or vice versa.

The calculations are the same for all models, but the expected mean squares and the components of variance are different. Furthermore, the appropriate F test has to be used. For a detailed treatment of these aspects, the reader is referred to a good standard statistical book like Brownlee's or Snedecor and Cochran's.

The present problem is for a model with *one observation per cell.* No "within cell" variation exists, only interaction mean square. The latter is used to test the main effects in a Model II problem. It may be a poor test for Model I main effects if the interaction is large, but is satisfactory if a significant test is obtained.

Equations.

1. Sum of squares (ss):

$$ss_{col} = \frac{\sum_j \left(\sum_i x_{ij}\right)^2}{k_2} - C$$

$$ss_{row} = \frac{\sum_i \left(\sum_j x_{ij}\right)^2}{k_1} - C$$

$$ss_{inter} = \sum_i \sum_j (x_{ij} + x - \bar{x}_i - \bar{x}_j)^2$$

where x_{ij} = cell value, x = grand mean, $\bar{x}_i$ = row mean, and $\bar{x}_j$ = column mean.

$$C = \frac{\left(\sum\sum x_{ij}\right)^2}{n},$$

where $n = k_1k_2$ = total sample size, k_1 = number of columns, and k_2 = number of rows.

$$ss_{total} = \sum\sum x_{ij}^2 - C.$$

2. Mean squares (ms):

$$ms_{col} = \frac{ss_{col}}{k_1 - 1}$$

$$ms_{row} = \frac{ss_{row}}{k_2 - 1}$$

$$ms_{inter} = \frac{ss_{inter}}{(k_1 - 1)(k_2 - 1)}.$$

The interaction sum of squares is calculated directly, not obtained by subtraction, with exceptions.

Example: The data on the next page represent 15 purchases in chain stores and 15 purchases of the same amounts of commodities in independent stores. The purpose was to test the charge of short-weighing in the chain stores. The entry is 1/16 ounce; (−) means the actual weight is less than the asked weight; otherwise it is more.

	Sugar		Green beans		Cheese		Total	
	ind.	chain	ind.	chain	ind.	chain	ind.	chain
	-7	1	9	44	45	11		
	18	6	12	-16	60	-1		
	-1	2	-3	19	36	0		
	16	14	19	15	-9	-21		
	26	0	-23	7	9	0		
Sum	52	23	14	69	141	-11	207	81
Sum	75		83		130		288	

A Bartlett test of the six variances is not significant.[9] Calculations for analysis of variance are: $n = 30$; $\Sigma x^2 = 13{,}140$; $\Sigma x = 288$; $C = 288^2/30 = 2764.8$.

		ss	df	ms	F
Between two store types:	$\frac{207^2 + 81^2}{15} - C$	529.2	1	529.2	
Between three commodities:	$\frac{75^2 + 83^2 + 130^2}{10} - C$	176.6	2	88.3	
Interaction: stores versus commodities		2167.8	2	1083.9	3.47
Within six classes (columns)		7501.6	24	312.6	
Total	13,140 - C	10,375.2	29		

Within sum of squares for first column: $-7^2 + 18^2 + (-1)^2 + 16^2 + 26^2 - \frac{52^2}{5} = 765.2$; the other five columns are calculated similarly. The six values added give 7501.6.

The interaction sum of squares is found by subtraction since the four other values are calculated. It can also be calculated independently. The interaction mean square is significant at the five percent level since $F = \frac{1083.9}{312.6} = 3.47$, showing inconsistent behavior of store clerks in the two types of stores selling the same weights of the same commodities. Based on these data and analysis, the null hypothesis of no difference between types of stores relative to short-weighing is not rejected. There is no evidence that chain stores are short-weighing and the others are not. The chain stores tended to give a weight closer to what was asked; for $x = 0$ to 15 there are 10 purchases for chain and three for independents.

Interaction. Interaction measures the extent to which *differences* in the successive classes or levels of one factor are maintained in the successive classes or levels of the other factor. The different levels follow a similar or consistent behavior. Interaction is important for at least four reasons: if it is zero or near zero, the main effect contains all of the information; if one or more interactions are significant, the interpretation of main effects is complicated or modified, its magnitude shows the internal consistency of the data, and the highest order interaction can be used as an estimate of the experimental error.

The interaction due to a single value in an mxk table is $x_{ij} - \bar{x}_i - \bar{x}_j + \bar{x}$.

		Factor A	
		1 2.........j......k	row mean
factor B	1		
	2		
	⋮		
	i	x_{ij}	$\bar{x}_i$
	⋮		
	m		
Column mean		$\bar{x}_j$	$\bar{x}$

A cell value is x_{ij}, a row mean is $\bar{x}_i$, and a column mean is $\bar{x}_j$; the grand mean for the table is $\bar{x}$ for km values of x.

Conditions for zero interaction

1. every $x_{ij} - \bar{x}_i - \bar{x}_j + \bar{x} = 0$.

2. $x_{ij} - \bar{x}_j = \bar{x}_i - \bar{x}$.

3. $x_{ij} - \bar{x}_i - \bar{x}_j + \bar{x}$ summed over entire table = 0.

4. If column values are designated $x_1, x_2, \ldots, x_k$, then zero interaction exists if $x_2 - x_1 = a$, $x_3 - x_2 = b$, $x_4 - x_3 = c$, etc. where a, b, c, etc. are constants not necessarily equal. The following table has zero interaction: column 2 differs from column 1 by three, and column 3 differs

from column 2 by six. A similar relation has to hold for the rows, but the constants can be different from those for columns.

Factor A levels

		1	2	3	Sum	Mean
Factor B levels	1	1	4	10	15	5
	2	3	6	12	21	7
	3	8	11	17	36	12
	Sum	12	21	39	72	
	Mean	4	7	13		8

Entries in column 2 differ from column 1 by three, and from column 3 by six; row 2 differs from row 1 by two and from row 3 by five. The interaction of every cell is zero. The differences from column to column are the same but not identical, and the same holds true for the rows. No change occurs if rows 2 and 3 or columns 2 and 3 are interchanged.

Maximum interaction. The condition for maximum interaction is that all major effects are zero; this means $\bar{x}_i = \bar{x}_j$ for all i and j.

Factor A levels

		1	2	3	Sum
	1	15	14	17	46
Factor B levels	2	20	15	11	46
	3	11	17	18	46
	Sum	46	46	46	

The sum of each row and column is the same, so the means are the same and both A and B take out zero variation in the measurements. All variation is due to interaction. Contradictory relationships are seen by examining the row and column values. Row 1 decreases at level 2 of A, row 2 decreases with levels of A, and row 3 increases with levels of A. Columns 1 and 2 are high at the second level of B, while column 3 is low at this level. This mixture of different relationships by columns and by rows means that interaction is going to be high, in this case a maximum.

Example. The data used in a previous example are given in Table 1 with column and row means and totals, the interaction of each cell, and the interaction sum of squares by column and for all of the data. The row data which are from different batches are now considered a separate factor or classification. Since there is but one value per cell, the interaction mean square is used as a measure of the experimental error. The individual cell interactions are calculated and measure how consistent column behavior is from row to row. In calculating the interaction sum of squares, we avoid the common practice of obtaining this figure by subtraction and verify the accuracy of the calculations of sums of squares.

The row and interaction effects in the analysis of variance table given below comprise the "within sum of squares" with 20 degrees of freedom in the previous example. Not much change took place by removing row (batch) effects: the within mean square is 100.9, while the interaction mean square is 98.7. With five and 15 degrees

Table 1
Measurements, Sums, and Means

	j columns (fats)				Total	
Batches	1	2	3	4	Sum	Mean $\bar{x}_i$
1	64	78	75	55	272	68
2	72	91	93	66	322	80.5
i rows 3	68	97	78	49	292	73
4	77	82	71	64	294	73.5
5	56	85	63	70	274	68.5
6	95	77	76	68	316	79
Sum	432	510	456	372	1770	
Mean $\bar{x}_j$	72	85	76	62		$73.75 = \bar{x}$

Cell Interactions and Sums of Squares*

					ss
1	− 2.25	− 1.25	4.75	− 1.25	30.75
2	− 6.75	− 0.75	10.25	− 2.75	158.75
rows 3	− 3.25	12.75	2.75	− 12.25	330.75
4	5.25	− 2.75	− 4.75	2.25	62.75
5	− 10.75	5.25	− 7.75	13.25	378.75
6	17.75	− 13.25	− 5.25	0.75	518.75
Sum x^2(int)	519.375	375.375	245.375	340.375	1480.50

*A cell interaction $= (x_{ij} + \bar{x} - \bar{x}_i - \bar{x}_j)$. The sum of the square of each of these terms is the interaction sum of squares x^2(int).

Examples: $x_{11} = 64 + 73.75 - 68 - 72 - 2.25$;
$x_{64} = 68 + 73.75 - 79 - 62 = 0.75$.

of freedom, the row mean square would have to be 2.9 times the interaction mean square to be significant at the five percent level.

Analysis of Variance Summary

component	df	ss	ms	F	remarks
between columns	3	1636.5	545.5	5.53	F(0.01) = 5.42
between rows	5	537.5	107.5	1.09	
interaction	15	1480.5	98.7		
Sum	23	3654.5			

Analysis of variance for a nested classification. In a nested classification, or hierarchical structure, each succeeding set of data is a subclass of the preceding class.[10] In the present problem, the three levels are identified as follows:

1st level: lots
2nd level: samples selected from lots
3rd level: measurements made on each sample

These may be shown in a tabular layout:

	Lots			
	1	2	3	k
samples 1	x_{111}			
	x_{112}			
	x_{113}			
	⋮			
	x_{11m}			
	⋮			
2				
	⋮			
n				

where lots = j = 1,2,3, . . . , k: samples = i = 1,2,3, . . . , n: measurements = h = 1,2,3, . . . , m; and measurement x_{ijh} = hth measurement in the ith sample from the jth lot.

Equations:

1. The model: $x_{ijh} = u + v_j + v_{ij} + z_{ijh}$,

where u = grand mean: v_j = lot or batch or group effect $N(O,\sigma_2^2)$: y_{ij} = sample effect within lots, batches, groups $N(O,\sigma_1^2)$: z_{ijh} = random variation in determinations within samples $N(O,\sigma^2)$.

Lots, samples, and measurements are randomly selected and normally distributed with zero mean and variances as indicated.

3. Components of variance:

lots: $\sigma_2^2 \text{ est} = \dfrac{s_2^2 - s_1^2}{mn}$,

samples: $\sigma_1^2 \text{ est} = \dfrac{s_1^2 - s^2}{m}$,

measurements: $\sigma^2 \text{ est} = s^2$.

4. Variance of the grand mean:

$$v_{\bar{x}} = \frac{s_2^2}{kmn} \text{ and } s_{\bar{x}} = (v_{\bar{x}})^{1/2}$$

5. Confidence limits of the grand mean:

$$\frac{\bar{x} - u}{(v_{\bar{x}})^{1/2}} = t, \text{ with k-1 degrees of freedom,}$$

$$u(\text{lower}) = \bar{x} - ts_{\bar{x}},$$

$$u(\text{upper}) = \bar{x} + ts_{\bar{x}}.$$

Example: The data in Table 2 consist of measurements for six lots within which four samples were taken with each sample consisting of three measurements: k=6, n=4, and m=3 for a total of 72 measurements. The calculation of the sums of squares, the analysis of variance table, and the components of variance are given below.

Sum of squares (ss):

1. Between lots:

$$\frac{(287^2 + 385^2 + 413^2 + 338^2 + 445^2 + 434^2)}{12} - \frac{2312^2}{72}$$

$= 1658.11.$

2. Between samples within lots:

$29.58 + 74.92 + 128.25 + 71.00 + 82.25 + 65.00$

$= 451.00.$

3. Measurements within 24 cells:

$25.33 + 18.00 + 18.67 + 16.67 + 24.66 + 18.67$

$= 122.00.$

4. Total: $\sum x_{ij}^2 - 2312^2/72 = 2231.11.$

The sum of squares between samples within lots is the square of the deviation of each sample mean within a lot

2. Analysis of variance:

Source of variation	df	ss	ms	Expected ms
between k lots	k-1	$mn\sum(\bar{x}_j-\bar{x})^2$	s_2^2	$\sigma^2 + m\sigma_1^2 + mn\sigma_2^2$
between n samples	(n-1)k	$m\sum\sum(x_{ij}-\bar{x}_j)^2$	s_1^2	$\sigma^2 + m\sigma_1^2$
between m measurements	(m-1)nk	$\sum\sum\sum(x_{ijh}-\bar{x}_{ij})^2$	s^2	σ^2
Total	kmn-1	$\sum\sum\sum(x_{ijh} - \bar{x})^2$		

from the lot mean, multiplied by the number of observations. For example, for the first lot $29.58 = 3\Sigma(\bar{x}_i - 23.917)^2$, where the four sample means are 26.333, 23.333, 22.000, and 24.000. The other lot values are calculated similarly. The sum of squares due to measurements is the sum of the squares of deviations from the mean of each sample of three, shown in the table summed by columns for presentation purposes. Actually, the calculations are made for each of 24 samples of three and then added. The analysis of variance table is:

component	df	ss	ms
lots	5	1658.11	331.622
samples	18	451.00	25.056
measurements	48	122.00	2.542
Sum	71	2231.11	

Components of variance:

$$\text{lot variance} = \frac{331.622 - 25.056}{12} = 25.547$$

$$\text{sample variance} = \frac{25.056 - 2.542}{3} = 7.505$$

measurement variance = 2.542

Standard error of the grand mean and confidence limits of the mean u: mean = 32.111; variance of the mean = 331.622/72 = 4.606; standard error of mean = 2.146 = $s_{\bar{x}}$. Using t = 2.571 for df = 5 and a two tailed test at the 95 percent level gives $\bar{x} \pm ts_{\bar{x}} = 32.11 \pm 5.52 = 26.59$ and 37.63. One lot mean is below 26.59 and one lot mean is above 37.63.

Table 2
Data for Nested Classification
(6×4×3 measurements)

n samples	k lots: 1	2	3	4	5	6	
1	25	30	30	24	35	31	
	28	33	31	27	36	34	
	26	31	28	25	36	32	
2	22	29	36	30	40	36	
	25	31	35	28	42	40	
	23	32	38	31	38	37	
3	22	36	35	32	35	35	
	20	37	32	32	36	38	
	24	36	33	30	34	36	
4	26	30	38	26	44	38	
	24	32	37	25	40	38	
	22	28	40	28	39	39	
Sum Σx	287	385	413	338	455	434	2312
Mean $\bar{x}_j$	23.917	32.083	34.417	28.167	37.917	36.167	32.111
Σx^2	6,919	12,445	14,361	9,608	17,359	15,780	76,472

Notes

[1]W. A. Shewhart, *Statistical Method from the Viewpoint of Quality Control,* edited by W. E. Deming, Graduate School, U.S. Department of Agriculture, Washington, 1939, p. 66.

[2]A. C. Rosander, *Elementary Principles of Statistics,* Van Nostrand, New York, 1951. Copyright by Wadsworth, Inc. Used by permission of Brooks Cole Publishing Co., pp. 104, 105, 220. Speed of light is from *The World Almanac,* 1976, p. 760. A NOVA television program on May 12, 1978 gave the velocity of light by the laser method as 299,793.0 kilometers per second.

[3]*Rocky Mountain News,* June 4, 1974.

[4]*The Denver Post,* February, 1974.

[5]*Mechanics Illustrated,* August, 1971.

[6]*Rocky Mountain News,* June 27, 1974.

[7]For a more detailed treatment of analysis of variance, see standard books on statistics such as G. W. Snedecor and W. G. Cochran, *Statistical Methods,* 7th edition, Iowa State University Press, Ames, Iowa, 1980.

[8]Adopted from G. W. Snedecor and W. G. Cochran, *Statistical Methods,* Iowa State University Press, Ames, Iowa, 1967, p. 259.

[9]A. C. Rosander, *Elementary Principles of Statistics,* Van Nostrand, New York, 1951. Copyright by Wadsworth, Inc. Used by permission of Brooks Cole Publishing Co., pp. 557-559. See pp. 500-501 for Bartlett test. The test is illustrated later in Chapter 25.

[10]For a detailed treatment of analysis of variance applied to nested classifications, see K. A. Brownlee, *Statistical Theory and Methodology in Science and Engineering,* Wiley, New York, 1960, pp. 389-395.

Part 3
Techniques

18

Quality: Its Nature, Measurement, and Attainment

Quality as conformance to specifications. The concept of "quality" is described in many different ways reflecting the complexity of the concept, its multidimensional nature, and its numerous aspects. One of these is conformance to specifications.

Specifications are the numerical values of key characteristics within a given range of variation (tolerance) which are judged necessary for a functioning product and descriptions of defects that should not exist such as scratches, flaws, and blemishes. Specifications include drawings, blueprints, and descriptions; they may also include standards. These are laid out by the manufacturer based on past experience.

Specifications include such key characteristics as diameter, hardness, density, length, tensile strength, thickness, resistance, smoothness, and temperature. They are aimed to obtain a proper fit, a smooth and silent operation of an assembly, an adequate length of life, or an acceptable mean time to failure.

In a service industry or operation, specifications are just as important for purchased parts and equipment and are used for the same purpose. They are a guide to what is considered acceptable and what is not. In dealing with the mass buyer who characterizes service industries, conformance to specifications may not be enough unless the specifications are based on findings from adequate market and customer studies. Specifications laid down by the manufacturer and service company may fall far short of meeting the requirements of the individual mass buyer. This means millions of buyers may reject the seller's specifications and that no amount of pressure from the sales force and no amount of sales psychology, gimmicks, and inducements will succeed. This is what has happened in the automobile industry.

In many service industries, the seller, not the buyer, sets the specifications — the framework of rules under which the buyer must purchase. A rare exception is automobile repair where the buyer can specify what he or she wants done and the repair shop will do it, and no more.

Quality as fitness for use. Fitness for use can exist under a wide range of quality, cost, and performance characteristics. A Cadillac is fit for use, but so is a Jeep, a Volkswagen, and a second-hand Chevrolet. Some buyers are interested in costs — original, operating, maintenance, and repair — and not in power, style, luxury, and prestige that means "fitness for use" to others. Different groups of mass buyers have very different preferences and very different ideas about what constitutes fitness for use.

Since the expression "fitness for use" is general and ambiguous, it needs to be analyzed before it can be used by the buyer to make a decision. This is because even a "lemon" is fit for use. Questions to ask are:

- Used for what?
- How often is it used?
- At what cost?
- How reliable is it?
- How often does it fail?
- What is required to correct a failure?
- How satisfactory does it perform?

Quite clearly "fitness for use" like "conformance to specifications" is not enough when we consider quality in the service industries and service operations.

Quality as buyer satisfaction. This view of quality puts the buyer of goods and services at the heart of the quality concern. This is most appropriate for service industries depending upon markets of mass buyers. This contrasts with the approach which centers on the seller, the specifications, or the use of a sales force and advertising to sell a product or service to the public. This gives the buyer a choice if real competition exists. The buyer has no choice at all in the case of a public utility or other monopoly and very little choice in case of a regulated service industry like banking and insurance where operating rules are set not by the buyer, but by laws and regulations, by legislatures, and by regulatory agencies.

In negative terms, buyer satisfaction means absence of errors, defects, delays, failures, and high costs. In positive terms it means: Zero Defects + Zero Errors + Minimum time delays + High reliability + Low life costs.

To the seller, buyer satisfaction means:

- Following market trends continuously and acting on them.
- Surveying customer attitudes and preferences continuously and acting on them as required.
- Measuring the quality of competitive products and services and seeing that competition is being met or excelled.
- Offering goods and services the mass markets will buy. This will result from adequate and continuous market and customer studies.

Quality as acceptability at an affordable price. This definition of quality combines acceptable quality characteristics with an affordable price. On the one hand, these buyers want such acceptable quality characteristics as Zero Errors and Defects, minimum time delays, high reliability, and safe operation at a price they can afford to pay. On the other hand, there are those buyers who are interested in these same quality characteristics, and in addition luxury, style, and prestige, but have such high incomes that they are not interested in what a product or service costs. In the service industries, it is well for companies to keep in mind that there are those buyers who buy the quality they can afford; they will tend to buy from the firm that gives them the best quality goods and services at the prices they can afford. While in many cases, the higher the price the better the quality of the goods and services, this is by no means always true. The rich have their troubles obtaining acceptable quality services, the same as those with lower incomes.

Quality as life benefits that are worth life costs. Quality may be interpreted to mean that the benefits from the product or service are worth the costs the buyer has to pay. Costs are easy to isolate since products and services are always sold at a price. Benefits, on the other hand, are real but intangible so they are rarely quantifiable. Hence, whether benefits are worth the costs has to be a subjective judgment made by the buyer. There is no money value for benefits to compare with the costs. Examples of benefits of products and services to the buyer are the following:

- Automobile: transportation: convenience, comfort, time saving, low cost, reliability, speed.
- Telephone: communication: convenience, time saving, reliability, low cost, speed, efficiency.
- Electricity: lighting, heating: convenience, time saving, low cost, reliability, speed, efficiency.

So the basic question for many is: Can I afford the benefits of various products and services at the prices being charged? This often reduces to the question of "to what extent"?:

- What kind of automobile can I afford?
- What telephone service can I afford?
- How much electrical service can I afford?
- How much and what kind of insurance can I pay for?
- Do I have enough money to open a checking account in a bank?

This aspect raises the question of what is called cost-benefit analysis. If the benefit is a profit as in a business, then costs can be compared with earnings. But even here cost-benefit analysis needs to be used with extreme caution.

Cost-benefit analysis does not apply to a situation where benefits are intangible but still very great, exceeding even the highest cost. Examples are polio vaccines, antibiotic drugs, and other life-saving medical advances. How does one estimate the money values of these benefits? Life-saving costs have no limits put on them. Billions of dollars have already been spent and billions more will be spent before cancer is conquered. In a case like this, cost-benefit analysis fails.

Another example is the case where it is more important to obtain accurate information than it is to consider cost and save money:

1. Obtaining an accurate diagnosis of a serious illness.
2. Obtaining an accurate count of the population of the United States.
3. Obtaining an accurate estimate of the number of unemployed in the United States.
4. Collecting accurate information when a large damage claim or product liability suit may be involved.
5. Obtaining an accurate estimate of the amount of tax error made on federal individual income tax returns.

It is more important to spend the necessary funds for a properly designed and managed test, experiment, or sample study than it is to economize and run the risk of obtaining inadequate, if not useless, data.

The same holds true for a product or service. Trying to save costs may lead to using poor quality materials which reduce the length of life, reduce safety, reduce quality, and increase the chance of failure, dissatisfaction, and loss of customers. If the bottom line is profit, the top line is customer satisfaction.

If management wants acceptable quality data, acceptable quality performance, and acceptable quality products and services, it will have to pay the price. A cost-benefit analysis may result in a short-term decision that is fatal in the long run, as illustrated by what happened to the automobile industry. Making large, more profitable cars ($1500) instead of smaller, more efficient cars ($400) worked in the short run, but led to disaster in the long run.

The market and customer survey or studying the nature and preferences of buyers. Monopolies, like public utilities, are not particularly concerned about individual buyers because they have a guaranteed market. Regulated industries, like banks and insurance companies, are concerned only to the extent that they have to compete for business, even though the buyer may have very little choice.

Market and customer surveys are important in service industries where real competition exists and where a buyer does have a choice:

- Wholesale and retail stores
- Doctors and dentists
- Restaurants and cafeterias
- Transportation and travel
- Automobile, radio, and television repair shops
- Entertainment, theaters, movies, concerts
- Cleaning and laundry
- Barbers and hair dressers
- Computers and data processing
- Lawyers
- Hotels and motels

In many of these services, such as medical and legal services, there is a choice for the buyer only in larger cities and communities; in a small town, there may be little or no choice.

The larger companies which are faced with competition will pay close attention to market trends, to population changes, to available sales figures, to their share of the industry's business, and to buying habits and trends. The smart seller will maintain the names and addresses of customers and make periodic customer surveys to determine not only what they prefer in the nature of services, but what they have found unsatisfactory. The seller will give them first choice of sales, reservations, reduced rates, and new accommodations. This practice of holding on to customers is one way to develop a successful business. This is found in motels, travel agencies, restaurants, and auto repair shops.

The Time Factor

Demand for service on a time basis. The demand for services is very highly correlated with time. Services required continuously or daily include water, electricity, gas, telephone, newspaper, mail, and bus or other transportation. Those on a weekly basis include buying groceries and gasoline. Those on a periodic basis (usually one or more times per year) include medical service, dental service, buying birthday and holiday gifts, vacation travel, auto repair, and auto maintenance. Those on a once-per-year basis include tax assistance, auto safety inspection, auto license plates, auto emissions inspection, and furnace check-up. Services required periodically, but usually only rarely, include an auto insurance claim, fire or other home damage claim, or household repair of appliances like a washer, dryer, refrigerator, or electric switches.

Components of time for customer. The customer is concerned with several components of total service time:

1. Time to order service: appointment, reservation, order.
2. Waiting or delay time before service can be performed.

3. Service time.
4. Post-service time, if any.
5. Repeated service, if any.

The service time sequence starts with an order for service in the form of making an appointment for some future date, making a reservation for some future date, submitting an order for one or more items to be delivered at some future date, or some other request for service. Usually the customer wants the future date to be as soon as possible or at some prescribed date which may have to be varied in order to obtain the required service. The customer may have to take a time when service is available: examples are doctor's and dentist's appointments, motel reservations during the vacation season, low cost housing, and a nursing home. The automobile driver may have to line up and wait his turn to obtain gasoline at a service station; a customer may have to wait in line at the supermarket check-out station, at the post office, and at the bank. Waiting and delay time have become part of modern living, but such time should be reasonable. Waiting in a doctor's office increases his income, but does not increase the quality of service.

The time factor for a company or agency. Most services are organized around the clock or the calendar. In other chapters, the role of time tables and time schedules in exerting control is described. The company or agency is faced with weighing the effect of waiting or delay time on customer buying against the cost of increased capability to reduce waiting time. Waiting time varies inversely with capability: the larger the capability, the shorter the waiting time, and vice versa.

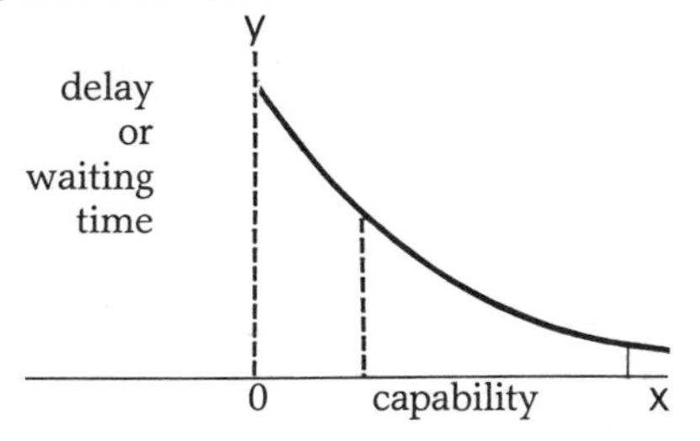

Service is limited by capacity to serve. This automatically creates waiting or delay time. This is because the company or business will keep capability limited and hence customers waiting to insure a profit — often a high profit — as in medical and dental services. Excessively large staffs would be needed to eliminate delay time. Hence, some delay or waiting time is an inherent price the customer has to pay for many services, whether in banking, postal service, medical service, dental service, buying gasoline at a service station, or buying groceries at a supermarket.

In some service industries, time is important because it determines the distribution of the workload, the load factor, and the peak load. In passenger transportation, the time schedule shows the departure and arrival times, the nature of the service, and the connections and waiting times when transfers are made. Traffic is heavier at certain times of the day than at other times. In freight transportation, the shipper orders the number and type of freight cars needed stating when and where he needs them. The demand for certain kinds of freight cars is much greater at certain times of the year, such as grain harvest time in the Midwest, than at other times. In connection with electric and gas service, meters have to be read monthly, individual bills calculated, taxes listed and added, the bills prepared and mailed, delinquencies recorded and accounted for, and action taken on delinquent accounts. The company has to lay out a time sequence of operations and see that the sequence is implemented monthly. In those areas where a very large volume of bills exist, the entire sequence may be staggered in terms of customer names so as to smooth out the work load over the entire month and year.

Total service time. Total service time measures how fast the service is rendered as well as how much it costs. This is because services are labor intensive and therefore are paid for at labor union or professional rates. Furthermore, the customer has to pay whether the services are satisfactory or not. One has to pay for a doctor's visit whether you are helped or not and whether your complaint is treated or the doctor elects to do something else, even though unnecessary. You have to pay for every visit of a repairman whether he fixes the trouble or not. The same is true when you take your automobile to a garage. You pay if they fix nothing, you pay even if they make your car worse, and you pay if they fix it.

Control of time by the seller. In some service industries, the time that service takes place and its length may be controlled by the firm in order to smooth out the work load and avoid having extra employees at peak loads. Several methods are described. Some may be made to aid the firm internally; the buyer is not affected.

1. Control by appointments. This method is used by doctors, dentists, lawyers, auto repair shops, and beauty parlors.

2. Control by reservations. This is used by hotels, motels, airlines, theaters, and restaurants. This measures demand ahead of time, prevents overbooking or it should, and acts as a guide for advance preparation as in a restaurant.

3. Incentives by special low rates. Low fares may be used during the off-rush hours as in the case of buses, airlines, and telephone companies.

4. Staggered billing. Retail stores and public utilities may use monthly billing which is distributed fairly evenly over the month, by dividing the alphabet or by some other method. This may be more of a benefit to the seller than to the buyer.

5. Special tours and excursions. Special rates may be set, not only for an individual but for groups by a company engaged in tours or excursions. Reduced fairs or packages of fares are offered only during certain times favorable to the company. The individual has to adjust his or her schedule to what is available.

6. Time schedules. Time schedules are everywhere controlling our lives. They control both the buyer and non-buyer of services. Time schedules control factory work, office work, transportation, (bus, rail, air, ship), medical service, schools, churches, restaurants, theaters and movies, and many more.

The buyer of services is forced to accept these schedules and adjust his living to them. It is imperative for the company or agency to adhere as strictly as possible to these schedules, so that waiting and delay times are minimized. This multi-dimension, time, is a quality characteristic of critical importance.

The Cost Factor

The cost factor for both customer and company is directly related to the time factor, but there are other costs not directly related to a time sequence. The costs to the customer in the service time sequence are:

1. Cost to order the service.
2. Cost of waiting or delay.
3. Cost of the service.
4. Cost of post service, if any.

Costs to the customer that are not directly related to time include:

1. Cost of correcting one or more errors.
2. Cost when company shifts responsibility to customer: "Call us".
3. Cost to the customer of company failing to do as promised.

The cost of correcting one or more errors includes the cost of phoning, writing letters, duplicating cancelled checks, losing working time, travel time, its effect on other activities, and inconvenience. The cost of the "Call us" tactic is not only the cost of telephoning, but also the cost of travelling to the store or establishment to determine what the situation is. The same costs are incurred if the company does not do what it promises.

The costs to the company involve a whole series of operational costs: training personnel; salaries and wages; computer and other equipment costs; processing operations such as billing, crediting, listing, filing, accounting; service and repair operations; guaranties and warranties; customer complaints; legal aspects; government regulations; and expanding area of service, a situation faced by public utilities.

Errors and Error Rates

A major determinant of quality is human error, both in actual value and in relative terms, which in importance may be minor, major, massive, dangerous, or fatal. *Individual errors* that may or may not be serious include:

- typing
- transcribing
- coding
- tabulating
- listing
- filing
- counting
- calculating
- recording
- observing
- measuring
- classifying
- printing
- interpreting words
- interpreting data
- inspecting
- verifying
- auditing
- failure to follow instructions
- failure to follow procedures
- failure to inspect or verify
- taking something for granted

Examples of *operations* that can be in error where management has a major responsibility include:

- planning
- designing
- billing
- accounting
- inspection
- computer programming
- instrumental failure
- sample selection
- sales transactions
- checking accounts
- savings accounts
- insurance claims
- preparing tax return
- making change
- charging wrong price
- stopping or starting a process
- repairing
- shipping
- servicing

Service industries where human error can be *dangerous, if not fatal,* include transportation, health services, and power plants. In transportation, errors can be made relative to weather conditions, road conditions, state of equipment, and other operating conditions. In health services, errors can be made in administering gases, medicines, and drugs; and in labeling and reading labels or not reading labels. In power plants, switches can be opened or closed accidentally or at the wrong time; equipment that should be turned on is turned off and vice versa.

The goal is error-free performance. The practice should be a continuous alert to inspect, verify, check, and otherwise take steps to prevent and avoid errors. This is because errors cost money, raise prices to the consumer, reduce productivity, fuel inflation, and may endanger the lives and safety of people.

Human error was a major cause of the Three Mile Island accident. Even though the danger from radiation was real, it was exaggerated. For example, Denver residents absorb *every year* 200 millirems of natural radiation which is much more than was released at Three Mile Island.[1]

The four kinds of errors are: esthetic and irritation errors, minor numerical errors, major and serious errors, and massive errors which were described in Chapter 1 and will not be described here.

Esthetic and irritation errors are not earth-shaking errors, but are of psychological importance because to some managers and executives they are just as serious as a major type of error. This type of error includes misspelled words, faulty grammar, erasures, punctuation or lack of it, poor margins, legibility, dim transcript, or uneven print. Neatness and language standards are required to avoid this type of error.

Minor numerical errors include the following: errors in transcribing, errors in reading an instrument, errors in original data record, transposition of digits, typographical errors, the sum of percentages is not exactly 100 percent, and the sum of vertical and horizontal entries in a two-way table do not add to the total. This statement should be qualified because it is possible for some of the above to result in serious errors.

Major sources of errors that can be serious in the sense that the error leads to wrong estimates and wrong inferences include the following:

1. Defective records
2. Wrong decimal point
3. Calculating error
4. Programming error
5. Tabulating error
6. Wrong unit of measurement (in 1,000's not units)
7. Biased sample
8. Faulty experiment or test
9. Errors in data at the source
10. Interview error
11. Observer error
12. Faulty instrument, faulty equipment.

The customer may make errors or mistakes because of faulty records or no records at all, because of misuse due to ignorance, because of misuse due to carelessness, and because of lack of technical and other relevant knowledge.

Management errors include ignoring or neglecting market trends and customer satisfaction, ignoring new management developments including quality control, neglecting new and improved computer and statistical techniques, faulty policy making and planning, ignoring human factors, inadequate controls, and misuse of cost-benefit analysis.

Measures of Quality and Its Acceptability

Specifications or their equivalent enumerate measures of quality and specify in operational, if not quantitative terms, what constitutes acceptable quality. The following tabular arrangement lists a large number of quality measures or factors and indicates what constitutes acceptable quality. The latter are qualitative, but would be made more precise in the specifications or their equivalent.

Quality measure or factor	Acceptable quality
1. Delay time	1. Reasonable, if not minimal
2. Service time	2. Reasonable, if not minimal
3. Initial cost	3. Affordable
4. Operating cost	4. Affordable
5. Maintenance cost	5. Affordable
6. Repair cost	6. Affordable
7. Life cost	7. Affordable
8. Trouble	8. Corrected, eliminated
9. Errors	9. No major or minor, no serious ones; error-free performance
10. Error already made	10. Corrected without delay and at no cost to customer
11. Reliability	11. High enough for all parts and systems to yield affordable life cost
12. Customer attitudes	12. Satisfied with service; civil, cooperative
13. Treatment of customer	13. Civil, helpful, honest, no abuse, courtesy; no age bigotry in doctor's offices, nursing homes, hospitals, other agencies; no taking advantage of customer's ignorance
14. Employee competence	14. Capable, competent, reliable
15. Employee attitude	15. Cooperative, helpful, honest, no abuse
16. Promises, guarantees	16. Honored, lived up to, kept
17. Standards of quality for air, water, radioactivity	17. Safe for health at all times
18. Standards for drugs, food, etc.	18. Effective, safe at all times
19. Data — survey, experimental, test	19. Sound sampling, testing, and experimentation, with proper controls and management so data are accurate, reliable, valid, and trustworthy.

Quality means acceptability. Quality means acceptability of the service rendered, and specifications define what constitutes acceptability:

1. Acceptable dimensions, sizes
2. Acceptable operating characteristics such as reliability, efficiency, cost
3. Acceptable number of errors or defects
4. Acceptable proportion of errors or defects
5. Acceptable length of life
6. Acceptable amount of total service time: order, service, delay, post service
7. Acceptable initial cost
8. Acceptable operating cost
9. Acceptable life cost
10. Acceptable reliability for all major parts
11. Acceptable psychological factors: convenience, treatment, attitudes, courtesy

Indicators of trouble or unacceptability. There are many indicators of trouble which can lead directly to a decision of unacceptability of the performance, service, or product. These involve the use of various criteria to determine whether "trouble" exists or not; obviously, one of the major problems is to make sure that trouble exists and that time is not being spent hunting for trouble that does not exist:

- Number of errors is too large
- Error rate is too high
- Delay time is excessive
- Values deviate too far from standard
- Cost is too high
- Reliability is too low
- Too many failures
- Failure rate is too high
- Length of life is too short
- Too much down time
- Deviation from standard procedure
- Massive error is found
- Too many serious complaints
- Value is outside critical limits
- Production is too low
- Idle time is too high
- Unethical practice found
- Illegal action found
- Conflicts over jurisdiction or responsibility found

Attaining Quality

What is necessary to attain quality in a service operation? What is needed to obtain error-free performance? To perform in a timely manner at an affordable price? To render service satisfactory to the buyer and consumer? To rectify a mistake once made? To correct a source of trouble? To prevent errors and mistakes and trouble from arising in the first place? There are at least seven major needs:

1. Managerial capability
2. Professional technical capability
3. Employee capability
4. Motivation
5. Adequate information
6. Appropriate and timely decisions
7. Appropriate and timely actions.

It is not enough to say "Quality is everybody's business". What constitutes quality must be made specific and operational and must be defined for every job, operation, function. Every individual must know what constitutes "quality" in every job he or she performs. The factors of quality that management is responsible for must be made clear. Examples are: facilities, equipment, conditions of work, standards, and quality goals. What the individual employee is responsible for must also be made clear, such as carrying out specified procedures carefully, completely, and accurately. The employee should not be blamed for poor quality which is due to factors beyond his or her control or which lie outside the authority inherent in the job or position.

Management capability. This capability includes that of the policy-making executive, planning director, operating manager, service foreman, or office supervisor. Except for top level policy-making, the basic supervisory functions are involved at every level: planning, directing, controlling, and appraising.

A basic quality policy for services needs to be laid down for the entire company or agency. This policy should be a guide and compass for everyone. This policy should include both customer and company:

1. Quality service that satisfies customers and which they can afford.
2. Quality service that eliminates waste, keeps down costs, but earns a profit for the company.
3. Expeditious and satisfactory resolution of all customer complaints.

Quality policy needs to be analyzed into what it means in operational terms for each part of the company and put into writing to provide specific guidance and direction.

Quality policy needs to take into consideration recent developments such as:

1. Growth of consumer oriented technology.
2. Growth of computer systems.
3. Quality of environment.
4. Product safety.
5. Relative ignorance of user and consumer relative to product science, engineering, and technology.
6. Increase in damage suits brought by consumers.
7. Difficulty of an individual to obtain redress from a large company or agency.
8. Increase in both Federal and State government regulations.
9. Shifts in mass markets and in consumer demand.

Management capability must include an adequate understanding of the human factors involved in a quality policy and in quality operations. It must also include competence to deal with these factors. This means an understanding of the psychology of both the individual and the group and of the important role of such factors as individual differences, motivation, and compatibility so necessary for team work. Where required, the factors of ambition and incentive must also be understood.

More specifically, this means that problems cannot be divorced from people. People create problems directly or indirectly, and people must solve or ameliorate them. Problems and people interact, and nothing is gained by refusing to face this reality. Individuals are responsible for quality. Who is responsible for quality at various levels, and what that quality should be, should be made very clear both verbally and in writing. This working rule applies to everyone from a top level executive to the lowest level worker.

A major challenge to management, whether at the middle or higher levels, is the ability to successfully run a team of specialists. This is because in any company or agency of any size there is specialization of knowledge and skills and abilities which requires scores, if not hundreds or thousands, of specialists. In one task force, there were one or more lawyers, economists, accountants, statisticians, computer specialists, rail traffic specialists, managers, and budget officers.

In connection with these specialists, management is faced with at least four tasks: How to hire a competent and capable specialist in the first place? How to get the most benefit from each of these specialists? How to get specialists who know only their own fields to work effectively as a team with a minimum of conflict and friction? How to get these specialists to reconcile their interests with those of the company or agency?

What top level management needs to know is what special knowledge and skills are needed for the successful operation of the company or agency. Then they must hire the specialists who meet these requirements. Management does not need to be knowledgeable in any of these specialties, but it does need to appreciate the value of each. It needs to keep in touch with new developments and their applications. New fields which have developed during the past 40 years include the computer field, operations research, management science, and the statistical field including sampling and quality control.

Many managers act as though they should be experts in all of these fields; on the other hand, there are those who berate the specialist for his or her narrow views. Some refer to the "lingo" or the "jargon" of the specialist. The manager should recognize that his job is to direct and coordinate the work of these specialists, not match their knowledge.[2] The manager should also recognize that every specialty has its own special vocabulary.

Management practices which create difficulties not only for managers in general but for managers of quality programs, and those aiming to become managers of quality programs, are the following:[3]

- Officials over-ride, over-rule, or otherwise interfere with the technical work of technical specialists. Those with power try to over-rule those with know-how. For example, they want to tell sampling experts how to design a probability sample.
- Officials oppose the use of improved techniques and procedures such as probability sampling, quality control, design of tests and experiments, and other aspects of statistical science. They have all kinds of "reasons" why changes should not be made.
- Policy-makers interfere with operations which they do not understand and in which they have no experience, or they actually try to run operations.
- Officials are misled by cost-benefit analysis. The long-run market may be much more important than short-run profits.
- A technical planning staff may do very little planning, especially important long-range planning.
- A quality assurance staff tries to do statistical work, but has no statisticians, nor anyone familiar with statistical techniques applied to quality control. Hence, its reports are crude, misleading, and ineffectual. Management does not know how to set up an effective quality assurance department.
- Officials want to force on to technical staffs persons who are not qualified or competent.
- High level officials do not know how to direct and manage a technical team.
- Officials confuse superficial or cosmetic changes with real improvements.
- The time of technical specialists and experts is wasted in conferences, talks, and activities of little or no importance.

The Management of Quality

The key factor in the initiation and implementation of a quality program, aimed at improving productivity and reducing costs, is management. Experience has shown that in both government and private industry the major obstacle to the introduction of new statistical and related techniques is high level managers and professionals. It is not the lower level employees who object and prevent change so much as it is top level managers and executives. They do not understand the power and versatility of the science of statistics or the important role the statistical practitioner can play in helping solve numerous management problems. Consequently, many applications which could increase productivity, improve quality, and reduce costs are ignored, rejected, or adopted half heartedly.

Examples of where the system is at fault, where management objects to substantial improvements, are the following:

- Judgment sampling is used to collect management operating data when a probability sample would be much better.
- A census, 100 percent coverage, or a 100 percent tabulation is used, or a 100 percent computer run is used when a properly designed probability sample could be much better.
- A computer edit is used on incoming sample documents being collected for the first time where the error rate is 30 percent or more. A manual edit by specialists familiar with the form and data would be much better, eliminating the high degree of wasted operations on the computer. The computer was used for about three months and then abandoned in favor of the more expeditious manual method. The error rate was too high for the computer to be a cost-benefit operation.
- Management lacks key operating information and is content with using guesstimates, hunches, and "expert" judgments. An example is the case of the stolen nylon hose. Control charts or a simple in-factory experiment of the relationship between input and output would have readily shown management that no hose had been stolen. Other areas where high level officials lack key information include tax errors on federal income tax returns, and marketing studies on consumer demand for various models of the products such as an automobile.
- Management refuses to act on the information available or draws false conclusions from current data such as a drop in sales being blamed on salesmen or has false data to start with such as the highly biased estimate of the output of a semi-automated textile machine.
- A computer post mortem of the output of a production line is mistaken for the quality control of the product.
- Spot checking of the product, or 100 percent inspection of the product, is confused with quality control.
- The product was not properly designed in the first place: wheels ran off a tricycle, sharp edges cut the fabric of a baby carriage whenever the top was put up or lowered, or a portable ditto machine had no stop making it difficult to pull a finished sheet off of the mould.
- Sampling inspection of a measurement is based upon a sample plan from MIL-STD-105 which is based on attributes: 100 pieces were inspected; when half that number could give the same protection, and half again would give the same protection if the vendor had a quality control chart on the dimensions. This chart would show that it was being produced under conditions of control and the population variation could be very closely approximated. Then the sample could be only 25.
- Variations exist throughout the plant, organization, or agency. These should be explored to determine which are essentially random and can be left alone and those that are due to assignable causes that should be eliminated.[4]

Quality circles. A quality circle is a management technique adopted from the Japanese; although it represents only a minor part of their total quality program. A detailed application of quality circles applied to the operations of a large American bank is described in Chapter 2. They are characterized by the following:

- Employees form a group (circle) to work collectively on quality problems they encounter in their work.
- Membership is voluntary.
- Emphasis is on employee participation in all aspects of quality.
- Employees work on quality problems in their area.
- They use problem-solving techniques which are mostly statistical.
- They receive management support and encouragement.
- Emphasis is on team work with the supervisor as leader.
- The assumption is that employees can improve the quality of their work by intimate participation in discovering and solving problems in their area.

To many these are aspects of good supervision: team work, cooperation, employee suggestions and contributions, supervisory guidance and assistance, and improved quality of performance.

It is easy to encounter technical and jurisdictional aspects which are beyond the capabilities and domain of the circle.

The circle may create personal and supervisory problems because some employees are in the circle and may receive special attention, while those outside the circle do not. Also, how does the supervisor handle the quality problem for those who are not in the circle?

Identifying statistical techniques as "problem-solving techniques" is misleading. In the main, these techniques discover problems and indicate the possible existence of problems; they do not solve them. Other persons — engineers, specialists, etc. — have to pin-point the problem, if it exists, and solve it.

An alternative to quality circles is to develop, if not hire, better supervisors, on the one hand, and more alert, concerned, and capable employees, on the other. The better supervisor will promote team work, participation, guidance, and challenging tasks.

Seven basic concerns in quality. Management aimed at fostering, developing, promoting, and improving quality faces seven basic concerns. All of them are of great importance; thus, the danger lies in ignoring or neglecting one or more of these factors in favor of one or two others. The challenge to management is to obtain an effective balance of all seven:

1. **Psychology** because quality involves people and human behavior.
2. **Statistics** because quality involves variability, quantitative data, and the effective use of such data.
3. **Management and supervision** because quality involves the directing of human efforts toward specific goals.
4. **Specialities** because quality involves knowledge of a wide variety of subject matters and the ability to use know-how effectively.

5. **Time** because quality involves trends, productivity, life cycle costs, delay, and synchronization.
6. **Techniques** because quality involves processes, methods, procedures, efficiency, and operations.
7. **Reliability** because quality involves trustworthiness of performance, continuity of operations, and dependability of people and equipment.

All quality planning should include an adequate provision for all seven; imbalance may lead to improvement in quality, but may fall far short of the maximum amount of improvement that is feasible.

Professional technical capability. Professional technical capability means all of the special knowledge needed in order to run an acceptable quality service operation. This includes not only those competent in special fields of knowledge, but also those competent in techniques which are now standard in quality control and quality assurance or can be used for solving quality problems in service operations.

Special fields include accounting, auditing, engineering, economics, statistics and probability, computer science, regulatory laws, sales engineering, service engineering, safety engineering, laboratory sciences, applied mathematics, operations research, and field tests and experiments.

The techniques used directly in quality control work include:

1. Probability sampling for customer surveys, marketing surveys, sample audits, laboratory tests, and field tests.
2. Process control for production work using Shewhartian techniques.
3. Acceptance-rejection sampling for control of lots, batches, or work units.
4. Statistical techniques for data analysis, regression analysis, and analysis of variance.
5. Random time and random area sampling for estimating costs, decomposing joint costs, estimating machine and employee utilization, and detecting and measuring down time.
6. Computerization for storage, tabulating, listing, calculating, printing, and retrieval.
7. Design of experiments for field tests and laboratory tests.

The basic problem here is to insure that the technical capability of every specialist is brought to bear, continually and persistently, on problems dealing with quality to prevent wandering off into areas that may be interesting but are not related to improving the quality function. These include the trouble-shooters.

Employee capability. The success of any program, whether dealing with quality or not, depends upon the employees. They are the ones who have to implement policies, plans, programs, and projects. They can be the difference between success and failure, and between an acceptable quality job and an inferior type of performance. They can thwart even the best laid policies and plans of executives, managers, and technical experts. Like a manager or specialist, a lower level employee should be hired for his or her competence and capability to perform a specified job, including the ability to meet quality standards.

At the time of hiring, and thereafter as needed, an employee should be told why his or her job is important to the company or agency and how it relates to other jobs. The reasons and explanations should be real, not phony, or otherwise the job should never have been created in the first place. In this connection, high level officials would find it helpful in improving and maintaining employee performance to leave their offices periodically to reassure employees *individually* of how important their jobs are. This would include all kinds of employees such as file clerks, secretaries, typists, computer librarians, control clerks, statistical clerks, and supervisors. These contacts should be made only if the official knows what he is talking about; it should not be a high pressured or superficial pep talk.

A strong comprehensive training program should be established for all parts of the company or agency. The purpose is to update or improve knowledge, skills, and abilities so as to improve both present performance and the chances for promotion.

Management should have a sound and effective employee policy:

1. Utilize individual talents to the maximum.
2. Provide opportunities for every employee to show his or her capabilities.
3. Recognize and develop individual differences.
4. Develop superior talents.
5. Provide appropriate rewards for superior and high quality performance.
6. Hire those who are promotable and promote them.
7. Foster an atmosphere acceptable to new ideas and new techniques that improve operations.
8. Promote team work.

Persons who are hired should always be interviewed; their education and experience should be examined relative to their potential for promotion. If promotability is not given considerable weight in hiring, the personnel situation may become rigid, inflexible, restrict promotions, create morale problems, and lead to unhealthy working conditions and poor production and performance. Promotability is reflected in an individual's:

1. Ability to learn, be accurate, and be alert.
2. Ability to follow instructions.
3. Desire to learn.
4. Desire to tackle difficult situations.
5. Acceptance of challenges.
6. Progressive advancement in jobs held.
7. Flexibility: ability to adjust to new situations, new conditions, new problems, new assignments.
8. Emotional stability: coolness under pressure and the ability to handle stress.

Job descriptions are highly desirable, if not a "must", so that every employee knows specifically what he or she is to do and just as necessary what *not* to do. Job descriptions provide a way for management to insure that all important functions and tasks are being covered, that duplications that could lead to friction are avoided, that relationships between jobs are spelled out, and that the most important tasks are emphasized. In this way, each employee is held responsible for what is on the job description or its revision, so there can be no misunderstanding as to what each person is to do.

Nothing is more annoying than to have some officious front-office employee answer a telephone call or an inquiry and try to make a decision another should make. I have encountered these persons in doctor's offices, banks, insurance companies, government offices, and libraries. Trying to get through this office "iron curtain" is often a difficult job; it is a situation management could afford to correct. Obviously, these employees need to be instructed as to how to handle inquiries, what kinds of questions they are competent to answer, and what kinds of inquiries they are to pass on to those competent to answer. They must stop being an obstacle to quality service. All this could be avoided if these persons had a written job description explained to them by their supervisor when they were hired.

Inter-person communication is a "must". "Communication" can be very ambiguous unless carefully defined. It means the exchange of information, the exchange of meanings between individuals, and the expressing of ideas effectively so that there is mutual understanding. All three of these definitions really mean the same thing. The emphasis is on expressing ideas or conveying meanings or information so that there is an understanding of what is being said or written.

How many times have you gone to a store or an office, or written to them, and found a new person unfamiliar with what has been said or written before. I recall encounters like this in a library, retail store, insurance company, and motel. I make a motel reservation over 1,000 miles away, by telephone. I give the man answering the dates desired. A confirming letter, in error, is sent by the reservation clerk, a woman. I send a deposit; confirmation of this from a third party is also in error. I have to send two letters before the error is corrected. I made a mistake in not having the first party repeat over the telephone what I had requested.

Motivation. Motivation is a psychological term referring to what impels, induces, or influences the actions of a person. The factors underlying motivation are numerous:

1. Successful performance
2. Recognition
3. Rewards: money, medals, honors, certificates
4. Approval of others
5. Power
6. Wealth
7. Leadership
8. Prestige
9. Authority
10. Independence
11. Promotion
12. Achievement, successful experience, new knowledge
13. Superiority
14. Desire to excel
15. Competition
16. Originality: creating, inventing, original contribution
17. Challenge
18. Desire to help others

A major task at every level of management is to use motivating factors that are applicable, appropriate, and effective. A careful study of individuals is necessary to determine the best motivating factors. What will work for one person may not work for another. A vice president of a data processing firm thought introducing incentive pay would speed up production of standard data formats. Inquiries by the supervisor revealed that only a few were interested in incentive pay, but that several were interested in gaining new and desirable experiences that would help them in their careers. Consequently, the idea was dropped. Progress made in production during the next two months, revealed by individual and composite learning curves, showed that the monetary incentive system was unnecessary; individuals were increasing their productivity as much as could reasonably be expected.

The negative side of motivation consists of those factors which upset the individual, cause resentment or even anger, turn persons off, and make rational performance difficult, if not impossible. Every attempt should be made to avoid and prevent these negative factors because of their destructive effects on the performance of the individual. The following terms give an idea of the nature of these factors:

injustices
partiality
false accusations
brow beating
unjust criticism
undeserved rebuke
refusal to give credit where credit is due
discrimination
favoritism
innuendo
psychological warfare
undeserved censure
undeserved reprimand
exploitation
unfair treatment
harassment
threats
envy
jealousy
intrigue

Far too many supervisors, foremen, managers, and executives exhibit at one time or another one or more of these moral crimes against an individual. I have seen or been subject to all of them. These tactics are often counterproductive; they are always harmful to the individual, although this may be internal and never show in connection with the individual's performance. They are much more common than we are led to believe because they are committed by those in power and the victim is silenced. He or she has to accept much of this in order to keep his or her job. Anyone who is subject to one or more of these tactics never forgets. If the situation continues, or is considered severe enough, there is nothing for the employee to do but to ask for a transfer or shift, arrange on the quiet for a transfer within the company or agency, or quit and find work elsewhere.

Motivation to some employees is related to the work schedule. Here the desires of the individual may quickly come into conflict with the discipline and requirements of the company or agency. The individual has no right to dictate to the company or the agency, nor has the company or agency the right to run rough shod over the individual. Conditions have to be arranged so as to meet the needs of each. One change suggested is the flexischedule. The individual is allowed to set his or her own hours of work, as long as the working day and working week is met. This type of scheduling is non-productive and excessively costly in common work situations because of the extra supervisory load:

1. Where functions, projects, and jobs must be performed on a *team* basis and all of the team has to be there.
2. Where the work input of one group is the work output of another and where production is sequential.
3. Where the boss relies on specialists for assistance which he may require at anytime of the working day. The boss may receive a call from top-side and need technical advice: from a computer specialist, from a statistician on a sampling problem or data analysis, from an accountant on a cost problem, from a lawyer on a legal question, or from an economist on an economic problem.

The flexi-schedule is for the "lone worker", for laboratory work, for library research, and for individual experimentation. It is for those individuals, especially professionals, whom management can trust to be self starters and self motivators. Otherwise, it raises a difficult problem of supervision and accountability.

The production of goods and services requires certain abilities, skills, knowledges, attitudes, and other forms of behavior which the worker must master and follow. The individual always has the freedom that he is not required to work in any fixed job or in any fixed place; he or she is free to work in the most congenial and satisfactory job although it may take time to accomplish this. In return, acceptable quality work is advantageous to the individual because:

1. Increased productivity keeps down inflation.
2. Acceptable quality work means lower prices.
3. Acceptable quality work helps meet foreign competition.
4. Acceptable quality work helps gain promotions.
5. Acceptable quality work means satisfied customers.

Participative management is another movement aimed at improving employee-management relations and productivity. It has gained attention because of the success of the Japanese quality circles. Participative work is that in which employees and supervisors or foremen work together to solve operating problems in the factory or office. This arrangement is not new; the question is how far can it be carried.

It will succeed if: 1) employees have the necessary knowledge, 2) employees are willing to learn, to study, and to think, 3) they are willing to work as a team, 4) they are willing to "pull their weight in the boat" and do their fair share or more, 5) they like solving problems facing management, 6) they analyze other's views and judgments on their merits, 7) they are willing to operate on the basis of evidence and facts, and 8) they accept accuracy and alertness.

It will fail if: 1) employees are ignorant, opinionated, self-righteous, and dictatorial, 2) employees are aggressive talkers but lack the knowledge, ability, and inclination to solve problems, 3) employees have anti-business and anti-corporation attitudes, 4) employees want to substitute their own world of fantasy for the real world of production and employment, 5) employees want to "do their thing" and ride rough shod over others whose "thing" does not coincide with theirs, 6) an employee thinks he or she is of the $20,000 caliber when he or she is only of the $10,000 caliber, 7) an employee lacks the patience and self control to work with others where differences with regard to problems, procedures, improvements, diagnoses, corrective action, and the like are bound to arise.

Management, from the supervisors on up, have a real responsibility to take all necessary steps to motivate those below them. Employees at all levels want to feel important and be recognized. The great majority wants to do a good job. Many times employees need to be reassured that their jobs are important and that much depends upon them doing an acceptable quality job. Many steps can be taken to promote motivation:

1. Managers need to explain personally to each worker why his or her job is important and how others are affected by what he or she does.

2. Managers need to have weekly staff meetings with professionals and supervisors to hear progress reports, to discuss problems, to resolve differences, to keep everyone informed about what is going on, to give assistance, and to give assignments or modify them.

3. Supervisors need to have weekly meetings with their employees for the same purpose: to discuss problems, to resolve differences, to keep everyone informed about what is going on, and to give advice and assistance.

4. Managers should maintain personal contact with the individuals who are doing the work. Taking a tour at least weekly through the various offices and operations is recommended.

5. Managers need to have personal contacts with key positions. An example is the chief of a section who goes to the files, sits down with the file clerk at her desk, asks about her work, and explains why careful error-free filing is so important. At another time, this same chief sits down with the clerk at the sample receipts control desk and explains why her desk is one of the most important in the section which it really is. The manager and supervisor must level with employees at all times. If they cannot talk with their employees frankly, honestly, and sincerely, they are in the wrong job.

Adequate Information

Adequate information is the heart of quality control, the basis for appropriate and timely decisions, and the basis for appropriate and timely actions. Information means knowledge derived from analyzed data collected for the purpose of answering an important question or solving an important problem. *The starting point is a problem facing management;* it is *not* data. Data grow out of a careful analysis of the problem so that adequate amounts and kinds of data are collected.

If the data are ambiguous no manipulations, however complex, can make them clear; if the data are irrelevant, no adjustments, however sophisticated, can make them sound; if the data are erroneous, no calculations, however fast, can make them accurate; and if the data are biased, no analysis, however elaborate, can make them correct.

Data are a means to an end, and that end is accurate, relevant, and timely knowledge or information which can be used to make an appropriate decision and take appropriate action. If data are not oriented toward real world problem-solving, then there is no end to the amount of data that can be collected and stored in computer data banks and no end to the amount of resources that can be wasted.

For purposes of discussion, three kinds of data can be isolated: descriptive data, problem-raising data, and problem-solving data. Descriptive data refer to data on kind, condition, characteristics, and qualities. These are non-normative characteristics such as heights, lengths, densities, areas, altitudes, amounts, and colors. Then there are numerical facts such as those given in the *Statistical Abstract of the United States* or *The World Almanac*. These range all the way from the number of votes a presidential candidate received in each state to the latest baseball scores and performances to the latest Olympic track and swimming records. Numerical facts may be arranged into trends over time, as comparisons by sex, race and age, and as percentage compositions by some classification to give them more meaning.

Many studies, whether based on sampling or not, are really problem-raising or problem-defining studies not problem-solving studies. When these studies are completed, one is ready to begin the real study. This is because initially the problem or problems are not clearly defined or analyzed; a study is necessary to obtain specific ideas about and to isolate the type and nature of the basic questions and problems. Again, it may be that the methods used preclude problem-solving: a biased sample, a judgment sample, or a very limited scope or coverage biased data at the source.

These problem-raising studies may be based on personal or professional interests. One may be interested in poverty, youth, minorities, unemployment, quality, inspection, customer complaints, or employee morale. They may be based on personal experiences such as high prices, low quality, poor service, or high taxes. They may be based on observations of unemployment, speeding autos, reckless driving, a "brown cloud" (air pollution), urban sprawl, rough roads, or dirty streets. They may be based on published reports or press accounts on vandalism, drunken driver accidents, freight car shortages, taxes wasted on government programs, errors on federal tax returns, graft and fraud in tax supported health and welfare and education assistance, recall of millions of defective automobiles, enormous amounts of goods stolen by shoplifters and employees, or human errors made in computer operations. They may be based on customer complaints, a high employee error rate, or an excessively high percentage of poor quality services such as delivery, installation and repair.

Problem-solving studies require both an intensive and an extensive approach. This means careful planning, analysis, and implementation:

1. Analysis of the problem until specifics emerge
2. Defining the scope of the study
3. Isolating the key characteristics
4. Isolating important groups and classes
5. Determining kinds and amounts of data needed
6. Selecting applicable techniques; this often requires exploratory work
7. Finding an adequate frame for sampling or census
8. Adequate sample design and specifications
9. Careful management
10. Control of sources of bias and error, and their elimination
11. Appropriate methods of analysis; these are in the specifications, although additional analysis may be thought necessary as analysis continues

Problem-solving calls for appropriate techniques as well as careful management at all stages. Techniques include one or more of the following:

1. Probability sample surveys
2. Designed experiments
3. Designed field tests
4. Laboratory tests
5. Sample audits
6. Acceptance-rejection sampling
7. Sampling for process control
8. Random time and space sampling
9. Regression analysis
10. Control charts
11. Input-output analysis
12. Learning curve analysis
12. Failure analysis
13. Error analysis

This list is illustrative. Techniques must be appropriate for the various aspects of the problem.

Two developments, both data oriented rather than problem oriented, need special attention: the computer and exploratory data analysis. Those who have made a profession out of the computer have developed a special vocabulary without regard to what has already been developed and used in other fields especially probability, mathematics, and statistics. Terms like "data elements," "data bank," and "data management" have been coined to try to build a science about a machine. These terms are not used, nor have they any use, in the field of statistical practice which deals exclusively with the collection, analysis, and use of acceptable quality data for management and research purposes. Computer science, with emphasis on data of the census type such as inventory, personnel, and finances, ignores completely some very important aspects: the role of problem analysis, the quality of data, the use of sampling, the pre-designed sample study, the dependence of the computer on the human beings who program and run it, and the waste inherent in storing and analyzing large masses of data of little or no use or of such poor quality as to be of no use whatever. Example: a company official runs 3,250,000 documents through the computer when a master probability sample of only 10,000 is all that is needed to make the several studies required. The high speed of the computer has turned many away from the use of sampling, but has resulted in wasting a lot of resources, especially where large volumes of documents are involved.

Exploratory data analysis is receiving attention largely because of a book by the same name written by Prof. John W. Tukey.[5] In this book, Tukey:

- Creates a new statistical language,
- Stresses analysis of data, mostly non-sample data,
- Draws heavily on data in *The World Almanac,*
- Gives quick methods for calculating squares and logarithms,
- Stresses the use of the median rather than the mean, and
- Ignores estimation.

He invents such terms as stem and leaf, dots and boxes, bins, hinges, box and whisker plots, fences, re-expression, H spread, froots, and flogs. "Table", "chart", and "graph" are rejected in favor of the lawyer's "exhibit".

Such standard expressions as the following are not indexed, nor could they be found: frequency distribution, class interval, mean, arithmetic mean, aggregate, sampling, probability, bias, error, estimation, and random. It is as though the reader, familiar with the terminology of statistics and it is far from standard, was faced with a foreign language.

Other questions raised by statistical practice include the following:

- Why spend time on quick logs, quick squares, the slide rule, and other quick methods when the *new desk and hand-held calculators have revolutionized calculation?* The tedium of calculating squares, square roots, logarithms, reciprocals, and other mathematical functions has been eliminated. It is admitted that a good hand-held calculator is easier to use.[6]
- The significance of *quality of the data* in analysis is ignored. Analysis cannot be sound if the analyst is trying to find meaning in a mass of data whose quality is unknown, whose method of collection is ignored, and whose biases are brushed aside. There may not be any meaning one can trust.

- The starting point is not data which are the means, not the ends. *The starting point is problems to be solved, questions to be answered, and hypotheses to be tested.* The data and meaningful information come from an analysis of these problems, questions, and hypotheses. Stressing the analysis of large masses of data is setting the clock back 50 years.
- Various graphic, tabular, and computational methods for preliminary exploratory analysis and interpretation of data have been in use for years. It is not that they have not been available: the problem is that they have not been used to the extent they should have been. It is not unlikely that many practitioners will find this new and strange vocabulary too big a change to make for the few benefits to be derived.

Quality of data. The concept of quality of data has not received the attention it deserves, although it is the critical factor in estimation and inference whether from samples or from 100 percent enumerations. In fact, one of the advantages of probability sample studies is that due to their reduced size, a much more effective control can be exerted over the collection and processing of data. Quality is measured in terms of relevance or validity, sampling errors, sampling bias, non-sampling errors, response errors, errors in the original data, errors in the estimates, errors in the source data, errors in observation and interviewing, errors in processing and tabulation, and errors in calculation and analysis. In statistical practice, the major source of error in the data usually is the data source itself including the respondent, interviewer, observer, or inspector. Sampling error is easier to control, and in numerous instances, it is smaller. This finding has been documented by practitioners for decades.

The wide concern of persons who are not specialists in statistics or data collection and analysis over the need for acceptable quality data or information is reflected in the following examples. Robert W. Sarnoff, president of RCA, is quoted as saying that business is showing "a greater preoccupation with quantity than quality — more concern with amassing new facts than with developing the structure and relationships that will convert them to meaningful information. . . . The evidence so far indicates widespread unpreparedness to absorb more than a fraction of the accumulated data. . . ."[7]

Edward L. Brady and Lewis M. Branscomb of the National Bureau of Standards cite from a report under the auspices of the Organization for Economic Cooperation and Development (OECD) with regards to uses and needs for scientific and technological information: ". . . quality of information — that is its reliability and credibility — is more important than access to large masses of raw data."[8]

J. Ross MacDonald of the National Research Council responds to the question — Who should own the data? — as follows: ". . . Are the data even worth owning? Unfortunately the answer is usually and embarrassingly 'No' across the entire spectrum of the research. The problem usually lies in lack of knowledge about the trustworthiness of the data. . . . Lide of the National Bureau of Standards . . . estimates that from 50 to over 90 percent of the published raw data available for producing trustworthy, evaluated results from the properties of scientific materials cannot in fact be used for this important purpose. . . . Youden . . . states that of 15 observations of the mean distance to the sun published from 1895 to 1961 each worker's estimated value is outside the uncertainty limits set by his immediate predecessor. . . ."[9]

J. Y. McClure of General Dynamics is quoted as follows:[10] "Second and third generation computer capabilities have progressively simplified the data collection process. Unfortunately, with this capacity exists *the temptation to collect all types and bits of information that 'might be needed by someone, somewhere, sometime, for some reason'.* Computer time is expensive. If a given system is not carefully planned in terms of minimum data requirements consistent with maximum use, a gross imbalance will exist between data acquisition costs and data utility." (Emphasis added).

That the quality of data poses a continuous challenge of top priority is described by Lide in a recent paper.[11] Lide discusses three classes of data: repeatable measurements as in physics and chemistry, observational data dependent on time and space as in geosciences, and statistical data. He discusses the growing need for *good* data, the need for *quality control of data bases,* and the need for *reliable* data. (Emphasis added). He states that "unfortunately the quality of the data preserved in the literature leaves much to be desired".

All of these remarks refer to data in fields of science where accurate measurement has a long history. He cites the wide variation in the scatter of 200 reported measurements of the thermal conductivity of copper as a function of temperature and the pitfalls of relying on a single value retrieved from the literature. If this is true of fields where precise laboratory measurement is common, how much more applicable is it to the counts and measurements in fields of application.

Lide does not describe how one obtains "good" data, does not explain how to produce "reliable" data, and does not outline how to apply "quality control" to a data bank. He does not mention the need for technical know-how and managerial capability to control sources of variation and prevent errors, whether human or instrumental.

There is no mention of the work over the past 50 years of Shewhart, Deming, and others in applying quality control techniques; nor is there any mention of the work of Fisher, Yates, Youden, and others in the design of experiments; nor is there any mention of the work of Hansen, Hurwitz, Madow, Deming, Cochran, and others in the field of probability sample studies — all of which were aimed at producing high quality data whose degree of uncertainty was known.

There is no mention of the fact derived from considerable experience that the quality of data is very often determined at the source — observation, measurement, interview, questionnaire, etc. — as well as by the sample or study design and how the sample — materials, people, time, area, etc. — is selected.

Appropriate and timely decisions. In the final analysis, the purpose of good management, technical competence, employee capability, motivation, and an effective information system is to insure appropriate, timely decisions and actions to render and improve services of acceptable quality. Everyone makes decisions because all jobs require them. Some of the most important decisions are made by lower-level employees because they are the ones most intimately associated with implementation of policies, plans, and procedures. Example of tasks where decisions are made include:

1. Selecting a sample	9. Inspecting
2. Coding	10. Computer library control
3. Editing	11. Typing
4. Filing	12. Reviewing
5. Transcribing	13. Recording
6. Sample receipts control	14. Answering a letter
7. Calculating	15. Mailing
8. Verifying	16. Routing

This is why it is imperative to apply quality control to all operations.

High-level and even middle-level managers hold the key to improved operations. For years, these managers opposed the application of effective techniques such as probability sampling, statistical quality control, statistical techniques, and design of tests and experiments. Even today, there are numerous agencies, companies, and areas where these techniques are not being applied or applied most effectively. Examples of the above are given below.

Computer science, although accepted too enthusiastically in many instances, brought dire consequences which are still far too common. It was not lower-level employees, it was high-level officials who refused to accept these very important innovations. I heard all kinds of "reasons" why probability sampling and statistical quality control techniques should not be used. These "reasons" were simply excuses because they did not want to accept improvements proposed by a lower-level manager or professional. A decade or two later all of these "reasons" had gone with the wind.

In addition to overcoming higher-level opposition, there is the persistent problem of convincing officials and professionals not to make technical decisions they have no business making but to leave them to the competent specialist, such as the statistician or quality specialist. Many times they do not realize what they are doing. The following are some real cases:

1. 72 sample documents, selected by the computer, could not be found in the files. The missing documents, about three percent of the sample, were going to be ignored, but fortunately were reported to the chief statistician as he had requested. He immediately asked for a recheck of the files by a professional. Results: 70 were found and two were duplicated from computer records. Cause: Unacceptable quality work done in filing. If the chief statistician had not requested a report on any sampling problem encountered, the sample would have been accepted and processed with the 72 sample documents ignored. This would have introduced an unknown bias into all sample results and would have greatly reduced or invalidated the effectiveness of the sample. One does not leave a source of bias in a sample when the bias can be eliminated. Every sample study is based on the assumption that all major sources of bias are eliminated or reduced to the point that the effects are negligible.

2. 427 sample records were not returned in a nationwide stratified sample study. These represented 1.5 percent of the sample, but included some large units sampled 100 percent. None of the 427 was ever obtained, although the persons in charge were clearly instructed to follow up and obtain as many as possible — especially those in the 100 percent stratum. Despite instructions from the sampling statistician, they were ignored by an accountant and a computer manager who decided since over 98 percent of the total sample had been obtained, the 427 could be ignored. This was a clear case of professionals making technical decisions they had no business making; they substituted their judgment for that of the sampling statistician who was really responsible for the sample design and its effective implementation.

3. Since a certain nationwide sample study was for the purpose of measuring the extent of compliance with the laws and regulations, the lawyers in charge decided that this was a legal study, not a statistical study, and so ignored the sampling experts in the agency. Result: A messy expensive study that wasted $250,000 and obtained nothing of any value. This was another case of high-level officials making the wrong decision, making a technical decision for which they had no competence.

4. A computer programmer had to program, for a nationwide sample study, ratio estimates of subclass aggregates and their standard errors (sampling variability). He thought he could find the appropriate equations in a statistics textbook and ignored the detailed computation instructions of the consulting statistician. Result: About 26,000 standard errors are wrong!

In this same study, there was evidence that in the final tabulation the classes that were subsampled were not properly handled and that the corrected computed edits which amounted to about 5,500 sample units were not included.

Jobs at the professional level where key decisions are made include:

1. Hiring
2. Promoting
3. Writing procedural manuals
4. Preparing training courses
5. Auditing
6. Accounting
7. Cost accounting
8. Work sampling
9. Isolating trouble spots
10. Sample design
11. Computer programming
12. Laboratory testing
13. Design of an experiment
14. Data analysis
15. Statistical analysis
16. Computer operations
17. Interpreting a quality control chart
18. Resolving sampling problems

Appropriate and timely actions. Actions may or may not grow out of decisions, but they usually do. In the preceding, the decision on the 72 cases was corrected by the statistician so proper action was taken. The decision relative to the missing 427 sample units stood and this action introduced unnecessary bias into the data and estimates. The compliance sample ended in no action at all because the data were worthless. Neither was any action taken to prevent this from recurring. In the case of the wrong computer program, the false data were published and nothing was ever done about it. Here again, no action was taken to insure that these errors were not repeated.

Management believes, or should believe, that the product produced or the service rendered is of such quality that it is both profitable to the company and affordable to the customer. Actions should be taken to foster, promote, and achieve this goal.

Responsibility for quality. The indicators of acceptable quality listed earlier show a departure from a level, a standard, a range, a procedure, a practice, or other actions or characteristics which may mean trouble. The sources of trouble may be *general, specific,* or a combination of the two. Any person or a group may be responsible for poor quality: from the top executives to the lowest paid workers.

General causes are those that are common, widespread, and outside the control of the individual worker and sometimes lower level supervisors and managers. They affect systems and operations. These causes are due to, and have to be corrected by, high level executives, managers, and professionals. Data or information indicating trouble must be fed back to them to take corrective action. Examples are:

1. Faulty decisions in planning, designing, buying, testing, marketing, and purchasing.
2. Poor marketable product, ignoring market trends and customer complaints.
3. Faulty system, faulty equipment and machinery, and faulty materials.
4. Poor design, poor planning, poor instructions, and poor maintenance.
5. Inefficient processes, procedures, scheduling, testing, auditing, and operations.
6. Conditions of work, inadequate training, and inadequate supervision.

Specific causes are individual errors, mistakes, and shortcomings; misreading or misunderstanding directions and instructions; ignoring directions and instructions; changing the processes or procedures to be followed; carelessness; indifference; fatigue; personal upset; lack of capability; natural causes such as weather; and erratic factors such as accidents and emergencies. These causes require that appropriate actions be taken with regard to individuals, groups of individuals, or whatever the cause is.

Both kinds of causes can create costly situations: management making a mistake in planning, designing, and selecting products to be produced, or an operative making a mistake that shuts down a system or a plant.

The emphasis needs to be on an AQJ (acceptable quality job) at all levels from the top executive to the lowest paid employee. If the AQJ goal is met, then the AQL (acceptable quality level) and AOQ (average outgoing quality) follow almost immediately.

Quality Costs in Service Industries

The four cost segments proposed by the Cost Effectiveness Committee of the American Society for Quality Control can be used, with adaptations, for costing the quality function in a company or agency in a service industry.[12] The four segments or components are: prevention, appraisal, internal failures, and external failures. Prevention costs include those associated with the personnel engaged in designing, implementing, and maintaining the quality system. Included will be those assigned to a quality unit, section, or division including quality engineers, quality technicians, quality specialists, and consultants hired to assist quality professionals. Included is an overall auditing of the system.

Appraisal includes costs associated with measuring, evaluating, or auditing related services or products, as well as purchased services, to assure conformance with quality standards and performance requirements. Presumably this includes specific quality audits, customer surveys to determine the attitude toward services rendered, cost of using quality control charts and acceptance-rejection sampling, and cost of computer and statistical analysis whose purpose is to measure and evaluate quality.

Internal failures refer to costs associated with correcting mistakes and errors or otherwise re-doing work that fails to meet quality requirements, and other performance that leads to a loss by the company or agency.

External failures refer to costs that are generated by defective or poor quality services to customers. This includes the cost of customer complaints, the cost of warranties or guaranties, and the cost of law suits.

In planning and implementing a program of quality costs applied to service operations, it is necessary first to define and isolate what constitutes quality tasks, jobs, and functions; then distinguish them from those not connected with the quality function. It will be necessary to sharply separate those who work on quality jobs full-time from those who work part of the time on quality jobs and part of the time on non-quality jobs. In connection with the latter group, it will be necessary to apply random time sampling (work sampling) in order to separate the quality cost from the non-quality cost. Random time sampling is the best available method to apply to decompose joint costs. It has been applied successfully to a wide variety of problems for this purpose.[13]

The method of random time sampling should be given careful consideration in quality costing despite the fact that the ASQC booklet on *Quality Costs — What and How* does not mention it. In costing service operations, some additional problems arise that are not as prominent in factory work. What should be done about the cost to a customer for errors the company makes? Should this be included in the total quality cost? It certainly appears as though it should if one wants to obtain the total quality cost created by the company.

Methods that can be used in connection with the determination of costs of quality in service operations include the following which may be used separately or in combination:

1. The ASQC booklet on quality costs. Some parts are not applicable to services, but will have to be adapted or new methods will have to be developed. Some parts are applicable as is, such as salaries of quality specialists and quality training.

2. Standard costs as used by a cost accountant. This expertise and this type of record may not be available. Furthermore, many may think it unnecessary and prefer instead to use actual costs and their trends. Learning curve analysis can be used to determine the capability of a group of workers doing a specific job.

3. Regular salary and wage records plus random time sampling using the minute model. The latter is used to separate costs of non-quality activities from the costs of quality activities. This calls for a detailed description of the various activities so that the former can be distinguished from the latter.

4. Learning curve analysis plus acceptance sampling of lots plus daily salary records. Sampling acceptance is used with immediate feedback of all rejected lots to the individual responsible for the work requiring a 100 percent re-do of the lot. Daily production of acceptable quality work is recorded to show learning curve effects. The unit cost of an acceptable quality document is calculated daily for the entire group. This combines control over quantity as well as quality and shows productivity as well as cost improvement over time.

The advantages of using learning curve analysis are:

- Shows the extent of improvement over time due to learning.
- The capability of a group is measured relative to both quantity and quality.
- Enables management to set realistic production and quality goals that fit the capability of the group.
- Shows the cost of acceptable quality performance and the reduction in unit cost due to learning effects.

Pareto Analysis in Service Operations

Pareto analysis is based on the fact that the frequency distributions of errors, failures, defects, or deficiencies, by types or sources, are skewed often highly skewed, making it efficient to concentrate on those that occur most frequently and therefore tend to be most costly. Juran refers to the "vital few and the trivial many," suggesting that most of the trouble can be eliminated by concentrating on the vital few. While this situation does occur in service operations, there are significant variations and even a reverse Pareto situation. Hence, these situations need to be analyzed with great care. What applies in the factory does not necessarily apply in a service operation.

Examples of a reverse Pareto, where low frequencies of occurrence are significant, are the following:

- The Internal Revenue Service makes a computer check of arithmetic errors on the face of the return for all individual income tax returns. It is worth doing because of the tens of millions of dollars of extra taxes disclosed. The general situation is $N\bar{x}=T$, where N is the total number, $\bar{x}$ is the mean or average, and T is the total amount. Pareto assumes $\bar{x}$ is large and N is small. The reverse, however, may be worth doing, where $\bar{x}$ is small and N is large.
- An example of where the skewed low frequency portion of the distribution defines the problem is the distribution of delay time in delivering freight cars to shippers. The skewed part measures the magnitude of the delay; this is the problem that has to be solved. The high frequency portion lies in the area of satisfactory service.[14]
- The skewed upper portion of any delay or service time distribution is the suspect area, suggesting possible use of excessive performance time.
- An error made by a company or poor service performed by a company, affects the pocketbook of the consumer. Hence, it cannot be ignored because it affects only one person or at most only a few. Where a person's property, money, financial welfare, or safety is involved, no errors can be accepted. The customer does not want to be penalized in any way whatever, nor should he or she, because of some error or poor performance made by a company or agency.

There are service operations and industries where Pareto analysis and the 85 percent division do not apply. These are industries like transportation, health services, and power plants where *one employee error can be dangerour, if not fatal.* Zero Error as a goal in these jobs is not a dream or a delusion; it is a necessity. In transportation, there are train wrecks, airplane crashes, and bus accidents. In the medical service, a cylinder of carbon dioxide is substituted for a cylinder of oxygen or a bottle of cocaine is mistaken for a bottle of phenobarbital. In a power plant, a switch that should be closed is opened or cooling systems that should be on are turned off. In all of these cases, operating personnel made a dangerous, if not fatal, mistake. In some, a case can be made that management was at least partly to blame, but in no case with a frequency as high as 80 or 85 percent.

A further complication of Pareto analysis is that it does not show any interaction among the various sources of error. A high frequency error may be correlated with a low frequency error so that the latter has to be eliminated in order to eliminate the former. Pareto analysis assumes that the various errors are independent which may or may not be true.

Quality control (QC) and quality assurance (QA). The field of quality control started with "statistical quality control" because Shewhart used probability and statistics to solve a manufacturing problem faced by Western Electric. Engineering and management failed because they were faced with a quality problem involving variability and needed the science of statistics which deals with the nature and characteristics of variability.

A new term "quality assurance" (QA) is now being widely used, although its relationship to quality control (QC) is never explained. In fact the term was used by Shewhart:[15]

> "To obtain economic control and maximum quality assurance, statistical theory and techniques must enter every one of the three steps (specification, production, and inspection) in the control of quality."

The word "maximum" clearly implies that quality assurance is measured on a continuum, that it exists in degrees, and that quality control, if properly applied, tends to maximize the assurance of acceptable quality. This is not the way this term and related terms are defined in Standard A3. The definitions of "quality system", "quality assurance", and "quality control" in this standard are:[16]

> "Quality System: The collective plans, activities, and events that are provided to ensure that a product, process, or service will satisfy given needs."
>
> "Quality Assurance: All those planned or systematic actions necessary to provide adequate confidence that a product or service will satisfy given needs."
>
> "Quality Control: The operational techniques and the activities which sustain a quality of product or service that will satisfy given needs; also, the use of such techniques and activities."

Some comments on these definitions are in order. Quality control includes specification, design, production, inspection, and review of usage. The heart of Shewhartian quality control — feedback of information indicating possible trouble and taking corrective action — is omitted from quality control, but included under quality assurance. Quality control is written in language that does not reflect the innovations that Shewhart introduced.

Under quality control the word "sustain" seems inappropriate. Since it means to support, confirm, or nourish, this implies that quality control supports the quality that is already built into the product or service. Of course, such an interpretation is misleading.

The difference between quality assurance and quality control is not made clear. For example:

- Quality assurance makes sure quality is what it should be. So does quality control.
- Quality assurance is based on plans and systematic actions. So is quality control.
- Quality assurance provides adequate confidence. Quality control is supposed to do the same.

The standard seems to say that quality assurance is an audit and evaluation function to see that all aspects of quality, including quality control, are what they should be.

Changes and developments in quality control. Attempts to apply the principles and charts of Shewhart to problems in manufacturing industries soon led to the development of and emphasis on new concepts, working rules, and even areas of application.

It was soon discovered that quality control need not be limited to the mass production of "identical" objects, such as telephone piece parts, but could be applied to quality problems in a wide variety of industries. Examples were the chemical industry, the drug industry, the food canning industry, and even the job shop.

Quality had to be built in, it was not inspected into the product. Sampling and charts in themselves did not improve quality; it was the elimination of sources of trouble that improved quality. This led to the axiom that quality should be built into the product (or service) in the first place.

Management can be a major source of poor quality. This significant role of management was implied, but never made explicit. It was assumed that engineers and foremen would learn statistics, how to apply it to quality control, and they would discover quality problems and have the ability and authority to implement the steps needed to eliminate the trouble. Nothing was ever stated as to how this could be done without the understanding and approval of higher level management.

The problem is not only one of stabilizing variability in counts and measurements, especially where defects and errors are involved, but one of reducing their variation. Statistical control can be obtained at much too high an error rate. We can move closer to the goal of Zero Defects, Zero Errors, and Zero Delay time than is commonly thought.

It was found that it was not only possible to control the variations in counts and measurements and to control the number of defects, but it was possible to control the number of errors and error rates outside the factory in the service industries. Early examples were the error rates made in key punching of IBM punch cards (Deming) and the error rates made by a group filling orders in a mail order house (Ballowe).

A broader approach to quality control than existed in the manufacturing industries was needed in the service industries. This broader approach called for the adaptation of Shewhartian charts and techniques to the service industries, but also for the application of a wide variety of techniques not used in the factory. This is because service industries are much more people-oriented, because individual buying usually predominates, because various mass markets may be involved, and because human factors and human preferences are of vital importance.

Shewhart deliberately avoided the subjective aspects and the subjective approach to quality.[17] In the service industries, where the mass buyer predominates, it is very necessary to measure buyer attitudes and perceptions of quality and to determine what satisfies and what does not satisfy various groups of mass buyers. The buyer has a subjective attitude toward errors, error rates, failures, delay time, service time, prices that have to be paid, and the seller's mistakes; the smart seller never ignores them. Some of these subjective attitudes lead to objective values:

1. A journalist and radio station owner stated publicly that he will spend no more than 20 minutes waiting in a doctor's office. After 20 minutes, he will walk out. This is a buyer setting a specification, setting a limit on waiting time.

2. A very large bank makes an error in a deposit: it records $18.45, when it should have recorded $1,845. As a result, the business man is penalized for overdrawing his bank account through no fault of his. He takes the bank into court and sues them for wasting his time and wins. This buyer was not going to assume a cost created by the seller.

3. Millions of persons buy smaller foreign cars because they are more efficient, less costly to operate and maintain, and are easier on the pocketbook in times of high prices and high costs of living. Their subjective attitudes about quality are converted into objective behavior relative to what car they will buy and continue to buy.

Summary of the requirements for success. Successful attainment of quality throughout the company or agency places severe requirements on personnel at all levels from the top executive to the lowest paid employee.

1. Management is responsible for quality policies; for quality planning; and for providing leadership, staffing, facilities, and other resources needed to put the quality program into effect at all levels. Responsibility for quality products and services that customers will buy lies more with top level management than with employees at lower levels. Dr. Deming states that 85% of the quality problems lie with management.
2. All levels of management have to agree on quality goals and cooperate to reach them.
3. Policies, plans, programs, goals, and procedures should be in writing.
4. The major goal is to prevent human errors from occurring, to do the job right the first time, and to aim at Zero Defects and Zero Errors and minimum delay times.
5. Urgency in detecting and eliminating errors is imperative.
6. Human factors must be given top priority; these include both the customer and the employees.
7. Capable and careful supervision should be stressed, including the ability and skill to manage large as well as small projects.
8. Professionals and technicians should not only have a mastery of the subject matter, but an ability and skill to successfully apply this knowledge to quality problems.
9. A quality information system is needed to insure that policies are carried out, goals are attained, urgency is practiced, and prevention of trouble succeeds.

Some working rules experience has shown vital are: Give careful attention to important details, make a habit of verifying and checking, assume nothing, take nothing for granted, make sure, be alert to what is going on, develop a sense of urgency, try to stop sources of trouble from arising, and try to prevent poor performance.

It should be noted that quality is knowledge intensive: it requires no capital investment, no large equipment outlays, no large additional space, and no new buildings. This is true of many service industries. It may not be true where products are involved whose quality can be improved only by re-design, new equipment, and large capital investments.

Finally, it cannot be emphasized too strongly that quality in services differs very markedly from quality in products. Several aspects commonly accepted in manufacturing, but requiring considerable modification if not a drastic change, are summarized:

1. Conformance to specifications is not enough. The buyer's specifications, preferences, and desires may be quite different from the seller's specifications.

2. Fitness for use is not enough. The expression is ambiguous. The criterion to meet is: Acceptable quality at an affordable price.

3. Correcting errors is not enough. It is necessary to eliminate sources of error as quickly as possible and stop them from occurring in the first place.

4. Profit on the bottom line, in itself, is not enough. There must be buyer satisfaction. Market trends must be faced and buyer preferences recognized.

5. Cost-benefit analysis may be very misleading. Cost-benefit must be reconciled with the buyer's preferences and with the market's trend or else short-term gains may turn into long-term losses.

6. The common practice of the producer setting specifications and rules the firm likes, and then expecting salesmanship including advertising to sell the product or service to a reluctant public, needs to be critically evaluated.

7. Management changes, such as "streamlining" management, participative management, flexi-scheduling, and the like, may not prove effective or may even prove counter-productive unless careful consideration is given to both internal and external factors. Internal factors include employee competence, team work, effective use of technical advances and improved techniques, and productivity with quality. External factors include buyer-customer satisfaction, meeting competition, and analyzing and facing market trends.

Notes

[1]See *Science,* vol. 204, April 20, 1979, pp. 280-281 for summary of the Nuclear Regulatory Commission's Report. For Denver figures, see statement by Al Hazle of the Colorado Department of Health in the *Rocky Mountain News,* April 4, 1979, p. 46.

[2]The question of generalists versus specialists, as well as other pertinent questions, are discussed by Clarence Randall, *The Folklore of Management,* Dunn and Bradstreet, New York, and the Atlantic Monthly Press, 1961.

[3]Many of these are illustrated in J. Frank Wright, *On a Clear Day You Can See General Motors,* Avon Books, New York, 1979. Experiences in both government and private industry show that these practices are widespread and worthy of careful study.

[4]Dr. W. Edwards Deming estimates that 85 percent of the quality problems in manufacturing have to be solved by management, not by the individual worker. In some service industries, such as transportation, health services, and electric power plants, where even one error can be dangerous, if not fatal, the individual operator is responsible for much more than 15 percent of the serious operating errors. Others use an 80-20 division.

[5]John W. Tukey, *Exploratory Data Analysis,* Addison-Wesley, Reading, Mass., 1977.

[6]*Ibid.,* p. 521.

[7]*The Washington, D.C. Sunday Star,* September 28, 1968.

[8]*Science,* vol. 175, (1972), p. 962.

[9]*Science,* vol. 176, (1972), p. 1377.

[10]*Quality Progress,* VIII, (1975), p. 14.

[11]David R. Lide, Jr., "Critical Data for Critical Needs," *Science,* vol. 212, (1981), pp. 1343-1349. Lide is Chief, Office of Standard Reference Data, National Bureau of Standards, Washington, D.C.

[12]*Quality Costs — What and How,* Second edition, 1971, American Society for Quality Control, Milwaukee, Wisconsin.

[13]A. C. Rosander, *Case Studies in Sample Design,* Marcel Dekker, Inc., New York, 1977, Chapters 14, 15, and 16, reprinted by courtesy of Marcel Dekker, Inc.

[14]*Ibid.,* Chapter 17 contains some actual delay time distributions. Also see Chapter 7 of this book.

[15]W. A. Shewhart and W. E. Deming (editor), *Statistical Method from the Viewpoint of Quality Control,* U.S. Department of Agriculture Graduate School, Washington, 1939, p. 46.

[16]ANSI/ASQC Standard A3-1978; items 2.4.1, 2.4.3, and 2.4.4.

[17]W. A. Shewhart, *Economic Control of Quality of Manufactured Product,* Van Nostrand, New York, 1931, p. 54. Republished in 1980 by American Society for Quality Control, Milwaukee, Wisconsin.

19

Exploratory Control by Time Plots

The standard method of control by time plots from samples is the Shewhartian charts which are based on sampling, probability, and statistics. These charts are described in detail in the next chapter.

The heart of control is to obtain stability in a characteristic by keeping it within tolerated limits or bounds about a specified level. Shewhartian charts obtain and maintain stability by a process of detection. Detection leads to the elimination of assignable causes — factors that shift the levels and bounds beyond what are specified. All charts, however, start with a time plot of a key characteristic estimated from a sample and plotted in time sequence from successive samples.

Stability of a characteristic, however, may exist without the use of any formal control chart due to a wide variety of factors which operate to fix a level or restrict variation. Sometimes a time plot shows a degree of stability which is hard to explain; an example is given later in this chapter. The value of a time plot of a key characteristic is that it may throw light on stability without using a formal chart, and it may be a preliminary step to designing a formal Shewhartian control chart.

In this chapter, we give five examples of actual time plots in service operations. In these examples, considerable stability existed over time without the use of any formal control charts and without the elimination of any assignable causes of out-of-control or instability.

The following are common measures used in exploratory time plots. These are measured vertically (y axis); time or time order is measured horizontally (x axis).

1. Individual measurements, written "x" or "y".
2. The average or arithmetic mean, written "$\bar{x}$" and read "x bar".
3. A range of sample values, written "R", which is the largest value minus the smallest value in the sample.
4. A count or frequency, written "c", derived from a comparable base such as an hour or a day, or from the same area.
5. A proportion or percentage, written "p".
6. A ratio of two quantities, written "f" or "w".

Other measures include actual time of day or night with points plotted in time sequence or the deviation from some standard such as late or early arrival of an airplane measured in minutes.

Examples of characteristics which can be time plotted. All kinds of characteristics of service operations can be plotted on an exploratory basis as a preliminary step to designing a formal control chart or constructing learning curves for individuals on a new job or for a group such as a unit, section, or division:

1. Error rate of an individual or group: daily.
2. Waiting time and service time in getting car inspected: individual instances.
3. Percent of drivers exceeding 55 mph: daily at point A on an interstate road for one hour in the morning and one hour in the afternoon.
4. Pollution index at point A: daily 5 pm.
5. Carbon monoxide measure at point A: daily at 5 pm.
6. Miles per gallon of gasoline for an automobile: miles for every 10 gallons.
7. Time required to get to work: daily for work days.
8. Number of long distance telephone calls made: every four weeks.
9. Kilowatt hours of electricity used monthly: for two years.
10. Time for a bus ride between points A and B during rush hour, during other times.
11. Percent of vacant places in an airplane between cities A and B: daily.
12. Minutes in 480, 720, or 1440 minutes usable equipment is not being used, unusable equipment is down.
13. Percent of total documents processed or read out by the computer: daily by type of document.
14. Percent of computer readouts which were corrected and action was taken: daily by type of document.
15. Percent of paid time employees are at work: weekly basis or monthly basis.
16. Number sunny or partly sunny days: by weeks with count 0 through 7.

Types of stability. Most sample information is used for the purpose of prediction — that is, the data are applied to the population in the future, not only to the present population from which the sample is drawn. Otherwise, the sample is being used only for a post mortem. Prediction from the sample will be effective only to the extent that the population and the population characteristic estimated from the sample are stable and can be projected into the future. A similar statement applies to any data, whether derived from a sample or not.

We describe and illustrate at least three types of stability: the stability of a constant level, the stability of a relationship, and the stability of time-related activities.

The first type of stability is basic to the Shewhartian quality control chart. An average or aimed-at value is established and upper and lower control limits are set about this level. Then, whenever an estimate derived from a sample falls *outside* these limits, it indicates an assignable cause. Appropriate action is then taken to eliminate the cause (if it exists) and keep the characteristic at the established level, that is, maintain stability. There is variation about the constant level, but this variation is random or equivalent to random variation, so that the level is maintained.

Stability of counts or measurements may exist at a fairly well-defined level even though no attempt is made to control the operation and no attempt is made to eliminate assignable causes using a formal chart. The time plot shows variations about a level. Extreme points are rare and not serious or can be identified with causes about which nothing can be done. Examples are given later in this chapter.

Stability Represented by Random Variations About a Constant Level y = a

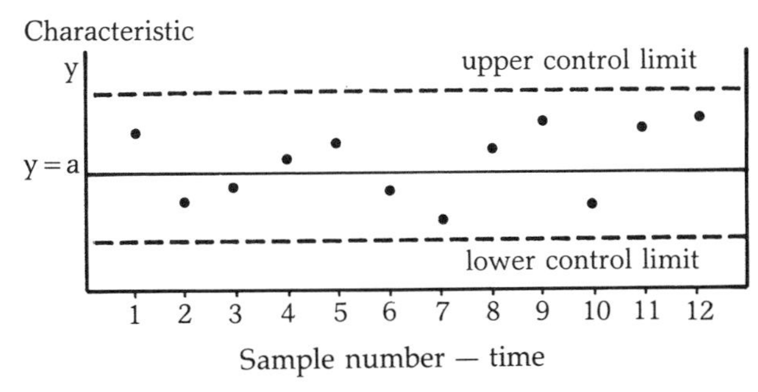

A second type of stability is the relationship between two characteristics x and y. If this relationship can be expressed by means of a straight line, then the equation relating the two quantities is $y = a + bx$, where a is the intercept on the y axis and b is the slope of the line $(y_2 - y_1)/(x_2 - x_1)$. Variations exist about the line, but they are random or essentially so so the line tends to remain fixed or changes very slowly. Control limits for random variation may be set about the line — the same as in the more common types of quality control charts.

Very often this type of stability exists without any conscious control over either x or y. Some of the factors[1] which give rise to this type of stability are the following:

1. x is part of y, or vice versa.
2. x and y are equivalent scales of measurement.
3. y is a direct effect of x.
4. y is a characteristic at a later sequence (in processing), where x is a characteristic at an earlier sequence.
5. y is related to levels of x or to changes in x.

Stability Represented by Variation About a Straight Line

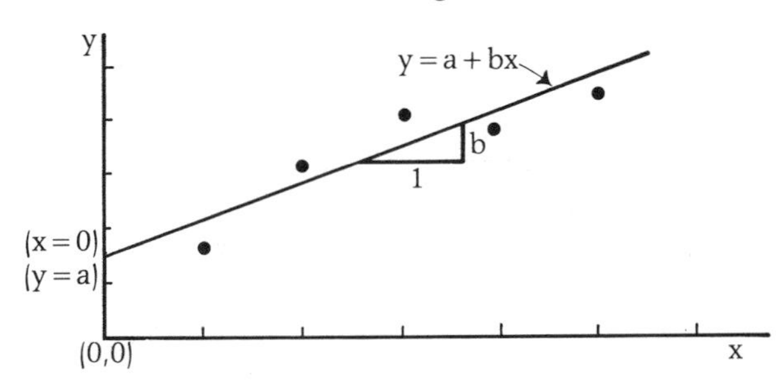

A third type of stability involves time-related activities which may be a recurring pattern or cycle. The work of a group of persons or service operation may follow a time pattern which tends to repeat itself. The length of this cycle may be short or long; it may be hours, a day, a month, or a year. It may be a job, it may be an operation, or it may be an activity. There are at least four bases for assuming some kind of activity cycle.

The service operation itself may be directly related to time such as the time schedules used in transportation: timetables of buses, airplanes, railroads, and ship lines are examples. These timetables reflect convenience, speed, and various on-board services which all enter into the quality of the service to the passenger. Charges for long-distance telephone service are directly related to time-of-day and days of the week. Hence, cost of the service, which is a quality characteristic, is time-related.

The legal basis creates time-related activities. Many laws are passed which call for certain actions at certain times of the year. Tax laws require certain actions by individuals, businesses, and government agencies. Laws require an estimate of nationwide unemployment every month, a business census every two years, and population and related censuses every ten years. Many government agencies and most private companies are required by law to make reports periodically, such as monthly, quarterly, and annually.

There is an administrative or managerial basis for a pattern or cycle. Once an organization or agency is set up with functions, procedures, and personnel, it may change very slowly. Budgets this year tend to be about the same as they were last year, except for adjustments required by growth or retrenchment. Procedures, as well as functions, are usually not drastically changed from year to year; if so, the organization moves from one time-pattern of activity to another — say from clerical, manual processing to machine or computer processing. Individual performance becomes routinized and stabilized by habit. Business practices such as mailing out periodicals, monthly statements, and monthly bills contribute to this time pattern.

Work scheduling tends to be time-oriented and work instructions tend to result in similar job performances regardless of who is doing the job. Work scheduling involves estimating the length of time required to do not only the total job, but each of the important parts. Estimating times required to perform a service such as repair, medical service, or dental service requires an appointment time, a waiting time, an actual service time, and what the service price is as set by the seller. This service price does not include any cost to the customer, patient, or buyer.

Much of our personal behavior is time-oriented: personal habits such as eating, sleeping, and working; time-scheduled activities such as store hours, paying bills, paying taxes, watching TV, and attending movies or the theater; seasonal activities such as holidays, festivals, and vacations; and activities such as conventions, society meetings, school and college attendance, professional meetings, shopping, and attending church.

Stability without direct control: computer readouts.[2] It is quite possible, therefore, to have stability in behavior, operations, or characteristics even though direct methods of control, as illustrated by the Shewhartian quality control chart, are not being applied. We cite five illustrations where actual data pertaining to a service operation were time plotted to gain a better understanding of the operation, to show where possible improvements might be made, and to give management an idea of how these operations were being controlled.

Figure 1 shows the proportion of total number of documents read-out by the computer in a computer edit and the proportion of these read-outs on which action was taken for 25 consecutive months from July 1960 through July 1963. The actual numbers and percents are given in the following table: "action taken" means documents in error on which some action was taken, "x" refers to the number of documents read-out by the computer, and "y" refers to the number of read-outs on which action was taken:

Month	Number of documents			Percent		Percent action of read-outs
	total received	read-out by computer x	action taken y	read-out	action taken	
1	14,295	677	210	4.74	1.47	31.0
2	24,772	1,426	337	5.76	1.36	23.6
3	20,995	1,053	303	5.01	1.44	28.8
4	17,529	818	371	4.67	2.12 maximum	45.2 maximum
5	16,935	822	211	4.85	1.25	25.7
6	23,724	1,110	341	4.68	1.44	30.7
7	19,162	845	275	4.41 minimum	1.44	32.5
8	17,602	901	272	5.12	1.55	30.2
9	23,864	1,347	371	5.64	1.55	27.5
10	16,751	824	225	4.92	1.34	27.3
11	18,837	959	280	5.09	1.49	29.2
12	20,229	963	289	4.76	1.43	30.0
13	18,438	925	314	5.02	1.70	33.9
14	20,276	980	295	4.83	1.45	30.1
15	18,211	926	301	5.08	1.65	32.5
16	18,860	1,039	332	5.51	1.76	32.0
17	19,460	1,050	287	5.40	1.47	27.3
18	24,075	1,259	345	5.23	1.43	27.4
19	19,164	1,002	265	5.23	1.38	26.4
20	17,329	965	233	5.57	1.34	24.1 minimum
21	21,938	1,389	439	6.33 maximum	2.00	31.6
22	25,277	1,206	327	4.77	1.29	27.1
23	20,105	1,000	311	4.97	1.55	31.1
24	20,134	1,106	289	5.49	1.44	26.1
25	18,518	890	220	4.81	1.19 minimum	24.7
Total	496,480	25,482	7,443	5.13	1.50	29.2
Average monthly	19,859	1,019	298			

1. Standard deviation of read-outs: $s_x = 187.7$.
2. Standard deviation of action taken: $s_y = 54.9$.
3. Coefficient of variation of total: $\frac{2770}{496,480} = 0.14$.
4. Coefficient of variation of read-outs: $cv_x = \frac{187.7}{1019.2} = 0.18$.
5. Coefficient of variation of action taken: $cv_y = \frac{54.9}{297.7} = 0.18$.
6. Action taken $y = 80.7 + 0.2129x$ (read-outs).
7. Correlation coefficient between x and y is 0.73, where $r^2 = 0.53$.

Figure 1 shows the 25 monthly percentages of computer read-outs for the consistency checks (Figure 1A) and the corresponding percentages of records on which errors were found and corrected (Figure 1B). The base for both sets of percentages is the number of records processed for each month. First we discuss Figure 1A.

In Figure 1A, the totals for 25 consecutive months were 25,482 computer read-outs from a grand total of 496,480 records processed, or slightly more than five percent. The monthly average processed was 19,859; the monthly average read-out was 1,019.

The monthly percents ranged from 4.41 for the seventh month to 6.63 for the twenty-first month. Three peaks are at the second, seventh, and twenty-first months. No unusual errors were found for the second and seventh months, but errors were found for the fourth and twenty-first months. The corrected values as well as the uncorrected values shown on the chart are as follows:

month	chart values	correct values
fourth	4.67	5.0
twenty-first	6.33	5.53

The unusual source of error found by the reviewer for the fourth month is not recorded. For the twenty-first month, however, the reviewer found 176 records read-out in error by the computer. When corrected, this reduced the 6.33 percent to 5.53 percent which put it more in line with how values had been varying in the past. Actually, this large error was discovered before the percentages were calculated; the same was true for the fourth month. Why the computer made these errors is not available. *The highly significant point is that an expert reviewer (a transportation specialist), not connected with the computer unit, discovered the computer errors; the computer personnel did not.*

An examination of Figure 1A shows that no time trend exists in the percentages, especially when the percent for

FIGURE 1
Computer Read-outs and Action Taken (Errors Corrected) in Consistency Checks

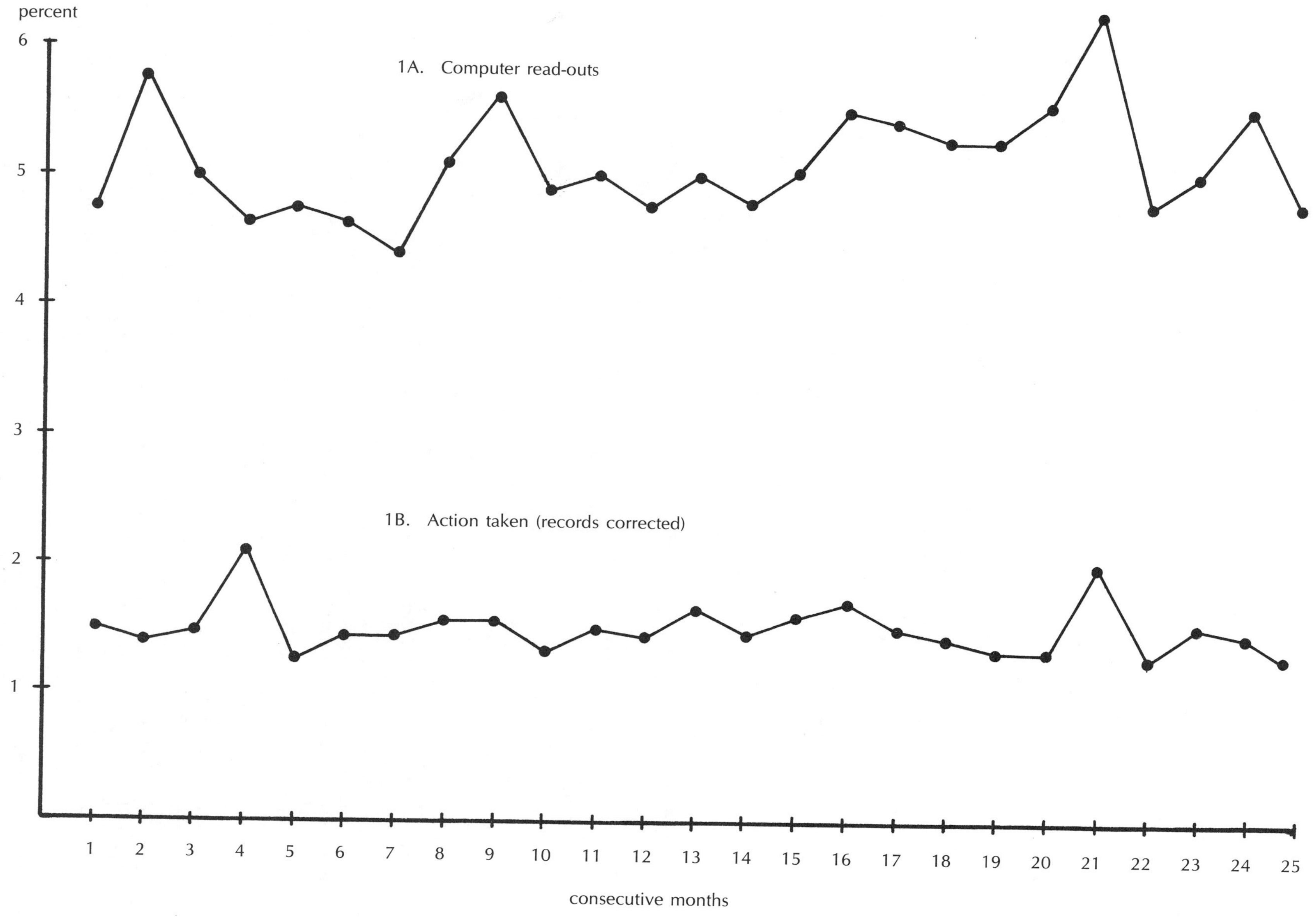

the twenty-first month is corrected. This stability is really surprising when one considers that thousands of people all over the United States selected and shipped the documents or records in question and that these documents followed no standard format. Furthermore, about 25 persons were connected with the manual and computer processing; there was not any change in the computer program for consistency checks during these 25 months.

No formal Shewhartian chart, no Tukey exploratory data analysis, and no sophisticated statistics were used — just a time plot with some simple calculations and a competent, alert reviewer. This statement also applies to Figure 1B, to which we now turn.

Stability without direct control: correcting computer read-outs. Computer read-outs were printed on the usual computer paper with one record per line. The various characteristics were checked as columns. These read-outs went to a specialist on the subject matter of the document or record who examined each characteristic of each line for possible errors. When an error was found, it was corrected on the print-out so that when it went back to the computer room, they could correct their tapes. Figure 1B shows the percent of the records processed that had one or more errors on them in the computer read-outs. The total number of read-out records that had errors on them during the 25 months was 7,443 or about 1.5 percent of the total number of records processed. The computer errors for months four and 21 had to be corrected. The uncorrected chart values, as well as the corrected values, are:

month	chart	corrected
fourth	2.12	1.7
twenty-first	2.00	1.2

When these corrections are made, the lowest percentage is 1.19 for the twenty-fifth month and 1.76 for the sixteenth month.

This line shows a surprising degree of stability, despite the fact that in about the middle of the time period under consideration the work making the consistency checks of the computer read-outs was shifted from one person to another. No changes were made in the correction rules during this time.

The variability relative to the average on a monthly basis was the same for error corrections where only two persons were involved, as it was for the computer read-outs where thousands of people were involved. This finding is certainly worthy of considerable thought although there seems to be no simple explanation for it in the present case.

Another very significant finding was the percentage of read-outs that were corrected. The average for the 25 months was 29 percent (7,443/25,482), with a range of 24 to 34 when correction is made for the computer error in the fourth month. These percentages are not plotted, but it is quite evident from a glance at the above table (last column), that an average close to 29 or 30 percent existed during the first six months and certainly for the first 12 months. In other words, nothing was ever done with slightly over 70 percent of the read-outs. Obviously, this value was excessive, but it was not discovered until these time plots were made. When this situation was discovered, a study was initiated to determine to what extent unnecessary read-outs could be eliminated and to what extent a more optimum relation between read-outs and corrected records could be established. If 10,000 read-outs could be eliminated, an average of 400 per month, this would not save too much computer time but it would reduce the number of records the reviewer examined by 40 percent — a very large savings.

Stability without direct control: effective working time. A third example shows the proportion of hours worked by about 350 employees including everyone from the messengers to the director in the Statistics Division of the Internal Revenue Service for five months during 1957. (See Figure 2.) These proportions were estimated from a continuous random time sample, in which random times were called on employees two or three times weekly depending upon the type of work, using 510 minutes daily (480 plus 30 minutes for the lunch hour). The proportion is the total hours worked divided by the total hours paid for. Since the lunch hour was included in the random time sample, an adjustment factor of 0.94 was used to eliminate it from the proportions. This factor was justified both by the prescribed time and by the sample data.

The average value of the proportion of paid-for time employees were at work during these five months was 75.7 percent. The monthly variations were between 74 and 77 percent. This certainly is a range which is quite satisfactory for management purposes. As a matter of fact, many figures or estimates used by management are not pinpointed to the extent indicated by this chart (Figure 2). Management would not be far off in estimating that 75 or 76 percent of paid-for time is worked, and that 24 or 25 percent of paid-for time is used for leave, personal, coffee breaks, and not at work. Actually the 12 month figure for the period July 1956-June 1957 was 75 percent.[3]

There is a good deal of stability in these estimates even though no direct control was placed on either groups or individuals. Of course, there is a broad type of control exerted through supervision, procedures, work routines, work schedules, leave routines, vacation practices, and other practices that tend to give stability to these overall proportions month after month. In planning work schedules and staffing for the future, a figure of 75 percent would be appropriate until some drastic change took place. This means four persons must be hired to have three working full time.

Stability without direct control: bus line schedule. A fourth example is illustrated by Figure 3. This chart shows the number of minutes after 7 am an express bus arrived at a downtown intersection which was about 10 miles from the origin stop. This was a scheduled bus line which was supposed to leave the origin at 6:58 am. The point at which the time was read was an intersection with a large electric clock which was read when the bus arrived at, or passed, this intersection. The clock was checked against a watch periodically for accuracy so that the times recorded are comparable.

Times were recorded for 38 consecutive working days during October and November 1962. Driver and weather information were also recorded. Two drivers were involved, one for the first 13 days and the other for the remaining time. The graph shows a break between the thirteenth and fourteenth days to indicate a change in drivers.

A simple analysis of the times for the two drivers is as follows. It shows that although a formal control chart could be constructed and used for control, it is really unnecessary since all extreme values recorded for both drivers can be readily explained in terms of the weather

FIGURE 2
Percent of Paid-for Time Worked, for 5 Months, 1957
(percents derived from a continuous random time sample)

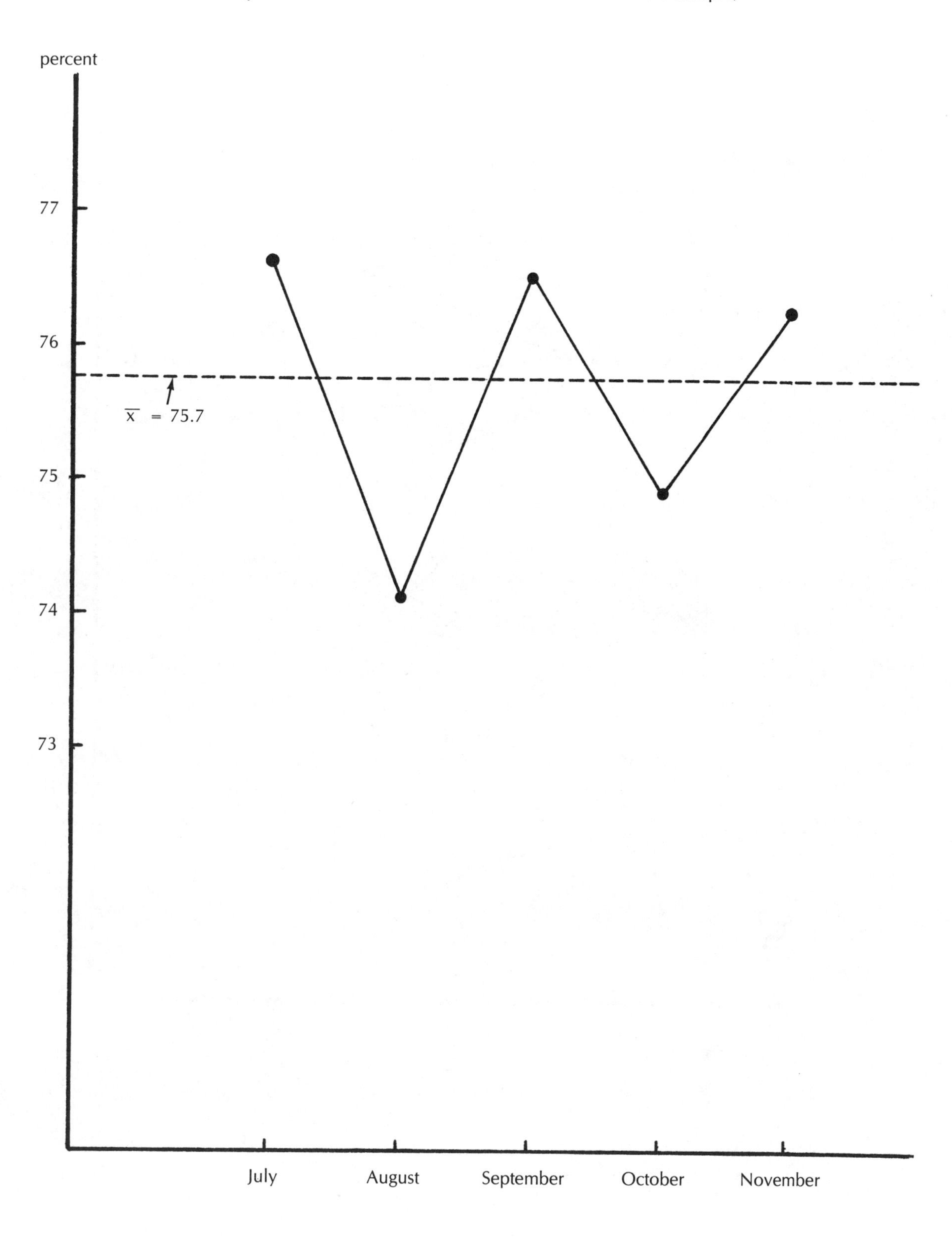

FIGURE 3
Arrival Time of a City Express Bus at Specified Point, at Same Origin, for 38 days, October-November 1962

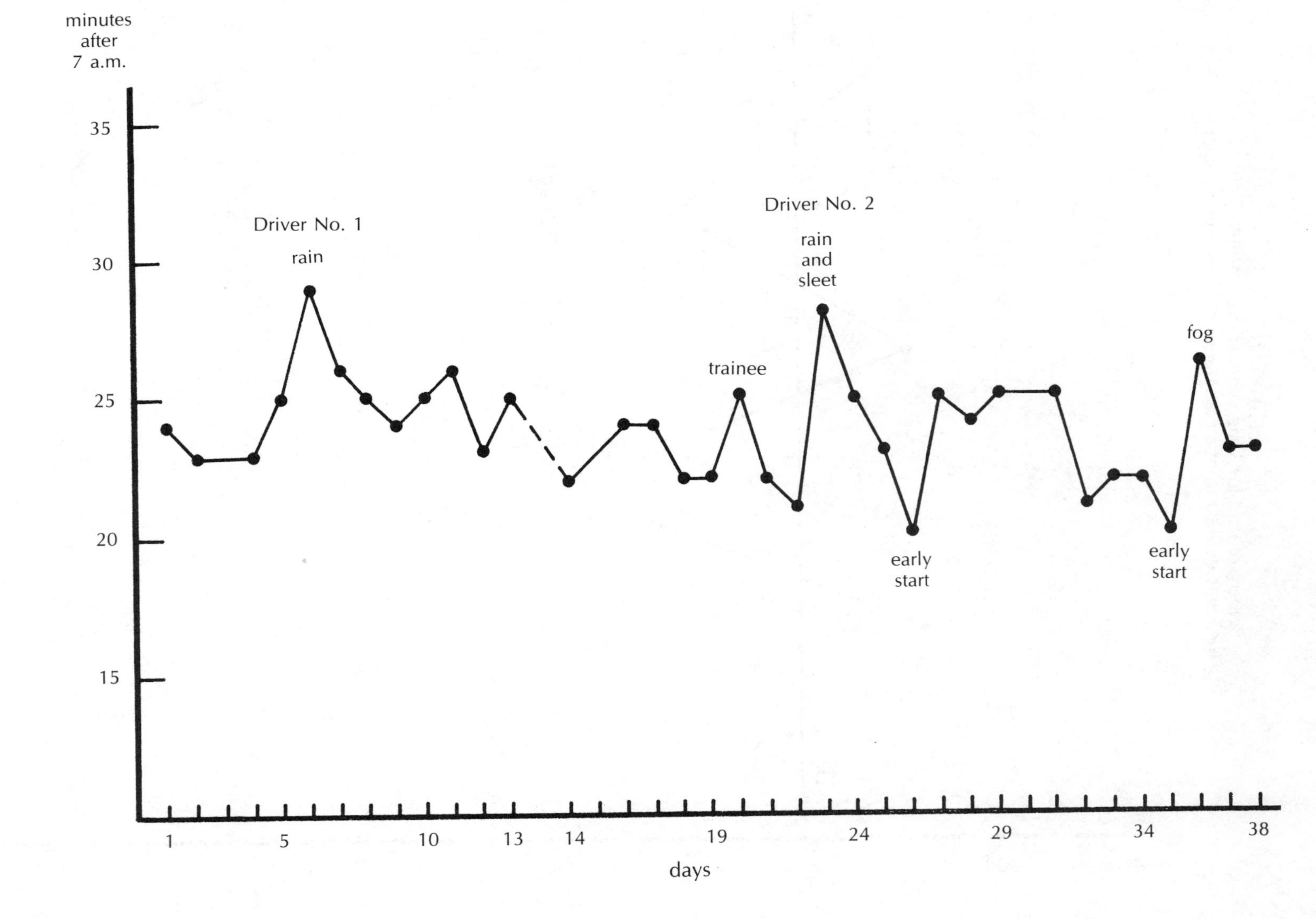

or other relevant factors. It also shows that the second driver made better time, taking on the average 1.4 minutes less, than the first. Ten trips (40% of 25) were driven in less time than the first driver's minimum of 23 minutes.

Item	Driver No. 1	Driver No. 2
Number of trips	13	25
Maximum time	29 minutes	28 minutes
Minimum time	23 minutes	20 minutes
Range (difference)	6 minutes	8 minutes
Average time	24.7 minutes	23.3 minutes

Frequency distribution			Frequency distribution			
x_i	n_i	$n_i x_i$	x_i	n_i	$n_i x_i$	
23	4	92	20	2	40	early start
24	2	48	21	2	42	
25	4	100	22	6	132	
26	2	52	23	4	92	
29	1	29 rain	24	3	72	
Sum	13	321	25	6	150	
			26	1	26	fog
			28	1	28	rain and sleet
			Sum	25	582	

$$\bar{x}_1 = \frac{321}{13} = 24.7 \text{ minutes.}$$

$$\bar{x}_2 = \frac{582}{25} = 23.3 \text{ minutes.}$$

Stability without direct control: employee utilization. A fifth example shows the number of idle workers in a maintenance crew of seven men at each of 10 instants (random minutes) selected during a working day. The basic information from which this sample was selected is given by MacNiece[4] and Hansen[5]. An observer followed this crew all day and recorded continuously how many men were working and including what each was actually doing.

The working day extended from 8:30 am to 4:30 pm. A lunch hour of 30 minutes was taken, giving a working day of 450 minutes. Using a table of random numbers, 10 random minutes were selected in the ranges 0830 through 1229, and 0100 through 0429, omitting the lunch hour from 12:30 to 1:00. The 10 sample values showing the number of men idle at each instant are given in Figure 4. The sample gives an estimate of 81 percent idle derived as follows:

Number idle	Number of minutes	Product
5	5	25
6	3	18
7	2	14
	10	57

The total number of employee-minutes in the sample is $7 \times 10 = 70$; 57 of these were idle man-minutes. Hence, the percent idle $= 57/70 = 0.81$ or 81 percent. By chance, 81 percent is what 100 percent coverage of the 450 minutes and the seven men obtained. A random time sample of 10 instants (minutes), however, gives a very close estimate of the 100 percent coverage as the results from 10 different random samples of 10 instants (minutes) show: one sample gave 79 percent, four gave 80 percent, three gave 81 percent, and two gave 84 percent. This example illustrates a principle in sampling: if the activity is stable over time, then a relatively small random time sample will give an estimate very close to the 100 percent coverage value and at considerably less cost. In this case, a random sample of 10 out of 450, or a 2.2 percent sample, was all that was needed. This was a simple unrestricted random sample; there was no stratification by hours although this could be used to increase the precision of an estimate still further if needed.

Other examples of stability. The concept of stability gets top priority, not only in quality control charts, but also in statistical aspects closely related to quality control. Three examples can be cited: stability of level, stability of relationship, and stability of size.

An excellent example of stability of level is the coefficient of variation (cv) of some measurement or count, x, which is the ratio of the variability to the average or more specifically, the ratio of the standard deviation to the mean of x. This value is related to the shape of the frequency distribution of x and to the value of the mean relative to the variation of x.[6] For example, the exponential distribution has a cv of 1.0, the right angled triangle has a cv of 0.72, and a normal distribution has a cv of 0.34.

The coefficient of variation of empirical distributions varies according to the nature of the characteristic:

1. Measurements: the weight of full whiskey barrels is between 0.010 and 0.015.
2. Economic characteristics: 0.50 to 1.50.
3. Financial characteristics: 0.50 to 1.50.
4. Tax errors on federal individual tax returns: 1.0 to 5.0.

The stability of a relationship is illustrated by the relationship between two and more variates. During World War II, for example, the War Production Board used these relationships very extensively in connection with estimating civilian requirements. Much to the surprise of economists, it was discovered that many of the pre-war relationships between economic characteristics did not change substantially, but could be used for wartime predictions one year in advance.[7] The relationship simply moved to a different, and usually higher level, without any substantial change of pattern. In a straight line relationship both x and y increased so that the past relationship was maintained.

Another example of stability is the stability of size, whether we are concerned with the size of cities, size of a business, size of a manufacturer in a certain industry, size of income, volume of sales, number of employees, or any number of other characteristics. The large tend to stay large, the middle-sized tend to stay middle-sized, and the small tend to stay small — at least over a short period of time and often over a long period of time. This stability is used in the design of probability samples using stratification of the population or frame into two or more strata based upon some measure of size.

Summary. Time plots can be exploratory, analytical, or a way to discover something significant that was not suspected. From the above examples:

• The computer runs were in error and were discovered, not by computer people, but by an outside reviewer.

• Read-out rules needed changing — 71 percent were never corrected or acted upon.

• 75 percent of the employes were time effective: the director used this figure for budget planning and work scheduling.

• The bus line covered 10 miles in 26 to 28 minutes. This time was as good or better than that of the private automobile driver. This service was very satisfactory.

• The random time sample of a maintenance crew showed a tremendous waste of time and money. It showed that a sample of 10 random minutes out of a total of 450 minutes, or a 2.2 percent sample, gave a very satisfactory estimate due to the stable pattern of work behavior. It is imperative that the time sample be random and selected from all minutes of the working day.

FIGURE 4

Number of Employees Idle in a Maintenance Crew of 7, Estimated from 10 Random Minutes from One Day

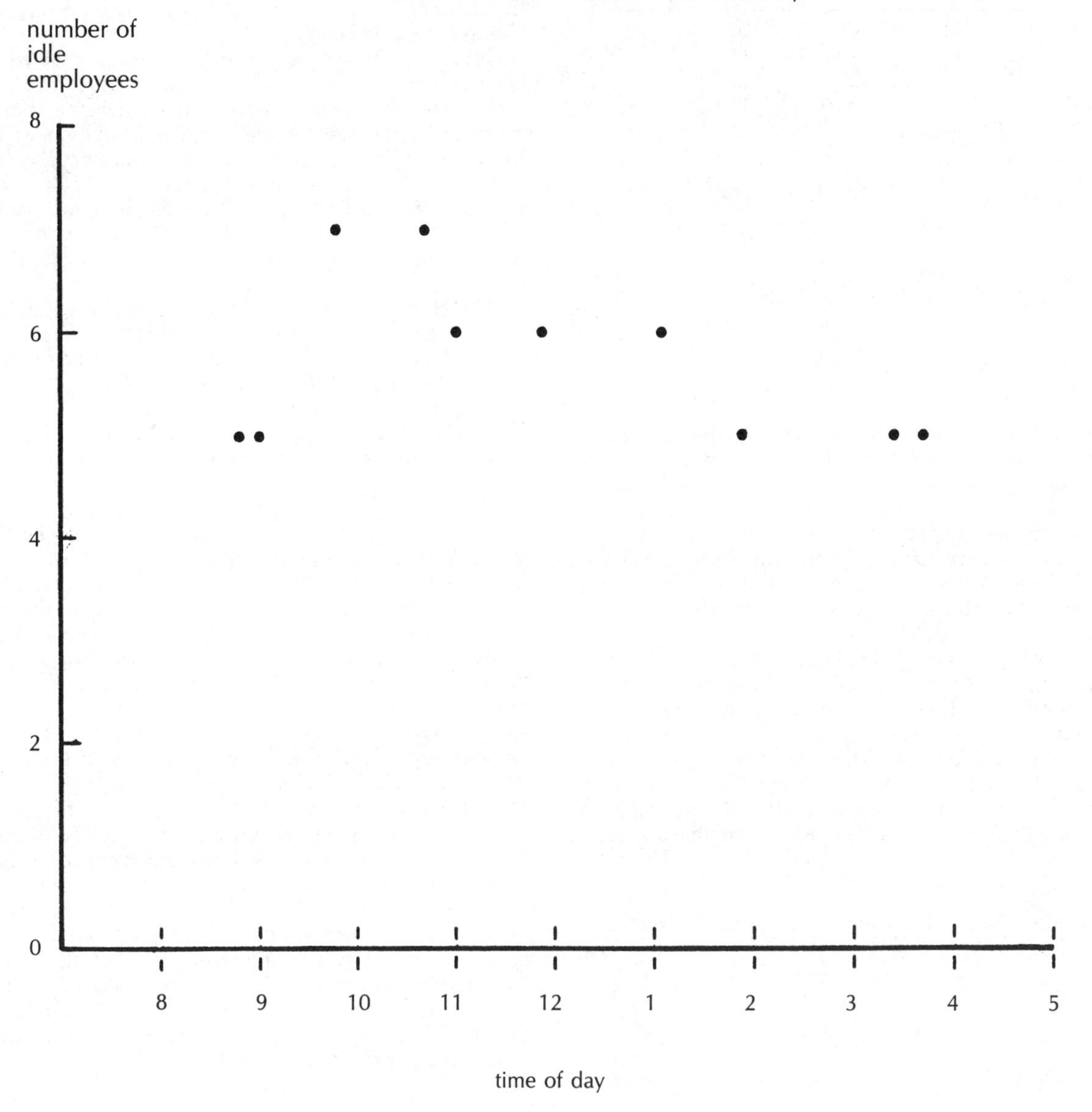

Notes

[1]A. C. Rosander, *Elementary Principles of Statistics,* Van Nostrand, New York, 1951. Copyright by Wadsworth, Inc. Used by permission of Brooks Cole Publishing Co., pp. 325-340.

[2]These examples are not to be taken as arguments against the use of formal control charts, but rather as examples of where time plots can be very helpful in showing what is going on, and as a possible preliminary step to a formal control chart.

[3]A. C. Rosander, *Case Studies in Sample Design,* Marcel Dekker Inc., New York, 1977, Chapter 14. Reprinted by courtesy of Marcel Dekker, Inc.

[4]Adapted from *Time,* December 20, 1952, in E. J. MacNiece, "Work Sampling," *Factory Management and Maintenance,* July 1953.

[5]B. L. Hansen, *Work Sampling,* Prentice Hall, Englewood Cliffs, N.J., 1960, p. 17.

[6]Deming gives the standard deviation for six distributions with known geometric shapes. The mean is $\frac{1}{2}h$ except for the right angle triangle where the mean is $\frac{1}{3}h$. W. Edwards Deming, *Some Theory of Sampling,* New York, Wiley, 1950, p. 62. This table, expanded and rearranged is Table 4 of Chapter 15.

[7]A. C. Rosander, *Elementary Principles of Statistics,* Van Nostrand, New York, 1951. Copyright by Wadsworth, Inc. Used by permission of Brooks Cole Publishing Co., pp. 414-416.

20

Control by Shewhartian Charts

Some of Shewhart's basic ideas and procedures. Shewhart described in detail in two books the basic theory of statistical quality control and its application to the manufacturing of products.[1] Some of Shewhart's basic ideas and procedures are presented as an introduction to this chapter, especially those which bear most directly on quality control practice:

1. The three steps in the quality control process are *specification* of what is wanted, *production* of what is specified, and *inspection* to determine if production meets specifications. These three operations are a circular interrelated cycle of a progressive nature.
2. Statistical control is necessary to establish a population or universe whose characteristics are predictable.
3. The control chart provides for the progressive detection and elimination of assignable causes so statistical control can be established. This control applies not only to a process, but also to research and measurement.
4. Points on the control chart are derived from small samples since they are quick and sensitive detectors of unusual variations or assignable causes.
5. It is imperative to plot sample points on the control chart in the time order in which they are selected.
6. Not less than 25 samples of four, with the elimination of assignable causes, are needed to obtain statistical control and are recommended before any control limits are set.
7. Samples are to be identified by rational sub-groups, strata, or factors which may be assignable causes or related to such causes.
8. Setting three sigma limits of variation detects assignable causes very well in practice, but sets a probability of 3/1000 for looking for non-existent trouble.

The principle of the control chart. Even though it was originally designed to solve an engineering problem in a factory, the control chart is an application of the theory of probability and mathematical statistics. It is assumed that the variation in values of a characteristic from repeated samples, such as the error rate made by a typist on the same kind of material, the percent of readouts of a computer for a consistency check, the bus time between two points, or the percent of errors corrected on computer read-outs, arises from two sources: random variation and non-random variation due to assignable causes.

If in some work operation or process the sources of non-random variation can be eliminated or identified, then the remaining variation is random. The theory of statistics can be applied to determine with a specified probability the limits of this variation about a desired level. Shewhart recommends using 25 samples to eliminate assignable causes before these limits are calculated. This applies to mass production; it does not necessarily hold for service operations.

This procedure provides a way of constructing a control chart with a specified or aimed-at level which is the center line, an upper control limit line, and a lower control limit line.

Periodic samples are taken from the operation or process, and a calculation of a characteristic is made and plotted in time order on the chart. If the points stay within the two limits, a state of statistical control exists, a stable population exists, and the chart has predictive value. Otherwise, we do not know what the process is producing. An exception is the stable time plot described in the preceding chapter.

It should be noted that an out-of-control situation can be indicated by three other situations: a trend within the two limits, a cyclical or oscillatory trend within the two limits, and a constant bias that a control chart does not detect.

Functions of the control chart. The Shewhartian control chart is deceptively simple, as can be seen by the long list of functions it performs. This simplicity has led many to greatly underestimate its importance and value.

1. It gives a history of the process.
2. It indicates trouble or the lack of it.
3. It measures quality progress and improvement.
4. It is an effective presentation device.
5. It is a control device.
6. It is an information system.
7. It is an action device.
8. It is an analytical device.
9. It is a test of a statistical hypothesis — equivalent to the use of analysis of variance with an F test for the homogeneity of the means of k groups.[2]
10. It shows when a process is shifting from one level to another, from one population or universe to another, or when the process is moving toward out-of-control.

To be valid, the sample values must be plotted *in the time order in which the samples are selected.* The samples cannot be too large or they will hide the very deviations one is trying to detect. Any out-of-control point (either above the upper limit or below the lower limit) must be investigated and fed back to the source *immediately* to insure that any assignable cause is eliminated. Otherwise, control may be lost, a large volume of defective material is produced before the cause is found and eliminated, and production costs are increased. The basic purpose of the chart is to prevent this waste from occurring. Identification of assignable causes is greatly helped by selecting samples from *rational sub-groups,* significant strata, or groups into which the data can be classified.

Eight charts. The basic principle on which control charts are based is that for each key estimate derived from a sample, a mathematical expression exists which gives the variability about the estimate due to random fluctuations. This expression is different for each type of estimate, but it can be used to set limits to the random variation with high probability. In this way, random variation can be separated from non-random variation due to assignable causes which can be eliminated.

Eight estimates are described and illustrated by an actual problem showing how the chart is constructed and used in each case. The eight are:

1. The p chart for a *proportion or percentage* applicable to the "heads or tails" situation or model (the binomial) with a fixed sample size n, where $p = x/n$ and x is a count or frequency and n is a constant.

2. The c or np chart for a *count* from comparable units or a characteristic approximating the Poisson distribution.

3. The $\bar{x}$ chart for the *mean* (arithmetic mean) of a measurement, such as tax error, delay time, service time, or carbon dioxide in air.

4. The R chart for the sample *ranges* of measurements obtained in item 3 above.

5. The σ chart for the *standard deviation* of measurements within samples, such as those obtained in item 3 above.

6. The s chart for the *standard deviation* using analysis of variance to obtain an estimate of the variance within samples.

7. The f chart for *ratios* of the form x/y, where both x and y are random variables. It may or may not be a proportion; if so, $p_i = x_i/y_i$.

8. The *regression* chart showing the straight line relationship $y = a + bx$ fitted to the sample data by the method of least squares, where a is the y intercept and b is the slope of the line.

There are now desk or hand-held calculators that make all of these estimates easy to calculate. A calculator with $\bar{x}$, the mean, and s, the standard deviation, programmed into the machine make it easy to make the necessary calculations for an $\bar{x}$ and s chart for any sample size. It is no longer necessary to limit oneself to an R chart which is efficient for small samples but not for larger samples where the standard deviation is better.

Remarks on sampling for control charts. A warning about sampling is needed at this point. Shewhart's samples of four with sample estimates plotted in time sequence as they are drawn for a measurement obtained from the work or production process apply especially to the $\bar{x}$ (mean) and R (range) charts. *There is no random sampling here.* The elimination of assignable causes using either chart is for the purpose of moving toward a condition of statistical control in which variation is entirely random, or predominantly random, about some desired level. If the process is going out of control, this will be most rapidly detected by taking the last items off the line, or the last ones produced, not by taking a random sample of what was produced during the last hour, the last minute, or the last half day.

If work outcomes or results are being inspected in lots accumulated from the work process, they should be inspected in time order and by identification of classification of service or rational sub-group. *A random sample of proper size* should be selected to detect or measure the characteristic in question. Here time order is combined with random sampling. This applies to percentages, proportions, and counts for comparable units, such as equal sample sizes, equal lengths of time, equal lengths, or equal areas. Large sample sizes are ordinarily required, such as 50, 100, or even more, depending upon the frequency of occurrence of the defect, error, or other characteristic being estimated, detected, or controlled.

Sample sizes of 25 or more are also required for charts for the standard deviation, a ratio, and a linear regression line $y = a + bx$.

Sampling by lots for receiving inspection, say of purchased materials and components, also requires random sampling for each lot since the theory of acceptance sampling is based on random selection applied to the entire lot. Usually, in this operation, nothing is known about the order in which the lots were produced or even if a lot is an aggregation of production at different times. This topic is treated in the next chapter.

The size and nature of the sampling not only depends upon the type of characteristic to be determined from each sample, but it also depends upon what the sample is going to be used for. Each of the following purposes usually requires a different sample design:

1. Sampling for detection.
2. Sampling for estimation.
3. Sampling for comparison.
4. Sampling for measuring effectiveness.
5. Sampling to test a mathematical model.

The p chart. The p chart is used to control fraction defective, proportion defective, or percent defective, all of which mean the same thing. "Defective" includes mistakes, errors, faults, imperfections, and blemishes — anything that makes for an unacceptable performance, outcome, or service. Common uses of the p chart in service operations include the following actual applications:

1. Typing errors; an example is given in Chapter 2 on Banking.

2. Error rate of a group filling mail orders; used by Alden's mail order house of Chicago.

3. Error rate in statistical coding for key punch operation.

4. Error rate in key punch operation.

5. Error rate in transcribing data from a document to a standard format for computer operation.

In a factory, defects are often connected with a machine operation or with an inspector using instruments or other aids to discover defective parts in a lot being inspected. In some service operations involving testing, such as laboratory testing of materials, such equipment may be available to aid the operator in measuring quality. But in many service operations, no such assistance from special equipment exists. In these operations, employee performance tends to stabilize at some level once the employee is proficient on the job. Then, when an unusual situation arises or some assignable cause appears, a p chart will show deviation from established control limits.

Experience shows that the continuous use of the p chart, the same as many of the other charts, often brings about a sharp reduction in the error rate, sometimes in a very short time. Alden's found this to be true and so did Blue Cross.[3]

If a p chart for every employee is too difficult to administer or is objectionable, a chart can be applied to a meaningful working group using a random sample of 100 items or units selected daily to calculate the error rate. The error rate is plotted daily on a large wall chart for all to see; this avoids singling out any individual. Alden's found the method very effective.

An alternative to the p chart for control is to use control by acceptance-rejection sampling of lots of work completed by each employee, with 100 percent verification of all rejected lots by those who performed the work originally. This method and application are described in detail in this chapter and in Chapter 9.

A sharp distinction needs to be made between two proportions. In the first case:

$$p = x/n,$$

where p = proportion x items is of total sample; n = total sample size and is *fixed;* and x = number of elements in the sample n with errors or defects, or other undesirable characteristics. This is the binomial "heads or tails" situation or model. It is assumed that the population proportion is constant or nearly so, the values of x are independent, and sample elements are selected at random from the entire population or universe. When a p chart is referred to, this is the type of proportion that is meant.

In the second case:

$$p_j = x_j / y_j$$

where both x_j and y_j can vary; x_j is a part, or a component, of y_j, but cannot exceed y_j; and $j = 1,2,\ldots,k$ samples. This proportion is a *ratio estimate* and calls for a ratio control chart described later in this chapter, *not* the p chart referred to above.

To start a p control chart select at least 20 random samples of 50, but preferably 100 or more items or units, from lots of work, calculate the proportion desired for each sample, and plot each in time sequence on a chart. If wide variations exist, assignable causes may be indicated which need to be eliminated. Once the proportions appear to be stable, calculate a center line and the two limits as follows:

1. The center line:

$$\bar{p} = \frac{1}{k}\sum_{1}^{k} x_j/n = \frac{\sum_{1}^{k} x_j}{kn},$$

where $\bar{p}$ = proportion estimated from all the samples eliminating all samples out-of-control (assignable causes found); k = number of samples of size n; $j = 1,2,\ldots,k$ (summation is over all of the k samples); and n = sample size (number of elements selected at random).

2. Measure of random variation about $\bar{p}$:

$$s = \sqrt{\frac{\bar{p}\,(1 - \bar{p})}{n}},$$

where $\bar{p}$ and n have the same meanings as given above.

3. Upper control limit:

$$UCL = \bar{p} + 3s.$$

4. Lower control limit:

$$LCL = \bar{p} - 3s.$$

If this value is negative, set a lower control limit equal to zero.

Example. A p chart is constructed for the computer read-outs for a consistency test as a proportion of the total sample received monthly. Corrected data prepared by the reviewer for the fourth and twenty-first months are used. The data are from Chapter 19 and the basic calculations are:

1. The total number of samples was $25 = k$.
2. The average sample size, since n varied from month to month is:

$$\bar{n} = 19{,}859.$$

3. The average (mean) proportion is:

$$\bar{p} = \frac{25{,}364}{496{,}480} = 0.051 \text{ or } 5.1 \text{ percent.}$$

4. $1 - \bar{p} = 1 - 0.051 = 0.949.$
5. The standard deviation of the average proportion $\bar{p}$ (standard error) is:

$$s = \sqrt{\frac{0.051(0.949)}{19{,}859}} = 0.00156.$$

6. Hence, $3s = 3(0.00156) = 0.00468.$

FIGURE 1
Control Chart for Monthly Percentages of Documents
Computer Read Out for Consistency Checks

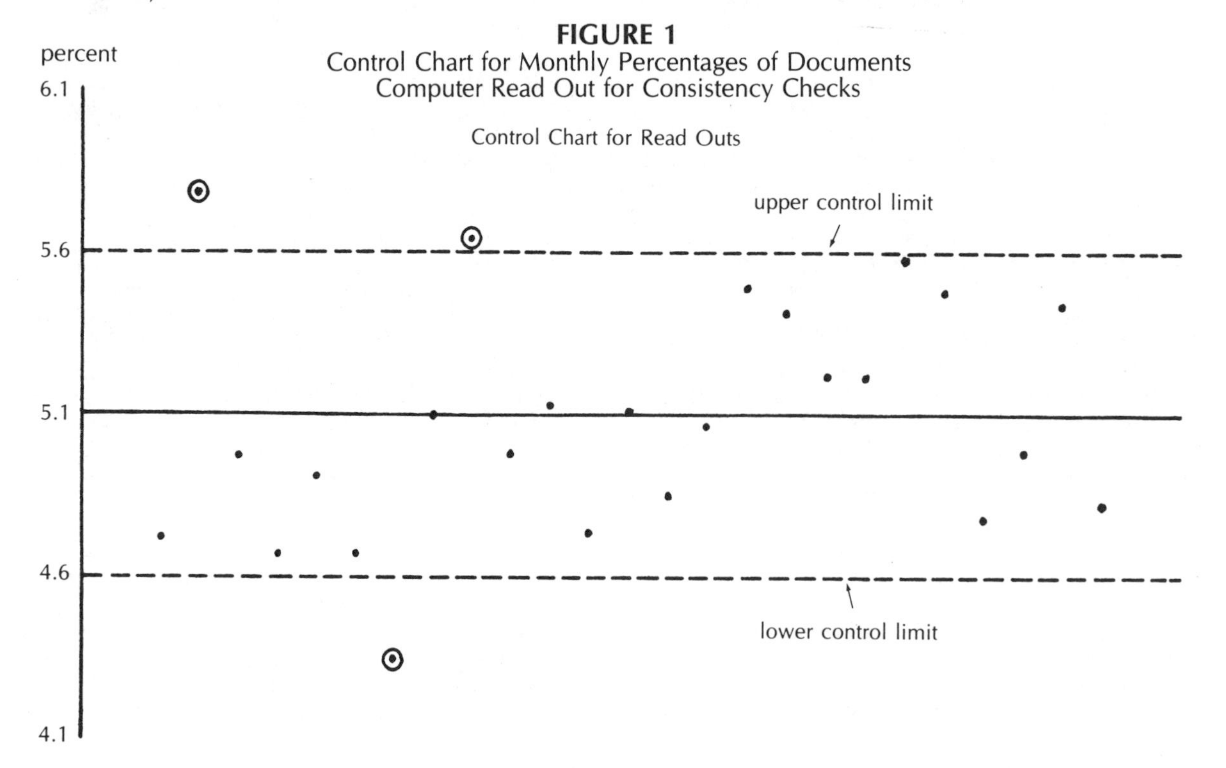

The three lines can now be calculated:

1. The center line is $\bar{p}=0.051$.
2. The upper control limit (UCL) $=0.051+0.00468=0.0557$ or 5.57 percent.
3. The lower control limit (LCL) $=0.051-0.00468=0.0463$ or 4.63 percent.

Examination of the monthly data shows that three months fall outside these limits:

second month: 5.76 high.
seventh month: 4.41 low.
ninth month: 5.64 high.

These are the high and low points in Figure 1. Even if a lower n of 16,000 had been used, exceeded by every month but one, the limits would have been wider but not enough — the three points would still have fallen outside the limits. The record does not show what happened during these months, but the probability is very high that something did go wrong, the same as during the fourth and twenty-first months.

The c or np chart. The c or np chart is used to control a count or frequency that occurs in comparable or equal units, such as equal sample sizes, equal areas, equal times, or equal lengths.

The errors are assumed to be independent, not bunched, relatively rare or of low frequency, and distributed as the Poisson probability distribution. This distribution has a mean equal to the variance, so that the standard deviation is equal to the square root of the mean.

If c is an individual count and $\bar{c}$ is the mean count, then the standard deviation (standard error) of $\bar{c}$ is:

$$s_{\bar{c}} = \sqrt{\bar{c}}$$

and

$$3s_{\bar{c}} = 3\sqrt{\bar{c}}\,.$$

The value of $\bar{c}$ may be derived from k samples, which are under control after assignable causes are removed, or it may be based upon an acceptable proportion p for the error rate or fraction defective multiplied by the fixed sample size n, so that:

$$\bar{c}=np.$$

Example: In a large nationwide sample of waybills of rail freight shipments, the documents had to be transcribed to a standard format for computer processing. To facilitate this work, lots of 300 documents were formed and processed by transcribers. A sample of 50 was drawn from each lot by reviewers who carefully examined them and counted the number of documents in error in the sample of 50. If more than five documents were found in error, the transcriber was required to re-do the entire lot of 300. The person who transcribed each lot of 300 was identified.

The number of documents in error in each of 21 samples of 50 selected with random start from lots of 300 documents, transcribed by three transcribers, are given below and plotted in Figure 2.

Sample number	Documents in error	Sample number	Documents in error	Sample number	Documents in error
1	2	8	2	15	1
2	5	9	1	16	2
3	9	10	1	17	4
4	0	11	0	18	1
5	4	12	4	19	0
6	10	13	11	20	3
7	13	14	0	21	1

Control equivalent to what is shown in Figure 2 was used. A sample count of six, the upper control limit, was considered out of control. The plan was equivalent to the following:

1. The desired p on the average was set at four percent.
2. The value of np for a sample of 50 was $50(0.04)=2$ on the average.
3. Hence, $\bar{c}=2$.
4. The sampling variation is $s_{\bar{c}} = \sqrt{2} = 1.4$.
5. $3s_{\bar{c}} = 4.2$.
6. The limits, therefore, were $\bar{c} \pm 3s_{\bar{c}} = 2 \pm 4$.

Therefore:

The upper control limit (UCL) $=6$.
The lower control limit (LCL) $=0$.

In the 21 samples, four samples (3, 6, 7, and 13) fell outside the upper control limit of six and the corresponding lots were rejected and re-done 100 percent by the employee doing the original work. This fed back unacceptable production directly to the person responsible for it. No more than one day elapsed between completion of a lot and its acceptance or rejection. Records for a period of about 40 days showed that about one in seven lots was rejected (14 percent).

The effectiveness of rejecting unacceptable quality lots is seen by the sharp drop in errors made once a lot is rejected. Sometimes this was due not only to the effect of rejection, but to the supervisor giving an employee extra attention on the kinds of errors made so as to avoid making them in the future. The actual acceptance-rejection plan used on this project, equivalent to the above, is described in Chapter 9.

The $\bar{x}$ and R charts. These two charts are described together because this is the way they are used in practice. The $\bar{x}$ chart, a mean chart, controls the average or aimed-at *level.* The R chart, a range chart, controls the *variation* about this level.[4] Indeed, the two have to be used together because a work process can go out-of-control when *either* the mean $\bar{x}$ or the range R shifts or *both* shift. Furthermore, the mean range $\bar{R}$ is used in constructing both charts.

The standard deviation is used instead of R to control variability, but for small samples R is simpler and just as effective; hence, its more common use. The standard deviation chart is explained in the next section.

Usually 10 to 20 consecutive samples of four or five each (Shewhart recommends 25) are collected, then the mean and range are calculated for each sample and plotted in time sequence on their respective charts. If wide fluctuations exist in either, or both, a hunt is made for assignable causes which are eliminated if found. Then the center line and the two control limits are calculated after eliminating all points affected by assignable causes; the lines are drawn and the chart used for control. The calculations required for an $\bar{x}$ and an R chart, where no standard is used or known, are as follows (this is the most common case found in practice):

$\bar{x}$ chart

1. The center line $x = \dfrac{\Sigma x}{kn} = \dfrac{\Sigma \bar{x}}{k}$,

   where n = sample size of four or five; k = number of samples of equal size (summation is over kn sampling units or k samples); x = value of the characteristic; and $\bar{x}$ = mean of x values in a single sample.

FIGURE 2

c Chart for Errors Made by 3 Transcribers in Samples of 50 Selected from Lots of 300 Transcribed from Documents to a Standard Format

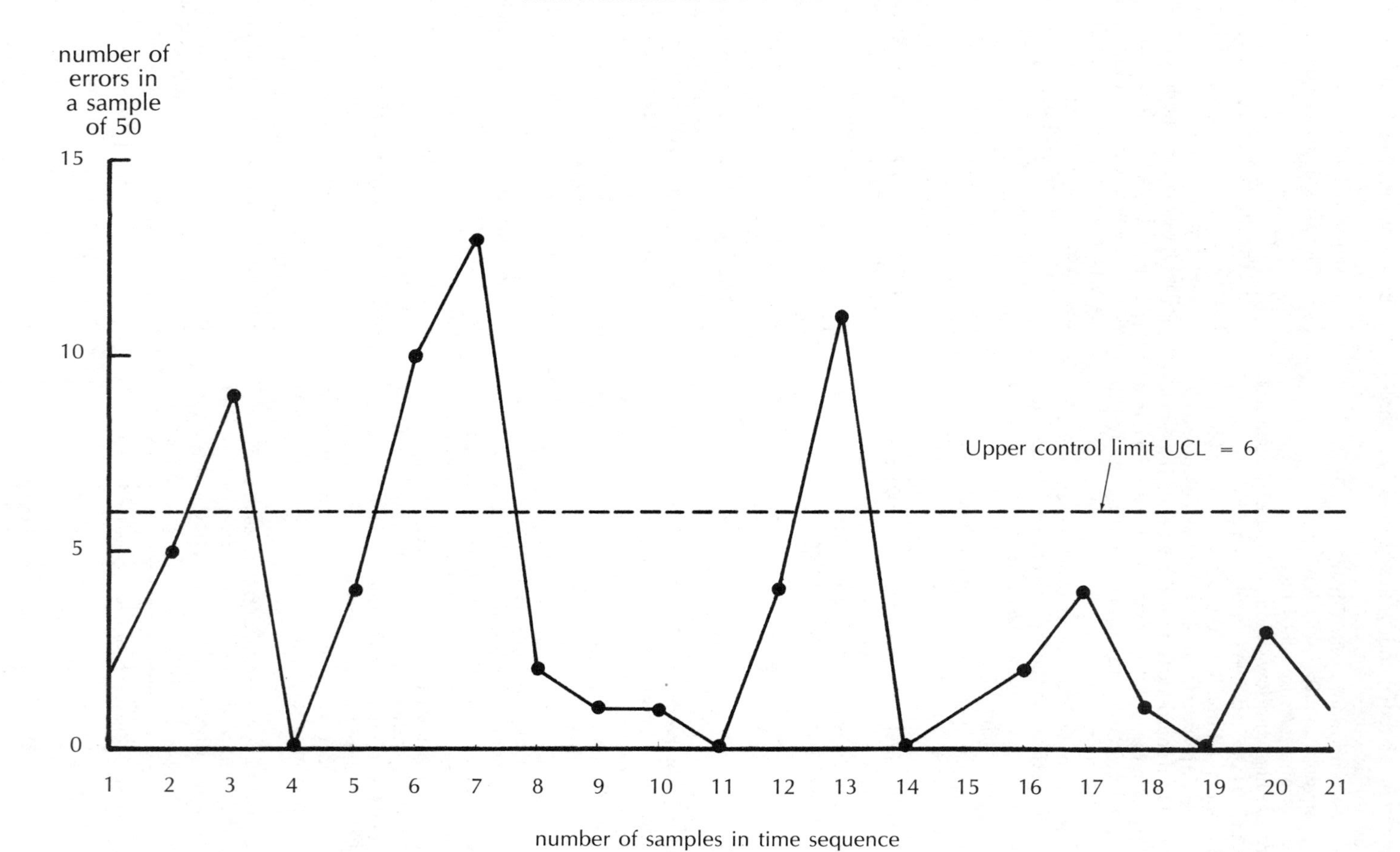

2. Mean range $= \bar{R} = \frac{\Sigma R}{k}$,

where R is summed over k samples.

3. Upper control limit (UCL) for $\bar{x}$:

$UCL = x + A_2\bar{R}$.

4. Lower control limit (LCL) for x:

$LCL = x - A_2\bar{R}$,

where A_2 is a constant depending upon the sample size n.

Some values of this constant and other constants are given in Table 1; a more complete table is found in the Appendix. The source of these constants is reference 4.

Table 1
Constants for Control Charts — No Standard Given

n	A_2	Range		σ	
		D_3	D_4	B_3	B_4
2	1.880	0	3.267	0	3.658
3	1.033	0	2.575	0	2.692
4	0.729	0	2.282	0	2.330
5	0.577	0	2.115	0	2.128
6	0.483	0	2.004	.003	1.997
7	0.419	.076	1.924	.097	1.903
8	0.373	.136	1.864	.169	1.831
9	0.337	.184	1.816	.227	1.774
10	0.308	.223	1.777	.273	1.727

Example. The $\bar{x}$ and R charts are illustrated by test measurements on a three minute granular timer for household use to determine at what level it is stable and how close this is to three minutes. The test measurements are for one end of the timer, since tests of several timers show that there is a significant difference statistically between the two ends in every case. They simply do not represent the same time distribution despite the fact that in some instances the difference is only a few seconds. The difference is so consistent for so many samples that no statistical test is needed to show that a real difference exists.

The data consist of 10 consecutive samples of four measurements each (Table 2). The actual measurements, made by a stop watch which read to tenths of a second, and the sample sums, means, and ranges for the 10 samples are given below. The data are followed by the calculations necessary for the center line and the two control limits. Figure 3 shows the $\bar{x}$ control chart; Figure 4 shows the R chart.

All of the 10 sample points for both the mean $\bar{x}$ and the range R fell within the control limits of the corresponding chart. This shows that for these 10 samples the timer is operating in a stable fashion and under statistical control. The average time is 3 minutes 2.7 seconds; the average range is 7.7 seconds.

For practical uses, such as in cooking or in telephoning, tenths of a second are unnecessary. Since the telephone company allows a 15 second interval to terminate a three minute call, no difficulty would arise since the highest individual value is 3 minutes 10.5 seconds (a value of 20.5).

This is an example of a stable timer; tests on two others showed gradual increases in time as the number of measurements increased. In one case, reading to tenths of a second showed a very definite upward trend over 10 samples of four, but if rounded to seconds, the trend was eliminated: 1, 2, 2, 2, 2, 1, 2, 2, 2, 2. So whether control exists or not depends upon how precise the measurements are. Some evidence exists to show that if the temperature was held constant, say at 68° F, the operation may have been more stable.

Table 2
Timer Data in Seconds Beyond 2:50 Multiplied by 10

Readings	Samples 1	2	3	4	5	6	7	8	9	10	Sum
1	92	197	65	154	119	141	132	111	149	139	
2	123	152	88	165	162	122	100	205	83	104	
3	100	131	139	183	95	65	106	104	132	142	
4	62	161	156	94	118	195	83	144	105	150	
Sum	377	641	448	596	494	523	421	564	469	535	5068
$\bar{x}$	94	160	112	149	123	131	105	141	117	134	1266
R	61	66	91	89	67	130	49	101	66	46	766

Calculations for the $\bar{x}$ and R charts using actual values from above expressed in tenths of a second:

$\bar{\bar{x}} = \frac{506.8}{40} = \frac{126.6}{10} = 12.7$. Hence, average time $= 2{:}50 + 12.7$ seconds
$= 3$ minutes 2.7 seconds.

$\bar{R} = \frac{76.6}{10} = 7.7$.

For $\bar{x}$ chart: Center line = 12.7 seconds.
Upper control limit (UCL) = 12.7 + (0.729)7.7 = 18.3 seconds.
Lower control limit (LCL) = 12.7 − (0.729)7.7 = 7.1 seconds.

0.729 is A_2 for n = 4 (Table 1)

For R chart: $\bar{R} = 7.7$ seconds.
UCL = 2.282(7.7) = 17.6 seconds.
LCL = 0(7.7) = 0 seconds.

2.282 and 0 are from Table 1 for n = 4.

FIGURE 3
$\bar{x}$ Chart Testing a 3-Minute Timer Using 10 Consecutive Samples of 4 Measurements Each

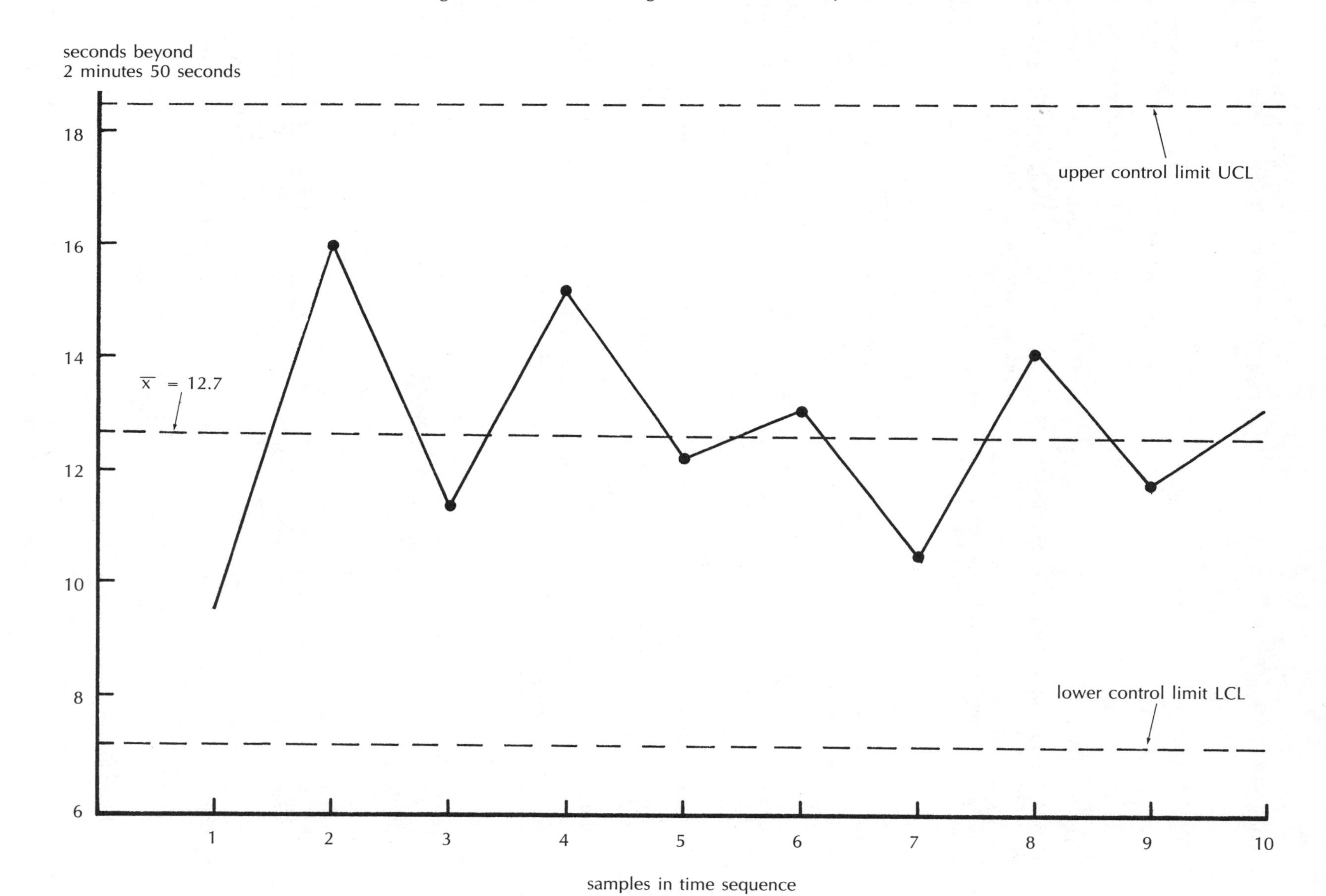

FIGURE 4
R Chart Testing a 3-Minute Timer Using 10 Consecutive Samples of 4 Measurements Each

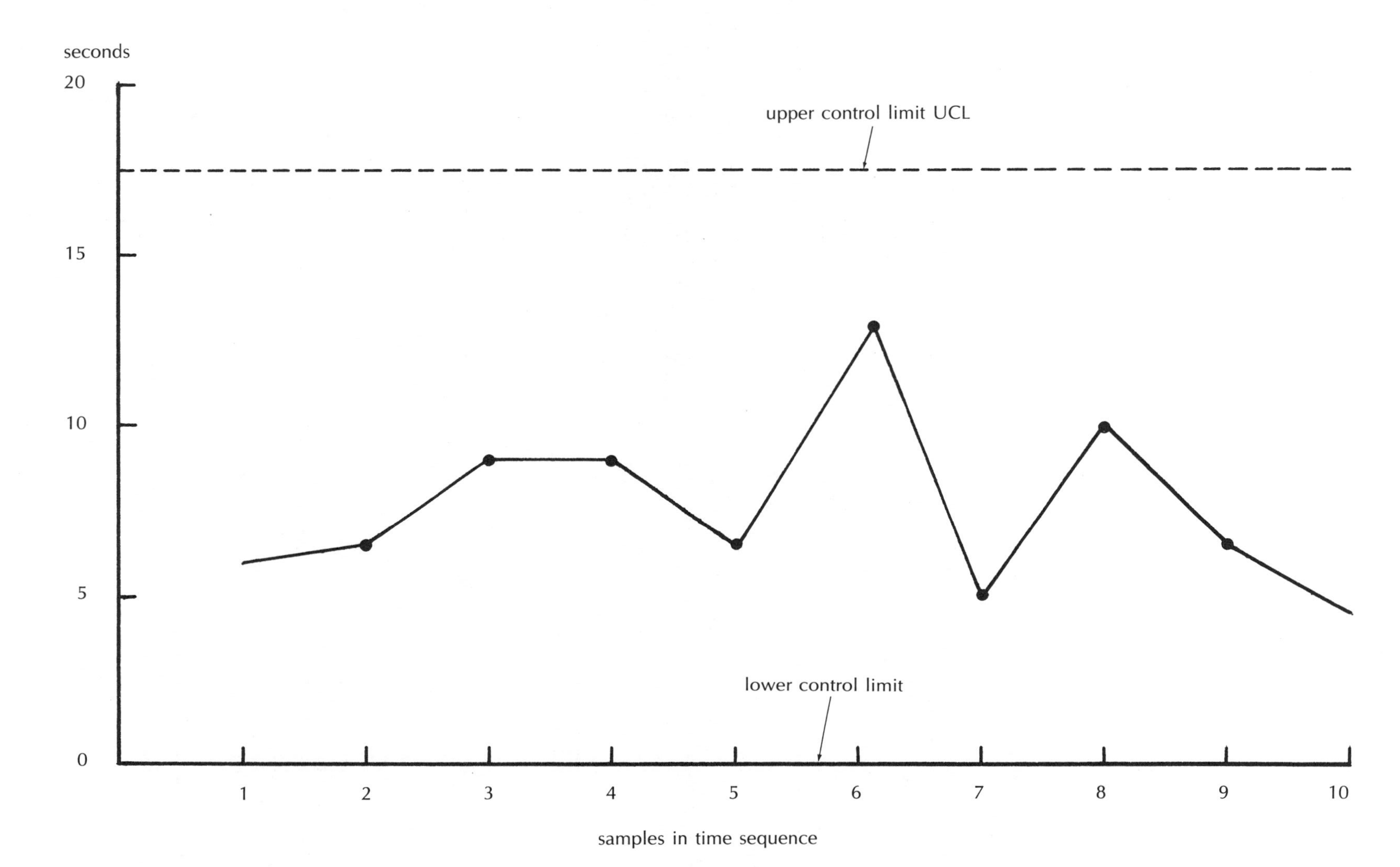

The σ chart. The σ chart, like the R chart, is for the purpose of controlling variability. At this point, it is necessary to explain the different meanings attached to σ in order to avoid confusion.

1. In quality control practice, σ means the *sample* standard deviation:

$$\sigma^2 = \frac{\Sigma(x - \bar{x})^2}{n}; \sigma = \sqrt{\sigma^2},$$

where n = sample size; x = a characteristic of a sample element (summation is over all n elements); and $\bar{x}$ = mean of the sample.

2. In statistics, σ stands for the *population* or *known* standard deviation:

$$\sigma^2 = \frac{\Sigma(x - u)^2}{N}; \sigma = \sqrt{\sigma^2},$$

where N = the population or universe size; and u = the population mean (summation is over all N elements in the population or universe).

3. In quality control:

$$\bar{\sigma} = \frac{\Sigma\sigma_j}{k},$$

where σ_j = standard deviation in the jth sample; and j = 1,2,3...,k samples.

The use of the mean sample value of σ is peculiar to quality control work. It is not described in statistics textbooks.

4. In statistics, s^2 is an unbiased *estimate* of the unknown population variance σ^2:

$$s^2 = \frac{\Sigma(x - \bar{x})^2}{n - 1}, s = \sqrt{s^2},$$

where x, $\bar{x}$, and n mean the same as under 1. above. Note that division is by n − 1, not by n.

5. From 1. and 4. above, it is seen that the sample σ and the estimates are related:

$$\sigma^2 = \frac{n-1}{n} s^2,$$

$$\sigma = s\sqrt{\frac{n-1}{n}}.$$

For n = 4, $\sigma = 0.866s$.

6. Analysis of variance can be applied to the k samples to test whether the k means could have come from the same population. This is done by comparing two independent estimates of the population variance — the variance between samples (s_b^2) with the variance within samples (s_w^2). The latter is an unbiased estimate of the population variance and is used to make an F test for homogeneity of the sample means. It is also used as a basis of an s chart.

The calculations for the σ chart.

1. For each sample, compute σ_j from:

$$\sigma_j^2 = \frac{\Sigma(x - \bar{x})^2}{n},$$

where summation is over k samples.

2. Compute:

$$\bar{\sigma} = \frac{\Sigma\sigma_j}{k},$$

where summation is over k samples. This is the mean value of σ for the k samples.

3. Calculate limits about the mean value $\bar{\sigma}$:

$$\text{Upper control limit (UCL)} = B_4\bar{\sigma},$$

$$\text{Lower control limit (LCL)} = B_3\bar{\sigma},$$

where B_3 and B_4 are the constants from Table 1 corresponding to the size of the sample n.

Draw lines for the UCL and the LCL; plot the k values of σ according to the order of the k samples; and see if any values lie outside these limits, thereby indicating lack of control.

The calculations for the $\bar{x}$ chart to correspond to the σ chart:

1. Center line = $\bar{\bar{x}}$.
2. Control limits = $\bar{\bar{x}} \pm A_1\bar{\sigma}$.
   Note that a new constant A_1 is used. See Appendix Table 1.
3. UCL = $\bar{\bar{x}} + A_1\bar{\sigma}$.
4. LCL = $\bar{\bar{x}} - A_1\bar{\sigma}$.
5. The value of A_1 for n = 4 is 1.880.

Example: The data are the same 10 samples of four measurements each used in the preceding $\bar{x}$ and R charts. The values of σ_j, s_j, and the sum of squares of deviations from each sample mean, ss_j, are shown in Table 3. The sum of the σ_j's is needed to obtain the mean $\bar{\sigma}$ for the 10 samples; the ss's are summed for use in an analysis of variance to be explained later.

Calculations for the σ chart. From the foregoing table of sample data and the previous table of constants, B_3 and B_4, the following calculations are made for the σ chart (Figure 5):

1. $\bar{\sigma} = \frac{28.70}{10} = 2.87$.
2. For n = 4, $B_3 = 0$, $B_4 = 2.330$.
3. UCL = $2.87B_4 = 2.87(2.330) = 6.69$.
4. LCL = $2.87B_3 = 2.87(0) = 0$.

Since values of σ_j range from 1.76 to 4.65, all fall within these limits, and lead to the same conclusion drawn from the R chart.

Table 3
Standard Deviations σ and s with Sums of Squares for Each of 10 Samples*

Sample j	σ_j	s_j	Sum of squares ss_j
1	2.18	2.52	19.05
2	2.38	2.75	22.75
3	3.69	4.26	54.50
4	3.34	3.86	44.62
5	2.42	2.80	23.45
6	4.65	5.36	86.33
7	1.76	2.03	12.39
8	3.99	4.61	63.74
9	2.52	2.91	25.49
10	1.77	2.04	12.45
Sum	28.70		364.77

*Since $\sigma = (n-1)/n(\sqrt{s})$, $\sigma = 0.866s$, or $s = 1.155\sigma$, when n = 4.

FIGURE 5
A σ and Corresponding s Chart

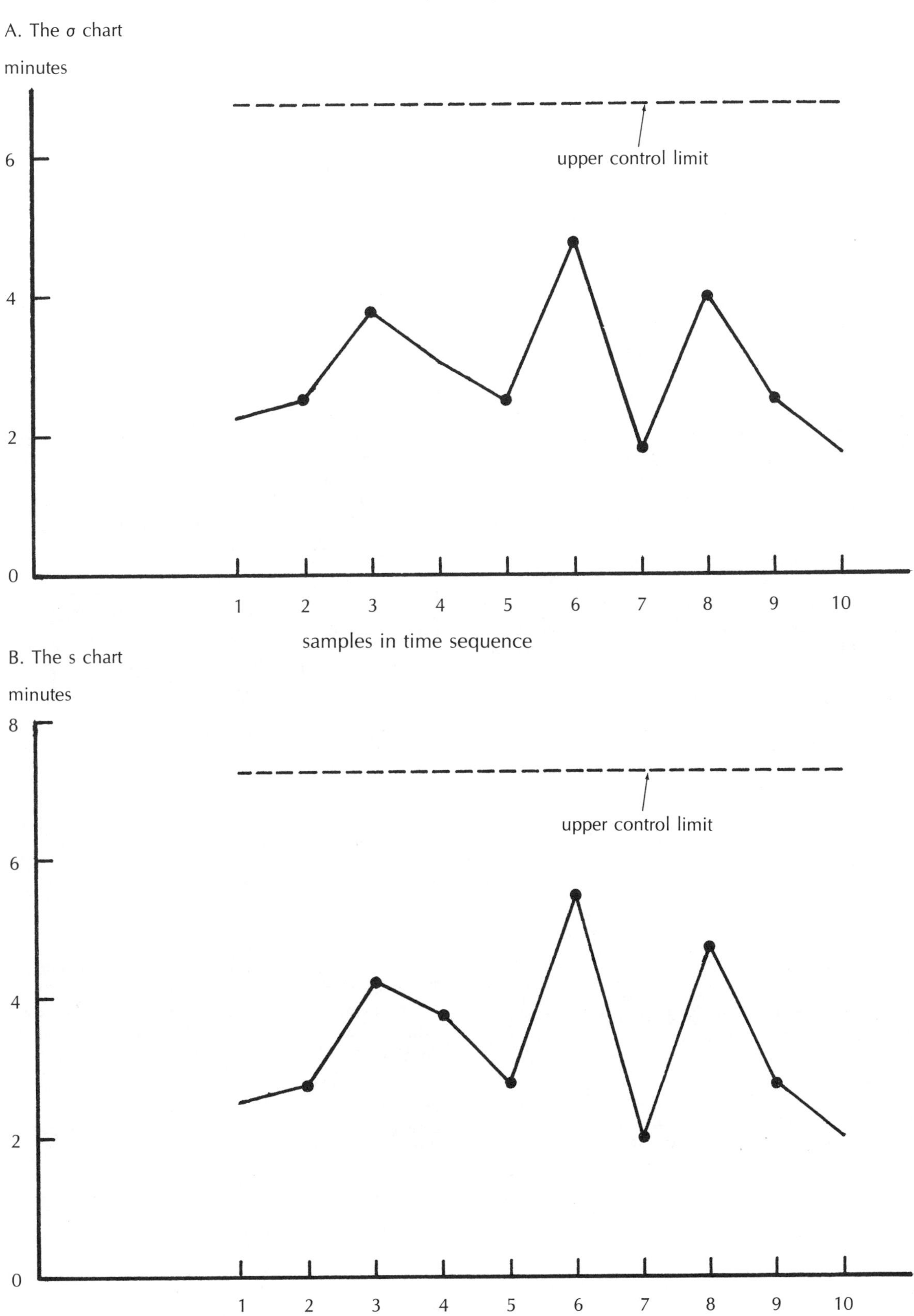

Calculations for the $\bar{x}$ chart corresponding to the σ chart:

1. Center line $\bar{\bar{x}} = \frac{\Sigma x}{kn} = \frac{506.8}{40} = 12.7.$
2. $\bar{\sigma} = 2.870.$
3. Limits: $\bar{\bar{x}} \pm A_1\bar{\sigma}$, where $A_1 = 1.880$ for $n = 4$.
4. $UCL = 12.7 + 1.880(2.870) = 18.1.$
5. $LCL = 12.7 - 1.880(2.870) = 7.3.$

The sample means range from 9.4 to 16.0, so that all of them fall within the limits of 7.3 and 18.1.

Analysis of variance. The technical aspects of the techniques called analysis of variance are described in detail in a section on statistics (Chapter 17). It is a very powerful analytical method. The calculations are not hard to perform using desk type programmable electronic calculators, or even hand-held calculators, while the simple model dealing with the sum or squares within and between samples is equivalent to the $\bar{x}$ chart. It can also be used to analyze observations within control limits to determine if an out-of-control situation exists.

Analysis of variance applied to data from several samples is used for two purposes: to make an F test to determine whether the k sample means could come from the same population indicating a condition of statistical control or not, the same as the $\bar{x}$ chart and to use the mean square within samples as an unbiased estimate of the population variance s_w^2 and then use s_w as the basis of a control chart similar to the one based on σ_j. In this connection, note that the F test is used to test means, *not* variances.

Calculations for an s chart:

1. Center line $= s_w$.
2. $UCL = s_w + \frac{3s_w}{\sqrt{2n}}$.
3. $LCL = s_w - \frac{3s_w}{\sqrt{2n}}$.

The sample sizes should be preferably 25 to 30 or larger. In using analysis of variance, the sample sizes do not have to be the same.

Example: The 10 samples of four measurements each, as presented in previous examples, are analyzed by the use of analysis of variance with an F test. The calculation of s_w and the construction of an s chart is illustrated even though the sample size is not 25 or 30. First, the analysis of variance is presented: total ss = between ss + within ss, where ss means sums of squares:

Analysis of variance, total sum of squares:

$$ss = \sum_1^{40} x_i^2 - \left(\sum_1^{40} x\right)^2 / nk$$

$$= 6934.8 - \frac{(506.8)^2}{40} = 513.64.$$

Between 10 samples:

$$ss = \frac{\sum_1^{10}\left(\sum_1^{4} x_i\right)^2}{4} - \frac{(506.8)^2}{40}$$

$$ss = \frac{26{,}280.18}{4} - 6421.16 = 148.87.$$

Within 10 samples:

$$= \sum_1^{10}\left[\sum_1^{4} x_i^2 - \left(\sum_1^{4} x_i\right)^2/4\right] = 364.77.$$

The *total* ss is the total sum of squares of deviations of the 40 measurements from the grand mean of the 40 values. The *between* ss is the sum of squares of the deviations of 10 sample means from the grand mean. The *within* ss is the sum of squares of each set of four sample values from its own sample mean, calculated for the 10 samples separately, and then added. The summary table is as follows: df means degrees of freedom which are $kn - 1$ for the total, $k - 1$ for the number of samples, and $k(n - 1)$ for within samples, where $k = 10$ and $n = 4$:

df	ss	Mean square	F
between 9 within 30	148.87 364.77	16.54 $12.16 = s_w^2$	1.36
Total 39	513.64		

The mean square is ss divided by df.

$$F = \frac{16.54}{12.16}, \quad s_w = \sqrt{12.16} = 3.49.$$

The value of F is the mean square between samples divided by the mean square within samples or s_w^2. The value is 1.36, but reference to an F table with nine degrees of freedom for the larger mean square and 30 degrees of freedom for the lesser mean square shows that to be significant at the one percent level, F has to have a value of 3.06. This level is equivalent to 2.58σ, so it is less severe than the usual 3σ limits of the Shewhartian chart. The F test is evidence, therefore, that the 10 sample means came from the same population; therefore, the measurements are in a state of statistical control. This is the same conclusion derived from the $\bar{x}$ chart.

The s chart. The calculations are illustrated for an s chart, even though it is recommended that this chart be constructed for sample sizes of 25 or 30 or even more (Figure 5).

1. Center line $= s_w = 3.49$.
2. $ULC = 3.49 + \frac{3(3.49)}{\sqrt{8}} = 7.19.$
3. $LCL = 3.49 - 3.70 = 0.$

The values of s range from 2.03 to 5.36; all of the values fall within these limits leading to the same conclusion that was derived from the R chart and the σ chart.

The ratio control chart. This chart is used for the purpose of controlling ratios, either in the form of proportions or in the form of ratios greater than one. The use of 3σ limits, which assumes the characteristic is distributed as the normal probability distribution, calls for fairly large samples, say those of size 25 to 30 or even larger. The example given below is based upon 20 different sets of observations.

A ratio is defined as:

$$f = \frac{\sum x}{\sum y} = \frac{\bar{x}}{\bar{y}}; \text{ Sum over n sample values.}$$

Since both x and y are obtained from each sample unit, the value of n is the same for both — which leads to the ratio being equal to the ratio of the two means.

The variance of f is:

$$s_f^2 = \frac{1}{n(n-1)\bar{y}^2}\left(\sum x_i^2 + f^2 \sum y_i^2 - 2f \sum x_i y_i\right),$$

where n = sample size or number of sets of observations; $i = 1,2,\ldots,n$; and $\bar{y}$ = mean of the y values.

To construct a control chart:

1. The center line = f.
2. $UCL = f + 3s_f$.
3. $LCL = f - 3s_f$.

The above form is used to calculate s_f^2 because it requires values of the squares of x and y, the cross products of x and y, and the mean of y. These can be obtained directly from the data or very quickly from an electronic programmable desk calculator or even a hand-held calculator which has a Stat Rom, or other built-in program which gives sums, squares, and cross products.

Example: The data show odometer miles, estimates number of gallons of gasoline consumed, and the ratio estimates miles per gallon, where x = miles and y = gallons. The miles, gallons, or miles per gallon were recorded and calculated for each of 20 purchases of gasoline made while driving on major highways, mainly interstate highways. The calculations from Table 4 are:

1. $f = \frac{3668}{132.1} = \frac{183.4}{6.605} = 27.77$ miles per gallon.
2. $s_f^2 = \frac{1}{20 \times 19 \times 6.605^2}$ $[707{,}800 + 27.77^2(908.69) - 2(27.77)25{,}337.2] = 0.080047$
3. $s_f = 0.2829$.
4. $3s_f = 0.85$.
5. $UCL = 27.77 + 0.85 = 28.6$.
6. $LCL = 27.77 - 0.85 = 26.9$.

The following points are outside these limits (Figure 6):

below	above
26.8	30.7
26.6	29.1
26.4	29.4
26.6	
24.3	
26.4	

These out-of-control points may arise from a number of causes. The major cause is, no doubt, the errors made in estimating the amount of gasoline consumed corresponding to the miles read on the odometer. Gasoline was matched with the recorded miles in a record book. Also, some varying traffic conditions may have been encountered which were not recorded. There is no doubt about the average figure of 27.8 miles per gallon, derived from 3,668 miles and 132 gallons, being very close to the correct value. Since the odometer over-read by four percent based on actual tests, a more accurate value is 3668(0.96)/132.1 = 26.7 miles per gallon.

Table 4
Miles, Gallons of Gasoline, and Miles per Gallon for 20 Purchases of Gasoline on Over-the-Highway Trips

Purchase number	Odometer miles x	Estimated gallons y	Estimated miles per gallon
1	171	6.2	27.6
2	203	7.2	28.2
3	198	7.4	26.8
4	192	6.9	27.8
5	198	7.1	27.9
6	192	7.0	27.4
7	117	4.4	26.6
8	221	8.0	27.6
9	145	5.5	26.4
10	113	4.2	26.9
11	204	7.2	28.3
12	226	8.5	26.6
13	97	4.0	24.3
14	233	7.6	30.7
15	227	7.8	29.1
16	244	8.3	29.4
17	140	5.3	26.4
18	145	5.1	28.4
19	199	7.1	28.0
20	203	7.3	27.8
Sum	3668	132.1	xxxxx
Mean	183.4	6.605	27.77

Another method that gives the sampling variance of the ratio x/y and has certain advantages, because it shows how much the correlation between x and y reduces this variance, is the following:

$$s_f^2 = \frac{f^2}{n}\left(c_x^2 + c_y^2 - 2r_{xy}c_x c_y\right),$$

where $f = \bar{x}/\bar{y}$; n = size of the sample or the number of observations; c_x = coefficient of variation of $x = s_x/\bar{x}$ in decimal form; c_y = coefficient of variation of $y = s_y/\bar{y}$ in decimal form; and r_{xy} = correlation coefficient between x and y. Substituting corresponding values in this equation gives:

$$s_f^2 = \frac{27.77^2}{20}\left(\frac{1846.78}{183.4^2} + \frac{1.9037}{6.605^2} - 2(0.9853)(0.2343)(0.2089)\right).$$

$s_f^2 = 0.080431$.
$s_f = 0.2836$.
$3s_f = 0.851$.
$UCL = 27.77 + 0.851 = 28.6$.
$LCL = 27.77 - 0.851 = 26.9$.

There are slight differences relative to the first method, but these are due to rounding; the three sigma limits are the same. It is seen that the high correlation between x and y (r = 0.9853) reduces the sampling variability tremendously due to the negative sign attached to the correlation term.

FIGURE 6
A Ratio Control Chart Applied to Miles per Gallon Based on 20 Purchases Totaling 132 Gallons

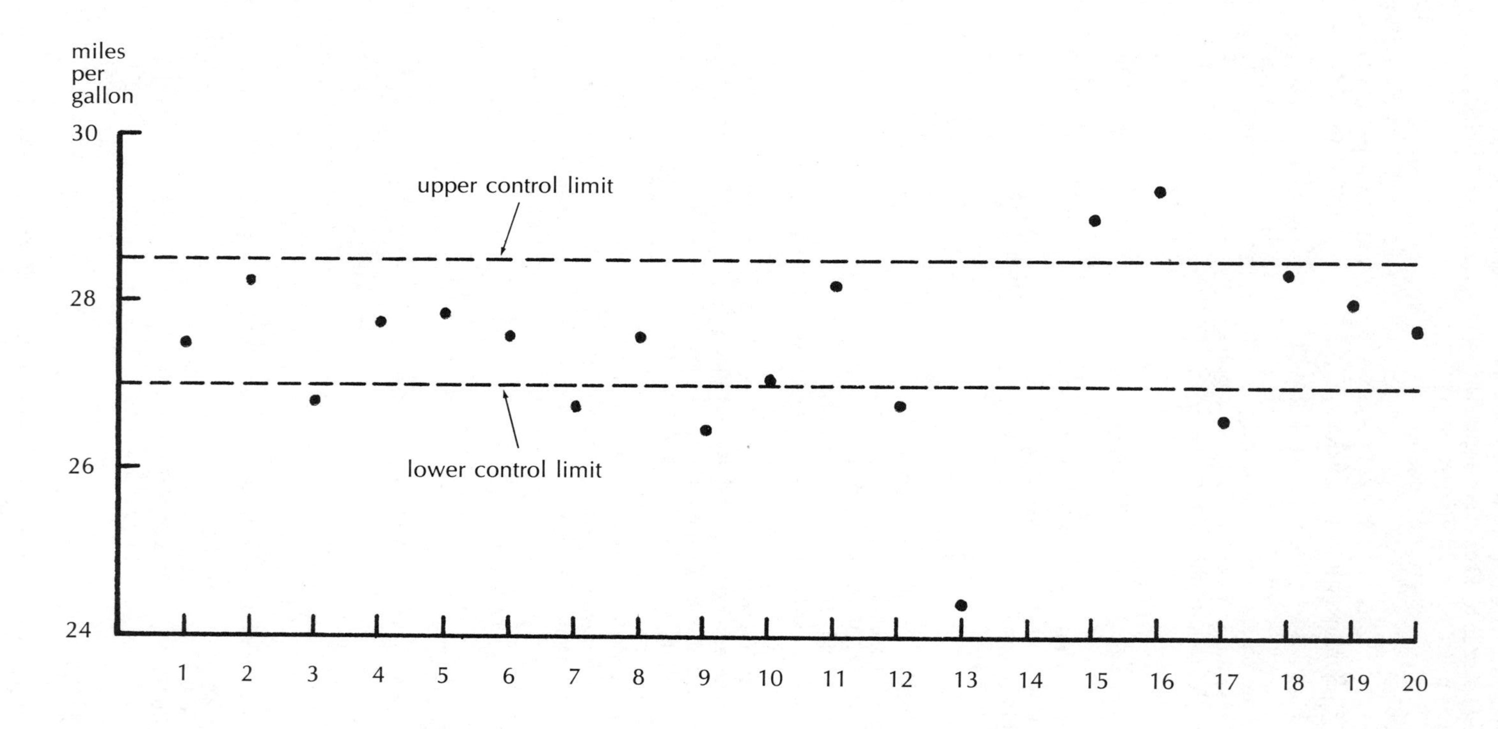

The regression chart. This chart is used to control the relationship between two or more variables or to use the relationship to show whether observed points appear to be consistent with this relationship or not. We limit the discussion and examples to the relationship between two variables, x and y, and more specifically a straight line of the form:

$$y = a + bx,$$

where the constants or parameters, a and b, are determined by the method of least squares from the n pairs (x_i, y_i) which preferably are selected from a large sample of 25 to 30 or more sampling units. Large samples are preferred so that the errors in a and b are negligible or at least very small compared with the other errors. Let P(x,y) be a point on the line and let $P(x_i, y_i)$ be an observed or sample point. Then, the residuals $v_i = y_i - y$ are assumed to be normally distributed if n is 25 to 30 or more; otherwise, it is assumed they are distributed as the t distribution. In any event the residual variance is:

$$s_v^2 = \frac{1}{n-2}\left(\Sigma y_i^2 - a\Sigma y_i - b\Sigma x_i y_i\right)$$

For the control chart:

1. Center line $y = a + bx$.
2. UCL $y + 3s_V$.
3. LCL $y - 3s_V$.

Using $3s_V$ will only reduce slightly the probability between the limits if the sample size is less than 30 and the t distribution is used:

Degrees of freedom	Probability between limits
13	0.99
14-29	0.99+
30	0.995
3σ	0.997 Normal curve

Example 1: In this example of wartime industry control, the past relationship between tons of alloy steel (y) and tons of carbon steel (x) for a specified industry shows how the allocation of these critical materials can be controlled on a quarterly basis. The data consist of 16 pairs of values and x and y, where x represents carbon steel in 100 tons and y alloy steel in tons. The results are:[5]

1. The least squares equation: $y = -55.13 + 26.86x$.
2. The residual variance:

$$s_v^2 = \frac{1}{n-2}(\Sigma y^2 - a\Sigma y - b\Sigma xy),$$

$$= \frac{1}{14}(5{,}913{,}004 + 55.131155 \times 8534 - 26.857098 \times 235{,}919.8),$$

$$= 3383.72.$$

3. $s_v = \sqrt{3383.72} = 58.2$.
4. $3s_v = 174.6$ (For 14 degrees of freedom, $t = 2.98$; the probability between the two control lines is 0.99).
5. Control equations: $-55.13 + 26.86x \pm 174.6$.
   Upper control limit (UCL): $119.47 + 26.86x$.
   Lower control limit (LCL): $-229.73 + 26.86x$. Applying these equations to the two points tested gives:

	x	y	UCL	LCL
Point A:	32.88	938	1003	654
Point B:	30.88	769	949	600

Remarks
A: x and y were allocations for one quarter
B: x and y were allocations for the following quarter

For each point, the x value is substituted in the control equations to obtain the y values on the UCL and LCL lines. The value of 938 for Point A falls between 654 and 1,003; while 769 for Point B falls between 600 and 949. This shows that the two allocations were well within the limits set. This shows that y is consistent with x: it does not show that x is right; x may be all right, or it may be too high or too low.

Example 2: This actual example shows how the regression control chart can be used to discover "out-of-line" cases which need to be audited — in this case a tax audit. It can also be used in those legal (court) cases where it is necessary to establish the average value of some important characteristic for a given size of business in a specified industry. This shows a way of solving these problems.

The straight line is a fit to the salaries of officers in relation to the volume of business of 514 large corporations. The least squares straight line is $y = 67 + 3.35x$, where x = volume of sales and receipts in *millions* of dollars and y = compensation of officers of the corporation in *thousands* of dollars.

Since this was a large sample, the following equation was used to obtain the residual variance. Since:

$$r_{xy}^2 = 1 - s_v^2/s_y^2,$$

then, the residual variance:

$$s_v^2 = s_y^2(1 - r_{xy}^2),$$

where s_v^2 = the residual variance of y about the line $y = a + bx$; s_y^2 = the variance of individual observed y values; and r_{xy} = the correlation coefficient between actual observed values of x and y.

The calculations gave: $r_{xy} = 0.90$; $s_y = 308$; $s_v = 134.25 = 308\sqrt{1 - 0.90^2}$.

The control limits are: $3s_v = 403$
Upper control limit (UCL) = $470 + 3.35x$.
Lower control limit (LCL) = $-336 + 3.35x$.

An actual corporation was represented by the point x = \$10 (million) and y = \$950 (thousand). For x = 10, the value for the upper control limit (UCL) = 470 + 33.5 = 503.5. But the actual value of y is 950, which is way above the UCL; therefore, this value is suspect and should be selected for audit.

Control chart based on individual values. The previous charts are based on *estimates* derived from original counts or measurements obtained from samples. It is possible to apply the control chart to individual counts or measurements, if the frequency distribution of such counts or measurements follows approximately the normal curve or normal probability distribution. One can use the actual frequency distribution, with extreme values eliminated, as in the first example below; or superimpose a normal distribution on the data based on past experience and a study of the distribution, as was done in the second example. To be an effective control chart, the limits should be set after extreme values that experience and analysis show clearly are way out of line are eliminated at both ends of the distribution.

In setting up this chart for individual values:

1. The center line is the mean of the distribution with extremes eliminated *or* a mean based on analysis of past experience.
2. The control limits are the mean plus or minus three standard deviations after extreme values have been eliminated *or* three standard deviations on a normal frequency distribution, which represents readily attainable performance.

Example 1: Murray and Blivens had 148 tax assessors in Iowa assess *independently the same residence* and published these figures with an analysis.[6] The present author obtained the individual values and made a frequency distribution of these assessments as well as a control chart.[7] The individual values are given in Chapter 25. The frequency distribution in intervals of $1,000 with the ends telescoped is as follows:

interval ($1000):	less than 9	9-	10-	11-	12-	13-	14-	15 and over
number:	3	8	32	63	22	9	7	4

The mean calculated from the 148 values is $11,460; the standard deviation is $1,830. The minimum value is $4,000; the maximum value is $21,000 for a range of $17,000. Clearly, this is an excessive range which indicates the need for corrective action.

Using the actual distribution gives limits as follows:

$$UCL = \bar{x} + 3s = 11{,}460 + 3(1830) = \$16{,}950.$$
$$LCL = \bar{x} - 3s = 11{,}460 - 3(1830) = \$5{,}970.$$

Four values fall outside these limits: one below and three above. However, if the seven extreme values are eliminated, a more reasonable control chart would have the following characteristics:

$$\bar{x} = \$11{,}500;\ s = 1{,}000;\ 3s = 3{,}000.$$

This gives a range for the random variation of 6,000 instead of 10,900:

$$UCL = 11{,}500 + 3{,}000 = \$14{,}500.$$
$$LCL = 11{,}500 - 3{,}000 = \$8{,}500.$$

No doubt substantial reduction in the variability of assessments would result from training the assessors in the use of objective standards, by a detailed analysis of the major components of the assessment and how to estimate them, by assistance to individual assessors, and by more practice in making assessments of a variety of houses. A written manual describing the various standards, rules, and criteria and how to apply them to specific examples would also be a great help in reducing variability.

Example 2: The Internal Revenue Service is required by law to collect a large amount of data relating to Federal taxes, as well as numerous factors and characteristics which relate to taxes such as income, deductions, exemptions, and business expenses. Before the introduction of the computer system, Form 1040A returns were sampled at the rate of 1/500 in each of the 61 district offices. This gave a multiplier (theoretically) of 500 for each of the 61 offices. The latter was the ratio of two quantities N_j/n_j, where N_j = total number of tax forms filed and n_j = total number of tax forms selected in the sample.

After careful study of the records a control chart for this tax return was constructed assuming the ratios approximate a normal distribution:

1. The mean or center line was set at 510;
2. The standard deviation was set at 10 so three sigma = 30;
3. Then, UCL = 540 and LCL = 480.

The 61 points were plotted on the chart. Both N_j and n_j were also to be verified for those falling outside the limits. In some cases N_j was in error, in others n_j; in a few both. This chart proved very effective in controlling sample receipts and ensuring that the proper number of sample returns was received from each of the 61 offices.

Notes

[1]W. A. Shewhart, *Economic Control of the Quality of Manufactured Product,* Van Nostrand, New York, 1931; republished by American Society for Quality Control, Milwaukee, 1980. See also W. A. Shewhart and W. E. Deming (editor) *Statistical Method from the Viewpoint of Quality Control.* The Graduate School, U.S. Department of Agriculture, Washington, D.C., 1939. The latter is based on four lectures given in Washington, D.C. during March 1938.

[2]Shewhart and Deming, *ibid.,* p. 33.

[3]For Alden's experience, see James Ballowe's report issued by the Office of Production Research and Development of the War Production Board, September 1945. For Blue Cross, see Connell's paper reproduced in A. C. Rosander, *Case Studies in Sample Design,* Marcel Dekker, Inc., New York, 1977, Chapter 10.

[4]For a detailed and authoritative treatment of this entire subject, with the various tables of constants required, see *ASTM Manual on Quality Control of Materials.* Special Technical Publication 15-C, January 1951, American Society for Testing Materials, Philadelphia, PA.

[5]A. C. Rosander, *Elementary Principles of Statistics,* Van Nostrand, New York, 1951. Copyright by Wadsworth, Inc. Used by permission of Brooks Cole Publishing Co., pp. 338-339, 346.

[6]W. G. Murray and G. E. Blivens, *The National Tax Journal,* V, (December, 1952), pp. 370-375.

[7]*Statistical Quality Control in Tax Operations.* Internal Revenue Service, Washington, September 1958, pp. 9, 19. This booklet was prepared by A. C. Rosander. The 148 individual values (collected by Murray and Blivens) are given and analyzed in Chapter 25.

21

Control by Lot Sampling

Background. Lot sampling was developed as a sound scientific technique by Dodge and Romig of the Bell Telephone Laboratories in the late 1920s. Prof. Walter Bartky of the University of Chicago proposed the use of multiple sampling at about the same time. Dodge and Romig not only developed the theory, but prepared detailed tables to use in sampling lots to obtain specified quality with a minimum amount of inspection.[1]

There are several reasons to use sampling for the purpose of accepting or rejecting a lot of product relative to some quality characteristic:

1. A destructive test must be used to measure the characteristic: strength of paper or plastic, strength of a container, chemical composition, taste of food or beverage, potency and efficacy of medicines and drugs, or various tests of human blood.
2. 100 percent inspection is prohibitive in both time and cost.
3. 100 percent inspection does not give 100 percent protection; a substantial percentage of defective items still remain due to fatigue and other reasons.
4. A huge volume of production precludes the use of 100 percent inspection.

This last reason is why World War II gave quality control, including both control charts and lot inspection by sampling, a big boost. Out of this wartime experience came the military standards relative to the use of sampling: MIL-STD-105 which has gone through A, B, C, and D versions since 1950 is for sampling inspection of lots based on attributes, a characteristic that is counted, not measured.[2] MIL-STD-414 is the military standard which was developed for use in sampling items whose quality characteristic is a measurement.

Source of lots. Lots may come from several different sources and have to be handled accordingly as described below:

1. Source is an *outside* vendor, supplier, manufacturer, or business.
2. Source is another plant of the *same company*.
3. Source is another department, division, or unit of the *same company*.
4. Source is another process, or part of a process, in the same department, division, or unit of the *same company*.

What is a lot? A lot may be an accumulation of finished product or it may be a partially completed set of items in a sequence of processing operations. In the latter instance, the lot may be tested for a different quality characteristic at each of several points in the operation or it may be tested for all of them at the end of the operation.

Examples of lots are the following:

1. A shipment which may contain a few items or hundreds.
2. A batch of work, e.g., 300 documents.
3. An accumulation of paper to be processed such as sales slips, orders, bills, claims, invoices, checks, tickets, or freight bills.
4. Purchased materials and supplies such as printed forms, bank checks, office supplies, towels, soap, or chemicals.
5. File drawers.
6. Record books.
7. A cargo such as a shipload of cocoa beans.
8. A carload or a truckload.

Two basic lot sampling situations. Two different sampling situations arise: external lot sampling and internal lot sampling. External lot sampling, or receiving inspection of purchased materials, is for the purpose of controlling the quality of purchases to see that they meet specifications furnished to the vendor. Internal lot inspection is within the business or office or plant and is for the purpose of controlling directly the performance of workers engaged in a process, work operation, or job.

Control of quality of purchased materials. In this situation, quality is controlled by using sample plans which accept lots of high quality with a high probability and reject lots of low quality with a high probability.

Rejecting lots made by the vendor penalizes the latter, increases his costs, and may call for replacing rejected lots with good material depending upon the vendee-vendor agreement Furthermore, a vendee may have more than one vendor or supplier. If one supplier does not meet satisfactorily the specifications of the vendee, the latter may shift his business to another vendor. Hence, the vendor is under pressure to turn out an acceptable quality service or product.

Usually the vendee or purchaser sets up a receiving department to test purchased items to determine if they meet specifications, negotiates vendee-vendor contracts for mutual agreement, and may rate vendors according to their reliability and trustworthiness in meeting specifications.

The goal: minimize or eliminate receiving inspection. In the early days of quality control, receiving inspection of purchased parts and components became an integral part of a quality control program. As time went on, a closer relationship developed between the vendee (buyer) and the vendor (seller), largely because vendors' products often did not meet the specifications of the buyer. Large companies like Ford Motor Company sent quality control specialists to the suppliers to try to reduce the percent defective in purchased parts. For example, Ford did this with Houdaille and it resulted in the percent of defective door locks being reduced from about 25 percent to about three percent.[3]

This situation led to closer vendee-vendor relations, vendor certification of the quality of its products, vendee-vendor negotiations, and vendor ratings by the vendee.

All of these steps led to a better receiving inspection function so that wasteful overhead costs could be reduced. Not only the vendor, but also the vendee needed to improve operations so as to reduce inspection costs as the following actual example shows:[4]

> In a large Midwest factory an inspector in the receiving department tested two diameters of a complex purchased part with a micrometer. He used a sample of 100 derived from a sample plan in MIL-STD-105, selected from each lot received. He did not know that he was applying a sample plan for attributes to measurements, and hence was using an excessively large sample. He was told that a sample of about 50 would give the same protection, if he used the standard deviation derived from the samples. Even better, if the vendor had these diameters under control using the proper control charts so that a stable value of the standard deviation existed, then the sample could be still further reduced to about 25 with the same protection. He was referred to Bowker and Good's book on *Acceptance Sampling by Variables.*

Obviously, this company was wasting time and money on its receiving inspection because no one seemed to know the most efficient sample plan to use. Furthermore, if the vendor had the two dimensions under control and certified to that effect, then receiving inspection of these dimensions could have been eliminated.

The responsibility for quality lies solely and entirely with the producer of the item or the renderer of the service. The vendor should meet and certify the specifications of the vendee. If this is done, then, in theory at least, receiving inspection can be eliminated. At most, receiving inspection by the vendee should be no more than an occasional audit. It certainly would pay a vendee to find a vendor that would do this so that receiving inspection could be minimized if not eliminated.

Control by internal lot sampling. Internal lot sampling can be used to control work performance in much the same way as a p chart or a c chart derived from the number of sample units in error in a sample selected from the lot. Lot sampling may be used, as described in an example previously, because it is simple, it is easier to administer, it may be more acceptable to management, it eliminates plotting time-consuming charts, it facilitates quick feedback to the source, and it is easy to relate quality to production and to measure improvement in productivity.

These advantages are not obtained without following a number of operating procedures:

1. Using work lots with equal numbers of units, or approximately equal numbers, so that only one sample plan is necessary and it is easy to estimate the over-all percentage of error.

2. Returning a rejected lot *immediately* to the person processing the lot for a 100 percent re-do. This tends to check, if not stop errors, and prevents them from accumulating. Supervisory assistance on an individual basis may be necessary in order to prevent recurrence of the same kinds of errors.

3. Keeping daily records of units processed by each employee separating new production from rejected lots corrected so that outgoing production is acceptable quality production that has passed through the lot acceptance stage and has been accepted.

4. Records are kept of the number of defective units (in error) both for the accepted lots and the rejected lots. This makes it possible to estimate the over-all percent error rate in outgoing production.

5. Daily production records mean learning curves can be plotted, both individually and collectively, to show group capability under existing conditions. This helps management to accept realistic quality and production goals; by knowing how long it takes individuals and groups to attain a given level of competence under competent supervision. This is especially applicable to an entirely new job or project, where all employees start from scratch and have to be put through an extended training course (2 to 6 weeks) before real production can begin.

Some principles of lot sampling. Reference has already been made to the fact that a sample may be designed for one of several different purposes. A design for one purpose is not usually effective for another. Sampling is not based on opinion, intuition, or common sense. It is based on probability and statistics; it is a complex technical subject requiring careful study to insure that the proper sample design is being used for a given purpose. In view of the widespread misunderstandings and misuses of sampling, this warning cannot be repeated too often. Some principles applicable to lot sampling are the following that constitute an introduction to the more technical material to follow:

1. A random sample of size n, selected from the entire lot of size N, tends to reflect the characteristics of the lot. All lot sampling is based on *random selection,* thus giving every unit in the lot an equal probability of being selected. This eliminates bias from the selection process, but not from other sources. The power of the sample lies in its ability to reflect the characteristics of the lot.

2. A risk is taken in using a random sample to make a decision about the lot because some acceptable lots will be rejected and some unacceptable lots will be accepted. Actually, this could happen even if 100 percent inspection were used.

3. An acceptance-rejection sample plan can be designed to keep these risks at desired levels, but the success of such a plan depends upon several factors including the training of inspectors and others so they can readily identify a defective unit.

4. As pointed out above, the sample plan must be appropriate for the quality characteristic being tested: a sample plan for an attribute (a count) is *not* appropriate for a measurement.

5. The sample selected from the lot to accept or reject the lot does not automatically build quality into the lots; it simply is an indicator as to whether quality is good or bad. Quality has to be built into the lots at the source.

6. A sample plan is designed by assuming a lot size N; then using sample size n and setting the maximum number of defective sample units c which is found in n so as to accept good lots with a high probability and bad lots with a low probability.

7. The best sample plan is *not* one in which a lot is accepted when the number of defective units c found in the sample is zero. This is illustrated later.

8. In order to determine the discriminating power of a sample plan, that is the power to discriminate sharply between lots with acceptable error rates and lots with unacceptable error rates, it is necessary to construct the

operating characteristic curve (OC curve) of the sampling plan. This curve shows for values of incoming quality (percent in error or defective) p, increasing from $p=0$, the probability of accepting the lot P_a, given N, n, and c.

For example, in a given situation, a discriminating sample plan may be one which accepts lots with $p=0.01$ or one percent, 0.02 or two percent, and 0.03 or three percent with a probability of 0.90 and above; and accepts lots with p = six percent, seven percent and eight percent with a probability of 0.10 and below.

9. Setting high quality levels in an acceptance-rejection sampling plan, such as 95 percent acceptance of lots with a two percent error rate, is futile if the incoming quality is always worse say at 10 percent or higher. In receiving inspection, this means protection from low quality items received from an *outside vendor* because all lots will be rejected. However, as pointed out above, this is a very wasteful practice since it increases the costs of the vendor and delays the operations of the vendee. If vendors cannot get below 10 percent, the buyer will have to screen out the defective items and make adjustments with the vendor.

10. If the incoming material is from *inside processing* and a similar situation exists, then sampling is a waste of time because all lots will be rejected. Re-training personnel or other remedial measures are in order immediately in order to bring down the error rate to a level where sampling will be effective.

11. Sampling lots for acceptance-rejection decisions works effectively only if the incoming quality is near or better than the quality level specified by the sampling plan. In fact, it is considered good practice to have the lot rejection rate at about 15 to 20 percent or even less, which means rejecting about one in five or one in seven lots. As stated above, just because a sample plan is set at a two percent error rate, this will not reduce an incoming error rate of 10 percent to two percent. What happens is that all lots will be rejected. Sampling does not work unless the error rate is close to that set in the sample plan or averages very close to that figure. Quality is built in, not sampled in.

12. Lots should be grouped, identified, tested, and sampled with regard to classifications directly related to the quality characteristic and corrective action to be taken, such as vendors, origins, time of production, employees, organizational unit, lot numbers, grades, types, work shift, shipment, and location. Identification of lots in this way speeds up feedback and corrective action, increases the effectiveness of the sample, and provides information that may lead to an improvement of the sampling plan.

Three kinds of sample plans for attributes. The three kinds of sample plans for attributes are single, double, and multiple. These terms refer to the number of samples which are, or may be, selected from the lot.

"Single" refers to the random selection of one sample from the lot. "Double" refers to the random selection of one sample from the lot; if this does not lead to a decision, a second sample of the same size is selected which when coupled with the first sample always leads to a decision. "Multiple" refers to a series of samples of the same size which when accumulated always result in a decision. The number of such samples is variable; it is not fixed in advance. In multiple sampling, it is possible to reach a decision with the first sample if the lot is highly defective or contains very few errors or defective units.

These terms are used to identify sample plans for attributes in Military Standard-105-D (MIL-STD-105-D), a very popular compilation of lot sampling plans. For a specified acceptable quality level (AQL) for several values up to 0.10 or 10 percent, the standard gives the sample size n and the acceptance and rejection numbers for several ranges of lot sizes N. Also included are OC curves and normal, tightened, and reduced inspection for single, double, and multiple sample plans. These three kinds of plans are described later for the same lot size and the same acceptable quality level (AQL).

The advantages and disadvantages of the three kinds of sampling are as follows:

1. **Single:** This plan is simple, requires only one sample selection, and is easy to administer. If the entire sample is inspected, not truncated because the acceptance or rejection number is reached before the entire sample n is inspected, the sample provides an estimate of the average quality level. This is a very important advantage. For the same AQL, the sample size is larger than for the other two plans.

2. **Double:** This plan reduces sampling, especially if the incoming quality is really good or really bad, because then only the first sample is needed. It is alleged that double sampling gives the lot two chances of being accepted. This leads to more complicated paper work, especially if the first sample is indeterminate.

3. **Multiple:** This plan may reduce the total sample required by as much as 50 percent. This is offset by the time consumed in selecting various samples and recording the number of defective items, accumulating sample sizes and number of defectives, and comparing them with the acceptance-rejection rules. This means that the plan is often difficult to administer under actual conditions. For example, there may be objections, if not obstacles, to going back and drawing samples from the lot at three or more different times.

Single Sampling Plans

In the following pages, we show how to calculate and draw the operating characteristic curve (OC curve) for three different single sampling plans. We show the effect of changing the values of c with N and n fixed. We show that using $c=0$ gives an inefficient rule because it rejects too high a proportion of acceptable quality lots contrary to intuition and "common sense". We show also that increasing the sample size, as well as c, gives the best sample plan of the three. We show also that while the easiest distribution to use to construct these OC curves is the Poisson distribution, the more exact hypergeometric distribution gives values of P_a in this case that are not significantly different for all practical purposes from those given by Poisson. This would not be the case for smaller lots and samples that include a large proportion of the lot.

The power of a sample plan to discriminate differences of incoming proportion defective p is illustrated by three single sample plans that show the effect of varying n and c on the probability of acceptance, P_a, for a range of p up to 0.08:

Plan	Lot size N	Sample size n	Accept lot if	Reject lot if
1	600	50	$c=0$	$c=1$ or more
2	600	50	$c=1$	$c=2$ or more
3	600	100	$c=3$	$c=4$ or more

For a selected value of p, the value of np was calculated and the Poisson tables[5] were used to obtain the probability of acceptance P_a for the specified value of c. The data are given in Table 1; the curves are shown in Figure 1. Each of the curves is called the *operating characteristic (OC) curve* of the sample plan. The OC curve shows, for a given submitted or incoming lot quality p, the probability of acceptance P_a of the lot under a given acceptance-rejection rule.

The first plan, where c = 0, shows large proportions of lots being rejected, of what is generally considered a good quality lot, p = 0.01 and p = 0.02, although only small proportions of relatively poor quality lots, p = 0.07 and p = 0.08, are being accepted. The second plan, which is identical to the first except that c = 1, shows some improvement especially in the area of accepting good lots at the lower values of p.

The third plan, with a larger sample and with c = 3, shows a much higher proportion of acceptance of good lots and a level of acceptance of poor lots between that of the other two plans. These plans show how, in a single sample plan, the values of n and c can be varied for a fixed lot size N to improve the power of the plan to accept good quality lots and reject unacceptable quality lots. Contrary to a common belief, a plan where c = 0 is *not* the best plan, as the OC curves clearly show. This is generally true.

FIGURE 1
Operating Characteristic Curves for Three Sample Plans

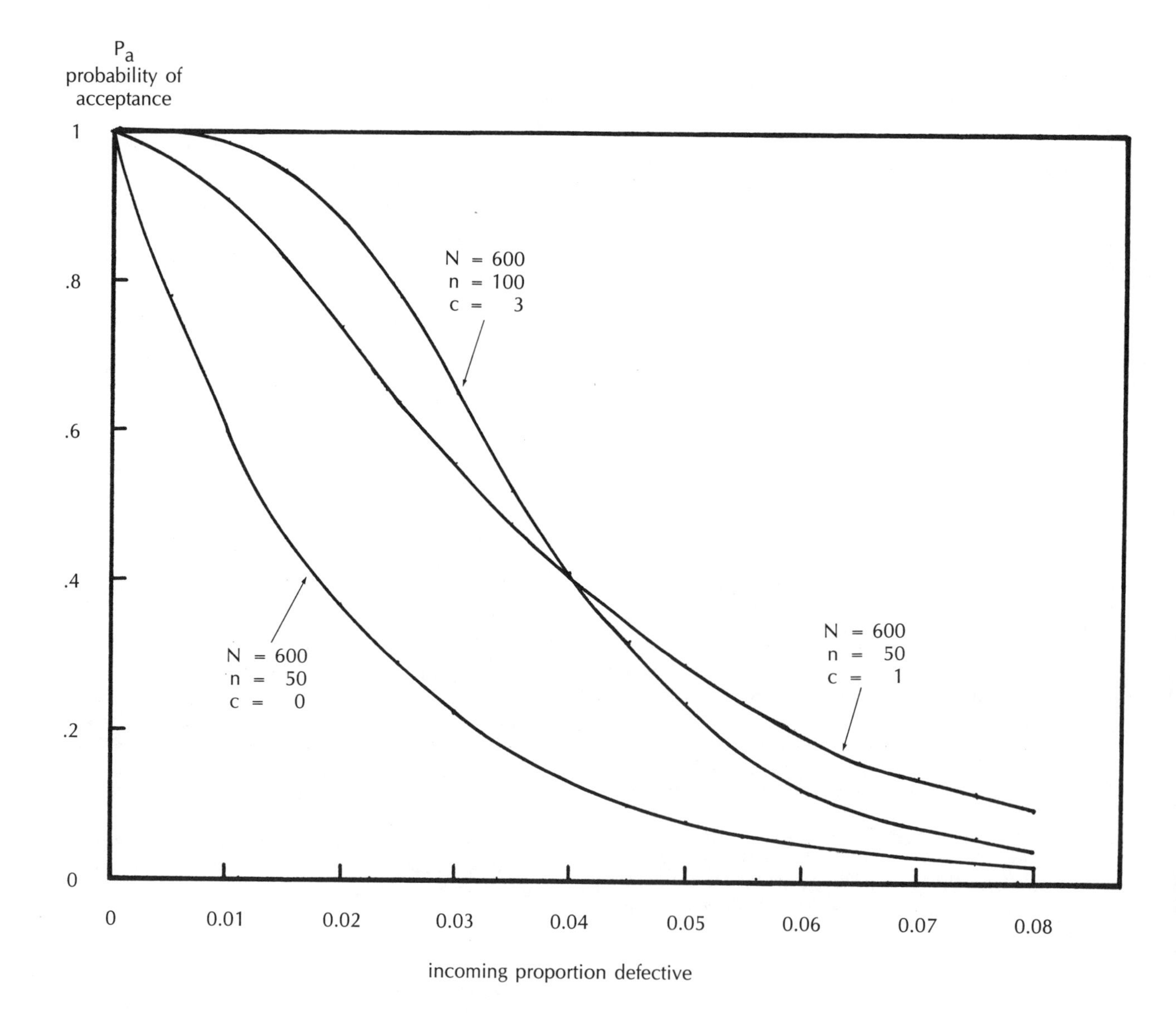

Table 1
Data for Three Operating Character (OC) Curves ($N = 600$)

Incoming p	np	n=50 c=0 P_a	n=50 c=1 P_a	n=100 c=3 np	(Poisson) P_a	(Hyper-geometric)[a] P_a
0.005	0.25	0.78	0.97	0.50	1.00	1.00
0.01	0.50	0.61	0.91	1.00	0.98	0.99
0.02	1.00	0.37	0.74	2.00	0.86	0.88
0.03	1.50	0.22	0.56	3.00	0.65	0.65
0.04	2.00	0.14	0.41	4.00	0.43	0.41
0.05	2.50	0.08	0.29	5.00	0.27	0.23
0.06	3.00	0.05	0.20	6.00	0.15	0.12
0.07	3.50	0.03	0.14	7.00	0.08	0.06
0.08	4.00	0.04	0.09	8.00	0.04	0.03

[a]This is the distribution plotted the Figure 1.

The calculations required to obtain the entries in Table 1 are illustrated below for an incoming p of 0.02; in all cases the lot size $N = 600$:

1. For the plan $n = 50$ and $c = 0$: $np = 50(0.02) = 1.00$. In the Poisson table for mean $= np = 1$ and $c = 0$, $P_a = 0.37$.
2. For the plan $n = 50$ and $c = 1$: In the Poisson table for mean $= np = 1$ and $c = 1$, $P_a = p_0 + p_1 = 0.74$.
3. For the plan $n = 100$ and $c = 3$ and using the Poisson distribution: $np = 100(0.02) = 2.00$. In the Poisson table for mean $= 2.00$ and $c = 3$, $P_a = p_0 + p_1 + p_2 + p_3 = 0.86$.
4. For the plan $n = 100$ and $c = 3$ and using the hypergeometric distribution and the MF approximation to p_0: $Np = 600(0.02) = 12$ defective items. The hypergeometric format is as follows:

Class	Population (lot)	Sample	Non-sample
defective	12	0	12
not defective	588	100	488
Total	600	100	500

By the MF method:

$$p_0 = \frac{(500 - 11/2)^{12}}{(600 - 11/2)^{12}} = \quad 0.110$$

$$p_1 = p_0(12)(100)/489 = \quad 0.269$$

$$p_2 = p_1(11)(99)/2(490) = \quad 0.299$$

$$p_3 = p_2(10)(98)/3(491) = \quad 0.199$$

$$0.877 \text{ or } 0.88$$

$P_a = 0.88$ is the value shown in Table 1 under hypergeometric.

For p_0, the MF method gives 0.109692; the exact figure is 0.109682. This small difference has no appreciable effect on the results. The MF method was used to obtain p_0 in all of the other calculations of P_a under the hypergeometric distribution.[6]

Comparison of three kinds of sampling plans for attributes. In this section, each of the three sampling plans — single, double, and multiple — is taken from MIL-STD-105-D, where the acceptable quality level (AQL) is assumed to be 1.5 percent, inspection is normal (designated by II), and the lot size is assumed to range from 281 to 500. The sample size code letter is H. Hence, we can compare sample sizes and acceptance-rejection rules for the three kinds of plans designed to meet the same specifications.[7] Those who desire more detailed information about the theory and design of these three kinds of sample plans are referred to standard textbooks on quality control as well as MIL-STD-105-D (the Burr reference given in footnote[8] contains much of this standard).

Sample plans for the above specification:

1. Single

$n = 50$
acceptance (acc.) 0, 1, 2
rejection (rej.) 3 or more

2. Double

	n	accepted	rejected
first sample:	32	0	3
second sample:	32	3	4

3. Multiple

sample	n	accepted	rejected
1	13	x	2
2	26	0	3
3	39	0	3
4	52	1	4
5	65	2	4
6	78	3	5
7	91	4	5

In the double sample, if one or two defectives are found in the first sample, select a second sample. If in both samples totaling 64 the sum of the number of defectives is three or less, accept the lot; if the number is four or more, reject the lot. In the multiple sample plan, seven cumulative samples of 13 each are shown with the acceptance and rejection numbers; the "x" means that a decision to accept the lot with a sample of 13 is not permitted.

Under the single sample plan, the lot requires a sample of 50 to be accepted or rejected. Under double sampling, the lot can be accepted or rejected with a first sample of only 32. Under multiple sampling, the lot can be rejected with a sample as small as 13 and accepted with a sample as small as 26.

Three basic curves for acceptance sampling by attributes. Equations and data are given for three related distributions and their curves: the operating characteristic (OC) curve where the probability of acceptance P_a is a function of the incoming proportion or fraction defective (p); the average number inspected (API) curve where the number of pieces inspected is a function of the incoming proportion p or fraction defective; and the average outgoing quality AOQ curve as a function of p. The basic equations are:

1. Probability of acceptance P_a:

$$P_a = e^{-m}\left(1 + m + \frac{m^2}{2!} + \ldots + \frac{m^c}{c!}\right).$$

2. Average number of units inspected per lot:

$$API = nP_a + N(1 - P_a) = N - P_a(N - n).$$

3. Average outgoing quality:

$$AOQ = pP_a \frac{(N - n)}{N} = pP_a (1 - \frac{n}{N}),$$

where $m = np$, c = acceptance number, n = sample size, N = lot size: P_a = probability of acceptance of a lot, and p = incoming proportion or fraction defective.

AOQL is the *average outgoing quality limit* and is the proportion defective p corresponding to the maximum point on the AOQ curve. This curve assumes that if the lot is accepted, only n units in the sample are inspected; if the lot is rejected, the entire lot is inspected and all defective units are corrected or replaced with non-defective units.

The factor $\frac{N-n}{N}$ or $1-\frac{n}{N}$ is the correction factor for sampling from a finite population. It is used when the sample size n is a substantial proportion of the lot size N, say four percent or more; otherwise, if the sample is only a very small fraction of the lot, it can be omitted. In the example, this factor is included since n is 10 percent of N.

Eleven values of p are used to construct the three curves; this number is sufficient to give a good idea of the nature of the curves. The data in Table 2 show that as p increases, P_a decreases, API increases, and AOQ increases then decreases to give a maximum (AOQL).

Table 2
Data for Operating Characteristic (OC) Curve, Average Number of Units Inspected (API) Curve, and Average Outgoing Quality (AOQ) Curve for N = 800, n = 80, and c = 2 for Values of Incoming Quality p from 0.005 through 0.08.

Incoming quality p	Mean np	Probability of acceptance P_a	Average no. units inspected (API)	Average outgoing quality (AOQ)
0.005	0.40	0.99	87	0.45
0.01	0.80	0.95	116	0.86
0.02	1.60	0.78	238	1.40
0.025	2.00	0.68	310	1.53
0.03	2.40	0.57	390	1.54 AOQL[a]
0.035	2.80	0.47	462	1.48
0.04	3.20	0.38	526	1.37
0.05	4.00	0.24	627	1.08
0.06	4.80	0.14	699	0.76
0.07	5.60	0.08	742	0.50
0.08	6.40	0.05	764	0.36

[a] p = 0.028 and 0.029 also round to 1.54.

Source: Knowler, Howell, Gold, Coleman, Moan, and Knowler, *Quality Control by Statistical Methods,* American Society for Quality Control, Milwaukee, Wisconsin. Used by permission.

Sample plan given p_1, p_2, α, and β. Single sample plans can be designed given an acceptable lot defective p_1 which will be rejected with a very small probability α and a larger lot defective p_2 which will be accepted only with a small probability β. Sample plans can be designed to meet these specifications by using the Poisson distribution. In what follows, it is assumed that $\alpha = 0.05$ and $\beta = 0.10$. An OC curve for a sample plan using these values is given in Figure 2. The same general method can be used to develop sample plans using other values of α and β. The steps in determining the sample size are:

1. For a specified value of c, find $m_1 = np_1$, where m_1 is the mean of a Poisson distribution and c corresponds to the cumulative probability which is nearest 0.95 for $\alpha = 0.05$.
2. For the same specified value of c, find $m_2 = np_2$, where m_2 is the mean of a Poisson distribution and c corresponds to the cumulative probability which is nearest $\beta = 0.10$.
3. Calculate the ratio $m_2/m_1 = np_2/np_1 = p_2/p_1$. These ratios for various values of c are given in the appendix.
4. Since numerous p_1's and p_2's have this ratio, select an acceptable p_1 and p_2 for the specified c which gives the needed protection.
5. Then the sample size: $n = m_1/p_1 = m_2/p_2$.
6. Repeat using other values of c until an appropriate plan is found relative to sample size, p_1, and p_2.

Example 1: For $c = 0$, $m_2 = 2.303$ and $m_1 = 0.0513$. The ratio is 44.893. Hence, if p_1 is assumed to be 0.0025, then $p_2 = 0.0025(44.893) = 0.1122$. The sample size $n = 0.0513/0.0025 = 2.303/0.1122 = 21$.

Example 2: For $c = 1$, $m_2 = 3.890$ and $m_1 = 0.355$. The ratio is 10.858. If p_1 is assumed to be 0.008, then $p_2 = 0.008 (10.958) = 0.088$. The sample size $n = .355/.008 = 3.890/.088 = 45$.

Example 3: For $c = 4$, $m_2 = 7.994$ and $m_1 = 1.970$. The ratio is 4.058. If $p_1 = 0.01$, then $p_2 = 0.0406$. The sample size $n = 1.970/0.01 = 7.994/0.0406 = 197$.

Figure 2 shows an operating characteristic (OC) curve meeting the following specification:

$p_1 = 0.01$; $\alpha = 0.05$; $1-\alpha = P_a = 0.95$;
$p_2 = 0.05$; $\beta = 0.10$; $\beta = P_a = 0.10$.

This plan requires a sample size of 133 for $c = 3$, a sample size which is derived as follows:

1. For $c = 3$, $m_2 = 6.681$, $m_1 = 1.316$, $m_2/m_1 = 5.077$ (see Appendix tables).

2. $p_1 = 0.01$ and $p_2 = 0.01(5.077) = 0.0508$.

3. Therefore: $n_1 = \frac{1.316}{0.01} = 132$ for $p_1 = 0.01$ and $p_2 = 0.0508$. Also, $n_2 = \frac{6.681}{0.05} = 134$ for $p_2 = 0.05$ exactly.

4. n is taken as 133 since n_1 and n_2 are not identical.

5. If $n_1 = n_2$, then $p_2/p_1 = n_2/n_1 = n_2p_2/n_1p_1$.
Then $n_1 = m_1/p_1$ and $n_2 = m_2/p_2$.

Values of m_1 and m_2 for values of c are given in the Appendix for $\alpha = 0.05$ and $\beta = 0.10$.

Additional points on the OC curve (P_a plotted against p) can be calculated from the Poisson distribution summing terms from $x = 0$ through $x = c = 3$ for $m = np$:

$$P_a = e^{-m}(1 + m + \frac{m^2}{2!} + \frac{m^3}{3!}).$$

For example, for a value of $p = 0.02$, $m = np = 133 (0.02) = 2.66$. Hence,

$$P_a = e^{-2.66}(1 + 2.66 + \frac{2.66^2}{2} + \frac{2.66^3}{6}) = 0.72.$$

FIGURE 2
Operating Characteristic Curve
$\alpha = 0.05\ \beta = 0.10,\ p_1 = 0.01,\ p_2 = 0.05$
$n = 133,\ c = 3$

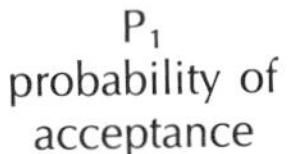

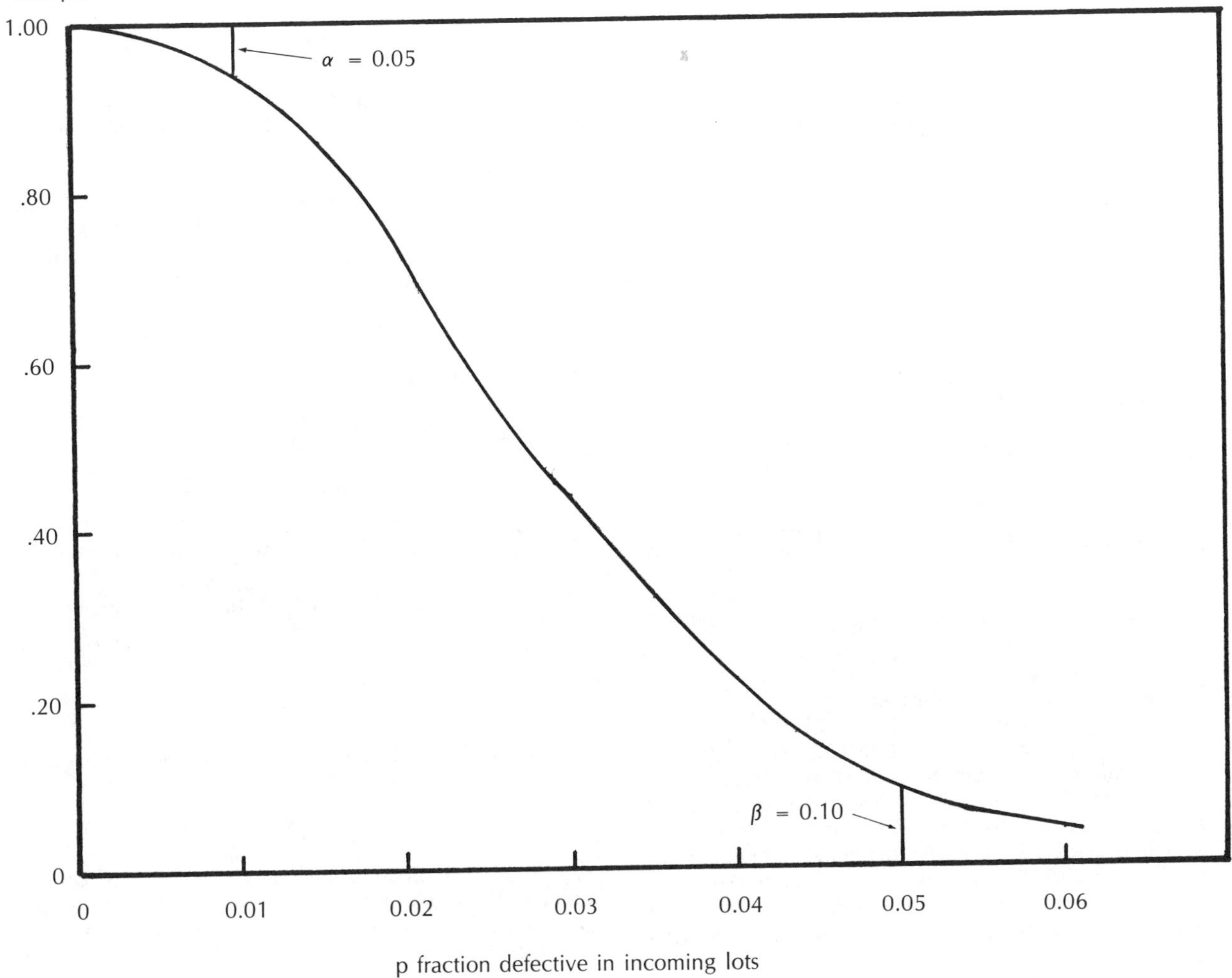

Hence, a point on the OC curve is $p = 0.02$ and $P_a = 0.72$. Other points on the OC curve are calculated in a similar manner: assume a value for p and solve for P_a. Tables to use to obtain a sample size n, given p_1, p_2, and c and assuming $\alpha = 0.05$ and $\beta = 0.10$ are given in the appendix.

Sequential acceptance sampling with control lines. Sample size is usually fixed, but in sequential sampling it is a variable — the selection of a random sample from a lot, either in the form of individual sample units or in groups of sample units, is continued until a decision is reached. The sequential sampling plan for lot acceptance or rejection described below assumes that each sample unit is either good or bad and is distributed as the binomial distribution. The specifications for the sample plan require the same four characteristics of the operating characteristic (OC) curve that were described in the previous section:

α = producer's risk of rejecting a good lot, corresponding to fraction defective p_1.

β = consumer's risk of accepting a poor quality lot, corresponding to fraction defective p_2.

In the previous statement about a multiple sample plan, samples were selected in groups and each group had a rejection number and an acceptance number. In the present method, two parallel straight lines serve as control lines establishing three decision areas: above the upper line is the zone of rejection, below the lower line is the zone of acceptance, and the zone between the two lines calls for a continuation of sampling as long as the value from the sample item or group falls in this area.

The two parallel lines are of the form $y = a + bx$, where x represents the individual sample units 1,2,3,... and y is the cumulated number of defective sample units. Four quantities are needed to determine these two lines:

$$c = \log \frac{1 - \beta}{\alpha}; \qquad g_1 = \log p_2/p_1;$$

$$d = \log \frac{1 - \alpha}{\beta}; \qquad g_2 = \log \frac{1 - p_1}{1 - p_2}.$$

The two parallel lines are:

$$\text{upper line } y_2 = \frac{c}{g_1 + g_2} + \frac{g_2}{g_1 + g_2} x.$$

$$\text{lower line } y_1 = \frac{-d}{g_1 + g_2} + \frac{g_2}{g_1 + g_2} x.$$

The slopes of the two lines are the same:

$$b = \frac{g_2}{g_1 + g_2}.$$

Example: The specifications are the same as those for the OC curve in Figure 2.

$p_1 = 0.01;\ 1 - p_1 = 0.99;\ \alpha = 0.05;$
$1 - \alpha = 0.95;\ p_2 = 0.05;\ 1 - p_2 = 0.95;$
$\beta = 0.01;\ 1 - \beta = 0.90;$

$$c = \log \frac{0.90}{0.05} = 1.25527; \quad g_1 = \log \frac{0.05}{0.01} = 0.69897;$$

$$d = \log \frac{0.95}{0.10} = 0.97772; \quad g_2 = \log \frac{0.99}{0.95} = 0.01791;$$

$$g_1 + g_2 = 0.71688.$$

The equations of the two lines are:

upper line: $y_2 = 1.75 + 0.025x$.
lower line: $y_1 = -1.36 + 0.025x$.

The smallest sample for acceptance is $1.36/0.025 = 55$. The average size of sample needed if the submitted lots are at p_2 is 58; if the submitted lots are at p_1 the average sample needed is 81. If $n = 50$, the upper limit is 3.00 and the lower limit is 0; hence, the lot can be rejected if $y = 4$ or more. Rejection can take place with a very small sample size. For example, if $n = x = 3$ and all three are defective (y), then $y_2 = 1.75 + 0.075 = 1.83$ so y falls above the line and calls for rejection. This illustrates how sequential sampling allows a decision to be made about the lot with a much smaller sample than that required by either single or double sampling; in general, samples are only about half as large. Some of the disadvantages of selecting sampling units individually and making a test on each can be avoided by using groups of sampling units, say groups of five or 10, and plotting points of cumulated sample size against cumulated defects (y). This may lead to some oversampling, but the administrative advantages may more than offset this loss.

A Poisson acceptance-rejection test. In a Poisson distribution, the mean m may be calculated from density per unit of volume and volume, from density per unit of time and time, from density per unit of length, from density per unit of area and area, or other similar situations. The problem described is one in which the mean is based on density per unit volume. The sample size is calculated and a sample rule determined which will accept or reject a batch from the production process to give the amount of protection required. The illustration used involves the control of the production of a vaccine so that the number of live viruses will be less than a specified mean number per unit of volume, although the method is obviously applicable to other similar situations.

A random sample of v milliliters (cubic centimeters) is selected from a batch of vaccine of V milliliters to determine whether to reject the batch or not. If x, the number of live viruses in the sample, is one or more the batch is rejected. The problem is to determine v, so that if n_p a specified number of live viruses exist per liter, the probability of rejection (of getting one or more in the sample) is very high, say 0.95 or more. The probability of getting no live viruses in the sample is very low, say 0.05 or less.

Assume x the number of viruses per liter is scattered uniformly and independently throughout the batch so that the proportion $P = v/V$ is small and constant or approximately so and x is a Poisson variate. From the Poisson distribution, where m is the mean:

$$p_x = e^{-m} \frac{m^x}{x!},$$

and the sum of p_x is:

$$F_x = e^{-m}\left(1 + m + \frac{m^2}{2!} + \frac{m^3}{3!} + \ldots + \frac{m^x}{x!}\right),$$

$$x = 0,1,2,\ldots,\text{infinity}.$$

If $c = x = 0$, set p_0 equal to some small probability B:

$$P_0 = e^{-m} = B.$$

This means:

$$m = \ln \frac{1}{B}.$$

The probability of x being greater than zero is $1 - e^{-m} = 1 - B$. This mean m becomes:

$$m = n_p v/1000,$$

where m is the mean number of viruses per liter or 1000 cubic centimeters, n_p is the number of live viruses per liter, and v is expressed in milliliters or cubic centimeters. Hence:

$$e^{-m} = e^{-n_p v/1000} = B.$$

Therefore, the sample size in milliliters or cubic centimeters is:

$$v = \frac{1000}{n_p} \ln \frac{1}{B} = \frac{1000m}{n_p}.$$

Values of v can be calculated given individual values of n_p and B. A table below shows values of v for five values of n_p and four values of B for each value of n_p. The table shows the large increases in sample sizes needed if the probability of B is increased from one in 20 to one in 200.

This problem assumes that the values of m or n_p are known or are closely approximated from prior tests or experiments. The first and most realistic problem in practice is that of designing a sample test or experiment that will yield satisfactory estimates of these two quantities.

That the value of n_p is very critical in determining the sample size in this type of problem is shown by the figures given in the table below. Obviously, a difference of only one in the value of n_p can make a large difference in the size of the sample v. These differences are greater at the one percent level than at the five percent level. The sample sizes are given in the following table for five values of n_p and four values of B for each value of n_p:

Sample Volume, v, in milliliters (ml)

n_p	B=0.005 m=5.2983	B=0.01 m=4.6052	B=0.02 m=3.9120	B=0.05 m=2.9957
3	1,767	1,536	1,304	999
4	1,325	1,152	978	749
5	1,060	921	783	600[9]
6	883	768	652	500
7	757	658	559	428

For the case in which c=x=1, in which rejection occurs if two or more objectionable particles are found:

$$F_x = e^{-m}(1 + m) = \beta,$$

$$v = \frac{1000m}{n_p}$$

It is not easy to solve for m in the first equation, but an approximate m can be found corresponding to B by using Poisson tables. Then, by using an electronic calculator with a key for e^x, a more refined value of m can be found. This was done in the table given below; first m was calculated and then substituted in the equation for v. For c=x=1 (acceptance number), the sample volume in milliliters and the mean for each of five values of B and four values of n_p are:

B	m	v in milliliters for n_p: 2	3	4	5
0.01	6.635	3318	2212	1659	1327
0.02	5.834	2917	1945	1459	1167
0.03	5.355	2678	1785	1339	1071
0.04	5.013	2507	1671	1253	1003
0.05	4.744	2372	1581	1186	949

It is possible to calculate m by using a programmable calculator. In calculating m for c=x=1, use:

$$e^{-m} = \frac{B}{m+1}.$$

For c=x=2, use:

$$e^{-m} = \frac{2B}{(m+1)^2 + 1}.$$

In both cases increment m until the equality is met — until the right hand side just exceeds the left hand side.

Notes

[1]H. F. Dodge and H. G. Romig, *Sample Inspection Tables*, Wiley, New York, 1959.

[2]Military Standard, MIL-STD-105-D, *Sampling Procedures and Tables for Inspection by Attributes*, U.S. Department of Defense, Washington. Sections of this Standard are given in the appendix.

[3]Reported in a talk before a Middle Atlantic Conference of the American Society for Quality Control, as heard and recorded by the author.

[4]From the author's actual observation of, and conversation with, a receiving inspector in this factory.

[5]General Electric Company, *Tables of Poisson Distribution*, Van Nostrand, Princeton, N.J., 1962. Also E. C. Molina, *Poisson's Exponential Binomial Limit*, Van Nostrand, Princeton, N.J., 1942.

[6]The MF method originated by the author is explained under the hypergeometric distribution in Chapter 14.

[7]Same as reference 2 above. Some of the material is reproduced in the appendix. A considerable amount of the basic material is reproduced in the Burr reference in footnote 8 below.

[8]Two recommended references are E. L. Grant and R. S. Leavenworth, *Statistical Quality Control*, 4th edition, McGraw Hill, New York, 1972 and I. W. Burr, *Statistical Quality Control Methods*, Marcel Dekker, New York, 1976.

[9]This problem, where n_p=5, is in K. A. Brownlee, *Statistical Theory and Methodology in Science and Engineering*, John Wiley, New York, 2nd edition, 1965, p. 185. Only the case for B=0.05 or five percent is discussed. See also, E. Parzen, *Modern Probability Theory*, John Wiley, New York, 1960, pp. 106-107, for a discussion of the same problem.

22

Detection Sampling

Sampling for detection. Situations occur where it is important to detect the existence of at least one anti-quality event or situation in a random sample selected from a population or work place, or to uncover a potentially dangerous situation in operations. Examples of the former are:

- One error may be costly, as in financial accounts and records.
- One defect may be serious, as in a manufactured product.
- One departure or deviation from instructions or procedures may be critical, as in laboratory testing, computer programming, health services, or power plant operation.
- One flaw may require changes or revisions, as in a data sheet or questionnaire.
- One case or condition may be harmful or unsafe, as in adulterated food, food additives, air and water pollution, or mislabeling of a drug.
- One machine or employee not working may be critical.

In the latter case, poor identification, labeling, packaging, storage, or failure to recognize dangerous operating conditions may lead to errors that are harmful or fatal. Examples are found in transportation and health services. The purpose is to *prevent* errors, not detect them.

Often the detection of the first type of situation is a warning that a serious situation, and not a single isolated event, actually exists. The same may be true of the second type of situation.

It should be noted that detection throws no light on the actual magnitude of the problem, frequency of occurrence, or the amount of sampling and other errors.

One serious error in a financial account or record, one serious error in the data collection system, one deviation from instructions or procedures, one instance of misunderstanding of a data sheet or questionnaire, one error in a computer program, or one error made in an operation may call for immediate corrective action without any further data collection.

In most, if not all cases, *detection should lead to a prevention program,* if such a program does not already exist.

Sampling for detection is directly related to *escape probability* — the probability of an event either never being detected or being detected very rarely. Sampling for detection is aimed at preventing high escape probabilities for a wide variety of anti-quality situations such as critical defects, serious errors, unsafe products, hazardous conditions, serious deviations from procedures and processes, tax evasion, violation of financial and other accounting rules and controls, violations of eligibility rules in government financial and other aid programs, and other highly significant events whose frequencies of occurrence may vary widely.

This requires that a sample study be designed, or a data collection system or monitoring system be established, that will reduce the probability of escape to a very low level, say one percent or less.

Probability distributions for detection. Three probability distributions are described that can be applied to this problem: binomial, Poisson, and hypergeometric. Interest lies in the probability p_o of a random sample of size n detecting none of the critical events in the population and in the probability $1-p_o$ that the sample will contain at least one such event.

Binomial. The probability of getting no event of a specified type in a random sample of size n is:

$$p_o=(1-p)^n=q^n.$$

Therefore, the probability of getting at least one event in this sample is:

$$(1-p_o)=(1-q^n)=1-(1-p)^n,$$

where p = probability of occurrence of the event and is constant or nearly so; q = probability of the non-occurrence of the event or $1-p$; n = size of the random sample; and $p+q=1$.

Equate $(1-q^n)$ to some high probability such as $(1-a)$ where a is some small value such as 0.05, 0.01, or 0.001. Then, $(1-q^n)=(1-a)$. Solving for n gives the size of sample needed to detect at least one event with probability $(1-a)$:[1]

$$n=\frac{\log a}{\log q}.$$

This expression may be expressed in terms of integers x. Let $p=\frac{1}{x}$ and $q=1-\frac{1}{x}=\frac{(x-1)}{x}$. Then the probability of getting at least one event in a random sample of n is:

$$1-\frac{(x-1)^n}{x^n}.$$

Equate this to a high probability $1-\frac{1}{b}$, where b is an integer such as 20 for 0.05, 50 for 0.02, 100 for 0.01, and 1000 for 0.001. Then the sample size to detect at least one event with probability $1-\frac{1}{b}$ is:

$$n=\frac{\log b}{\log x-\log(x-1)}=\frac{\log a}{\log q}.$$

This is exactly the same as the preceding equation, but is based on integers instead of on decimals.

Poisson. In the Poisson distribution, the probability of getting exactly x specified events is:

$$p(x)=e^{-m}\frac{m^x}{x!},$$

where m is the mean. The probability of getting $x = 0$ is $p_o = e^{-m}$, so that the probability of getting at least one event is:

$$1 - p_o = 1 - e^{-m}.$$

Equate this quantity to a high probability such as $(1-a)$, where a is some small value such as 0.05, 0.01, or 0.001. But the mean $m = np$, where p is the probability of occurrence of the event and n is the random sample size. Therefore, $1 - e^{-np} = (1-a)$ and the sample size required to obtain at least one event with probability $(1-a)$ is:

$$n = \frac{1}{p} \ln \frac{1}{a},$$

where logarithms are taken to the base e because this quantity appears in the Poisson distribution.

Example 1: Find the size of sample needed if $p = 0.05$, $q = 0.95$ and $a = 0.05$ so $(1-a) = 0.95$:

$$\text{binomial: } n = \frac{\log 0.05}{\log 0.95} = 59.$$

$$\text{Poisson: } n = \frac{1}{0.05} \ln \frac{1}{0.05} = 60.$$

Hypergeometric. The hypergeometric distribution, described in detail in Chapter 14, gives the probability of obtaining exactly x units in a specified class in a random sample of size n. Selection is made without replacement of sampling units. In practice, it matches real situations better because:

- Sampling without replacement is the only feasible method or the only possible method.
- The population (lot) is finite in size.
- The sample may be a fairly large proportion of the population; if so, the sample size will be less than that obtained from the binomial or Poisson.

The following example illustrates two points: 1) determining sample sizes by rules of thumb such as 10 percent or $n = 1/p$ leads to a high escape probability and 2) the hypergeometric gives a smaller adequate size sample than does the binomial or the Poisson.

Example 2: In the following problem, $p = 10/200 = 0.05$ and $q = 0.95$. The sample size is taken as $1/p = 20$.

Class	Population or lot	Sample
A	10	0 = x
not A	190	20
	200 = N	20 = n

The exact probability of obtaining $x = 0$ is:

$$p_o = \frac{190(189)\ldots(171)}{200(199)\ldots(181)} = 0.3398.$$

$$\text{By the MF method: } p_o = \frac{(180.5)^{20}}{(190.5)^{20}} = 0.3401.$$

$$\text{Ave. binominal} = \tfrac{1}{2}\,(.3585 + .3209) = 0.3397$$

The binomial gives: $p_o = q^n = 0.95^{20} = 0.3585$.
The Poisson gives: $p_o = e^{-m} = e^{-1} = 0.3679$.
The probability of escape, getting $x = 0$, is very high with a sample of 20.

Example 3: What sample size is necessary to reduce the escape probability to five percent? Using trial and error to obtain five percent we have:

$$\text{By the MF method: } p_o = \frac{(190 - 25)^{51}}{(200 - 25)^{51}} = 0.0497.$$

Hence, $n = 51$.

binomial: $p_o = 0.95^{58} = 0.0510$.

Hence, $n = 58$.

Poisson: $p_o = e^{-3} = 0.0498$.

Hence, $n = 60$.

The MF approximation to the hypergeometric gives a sample size substantially lower than the binomial or Poisson, as it should since it takes into consideration the fact that the sample is a large proportion of the population or lot.

Problems

1. How many throws of a single die are required to obtain a six if $p = \tfrac{1}{2}$?

$$\text{Given: } p = \frac{1}{6},\ q = \frac{5}{6},\ a = \tfrac{1}{2};\ x = 6,\ b = 2.$$

Solution:

$$n = \frac{\ln 0.5}{\ln 0.8333} = \frac{\ln 2}{\ln 6 - \ln 5} = 3.8 = 4 \text{ rounded.}$$

2. How many throws of two dice are required to obtain a pair of sixes if $p = \tfrac{1}{2}$?

$$\text{Given: } p = \frac{1}{36},\ q = \frac{35}{36},\ a = \tfrac{1}{2};\ x = 36,\ b = 2.$$

Solution:

$$n = \frac{\ln 0.5}{\ln 0.972222} = \frac{\ln 2}{\ln 36 - \ln 35} = 24.6 = 25 \text{ rounded.}$$

3. How many throws of three dice are required to obtain three sixes if $p = \tfrac{1}{2}$?

$$\text{Given: } p = \frac{1}{216},\ q = \frac{215}{216},\ a = \tfrac{1}{2};\ x = 216,\ b = 2.$$

Solution:

$$n = \frac{\ln 0.5}{\ln 0.995370} = \frac{\ln 2}{\ln 216 - \ln 215} = 150.$$

4. How many throws of four dice are required to obtain four sixes if $p = 1/2$?

Given: $p = \frac{1}{1296}$, $q = \frac{1295}{1296}$, $a = 1/2$; $x = 1296$, $b = 2$.

Solution:

$$n = \frac{\ln 0.5}{\ln 0.999228} = \frac{\ln 2}{\ln 1296 - \ln 1295} = 898.$$

The sample sizes 4, 25, 150, and 898 increase in almost the identical ratio as x: 6, 36, 216, and 1,296; that is, each value after the first is six times as large as the preceding value.

The first two problems are of historical interest because it was the slight difference between 25/4 and 36/6 that led Chevalier de Méré, a French nobleman greatly interested in gambling, to send this problem in 1654 to Pascal who solved it. This event is usually considered the beginning of the science of probability. Pascal and another Frenchman, Fermat, were the first to lay the foundations of the theory of probability.

5. How many throws of two dice are required to obtain two sixes with a probability of 0.90? 0.95? 0.99? 0.999?

Given (common to all): $p = \frac{1}{36}$,

$$q = \frac{35}{36} = 0.972222, \; x = 36.$$

1. 0.90

Given: $a = 0.10$, $b = 10$.

Solution:

$$n = \frac{\ln 0.10}{\ln 0.972222} = \frac{\ln 10}{\ln 36 - \ln 35} = 82.$$

2. 0.95

Given: $a = 0.05$, $b = 20$.

Solution:

$$n = \frac{\ln 0.05}{\ln 0.972222} = \frac{\ln 20}{\ln 36 - \ln 35} = 107.$$

3. 0.99

Given: $a = 0.01$, $b = 100$.

Solution:

$$n = \frac{\ln 0.01}{\ln 0.972222} = \frac{\ln 100}{\ln 36 - \ln 35} = 164.$$

4. 0.999

Given: $a = 0.001$, $b = 1000$.

Solution:

$$n = \frac{\ln 0.001}{\ln 0.972222} = \frac{\ln 1000}{\ln 36 - \ln 35} = 246.$$

The sample sizes are proportional to log of b to base 10; the four values of b are 10, 20, 100, and 1000. Their logs to base 10 are 1, 1.3, 2, and 3. The sample sizes are in these exact ratios:

For 0.90: $b = 10$, $\log 10 = 1$, $n = 82$.
0.95: $n = 82 \log 20 = 82(1.30) = 107$.
0.99: $n = 82 \log 100 = 82(2) = 164$.
0.999: $n = 82 \log 1000 = 82(3) = 246$.

These problems illustrate how the sample size has to be increased if we want to increase the probability of detection and the probability of obtaining at least one of the specified events.

6. How large a sample is needed to detect at least one event that occurs one percent of the time ($p = 0.01$) with a probability of 0.95?

Given: $a = 0.05$, $q = 0.99$; $b = 20$, $x = 100$.

Solution:

$$n = \frac{\ln 0.05}{\ln 0.99} = \frac{\ln 20}{\ln 100 - \ln 99} = 298.$$

7. How large a sample is needed to detect at least one event with a probability of 0.99 if it occurs 1/10 of the time? 1/100? 1/1000? 1/10,000?

Given the same for all: $a = 0.01$, $b = 100$.

1. $p = 1/10 = 0.10$.

Given: $q = 0.9$, $x = 10$.

Solution:

$$n = \frac{\ln 0.01}{\ln 0.9} = \frac{\ln 100}{\ln 10 - \ln 9} = 44.$$

2. $p = 1/100 = 0.01$.

Given: $q = 0.99$, $x = 100$.

Solution:

$$n = \frac{\ln 0.01}{\ln 0.99} = \frac{\ln 100}{\ln 100 - \ln 99} = 459.$$

3. $p = 1/1000 = 0.001$.

Given: $q = 0.999$, $x = 1000$.

Solution:

$$n = \frac{\ln 0.01}{\ln 0.999} = \frac{\ln 100}{\ln 1000 - \ln 999} = 4{,}603.$$

4. $p = 1/10{,}000 = 0.0001$.

Given: $q = 0.9999$, $x = 10{,}000$.

Solution:

$$n = \frac{\ln 0.01}{\ln 0.9999} = \frac{\ln 100}{\ln 10{,}000 - \ln 9{,}999} = 46{,}050.$$

The sample is roughly increased 10 times as p is reduced by 1/10.

Problem 7 illustrates the difficult position faced by those in cancer research, and other similar fields and situations, involving the detection of rare events — events that occur 1/100 times, 1/1,000 times, or 1/10,000 times. Biologists, geneticists, and others accustomed to using 10 to 50 test animals in an experiment find that this situation requires very large samples if at least one case is to be detected with high probability. Problem 7 shows that if $p = 1/100$ in the population and the probability of detection is set at 0.99, then the sample size is 459. For the same probability of detection, an event occurring 1/1000 requires a sample of at least 4600; if the event occurs only 1/10,000, then the sample size must be at least 46,000. Even if the probability of detection is reduced to 0.95 and the event occurs 1/100, the sample still has to be 298 as shown in problem 6.

These very large sample sizes are one of the reasons, if not the major reason, why experimenters in these fields are using the massive dosage techniques, such as 50 times normal, as well as other accelerated techniques, to keep the sample small and obtain results in a hurry. Suffice it to say that some scientists, if not a large number, question the validity of an experimental technique that uses 50 times normal dosages — the technique, incidentally, that put cyclamates off the market. Apparently no tests have been made on common foods, drugs, drinks, and chemicals to determine if 50 times normal intake in test animals or human beings create pathological conditions, e.g., carcinogenic effects. It is only fair to state that consideration is being given by some scientists to the use of much larger samples of test animals in detecting rare characteristics.

These large samples also explain, in part, why the 1955 nationwide test of Salk vaccine for polio, with a rate of occurrence of about 1/2000, required a sample of more than a million children. This study is described in detail in the chapter on control through testing effectiveness.[2]

8. This is an actual problem faced by the Internal Revenue Service. They wanted to know if a sample could be used to detect tampered whiskey barrels in a bonded warehouse containing about 100,000 barrels, if it was assumed that about 1/4000 barrels had been tampered?

Given: $a = 0.01,\ q = \frac{3999}{4000} = 0.999750;$

$b = 100,\ x = 4000.$

Solution: $n = \frac{\ln 0.01}{\ln 0.999750} = \frac{\ln 100}{\ln 4000 - \ln 3999} = 18{,}419$

Actually, this was the wrong question because they already knew barrels were being tampered. They really wanted to know how extensive the tampering was. This called for determining whether sampling could be used to *estimate* the extent of tampering. The problem was *not* to discover whether one or more barrels were being tampered. There were also two economic problems: Was it worth the cost of trying to determine the magnitude of the problem? Once the problem was defined, was it worth the price of trying to reduce or eliminate it? It might have been because the federal tax was then $10.50 per gallon.

Although control by estimation is discussed in the next chapter, the calculation of the sample sizes needed for estimation in this problem are given below, using the hypergeometric distribution format:

Class	Population	Sample
tampered	25	x
not tampered	99,975	n − x
	100,000	n

The expected value of $x = \frac{25n}{100{,}000}$. If we allow a tolerance of two barrels so expected $x = 23$, then $n = 92{,}000$. If the tolerance is five barrels, $n = 80{,}000$. If the correct value is 50, instead of 25, then a tolerance of five will still give a sample size of 90,000. In other words, this means 100 percent inspection, not sample inspection.

Use of probability tables for stop-or-go sampling.[3] These tables give the probability that the error rate in a known or approximated population or lot of size N is less than p percent. They allow an inspector, auditor, or other person to accumulate a sample size n and number of errors x until a satisfactory maximum value of p is obtained with an acceptable probability; or until the number of errors in n is so large as to reject the lot or population or conclude that error rate is so high as to warrant conclusion controls or other procedures are not operating properly. Examples to follow illustrate the uses of tables. Tables are based on random selection of samples, preferably using tables of random numbers. Table 1 reproduces probabilities rounded to nearest 0.1 for $N = 1000$, $n = 50$, and population proportions from 0.01-0.10.

Table 1
Probability that Error Rate for Population N = 1000 Is Less than Given Proportion p for n = 50

Number of errors x	Proportion p 0.01	0.02	0.03	0.04	0.05	0.06	0.07	0.08	0.09	0.10
0	40.3	64.5	79.0	87.7	92.8	95.8	97.6	98.6	99.2	99.6
1	8.5	26.4	44.9	60.6	72.8	81.7	88.0	92.3	95.1	96.9
2	1.1	7.4	18.6	32.3	46.3	58.9	69.6	78.1	84.6	89.4
3	0.1	1.5	5.8	13.4	23.6	35.2	47.0	58.0	67.6	75.7
4		0.2	1.4	4.5	9.8	17.5	26.8	37.1	47.4	57.3
5			0.3	1.2	3.4	7.3	13.0	20.4	29.0	38.3
6				0.3	1.0	2.5	5.4	9.6	15.4	22.6
Binomial (x = 0) $p_o = q^n$	39.5	63.6	78.2	87.0	92.3	95.5	97.3	98.5	99.1	99.5
Poisson (x = 0) m = 50p	39.3	63.2	77.7	86.5	91.8	95.0	97.0	98.2	98.9	99.3

Example: A random sample of 50 records is selected at random from a total lot or population of 600 records and one error is found in the 50. The table for N = 600 gives a probability of 73.4 percent (x = one error) that the population error rate is less than five percent — that is, at least 95 percent of the records are correct. The probability is 45.2 percent that the population error rate is less than three percent.

Example: If the probability of 73.4 percent is too low in the previous example, increasing the sample size to 100 by taking an additional random sample of 50 with an additional error giving a total of two errors, increases the probability to 90.3 percent that the population error rate is less than five percent.

Example: If N = 1000 and the error rate in the lot or population is less than two percent due to the nature of the records, a random sample of 210 gives a probability of 94.5 with one error and 82.6 with two errors. A random sample of 300 is required to give a probability of 99.3 percent for one error, 96.6 percent for two errors, and 89.5 for three errors that the error rate is less than two percent.

Example: What is the smallest sample size n with x = 0 (no errors in sample), if N = 800 and at least 95 percent probability is desired for a maximum error rate of one percent? two percent? three percent? four percent? five percent?

one percent; x = 0; 95.0 percent; n = 249
two percent; x = 0; 96.5 percent; n = 150
three percent; x = 0; 96.1 percent; n = 100
four percent; x = 0; 95.0 percent; n = 70
five percent; x = 0; 95.9 percent; n = 60

Example: Suppose in the first example, where N = 600 and n = 50, the number of errors is three, not one. The probability now that the error rate is less than five percent is only 23.4 percent. The probability that it is less than 16 percent is 97.4 percent and the probability it is less than seven percent is 47.1 percent — a 50 percent difference. The probability is now 50 percent that the error rate lies between seven and 15 percent inclusive, and 50 percent that it lies outside this range. This is a high risk to take as a usual rule. The answer is to decide that the error rate is too high and take appropriate action or increase the sample by another 100 so n = 150, in order to obtain better protection. Even if x = 5 for n = 150, the probability of an error rate less than six percent is 92.3 percent.

Sample sizes for 95 percent detection. Table 2 gives the sample sizes required to detect at least one unit in a specified class A for population proportions from 0.005 to 0.10 and for population or lot sizes from 200 through 5,000. These were calculated by the method described in detail below using the hypergeometric distribution and the median fraction MF method to obtain p_o's. Sample sizes vary most for the small proportions and very little for the larger proportions. The value of n is that giving the least deviation from 0.050; 134 values lie between 0.049 and 0.051, while nine lie outside these limits, but all round to 0.05. In practice the population proportion P is unknown, so an estimate p is used.

In a previous chapter the probability of getting no units in a specified class of size $N_a = Np$, where p is the population proportion of occurrence and sampling is without replacement, is of the form:

$$p_o = \frac{v(v-1)\ldots(v+k-1)}{y(y-1)\ldots(y+k-1)},$$

where the first terms y = N and v = N − n. We use the MF approximation to obtain p_o:

$$p_m = \frac{[v - (k-1)/2]^k}{[y - (k-1)/2]^k} = (v_m/y_m)^k.$$

Set this equal to "a". For any given problem only v_m is unknown, since y_m is determined from the problem, k is the number of terms, (1 − a) is the probability of getting at least one unit in the specified class and is set equal at the desired level, such as 0.90, 0.95, or 0.99 corresponding to values of a of 0.10, 0.05, and 0.01, respectively. Solve for v_m from:

$$\log v_m = \frac{\log a}{k} + \log y_m.$$

Find the value of v_m which gives $(v_m/y_m)^k$ nearest to the value of a. Solve $k \log (v_m/y_m) = \log a$; $a = \exp k \, 1n\, (v_m/y_m)$ if base e is used, or $a = 10 \exp k \log (v_m/y_m)$ if base 10 is used. At most, only two values of v_m need to be tested to determine which gives a value nearest to a: one when the result exceeds a and one when the result is less than a. Finally the sample size:

$$n = y_m - v_m.$$

Two examples illustrate the calculations: one where Np is an even number and one where it is odd. The method is general and can be applied to any values of N, p, k, and a met in practice.

Example: N = 4000, p = 0.005, Np = k = 20. Set a = 0.05 for 95 percent detection. The format is:

Class	Population	Sample
A	20	0
not A	3980	n
	4000	n

The problem is to calculate n. Set up the denominator to find y_m. Since k = Np = 20, 10 values are on each side of the median line y_m. Set up a decreasing series for the denominator beginning with N; the median lies halfway between the tenth and eleventh values:

	median line
	3435 = v
(4000)(3999) ... (3991)	3990

Hence, median y_m = 3990.5. Solve for v_m:

$$\ln v = \frac{\ln 0.05}{20} + \ln 3990.5; \; v = 3435.4$$

Use v = 3435 to correspond to 3990. Calculate:

$$p_m = p_o = \frac{(3435.5)^{20}}{(3990.5)} = 0.050033.$$

This meets the specification a = 0.05. No further testing is necessary with this value of p_o. Hence, the sample size is:

$$n = 3990 - 3435 = 555.$$

All of the values needed are given above. Completing the series in both the denominator and numerator is unnecessary since every value in the numerator is 555 less than the corresponding value in the denominator.

Example: N = 700, p = 0.03, Np = 21 and a = 0.05. Since k = Np = 21, the median fraction consists of the eleventh value in both numerator and denominator.

$$\frac{\quad\quad \Big| \text{median } 598 \Big|}{(700)(699)\ldots(691)\ \Big| \ 690 \Big|} = v = v_m$$

$$\ln v = \frac{\ln 0.05}{21} + \ln 690;\ v = 598.3.$$

Test 598 and 599:

$$\frac{(598)^{21}}{(690)} = 0.049533 \text{ (This is nearest to a = 0.05.)}$$

$$\frac{(599)^{21}}{(690)} = 0.051302.$$

We reject the second and accept the first. The sample size n = 690 − 598 = 92.

Sample size for 95 percent detection assuming the Poisson distribution. The Poisson distribution can be used to give sample sizes which give a probability of 95 percent of detecting one or more events in a random sample of size n. This assumes that the events, such as errors or defects, are distributed in the population or lots according to the Poisson distribution and that the lots are very large relative to the sample, say at least 20 times as large as the sample.

This means a Poisson distribution has to be found that gives a probability of 0.05 of getting x = 0 (Zero Errors or Defects) and a probability of 0.95 of getting one or more errors or defects, that is x = 1 or more. The distribution with this characteristic has a mean of three, or m = 3. Therefore:

$$np = 3 \text{ or } n = \frac{3}{p}.$$

Various values are assigned to the population proportions p and the corresponding sample sizes n are calculated. The value of p from 0.05 to 0.10 and the corresponding sample sizes are given in the table below:

p	n	p	n
0.005	600	0.04	75
0.006	500	0.05	60
0.0075	400	0.06	50
0.010	300	0.07	43
0.015	200	0.08	38
0.020	150	0.09	33
0.025	120	0.10	30
0.030	100		

These values are fairly close to, but larger than, the values given for the hypergeometric distribution. The table based on the latter distribution, however, gives smaller sample sizes because it takes into consideration the size of the lot as well as the fact that random sampling is being used without replacement.

Table 2
Sample Size for 95 Percent Detection of One or More Units in Class A

Population size N units	Population proportion in class A										
	0.005	**0.01**	**0.015**	**0.02**	**0.025**	**0.03**	**0.04**	**0.05**	**0.06**	**0.08**	**0.10**
200	190	155	126	105	89	78	61	51	43	33	27
300	259	189	145	117	98	84	65	53	45	34	27
400	310	210	156	124	102	87	67	54	46	34	27
500	348	224	164	128	105	89	68	55	46	35	28
600	378	235	169	131	107	91	69	56	47	35	28
700	402	243	172	134	109	92	70	56	47	35	28
800	421	249	176	135	110	93	70	56	47	35	28
900	437	254	178	137	111	93	70	57	47	35	28
1,000	450	258	180	138	112	94	71	57	47	35	28
2,000	517	277	189	143	115	96	72	58	48	36	28
3,000	542	284	192	145	116	97	73	58	48	36	28
4,000	555	287	193	146	117	97	73	58	48	36	28
5,000	564	289	194	146	117	97	73	58	48	36	28

The calculations are easy with a minicomputer or electronic calculator, if the machine has keys for ln x and e^x, or log x and 10^x or y^x, so logarithms to the base 10 can be used. Divide $(v_m/y_m) = F$, take ln F, multiply this value by k which gives the exponent x, and then use the e^x key to get p_o. The MF method is simple, accurate, and eliminates the use of logarithms of factorials and has general application.

How many is "at least one" or "one or more"? When a random sample of size n is used to detect at least one event with a probability of 95 percent, how many does "at least one" really mean? This question was raised earlier, but now we apply it to Table 2 using the Poisson probabilities as approximations. Note, in the data given below, where p_o rounds to 0.05, that the Poisson gives somewhat higher values; this means that probabilities given for x = 1,2,3, etc. tend to be slightly lower than they really are.

Poisson probabilities for x events

N	p	n	np	0	1	2	3	4	5
2,000	0.02	143	2.86	0.057	0.165	0.234	0.223	0.160	0.091
2,000	0.01	277	2.77	0.063	0.170	0.238	0.222	0.156	0.087
1,000	0.02	138	2.76	0.063	0.170	0.238	0.222	0.156	0.087
1,000	0.01	258	2.58	0.076	0.193	0.251	0.217	0.141	0.073
3,000	0.02	145	2.90	0.055	0.160	0.231	0.224	0.162	0.094
	0.01	284	2.84	0.058	0.165	0.234	0.223	0.159	0.091
5,000	0.02	146	2.92	0.054	0.158	0.230	0.224	0.163	0.095
	0.01	289	2.89	0.056	0.162	0.232	0.223	0.162	0.093

An examination of these values shows that the maximum probabilities are associated with x=2 and x=3, which together have a probability of occurring of about 0.45. The probability of one occurring is about the same as the probability of four occurring. Hence, the probability of getting at least one means that the chances are about 3.5 or 4.0 to one that x will equal two or more. This is due to the fact that these probability distributions are all skewed toward the right — toward the higher values of x. If six or more events occur in n, this is reason to suspect that p is higher than the specified value with a probability of about 90 percent. If zero or even one event occurs in n, this is reason to suspect that p is lower than specified with a probability of about 80 percent.

Audit for internal control. The auditor examines those records, entries and transactions, which in his judgment and experience, produce evidence that the internal controls are or are not effective. Test checks or samples are used for this purpose. Lack of control is evidenced by serious clerical errors, violations of accounting principles, failure to follow established rules or procedures, failure to correct known deficiencies, and other irregularities or defects. An isolated or rare violation does not mean a lack of control unless something very serious is indicated such as fraud, diversion, theft, or falsified entries. The discovery or detection of only one such instance is enough to take appropriate action. The problem of designing a sample to give adequate estimates of quantities such as amounts of money error in transactions is described in Chapter 9 and 15.

Random sampling is applied to a defined frame or population to decide whether or not internal controls are satisfactory. Two kinds of errors can be made by the auditor, whether using test checks or random samples: a satisfactory situation can be asserted to be unsatisfactory and an unsatisfactory situation can be asserted to be satisfactory. Of the two, the latter is potentially more serious, although in the first situation it would be possible, by taking various actions, to make a satisfactory situation quite unsatisfactory. These two kinds of errors can be controlled using sample plans as described below. These errors can exist whether random sampling is used or not, but random probability sampling eliminates the bias that may easily exist in a judgment sample and allows a sample to be designed based upon a specified margin of error and a specified probability of getting that error or something worse.[4]

The auditor proceeds as follows to answer the question "Do the sample data warrant accepting present controls as satisfactory?":

1. Sets the upper precision limit (maximum allowable error rate.)
2. Determines risks to be taken in making the two types of errors.
3. Formulates decision rule based on sample findings; this may be modified by other information and experience.
4. Determines population or lot or frame from which the sample is to be selected.
5. Determines the sample size.
6. Determines the number and percentage of errors, defects, and deficiencies in the sample.
7. If the sample shows that the upper limit is not exceeded decides controls are satisfactory, otherwise, takes appropriate audit action.

The examples below illustrate the risks involved in using random samples of various sizes and how these risks can be controlled.

Example: Assume a population error rate is four percent and take a random sample of 75; then, np = 3 and the probability of getting no defects is 0.0498 or five percent. Decide if x = 0 in a sample of 75 control is satisfactory, otherwise it is not. The upper limit is four percent and this will be exceeded five percent of the time; 95 percent of the time the maximum rate will be four percent or less. The Poisson distribution is being used.

The question arises: What is the probability of getting one or more defects, x equal or greater than one, calling for a decision of unsatisfactory controls when actually they are satisfactory; that is, the population rate is three percent or less.

Population p	n = 75 np	Probability x = one or more
0.01	0.75	0.528
0.02	1.50	0.777
0.03	2.25	0.895

When n = 75, the probability of finding one or more defects is high and so is the probability of deciding unsatisfactory compliance even though the proportions in the population are acceptable — three percent or less. Hence, using a decision rule based on x = 0 will lead to the error of rejecting acceptable quality situations.

Example: The way to reduce the high risk in the previous example is to increase the sample size and change the decision rule. Let n = 225; p, the maximum error rate is still four percent, np = 9, and the probability of x being equal to or less than four is 0.055. Hence, the decision rule is to accept the situation as satisfactory if four or fewer defects are found in a sample of 225; otherwise, take necessary audit action. The probability now of rejecting a satisfactory population situation is greatly reduced.

Population p	np	Probability x = 5 or more
0.01	2.25	0.072
0.02	4.50	0.297
0.03	6.75	0.665

This increased protection was obtained by increasing the size of the sample — in this case by three times. It illustrates that protection against both kinds of errors requires samples of substantial sizes.

The second type of error of accepting an unsatisfactory situation can easily arise if the sample is too small. The following table shows the probability of getting x = 0 when the sample sizes range from 10 to 75 and the population proportions range from 0.04 to 0.10, levels considered unsatisfactory:

Population p	Sample size						
	10	20	30	40	50	60	75
0.04	0.665	0.442	0.294	0.195	0.130	0.086	0.047
0.05	0.599	0.358	0.215	0.129	0.077	0.046	0.021
0.06	0.539	0.290	0.156	0.084	0.045	0.024	0.010
0.07	0.484	0.234	0.113	0.055	0.027	0.013	0.004
0.08	0.434	0.189	0.082	0.036	0.015	0.007	0.002
0.09	0.389	0.152	0.059	0.023	0.009	0.003	0.001
0.10	0.349	0.122	0.042	0.015	0.005	0.002	0.0004

The table shows that samples under 30 give high probabilities of non-detection in populations with four to 10 percent defective, or unsatisfactory. Only samples of 75 or above give probabilities of 0.95 or more of detecting at least one defect or error for all seven percentages, and therefore deciding that an unsatisfactory condition exists. However, as shown above, a sample of 75 is too risky for making correct decisions about satisfactory conditions where the maximum error rate is three percent or less. A much larger sample is needed to reduce the risk of rejecting a satisfactory situation.

Random time sampling to detect certain work and other situations. Random time sampling can be used to detect unsatisfactory conditions in a factory, office, work location, or other site. Random time sampling means selecting units of time from a time frame such as minutes or days from a working year. The minute model has the advantage that it can either be selected as a discrete instant to be counted or as a duration for measuring variables. In the present applications, the minute model represents instants which are identified at the beginning of the minute. The hypergeometric distribution is used assuming a working day of 480 minutes and N as the daily frame or population during which N_a are associated with activity A which is of interest:

Class	Population	Sample	Non-sample
A	N_a	x	$N_a - x$
not A	$480 - N_a$	n − x	$480 - N_a - n + x$
Total	480	n	480 − n

Interest lies in determining the sample size that will detect at least one situation in class A with high probability; this probability is $p = 1 - p_o$, where p_o is the probability of getting x = 0 in a random sample of n instants (minutes). The above reduces to the following when x = 0:

Class	Population	Sample	Non-sample
A	N_a	0	N_a
not A	$480 - N_a$	n	$480 - N_a - n$
Total	480	n	480 − n

The probability of getting x = 0 is:

$$p_o = \frac{(480 - N_a)!(480 - n)!}{(480 - N_a - n)!480!}$$

Example: In an actual case, a graphic arts section was without work daily for about 60 minutes due to poor work scheduling. How many random minutes (instants) are required in one day to detect this situation with a probability of at least 95 percent? Using the MF approximation:

$$p_o = \frac{(409.5)^{22}}{(469.5)^{22}} = 0.049;$$

hence, $1 - p_o = 0.951$ for n = 22.

It requires three random minutes for each of eight hours to detect this situation in one day.

If eight random minutes, instead of 22, are selected per day, how many days before the probability of detection is 95 percent? For n = 8,

$$p_o = \frac{(416.5)^8}{(476.5)^8}$$

= 0.3407. Hence, $1 - 0.3407^m = 0.95$; m = 2.8 or three days. So the number of minute days is 24, the same as before. What is the probability of discovery after five days? $1 - 0.3407^5 = (1 - a)$ so probability a = 0.005. The actual situation was discovered after five days although the situation would have been evident after three days. After five days, the probability of detection is (1 − a) or 0.995 or 99.5 percent. When this situation was discovered, the director had the supervisor take immediate corrective action.

Detecting deficiencies in questionnaires and data sheets. Data sheets received from freight stations in connection with a nationwide sample study of freight car supply and demand were in error. During the initial stages of the study, the error rate was between 30 and 40 percent. This was because the data sheet and instructions were misunderstood, the data were being collected for the first time, and some of the data had to be compiled since they were not in established reports or summaries. How many freight stations selected at random in a pre-test would have discovered an error rate of 30 percent with a 99 percent probability?

$$p = 1 - q^n = 1 - 0.70^n = (1 - a) = 1 - 0.01;$$
$$q = 0.70: (1 - a) = 0.99$$
$$0.70^n = 0.01 \text{ giving } n = 13.$$

A random sample of 39: 13 from the larger offices, 13 from the medium size offices, and 13 from the small offices would have been sufficient to reveal how the items on the data sheet and the instructions should be changed. It would have paid off handsomely in view of the high cost required to reduce the 30 percent error rate to an acceptable level of five percent or below. The 30 percent error rate meant that a manual review had to be used and that three out of 10 data sheets had to be mailed back to the freight station with detailed explanations and instructions as to how to eliminate the errors. Computer edit was out of the question since it would have increased costs tremendously; actually, it was tried and abandoned.

If a question on a questionnaire is misunderstood by 20 percent of the potential respondents in the frame or population, how large a random sample of respondents is required to detect at least one such case with a probability of 0.99?

$$p = 1 - 0.80^n = (1 - a) = 1 - 0.01 = 0.99;$$
$$0.80^n = 0.01. \text{ Hence, } n = 21.$$

A random sample of 21 should detect this situation. The sample sizes required to detect cases of this kind, when the proportions who misunderstood are as given, are:

p in error	sample size
0.50	7
0.40	9
0.30	13
0.20	21
0.10	44

Data sheets and questionnaires usually can be adequately pre-tested for serious deficiencies by using a random sample of no more than 100 units selected from the frame or population. Effectiveness is greatly increased by using stratification of the sample by factors which are likely to correlate highly with misunderstandings and discovery of defects; four or five strata with 20 to 25 allocated to each stratum may be necessary. Examples are level of schooling: college graduate, high school graduate, some high school, or elementary school; size of business such as small, medium, and large; area of residence such as farm, rural town, small cities, medium cities, large cities suburban, and large cities core area including racial and ethnic areas where language difficulties are very important; and geographical areas with ethnic and regional language differences.

Notes

[1]Logarithms can be taken either to base 10 (written "log") or to base e (written "ln").

[2]Paul Meier, "The biggest public health experiment ever," (1972), *Statistical Guide to the Unknown,* F. Mosteller, W. Kruskal, R. Link, R. Pieters, G. Rising, and Judith Panur, Holden Day, San Francisco, 1972, pp. 2-13.

[3]Tables for Stop and Go Sampling, U.S. Department of the Air Force, Auditor General, 1961, Government Printing Office, Washington, D.C. 75 cents. 72 pages plus appendix. Tables are based on the hypergeometric model.

[4]Donald M. Roberts, "A Statistical Interpretation of SAP No. 54," *Journal of Accountancy,* vol. 136, 1974, pp. 47-53. "SAP" means Statement on Auditing Procedure. An adaptation of material copyrighted 1974 by the American Institute of Certified Public Accountants, Inc.

23

Control by Estimates

Control by estimates means direct or indirect control over activities, conditions, operations, or environment by using estimates of important population characteristics. The common methods used to obtain these estimates are probability samples in tests, surveys, or studies, laboratory or field experiments, or some form of 100 percent coverage over a limited time, area, or space. Probability samples are preferable to spot checks or judgment samples because the power of probability and statistics can be brought to bear on the former, but not on the latter.

Control is exerted because these estimates furnish key or critical information that is required for important decisions, improved operations, better quality service, more effective supervision and management, corrective action, effective planning, and better work scheduling. The need for estimates grows out of several situations.

1. It is the only way to define many situations that need to be controlled because of their very nature, such as tax error, unemployment, air and water pollution, freight car shortages, delay time, down time, idle time, re-work, failure error rates and magnitudes, losses, poor service, customer complaints, defective purchases, and computer programming of mathematical and statistical formulas and equations. Management needs to know the level and extent of each key characteristic so that timely, effective and valid decisions are made.

2. Management needs sound information to define the problem. Any attempt to control is futile and wasteful if management tries to solve the worng problem.

Example: The president of a company blamed the sales force for declining sales, but a survey of the ex-customers showed that the cause of declining sales was a reduction in the quality of the product due to changing to a new vendor in order to save money. The situation was corrected by returning to the old vendor.

Example: A top manager has a test run made by one of the best machine operators. Using this information, he concludes that output should be a million dollars more than it is. He accuses the employees of stealing this amount of merchandise, and hires detectives and psychologists to study the situation. The test run was very biased, so the estimate was excessively high. A million dollars worth of goods had not been stolen; they had never been produced! He had an absurd estimate, but did not know it. His problem was neither legal nor psychological, it was statistical.

3. Management does not know how to hire the technical capability needed to obtain adequate estimates. The manager in the preceding example did not have a technical specialist who knew how to 1) use quality control charts, 2) use input-output analysis, and 3) design a sound test or experiment. Neither did he understand the situation well enough to hire a technical consultant who could have resolved his "problem" in short order. Actually, he was the problem.

Top level managers cannot solve statistical problems if they do not know what a statistical problem is. And they do not. This is why it pays to have competent technical specialists in statistics, quality control, and the design of experiments.

4. Information that management has about the problem, operation, or situation is incomplete, inaccurate, and misleading. This is not uncommon. The information is not in the right form, nor in sufficient amounts, due to conceptual differences. A source of data may appear to be applicable when it is not. If data are not applicable, no amount of adjusting, manipulating, or calculating will make them so.

Adequate data can be obtained in many instances, as ample experience shows, without requiring a costly time-consuming study depending, of course, on the nature of the problem and what information is required. Well-designed and well-managed probability sample studies may be the answer. However, if management wants acceptable quality data for numerous classes and subclasses, it should be prepared to pay the price.

Actual examples of where estimates were used for control purposes. Several examples are listed below to show that estimates are used for the purpose of control — indeed, they are absolutely necessary for many kinds of control.

1. Errors on federal individual income taxes reported on Form 1040 — kinds, frequency, totals, averages, and classification by types of taxpayers.
2. Amounts of carbon steel allotted to an industry during wartime — War Production Board.
3. Freight car shortages, especially shortages of box cars in the Midwest grain states — Interstate Commerce Commission.
4. Amount of unemployment in states and in United States, by sex, race, residence, and locality.
5. Characteristics of automobile traffic moving through toll booths at a bridge or terminal.
6. Amount of carbon monoxide in the air on a city street.
7. Amount of infestation in imported cocoa beans, by country of origin.
8. Whiskey output, as a function of corn input, in a distillery.
9. Daily number of acceptable quality documents transcribed, as a function of time.
10. Amount of plutonium and other radioactive substances in air, water, and soil in a specified area.
11. Waste water being discharged from a manufacturing, utility, or other plant or facility.
12. Testing drinking water source for a city or community.
13. Making an inventory check to determine the types of commodities, and their money value, which are being stolen by shoplifters and employees.

Estimates for action. Many times management needs estimates of key characteristics as a basis for needed action. Examples are estimates of cost, error rates, percent defective, failure rates, or customer complaints. More specific examples are daily volumes of acceptable quality product, sample audit findings, shortages of various types of freight cars, number of unemployed by age, sex, and occupation, and level of critical pollutants in air and water at key points or areas. One of the best examples of the use of estimates to lay out a comprehensive course of action is the audit control program of the Internal Revenue Service based on a nationwide sample of about 160,000 Forms 1040 (Case 5 below). The major purpose of this sample study was to define the problem of non-compliance in quantitative terms and to determine the major sources of tax error for better allocation of audit resources.

The problem called for the sample to provide estimates of the number of taxpayers in non-compliance, the total amount of tax error, and the average amount of tax error per return for a wide variety of subdivisions of taxpayers, such as business and non-business, various types of business, or size of income. Estimates were obtained for the major sources of error: unreported and under-reported income, personal exemptions, deductions, and arithmetical calculations on the tax returns, the latter being a very minor source of error compared with the other three.

An example of a taxpayer error, about which corrective action was taken, was that of claiming one or more aliens as personal exemptions. This is illegal, even though the taxpayer provides full support. Agents already knew about this situation, but the sample audit showed where it was and how large it was. Actions taken included a more careful listing and explanation of the conditions that had to be met to claim an individual as a personal exemption. A special schedule was added to Form 1040 to obtain relevant information on dependents.

Other specific actions taken included a provision for a two-year cycle of audit on high income returns, the segregation of business returns for a more careful scrutiny, provision for itemization of dividend and interest income on Form 1040, and proposed regulations to require corporations to file information returns on all dividend payments.

This sample study was not a mere compilation of numbers like a census. The estimates were made in such a form that action could be taken to improve taxpayer response, at least in a limited area; to allocate audit resources more effectively; and to give high-level management, for the first time, an accurate picture of the magnitude of non-compliance which could be used for improved policies, planning, and budgeting as well as operations. Estimates from a sample study were used to improve operations and actions in a significant aspect of tax administration.

Types of estimates. Estimates refer to estimates of population values which are unknown and are usually derived from probability samples, field tests, experiments, or 100 percent tabulations. The most common are the following:

1. Proportions or percents of the binomial or multinominal type.
2. Averages such as the arithmetic mean.
3. Aggregates or totals.
4. Counts or frequencies of occurrence.
5. Ratios, including proportion-type ratios.
6. Measures of variability, such as range, variance, standard deviation, and coefficient of variation.
7. Measures of relationship, such as correlation coefficient and regression coefficients, including the slope and intercept of a straight line.

An efficient sample design for one of these estimates is not necessarily the best sample design for another estimate. If a study or problem involves more than one kind of estimate, it is necessary to design an adequate sample for each estimate and then use the largest sample size obtained. This insures that all estimates meet or better the required specifications.

Sampling error or sampling variation. In repeated random samples, each estimate varies and, hence, forms a frequency or sampling distribution with a mean and standard deviation — the latter is called the *standard error.* The former tends to be the population mean; the latter is expressed by a known mathematical equation. If these two quantities are designated $\bar{x}$ and $s_{\bar{x}}$, then the *relative sampling error* $c_{\bar{x}} = s_{\bar{x}}/\bar{x}$, which is the proportion the standard error is of the mean.

Every estimate derived from a probability random sample has to be evaluated in terms of its sampling error or standard error. Most estimates derived from samples should have a relative sampling error of five percent or less; this does not mean that estimates with larger relative errors may not be useful. It depends upon many factors, one of which is the magnitude of the *non-sampling or biased* error in the basic data. There is no point in worrying about a 10 percent sampling error when the bias in the basic data is 20 percent. Many examples can be cited of instances like this one, or even worse. In these situations the problem is to reduce the bias and stop worrying about the sample.

Eight cases. Eight cases are described in some detail to illustrate how estimates are derived and used in some aspect of managerial or operational control. The name of the study, the estimates used, and the nature of the control are described below:

Case 1. Consumer preference study. This was a market study based on a nationwide sample study designed to estimate a proportion so that action could be taken relative to an important consumer preference.

Case 2. Estimating the number of fish in a lake or animals in an area. The capture-recapture method is used to estimate the number of fish in a lake or animals in an area. The purpose is environmental control.

Case 3. Distillery characteristics. A sample was used to estimate aggregates, averages, and relationships to control weights, alcohol content, and input for purposes of government regulation.

Case 4. Industry shipments. A stratified random sample of companies comprising an industry was used to estimate aggregate dollar shipments and shipments of critical materials for wartime control by the War Production Board.

Case 5. Tax audit control program. A nationwide sample of individual income tax returns was used by the Internal Revenue Service to estimate aggregate and average amounts of tax error for the purpose of defining the audit problem, defining how big the non-compliance problem was, and improving allocation of audit resources to reduce errors on tax returns.

Case 6. Railroad diversion traffic study. A replicated sample was used by a railroad to obtain the estimated amount of money to be gained by a proposed merger, the data to be used in a legal proceeding.

Case 7. Rail traffic characteristics. This is a layout of calculations for computer programming control to obtain accurate estimates of ratios and aggregates.

Case 8. Motor carrier traffic study. This sample was used to obtain totals and ratios on traffic and costs for use in a legal proceeding.

CASE 1

National consumer survey. This was a market study. A nationwide sample survey of women in households was conducted to estimate the proportion answering "yes" to the question "Do you make your own cake frosting at home?" The sample consisted of 400 area segments divided into two independent parts or replicates of 200 segments each. All women in households in these sample segments were contacted. The use of dwelling units made it possible to use the United States Bureau of the Census figures to make ratio estimates of the total number of women.[1]

Each replicate was also divided into eight geographical regions, four for metropolitan areas and four for non-metropolitan areas. The total number of interviews was about 1,200; the total number of dwelling units in the 400 segments was 1,961.

The estimated proportion of women preferring to make cake frosting at home was 0.40, with a standard error of 0.01. Assuming p/s_p is distributed as the t distribution with seven degrees of freedom gives confidence limits at the 99 percent level of 0.365 and 0.435. The basic data with the calculation of p and its standard error are shown in Table 1.

In Table 1, subscript "1" stands for replicate one and subscript "2" stands for replicate two; x_{i1} and x_{i2} are the numbers of housewives answering "yes" in replicates one and two respectively; y_{i1} and y_{i2} are the corresponding numbers of households; and the index "i" is summed over the eight geographical regions. For one such area the proportion:

$$p_i = \frac{x_{i1} + x_{i2}}{y_{i1} + y_{i2}}.$$

The estimate of the proportion p for the entire sample is:

$$p = \frac{\sum_i (x_{i1} + x_{i2})}{\sum_i (y_{i1} + y_{i2})}.$$

For the ratio estimate $p = x/y$ without correction for finite population:

$$s_p^2 = \frac{1}{n(n-1)\bar{y}^2} \sum\left[(x_i - \bar{x}) - p(y_i - \bar{y})\right]^2.$$

Equating $n(n-1)$ to n^2 and using the differences between pairs of x's and y's gives:

$$s_p^2 = \frac{1}{n^2\bar{y}^2} \sum\left[(x_{i1} - x_{i2}) - p_i(y_{i1} - y_{i2})\right]^2 = \frac{S^2}{y^2},$$

where $y = n\bar{y}$; $i = 1,2,\ldots,8$; and $\sum h_i^2 = S^2$.

Also, $h_i = (x_{i1} - x_{i2}) - p_i(y_{i1} - y_{i2})$. Therefore: $s_p = \frac{S}{y}$.

These symbols correspond to those given in Table 1.

Table 1

Data and Calculations for Consumer Survey

Item	Metropolitan NE	Metropolitan NC	Metropolitan S	Metropolitan W	Non-metropolitan NE	Non-metropolitan NC	Non-metropolitan S	Non-metropolitan W	U.S.
i	1	2	3	4	5	6	7	8	1-8
x_1	98	100	40	38	35	33	47	33	424
x_2	79	69	38	45	17	38	52	29	367
Sum x_j	177	169	78	83	52	71	99	62	791
Difference	19	31	2	−7	18	−5	−5	4	57
y_1	213	203	123	87	64	94	174	62	1020
y_2	176	162	110	89	55	128	165	56	941
Sum y_i	389	365	233	176	119	222	339	118	1961
Difference	37	41	13	−2	9	−34	9	6	79
$p_i = \frac{\Sigma x_i}{\Sigma y_i}$	.455	.463	.335	.472	.437	.320	.292	.525	.403
$h_i = (x_{i1} - x_{i2}) - p_i(y_{i1} - y_{i2})$	2.165	12.017	−2.355	−6.056	14.067	5.880	−7.628	.850	25.163
$S^2 = \Sigma h_i^2$									482.680
S									21.970
$s_p = S/\Sigma y_i = S/y = 21.97/1961$									.011

Every area gives two independent estimates of the proportion in question, one for each replicate. From these two estimates, a range may be obtained and the average of the eight ranges used to estimate the standard error of the over-all estimate p. This method gives the same standard error as that described above.

Area	p_1	p_2	Difference (range)	
1	0.460	0.449	0.011	
2	0.493	0.426	0.067	
3	0.325	0.345	0.020	$p_1 = x_1/y_1$
4	0.437	0.506	0.069	
5	0.547	0.309	0.238	$p_2 = x_2/y_2$
6	0.351	0.297	0.054	
7	0.270	0.315	0.045	
8	0.532	0.518	0.014	
Sum			0.518	
Mean Range $\bar{R}$			0.0648	

Using the range to estimate the standard error:[2]

$$s = \frac{\bar{R}}{d_2(n)^{1/2}}.$$

For $n = 8$, $d_2 = 2.847$:

$$s = \frac{0.0648}{2.847(8)^{1/2}} = 0.01.$$

Area five contributes 41 percent to S^2 (as shown in Table 1); it contributes 46 percent of the sum on which $\bar{R}$ is calculated.

It may be observed that the expression for a proportion derived from a binomial model $[(0.40 \times 0.60)^{1/2}/1961^{1/2}]$ also gives a standard error of 0.01, but this is merely happenstance. The appropriate model obviously is the ratio type estimate.

CASE 2

Estimating the number of fish in a lake. A method used in environmental control is the capture-recapture technique to obtain counts of fish in a lake or animals in an area. First, a sample of fish is caught, marked in some way, and then returned to the lake. After a time interval, a second sample is caught and the number marked recorded. The hypergeometric distribution applies to this situation:

Class	Population	Sample
A (marked)	n_a	m
not A	$x - n_a$	$n - m$
Total	x	n

The only unknown is the population count x, n_a is the first sample which is marked 100 percent and is a population value, n is the second sample, and m is the number marked found in the sample n. The total x is estimated by assuming $m/n = n_a/x$ — the proportion found in the sample is a good estimate of the same proportion in the population. This is true only if those marked are scattered through the population area, there is no clustering, and n is selected at random from the entire population and is large.

Hence $x = nn_a/m$ or, since n and n_a are constants, $x = (nn_a)\frac{1}{m}$ Then the variance of x is $s_x^2 = (nn_a)^2 s_{1/m}^2$ which is approximated by:

$$s_x^2 = (nn_a)^2 \frac{n}{m^4}\left(\frac{m}{n}\right)\left(1 - \frac{m}{n}\right)(1 - f)$$

$$= \frac{n}{m^3} n_a^2 (n - m)(1 - f),$$

where $(1 - f)$ is the correction for finite population sampling. Hence, the standard error:

$$s_x = \frac{nn_a}{m^2}\left[\frac{m}{n}(n - m)(1 - f)\right]^{1/2}.$$

The relative sample error s_x/x is a simple expression:

$$c = \left(\frac{1}{m} + \frac{1}{x} - \frac{1}{n_a} - \frac{1}{n}\right)^{1/2}.$$

All three values m, n, and n_a need to be large to obtain a very good relative error. If $m = 10$ and $n = n_a = 100$, then $x = 1{,}000$ and $c = 28$ percent, a value much too high. If all values are increased 10 times so $m = 100$, $n = n_a = 1{,}000$, and $x = 10{,}000$, then c is 9 percent, which is more satisfactory.

Considering m a binomial variate where $s^2 = npq(1 - f)$; $p = m/n$ and $q = 1 - p$:

$$s_m^2 = n\left(\frac{m}{n}\right)\left(1 - \frac{m}{n}\right)(1 - f).$$

Hence,

$$s_m = \left[\frac{m}{n}(n - m)(1 - f)\right]^{1/2}.$$

Use $m - s_m$ to estimate x_1 and $m + s_m$ to estimate x_2. Then, an estimate of the standard error of x based on m and its standard error is:

$$s_x = (x_1 - x_2)/2.$$

This gives a range of x equivalent to two standard errors of m, but gives the same relative sampling error that exists for m.

Example: This method was used to estimate the fish population in Dryden Lake in New York State.[3] One type of fish studied was the large mouth bass. The first catch (sample) was made between mid-September and mid-November, 1970. Identifying marks were made on the fins before the fish were returned to the lake. There was no migration into or out of the lake, and the marked fish were assumed to have the same death rate as the unmarked fish and to be thoroughly mixed with the other fish before the second sample was selected. It should be observed that the marked fish returned to the lake constitute a class of the entire population of fish in the lake. The data for the bass were as follows:

Class	Population	Sample
A (marked)	$213 = n_a$	$13 = m$
not A (not marked)	$x - 213$	$91 = n - m$
	x	$104 = n$

The estimate of the fish population is:

$$x = \frac{213\,(104)}{13} = 1704.$$

The correction for finite population sampling is

$$(1 - f) = 1 - \frac{104}{1704} = 0.939.$$

The standard error of the estimate of 1,704 is:

$$s_x = \frac{104(213)}{13^2}\left[\frac{13}{104}(91)(0.939)\right]^{1/2} = 428.$$

The relative error is:

$$c_x = \frac{428}{1704} = 0.25 \text{ or } 25 \text{ percent.}$$

Assuming m is a binomial variate:

$$s_m^2 = \frac{13}{104}(91)(0.939); \text{ hence, } s_m = 3.268,$$

and

$c_m = \dfrac{3.268}{13} = 0.25$ or 25 percent (the same as for the population estimate). This means that:

$$m \pm s_m = 13 \pm 3.268 \text{ or } 16.268 \text{ and } 9.732.$$

Substituting these two values of m in $x = nn_a/m$ gives $x_2 = 1362$ and $x_1 = 2276$.

Hence:

$$s_x = \frac{2276 - 1362}{2} = 457$$

which compares with 428 above.

Assuming m is a Poisson variate gives:

$$s_m^2 = (m)(1-f) = 13(0.939), \text{ so } s_m = 3.494,$$

which compares with 3.268 for the binomial assumption.

Regardless of what method is used, the relative sampling error is m = 13 and x = 1,704 is about 25 percent which is much too high for any major point estimate required by management in other areas of government and business. This high relative error can be reduced by using larger samples, as the following examples show: However, it appears very difficult to get a relative error below 10 percent in any realistic problem.

Example 1

Class	Population	Sample
marked not marked	300 x − 300	40 160
	x = 1500	200

Example 2

Class	Population	Sample
marked not marked	300 x − 300	60 240
	x = 1500	300

In example 1, $x = \dfrac{300(200)}{40} = 1500.$

In example 2, $x = \dfrac{300(300)}{60} = 1500.$

In example 1, $(1\text{-}f) = 1 - \dfrac{200}{1500} = 0.867.$

In example 2, $(1\text{-}f) = 1 - 0.20 = 0.80.$

Example 1:

Standard error of x:

$$s_x = \frac{300(200)}{40^2}\left(\frac{40}{200}(160)(0.867)\right)^{1/2} = 197.$$

Relative error:

$$c_x = \frac{197}{1500} = 0.131 \text{ or } 13 \text{ percent;}$$

$$c_x = \left(\frac{1}{40} + \frac{1}{1500} - \frac{1}{200} - \frac{1}{300}\right)^{1/2} = 0.132.$$

Standard error of m:

$$s_m = \left(\frac{40}{200}(160)(0.867)\right)^{1/2} = 5.267.$$

$$m_1 = 34.733 \text{ and } m_2 = 45.267;$$

Range $m \pm s_m$ gives:

$$x_1 = 1{,}727 \text{ and } x_2 = 1{,}325.$$

Example 2:

Standard error of x:

$$s_x = \frac{300(300)}{60^2}\left(\frac{60}{300}(240)(0.80)\right)^{1/2} = 155.$$

Relative error of x:

$$c_x = \frac{155}{1500} = 0.103 \text{ or } 10 \text{ percent;}$$

$$c_x = \left(\frac{1}{60} + \frac{1}{1500} - \frac{1}{300} - \frac{1}{300}\right) = 0.103.$$

In these examples n_a, the number of marked fish; n, the size of the random sample selected; and m, the number of marked fish found in n, were all increased with very substantial reductions in the relative sampling error in the estimated population total. Unless large samples are used at both stages the relative error is going to be large. Even with the larger values used in example 2 the relative sampling error is still 10 percent and this required a 20 percent sample of the estimated population.

The reason for this is that the inverse of an estimated proportion is used as a weight to obtain the population estimate. It is well known that it requires relatively large samples to obtain estimates of proportions with relatively small sampling errors. Since $x = nn_a/m$, this can be written as n_a/p_m, where p_m is m/n and is a variable (not a fixed rate or proportion). Furthermore, the sampling variation in p_m is what creates the sampling variation in the estimate x.

Neyman[4] cites two examples of this method; in which large samples were used to estimate the number of salmon, but did not solve them. A sample of salmon swimming into Cotus Lake, Canada was marked and released. Later, it was found that in a sample of those that died a certain number was marked; the problem was to estimate the number that spawned and died. The data were as follows:

Example:

Class	Population	Sample
A (marked)	$7{,}809 = n_a$	1,529 died, marked = m
not marked	$x - 7{,}809$	12,050
	x	13,579 died = n

The estimates are:

1. Estimated population

$$x = \frac{7{,}809(13{,}579)}{1{,}529} = 69{,}351.$$

2. Relative error $c_x = (1/1509 + 1/69{,}351 - 1/13{,}579 - 1/7{,}809)^{1/2} = 0.021603$.

3. Sampling error $s_x = c_x x = 0.021603(69{,}351) = 1{,}498$.

Assuming m/n is a binomial proportion, results very close to the above are obtained. $p = m/n = 1{,}529/13{,}579 = 0.112600$; $s_p^2 = p(1-p)(1-f)/n$; $s_p = 0.002433$; $c_p = 0.021604$; and $f = n/x$. Therefore, $p \pm s_p = 0.110167$ and 0.115033. Substitute these two values into $x = n_a/p$ to obtain two values of x, two standard errors apart: $x_1 = 70{,}883$ and $x_2 = 67{,}885$. Hence, $s_x = (70{,}883 - 67{,}885)/2 = 1{,}499$, which compares with 1,498 above; $x = (70{,}883 + 67{,}885)/2 = 69{,}384$, which compares with 69,351. The final estimate is $69{,}400 \pm 1{,}500$.

Example: The total number of animals, such as bears, in a specified area may be estimated in a similar manner. In a national park, 40 bears are captured and colored collars are attached to them. They are then released throughout the park. After some time, a survey of the area is made by helicopter; 50 bears are sighted of which 10 have collars. Then, $n_a = 40$, $n = 50$ and $m = 10$. The population estimate is:

$$x = \frac{40 \times 50}{10} = 200,$$

but the relative error is 24 percent:

$$c_x = (1/10 + 1/200 - 1/40 - 1/50)^{1/2} = 0.24.$$

This may be, however, about the best that can be done under the circumstances.

CASE 3

Estimating distillery characteristics. Tests were made for regulatory purposes using data from four distilleries to show that random sampling could be substituted, due to very small standard deviations and coefficients of variation, for 100 percent weighing or measuring of whiskey in barrels. A comparison was made between average *taxable gallons* per barrel, based on complete coverage, and on a random sample with the following results:

Distillery number	Total barrels N	Random sample n	Random sample n/N	Tax gallons per bbl. N	Tax gallons per bbl. n	Tax gallons per bbl. diff.	Std. dev. s_x	Coeff. of var. c_x
1	2800	200	0.071	50.68	50.65	−0.03	0.61	0.0120
2	3950	200	0.051	51.20	51.25	0.05	0.90	0.0176
3	4800	76	0.016	54.10	54.18	0.08	0.63	0.0116
4	3550	142	0.040	55.38	55.39	0.01	0.66	0.0119

The taxable gallons estimated from the sample deviate very little from the values obtained from the entire lot. The largest deviation is only 0.15 percent and this is from the smallest sample. If the sample of 76 was increased to 200, the difference would be much smaller. Sampling is very effective because of the very small standard deviations or very small coefficients of variation, even though the sampling rates are two, four, five, and seven percent. Furthermore, the random differences will tend over time to zero.

The standard deviations and coefficients of variation are also very small for gross weight and net weight per barrel. The latter vary from 1.2 percent to 1.8 percent for the same four distilleries. This means that the *net weight per barrel* can also be estimated from relatively small samples. Consider distillery number three above, where 4,800 barrels may be weighed daily. Estimates from a random sample give: $\bar{x} = 382.9$ pounds, $s_x = 4.75$ pounds, and $c_x = 0.0124$. If the absolute deviation in the mean is set at 1.77 pounds at the 99 percent level, then the sample size is:

$$n = \frac{s^2 z^2}{d^2} = \frac{4.75^2 (2.58)^2}{1.77^2} = 48.$$

With this sample size, what is the expected sample variation in the estimate of the total weight, X, of 4,800 barrels? The number of barrels, N, is known since this count must be made. Assume the mean and standard deviation given above. Then:

$$X = 4800(382.9) = 1{,}837{,}920 \text{ pounds}.$$

The standard error of X:

$$S_X = N s_{\bar{x}} = \frac{4800(4.75)(1-0.01)^{1/2}}{(48)^{1/2}} = 3274 \text{ pounds}.$$

The relative error in X:

$$c_X = \frac{3274}{1{,}837{,}920} = 0.0018 \text{ or } 0.18 \text{ percent},$$

which is less than 1/5 of one percent.

A sample of 48 barrels, selected at random daily from 4,800 barrels is all that is needed to obtain a highly precise estimate of the total net weight. Weighing all 4,800 barrels is a tremendous waste of time, money, and other resources. If 3,274 is too large an absolute figure, then n=400, a sample of one in 12, reduces it to 1,091 pounds. This latter sample is really unnecessary on a daily basis, since the estimate is subject to random variations about the "true" figure which tends to zero in repeated sampling (providing the basic characteristics used above tend to remain stable).

A random sample of 48, selected daily from 4,800, can be selected in the form of eight replicates of six samples each. The eight replicates are selected with random starts from blocks or zones of 800 barrels each. Sample barrels are identified by counting or from an official record as the barrels are being handled. This replication provides for the direct calculation of the standard error of both means and totals, as well as providing a means for detecting any abnormalities or unusual measurements. Replication with eight random starts takes the following form:

Zone or block		Replicate 1	2	3	4	5	6	7	8
random start	1-800	362	468	49	106	553	184	390	707
(add 800)	801-1600	1162	1268	849	906	1353	984	1190	1507
etc.									

The first row of three digit random numbers is the first eight found in the range between 001 and 800 in a table of random numbers. They are *not* ordered, but assigned to the eight replicates in the order found in the table.

CASE 4

Estimating an industry characteristic. The problem was to estimate for wartime control the anticipated dollar shipments in an industry for the first quarter of a year, given the actual shipments and the sample shipments for the third quarter of the past year and the projected shipments for the first quarter for a stratified random sample of 50 companies. A quick estimate was needed because a final figure would not be available for months. These sample companies were allocated on an approximate optimum allocation basis to 15 strata into which the population of 381 companies was divided by size of their third quarter shipments. The third quarter distribution of shipments, by money value, was highly skewed with the maximum frequency in the interval below \$70,000. It was also steadily decreasing with an increasing volume of shipments to a maximum value of \$6,824,000. The five largest companies with shipments above \$2,700,000 were included 100 percent in a separate stratum. The same 50 companies were used in the sample for both quarters. (See Table 2.)

The results for the first quarter were:

1. Estimate of total volume: $X = \$161,061,000$.
2. Estimate of standard error: $s_X = 2,011,200$.
3. Coefficient of variation: $c_X = 0.0125$ or 1.25 percent.
4. Population value (obtained later) = \$162,267,000.
5. Estimate X was too low by \$1,206,000 or by about three-fourths of one percent.
6. 95 percent confidence limits, 161,061,000 ± 1.96(2,011,200), obviously includes the population value. The lower limit is \$157,119,000.

A ratio estimate was applied to each stratum expressed in the form:

$$X_j = \frac{N_j\bar{Y}_j}{n_j\bar{y}} n_j\bar{x}_j,$$

where $j = 1,2,3, \ldots, 15$ strata; $N_j, \bar{Y}_j$ = known third quarter stratum population; $n_j\bar{y}_j$ = known third quarter sample total for jth stratum; $n_j\bar{x}_j$ = aggregate or total from first quarter sample; these are the same sample companies used in the third quarter sample; and $w_j = (N_j\bar{Y}_j)/n_j\bar{y}_j$ (known values from the third quarter). This is the same form as $X_j = Y_j(x_j/y_j) = f_jY_j$, where x_j and y_j are sample totals whose ratio is f_j.

The standard error of the estimated aggregate and its coefficient of variation were calculated for the third quarter data, but not for the first quarter. It is assumed that the coefficient of variation for the first quarter would not be very different. Actually, these calculations should also have been made for the first quarter. The expression for the standard error squared is as follows (since s_j is in \$1,000):

$$s^2_{X3} = 1000^2 \left(\sum \frac{(N_js_j)^2}{n_j} - \sum N_js_j^2 \right),$$

where the second term inside parentheses is the correction for sampling from a finite population.

$$\begin{array}{r} 3,010,442 \\ -465,683 \\ \hline 2,544,759 \end{array}$$

Hence: $s^2_{X3} = 1000^2\ (2,544,759)$.

The standard error is: $s_{X3} = 1,595,230$.

The relative error is:

$$c_{X3} = \frac{1,595,230}{127,747,000} = 0.0124874 \text{ or } 1.25 \text{ percent},$$

where the denominator is the known population value for the third quarter. (See Table 2.) Assuming that c_{x3} holds for the first quarter, then the standard error of X for the first quarter is:

$$s_X = 161,061,000(0.0124874) = 2,011,200.$$

Features of this design were:

1. Deep stratification with 15 strata averaging about $n_j = 3$ per stratum.
2. Random selection of sampling units (companies) within each stratum using tables of random numbers.
3. Ratio estimates in each stratum using past data from population and sample.

Table 2

Basic data and calculations for stratified random sample of manufacturers to estimate shipments[5]

Strata	Population N_j	Std. dev. $s_j/1000$	Sample n_j	Third quarter shipments \$1000 Popul.	Sample	weight w_j	First quarter shipments \$1000 Sample	Estimate	Actual	Diff.
1	6	0	0	0	0	1	0	0	362	− 362
2	146	20.4	8	4064	209	19.445	271	5270	7606	−2336
3	69	20.4	4	7116	408	17.441	517	9017	10187	−1170
4	32	17.1	4	5389	690	7.810	1000	7810	7387	423
5	30	34.7	4	7589	965	7.864	1235	9712	9589	123
6	21	39.5	4	7509	1424	5.273	1675	8832	11019	−2187
7	17	27.2	3	8335	1521	5.480	2135	11700	9622	2078
8	15	32.4	3	8644	1784	4.845	2275	11022	10657	365
9	11	81.7	3	9059	2576	3.517	3815	13417	11023	2394
10	9	59.4	3	9838	3230	3.046	4550	13859	12996	863
11	8	91.9	3	10738	4183	2.567	4651	11939	12031	− 92
12	4	63.2	2	7394	3757	1.968	3935	7744	8609	− 865
13	4	107.7	2	8827	4310	2.048	5150	10547	11780	−1233
14	4	92.9	2	10049	4843	2.075	6319	13112	12319	793
15	5	1273.1	5	23196	23196	1.000	27080	27080	27080	0
Sum	381		50	127747				161061	162267	−1206

Source: Records of the War Production Board for 381 companies.

Notes on Table 2:

1. The two methods of weighting are illustrated by the data in stratum two:

$$X_2 = \frac{4064}{209}(271) = 19.445(271) = 5270. \qquad X_2 = (4064)\frac{271}{209} = 1.2967(4064) = 5270.$$

2. The weights, w_j, are larger than N_j/n_j in seven strata and smaller in six strata; in some cases, the weights are larger but the estimate is still too small.
3. The lower bounds of the 15 strata are: 0; 1; 70,000; 140,000; 205,000; 305,000; 450,000; 535,000; 635,000; 1,000,000; 1,200,000; 1,550,000; 2,000,000; 2,400,000; 2,700,000. These intervals are irregular to take advantage of natural clusters.
4. Six companies not in the population in the third quarter appeared in the first quarter and hence were not subject to sampling. Their shipments are, therefore, treated as a negative bias, the same as an under-estimate.

---

4. Approximate optimum allocation.
5. Cutoff of five largest companies including them 100 percent.
6. Use of past shipments for stratification and weights.
7. Estimates could be compared with the final population values by each stratum as well as for entire population.
8. In this way, showing offsetting sampling variations in 14 strata: seven estimates were too high and seven were too low.

CASE 5

Estimating tax return characteristics.[6] The Internal Revenue Service published the results of a nationwide probability sample study of 160,000 federal individual income tax returns, selected from 52 million tax returns filed for 1948, as part of an audit program. It was of historic significance because, for the first time, the agency had firm nationwide estimates of the basic error characteristics of individual tax returns by a large number of classes and subclasses. This sample was a sub-sample selected from the regular statistics of an income sample selected from nine strata — business and non-business within four gross income classes and Form 1040A. The 160,000 returns were sent to the field offices where they were audited by revenue agents in the usual manner. This nationwide sample furnished management with nationwide point estimates of tax error and other characteristics and, hence, defined the dimensions of the tax audit problem.

Major findings were:

1. About 25 percent or about 13 million tax returns had a tax error of \$2 or more.

2. The total tax error was \$1.5 billion.

3. Tax errors were not random — they favored the taxpayer nine to one.

4. Business returns were twice as likely to contain errors as were non-business returns.

5. The major sources of error were unreported income or under-reported income. Personal deductions and exemptions were also major sources of error.

6. Arithmetic errors on the face of the return were a distinctly minor source of error, despite a common belief that the arithmetic per se was very hard.

7. It was found that many taxpayers were erroneously taking alien relatives and others as personal exemptions. As a result, more detailed instructions were issued on this point and a new section was added to Form 1040. This

feedback of sample estimates and taking action to try to reduce the error rate at the source is an example of real quality control.

8. Business returns and higher income returns accounted for most of the tax error.

9. Some occupations, businesses, and industries are more prone to tax error than others.

10. It furnished information about various types of tax returns, where the most tax error could be corrected on the average per man hour of audit effort — an example of cost benefit analysis.

What this sample study illustrates is that top management cannot make adequate decisions and plan operations without having acceptable quality point estimates of the key characteristics. How, for example, can an effective tax audit program be planned and executed without knowing the incidence, types, magnitudes, the concentrations, and the locations of tax errors?

Top managers were able to make sound decisions once they had point estimates on key characteristics. Point estimates are stressed because management has to operate with one figure in so many problems, the best estimate, not with a range within which the "population" value falls (the confidence interval).

CASE 6

Estimates to control rail traffic diversion. When a railroad purchase or merger plan is proposed by the railroads or by the Interstate Commerce Commission, or both, the railroads usually make a study to determine how the plan will effect the control of their freight revenues. In the Rock Island case, the Santa Fe Railroad made such a study because millions of dollars were involved.[7] The procedure is to select a sample of carloads and have freight traffic specialists familiar with the sample movements make judgments of the extent to which a carload may be diverted off, or diverted to, their line. The study made by Santa Fe, and used in hearings before the Interstate Commerce Commission, is summarized below:

1. Population: Car movements for 12 months October 1963-September 1964, including local traffic and other movements out of scope which were eliminated from the sample.
2. Frame: 1,685,100 car movement records on computer tape.
3. Size of sample: one percent of frame or 16,851 movements.
4. Number of replications: 10.
5. Zone interval: 1,000 [$Z = kN/n = 10(100) = 1000$].
6. Random starts: 042, 175, 213, 353, 418, 597, 651, 736, 815, 937.
7. Number of carloads studied for diversion: 2,177.
8. Total money gain to Santa Fe: \$10,770,000 with a standard error of \$276,000.
9. Relative standard error: 2.56 percent.
10. 90 percent confidence limits using t and 9 df: \$10,770,000 ± 1.83(276,000) = \$10,265,000 and \$11.275,000.

The variability in the final estimate, due to differences in the judgments of independent traffic experts, is unknown.

CASE 7

Layout of subclass equations and calculations for computer control. An actual set of specifications is presented in detail for the major purpose of exerting control over computer programming, computer calculations, and computer tabulation. The data are from one subclass of a nationwide sample study. The specifications contain:

1. 12 different estimates.
2. 12 estimating equations.
3. The 12 equations for the standard errors.
4. The 12 equations for the relative sampling errors.
5. An actual subclass of data to which the 36 equations are applied.
6. A table showing the 36 correct answers.

The 36 equations are given for the purpose of showing the computer programmer what equations are needed to obtain the estimates, their standard errors, and their relative sampling errors. This is necessary because a programmer is not expected to understand the technical mathematics required for complex statistical situations, problems, and calculations. Assistance and guidance from a competent statistician is necessary. Also, by furnishing a solved problem with intermediate values as well as the final results, the statistician enables the programmer, the computer manager, and the operator to test the program and avoid time-consuming errors and de-bugging. In an actual study, neglect to follow these specifications led to the calculation and publishing of 26,000 values in error.

These specifications serve other important purposes: they show the methods used to calculate the results, they show the magnitudes of the sampling errors so that users can make the proper allowances in interpreting the data and the estimates, they show the role of correlation in applying ratio estimates, and they show the computer people the content and nature of the tabulations desired.

In this example, a nationwide sample of carload waybills was used to obtain data on rail traffic movement characteristics by commodities and by movement within and between geographical regions. These categories created more than 25,000 classes and subclasses for which estimates had to be calculated. The sampling unit is the waybill, although in over 90 percent of the shipments one waybill is also the same as one carload. The 12 estimates are as follows:

1. Means or ratios; these are means if one waybill equals one carload:

   $\bar{x}'$ tons per car
   $\bar{y}'$ miles per car
   $\bar{z}'$ dollars per car

2. Aggregates or total amounts or total counts:

M	cars	$U = XY$	ton miles
Z	dollars	$V = MY$	car miles
X	tons		

3. Ratios or ratio estimates:

   f_1 miles per ton
   f_2 revenue per cwt (hundredweight); 20 cwt = one ton of 2000 lbs.
   f_3 revenue per car mile
   f_4 revenue per ton mile

Explanation of equations

1. Symbols

   i = index
   n_i = a single sampling unit (waybill)
   m_i = number of carloads in a single sampling unit
   x_i = net tons of 2,000 pounds
   y_i = short line miles
   z_i = dollars of revenue for the shipment (waybill)
   $u_i = x_i y_i$ = ton miles for the shipment (waybill)
   $v_i = m_i y_i$ car miles

2. 100 $\bar{z}/\bar{u}$ cents per ton mile

   $5\bar{z}/\bar{x}$ cents per cwt (hundredweight) $= \frac{100}{20}\ \bar{z}/\bar{x}$
3. Seven pairs of variates (index omitted)
   - mx car tons per car
   - mv car miles per car
   - mz car dollars per car
   - ux tons miles per ton
   - xz dollars per ton (cents per cwt)
   - vz dollars (cents) per car mile
   - uz dollars (cents) per ton mile
4. c relative sampling error of the estimate

   w multiplier to apply to sample to get population or frame estimate
5. The relative standard error squared for a subclass aggregate is $c^2=(c_a^2+q_a)/n_a$, where a stands for a specified cell or subclass. If the cell size, n_a, is less than two percent of the total sample, this equation holds; otherwise, the correction factor for sampling from a finite population (1-f) should be used, where f is the sampling fraction or sampling rate n_a/n. If n_a is very small compared with the total sample, as in the present case, then q is very close to one and $c^2=(c_a^2+1)/n_a$.
6. The factor w is known because it is the inverse of the sampling rate or 1/f or n/N, where N is the total frame count. In the example used, w is assumed to be 110, which is very close to reality or 1/.90.
7. The coefficient of variation and the correlation coefficient are used to obtain the standard error of a ratio estimate, although a simpler form can be used because: 1) these tend to be stable characteristics of the data; 2) knowledge of these coefficient facilitates sample design; especially with regard to determining an efficient sample size; and 3) they show the correlation between variates reduces the standard error.

Equations. The 36 basic equations for estimating the ratio or aggregate, its standard error s, and its relative standard error c are as follows:

1. $\bar{x}' = \frac{\Sigma x_i}{\Sigma m_i}$; $c^2 = (c_m^2 + c_x^2 - 2r_{mx}c_mc_x)/n_a$; $s = c\bar{x}'$.
2. $\bar{y}' = \frac{\Sigma v_i}{\Sigma m_i}$; $c^2 = (c_m^2 + c_v^2 - 2r_{mv}c_mc_v)/n_a$; $s = c\bar{y}'$.
3. $\bar{z}' = \frac{\Sigma z_i}{\Sigma m_i}$; $c^2 = (c_m^2 + c_z^2 - 2r_{mz}c_mc_z)/n_a$; $s = c\bar{z}'$.
4. $M = w\,\Sigma\, m_i$; $c^2 = (c_m^2 + 1)/6$; $s = cM$, where $w = 110$, $n_a = 6$.
5. $Z = w\,\Sigma\, z_i$; $c^2 = (c_z^2 + 1)/6$; $s = cZ$.
6. $X = w\,\Sigma\, x_i$; $c^2 = (c_x^2 + 1)/6$; $s = cX$.
7. $U = XY = w\,\Sigma\, u_i$; $c^2 = (c_u^2 + 1)/6$; $s = cU = cXY$.
8. $V = MY = w\,\Sigma\, m_iy_i = w\,\Sigma\, v_i$; $c^2 = (c_v^2 + 1)/6$; $s = cV = cMY$.
9. $f_1 = \Sigma\, u_i/\Sigma x_i$; $c^2 = (c_u^2 + c_x^2 - 2r_{ux}c_uc_x)/6$; $s = cf_1$.
10. $f_2 = 5\,\Sigma\, z_i/\Sigma x_i$; $c^2 = (c_x^2 + c_z^2 - 2r_{xz}c_zc_z)/6$; $s = 5cf_2$.
11. $f_3 = \Sigma z_i/\Sigma v_i$; $c^2 = (c_v^2 + c_z^2 - 2r_{vz}c_vc_z)/6$; $s = cf_3$.
12. $f_4 = 100\,\Sigma z_i/\Sigma u_i$; $c^2 = (c_u^2 + c_z^2 - 2r_{uz}c_uc_z)/6$; $s = cf_4$.

The original data, consisting of number of carloads (cars), dollars, miles, and tons, are given below. All other values shown are derived from these basic figures:

Sample unit	Cars m_i	Dollars z_i	Miles y_i	Tons x_i	Ton miles u_i	Car miles v_i
1	1	$911	498	56.95	28,361	498
2	1	555	217	50.00	10,850	217
3	1	761	381	58.05	22,117	381
4	1	898	476	60.70	28,893	476
5	2	538	96	88.30	8,477	192
6	2	952	135	99.40	13,419	270
Sum	8	4615	1803	413.40	112,117	2034
Mean	1.3333	769.17	300.5	68.90	18,686.17	339.00
S	0.5164	184.1276		19.9579	8,976.95	131.9606

The pairs of variates, together with the linear correlation coefficient r_{12}, are:

1	2	r_{12}
m	x	0.9683
m	v	0.6340
m	z	0.1017
x	u	0.4991
x	z	0.1348
v	z	0.7190
u	z	0.6939

The values of the coefficient of variation c are:

	c
m	0.3873
x	0.2897
v	0.3893
z	0.2394
u	0.4804

The 36 estimates are as follows:

Estimate name	Symbol	Estimate	Standard error	Relative error
1. tons per car	$\bar{x}'$	51.675	2.7208	0.0527
2. tons	X	45,474	19,328	0.4250
3. carloads	M	880	385	0.4378
4. miles per car	$\bar{y}'$	254.3	34.48	0.1356
5. car miles	V = MY	223,740	98,018	0.4381
6. dollars per car	$\bar{z}'$	$577	$102	0.1772
7. dollars	Z	$507,650	$213,103	0.4198
8. miles per ton	f_1	271.2	46.42	0.1712
9. ton miles	U = XY	12,333,000	5,586,000	0.4529
10. revenue per cwt	f_2 (cents)	55.82	7.976	0.1429
revenue per ton	(dollars)	11.164	1.5952	0.1429
11. revenue per car mile	f_3 (cents)	226.9	25.34	0.1117
12. revenue per ton mile	f_4 (cents)	4.116	0.602	0.1463
revenue per ton mile	(dollars)	0.04116	0.00602	0.1463

In item 10, there are 20 hundredweight (cwt) in a ton of 2,000 pounds. Hence, 55.82 cents per cwt = 55.82(20) = 1116.64 cents or $11.164 per ton.

CASE 8

Maintaining quality of data by reduced sampling. This case parallels an actual situation in which a nationwide probability sample is used to estimate motor carrier traffic characteristics such as shipments, weights, revenues, rates, origin and destination, and commodities. It is possible to maintain the quality of the basic data, but at the same time use a more efficient sample design in which the sample size is reduced as much as 25 percent. This is accomplished by simply using known technical knowledge about the design of probability sampling.

The frame was divided into four strata based upon the weight of a shipment due to the wide range and variability of the weights of shipments. The frame and the sample had the following characteristics; values for the frame are the best estimates available:

Strata	Estimated frame size N_j	Estimated std. dev. s_j	Actual sample	Weights	Mean $\bar{x}_j$
1	6,500,000	125	3,250	2000	250
2	2,400,000	600	2,400	1000	1,000
3	900,000	3,000	4,500	200	4,000
4	200,000	36,000	5,000	40	40,000
Total	10,000,000 = N		15,150 = n		1,562.5 = $\bar{x}$

The standard errors of the mean $\bar{x}$ are found to differ greatly depending upon how the total sample size of 15,150 is allocated to the four strata:

	$s_{\bar{x}}$
1. Allocation as given above	11.31
2. Allocation proportional to N_j	42.05
3. Optimum proportional to $N_j s_j$	9.74

Since a standard error of 11.31 was satisfactory — a relative error of 0.7 percent — the sample could be reduced, by using optimum allocation, by $9.74^2/11.31^2$ or 74.2 percent of 15,150 to 11,241.

The allocation of the reduced sample to the four strata is as follows with the new weights:

Strata	Optimum allocation n_j	Weight w_j	Rounded w_j
1	752	8643	8000
2	1,332	1801	2000
3	2,497	360	400
4	6,660	30	30
Total	11,241		

The standard error of the mean, using this allocation, gives the actual as the optimum. Both are equal to 11.34, which for rounding variations is the same as 11.31 for the larger sample. Using the new rounded weights, w_j, gives the expected sample size as 11,200 as follows:

Strata	Weight w_j	$n_j = N_j/w_j$
1	8,000	800
2	2,000	1,200
3	400	2,200
4	30	7,000
	Total	11,200

Notes

[1]W. E. Deming, "An Application of a Replicated National Sample to Consumer Research," *New Frontiers in Administrative and Engineering Quality Control,* 1962 Middle Atlantic Conference, American Society for Quality Control, Washington, D.C. Used by permission.

[2]*ASTM Manual on Quality Control of Materials (1951),* S.T. Publication 15-C, American Society for Testing Materials, Philadelphia, PA., p. 63. See the appendix for factors like d_2 to use for control charts.

[3]Data from S. Chatterjee, "Estimating Wildlife Populations by the Capture-Recapture Method," Annual Meeting of the *National Council of Teachers of Mathematics,* Denver, April, 1975.

[4]From *First Course in Probability and Statistics* by J. Neyman. Copyright (c) 1950 by Henry Holt and Co. Reprinted by permission of Holt, Rinehart, and Winston, CBS College Publishing, p. 338.

[5]A. C. Rosander, *Elementary Principles of Statistics,* Van Nostrand, New York, 1951. Copyright by Wadsworth, Inc. Used by permission of Brooks Cole Publishing Co., pp. 282-285.

[6]*The Audit Control Program,* Bureau of Internal Revenue, U.S. Treasury Department, Washington, D.C., May 1951.

[7]Summarized by the author from ICC I and S Docket M 22688. Robert Keyes of the Santa Fe Railway Company, Chicago has kindly furnished clarifying information in a personal communication.

24
Control by Work Sampling

Work sampling: background. The technique now called work sampling originated with L.H.C. Tippett, an English statistician, who applied what he called "a snap-reading method" to employees and machines in the English textile industry during the 1930s. The method required an observer to tour the factory at random intervals. The purpose was to estimate the percent of machine down time and its causes, the percentage distribution of work activities, and the percentage of time spent on various products. Tippett pointed out some major advantages which are still true: it costs less than 100 percent production records, it eliminates stop watches and, hence, is more acceptable to workers, the data are more accurate, and major sources of variation can be measured. To obtain these advantages, certain conditions have to be met.[1]

In this country, Morrow applied Tippett's method to a number of industries and came to about the same conclusions. He called the method "ratio delay study," because it could be used to estimate delay time as well as the percentage of time employees and machines were or were not working. He noted that the observations should be distributed over all hours of the day and week, and that the results could be used to evaluate departmental operations.[2]

Waddell introduced the term "work sampling" to show that the method need not be limited to a study of delay time. Brisley showed that an estimate from a random time sample approximated the 100 percent value obtained from a motion picture of a person at work. He applied the method to nurses, professional engineers, and executives.[3]

Abruzzi described how the quality control chart could be used in connection with work sampling to show whether the estimates obtained were stable over time. Like Tippett, he found that stability could exist.[4] Barnes described several papers on applications of work sampling to factory operations. He went further and showed how work sampling data could be used in work measurement, performance ratings, and wage incentive planning.[5]

Rosander, with Guterman and McKeon, in 1955 put work sampling on a completely random time sampling basis by applying the theory and practice of probability sampling. They made use of the many technical advances which had been made in probability sampling during the 20 years since Tippett had introduced his snap reading method.[6]

They used the minute model, a complete time frame of both instants and durations, random selection of minutes from the time frame, cluster sampling units, estimation equations from the sample design, standard error equations appropriate for the sampling units used, and costs per minute for every employee and machine. This enabled them to make the following highly significant improvements:

1. Estimates were no longer limited to proportions.
2. Aggregates and totals for money and time were estimated directly from the estimating equations in the sampling plan. Intuitive estimation of totals and subtotals was eliminated.
3. Biases inherent in the tour method were reduced, if not eliminated.
4. Valid estimates of the standard error could be obtained for all estimates.

They found that the planning and management aspects were just as important, if not more so, than the technical sampling aspects. Implementation called for such steps as explanatory talks to employees, a simple data sheet, an employee instruction booklet, an explanation of how the random times were to be selected, and a trouble shooting operation.

Later, Rosander developed the standard month for constant probabilities, the balanced replicated design for time and areas and replicates, and prepared tables of 4,000 random days and 4,000 random minutes.[7]

Probability sampling of work activities. When probability sampling is combined with random time selection to estimate and control characteristics of work activities, the power and versatility of probability and statistics are applicable to the design, collection, and analytical phases of the study. Science is substituted for intuition and opinion.

Three techniques are described and appraised: the tour method, the pseudo time sampling method, and the comprehensive random time sample. In the *tour* method, an observer walks through the factory or parts under study and records whether each worker is working or not, and if so, what each is doing. Similar observations are made on machines. Tally counts are made and proportions calculated. The binomial, heads or tails model, is used to analyze the data and calculate the variation due to sampling. This assumption is highly questionable because there is nothing in the physical aspects of the situation that suggests the binomial. It is assumed that the sample is random; the observations are not necessarily independent because workers may see the observer coming and change their work behavior; and a constant proportion, required by the binomial model, may not exist because, as Tippett warned, of different machines, times, operations, and employees. Few are aware that if these differences exist, the formula PQ/n can greatly underestimate the true sampling variance. A proportion came out of the data and the textbooks said PQ/n applied to a proportion. Very few tested to see whether the model applies to a particular situation, as Tippett did. Another real shortcoming of the tour method is that only proportions can be estimated from the sample, thus greatly limiting its usefulness.

In the *pseudo time sample,* attempts are made to improve on the tour method. Samples are taken at 20 or 30 minute intervals or whatever is the duration of a "tour". Sampling is concentrated in time so that 1,000 or

2,000 observations are made in a day or two, thus giving the false notion of a "large" random sample over time. In conducting the study or analyzing the data, time is eliminated from the work day, or certain observations are eliminated from the results, because it is alleged they are not "typical" or "representative". These decisions increase bias; they do not eliminate it. A knowledge of probability sampling theory and practice enables one to avoid these pitfalls.

Comprehensive random time sampling accepts the fact that work sampling is sampling and applies the best knowledge available to the problem. This means applying the theory and practice of probability sampling which has been developed over the past 50 years. Actually four basic books on probability sampling appeared between 1949 and 1953: those by Yates (1949), Deming (1950), Cochran (1953), and Hansen, Hurwitz, and Madow (1953). There is no reason why anyone dealing with sampling problems by 1960 and later should not have known about this new science. This means abandoning the tour method and the pseudo sampling methods for more efficient techniques which will yield more and better data in a timely fashion and at minimum cost. But two conditions must be met: 1) a sound technical design, and 2) a careful management and implementation of the entire sample plan. In this chapter, we present applications of this kind of work sampling, including some recent developments.[8]

Why random time sampling works. In random time sampling, a continuous variable time is replaced by a discrete variable, instants at minute intervals. A random sample of these instants is selected and the number counted which hit a specified activity. From these two counts, a proportion is calculated. Using the minute as a duration (the minute model) and knowing the cost per minute (wages, salaries, rent, etc.), an aggregate cost is calculated. This latter model is valid since practically all work activities of interest last more than one minute. Why random time sampling works and gives very good estimates of population parameters is illustrated by Figure 1 and its analysis. This diagram shows six time lines: the continuous time variable T, the duration of activity A by T_a, the time T divided into M minutes (instants), the number of instants M_a that hit activity A, a random sample of m instants, and the number of sample instants that hit activity A or m_a.

In Figure 1, the *population values* for time as a continuous variable are expressed as follows:

1. The population proportion:

$$P_a = \frac{T_a}{T}.$$

2. The population aggregate cost: $C_a = c_a T_a$, where c_a is the cost per minute for activity A.

In this diagram, the population values for time as a *discrete variable (instants)* are expressed as follows; as M increases and gets very large $P_m = P_a$ for all practical purposes.

1. The population proportion:

$$P_m = \frac{M_a}{M}.$$

2. The population aggregate cost: $C_a = c_a M_a$.

In the above equations, T_a and M_a are unknown, so a random time sample is used to obtain *estimates* of P_a and C_a:

1. The estimate of P_a or P_m:

$$p = \frac{m_a}{m}.$$

2. The estimate of aggregate cost C:

$$C_{est} = pMc_a = \frac{M}{m} m_a c_a.$$

3. The estimate of aggregate time T_a:

$$T_{a_{est}} = pM = \frac{M}{m} m_a.$$

Every employee working at activity A may not have the same cost per minute; furthermore, the number of persons at each cost per minute may differ. Hence, there may be k groups with different numbers of random minutes and different minute costs. Therefore:

1. The estimate of aggregate cost C_{est} for activity A is:

$$C_{est} = \frac{M}{m}\left(m_1c_1 + m_2c_2 + \ldots + m_kc_k\right) = \frac{M}{m}\sum_{1}^{k} m_j c_j.$$

The aggregate costs for other activities are calculated in the same way.

FIGURE 1

The Random Time Sample Model C

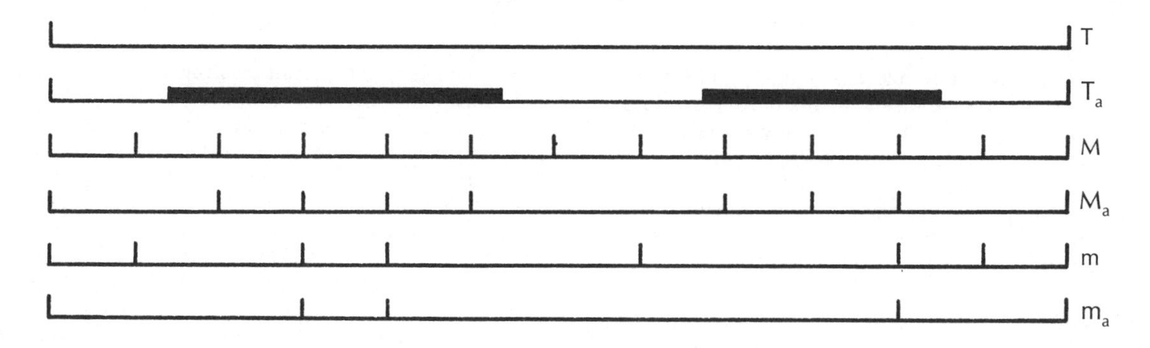

Advantages of random time sampling applied to work activities. First, we need to explain why work sampling is preferred to the use of some kind of 100 percent time production records where each employee keeps a record of how he spends his time each day. The limitations of 100 percent time reporting as reported by management are as follows:[9]

1. It is difficult to apply this method to professional workers or to those with several daily job assignments.
2. Inaccurate records result due to faulty memory, carelessness, and forgetfulness.
3. Adjustments for the costs of leave, training, supervision, and other overhead expenses are difficult to make.
4. Possible objections by some employees to the extra work required to prepare these reports.
5. It is more costly than work sampling to collect, process, and interpret the data.

We do have one very good test: for several months employees in the accounting division of a telephone company kept 100 percent time sheets, but at the same time a random time sample test was made. McMurdo reported that random time sampling had six advantages over time sheets: it produced more accurate information, there was less interruption of valuable machine time, operators were not distracted, paperwork by operators was reduced, reports were not biased by human judgment, and about 40 percent of the time required to summarize reports was saved.[10]

McMurdo pointed out the advantages of random time sample in combined machine and clerical operations in an accounting department. But random time sampling can be applied effectively to a broad range of situations to which ordinary methods of accounting and management do not apply or application calls for making questionable assumptions. These broader applications are:

1. Evaluates performance of employees and machines.
2. Aids in budget planning and in making allotments.
3. Used to allocate resources to functions and projects.
4. Used to set standards for time, money, and unit costs.
5. Provides a way of making reimbursable charges.
6. It is the only effective way we have to decompose joint costs.
7. It can be used to estimate costs of new jobs and new projects.
8. It can be used to estimate costs of "outside" work (work a department is forced to absorb).
9. It can be used to classify job costs according to job funds, where they exist.
10. It can be used to cost complex and difficult work situations, where workers are mobile and cover large areas.
11. It has built into it a sound method of estimation.
12. It provides valid sampling errors of all estimates.
13. It provides a basis for initiating or improving a quality control program.

The fact that this list is lengthy, but far from complete, is evidence of the power and effectiveness of random time sampling. There is no question that management and professionals alike have neglected a very powerful and effective technique.

Time frames. The time frame is basic to random time sampling. Time frames have a unique characteristic very rare in the real world: they do not change, they are known exactly, they are known beforehand, and they are known for the future.

The time frame provides the domain to which the sample estimates and the sample inferences apply. There is no justification for taking a large time sample in two or three days, and then generalizing to 12 months or even to one month, unless one shows that the characteristics in question are not time related. But to obtain this information requires a valid study over the time period involved. So the random time sample might just as well be designed from the start to measure stability of the characteristics over time including 12 months. Such a sample will estimate weekly, monthly, and seasonal variations and differences and insure that variations between weeks and months are properly represented in the annual estimates. Otherwise, the latter may be highly biased.

Examples of time frames. Examples of time frames are the following:

1. One day a week balanced and equalized:

$7 \text{ days} \times 7 \text{ weeks} = 49 \text{ weeks.}$

$6 \text{ days} \times 8 \text{ weeks} = 48 \text{ weeks.}$

$5 \text{ days} \times 10 \text{ weeks} = 50 \text{ weeks.}$

In this arrangement each of seven days occurs seven times, each of the six days eight times, and each of the five days 10 times.

To select one day a week for a seven day week, divide a year of 49 weeks into seven blocks of seven days each. Then one can assign days to weeks using 7×7 Latin square design[11] or construct a table using a table of random numbers. The first two assignments of days to weeks for the first two blocks of seven weeks each is given below. To construct a table, draw seven random digits from one to seven and record in *order* as drawn for each of the seven blocks. In the first line in the table below, "03" was the first random number drawn, but "3" stands for Tuesday, so Tuesday is assigned to the first week of the first block; "6" was the first random digit drawn for the second block, so Friday is assigned to the first week in the second block of seven weeks.

	Identification number						
	1	2	3	4	5	6	7
block	S	M	T	W	T	F	S
1	7	6	1	4	2	5	3
2	3	5	6	7	4	1	2
3			etc.				
4							
5							
6							
7							

2. Yearly frames of minutes: examples:
   1. 60 minutes × 12 hours × 365 days = 262,800 minutes.
   2. 60 minutes × 40 hours × 52 weeks = 124,800 minutes.
   3. 60 minutes × 8 hours × 250 working days = 120,000 minutes.
   4. 60 minutes × 8 hours × 220 paid days worked = 105,600 minutes.

Clearly, the time frame constructed depends upon the purpose and the problem, on the length of the daily and weekly working week, and on what is to be included in the working year.

In some studies, including at least two which have already been reported, it was easier, due to staggered and varying lunch hours, to include the lunch period in the time frame and then reject it from the sample. This avoided a very difficult administrative problem. In cases like this, the daily work time frame would be not 480 minutes but 510 minutes for a 30 minute lunch period and 540 minutes for an hour lunch period.

The major reason for using random time sampling studies is that they provide important information for management use that is not provided by the established data-collection system. If the data collected by the random time sample is so important and management is willing to have the information collected on data sheets on a regular basis (daily, weekly, monthly), then random time sampling can be dispensed with and sampling can be applied to records collected over a period of a year. Records such as tax returns, freight bills, and automobile driver's licenses, which are collected on an annual basis, can be sampled directly because the time factor is automatically taken care of.

A random time sample, such as a balanced sample of days covering a year, may tremendously reduce the frame needed to sample for basic characteristics. For example, interest may lie in characteristics of pick up and delivery from a truck terminal or over-the-highway trips from the same terminal. In either case, the trip is an effective and convenient unit to use for sampling, but *a trip frame needs to be constructed only for the days in the random time sample.* This method has been used very successfully.[12]

Random time sampling situations. Examples of the kinds of sampling situations encountered in random time sampling practice are the following:

1. Random time sampling; 100 percent coverage of employees and machines at each random time.
2. Random time sampling; clusters of employees and machines with 100 percent coverage, but at different random times.
3. Random time sampling; random sample of employees.
4. Random time sampling; stratified random sample of employees.
5. Random time sampling; random sample of work areas with 100 percent coverage of each sample area.
6. Random time sampling; random sample from a frame constructed only for the random time, e.g., random day.
7. Replicated sample for each of the above.
8. Balanced replicated design by days: areas, shifts, replicates.

Henceforth, random time sampling will be designated by RTS and random area sampling by RAS.

Sample designs with formulas for estimating proportions, aggregates and their standard errors.

Case 1. m minutes are selected at random from a population of M minutes. All N persons are included in each random minute. This is RTS with a group or cluster of persons included 100 percent.

Proportion p

1. $p = \frac{1}{m}\sum_{1}^{m} p_j;\ j = 1,2,3,\ldots,m.$

2. $s_p^2 = \left(1 - \frac{m}{M}\right) \frac{\sum_{1}^{m}(p_j - p)^2}{m(m-1)},$

where p is an estimate of the population proportion; p_j = proportion of N engaged in activity A at the jth random minute; and j is summed over the m random minutes.

3. $s_p = \sqrt{s_p^2}$ is the standard error of p.

4. The factor $1 - \frac{m}{M}$ can be ignored if m is very small (less than three percent) compared with the population size M.

Aggregate cost X

1. $X = \frac{M}{m}\sum_{1}^{m} x_j = M\bar{x}$, where $\bar{x} = \frac{1}{m}\sum_{1}^{m} x_j$.

$\frac{M}{m}$ = w = the multiplier, weight, or inflating factor which is the inverse of the sampling rate m/M.

x_j = sample aggregate for the jth minute for all N individuals.

2. $s_x^2 = \frac{M^2}{m(m-1)}\sum_{1}^{m}(x_j - \bar{x})^2,$

where m is assumed small compared with M. For calculation, it is easier to use $\Sigma x_j^2 - (\Sigma x_j)^2/m$ than $\Sigma(x_j - \bar{x})^2$.

Example for estimating proportions. Ten random minutes were called on a group of 100 persons. It was found that the following numbers were engaged in activity A for each of the 10 random minutes: 21, 8, 13, 21, 10, 17, 27, 14, 22, and 11 for a total of 164. Hence:

$p = 1.64/10 = 0.164$ or 16.4 percent.

$$s^2 = \frac{\Sigma(p_j - 0.164)^2}{9 \times 10} = \frac{0.034440}{90} = 0.00383.$$

s = 0.0196 = 0.02 or two percent.

Check:

$$\Sigma x^2 - (\Sigma x)^2/m = \frac{344.40}{90} = 3.83$$

using whole numbers (percentages).

$$s = \frac{1}{100}\sqrt{3.83} = 0.0196 = 0.02 \text{ or two percent.}$$

Example for estimating aggregate cost. Ten random minutes were called on a group of eight employees. The cents per minute were recorded for each employee engaged in activity A. The results are shown in the following table (m = 10, N = 8, and assume M = 4,800):

Cents per Minute for Activity A and Total Cents

Employee	Random minute 1	2	3	4	5	6	7	8	9	10	Total cents
1				6							6
2	7	7				7					21
3	6							6		6	18
4	5				5				5		15
5		6		6		6	6				24
6					8				8		16
7											0
8		9		9						9	27
Sum	18	22	0	21	13	13	6	6	13	15	127

It is assumed that wages or salaries did not change during the sampling period, so the cost per minute assigned to an employee remains constant. Of the total of 80 employee minutes (8×10); 19 show activity A, so the proportion working on activity A is estimated at 19/80 or 0.24 or 24 percent.

The average cost per minute is:

$$\bar{c} = \frac{127}{10} = 12.7¢ = \$0.127.$$

Therefore, total cost is:

$$C_{est} = \frac{4800}{10}(1.27) = \$609.60.$$

$$s^2 = \frac{4800^2}{9 \times 10}\sum(c_j - 0.127)^2.$$

$$s = \$106.14.$$

$$cv = \frac{106.14}{609.60} = 0.17 \text{ or } 17 \text{ percent.}$$

Calculating (using cents) gives:

$$s^2 = \frac{4800^2}{9 \times 10}\left(\sum x_j^2 - \frac{(\sum x_j)^2}{10}\right)$$

$$= \frac{4800^2}{90}(440.10) = 10614 \text{ or } \$106.14,$$

where $\sum x_j = 127$.

The latter method is easier and more accurate because whole numbers are used; decimals are not used because they introduce a greater chance of making errors. Hence, in this problem, the estimated total cost of activity A is \$610 with a standard error of \$106 or 17 percent. This is an illustration; in an actual problem, a much larger sample would be used so that the relative sampling error would be below five percent.

Case 2. This is simple RTS, the same as Case 1, but with replication for ease in estimating the proportion p, the aggregate C, and their standard errors. We assume $M=480$, $N=8$, $k=4$ replicates, 1 stands for activity A, 0 for non-A, and $m=4$ random minutes as indicated. The observations with derived totals and calculations are given in the table:

Costing Activity A

Employees	Replicated random minutes: 3:45 pm 1	9:12 am 2	9:48 am 3	1:33 pm 4	Frequency f	Cents costs per min c	f x c
1	1	1	0	0	2	6	12
2	1	0	1	0	2	5	10
3	0	0	0	0	0	5	0
4	1	1	1	1	4	5	20
5	0	0	0	0	0	6	0
6	0	1	0	0	1	6	6
7	0	1	0	0	1	5	5
8	1	0	0	1	2	7	14
Sum	4	4	2	2	12		
p_j	1/3	1/3	1/6	1/6	3/8		
c_j	23	22	10	12			67
C_{est} \$	110.40	105.60	48.00	57.60			80.40

The cost per minute for each employee is shown in the next to the last column. These minute costs are used to obtain a total of 67 cents for the cost assignable to activity A for eight employees during four random minutes. The multiplier for each random minute is 480; for four minutes it is 120 or 480/4. For example, \$110.40 = 0.23(480) and \$80.40 = \$0.67(120); the latter is also (110.40 + 105.60 + 48.00 + 57.60)/4.

Proportion p

P = 12/32 = 0.375, where 32 = 8 × 4, the total employee minutes.

$$s^2 = \frac{1}{4 \times 3} \sum_1^4 (p_j - 0.375)^2.$$

$$s = 0.087.$$

Aggregate cost C

$$C = 0.67(120) = \$80.40$$

$$= \frac{110.40 + 105.60 + 48.00 + 57.60}{4}.$$

standard error s

Range method: $s = \dfrac{C\ max - C\ min}{4}$

$$= \frac{110.40 - 48.00}{4} = \$15.60.$$

Variance method:

$$s^2 = \frac{1}{4 \times 3} \sum_1^4 (C_j - 80.40)^2,$$

$$s = \$16.08.$$

Case 3. This is a combination of RTS and RAS. The frame consists of M = 480 minutes and N = 16 areas. The sample consists of m = 4 random minutes applied to n = 4 random areas. The weights are:

$$w_m = \frac{M}{m} = \frac{480}{4} = 120; \quad w_a = \frac{N}{n} = \frac{16}{4} = 4.$$

$$w = \frac{\text{frame}}{\text{sample}} = \frac{480 \times 16}{4 \times 4} = 480.$$

Random area number	Random minute and area-minute identification							
	1		2		3		4	
2	805	1	945	5	151	11	303	13
8	851	2	1111	7	136	10	423	16
11	904	3	924	4	107	9	346	15
15	1008	6	1152	8	239	12	322	14

The serial numbers from 1-16 identify the random minutes in their time order, an arrangement an observer would need in order to plan his time and area schedule. The problem is to estimate the proportion engaged in activity A and its aggregate salary or wage cost, together with their standard errors. Since the proportion in this design is a ratio estimate, the equation for the standard error of a ratio estimate, not that for a binomial proportion, must be used. The data for the 16 area-minutes are as follows:

Area-minute	Number in sample area (y)	Number engaged in activity A (x)	Total cents per minute activity A (z)
1	5	2	12
2	6	1	5
3	3	0	0
4	5	2	13
5	4	1	6
6	6	2	12
7	5	0	0
8	7	2	10
9	2	0	0
10	6	1	6
11	5	2	12
12	5	2	11
13	6	3	19
14	4	1	7
15	3	1	6
16	6	2	14
Sum	78	22	133

Proportion p

$$p = \frac{22}{78} = 0.2821 \text{ or } 28.2 \text{ percent.}$$

Standard error of p: basic values:

$$n = 16; \bar{y} = \frac{78}{16} = 4.875; \bar{y}^2 = 23.766.$$

$$2p = 0.5642; p^2 = 0.07958; \sum x^2 = 38;$$

$$\sum y^2 = 408; \sum xy = 119.$$

$$s^2 = \frac{1}{16 \times 15 \times 23.766}\left[38 + 0.07958(408) - 0.5642(119)\right]$$

$$= 0.000584$$

$$s = 0.024 \text{ or } 2.4 \text{ percent;}$$

$$cv = \frac{0.024}{0.282} = 0.085 \text{ or } 8.5 \text{ percent.}$$

Aggregate C

$$C = 480(1.33) = \$638.40.$$

$$s^2 = \frac{(480 \times 16)^2}{16 \times 15}\left(\sum_1^{16} z_1^2 - 133^2/16\right) \text{ (in cents)}$$

$$s = \$105.80.$$

$$cv = \frac{105.80}{638.40} = 0.166 \text{ or } 16.6 \text{ percent.}$$

A number of problems arise in connection with RTS and RAS. Here we illustrate a random sample of four areas for one day of 480 minutes. In practice, enough days would have to be included in order to measure any time effects, in order to stabilize the proportion and aggregate, and in order to reduce the sampling variations relative to the magnitude of the estimates. Furthermore, the sampling variations may not be as important as the biases that could be introduced by errors in recording the number of employees in a sample area and identifying the number engaged in activity A as well as identifying them individually so that their wage rates can be reduced to cents per minute. This means the observer should preferably be someone familiar with the work and the workers, as well as a careful and accurate observer.

Case 4. This is a balanced RTS and RAS design. Four random minutes are called per day on four random areas and are balanced daily and over three days. Each of the four replicates contains the same three days, the same 12 areas, has six random minutes in the am, and six random minutes in the pm. This arrangement assumes that the observer is able to observe daily four random minutes in four areas; if not, the number of random areas per day can be reduced to three, and four days are used to obtain balance. The basic values are as follows:

$R = 4$ replicates of random minutes.
$M = 480 \times 3 = 1440$ minutes.
$m = 4 \times 3 = 12$ minutes per replicate.
$m = 4 \times 12 = 48$ minutes in 4 replicates.
$n = N = 12$ areas.

$$w(\text{total}) = \frac{1440}{12} = \frac{480}{4} = 120; \text{ also, } \frac{480 \times 3 \times 4}{4 \times 12} = 120.$$

$$w_j(1 \text{ replicate}) = \frac{480}{1} = 480.$$

Design for Four Replicates in Each of 12 Areas for Three Days of 480 Minutes Each.

Day	Random area no.	Random minute (replicate) 1	2	3	4
1	3	1	2	1	2
	7	2	1	2	1
	1	1	1	2	2
	11	2	2	1	1
2	4	2	1	1	2
	6	1	2	2	1
	12	2	1	2	1
	9	1	2	1	2
3	5	2	2	1	1
	10	2	1	1	2
	8	1	1	2	2
	2	1	2	2	1

In the table "1" stands for a random minute in the am, "2" for a random minute in the pm.

Differences between areas are confounded with differences between days. This does not affect the total estimates. If variations between days must be separated from differences between areas, then one way is to cover all 12 areas every day.

Example: A data summary for the four replicates cumulated over three days, together with the necessary calculations, are given below. Total sample cents are derived by assuming an average salary and wage cost of eight cents per minute. Actually, the observer would record cents per minute or its equivalent for everyone performing activity A.

Proportion p

$$p = \frac{60}{240} = 0.25 \text{ or } 25 \text{ percent.}$$

$$s_p^2 = \frac{1}{4 \times 3} \sum_1^4 (p_j - 0.250)^2.$$

$$s_p = 0.031 \text{ or three percent.}$$

Aggregate C

$$C = \frac{480}{4}(57.60) = \frac{27{,}648}{4} = \$6912,$$

where $C_j = 480c_j/100$.

$$s^2 = \frac{1}{4 \times 3} \sum_1^4 (C_j - 6912)^2 = \frac{1}{12}\left(\sum C_j^2 - \frac{27{,}648^2}{4}\right).$$

$s = \$1,047$.

Relative sampling error cv

$$cv = \frac{1047}{6912} = 0.15 \text{ or } 15 \text{ percent.}$$

It should be emphasized that unless sources of bias are carefully controlled, such as observer counts and recording errors, they may be several times as great as the sampling variations, making the latter very misleading, if not useless.

Applications to Real Problems

Five applications of RTS to real problems are discussed and described in this section; four represent actual applications of RTS, while the other one is a case to which RTS can readily be applied. What these examples show is that the RTS principles and techniques, described in the foregoing pages, have been applied successfully to real-world problems.

Federal statistics division of 350 employees.[13] The Division, which had 24 organizational units, had about 20 major classes of projects and over 150 sub-

Replicate j	Total count	Activity A count	Proportion p_a	Total sample cents act. A (c_j)	Estimated total dollars act. A (C_j)
1	60	14	0.233	1344	$6,451
2	57	10	0.175	960	4,608
3	65	21	0.323	2016	9,677
4	58	15	0.259	1440	6,912
All Replicates	240	60	0.250	5760	27,648

projects and jobs in its regular work schedule. The director had been using 100 percent time records from certain sections of the division, but he was not satisfied with them. He asked the chief statistician about the possibility of using work sampling to obtain the data he wanted. The chief statistician said random work sampling was applicable, but recommended that a pilot study be run for a week of five days to test the sample plan, instructions, employee response, and specific procedures. This was done.

The work sample was to be used to determine time and money spent on regular jobs and projects for budget planning and hearings, for work control and work scheduling, and for costing special jobs and projects requested by outside agencies. The latter tended to be high priority jobs which the division was expected to "absorb" at the expense of delaying regular work. How much time and money were being diverted to these special jobs was an important purpose of the work sample. The work sample would provide unit costs for punch card work where detailed production records were kept for each operator, information on personnel utilization in each of the 24 organizational units, and a basis for costing out new jobs and new projects developed in or assigned to the division.

As a result of the pilot study, which produced some highly useful data, several changes were made in the procedures: the sample was revised, a different method of calling random times was used, the instruction manual was expanded and revised, and other procedures were improved. Steps were taken which successfully overcame some opposition which arose in connection with the plan.

The major features of the work sample, which ran for 26 consecutive months, were as follows:

1. RTS was applied to each of the 24 organizational units.

2. The daily time frame was 510 minutes, including a 30 minute lunch period.

3. All employees were included in the employee frame — everyone from messenger boys to the director: typists, clerks, secretaries, key punch operators, coders, supervisors, economists, statisticians, computer specialists, graphic artists, chiefs of sections, and branch chiefs.

4. Different random times were called by telephone from an administrative office to each of the 24 organizational units. One of the clerks in the chief statistician's office called random times on the administrative office.

5. No one ever knew when a random time was going to be called.

6. Anonymous self-reporting was used. At each random minute, each employee in the unit called filled out a data sheet giving the code number of the activity he or she was engaged in. Salary grade was checked.

7. Secretaries or supervisors verified every data sheet and filled out sheets for absentees. Absentees included those on paid leave, as well as those present, but on detail or out of the office or building for some reason.

8. Every employee was furnished a pad of data sheets, a book of instructions, the code number of every new job assigned, and the code numbers of all current jobs. Once the procedure was learned, it required only 10 to 15 seconds to fill out a data sheet.

Table 1

Distribution of Working and Not-Working Time (June 6-10, 1955)

Organizational unit	Personnel available for work	Performing assigned duties	Performing related duties	Waiting	Other
Statistics Division	100.0%	82.7	6.0	.9	10.4
1. Office of Director	100.0	89.9	7.0	.0	3.1
2. Administrative Office	100.0	94.6	1.1	.8	3.6
3. Office Chief Prog. Branch	100.0	95.8	2.8	.0	1.4
4. Statistics of Income	100.0	78.4	10.3	.7	10.6
5. Management Statistics	100.0	89.9	5.1	.0	5.0
6. Operation Reports	100.0	87.8	6.8	.0	5.4
7. Sampling & Est. Methods Sect.	100.0	97.7	.6	.0	1.7
8. Office Chief, Operation Br.	100.0	92.5	2.5	.0	5.0
9. Statistical Tech. Lab.	100.0	88.7	4.5	.8	6.0
10. Office Chief, Presentation	100.0	88.2	3.0	.0	8.8
11. Graphics	100.0	57.1	19.9	9.0	14.0
12. Typing and Proofing	100.0	83.5	4.1	5.9	6.5
13. Office Chief Manual Oper.	100.0	97.0	1.0	.0	2.0
14. Coding	100.0	86.1	3.0	.5	10.4
15. Statistical Audit	100.0	83.4	4.1	.2	12.3
16. SOI Compilation & Analysis	100.0	82.1	2.1	1.0	14.8
17. Production Statistics	100.0	68.5	17.6	1.2	12.7
18. Returns and Records Files	100.0	87.0	3.7	.1	9.2
19. Office Chief Mechanical Oper.	100.0	80.0	11.7	.7	7.6
20. Project Planning	100.0	70.7	10.7	.4	18.2
21. Transcribing	100.0	84.5	4.9	.4	10.2
22. Compiling	100.0	75.5	11.9	4.2	8.4
23. Punched Card File	100.0	61.4	.0	7.2	31.4
24. Machine Review	100.0	84.5	3.4	.3	11.8
1956 Division distribution	100.0	85.7	2.6	0.5	11.2
1957 Division distribution	100.0	87.3	2.4	0.3	10.0

9. In the pilot study, eight random minutes were called on each employee each day; in the final study only two random minutes per week were called, since only monthly and annual data were required.

10. The use of the minute model made it possible to estimate aggregate costs as well as proportions, using the techniques described under Case 1 above.

Table 1 shows the distribution of working and non-working time for the 24 organizational units during the test week. Even the pilot run was used for control when four units showed four percent or more of idle time (Table 1). Management was highly pleased with the entire project and pointed out several advantages to managers of a random time sample of the type used in this study.[14] As a result of this project, the author constructed two tables, one listing 4,000 random minutes and one listing 4,000 random days. Strangely enough, the latter is a sub-set of the former. This means a table of random minutes can also be used as a table of random days.[15]

After 26 consecutive months, it was found that the major estimates and factors developed were quite stable and highly satisfactory for management purposes. The project was then discontinued. Some employees were so accustomed to it that they did not want to stop! The plan, however, was available for re-activation in case it was found necessary to test for changes or obtain new information, but this was never done.

Estimating work time by a 10 replicate sample. The preceding work sampling problem can be solved by using a 10 replicate RTS. The assumptions are the same: two random minutes are selected per week for 50 weeks for each employee. The basic values are as follows:

1. Time frame: $M = 60 \times 40 \times 50 = 120{,}000$ minutes.
2. Employee frame: $N = 350$ and is assumed to be constant.
3. Time sample: $m = 2 \times 50 = 100$ for every employee.
4. Employee-minute frame: $MN = 120{,}000 \times 350$.
5. Employee-minute sample: $mN = 100 \times 350$.

For a 10 replicate sample the zone width Z in minutes is:

$$Z = \frac{kM}{m} = \frac{10 \times 120{,}000}{100} = 12{,}000 \text{ minutes}, = 480 \text{ minutes} \times 25 \text{ days}.$$

This means that in 50 weeks of five days each there are 10 zones of five weeks each. In each zone, 10 random minutes are selected on 350 employees for 350 employee minutes. In 50 weeks or 10 zones, there are a total of 35,000 employee minutes.

Select two random minutes per week of five days, which gives 10 random minutes in five weeks or 25 days. This totals to 100 random minutes in 50 weeks or 250 days. Balance each day of the week twice in each zone and do the same over 10 zones for each replicate. Within zones and replicates, equalize times as between am and pm. The format is as follows:

Week	50 weeks 10 zones (minutes)	Days assigned to 10 random minute replicates covering five weeks randomized										Number sum
		1	2	3	4	5	6	7	8	9	10	
1-5	1-12,000	W	T	M	F	Th	W	Th	F	T	M	10
6-10	12,001-24,000	F	W	Th	T	F	T	M	Th	M	W	10
'	'	T	'								'	
'	'	Th	'								'	
'	'	M	'								'	
'	'	W	'								'	
'	'	T	'								'	
'	'	F	'								'	
'	'	M	'								'	
46-50	108,001-120,000	Th	'								'	10
Sum (number)		10	10	10	10	10	10	10	10	10	10	100
Employee minutes		3,500										35,000

Table 2 shows the counts for each replicate for the 50 weeks. The actual counts are divided by 10 and rounded for each of four categories of work status: total count, number at work, number present but not at work, and number on paid leave. These are simulated counts, but the estimates for the three classes of work status very closely match those actually obtained.

Table 2 shows that 74.5 percent were at work, 9.2 percent were present but not working, and 16.3 percent were absent on paid leave. The actual sample gave 75 percent, 9 percent, and 16 percent, respectively. Table 2 shows that the effective work force is about 75 percent; that is, four persons have to be hired in order to have three working full time. At the time, the 75 percent effective compared very favorably with figures found by work sampling studies in private industry.[16]

Table 3 shows the percentage distribution of work status for each of the 10 replicates, the estimate for each of the work categories, and the standard error of each of these estimates. The latter are calculated by the minimum-maximum method as follows:

1. At work 74.5 percent: $s = (76.5 - 71.9)/10 = 0.46$ percent.
2. Present, but not working 9.2 percent: $s = (12.6 - 7.6)/10 = 0.50$ percent.
3. On paid leave 16.3 percent: $s = (19.7 - 13.6)/10 = 0.61$ percent.

Table 2

Work Status Counts by Replicates*

Replicate number	Total count	At work	Present, not at work	On paid leave
1	339	246	33	60
2	353	265	33	55
3	361	274	29	58
4	356	267	27	62
5	342	246	43	53
6	353	270	27	56
7	348	258	37	53
8	353	263	30	60
9	340	245	28	67
10	355	272	35	48
Sum	3,500	2,606	322	572
Percentage	100.0	74.5	9.2	16.3

*All entries = original value divided by 10 and rounded.

Table 3

Work Status Percentages by Replicates

Replicate number	Total count	At work	Present, not at work	On paid leave
1	100.0	72.7	9.7	17.6
2	100.0	75.1	9.4	15.5
3	100.0	75.8	8.0	16.2
4	100.0	74.9	7.6	17.5
5	100.0	71.9	12.6	15.5
6	100.0	76.4	7.6	16.0
7	100.0	74.0	10.7	15.3
8	100.0	74.5	8.6	16.9
9	100.0	72.1	8.2	19.7
10	100.0	76.5	9.9	13.6
Average	100.0	74.5	9.2	16.3
Standard error		0.46	0.50	0.61

Railroad accounting office. A large midwestern railroad asked the author to apply RTS to their accounting office which consisted of three separate offices with a total of 37 employees, including supervisors. The purpose of RTS was to determine the allocation of time and expense to 20 classes of activities, to furnish information on the utilization of employees, and to test the feasibility and effectiveness of RTS.

The plan was tested for 10 consecutive working days (two weeks) in May with four random minutes being called on each employee each day. The plan was applied to each of the three offices separately, and data were obtained for each and for all combined. Anonymous self-reporting was used, with a special data sheet showing the items required such as the code of the activity engaged in, hourly wage rate, whether on leave, at lunch, at work assigned, doing related work activities, waiting for work, present, but not working, and other, including personal and coffee break. The wage rate per hour was obtained so that the method of estimation inherent in the sample plan could be used to estimate both time and wages.

The estimating equation, used to obtain the sub-totals on salaries and wages by classes of activities, is as follows:

$$X_a = \frac{480}{4}\,\frac{1}{60}\sum m_{ia}c_{iah} = 2y_{ia},$$

where X_a = total dollars for activity A, m_{ia} = number of random minutes identified with activity A, c_{iah} = *hourly* wage rate for persons performing activity, $\frac{480}{4} = 120$ = inverse of the sampling rate (4/480), and 60 = factor converting hourly rates into dollars per minute. Summation is over all individuals engaged in activity A; $m_{ia}c_{iah}$ is the product of those with the same rate (c_{iah}) multiplied by the total number of random minutes associated with activity A. Hourly rates are easy to obtain. This equation avoids converting each worker's rate to cents per minute; the conversion is made in the equation by the factor 60.

The findings are summarized in Tables 4 and 5. The major findings were as follows:

1. 83 percent of the workers were at work, 9 percent were on leave, and 8 percent were waiting for work or on "other", such as personal or coffee breaks (Table 4).
2. The total cost for the 10 days was $9,218; of which $7,683 was expended for those at work, $808 for those on leave, and $726 for "other." The cost of those waiting for work was negligible, being only an estimated $50 or about 1/2 of one percent (Table 4).
3. The average hourly wage of those at work was 7683/2454 = $3.131.
4. The time distribution was about the same as that for the wages and salaries, except for activity 20 which was supervision. The rate here obviously

Table 4

Summary of Work Status from 10 Day Random Time Sample

Work status	Sample employee minutes	Percent	Estimated total	
			Employee hrs.	Dollars
At work	1,227	83.4	2,454	$7,683
Other	108	7.3	216	676
Waiting for work	8	0.5	16	50
On leave	129	8.8	258	808
Sum	1,472[a]	100.0	2,944[b]	9,218[c]

[a]This is 368 employee days = (37×9) + (35×1).
[b]This checks with the total frame of employee hours: 368×8 = 2,944. On one day, there were 35 employees, not 37.
[c]It is assumed that those not at work averaged the same rate as those at work.

Table 5

Percentage Distribution by Time and Wages by Activity Codes for Those at Work

Activity code	Employee minutes Number	Employee minutes Percent	Employee min. × hr. rate	Estimated dollars (frame)	Percent
3	561	45.7	1,768.0864	$3,536.17	46.0
3, 4	21	1.7	66.7700	133.54	1.7
3, 4, 5	182	14.8	564.1675	1,128.34	14.7
3, 5	1	0.1	3.3100	6.62	0.1
4	10	0.8	31.2725	62.55	0.8
4, 5	4	0.3	12.5800	25.16	0.3
5	132	10.8	408.27625	816.55	10.7
6	7	0.6	23.1525	46.30	0.6
8	4	0.3	12.2475	24.49	0.3
9, 10, 15	3	0.2	9.9300	19.86	0.3
11	1	0.1	3.3900	6.78	0.1
11, 13	32	2.6	108.4800	216.96	2.8
15	46	3.8	141.92875	283.86	3.7
15, 16, 17	75	6.1	164.6475	392.30	4.3
15, 17	1	0.1	3.0450	6.09	0.1
17	23	1.9	71.3475	142.69	1.9
19	50	4.1	159.0375	318.08	4.1
20	71	5.8	281.1405	562.28	7.3
no code	3	0.2	8.8225	17.64	0.2
Sum	1,227	100.0	3,841.63	$7,683.26	100.0

was higher than that for the other classes. A time distribution can be very misleading when the wage and salary rates vary greatly from one class of activities to another (Table 5).

No serious difficulties arose with regard to the sample plan, nor with its implementation. There was a minimum amount of interference with the work of the employees. The major problem encountered was not a sampling problem, nor a management problem, but an accounting problem. This involved the soundness of the 20 classes of activities. It was immediately discovered that many actual activities the clerks were performing did not fit into the classifications. Actual work tended to overlap many of the classes, as is shown by the combinations of activity codes shown in Table 5. As a result of this experience, a major recommendation was to revise the classification of activities so as to reduce to a minimum the number of overlapping classes.

Billing department of a telephone company.[17] The Chesapeake and Potomac Telephone Company of Virginia ran a test to determine whether a random time sampling method or a 100 percent time record method was better to determine the distribution of time used by clerks and machines on specified jobs in their revenue accounting center. The results of the test would determine what the four telephone companies would do about this problem.

Several hundred thousand customers had to be billed each month, so it was highly important to determine utilization of both clerks and machines. The billing process is fully mechanized using punch cards, accounting machines, and computers. A number of different work operations are performed on a variety of machines. Time reporting 100 percent on these units had always been difficult, not only because of the variety of work operations and machines used, but because of overlap operations. A clerk may operate simultaneously two or more machines which are performing different jobs. An operator is required to pro-rate, by some method, the total time over several jobs performed at the same time.

A group of accountants and statisticians planned the project. Five random minutes were selected per day from 510 minutes; the lunch period of 30 minutes was included. Since a machine unit consisted of about 10 clerks and 20 machines, it was estimated that observations for 100 random minutes per month would be adequate. The five random minutes were assigned to subsamples or replicates, one through five, so that the calculation of the estimates of the standard errors and confidence limits was greatly simplified. So far as is known, this is the first work sampling study to use replication of randomly selected minutes to estimate sampling errors.

The work activities were coded into 17 classes; the machines were coded into four classes and 10 types. Final results were obtained for each of these classes and types. Data at each random minute for each clerk and each machine were recorded on mark sense cards. The production (activity) code and the type of machine being operated was recorded on each clerk card. Each machine card was marked to indicate whether it was idle; out of order; working; and if working, the production code of the job. The date time of the observation, subsample or replicate number, clerk and machine identification number, and other information were also recorded. At the end of each month, the clerical hours were summarized and distributed by production codes; similarly, machine hours were tabulated showing productive hours, idle time, and out-of-order time according to the job codes.

Finally, the results from the 100 percent time reports were compared with those obtained from the random time sample including 90 percent confidence limits. In 12 instances, the time from the time report fell outside these limits; nine of these were in machine production, where the time report showed consistently lower figures. In clerical jobs, two time report figures were too high and one was too low. A total of 1,245 observations were made on clerical work; 3,903 were made on machine work. The machines were in production 55 percent of the time; the time sheets showed 51 percent, but this value fell below the 90 percent confidence limit suggesting that the time sheet gave too low a figure.

The committee concluded that RTS was better: it gave more accurate reporting, it interrupted valuable machine time the least, operators could devote full time to the machines, paper work for operators was reduced, the problem of overlap operation was solved, reports were unbiased by human judgment, pro-rating of time by operators was no longer required, and approximately 40 percent of the time formerly required to summarize reports was saved.

It should be observed that in this application, as in the preceding application, the classification and coding of activities and the accurate identification and recording of these codes are extremely important. Indeed, unless the original data are valid and accurate, the concepts of standard error and confidence limits collapse and are of little use because they are eclipsed by various sources of bias.

Deriving weighting factors in a Naval communications center.[18] RTS was used to determine how radiomen could be equitably allocated in a Naval communication center. The need grew out of the fact that the workload consisted of a wide variety of work units of varying difficulty. Originally the work was broken down into five broad functional areas and weighting factors were applied to the different work units, the factors being averages of judgments. Certain critical questions arose about this procedure. This led to the formation of a research team whose function was to find the reason for the weakness of the present system and formulate a new work measurement program that would be more acceptable to all personnel.

At first, the research team suggested that personnel keep logs, but this was objected to. So was the use of the stop watch. Finally, they laid out a work sampling plan based on random time sampling which would yield new weighting factors based on how much time, on the average, was required to process each message. Messages were classified into seven types, the average time required for each type of message was calculated using time spent on each type of message as derived from a random time sample, and the number of messages of each type readily available from operating records. These average times resulted in weights ranging from one to eight compared with a range of from one to five under the older system. Whereas employees were dissatisfied with the older weights, there was complete agreement and acceptance of the new factors derived from the random time sample.

In designing and implementing this RTS study, several problems arose which had to be resolved. These were:

1. A new classification of the functional areas and the work activities had to be developed.

2. Agreement on the meaning of terms had to be reached: terms such as "message center", "message section" and how "routing section", "broadcast section", and "wire room section" fit into them. Considerable differences of opinion existed as to what these terms included and excluded.

3. It was discovered that a brochure or booklet was needed to explain in clear language the need for the study and the procedures involved, so every individual would know how he would be affected.

4. At first, the supervisors objected to being observers because of the extra work, but several trial runs demonstrated that their fears were unfounded.

5. Supervisors also objected to the use of a table of random numbers since it interfered with their regular work. To resolve this, a clock was obtained which could be set so that a chime sounded at each random minute included in the RTS.

Some special points in this study need to be emphasized. Note especially that the workers and the supervisors did not object to RTS, but they did object to 100 percent time logs and stop watches. The acceptance of RTS by workers, including union members, has been observed by others, and is an advantage not to be taken lightly. Note also that a new classification of areas and work activities had to be developed. This shows that, as was pointed out in two previous applications, the formulation of a sound classification of work activities is very vital. Note also how some trial runs of the RTS allayed the fears of the supervisors, just as it did in the first application described above. Note how a clock with a chime was used to overcome other objections. This RTS study is an excellent example of how what appeared to be an almost impossible situation was finally resolved in a series of steps that led to a very successful implementation.

How an airline could have used RTS in a rate hearing. Delta Airlines had a joint cost operation which could have been easily decomposed by RTS. The problem arose in a rate hearing before the Civil Aeronautics Board. The actual, relevant testimony is reproduced below.[19]

Lawyer: "Is it possible for Delta to break down the $20.7 million figure for other cargo handlers which is far and away the largest item in this account, as between baggage handling and non-passenger cargo handling?"

Delta official: "A vast majority of our ramp service agents wear two or three hats. We cross-utilize baggage handlers. They may be handling baggage today and mail and express tomorrow."

Lawyer: "There is no way Delta could keep track of that that (sic) would permit an assignment of costs as between baggage handling costs and other cargo handling costs?"

Delta official: "There is a way. You can always come up with some allocation techniques. It is a matter of judgment. . . ."

Obviously, the official was not familiar with the fact that RTS applied to work is a made-to-order technique to solve the problem of joint costs, regardless of whether workers wear two, three, or a dozen hats. The lawyer asked a very fair question because the joint cost figure of $20.7 million was the largest item in the account under consideration.

It is *not* a matter of hunting for some allocation technique. It is *not* a matter of judgment. It is a matter of knowing that there is a powerful technique, RTS, which had been successfully applied and widely published by 1970, the time of this hearing.

What was needed, as the lawyer clearly stated, was a technique that would decompose $20.7 million into two classes — baggage handling and non-passenger cargo handling. A well designed and managed RTS study could easily have produced exactly this. Application of RTS would have allowed Delta to control these joint costs.

Notes

[1]L. H. C. Tippett, "A Snap-Reading Method for Making Time-Studies of Machines and Operators in Factory Surveys," *The Journal of the Textile Institute Transactions,* February 1935, pp. 51-70.

[2]R. L. Morrow, *Time Study and Motion Economy,* Ronald Press, New York, 1946, pp. 175-199; also *Mechanical Engineering,* April 1941, pp. 302-303.

[3]C. L. Brisley, "Work Sampling Technique," *Quality Control Conference Papers, 1953,* American Society for Quality Control, pp. 115-125; Brisley, "How to Put Work Sampling to Work," *Factory Management and Maintenance,* July 1952, pp. 84-89.

[4]A. Abruzzi, *Work Measurement,* Columbia University Press, New York, 1952, pp. 245-272.

[5]Ralph M. Barnes, *Work Sampling,* W. C. Brown Co., Dubuque, Iowa, 1956.

[6]A. C. Rosander, H. E. Guterman, and A. J. McKeon, "The Use of Random Work Sampling for Cost Analysis and Control," *Journal of the American Statistical Association,* vol. 53, June 1958, pp. 382-397.

[7]A. C. Rosander, *Case Studies in Sample Design,* Marcel Dekker Inc., New York, 1977. Tables on random minutes and days are in the appendix. Reprinted by courtesy of Marcel Dekker, Inc.

[8]Rosander, *ibid.,* Chapters 14, 15 and 16.

[9]E. J. Engquist, "How Management can use Work Sampling," *New Frontiers in Quality Control,* Proceedings of the Middle Atlantic Conference, American Society for Quality Control, Washington, D.C., March 1962.

[10]Rosander, *ibid.,* pp. 318-324.

[11]R. A. Fisher and F. Yates, *Statistical Tables,* 6th edition, Hafner Press, New York, 1963, p. 87.

[12]Rosander, *ibid.,* pp. 253-263.

[13]Rosander, Guterman, and McKeon, *ibid.,* pp. 382-397.

[14]E. J. Engquist, *ibid.*

[15]Rosander, *ibid.,* Appendix.

[16]R. E. Heiland and W. J. Richardson, *Work Sampling,* McGraw Hill, New York, 1957.

[17]C. E. McMurdo, "Work Sampling for Distributing Machine and Clerical Time," *Middle Atlantic Conference Proceedings,* American Society for Quality Control, Washington, D.C., March 1962, pp. 305-310. This study is described in detail in Rosander, *ibid.,* Chapter 16.

[18]*Navy Management Review,* April 1961, pp. 10-12; also Rosander, *ibid.,* Chapter 16.

[19]Civil Aeronautics Board, *Docket Nos. 21866-7,* p. 717.

25

Control by Comparisons

On the positive side, control by comparisons means finding out what is the best, the better, the superior, the successful, the safer, the most reliable, the most efficient, the most effective, and the most productive and taking action accordingly.

On the negative side, control by comparisons means finding out what is wrong, faulty, malfunctioning, in error, dangerous, defective, wasteful, below standard, undesirable, inferior, unreliable, and failing and taking *corrective* action.

Kinds of comparisons. There are several kinds of comparisons that can be made, but in all of them the question to answer is the same "Does a real difference exist?" If so, is the difference important in the real world? If so, what should be done about it?

1. Compare two elements A and B which may be materials, methods, machines, products, processes, procedures, or treatments.
2. Compare two elements A and B where 1) B is a standard, 2) B is some critical level, and 3) B is the method, machine, procedure, etc. now being used.
3. A and B are two test groups, where A is a control group and B is the test group. The same treatment, experimental, or test factor is applied to the B group, but not to the A group.
4. Compare event A or B in two areas or two groups at the same time or at several different times.
5. Analyze group C: Is it meeting a standard? Is it meeting good practice? Examples are machine operators, laboratories, chemists, tax assessors, and tax auditors.

Examples: Some specific examples of comparisons are the following: Some of these are described in detail in this chapter; the source is given in parenthesis.

1. Effect of two sleep-inducing drugs on the same 10 persons, measured in terms of the extra hours of sleep (R.A. Fisher).
2. Comparison of two methods: 100 percent time records and RTS (McMurdo in previous chapter).
3. Testing for extent of plutonium in the same areas by two different methods of sampling (Johnson and Colorado Department of Health).
4. Testing for the differences of two ends of a three minute timer with replication (Rosander).
5. Testing of two calculating machines by the same person on 10 different problems involving sums of squares (Cochran and Cox).
6. Using two instruments to test air composition at the same point, every hour, for 12 hours.
7. Testing a food additive by a control group of 10 animals versus an experimental group of 10 animals, paired or unpaired.
8. Testing two stores for differences in the price of several identical items.
9. Testing two kinds of shipping containers used to ship same item (Purcell).
10. Testing 148 tax assessors' appraisal of same property (Murray and Bivens).
11. Testing 29 laboratories which test the same material (Youden).
12. Testing four chemists on the same material (Noel and Brumbaugh).
13. Testing two different methods of printing one dollar bills (U.S. Bureau of Engraving).
14. Testing two different kinds of packaging to display the same kind of candy in a retail store (U.S. Department of Commerce).

A note on statistics used in this chapter. Several statistical techniques are used in the applications described in this chapter; they are explained in detail in Part 2 — Statistical Techniques. References to appropriate chapters, however, are given for the reader who desires these details. The following topics and concepts are referenced:

1. Linear regression $y = a + bx$, where a and b are calculated by the method of least squares.
2. Correlation coefficient r and r^2.
3. t test of paired differences.
4. t test of two independent means.
5. Test of the difference between two proportions.
6. Chi-squared test of independence.
7. Chi-squared test for variances (Bartlett's test of k variances).
8. Test of the difference between two counts assuming Poisson distribution.
9. Analysis of variance with F tests.
10. Standard error of a single measurement.
11. Circular probability.

Many of these calculations are greatly simplified by use of the new programmable desk calculators for several reasons: a program may be available either in the machine or in a "stat rom", a program may be written, or the problem may be worked directly from the keyboard. With these new electronic calculators, the problem of calculation is no longer the time-consuming chore it once was. This greatly increases by several times the productivity of anyone who learns how to operate the machine. One can use these machines most efficiently on statistical and mathematical problems only if one understands the basic assumptions and logic underlying the various mathematical and statistical calculations.

A review of some basic concepts and calculations. At this point, we present some of the basic concepts in statistics and the expression for each. For those familiar with them, this section can be skipped:

1. The mean of x:

$$\bar{x} = \frac{1}{n}\sum_{1}^{n} x_i, \text{ where } n = \text{sample size.}$$

2. The variance of x:

$$s_x^2 = \frac{1}{n-1}\sum_1^n (x_i-\bar{x})^2 = \frac{1}{n-1}\left[\sum_1^m x_i^2 - \frac{\left(\sum_1^n x_i\right)^2}{n}\right]$$

3. The standard deviation of x:

$$s_x = \sqrt{s_x^2}.$$

4. The coefficient of variation of x:

$$c_x = \frac{s_x}{\bar{x}}.$$

5. The standard error of a mean $\bar{x}$ calculated from *random sample* of size n:

$$s_{\bar{x}} = \frac{s_x}{\sqrt{n}}\sqrt{1-\frac{n}{N}},$$

where N = population size and $\frac{n}{N}$ = sampling rate.

6. The relative sampling error in the mean $\bar{x}$:

$$c_{\bar{x}} = \frac{s_{\bar{x}}}{\bar{x}}.$$

7. The estimate of a *binomial* proportion p_a:

$$p_a = \frac{n_a}{n}.$$

8. The standard error of p_a:

$$s_a = \sqrt{\frac{p_a(1-p_a)(1-n/N)}{n}}$$

9. Estimate of a *ratio* proportion:

$$f = \frac{\sum x}{\sum y} = \frac{\bar{x}}{\bar{y}}.$$

10. Standard error of a ratio proportion f:

$$s_f^2 = \frac{1}{n(n-1)\bar{y}^2}\left(\sum x^2 + f^2\sum y^2 - 2f\sum xy\right),\ s_f = \sqrt{s_f^2}.$$

Testing the difference between two calculating machines.[1] The speed of two standard calculating machines was tested by having the same person calculate 10 problems calling for the calculation of the sums of squares on each machine. The time in seconds required on each machine beyond two minutes for each of the 10 problems (replicates) are given below. The difference in each problem is also given:

Problem (replicate)	Machine A=y	Machine B=x	Difference (A−B): y−x=d
1	30	14	16
2	21	21	0
3	22	5	17
4	22	13	9
5	18	13	5
6	29	17	12
7	16	7	9
8	12	14	-2
9	23	8	15
10	23	24	-1
Sum	216	136	80
Mean	21.6	13.6	8.0
Std. Dev.	5.4	6.0	

Linear regression gives a = 19.0, b = 0.19, $r^2 = 0.05$, and r = 0.21. The equation of the line is y = 19.0 + 0.19x. The hypothesis to test is that the population difference is zero. The t test for this is as follows:

$$t = (\bar{x} - u)\sqrt{\frac{n(n-1)}{\sum (d_i - \bar{d})^2}} = (8.0 - 0)\sqrt{\frac{9 \times 10}{466}} = 3.52,$$

where u, the population mean is set equal to zero;

$$466 = \sum d^2 - \frac{(80)^2}{10}$$

For t = 3.52 and n − 1 or nine degrees of freedom, the probability of obtaining a larger t value is 0.003; or for a two tailed test, 0.006. This means that the difference of eight is significant at the one percent level. This is what the authors give, even though actually they used a crossover design where each machine was used first in five replications to correct for any advantage the second machine might receive. Pairing gives nine degrees of freedom; the two groups separately have 18, (10 + 10 − 2), but the positive correlation between x and y overcomes this advantage. In the next example, a high correlation tends to make even a smaller difference significant.

Testing a three minute timer. Using a stop watch that read to 0.1 second, the author tested the two ends of a three minute timer to determine if any significant difference existed in the two readings. Consecutive readings were paired so that if any time effects existed, both ends would be equally affected. Ten pairs of readings were recorded as follows (seconds beyond 2 minutes 50 seconds):

Trial	A (y)	B (x)	Difference y−x=d	Difference (d−1.75)
1	13.9	10.0	3.9	2.15
2	15.6	13.0	2.6	0.85
3	15.4	15.4	0.0	−1.75
4	16.5	15.6	0.9	−0.85
5	18.3	12.0	6.3	4.55
6	9.4	7.7	1.7	−0.05
7	11.9	13.7	−1.8	−3.55
8	16.2	14.9	1.3	−0.45
9	9.5	7.6	1.9	0.15
10	11.8	11.1	0.7	−1.05
Sum	138.5	121.0	17.5	15.40 absolute value
Mean	13.85	12.10	1.75	ignoring signs
Std. Dev.	3.06	2.97		

Linear regression gave a = 4.69, b = 0.76, $r^2 = 0.54$, and r = 0.73.

the t test gave $t = 1.75\sqrt{\frac{9 \times 10}{43.765}} = 2.51,$

where $43.765 = \sum(d-1.75)^2;\ 43.765 = \sum d^2 - \frac{17.5^2}{10.}$

For t = 2.51 and nine degrees of freedom, the difference is significant at the four percent level. This small difference is significant because r is high and positive.

For three minute timing, an additional question arises. To what extent, if any, do the timing ends deviate significantly from three minutes or 180 seconds? The standard deviation of a single value is needed. One can use either analysis of variance or the absolute value method; here the latter is used:

$$s' = \sqrt{\tfrac{1}{2}\pi}\sum\left|\frac{(d-1.75)}{n}\right| = 0.886\left(\frac{15.40}{10}\right) = 1.36;$$

At the 95 percent level, zs' = 1.96(1.36) = 2.7. Assuming a normal distribution, with a mean of 180, the allowable deviation at the 95 percent level is 2.7 seconds about 180 or 12.7 seconds about 170. The table above shows nine values within these limits and 11 outside.

$$s_d = \sqrt{\frac{2s^2}{n}} = 3\sqrt{0.20} = 1.34,$$

where s = 3, n = 10.

For timing a three minute operator-type telephone call, all 20 values are within the allowed tolerance of 15 seconds, but on direct dialing the computer allows six seconds so 186 seconds is the upper limit. Of the 20 values, three are above this limit, all at the high end of the timer. For boiling an egg, the mean difference of 1.75 seconds is of no practical significance even though the difference is significant statistically. Certainly, the 2.4 seconds short of 180 minutes will make no difference, nor is it likely that the largest excess, 8.3 seconds, will make any difference. Deviations from three minutes are not important unless a variation of from five to 10 seconds is critical.

The importance of the correlation coefficient in using paired differences can be shown quantitatively. The effect of correlation between x and y on the variance of the difference (y − x) can be seen by applying the expression of this variance when x and y are correlated:

$$s_d^2 = s_x^2 + s_y^2 - 2r_{xy}s_xs_y \text{ for large samples.}$$

In this problem, if r = 0, then:

$$s_d^2 = 3.06^2 + 2.97^2 = 18.1845;$$

$$s_d = 4.26.$$

But, r = 0.73. Hence:

$$s_d^2 = 3.06^2 + 2.97^2 - 2(0.73)(3.06)(2.97);$$

$$s_d = 2.22.$$

Correlation has reduced the standard deviation from 4.26 to 2.22.

Comparing two methods of testing plutonium. About 15 miles northwest of Denver is the Rocky Flats plutonium plant. It is run by a private contractor under the Department of Energy. Periodic tests are made by the contractor, as well as by government agencies, to measure plutonium 239 in the air, water, and soil in the immediate and surrounding areas, including Walnut Creek which is a source of drinking water.

Residential developers would like to build houses in four areas in the neighborhood, but authorities differ as to whether the areas are safe or not. Data have been collected by various agencies and by different methods leading to conflicting conclusions. For example, two different government agencies tested soil samples collected in the four proposed residential areas. Since the samples were collected in the same areas, it is possible to pair sample results in a total of 10 areas.

The major difference between the two sets of samples was the method of sampling. One set consisted of respirable dust collected by brushing the surface and collecting two pounds in each sample; in the other, soil was spooned from the topsoil. Both sets of samples were tested for plutonium 239 and the results were expressed in the number of disintegrations per minute per gram. The data are given in Table 1. The Jefferson County Health Department used the dust samples; the State Department of Health used the topsoil samples.

As Table 1 shows, the differences are so large that a statistical test is unnecessary to show the difference. On the average, plutonium 239 was about 10 times as concentrated in the 10 samples of respirable dust as in comparable samples of topsoil. The linear correlation coefficient r = 0.84, with $r^2 = 0.71$, shows a relatively high degree of relationship between the 10 pairs of values, but at widely different levels.

The data raise the basic question of what constitutes a valid sample of soil in a situation of this kind. The representative of one group, a medical doctor, says the sample should be selected from dust which persons are likely to breathe since protecting health is the real purpose of the sampling tests. To them, the collection of topsoil, including stones and gravel, does not yield a valid sample for health purpose testing.

Table 1

Comparison of Two Methods of Sampling Soil for Plutonium 239

Test area	Respirable dust ave. dpm/gram (y)	Topsoil ave. dpm/gram (x)	Difference y − x
1	59	13.5	45.5
2	120	14.1	105.9
3	36	0.2	35.8
4	24	0.14	23.86
5	26	2.96	23.04
6	40	0.14	39.86
7	1.3	0.23	1.07
8	1.0	0.05	0.95
9	7.7	0.72	6.98
10	6.9	0.72	6.18
Sum	321.9	32.76	289.14
Mean $\bar{x}$	32.19	3.28	28.91
Std. dev. s.	36.17	5.61	
$c = s/\bar{x}$	1.12	1.71	

straight line fit: y = a + bx
a = 14.429,
b = 5.422,
$r^2 = .71$,
r = .84.

Source: Carl J. Johnson, M.D., Director of Health, Jefferson County Colorado Health Department, *Remarks before State Board of Health,* January 21, 1976, Table 2. In areas where more than one sample measurement of respirable dust was made, the minimum measurement is used to compare with the topsoil value. "dpm" means disintegrations per minute. Used by permission of Dr. Johnson. Calculations by the author.

The Department of Health uses topsoil samples and samples from six air filters in the area. These filters are weighed before and after they are used to catch particulate materials which are tested for radioactivity. Air is drawn into the sampler at the rate of 40 to 60 cubic feet per minute with a velocity that does not differ markedly from velocities associated with normal breathing. The sampling unit is located about 44 inches from the surface of the ground. This approximates the height of a young child.

It has been found that land development results in a reduced concentration of the plutonium contaminant. Agricultural plowing of the land in the area of plutonium contamination may reduce the level by 90 percent (to 1/10 the former level).

This is an excellent example of the critical nature of the method of sampling including location and height above the ground — determining how to sample as well as where and when to sample are of crucial importance in obtaining valid data for purposes of environmental control. The extent to which the soil has been tilled is very important in testing soil. This is the same situation that can be raised relative to the sampling of air: Should air be sampled 15 feet above the street or at a four to six foot level where persons breathe? It is reported that a study in New York City did not show any differences in amounts of particulates at various heights above the street. The question is whether the findings of one study in one city should be generalized to the entire country.

Testing variation between traffic experts. A railroad merger, which must be approved by the Interstate Commerce Commission, may affect adversely one or more other railroads by diverting freight movements from their routes to those of the proposed merger. Usually railroads so affected appear as parties to the proceedings held by the Commission and present evidence, for example, that so much traffic will be diverted that the proposed merger should be denied. Tens of thousands, if not hundreds of thousands, of movements may be affected involving revenues in the millions of dollars.

An estimate of the amount and type of rail traffic to be diverted is derived from a sample of interline rail movements which might be affected by the merger. The frame for this sample may be interline waybills, abstracts, listings, or records on computer tapes. Local traffic originating and terminating on one railroad is ignored since this traffic is usually not affected by a merger.

After a random or probability sample is selected, it may be divided into regions, territories, or commodities and given to traffic experts familiar with these regions, territories, or commodities. These experts determine, on the basis of their experience and knowledge of rail traffic, whether each movement is likely to be diverted and, if so, what part of it. Stated another way, what is the probability of diversion? Hence, each sample movement is assigned a proportion from 0.00 to 1.00 which measures the extent of diversion. When this proportion is applied to the revenue earned by that movement, an estimate is obtained of the money loss assigned to that movement. Sometimes these proportions are derived by one person, by consensus of a group, or by two or more experts making independent judgments. This latter method is preferable because it shows to what extent the experts differ and to what extent they introduce variability into the final estimate.

We describe a sample diversion study which consisted of 12 replicates, three of which were assigned at random to three experts to make independent judgments. The problem was to determine whether any significant difference existed between these three experts.[2] Each replicate was divided into 27 subgroups or slices. The proportion judged divertible in each of the 27 subgroups and the variance within experts, that is the variance between the 27 subgroups for each expert, is given in Table 2. Analysis of the data showed no significant difference between experts. The same results are obtained from the usual analysis of variance table derived from the percentages shown in Table 2.

Component	df	ss	ms	F
Between experts	2	651.78	325.9	0.74
Within experts	78	34,484.99	442.1	
Total	80	35,136.77		

The calculations are:

Item	Expert A	Expert B	Expert C	Sum
number of subgroups	27	27	27	81
sum of proportions	1025.62	1180.39	1011.18	3217.19
mean $\bar{p}$	37.99	43.72	37.45	39.72
sum of squares (ss)	12,026	15,496	6,963	34,485
variance	462.53	596.00	267.81	1,326.24

The total sum of squares is 162,918.39

$$162{,}918.39 - \frac{3217.19^2}{81} = 35{,}136.77.$$

The sum of squares between experts is

$$\frac{(1025.62^2 + 1180.39^2 + 1011.18^2)}{27} - \frac{3217.19^2}{81} = 651.78.$$

Since the variation within experts exceeds that between experts, there is no statistical evidence of a difference between the judgments made by the experts. Two points are noted. Bartlett's test of the variances does not give a significant test.

$$\chi^2 = 2.3026(78 \times 2.6455 - 26 \times 7.8681) = 4.10.$$

For two degrees of freedom a chi-square of 4.10 gives $P = 0.13$; hence, the test is not significant.

The F value of 0.74 is not an unusual value to obtain if three independent random samples of 27 each were drawn from a very large normal population. Both the between sample variance and the within sample variance would give unbiased estimates of the population variance. Their ratio in a large number of repeated selections would vary about the expected value of one, with a deviation that would very often exceed 0.26.

Table 2

Variance of Percentage of Loss Within Experts

Slice		Subsample 2 Expert A		Subsample 8 Expert B		Subsample 9 Expert C	
		P_A	$P_A - \bar{P}_A$	P_B	$P_B - \bar{P}_B$	P_C	$P_C - \bar{P}_C$
$500 or over	1	58.38	20.39	56.49	12.77	21.20	-16.25
Under $500	2	3.40	-34.59	36.08	-7.64	24.67	-12.78
	3	41.27	3.28	27.63	-16.09	43.63	6.18
	4	0	-37.09	23.67	-20.05	19.88	-17.57
	5	0	-37.09	61.03	17.31	21.04	-16.41
	6	36.58	-1.41	0	-43.72	25.35	-12.10
	7	30.21	-7.78	52.24	8.52	29.60	-7.85
	8	19.17	-18.82	25.51	-18.21	18.12	-19.33
	9	38.93	0.94	18.64	-25.08	20.69	-16.76
	10	69.82	31.83	65.90	22.18	62.32	24.87
	11	19.98	-18.01	16.03	-27.69	20.61	-16.84
	12	50.27	12.28	26.84	-16.88	40.15	2.70
	13	40.73	2.74	61.06	17.34	62.05	24.60
	14	19.04	-18.95	9.60	-34.12	28.32	-9.13
	15	72.94	34.95	72.70	28.98	33.53	-3.92
	16	64.17	26.18	25.40	-18.32	46.05	8.60
	17	32.19	-5.80	16.03	-27.69	61.54	24.09
	18	21.77	-16.22	71.24	27.52	31.81	-5.64
	19	59.85	21.86	45.08	1.36	58.31	20.86
	20	37.21	-0.78	45.14	1.42	31.15	-6.30
	21	41.58	3.59	46.06	2.34	27.92	-9.53
	22	34.50	-3.49	22.94	-20.78	35.61	-1.84
	23	50.61	12.62	54.90	11.18	45.77	8.32
	24	73.53	35.54	71.78	28.06	61.14	23.69
	25	60.20	22.21	51.93	8.21	58.76	21.31
	26	13.13	-24.86	76.47	32.75	17.75	-19.70
	27	36.16	-1.83	100.00	56.28	64.21	26.76
Total		xxx		xxx		xxx	
Average		$37.99 = \bar{p}_A$		$43.72 = \bar{p}_B$		$37.44 = \bar{p}_C$	
SS		12,026		15,495		6,963	
SS/(27-1)		463		596		268	

$p_n = 0.41$, $s_p = 0.017$ based on revenues

Testing four analysts.[3] The purpose of this experiment was to study sources of variability of a chemical assay, especially that due to analysts. Samples of a chemical were prepared at two different potencies. Each sample was divided into 16 parts, eight parts for the first day and eight parts for the second day. Four analysts made two determinations each day on each sample, for two days for two samples, yielding eight determinations per analyst. The values of the 32 measurements are given below. Analysts are designated A, B, C, and D; the first day is D-1, the second day D-2.

	Sample 1		Sample 2	
	D-1	D-2	D-1	D-2
A	1053	1195	1305	1415
	1053	1258	1336	1415
B	1233	944	1417	1349
	1217	1142	1356	1158
C	1157	1163	1474	1163
	1062	1138	1348	834
D	1248	1209	1311	1319
	1147	1099	1358	1311

These data yield the following analysis of variance table:

Component	df	ss	ms	F	F 5%	F 1%
analysts A	3	38,305	12,768			
samples S	1	203,363	203,363			
days D	1	28,981	28,981			
interactions						
AS	3	25,374	8,458	1.11	3.24	
AD	3	127,534	42,511	5.60		5.29
SD	1	26,392	26,392	3.48	4.49	
ASD	3	81,049	27,016	3.56	3.24	
residual (within cells)	16	121,466	7,592			
Total	31	652,464				

The residual mean square is the variation arising from the 16 pairs of duplicates. It measures the extent to which analysts are able to make two independent determinations on the same sample on the same day. This mean square is used as the denominator in obtaining F from the other mean squares. The interactions are high: the interaction between analysts and days (AD) is significant at the one percent level; SD is high, but hardly significant while the highest interaction ASD is significant at the five percent level.

1. AS is low because every analyst had sample two larger than sample one, as it should be because of the different potencies. Despite this, analyst C had such a large residual mean square (analytical variance) that he was unable to detect this difference.
2. AD is high because analysts were not consistent from day one to day two; one analyst had a significant difference between days.
3. SD is high because the difference between samples on day one was much greater than the difference on day two.
4. ASD is significant because of the wide variation in SD between analysts, especially analyst C's high SD interaction.

As a result of the analysis of variance applied to each analyst, a residual variance measuring analytical error for each analyst was obtained. Limits were set about the mean of each analyst using $t=2.78$ for four degrees of freedom at the 95 percent level. This gave the following results:

Analyst	s^2	s	2.78s	$\bar{x}$	$\bar{x}+2.78s$
A	616	24.8	68.9	1254	1185 and 1323
B	9958	99.8	277.4	1227	950 and 1504
C	16,721	129.7	360.6	1167	860 and 1528
D	3072	55.4	154.0	1250	1096 and 1404

A had five points outside the 95 percent limits, two low and three high, B had one point low, C and D had no points outside the limits. However, C had such a high analytical error that no points fell outside the limits. The residual variance s^2 is a measure of the analytical variability based on four pairs of determinations. A had the lowest value; C had the highest. C had such a large variance, he could not detect the difference between samples that the other three detected. It was concluded that A was making some error in techniques from day to day; this was corrected. B and C were doing poor work and should be re-trained. D was capable of doing the work, but seemed careless. It was recommended that control charts be used by analysts to give a continuous record of their analytical variability.

Testing differences between laboratories.[4] Youden used paired materials, A and B, coupled with circular probability, to measure the precision and bias of 29 different laboratories testing one sample of each material. Each material was very much alike relative to the characteristic being measured. Each laboratory ran one test on each sample. These two measurements were plotted as one point on the usual rectangular grid; x represents material A and y represents material B. Youden does not explain why he uses two materials rather than two samples from the same material, nor how the two materials are selected so that they have about the same magnitude of the characteristic being tested. Examples of the idea, however, might be testing two different materials which have about the same electrical resistance, testing two fertilizers which have about the same proportion of nitrogen, or testing two chemicals which have about the same pH value.

The basic theory is that the points (laboratories) will tend to fall on a 45° line and be equally distributed within the four quadrants formed by the intersection of two perpendicular lines. These lines are represented by the two means or medians, providing that the laboratories differ only because of random variations and that biases or constant errors due to various causes are eliminated or are negligible. Furthermore, if x and y are independently and normally distributed, then the points will tend to fall within a circle drawn with a proper radius using the means as the center. A 45° line is not required.

Points distant from the 45° line indicate imprecision and not following the same procedures and controls in both samples. Points near the 45° line but far beyond the circle indicate constant errors leading to high values. Points near the 45° line below the circle indicate constant errors leading to low values. In general, points lying beyond the circle for the 95 percent level, for example, indicate the existence of one or more types of bias or constant errors in laboratory testing.

Table 3 gives the basic data for 25 laboratories; four outliers were eliminated from the analysis by Youden. The basic calculations, including a and b, from a linear least square fit, as well as r^2 and r, are as follows:

	x	y	x−y=d	$d-\bar{d}$
mean	0.2292	0.1340	0.0952	0.0530
s	0.07314	0.05965	0.06947	

Linear regression:

$a=0.04656$; $r^2=0.21881$;
$b=0.38151$; $r=0.468=0.47$.

A t test, that the population $\rho=0$ gives $t=2.54$ for 23 degrees of freedom where $t_{0.02}=2.50$ so the hypothesis correlation coefficient is zero, is rejected.

$$t=\frac{r(n-2)^{1/2}}{(1-r^2)^{1/2}} \text{ for n-2 degrees of freedom.}$$

The standard deviation of a single result, using the mean of the absolute values of the deviation of each difference d from its mean $\bar{d}$, is 0.047. This is the method used by Youden:

$$s' = \frac{1}{\sqrt{2}} \sqrt{\frac{\pi}{2}} \sum \frac{|d-\bar{d}|}{n} = 0.886 \frac{\sum |d-\bar{d}|}{n},$$

$$s' = 0.886(0.053) = 0.047.$$

The standard deviation, using analysis of variance, is 0.049:

	df	ss	ms	F	F(1%)
Between laboratories	24	0.155872	0.006495	2.69	1.98
Between A and B	1	0.113288	0.113288	46.95	4.26
Error	24	0.057912	0.002413		
Total	49	0.327072			

Therefore, the standard deviation or standard error $s'=(0.002413)^{1/2}=0.049$. It is the square root of the residual mean square, in this case the interaction between laboratories and the two samples A and B.

The sums of squares are calculated in the usual way.

A: $\Sigma x=5.73$; $\Sigma x^2=1.4417$; $C=9.08^2/50=1.648928$.
B: $\Sigma y=3.35$; $\Sigma y^2=.5343$.
sum 9.08 sum 1.9760

To obtain the sums of squares between A and B, square each sum, divide by the number of values each square is

$$\text{total ss}=1.9760-C=0.327072$$

$$\text{between laboratories } \frac{0.53^2+\ldots+0.24^2}{2}-C=0.155872$$

$$\text{between A and B } \frac{5.73^2+3.35^2}{25}-C=0.113288$$

The error is the remainder.

The value s′ is needed to determine the radius of a circle which will include the required percentage of observations, such as 95 percent or 99 percent. The center of the circle is the means $\bar{x},\bar{y}$; Youden used the median values of x and y. The equation of the circle is:

$$(x-\bar{x})^2+(y-\bar{y})^2=\lambda^2 s'^2=r^2,$$

where r is the radius of the circle. In this problem $\lambda=2.448$ for the circle to include 95 percent of the points. The circle is drawn by using a radius $\lambda s'$ with the center determined by the two means. In this problem:

1. The center is at $\bar{x}=0.229$, $\bar{y}=0.134$.
2. The radius $r=2.448(0.047)=0.1150$ or $2.448(0.049)=0.1200$.

Using $r=0.1150$ with the center at the *medians* places seven more points outside the circular limits in addition to the four outliers excluded. With the center at the *means* places six more points outside. Using $r=0.1200$ with the center at the *means* places five more points outside the circular limits. By using $s'=0.047$ and the medians, 11 laboratories are outside the circular limits; by using the means, 10 laboratories are outside the 95 percent limits. This shows the prevalence of constant errors or bias in these laboratories. This calls for a reexamination of the testing procedures and their execution. Youden states that there is no substitute for careful work in the laboratory. The plot of the 29 points and the circle using the means and $r=0.1150$ is shown in Figure 1.

Table 3

Data and Calculations on Percent Insoluble Residue in Cement Reported by 29 Laboratories

Laboratory[a]	Percent residue		Sum[b]	Difference	Deviation
	A=x	B=y	x+y	x−y=d	$d-\bar{d}$
1	0.31	0.22	0.53	0.09	−0.005
2	0.08	0.12	0.20	−0.04	−0.135
3	0.24	0.14	0.38	0.10	0.005
4	0.14	0.07	0.21	0.07	−0.025
5	0.52	0.37			
6	0.38	0.19	0.57	0.19	0.095
7	0.22	0.14	0.36	0.08	−0.015
8	0.46	0.23			
9	0.26	0.05	0.31	0.21	0.115
10	0.28	0.14	0.42	0.14	0.045
11	0.10	0.18	0.28	−0.08	−0.175
12	0.20	0.09	0.29	0.11	0.015
13	0.26	0.10	0.36	0.16	0.065
14	0.28	0.14	0.42	0.14	0.045
15	0.25	0.13	0.38	0.12	0.025
16	0.25	0.11	0.36	0.14	0.045
17	0.26	0.17	0.43	0.09	−0.005
18	0.26	0.18	0.44	0.08	−0.015
19	0.12	0.05	0.17	0.07	−0.025
20	0.29	0.14	0.43	0.15	0.055
21	0.22	0.11	0.33	0.11	0.015
22	0.13	0.10	0.23	0.03	−0.065
23	0.56	0.42			
24	0.30	0.30	0.60	0.00	−0.095
25	0.24	0.06	0.30	0.18	0.085
26	0.25	0.35			
27	0.24	0.09	0.33	0.15	0.055
28	0.28	0.23	0.51	0.05	−0.045
29	0.14	0.10	0.24	0.04	−0.055
Average	0.229	0.134		$\bar{d}=0.095$	0.053[c]

[a]Laboratories 5, 8, 23, and 26 were omitted from the analysis because they were considered outliers.
[b]This was added to the original table.
[c]This is the mean derived without regard to signs of the deviations.

Collaborative tests. Years ago considerable differences of opinion arose over whether the scientific methods used by the Food and Drug Administration were sound and reproducible. The collaborative tests developed out of this experience. Their purpose was to establish, on a broad scientific base, valid and efficient methods of chemical analysis. By testing a method at several laboratories, it was possible to develop methods that were scientifically and legally acceptable. It was possible, since all laboratories tested the same sample, to eliminate sampling variability and obtain measures of the amount of variation within and between laboratories. It was possible, by using samples "spiked" at various levels or concentrations, to determine the capability of laboratories to make an acceptable test at a critical or tolerance level, as well as reproduce known concentrations.

An actual case consisted of three food samples, each of which was submitted to nine laboratories in the collaborative testing program.[5] The first sample contained about 0.3 percent of the chemical under test, the second sample had 0.7 percent added, and the third sample had 1.5 percent added. Of the 27 measurements, two were omitted as outliers. These are starred and omitted from calculations of the mean and variance:

Laboratory	A (0.3%)	B (0.7%)	C (1.5%)
1	0.18	0.60	1.36
2	0.27	0.71	1.50
3 omitted	4.38* too high	0.79	1.42
4	0.20	0.62	1.52
5 omitted	0.28	0.73	0.78* too low
6	0.33	0.74	1.62
7	0.30	0.68	1.55
8	0.30	0.69	1.60
9	0.34	0.80	1.58
Mean	0.28	0.71	1.52
Std. dev. s	0.06	0.07	0.09

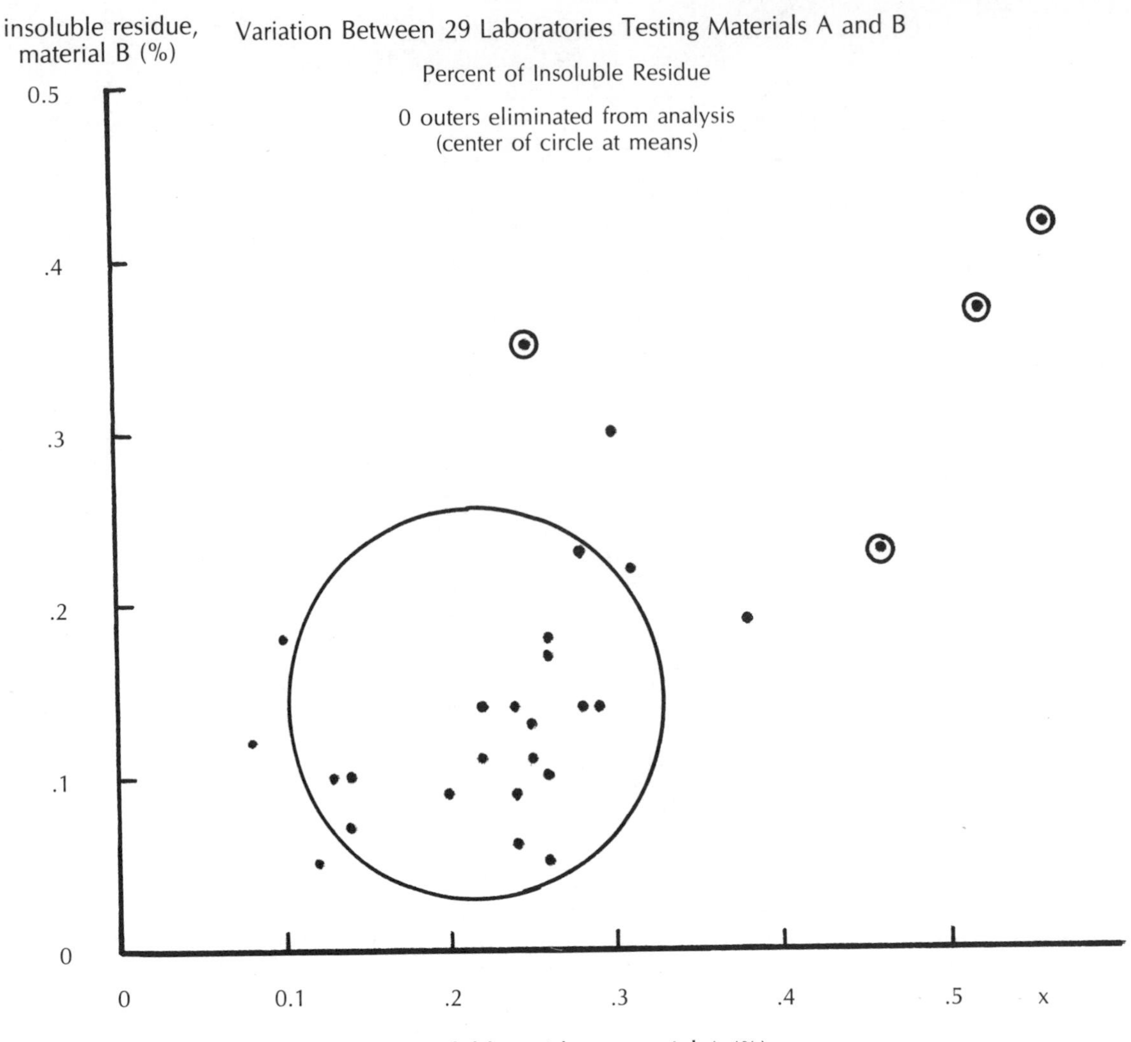

Analysis of variance was applied to these measurements; laboratories three and five were omitted because of the outliers. In both instances, some serious mistake was made so that one measurement was extremely high and the other was very low; both of these values would show "out-of-control" if an $\bar{x}$, R chart, was applied to each set of readings. Analysis of variance is used to test whether the laboratories are different assuming each set of values came from a population with a common variance. We know that the samples are different. The values used are 100x, where x is the original measurement.

	Samples			
Laboratory	A	B	C	Sum
1	18	60	136	214
2	27	71	150	248
4	20	62	152	234
6	33	74	162	269
7	30	68	155	253
8	30	69	160	259
9	34	80	158	272
Sum	192	484	1073	1749

Component	df	ss	ms	F	F (0.1%)
between samples	2	57,540	28,770		
between laboratories	6	830	138	11.8	8.38
error (interaction)	12	140	11.7		
Total	20	58,510			

There is a highly significant difference between laboratories for these three samples using the interaction mean square as a measure of random error. The low interaction mean square is due to the fact that a laboratory that produces a low, intermediate, or high value on one sample tends to be the same on the other samples. For example, laboratories one and four are low and laboratories six and nine are high on all three samples. This indicates a constant bias in laboratory performance — some negative and some positive with regard to the known value — and accounts for the significant test.

Comparing tax assessors.[6] A universal complaint of home owners is that they get unfair appraisals of their property for tax purposes. Complaints arise, for example, because of wide variations in the assessed valuation of similar houses in the block or in different parts of the city, or because of the same valuations given to houses quite different in size and condition.

Murray and Bivens of Iowa State College measured objectively the variation among 148 tax assessors from various parts of Iowa appraising independently the same residential property. The frequency distribution of these assessors by $1,000 intervals of assessed valuation shows a single peaked distribution, but the range is from $4,000 to $20,000. The values, to the nearest $100, are shown in Table 4.

Interval of assessed valuation in $1,000	Number of assessors
4-	1
5-	0
6-	1
7-	1
8-	3
9-	7
10-	32
11-	62
12-	22
13-	9
14-	6
15-	0
16-	1
17-	0
18-	1
19-	1
20-	1
Total	148

The arithmetic mean of the ungrouped values is $11,462 and the standard deviation is $1,847, giving a coefficient of variation of 0.16 or 16 percent. The distribution is fairly symmetrical in the range $8,000 to $15,000. If the seven extreme values are eliminated, three at the lower level and four at the higher level, the arithmetic mean is $11,375 and the standard deviation is $1,187, giving a coefficient of variation of 0.10 or 10 percent.

Assuming a normal model with mean $11,375 and standard deviation 1,187, then three sigma limits about the mean would eliminate six assessments; setting 95 percent limits would eliminate 16 out of the 148.

$11{,}375 \pm 3(1{,}187) = 7{,}814$ and $14{,}936$.

$11{,}375 \pm 1.96(1{,}187) = 9{,}048$ and $13{,}702$.

Murray and Bivens divided the assessors into 11 groups, so as to include in each group assessors from all parts of the state. Bartlett's test of the homogeneity of k variances shows these divide into two larger groups of sets (A and B) with different variances. Set A, including groups 1, 2, 3, 4, 8, 10, and 11, has a variance of 166.6 for 89 degrees of freedom; Set B, consisting of groups 5, 6, 7, and 9, has a variance of 669.1 with 48 degrees of freedom. The F ratio of $4.02 = 669.1/166.6$ is highly significant since $P(48,89) = 1.77$ at the one percent level. The variance tests are as follows:

$$\chi^2 = 2.30(n' \log s^2 - \Sigma n'_j \log s_j^2),$$

where n' is the total number of degrees of freedom, n'_j is the number of degrees of freedom in each group, s_j^2 is the variance of the jth group, and s^2 is the within-group mean square computed like the analysis of variance.

Set A:

$n' = 89$ for 96 assessors in seven groups; $s^2 = 166.6$;

$$\chi^2 = 2.30(89 \times 2.2217 - 195.3975) = 5.37.$$

For six degrees of freedom, $P_{0.50} = 5.348$, so the hypothesis of same variance is not rejected.

Set B:

$n' = 48$ for 52 assessors in four groups; $s^2 = 669.1$;

$$\chi^2 = 2.30(48 \times 2.8255 - 135.3532) = 0.623.$$

For three degrees of freedom, the probability P of exceeding this value is 0.89, so the hypothesis of same variance is not rejected although there is nothing in the data to indicate such a low value of chi-square.

Separate analysis of variance tables were prepared for A and B.

Set A

Component	df	ss	ms	F	F (5%)
Between 7 groups	6	1,786.3	297.7	1.79	2.20
Within 7 groups	89	14,828.4	166.6		
Total	95	16,614.7	174.9		

This is evidence that these seven groups of 96 assessors assigned valuations as a homogeneous group.

Set B

Component	df	ss	ms	F
Between 4 groups	3	1,383.7	461.2	0.69
Within 4 groups	48	32,117.3	669.1	
Total	51	33,501.0	656.9	

These four groups of 52 assessors assigned valuations as a homogeneous group, but with about four times the variance as Set A — 669 compared with 167. Five of the seven extreme values fell in this group of 52. These four groups were quite hetereogeneous internally. This could have resulted from the method of forming the groups and no doubt accounts for an F value less than one. Making an F test using 1/F with 48 and three degrees of freedom gives $F' = 1.45$, which is far from being significant since the five percent level is 8.58.

The analysis shows that two sets of assessors differed very significantly in the variability of their assessments, but between groups within sets no significant difference between means existed. No information is available of the extent to which assessor clinics would eliminate extreme values and reduce variability. No doubt training courses, based upon an established set of rules and standards to use in assessing residential property would go a long way toward reducing the variability in these assessed valuations.

Calculation of mean, standard deviation, and relative error after eliminating seven extreme values:

1. The seven extreme values are 4,000; 6,400; 7,900; 16,400; 18,000; 19,800; 20,000.

2. The mean $\bar{x} = \frac{1,603,900}{141} = \$11,375$ compared with $11,462.

3. The variance $s^2 = \frac{18,441,790,000 - 1,603,900^2/141}{140}$

   $= 1,408,165.$

   $s = \$1,187$ compared with $1,847.

4. The relative error $c = 1,187/11,375 = 0.10$ or 10 percent compared with 16 percent.

Testing two shipping containers.[7] The problem was to test whether a new container would afford as much shipping protection to a manufactured product as did the regular container. The test was important because the new container would save the company several thousands of dollars annually. A round trip transportation test was run with 1,000 inspected units packed in the new container and 1,000 inspected units packed in the regular container. At the end of the test run 10 defective units were found in the product packed in the new container; only one defective unit was found in the regular container.

Before a decision was reached in favor of the regular container, it was pointed out that the difference could be due to a weakness in the product rather than to differences in the protection of the two containers. What happened was that the one day's production was placed in the new container and another day's production was placed in the regular container. This meant that the quality of the product on the first day would be associated with the new container; the quality of the product on the second day would be associated with the regular container. If there was a difference in the quality of the output on these two days, the difference would appear as a difference between the two types of containers. Thus, the difference would be assigned to type of container rather than to daily variations in quality. That is precisely what happened; differences between days were transmitted to the two types of containers and it was erroneously asserted to be container differences. When these daily differences were eliminated, there was no significant difference between the two types of containers so the newer and cheaper one could be adopted. Daily production differences had been confounded with container differences.

Table 4

Assessed Valuations to the Nearest $100 Made on the Same House by 11 Groups of Tax Assessors

	1	2	3	4	5	6	7	8	9	10	11	Total
	139	117	112	112	125	115	124	112	83	116	110	
	107	98	120	134	116	110	87	100	143	97	110	
	106	103	112	113	111	123	121	108	119	115	136	
	110	115	108	103	40	115	110	121	108	110	116	
	138	115	96	137	105	124	127	120	121	90	119	
	117	107	112	136	113	110	110	111	102	110	113	
	121	110	115	110	122	111	113	116	111	110	122	
	104	92	130	105	122	109	200	113	64	144	105	
	103	121	115	143	105	115	123	118	80	121	111	
	117	111	131	115		113	107	127	115	91	105	
	142	100	135	140		91	107	164	121	100	101	
	106	111		111		120	109	112	113	110	114	
	113	107		114		198	121	105	180	126	106	
	111			114		107	108			104	79	
				114		114	144			103	110	
Sum	1,634	1,407	1,286	1,801	959	1,775	1,811	1,527	1,460	1,647	1,657	16,964
No.	14	13	11	15	9	15	15	13	13	15	15	148
Mean	116.7	108.2	116.9	120.1	106.6	118.3	120.7	117.5	112.3	109.8	110.5	114.62

Group	Sum of Squares
1	193,084
2	153,097
3	151,648
4	218,851
5	107,589
6	217,681
7	227,633
8	182,313
9	174,060
10	183,549
11	185,111
Total	1,994,616

Apparently the test was repeated using days as blocks or strata, the same number of units of product were assigned from each day to each container, and this assignment was made at random to equalize effects of all other factors that might affect the outcome. Ten packages of 100 units each were made up for each container; five packages were assigned to each date. The number of defective units found in the 20 packages were:

	Standard	Sum	Proposed	Sum	Total
Date I	0 2 1 1 3	7	2 0 2 1 2	7	14
Date II	0 0 0 1 0	1	1 0 0 1 1	3	4
Total	8		10		18

For a chi-square test of the container in relation to the data, the frequency table is:

	Standard	Proposed	Total	Differences	
				standard	proposed
Date I	7	7	14	2.5	2.5
Date II	1	3	4	−3.5	−1.5
Total	8	10	18		

The expected frequency in each cell, if no relation exists between containers and days, is 18/4=4.5. To correct for continuity, ½ is subtracted from the absolute value of each difference. Hence, the chi-square is:

$$\frac{2^2 + 2^2 + 3^2 + 1^2}{4.5} = 4.0,$$

which for one degree of freedom gives a probability of 0.0455. Hence, the null hypothesis of no relationship is rejected at the five percent level.

Just as container type and day can be confounded, so can the position the container is placed on the truck be confounded with the type of container if the new containers are placed in the front part of the truck and the regular containers are placed in the rear of the truck or vice versa. Similarly, if all of one type is placed on the left side of the truck and all of the other type is placed on the right side of the truck. Using a process of randomization, to place each type of container in each position on the truck, will eliminate any bias that might arise from this source.

It should be noted that selecting the two days at random would have given a broader base on which to select the product; by using a pair of such random days would have given an even broader base for testing because replication would have been used. Replication is just as important as blocking or stratifying, balancing, and randomizing because it gives a measure of the consistency of the results.

Testing two Poisson counts. Testing two Poisson counts is needed in a large number of situations where relatively rare events occur independently with constant probability. Examples are:

1. Deaths due to automobile accidents.
2. Violations of laws.
3. Certain crimes.
4. Errors in typing and printing.
5. Defective parts.
6. Defects of various kinds.
7. Accidents.
8. Failures of machines, equipment, and installations.
9. Blemishes.

The Poisson distribution has a density function:

$$p(x) = e^{-m}\frac{m^x}{x!},$$

where m = mean and x = count or number ranging from 0,1,2,..., infinity. In this distribution, the variance equals the mean:

mean $= m = np$, where n = sample size
p = proportion event A.
variance $= m$.
standard deviation $= \sqrt{m}$.

In the binomial distribution, $m=\bar{x}=np$ and variance $=npq$. As q goes to one or approaches one, then the variance $= np = m$, which is that of the Poisson distribution.

To test the difference between two counts that are distributed as the Poisson, the normal approximation is used. Under certain conditions, the square root of a Poisson variate tends to be normally distributed with variance ¼ or 0.25.[8] For two independent variates x_1 and x_2:

Sample count	$\sqrt{x}$	Variance
x_1	$\sqrt{x_1}$	1/4
x_2	$\sqrt{x_2}$	1/4
$x_1 - x_2$	$\sqrt{x_1} - \sqrt{x_2}$	1/2

standard deviation $= \sqrt{1/2}$.

The normal deviate z for testing the difference

$$z = \frac{\sqrt{x_1} - \sqrt{x_2}}{\sqrt{1/2}} = 2(\sqrt{x_1} - \sqrt{x_2}).$$

We set $z = 2$ for the two sigma level and 3 for the three sigma level.

Two sigma level or 95 percent level:

$x_2 = x_1 - 2\sqrt{2x_1} + 2$, where x_1 is given and x_2 is reduced count or frequency.

Three sigma level or 99.7 percent level:

$x_3 = x_1 - 3\sqrt{2x_1} + 4.5$, where x_1 is given and x_3 is reduced count or frequency.

Table 5

Poisson Counts at two sigma level (x_2) and three sigma level (x_3) to test deviation from x_1

x_1	x_2	x_3	2 sigma difference $x_1 - x_2$	3 sigma difference $x_1 - x_3$
8	2	0.5	6	7.5
18	8	4.5	10	13.5
32	18	12.5	14	19.5
50	32	24.5	18	25.5
72	50	40.5	22	31.5
98	72	60.5	26	37.5
128	98	84.5	30	43.5
162	128	112.5	34	49.5
200	162	144.5	38	55.5
242	200	180.5	42	61.5
288	242	220.5	46	67.5
338	288	264.5	50	73.5
392	338	312.5	54	79.5
450	392	364.5	58	85.5
.	.	.	.	.
.	.	.	.	.

Note: To extend this table: $x_{1i} = x_{2(i+1)}$; $(x_1 - x_2)$ increases by four and $(x_1 - x_3)$ increases by six.

$$x_3 = x_2 - \sqrt{2x_1} + 2.5, \text{ (three sigma level in terms of the two sigma level).}$$

Values of x_2 and x_3 for selected values of x_1 are given in Table 5 and the lines are plotted in Chapter 14. Table 5 shows the differences which have to exist between x_1 and x_2 for a difference to be significant at the two sigma level, and the differences which have to exist between x_1 and x_3 for a difference to be significant at the three sigma level. By subtraction, it is easy to find the change that is necessary to move from the two sigma level to the three sigma level. Increases can be tested by entering column x_2 or x_3 and noting whether the critical difference at the two sigma level or at the three sigma level is being met.

Other tests and experiments. These are real tests or experiments for which detailed data are not available or because they dealt with important comparisons, but did not use modern testing or experimental methods.

The Federal Trade Commission was requested to make a survey of shortweighing in chain and non-chain grocery stores, since there was considerable criticism of the weighing practices in the former type of store. Four cities with populations over 100,000 were selected in four regions and practically all stores in these cities were shopped. The sample consisted of purchases of five bulk commodities — beans, prunes, lima beans, sugar and crackers — made in 702 chain stores and 989 non-chain stores for a total of 6,640 purchases. Shoppers were instructed to avoid small, poorly stocked stores or stores in ethnic neighborhoods. The greatest precaution was taken *not* to shop the same store twice. Purchases were weighed to the nearest 1/8th ounce and compared with the asked weight. The study suffered from obvious weaknesses. Cities could have been selected at random by size strata throughout the country, the stores could have been selected at random, and the purchases within the stores could have been randomized and replicated. Replication was prevented so no measure of the variation within stores was possible. Hence, there was no basis to test the difference between types of stores. For all anyone knew, the variation within stores was as great as the variation between the two types of stores being tested.

During World War II an important decision had to be made by the Federal government with regard to the relative merits of cotton cord and rayon cord in the manufacture of synthetic rubber tires. This decision was very important because it would determine whether an expansion of facilities was to be made in cotton, in rayon, or in some combination of the two. The congressmen from the cotton states were very concerned because some of them saw it as a threat to the use of cotton. During the summer of 1942, the War Department set up a tire testing program at proving grounds in Texas and California — localities selected to simulate war conditions in North Africa. The testing method used was to mount a number of tires on one side of a test truck and drive through very rocky country where in some cases the stones were of considerable size. To be judged satisfactory, a tire had to pass this test.

A Senate investigating committee made a study of these tests and hired an independent expert to examine their validity. It issued a report which is the source of this description. Failure of a tire depended upon whether the driver of the test truck hit a rock or not. This meant unequal treatment of the several tires under test; it appeared that the test was more related to the driver's behavior than to the performance of the tire. It was found that 26 percent of the tires removed from the test truck because of failure did not fail because of the type of cord in the tire.

In addition to the highly questionable method used to test the tires, apparently no control was exerted over such basic factors as the make of the tire, size of the tire, the rubber content of the tire, the quality of the synthetic rubber used, and the time at which the synthetic tire was made (the process of making synthetic tires was reportedly being improved constantly). Nearly all of the tires came from the four largest tire manufacturers. The sample consisted of 100 tires, a relatively small sample considering the importance of the decision to be made, the number of variables involved, and the easy availability of tires. Finally, a tire company interested in the outcome supplied the personnel for the test at the Texas proving ground. This illustrates the complexity of the problem of testing the difference between two kinds of rubber tires that differ only in one major characteristic.

The question arose: In a retail store was it better to display candy in a single walled or double walled cellophane bag? The former displayed the contents clearer, but was weaker; the latter was stronger, but gave the product a different look. A retail grocer ran a test by exhibiting hard candy in double walled bags one week and the same kind of candy in single walled bags the next week. Since he sold four times as much candy the second week as the first, he concluded that the single walled cellophane bag was better.

Certain important factors were apparently not taken into consideration: time, number of each bag on the shelf, position of the bags on the shelf, and type of candy in the bags. The two weeks may have reflected a holiday or seasonal change in buying; the number of bags of each type may not have been the same, thereby giving an advantage to the type with the largest number displayed; the two different types may not have been equally accessible to the buyer since they may have been on different shelves or more of one type may have been out front; and the candy in the two kinds of bags may not have been exactly the same color or shape. The influence of factors of time, accessibility, exposure, color, and shape were not eliminated by randomizing, balancing, or equalizing. The test should have continued over several weeks, or even more, to eliminate time effects; the two types of bags should be exposed side by side, which tends to eliminate time effects; the same number of each bag should be on the shelves; their positions on the shelves should be balanced so that each type of the same number are readily accessible or their positions on the shelves should be randomized relative to the position so that neither type has an advantage; and the contents of each bag should consist of candy of the same color or equal mixtures of different colors and shapes so that preference by color is not confounded with type of container. A third method should be tested: Give the buyer a chance to select the same candy displayed in a large container and use a brown paper sack, but sell it at a lower price. This will test whether price is a significant factor and not the bag.

The Federal Bureau of Engraving in the Treasury Department wanted to test whether a new dry process was better than the present wet process for printing dollar bills. Physical measurements of margins and distortion were available as a basis of comparison. Bills were printed in sheets with each bill having a letter identification; 18 bills were printed on a sheet three wide and six deep with letters a to f, g to l and m to r. The Bureau told the testors that there were 1,000,000 dollar bills available for the test! No real experiment was ever designed, although this was discussed with high level officials. Nothing was ever done to set up a real experimental test. Measurements were made on bills produced by each process and these were sent to the statisticians for analysis. The analysis consisted mostly of calculating and comparing means and variances by position of the bill in the sheet.

Notes

[1]W. G. Cochran and Gertrude M. Cox, *Experimental Designs,* John Wiley, New York, 1950, pp. 2, 3, 113-116. The t test originated with W. S. Gosset, an English statistician, who wrote under the pseudonym of "Student."

[2]Verified statement and testimony of Dr. W. Edwards Deming for Chicago Great Northern Railway Company, *ICC Docket No. 22688 et. al.* (Rock Island merger and purchase case). The usual analysis of variance table is added by the author, but the same hypothesis is under test and the same results are obtained. The entire verified statement is reproduced in Chapter 19 of A. C. Rosander, *Case Studies in Sample Design,* Marcel Dekker, Inc., New York, 1977.

[3]R. H. Noel and M. A. Brumbaugh, "Application of Statistics to Drug Manufacturing," *Industrial Quality Control,* VII, (September 1950), pp. 11-14.

[4]W. J. Youden, "Graphical Diagnosis of Interlaboratory Test Results," *Industrial Quality Control,* XV, (May 1959), pp. 24-28.

[5]Data are from the Food and Drug Administration.

[6]W. G. Murray and G. E. Bivens, "Clinic, Bench Marks, and Improved Assessments," *National Tax Journal,* V, (December 1952), pp. 370-375. Professors Murray and Blivens furnished the author with the individual values which did not appear in this paper.

[7]W. H. Purcell, "Balancing and Randomizing in Experiments," *Industrial Quality Control,* VII, (1951), p. 7.

[8]W. Edwards Deming, *Some Theory of Sampling,* John Wiley, New York, 1950, pp. 420-421; also, Deming, *Sample Design in Business Research,* John Wiley, New York, 1960, pp. 461-462.

26
Control by Measuring Effectiveness

Several examples of control by measuring effectiveness have already been described:

1. Is supervision effective? Use RTS.
2. Are machines and employees being utilized effectively? Use RTS.
3. Are tax assessors effective? Test them independently on the same property.
4. Are chemists effective? Test them on identical samples daily for a few days.
5. Are testing laboratories effective? Test them on identical samples of paired materials or food chemicals.

Effectiveness in these cases was measured objectively: in terms of deviations from known values (items 4 and 5), in terms of variability (item 3), and in terms of the amount or percent of down time or idle time (items 1 and 2). In this chapter, we deal more directly with the effectiveness of changes or modifications made by intervention or introduction. Several examples of measuring the effectiveness of government programs or actions are given in Chapter 4.

Situations involving control by measuring effectiveness. Two different kinds of situations involve the need to measure effectiveness: those in which intervention is involved and those in which the introduction of new techniques is involved.

Examples of intervention are laws, legislation, social and economic programs, rules and regulations, ordinances, institutions, legal actions, and court decisions. Measuring effectiveness is for the purpose of determining whether or not these actions are accomplishing the function, objective, or purpose for which they were initiated. This is not a question of testing whether A is better than B, but a question of testing whether some established procedure or action is really effective in meeting the purpose or need for which it was created.

Specific examples of intervention include public housing, public welfare, food stamps, aid to families with dependent children, medicare, nursing homes, school busing for racial equality, no fault insurance, the 55 mile per hour speed limit, auto inspections, pollution standards, free flu shots, tax audit program of IRS, curfews, truancy laws, and bilingual education.

The second area, that of trying to improve how actions and operations are carried out, includes new techniques, new procedures, new materials, new machines, new equipment, new programs, new technologies, new systems, new products, new structures of organization and management, and new working rules and conditions.

Specific examples include new techniques, such as statistical quality control, probability sampling, design of experiments; new machines, equipment, and technology, such as those included in a computer center; new working arrangements, such as the flexi-schedule and the four day week; new drugs and vaccines, such as the polio vaccine; new machines such as the desk top programmable calculator and the hand-held calculator; and new managerial and operating arrangements, such as quality control circles.

In this area changes, often drastic changes, are being made. The situation is not only complex, but often comprehensive. Hence, the problem is quite different from a simple test showing whether A is better than B. The basic question is: A drastic change is being proposed or has been made. Is the new proposal so much more effective than the present procedure that it should be adopted or continued if already in operation? The ultimate goal is to obtain control over this situation by designing valid and efficient measures of effectiveness and changes in effectiveness. This leads to the question of how effectiveness is tested or measured.

Measures of effectiveness. Measures of effectiveness include 1) estimates by sample audits, 2) estimates by tests and experiments, 3) detecting and estimating violations, 4) determining errors and error rates, 5) use of cost-benefit analysis, 6) measuring improvements in production, reduction in errors, increase in productivity, reduction in paper work, reduction in costs, and better utilization of machines, equipment, and personnel, and 7) reduction in failures and customer and employee complaints.

Several problems arise in testing or measuring effectiveness. In some situations there is disagreement over what the intervention should accomplish. There is even opposition to it and, hence, there is disagreement over what "effectiveness" means and how it should be measured. Situations of disagreement include:

1. School busing for racial equality.
2. The 55 mile per hour speed limit.
3. Auto inspections for safety.
4. Nursing homes.
5. The flexi-schedule and the four day week as work arrangements.

In other situations, there is little or no disagreement largely because it is much easier to prove effectiveness. These situations include:

1. Polio vaccine.
2. Probability sampling.
3. Statistical quality control.
4. The electronic computer.
5. The electronic desk calculator.

Since many interventions are legal and legislative, and have strong political, social, and economic overtones, it is difficult 1) to conduct an adequate study or make a fair study and 2) to obtain acceptance of sound conclusions from careful studies, especially if they run contrary to original social and political biases, prejudices, and assumptions. People do not change strong, emotionally-loaded beliefs regardless of facts and realities.

Furthermore, studies are often made by those who want to "prove" that a certain program is effective because they believe in it or support it. Academic economists, sociologists, as well as politicians and special interest groups emotionally attached to some program or view, obviously cannot honestly and objectively study or appraise a program for its effectiveness. It has to be studied by those who will do a careful technical and managerial job, and who are willing to test hypotheses rather than "prove" that a view is correct.

Effectiveness of interventions. Box and Tiao[1] refer to "interventions" which are various kinds of actions taken to bring about a change. The effectiveness of these actions is measured in terms of observable and measurable changes. These interventions may be laws, social programs, administrative and governmental actions, or judicial decisions. They may be changes in policy, rules and regulations, procedures, methods, or organizations.

The problem is to find one or more measurable characteristics which is a valid reflector of the improvement that is supposed to take place. Obviously, this may be very difficult in the complex types listed above: there are many variables, complex interactions, and significant factors that are difficult, if not impossible, to measure. Also, it may be difficult to separate real factors from extraneous ones, such as random variation or "noise".

The interventions Box and Tiao discuss are complex types, where the data available for the measurement of improvement is a time series. There the standard assumptions of independence of observations, random samples, normal distributions, and a definable population are *not* met. They use an iterative procedure of model building to solve this problem; for example, the amount of ozone in the air measured hourly at some point in downtown Los Angeles.

Fortunately, other examples of intervention use standard statistical techniques, such as probability sampling, sample audits, and design of tests and experiments. Measuring the effectiveness of a vaccine, a social program, or a quality control program are examples.

In some instances, the improvement from an intervention is so pronounced in reduced paper work and improved quality that it is not necessary to conduct any tests to show that the intervention saves money and improves quality. Examples are the substitution of a well designed and managed probability sample for a judgment or 100 percent sample, the use of a designed test or experiment, or the use of a well planned quality control program in the operations of a factory or office.

In the present context "effectiveness" includes situations where there is no intervention to change the direction or nature of a certain course of events. The problem is to discover whether an on-going activity or operation that is supposed to be effective really is. Examples are:

1. How effective is an inspector in inspecting a manufactured product?
2. How effective is an inspector in inspecting a restaurant for health and sanitation violations?
3. How effective is an auditor in discovering errors or non-conforming practices?
4. How effective is an Internal Revenue agent in auditing individual income tax returns? Corporation income tax returns?
5. How effective are the police in solving crimes such as murder?
6. How effective are interviewers in getting accurate and valid data from persons interviewed?

In all of these cases, objective measures of effectiveness are available. With inspectors, it is the proportion of all defective parts or violations which were found and reported. With auditors and Revenue agents, it is not only the proportion of all errors found but whether they isolate or calculate the correct values. With police, it is the proportion of crimes successfully solved. With interviewers, it is the extent to which they obtained accurate and valid answers to questions; this can be determined by a repeat interview using an expert interviewer.

Two intervention examples are described: the effectiveness of a poliomyelitis vaccine and the effectiveness of services offered in connection with the national AFDC program (aid to families with dependent children) in terms of error rates by states. Examples of the non-intervention type include testing the effectiveness of a group of persons to make a sensory discriminal judgment, and testing the effectiveness of interviewers when measured against a repeat check interview of the same families with regard to certain financial items.

Testing effectiveness of poliomyelitis vaccine. In 1954 the National Foundation for Infantile Paralysis contracted with the University of Michigan to conduct a nationwide test of the effectiveness of the Salk vaccine for poliomyelitis (or polio). It was called infantile paralysis in earlier days because of its crippling effect and its concentration among children. It was thought that an extensive study was necessary before the vaccine was put into general use. The first vaccine that seemed effective against polio was developed by Salk and the purpose of this study was to test its effectiveness. Two reports are of interest: the summary report[2] and an independent appraisal of the techniques used in the study.[3]

At that time, the average rate of occurrence of polio was about 50 per 100,000 children. Thus, it was necessary to use large samples of children to obtain a difference that could be attributed to the vaccine and not to sampling fluctuations. Assuming the vaccine is 50 percent effective, a sample of 100,000 vaccinated children would have about 25 cases; a sample of 100,000 children not vaccinated would have about 50 cases. This reduction would be strong evidence the vaccine is effective. This can be shown by assuming 25 and 50 are Poisson counts. Then the normal deviate is:

$$z = \frac{50^{1/2} - 25^{1/2}}{0.5^{1/2}} = 2.93$$

which is significant at the 99.8 percent level. (See Table 5, Chapter 25).

Actually, slightly more than 200,000 children were used in the experimental group and about the same number were used in the placebo (no vaccine) group. See Table 1.

Two plans were used in the nationwide sample survey. In the observer-control plan, the vaccine was given to second graders in the study, but not to the first and third graders. The second graders were the experimental group; the other two grades were the control group. This method confounds the effects of the vaccine with age and other factors which may affect the outcome. Slightly more than a million children or 59 percent of the total sample were included in this plan.

About 750,000 additional children were subjected to the experimental-placebo plan. The three grades were combined and half the children received the vaccine and half did not (the placebo group). The latter groups received a similar-appearing solution which had no effect. Furthermore, the vaccinations were coded and reports were made on a concealed or "blindfolded" basis so neither children nor doctors knew who got the vaccine and who did not. This was done to eliminate a source of bias since medical judgment is involved. Furthermore, whether a child received the vaccine or the placebo was assigned at random so that a child, regardless of various factors affecting the outcome, had the same chance of being assigned to the one as to the other. Randomization tends to equalize the effects of extraneous factors on the two groups much better than if we try to equate the groups by judgment. Randomization also made it possible to interpret the results in terms of probability theory and provides an objective method of determining whether any observed difference could have arisen by chance or was so great that chance would be ruled out in favor of the experimental factor, the vaccine.

Brownlee cites two examples from the study to show why randomization was so important in assigning the vaccine to children. In Schenectady, New York, absenteeism was higher among the vaccinated than among the placebos. Furthermore, there was evidence that the placebo group came from a lower socio-economic class than did the vaccinated group. Brownlee concluded that testing the observer-control group was futile because it violated sound principles of experimentation.

Brownlee also pointed out that, while the vaccinated-placebo groups have about the same age distribution, there was medical evidence that a condition of low antibodies was more prevalent in the placebo group thereby increasing the probability of polo in this group. Finally, the study was deficient because no reference was made to the problems or methods of manufacturing the vaccine. In other words, nothing was done to ascertain whether the vaccine used in the study was subject to quality control which insured that various batches of the vaccine were equally potent.

Table 1 is abridged from the data reported in Table 2b of the Summary Report. Certain parts of the table, which do not bear directly on the effectiveness of the vaccine, are eliminated. The rates given in the table are rates per 100,000 children in the study.

The most convincing evidence of the effectiveness of the vaccine is the difference between the two rates for paralytic polio in the experimental-placebo group — a rate of 16 for those vaccinated and a rate of 57 for those not vaccinated corresponding to frequencies of 33 and 115, respectively. These latter frequencies are comparable since both totals, on which they are based, exceed 200,000 and differ by only 484. See Table 1. We can test whether these two frequencies are significantly different by assuming that they are Poisson counts, and using the square root approximation, in which $(x)^{1/2}$ with a variance of $1/4$ is normally distributed:

$$\begin{aligned} (115)^{1/2} &= 10.72 \\ (33)^{1/2} &= \underline{\ 5.74} \\ \text{difference} &= \ 4.98 \end{aligned}$$

The variance of this difference is $1/4 + 1/4 = 1/2$ and the standard error is $(1/2)^{1/2}$ or 0.707. Hence, the normal deviate:

$$z = \frac{4.98}{0.707} = 7.04,$$

which is a very rare value to obtain by chance. This is statistical evidence that there is a real difference between the counts of 33 and 115 which cannot be explained by random fluctuations due to sampling. To obtain this result, it is easier to use the chart for the difference of two Poisson variates (Chapter 14). Using $x_1 = 115$, it is seen immediately that a count of $x_2 = 33$ is way beyond the three sigma limits. The point $x_1 = 115$ and $x_2 = 33$ falls way below the three sigma line, indicating significance at a level exceeding the three sigma level on the normal probability distribution. Table 5, Chapter 25, shows the same thing: enter x_1 with 115 and see that three sigma difference has to be between 38 and 44. But the actual difference is 82, so significance is way beyond the three sigma level.

Table 1

Selected Data from Field Tests of Poliomyelitis Vaccine — 1954[a]

Group	Number of children in study group	Poliomyelitis cases: Total no.	Total rate	Paralytic no.	Paralytic rate	Non-paralytic no.	Non-paralytic rate
All areas total	1,829,916	863	47	685	37	178	10
Placebo areas total	749,236[b]	358	48	270	36	88	12
vaccinated	200,745	57	28	33[d]	16	24	12
placebo	201,229	142	71	115[d]	57	27	13
Observed areas total	1,080,680[b]	505	47	415	38	90	8
vaccinated	221,998	56	25	38	17	18	8
controls[c]	725,173	391	54	330	46	61	8

[a]Abridged from Table 2b of the Summary Report.
[b]Remainder of totals are children not inoculated or with incomplete inoculations.
[c]Total of first and third grade children.
[d]Key frequencies.

Effectiveness of a welfare program (AFDC). The welfare program of aid to families with dependent children (AFDC) is one of the major welfare programs of the Department of Health and Human Services (formerly HEW). There are a number of conditions which have to be met before a family is eligible for this relief. In making payments to families under this program, three kinds of errors are made: payments are made to ineligible families, payments are made in excess of what they should be, and eligible families receive no payment at all.

The Social Security Administration reports that the error rate made under this program, due to those ineligible and overpaid, increased from 9.4 percent to 10.4 percent in the six month period ending March 1979. This amounted to excess payments totaling $533,000,000, compared with excess payments of $479,000,000 during the previous six months. The total for 12 months was $1,012,000,000. During this time, about $10,500,000,000 was paid to about 3,500,000 families. The excess payment averages about $300 per family per year. If the ineligible families were like the average family, they received about $3,300 a year illegally from this program. These data were derived from an audit sample of about 40,000 cases, or nearly 800 on the average from each state, including the District of Columbia and Puerto Rico.

The error rates for the 52 governments and for the United States for the six months ending September 1978, the six months ending March 1979, and the difference with a minus sign showing a decrease and no sign showing an increase are given in Table 2. Only seven of the 52 had an error rate less than five percent for the six months ending March 1979: Indiana, Minnesota, Nebraska, Oklahoma, South Dakota, Utah, and Nevada.

State and local officials administer this program. These figures cast serious doubt on the ability and competence of not only these officials, but Federal officials as well, to carry out this program without a terrible amount of waste. A program that spends a billion dollars ($1,000,000,000) in error in 12 months is hardly effective. Those who manage it are not effective; those who operate it are not effective. As a starter, a reasonable level of error is 5 percent which would save $500,000,000. The final goal should be at least 1 percent, thus saving $900,000,000 annually.

In Table 2, the error rate increased in 31 areas and decreased in 21 areas. The increases came in the range from five to fifteen percent error; the bracket under 5 percent lost eight states, as the following shows:

Error rate range (percent)	March 1979	Sept. 1978
0-4	7	15
5-9	24	20
10-14	18	11
15-19	0	4
20-24	2	1
25-29	1	0
30-34	0	1
Sum	52	52

Some warnings are in order about Table 2. Figures are derived from a sample audit of about 40,000 cases (families) or about one in 85. Hence, the figures are subject to bias due to sampling, interviewing, and reporting; to random variations due to sampling; and to errors in estimating money payments from the sample.

Table 2

Percent Error Rate in Welfare Payments under AFDC for Six Months Ending, by States*

State	March 1979	Sept. 1978	Change
Alabama	7.4	9.5	−2.1
Alaska	28.8	31.2	−2.4
Arizona	6.4	8.0	−1.6
Arkansas	8.8	9.1	−0.3
California	7.2	3.7	3.5
Colorado	6.5	4.3	2.2
Connecticut	9.7	8.6	1.1
Delaware	9.8	16.1	−6.3
Washington, D.C.	23.8	22.4	1.4
Florida	6.2	5.6	0.6
Georgia	6.5	7.8	−1.3
Hawaii	8.6	9.2	−0.6
Idaho	7.5	4.4	3.1
Illinois	13.8	17.1	−3.3
Indiana	4.6	3.7	0.9
Iowa	10.1	7.8	2.3
Kansas	8.0	4.1	3.9
Kentucky	6.6	10.2	−3.6
Louisiana	8.5	11.1	−2.6
Maine	13.6	9.2	4.4
Maryland	14.7	13.6	1.1
Massachusetts	24.8	15.9	8.9
Michigan	10.3	9.2	1.1
Minnesota	2.4	3.4	−1.0
Mississippi	11.4	11.6	−0.2
Missouri	11.2	10.1	1.1
Montana	10.7	9.7	1.0
Nebraska	2.9	4.6	−1.7
Nevada	0.8	0.6	0.2
New Hampshire	12.2	11.0	1.2
New Jersey	10.0	9.3	0.7
New Mexico	5.3	4.8	0.5
New York	10.3	8.8	1.5
North Carolina	7.1	7.9	−0.8
North Dakota	5.4	1.6	3.8
Ohio	11.9	9.5	2.4
Oklahoma	4.1	3.2	0.9
Oregon	11.9	12.7	−0.8
Pennsylvania	11.9	16.3	−4.4
Puerto Rico	9.0	7.8	1.2
Rhode Island	9.9	12.7	−2.8
South Carolina	7.7	7.1	0.6
South Dakota	2.4	4.8	−2.4
Tennessee	6.3	7.0	−0.7
Texas	7.4	6.9	0.5
Utah	4.6	2.8	1.8
Vermont	12.5	4.5	8.0
Virginia	10.2	11.7	−1.5
Washington	9.6	6.7	2.9
West Virginia	10.4	11.3	−0.9
Wisconsin	11.7	11.1	0.6
Wyoming	6.0	4.0	2.0
U.S.	10.4	9.4	1.0

*Source: Social Security Administration, Washington, D.C. as reported in the *Rocky Mountain News,* March 13, 1980, p. 21.

Alaska and Vermont have high error rates, but each has less than 6,500 families so the total amount of money involved in payments is less than $2,000,000 annually in each state. Ten largest states — California, New York, Illinois, Pennsylvania, Ohio, Texas, New Jersey, Massachusetts, Michigan, and Florida — account for 59 percent of the families and 69 percent of the money payments.[4]

With an error rate averaging 11 percent, the amount of overpayment in these 10 states alone is over $600,000,000 annually. Immediate corrective actions need to be greatly increased in these states, not in states like Alaska and Vermont.

Discrimination of aromas.[5] How effective is a group of 200 persons in distinguishing three strengths of aroma of rose water: 1 percent, 10 percent, and 30 percent? Each of the three bottles of extract was colored so aroma could not be determined by sight. Subjects were asked to arrange the bottles in a ascending order according to the strength of the aroma: A = 1 percent, B = 10 percent, and C = 30 percent. The method of inversions is appropriate to test for departure from chance, since a standard order or permutation is known. The number of inversions of order for each of the 6 orders or permutations is as follows; the number of inversions is the number of positions a letter is out of its correct order:

Permutation (order)	Number of inversions (x)
A B C	0
A C B	1
B A C	1
C A B	2
B C A	2
C B A	3

The distribution of inversions is known and its mean and variance are functions only of m, the number of permutations or orders.[6] The mean $u = m(m-1)/4$ and the variance $\sigma^2 = m(m-1)(2m+5)/72$. This is the distribution expected on the basis of random order or a large random sample of all possible permutations of m orders. In this problem, it measures the lack of ability to discriminate one order from another. The greater the departure of an actual distribution from the expected distribution, the greater the ability to discriminate the actual order.

The frequency distribution of x obtained from 200 persons, the expected distribution derived from the hypothesis of no ability to discriminate, and the chi-square calculations are as follows:

Inversions x_i	Frequency n_i	n_ix_i	Expected relative no.
0	70	0	1
1	73	73	2
2	39	78	2
3	18	54	1
Sum	200	205	6

Inversions x_i	Expected actual no.	Actual less expected	Chi-square term
0	33.34	+36.66	40.31
1	66.66	+ 6.34	0.60
2	66.66	−27.66	11.48
3	33.34	−15.34	7.06
Sum	200.00	0	59.45

The 18 who completely reversed the order may have misunderstood the instructions. In tests of this kind, it is very necessary to check all subjects to insure that they understand the order in which the objects are to be arranged. This does not invalidate the chi-square test.

A chi-square of 59.45 for three degrees of freedom is highly significant, providing statistical evidence of the ability to discriminate. This is also revealed by examination of actual minus expected values: for $x_i = 0$ and 1 the differences are positive; for $x_i = 2$ and 3 the differences are negative.

The population mean is $(3)(2)/4 = 1.50$, the actual mean $\bar{x} = 1.03 = 205/200$, and the difference is 0.47. The population variance is $(3)(2)(11)/72 = 11/12 = 0.917$ or a standard deviation of 0.957. Assuming a normal curve, a mean of 1.50 and a standard deviation of 0.957 gives a standard error of 0.068 for a sample of 200:

$$s_x = \frac{0.957}{(200)^{1/2}} = 0.068.$$

The normal deviate corresponding to a difference of 0.47 is:

$$z = \frac{1.03 - 1.50}{0.068} = -6.9,$$

which is a very rare value to exceed on the normal distribution. The tests call for a rejection of the hypothesis that no discriminating ability existed. We conclude, on the basis of the data and the analysis, that the group was effective in discriminating the three aromas.

Effectiveness of interviewers. Interviewers obtained certain money information from a sample of families in an actual city. Some months later, a 10 percent sample was checked (re-interviewed) by another interviewer, either a selected superior interviewer or a supervisor. In the re-interview, all items and amounts were recorded so that the check interview values may be considered as close to the "true" values as one can expect in work of this kind. Table 3 and Figure 1 show the results of these two interviews. In Figure 1, the points represent x the first interview value and y the corresponding check interview value; the line y = x is shown for purposes of comparison. The number of families is 54.

A glance at either Table 3 or Figure 1 shows that the check interviewer obtained larger values than the first interviewer for most of the families. The first interviewer obtained a mean of 1,460, but the check interviewer obtained a mean of 1,627 — an increase of 167 or about a 10 percent bias:

	Number	Amount (dollars)
Original value increased	38	+258
Original value decreased	16	− 49
total number and net change	54	+167
Difference, $10 or less	13	
Difference, two low, $100 or more	19	
Difference, too high, $100 or more	4	

The net additional amount was $9,011, which becomes $90,110 considering the 10 percent sample. The linear regression $y = 198 + 0.978$ gives $r^2 = 0.81$ and a correlation coefficient $r = 0.90$. In view of the large average bias of +$167 (or about 10 percent) due to under-reporting by the first interviewer, an r of 0.90, even though usually considered high, obviously cannot be interpreted as indicating satisfactory interviewing.

If the 23 values deviating from the check interview by 100 or more are eliminated so $n = 31$, then $r^2 = 0.996$ and $r = 0.998$. Even with this high correlation, $\bar{x} = 1{,}550$ and $\bar{y} = 1{,}578$, a difference of +$28 — still almost a two percent bias.

In this case, the data were not corrected nor were the readers informed of it. Clearly, a sample of interviews should be examined at the start, and during, the field work. The interviewer errors should be explained and corrected, so further errors are reduced, if not eliminated. Quality control over interviewers should be built into the study right from the start.

Table 3

Values from Original Interview and Corresponding Check Interview for Individual Families (dollars)

Original x	Check y	Check minus original	Original x	Check y	Check minus original
1040	1409	369	1105	1144	39
3400	3379	−21	1924	1794	−130
2193	3620	1427	1768	1597	−171
1820	1900	80	2337	2334	−3
1172	1266	94	1796	2071	275
1300	2185	885	1449	1447	−2
300	309	9	1170	1560	390
1560	1749	189	1969	1971	2
1536	1291	−245	570	745	175
1352	1443	91	859	867	8
1300	1378	78	1340	1360	20
2105	2104	−1	1895	1900	5
2116	2701	585	468	517	49
1523	1992	469	728	726	−2
2236	2391	155	2099	2074	−25
1520	1516	−4	2086	2047	−39
1784	1862	78	452	724	272
1191	1828	637	520	546	26
946	842	−104	360	702	342
1092	1185	93	2366	2374	8
829	907	78	1444	1544	100
1576	1736	160	2211	2208	−3
730	1584	854	1340	1790	450
1649	1634	−15	1534	1604	70
700	946	246			
			Sum 78830	87841	+9011
1690	1684	−6			
1872	1924	52	Mean 1460	1627	+167
2080	2892	812			
1768	1758	−10	Standard Deviation 628	684	301
660	780	120			

Automobile driving intervention. Automobile driving appears to be subject to more intervention than any other common activity of the American people. Since all of this intervention is forced upon us by laws and legislation, usually by those who know what is best for us, the claims are long and loud that all these interventions are effective. We consider available evidence.

Methods of intervention take many forms including:

1. Use of seat belts, shoulder harnesses, padding, balloons, and special bumpers.
2. Periodic safety inspections, usually annually.
3. Driver's tests in order to obtain a license to drive.
4. 55 miles per hour speed limits on interstate highways.
5. Compulsory automobile insurance.

All of them are connected in some way with the desire to keep down the number of automobile accidents, especially those resulting in fatalities. The first two are based on the assumption that the major cause of these accidents is the automobile, not the driver. The other three recognize that the driver is involved. We now turn to some evidence and see what it shows.

Example: Some years ago Connecticut put on a drive to reduce fatal automobile accidents. The number of deaths for one quarter of one year was 73; the number of deaths for the same quarter for the next year was 64, a reduction of nine. This reduction was hailed as evidence that the drive was successful. Assuming that these counts follow a Poisson distribution, reference to the chart in Chapter 14 or Table 5, Chapter 25 shows that for $x_1 = 73$, $x_2 = 51$ on the two sigma line. Hence, 64 is not a significant reduction from 73; it has to be 51 to be significant at the 95 percent level. Only successive reductions in the 64 deaths in future quarters would be evidence that a real decrease is taking place.[7]

Politicians, journalists, and many others who never go beyond the apple arithmetic of the elementary school fail to understand that variability exists in counts and measurements. This is why it is very difficult to make an accurate test and interpretation of data dealing with effectiveness.

Example: Over a period of almost four months, the details of automobile accidents, resulting in 148 deaths, were reported by the Colorado traffic police.[8] The reasons

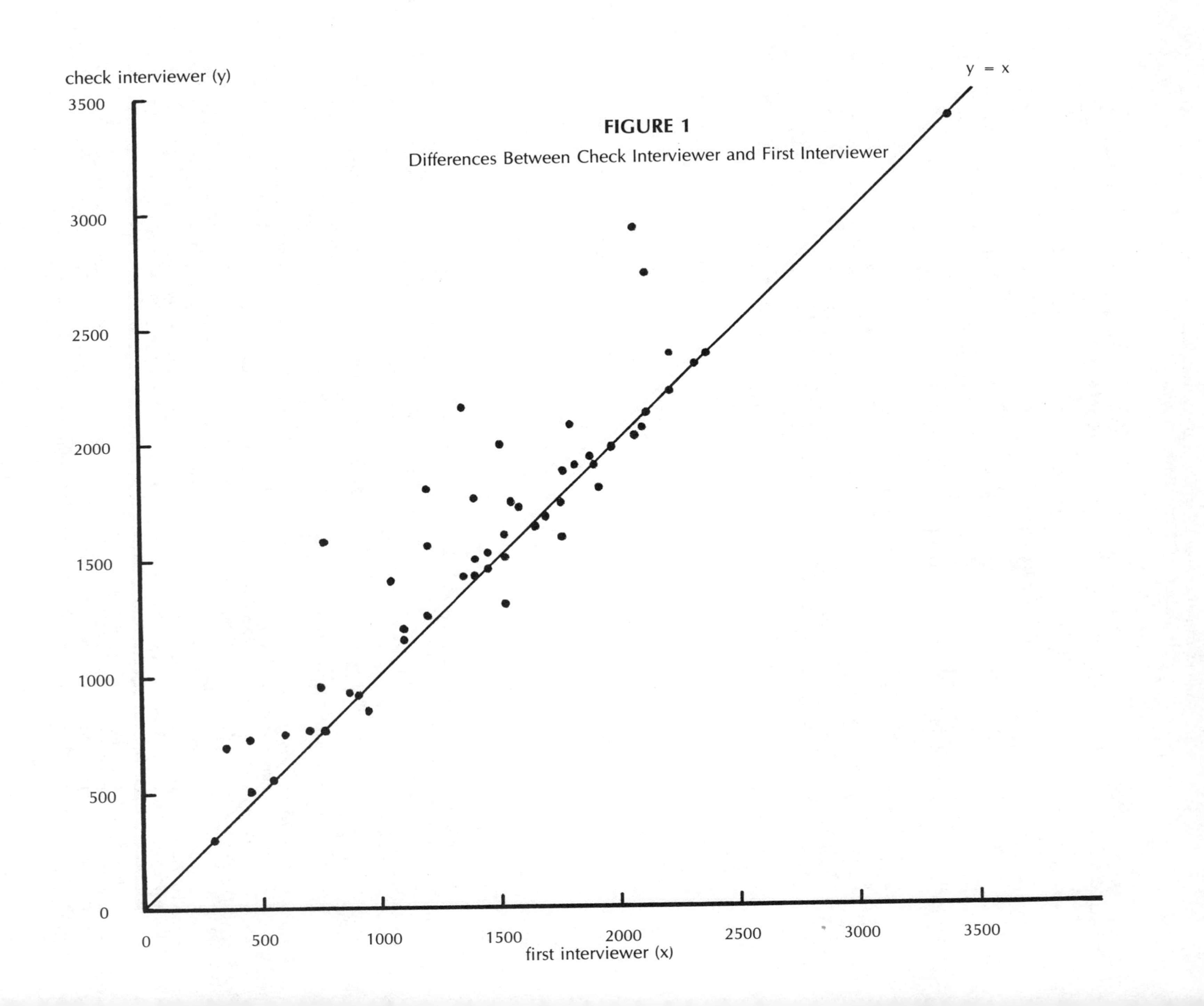

FIGURE 1

Differences Between Check Interviewer and First Interviewer

for these deaths, classified according to traffic officer reports, are in order of frequency:

Ran off road (one car)	74
Pedestrians (one car)	22
Intersection (two cars)	12
Wrong side of the road (two cars)	8
Railroad crossing (one car)	5
Other (one car)	5
Other (two cars)	22
	148

"Other" for one car includes skidding and hydroplaning; "other" for two cars includes passing, changing lanes, ramming or avoiding a stopped vehicle, hitting a car entering the road, and left turn off road or street. Pedestrians included both older persons and children.

There was no evidence in any of these cases that mechanical failure of the automobile was the cause. Quite the contrary, all the evidence pointed to the failure of the driver, except in some pedestrian cases where the driver could not avoid the accident. Driver failure was due to too high a speed on roads with curves, an inability to drive properly on wet, snowy, and icy roads, an inability to estimate accurately speeds and distances of oncoming cars, driving while under the influence of alcohol, and failure to give full attention to the conditions surrounding the automobile. A few accidents may have been due to persons falling asleep or driving while under the influence of drugs.

Example: The data on driving an automobile while under the influence of alcohol, as reported by traffic officers and city officials in various parts of the country, have remained about the same for at least three decades: 40 to 50 percent of the fatal automobile accidents are due to persons driving under the influence of alcohol. This means that at least 20,000 persons are killed each year by the alcoholic driver, of which 10,000 are those innocents murdered while driving or riding carefully, competently, and legally. This does not include the more than 100,000 innocent persons who are injured some permanently. Recently, many states have passed stricter laws against those driving while under the influence of alcohol. These measures include stiffer fines and sentences, including public service. It is doubtful that they will prove effective because they do not get at the heart of the problem — which is keeping the intoxicated person from driving a vehicle down a public street or highway. This, in itself, should be a crime because such a person is a potential killer.

Example: The 55 miles per hour speed limit on interstate highways was imposed to reduce the consumption of petroleum, from which gasoline is made, and to reduce fatal accidents. All evidence, based on actual observation and driving thousands of miles on the open road on Interstates 25, 70, and 80, in many parts of the country, shows that 75 to 90 percent of automobile drivers exceed this limit. As a specific example, the author, while driving a stretch of I-25 one morning at a speed between 55 and 60 miles per hour, was passed by over 100 automobiles, most of which were going over 65 miles per hour. The author passed only six. No one was in sight going 55 to 60 miles per hour.

In driving across Indiana, a distance of about 160 miles, at 60 miles per hour, the author by actual count passed 26 automobiles, but was passed by 82 showing that about 75 percent were going 65 miles per hour or more. This makes allowance for a few going 60 miles per hour.

Very few drivers on the open road ever drive 55 miles per hour, except when they sight a policeman or see someone being ticketed for speeding. This occurred before the limit was imposed and it still occurs.

Laws, rules, and regulations are effective interventions only to the extent that they are observed or enforced. It is very difficult to obtain an objective measure of the effectiveness of numerous social, economic, and political programs because of vested interests in these programs and the strong belief that they are justified on ideological grounds regardless of how much inefficiency and waste may be connected with them.

Effectiveness of probability sampling. This is an example of how the effectiveness of new research, operations, and management techniques can be demonstrated. An entire book could be filled with successful demonstrations and applications of these new techniques. Two examples are cited to show how effective probability sampling is in producing high quality data in a timely fashion at a greatly reduced cost.

Example: Motor carrier information on three forms was collected, for cost purposes on a one or two day basis annually from all carriers, by an accounting bureau of a Federal agency. One large carrier had to submit over 7,000 pieces of paper for one form alone.

When a probability sample was substituted for the one or two day judgment "sample," better quality data were collected at a greatly reduced cost both to the carriers and to the agency. The actual data given below show that the number of data forms required was *reduced 35 percent.*[9]

Data form number	Number of forms	
	Prob. sample	1 or 2 day study
4	22,600	24,960
7	21,200	44,530
10	2,740	1,670
Sum	46,540	71,160
Reduction	24,620	
	71,160	

The reduction in the amount of paper was 24,620/71,160 or about 35 percent. Management was impressed by this saving of paper work and this was one of the major reasons more than one top manager became convinced of the effectiveness of probability sampling.

Example: In this case, a highly efficient probability sample of 30, designed for purposes of estimation, turned out to be a very effective audit sample by uncovering a very serious error in a 100 percent listing.

For purposes of war control, a Federal agency collected data on quarterly shipments, from every company in an industry, and recorded these 100 percent in an accounting book. At the same time, a probability sample was designed to estimate the total shipments for the same industry for the same quarter, but to obtain this figure much sooner than the 100 percent tabulation figure.

For one quarter, the sample of 30 companies gave about $18,500,000; the accounting book gave $23,500,000. Since this difference was much greater than the sampling error in the $18,500,000, both the sample data and the accounting book data were carefully examined. As a result, the reason for the wide discrepancy was discovered. Shipments were to be reported in $1,000. On one form, 4,600 was entered instead of 4.6. The transcriber entered $4,600,000 in the accounting book, when it should have been $4,600 — an error of $4,595,400.

When this was substracted from $23,500,000, it gave $18,904,600 which compared with the $18,500,000 derived from the probability sample. This two percent difference was well within the sampling error. This two percent difference was much less than the 24 percent error made in the original so-called "true" 100 percent accounting record. Only the use of the probability sample and the controls associated with it provided the means of detecting this large error. This is but one of many examples, which have been reported in sampling practice, where a well designed and carefully managed probability sample is better than a 100 percent census, tabulation, or listing for purposes of estimation.

Effectiveness and cost-benefit analysis. Control by measuring effectiveness brings up the question of cost-benefit analysis. Is the intervention worth the cost? Put another way, is intervention beneficial and is it effective to an extent that justifies the cost?

1. There is no doubt about the polio vaccine. Whatever the cost, the benefits in preventing a crippling disease are well worth it because the benefits accruing over the years cannot be calculated.

2. Neither are there any serious doubts about the effectiveness of applications of probability sampling and quality control techniques. A wide variety of documented evidence has accumulated over the past 40 years: improving the quality of data, improving the quality of performance, improving the quality of decisions, improving the quality of products, the tremendous expansion in the number of professionals and technicians who are now employed in new occupations which grew out of these fields, and improving quality at the same time reducing costs.

3. There are, however, programs of intervention that raise questions of cost. Public welfare is necessary, but serious questions have arisen as to why the error rates which result in Federal excess payments of over $1 billion annually cannot be sharply reduced.

Intervention in the area of automobile driving also raises questions, especially the practice of requiring annual safety inspections. This practice is based on the assumption that automobile accidents are caused by mechanical defects in the auto and not by any fault of the driver. There is little or no evidence to support this claim. Indeed, studies show that inspections do not reduce the number of accidents. As a result, Colorado has suspended auto safety inspections for at least three years, if not indefinitely. The annual fee of $5.50, formerly paid, was sponsored in the state legislature by a legislator in the automobile business.

4. The effectiveness of interviewing, whether in person or by telephone, raises serious questions unless the questions are carefully framed and rapport is established with the person being interviewed. To improve effectiveness of people generally whether interviewers, tax auditors, tax assessors, inspectors, supervisors, and managers requires careful and adequate training as well as effective job experience.

Notes

[1]George E. P. Box and G. C. Tiao, "Intervention Analysis with Applications to Economic and Environmental Problems," *Journal of the American Statistical Association,* vol. 70, 1975, p. 70.

[2]Tabulation of 1954 field trials of poliomyelitis vaccine: Summary Report, *American Journal of Public Health,* vol. 45, part 1, (1955), pp. 1-63.

[3]K. A. Brownlee, "Statistics of the 1954 Polio Vaccine Trials," *Journal of the American Statistical Association,* vol. 50, (1955), pp. 1005-1013.

[4]*The World Almanac,* 1980, p. 202. Data are for the month of November, 1978.

[5]Reproduced from W. D. Baten, "Analysis of Scores from Smelling Tests," *Biometrics* 2:, 1946, pp. 11-14, 1946 with permission from the Biometric Society. Baten used the multinomial but the method of inversions is much simpler and gives the same answer.

[6]A. C. Rosander, *Elementary Principles of Statistics,* Van Nostrand, New York, 1951. Copyright by Wadsworth, Inc. Used by permission of Brooks Cole Publishing Co., p. 619.

[7]W. Edwards Deming, *Sample Design in Business Research,* John Wiley, New York, 1960, p. 463.

[8]Summarized from reports in the *Rocky Mountain News* from February 3, 1976 through May 29, 1976.

[9]A. C. Rosander, *Case Studies in Sample Design,* Marcel Dekker, Inc., New York, 1977, pp. 269-270, reprinted by courtesy of Marcel Dekker, Inc.

27

Control by Quantitative Models

Quantitative models. We have already described examples of how quantitative models are used in describing and controlling quality. One example of these models is the several probability distributions which are used in quality control work. Some of these applications, plus others, are listed below:

1. Use of the normal distribution in constructing and interpreting the Shewhartian $\bar{x}$ chart.
2. Use of the Poisson distribution in constructing and interpreting the Shewhartian c chart.
3. Use of the Poisson distribution in testing whether two counts, such as two counts of fatalities in automobile accidents, are significantly different or could have arisen from a constant system of chance causes.
4. The minute model is used in random time sampling (RTS) to estimate time and money aggregates.
5. The circular normal distribution was used by Youden to test variations among laboratories.
6. The normal distribution was applied to assessments made on the same property by 148 tax assessors.
7. The hypergeometric distribution is used in designing and interpreting random samples used for detection.
8. The binominal distribution is applied to events or conditions which match the "heads or tails" model.
9. Various models are postulated in formulating and calculating the analysis of variance appropriate to various problems.
10. The linear regression line (model) is used in input-output analysis.

These models provide effective ways of making estimates, determining whether differences are significant or not, determining whether a characteristic is under control or not, and determining when and where corrective action should be taken.

Formulating, testing, and using quantitative models. In previous chapters, simple quantitative models have been described and illustrated. These are largely theoretical probability distributions which have been proven sound and effective over many years of application to a wide variety of real-world problems.

In this chapter, we describe how a simple model is formulated and tested, and how important conclusions are drawn from it. Also, an example is described where considerable research and analysis of operations were required before a decision could be made as to what was the most effective model. Another example shows how a simple straight line relating output as a function of input was found to be a very simple yet effective control method.

The major steps in formulating, testing, and using a quantitative model are as follows:

1. Isolate and define the problem situation.
2. Analyze this problem situation.
3. Determine the major characteristic, y, and the characteristics related to it x_1, x_2, x_3,...,x_k.
4. Express quantitatively $y = f(x_1, x_3, \ldots, x_k)$.
5. Analyze the function to determine how the function is affected by varying values of parameters and to determine if the function makes sense and does not violate known principles, relationships, and experience.
6. Use available data, or collect relevant data, to estimate parameters in the function.
7. The function does not make sense or does not explain available or sample data; reject and revise model.
8. If the model appears satisfactory, apply it to a new set of data.
9. Apply the model to predict the function using various values of the parameters and determine to what extent it predicts the real-world situation.

Attention is called to the first example below because it is simple and illustrates all of the major aspects of formulating and using a quantitative model. This is an expression for predicting the number of automobiles in a single lane that can get through a green light with a specified duration. The steps in developing this model are as follows:

1. The physical situation is laid out on paper and the various key variables or factors are identified as: T the duration of the light, d the distance between cars, k the length of a car, v the speed of a car, and t the time required to get a car moving from a stopped position.
2. The time required for each car in sequence to get through the light is written out: time for the first car, time for the second car, and time for the nth car, making certain assumptions about k, v, t, and d.
3. These times are summed for the nth car and equated to T.
4. n is finally expressed in terms of T and average values of d, k, t, and v.
5. n is expressed as a function of v given T, d, k, and t.
6. Deductions are made about the size of n when two conditions exist: cars start from a stopped position and cars are moving, so $t = 0$.
7. These deductions are compared with principles of time, distance, speed, and motion and they are found to be consistent with them; no contradictions are found.
8. The conclusions drawn by simply studying and analyzing the model and its parameters are: in the stopped position, the starting time t dominates and determines how many cars get through the light; in a moving position, the speed v tends to dominate given certain values of k and d. Clearly, the average length of cars tends to be fixed and to a certain extent the distance between cars.

This simple example illustrates the power of a quantitative model in describing a situation and predicting what will happen when the basic characteristics or parameters are varied. If the model is sound, all of this can be done without making a single field test or experiment. Of course, there is a big "if" — if the model is a reflection of reality and gives a close approximation of the way the parameters actually vary. This is the most difficult part of using a quantitative model.

Automobiles at a traffic light. An example of a simple mathematical model and how sampling can be used to test it is found in the question "How many automobiles in one lane can get through a green light of specified duration, assuming the automobiles start from a stopped position and are in motion?" This can be used to study traffic control problems.

The physical model assumes that each car is k feet long, that they are d feet apart, and that the first car is d feet from the light. Each car starts from rest and has an average speed of v miles per hour. Each driver requires t seconds to start and the green light is on T seconds. The length of time required for the first, second, third, and nth automobiles to get through the light are as follows:

$$\text{first car: } t_1 + \frac{k + d}{v_1}.$$

$$\text{second car: } t_1 + t_2 + \frac{2(k + d)}{v_2}.$$

$$\text{third car: } t_1 + t_2 + t_3 + \frac{3(k + d)}{v_3}.$$

$$\text{nth car: } t_1 + t_2 + t_3 + \ldots + t_n + \frac{n(k + d)}{v_n}.$$

To reduce this to a usable mathematical form, it is assumed that the t's are independent and not overlapping, and that accurate averages can be obtained for t, k, d, and v. Then:

$$n\bar{t} + \frac{n(\bar{k} + \bar{d})}{\bar{v}} = T.$$

Therefore, the number of automobiles getting through the light is:

$$n = \frac{T}{\bar{t} + \dfrac{\bar{k} + \bar{d}}{\bar{v}}},$$

where it is assumed average values are obtained for t, k, d, and v.

This equation shows that with T fixed, n can be increased by decreasing the size of the denominator. Since the length of automobiles is fixed, this can be done by decreasing t, by decreasing d, or by increasing v. These relationships are consistent with experience. The starting time is very important since it includes not only driver reaction time, but also delay time due to inattention of the driver, a stalled engine, and the like. The speed v is important because it is limited by the speed of the slowest driver ahead of the nth driver.

Examination of the equation shows that even if $k+d$ approaches zero and v goes to infinity, n is still limited to the value T/t. This means that if $T = 25$, the upper limit of n is 25 if $t = 1$, the upper limit is 13 if $t = 2$, the upper limit is only 8 if $t = 3$ seconds. Since $k + d$, however tends to be fixed, as v increases the second term in the denominator decreases which makes the denominator less and the size of n greater. This is in harmony with experience: the greater the speed, the more cars that get through the light; the effect of v is greatly dampened if t is large, say two or more seconds.

If $t = 0$, then all reaction times are zero and the condition for moving vehicles exists. Hence, the number of automobiles getting through the light is:

$$n = \frac{vT}{k+d} = \frac{vT}{S}$$

where $S = k+d$, the linear space in a traffic lane "occupied" by the moving vehicle. The number of vehicles n varies directly as speed v and time T and inversely as S. The faster the vehicles move, the shorter they are, the closer they are together and the longer the duration of the green light, the greater is n. These deductions from the model are in harmony with the laws of time, space, and motion. The parameters v, k, and d affect n much more when vehicles are moving than when vehicles are starting from a stationary position. In the latter case, as shown above, the starting time t is of major importance.

The model was tested at a 50 foot intersection for both through traffic in one lane and for a left turn controlled by an arrow (see Table 1). The model tested was for automobiles starting from a stopped position at this intersection. The duration of the light T and the number of vehicles getting through the light were both very accurately measured, the former by means of a stop watch. The T values of 24 and 15 seconds were based on a total of about 20 measurements each including time of each traffic count. The left hand turn time of 15 seconds gave no problem since it was the time from the beginning of the green arrow through the amber light. The through time of 24 seconds was more difficult because repeated measurements showed the amber light was on for very close to three seconds. Since the amber light stops some and not others, it was decided, after considerable observation, to include half of the amber time in T which led to a value of 24 seconds. It was much more difficult to measure starting time except in the two cases where the first driver was slow in getting started, requiring about three seconds. The speeds v were estimated by actually driving in traffic at this intersection. The length of a car was based on measurements of both short and long cars ranging from about 13 feet to 16 feet. The distance d between cars was difficult to estimate in the through traffic where considerable variation existed, but it was not difficult to estimate for the left turn because the cars were close together. Furthermore, there were usually six or more cars waiting to make the left turn, but there were not always 10 cars bunched at the intersection to get through. Some were approaching the intersection and did not have to slow down to get through during the 24 seconds. Since it was difficult to separate those affected by starting time t and those not, two separate calculations for the two models were neither made nor added; instead the starting time was reduced to one second. Actually, it would have been better to have applied the two models to two portions of traffic.

This problem and model illustrates how a model is formulated, how it is tested, and the problems encountered in testing it, especially in a case like this one where four parameters were difficult to measure accurately: t, v, d, and k.

Table 1

Automobile Traffic Test Results[a]

Duration of light T	Estimated averages t sec.	Speed v ft. per sec.[b]	Feet apart d	Number n Predicted	Actual	Diff.	Remarks
Through traffic 24 seconds[c]	1.5	22	10	9	10	−1	cars close together
	1.5	22	10	9	8	1	
	1.5	29	20	9	10	−1	
	1.5	29	20	9	8	1	
	1.5	29	25	10	8	2	
	1	37	30	11	10	1	cars spread out
	1	37	25	12	10	2	
	1.5	29	20	9	9	0	
	1	37	25	12	10	2	cars spread out
	1	37	25	12	10	2	
Left turn 15 seconds	1.5	22	10	6	6	0	cars close together
	1.5	22	10	6	7	−1	
	1.5	22	10	6	6	0	
	1.5	22	10	6	6	0	
	3	22	10	4	4	0	slow start by first driver
	3	22	10	4	4	0	
	1.5	22	10	6	5	1	
	1.5	22	10	6	6	0	
	1.5	22	10	6	6	0	

[a]50 foot intersection; average length of car estimated to be 15 feet = k.
[b]22 is 15 miles per hour; 29 is 20 miles per hour; and 37 is 25 miles per hour.
[c]Autos come through on the amber light so half of the amber light time (three seconds) was added to the green light time of 22.5 seconds to obtain 24 seconds. On the left turn, total time is green light time plus amber light time, since turn was controlled by an arrow.

Traffic delays at toll booths. An effective use of the Poisson and normal models in solving the problems of traffic delays at toll booths at Lincoln Tunnel and the George Washington Bridge is described by Edie in a prize winning article.[1] Basic data collected included the number of vehicles arriving in each lane during 30 second intervals, backup of vehicles in each lane at 30 second intervals, and the toll transaction count at 30 minute intervals. These and other data were used to determine a better use of toll booths and the efficient assignment of personnel to them.

At each toll plaza an observer recorded the count and time of the vehicles arriving during 30 second intervals; a second observer took count of number of vehicles backed up in each lane at 30 second intervals; and the toll transaction count was recorded for each lane at 30 minute intervals. An example of the data is shown for 13 half minutes for each of the three lanes giving vehicles backed up in each lane, total backup, traffic arrivals, and number of lanes occupied at each observation.

Frequency distributions of arrivals at 30 second intervals were constructed for various volumes of traffic, using groups of 200 vehicles per hour. It is not explained how long, at what time of day, or how the frequency distribution data were collected. Apparently, a few hundred intervals were used for each distribution. Frequency distributions were constructed for 246, 480, 655, 865, 1265, and 1580 vehicles per hour.

Figures 1 and 2 show how the Normal and Poisson distributions fit the actual arrival frequency distributions at Lincoln Tunnel and George Washington Bridge for 655 and 1100 vehicles per hour, respectively. The basic values for traffic arrivals in these two cases where $\bar{x}$ and s apply to 30 second intervals:

Location	Number vehicles	Mean $\bar{x}$	Stand. dev. s	Variance s^2	Range	Remarks
Lincoln Tunnel	655	5.46	2.73	7.45	0-12	Poisson
George Washington Bridge	1,100	9.17	3.00	9.00	0-18	Normal slightly better

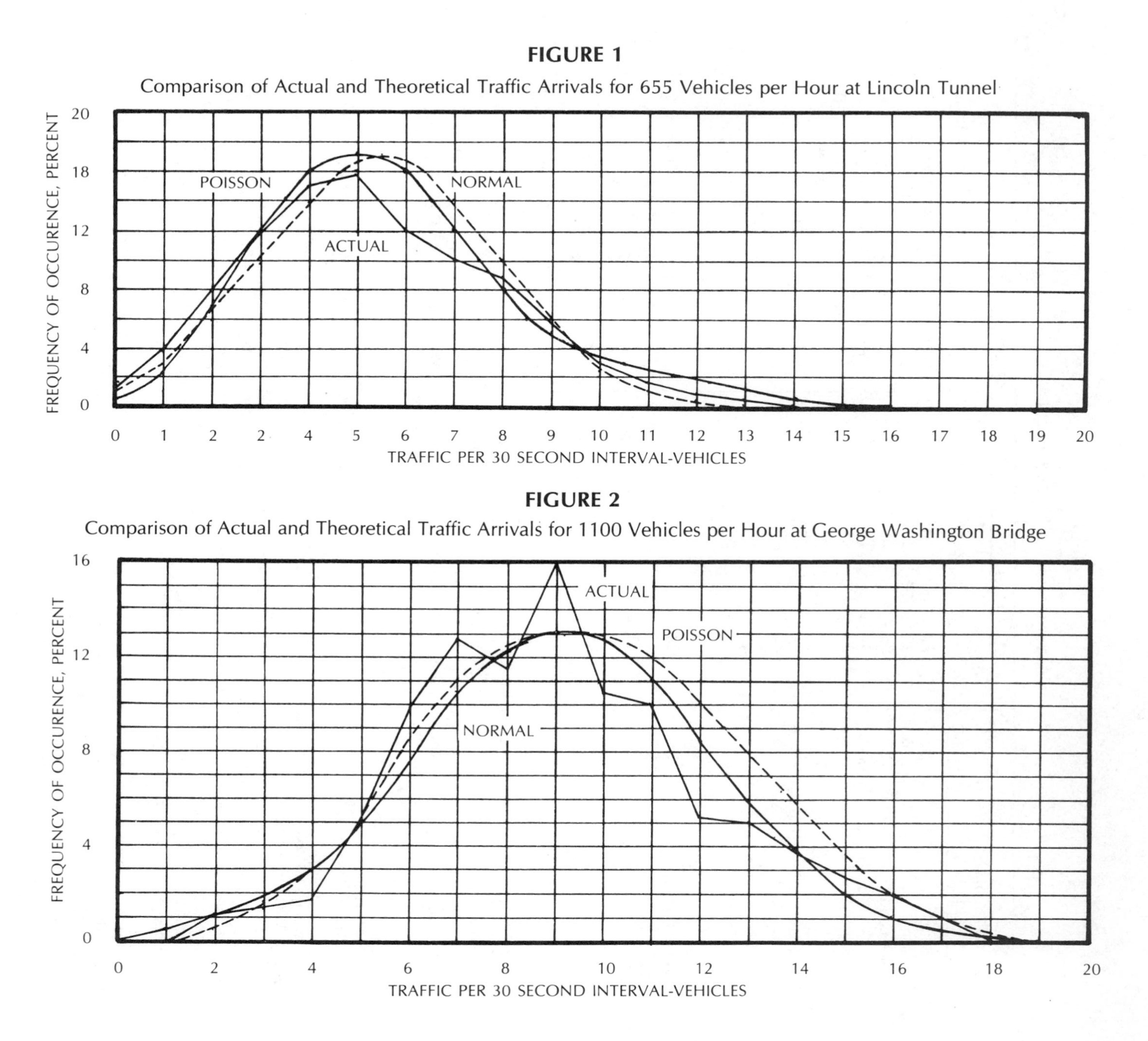
FIGURE 1
Comparison of Actual and Theoretical Traffic Arrivals for 655 Vehicles per Hour at Lincoln Tunnel
FREQUENCY OF OCCURENCE, PERCENT
20
18
12
8
4
0
POISSON
NORMAL
ACTUAL
0 1 2 2 4 5 6 7 8 9 10 11 12 13 14 15 16 17 18 19 20
TRAFFIC PER 30 SECOND INTERVAL-VEHICLES
FIGURE 2
Comparison of Actual and Theoretical Traffic Arrivals for 1100 Vehicles per Hour at George Washington Bridge
FREQUENCY OF OCCURENCE, PERCENT
16
12
8
4
0
ACTUAL
POISSON
NORMAL
0 2 4 6 8 10 12 14 16 18 20
TRAFFIC PER 30 SECOND INTERVAL-VEHICLES

It is noted that, for the George Washington Bridge, the normal distribution is considered a slightly better fit even though s^2 the variance is very close to the mean (much closer than in the case of the Lincoln Tunnel) even though $s^2 = \bar{x}$ is a characteristic of the Poisson distribution.

On traffic backup, samples of 20 minutes or 40 half minutes were used and the curves smoothed since changes in toll booth arrangements two or three times per hour required the smaller samples. The backup data for both the Lincoln Tunnel and the George Washington Bridge give frequency distributions closely approximated by the Poisson:

Location	No. vehicles per hr.	Mean $\bar{x}$	Std. dev. s	Variance s^2	Range	Remarks
Lincoln Tunnel	615	2.16	1.52	2.31	0-8	Poisson
George Washington Bridge	705	2.79	1.67	2.79	0-10	Poisson

The Poisson model breaks down at about 800 vehicles per hour due to congestion. At these higher densities, the traffic in one 30 second interval is not independent of adjacent 30 second intervals.

The analysis of the data included the following:

1. Distribution of traffic arrivals.
2. Delay time, which was backup time plus booth time.
3. Occupancy versus delay ratio, which was average delay per vehicle divided by average booth holding time.
4. Traffic volume in vehicles per hour per lane versus booth holding time.
5. Average delay curves.
6. Traffic backup.
7. Volume of traffic versus backup.
8. When to open a new toll booth.
9. Using the above for scheduling toll booths and manpower.

To verify this intensive and extensive analysis, a test was run for one week at the Lincoln Tunnel. The number of toll booths required every half hour was predicted. This required 512 predictions in both directions. A schedule was laid out for each of the toll collectors who were instructed to follow the schedule without deviation. The test was successful: no excessive backups developed, the manpower allocations worked satisfactorily, and the number of excessive booths open was less than with the former procedure. All this attested to the soundness of the analysis for this complex situation.

Input-output models. The simplest input-output model to use is the straight line, but more complex forms can be used following Leontief.[2] In the straight line $y = a + bx$, x is a measure of the input and y is a corresponding measure of output. The line is fitted to the data by the method of least squares. The values of x and y correspond to the same time period, which may be as short as a day or as long as a year. To be effective for control purposes, the correlation between x and y should be high, certainly above 0.90 so that y can be estimated or predicted very closely, given some value of x.

Example: A situation where a straight line is very effective exists in a distillery, where input x in units of 10,000 pounds of corn is plotted against output y in 1,000 gallons of whiskey. Values of A, b, and r^2 were calculated from monthly data. These were independent values, not cumulative, and totaled nine to 12 pairs of values:[3]

Distillery	a	b	Percent of y variation: Explained by x: r^2	Not explained: $1 - r^2$
A	– 14.552	1.0744	92.9	7.1
B	19.220	1.0259	96.4	3.6
C	0.207	1.1939	99.6	0.4
D	105.360	0.9417	98.0	2.0
E	– 128.57	1.2912	98.0	2.0

With one exception, more than 96 percent of the variation in y is explained by x. The straight line, therefore, can be used as a control over y knowing the amount of x, or in predicting the value of y for a given value of x.

Example: This is the case of a hosiery mill where the manager thought he had a serious problem which input-output analysis would have shown did not exist. Using a superior operator, he ran a test to find the amount of yarn used per pair of hose. From this low figure he obtained a very inflated estimate of output: actually $1 million too high! Simply plotting total yarn input against total hose output for the past five years would have quickly shown the number of hose per pound of yarn, or conversely, the amount of yarn per pair of hose. Instead, he spent thousands of dollars to discover the "problem" he thought he had did not exist.

Learning curves. These curves take four forms and are all directly related to quality performance over time:

- The error curve.
- The production curve.
- The productivity curve.
- The unit cost curve.

The *error curve* shows the reduction in number of errors, the error rate of an individual over time, or the reduction in the average error rate of a group over time.

The *production curve* shows the increase in volume of production over time or, better yet, the volume of acceptable quality production over time. This may be total production, with a constant work force, or the average daily production per employee, where the work force varies in number from day to day.

The *productivity curve* shows the increase in acceptable quality production per hour or day per employee. It may also be a decreasing curve, showing the reduction in time to produce a fixed amount of product, such as 100 units, or the reduction in time to do a repetitive job over time.

The *unit cost curve* shows the reduction in direct employee cost per unit of acceptable quality product over time.

The last three are described and illustrated in detail in Chapter 9 with actual data from a large nationwide sample study processed and analyzed on a computer. For example:

Day 2: 12 employees, 929 documents produced = 77/person; 21.8 cents each

Day 40: 11 employees, 2,495 documents produced = 227/person; 7.4 cents each

Overall production increased 2.7 times and productivity 2.9 times; unit cost was reduced 66 percent or two-thirds. Production of acceptable quality documents did not reach an average of 200 per day per employee until the 37th day. An unrealistic management goal of 2,000 per day total was not reached until the 18th day. The learning curve stops management from setting arbitrary and unrealistic goals because it measures the capability of the group under real-world conditions.

The learning curve situation assumes that workers in an office or factory are assigned to a new job or task, that they start from "scratch" with no previous training or experience in that job, and that they have to learn on the job or, in some cases, that some preliminary training has been given to everyone. In this situation, the rate at which acceptable quality production occurs depends almost entirely upon the rate at which learning takes place — the rate at which errors and mistakes are corrected, the rate at which false moves and wasted motion are eliminated, the rate at which mutual interference is prevented, and the rate at which management eliminates factors obstructing the learning process.

Example: Psychologists made early studies of the learning of typing and telegraphy, where practice time, error rates, and speed could be objectively measured. Although these were more like laboratory tests, they revealed some common properties of learning curves: plateaus and improvement at a decreasing rate. An example from typing is the following, where x is a measure of practice and y is the average number of words typed per minute by 51 persons based on four minutes of typing:[4]

Points on curve	Pages x	Words/min. y
1	10	9.7
2	30	15.0
3	50	18.2
4	70	22.5
5	90	24.7
6	110	27.5
7	130	28.2
8	150	30.0
9	170	32.5
10	190	33.2
11	210	34.5
12	230	36.2
13	250	37.0

Speed increases with more practice, but the biggest gain (two-thirds) is made in the first six steps. Thereafter, with smaller gains, the curve tends to flatten out — a characteristic of all learning curves.

Example: World War II furnished a wealth of significant data on learning curves in factory operations producing bombers, fighters, ships, dry cells, munitions, and other war materials. These data are significant for all industries because they show what was accomplished by the untrained: unemployed, housewives, students, and older men and women. Many of these learned the job while working or with a minimum amount of training. Three examples are described:[5]

B-24 bomber, Willow Run

Month	Employment	Monthly production	Employees per plane
June 1943	42,300 max	107	395
August 1944	26,600	428 max	62

Employment during this time decreased 37 percent, production increased four times, and productivity increased 6.4 times. They were reaping the benefits of learning, getting rid of excess employees, and the result was a tightened operation.

Dry cells, plant A

Month	Employees	Production
7th	4,700	38 million units
10th	4,700	50 million units

They reaped the benefits of learning that had already taken place. In three months, with the same number of workers, production increased 32 percent. A major factor was the increased volume of electronic equipment that required dry cells.

Liberty ship, Yard A. The increase in productivity in building Liberty ships in Yard A, 1941-1943, is given below:

Order of ship	1,000 man hrs. per ship	Ships per 10^6 man hrs.
5	1094	0.91
10	894	1.12
25	647	1.55
30	659	1.52
75	529	1.89
100	424	2.36
125	376	2.66
150	395	2.53
175	395	2.53
200	388	2.58
225	376	2.66
250	353	2.83
285	280	3.57

During this period, productivity increased nearly four times: from 1,094,000 man hours per ship to 280,000 man hours per ship. Plateaus existed in the neighborhood of 650,000 and 400,000 man hours, but these were false indicators of maximum attainable productivity.

Learning curves reflect not only increased skill but elimination of errors, wasted motion and interference. Plateaus should not be accepted as limits on learning or productivity even though they occur high on the curve.

Why formulating effective models is difficult. The formulation and testing of a mathematical model is an intensive as well as an extensive process; it is not simply applying known theory to a problem or situation. It requires training in mathematics, the subject matter fields or problem areas to which the mathematics is to be applied, an ability to apply mathematics to real-world problems, and an ability to select or create relevant mathematical expressions. It requires thinking on a complex and creative level. The formulation and development of an adequate model cannot be developed according to a time schedule.

There are several reasons for these demands and requirements. The problems are complex with a large number of variables. The phenomena are complex, as illustrated by weather control, air and water pollution control, and the operations of a business to maximize profits or return on investment. Situations very often involve the living rather than the non-living world. This means that the variability of living organisms, in general, and human behavior, in particular, are often involved. There may be a large number of probable or plausible models requiring the identification of one of the more desirable models. A promising model may be difficult to test because data in the forms and amounts required are not available, and a costly sample study or experiment is required to obtain them. This means management has to be convinced that such a data collection program is worth the cost.

This is why it is so much easier to carry out mathematical modeling in research organizations, research laboratories, and research units rather than in operating divisions where the real-world problems are. Most operating units do not spend half enough time analyzing and interpreting the data they collect in connection with their operations. Careful planning, implementation, and analysis are often short changed. This is why it is so much easier to do regression analysis which produces immediate results than to develop instead a more rational mathematical model.

General approaches to model building. Three general approaches to mathematical model building may be distinguished: the inductive, the deductive, and a combination of the two.

In the inductive model, the steps usually are: collect data such as a designed sample, analyze the data, formulate the model, test the model with more sample data, revise the model, test, etc. The data are analyzed and the mathematical expression which is considered the best fit to the data is the mathematical model. This is common in statistical analysis. Sometimes an attempt is made to develop a more rational model from a careful and intensive study of the underlying variables connected with the situation or phenomenon, especially in a laboratory experiment or a factory test. Statistical analysis could be the basis for better model building if more time and thought were given to the variables and their relationships in any given situation, especially in cooperation with experts in the subject matter fields involved. But this is not usually done.

In the deductive approach, the phenomenon under consideration is studied, or has been studied, rather intensively, the variables and their relationships have been postulated, and certain simplifications have been made before a mathematical equation or system of equations is developed to explain these variables and their relationships. In doing this, the situation may be abstracted on paper in the form of graphs, diagrams, and flow charts — aids in arriving at a mathematical model. In the initial stage, the inductive approach is based upon the analysis of data; the deductive approach is based upon an analysis of the phenomenon. Once the model is formulated, various deductions are made from it to see if they are consistent with experience and known principles. This is the initial way of validating the model. The second is to design and collect data, either from sample studies or experiments, of the various types and amounts, and under the conditions required by the model, to see how closely the model reflects real-world conditions. This testing by experiment, field tests, or sample studies may be left to others by the originator of the model.

Developing mathematical models. Draper and Smith[6] describe three types of mathematical models: *function, control,* and *predictive.* Functional models are rare, except in physical sciences where the mathematical relationships between a dependent variable (response) and the independent variables are known exactly. Control models are also difficult to formulate because of the possible existence of uncontrollable factors which affect the outcome. The common model, despite its shortcomings, is the predictive model, which to be effective has to isolate the major factors affecting the response or dependent variable.

The predictive model is often based on multiple regression, but it must be used with caution because of possible intercorrelations between the independent variables, the low quality data to which it is applied because better data are not available, and the difficulty of designing a proper experiment or sample study to obtain the acceptable quality data needed. The "predictive" model, however, may turn out to be a smoothing or descriptive model with little or no predictive value. It is not made clear whether "predictive" means ability to extrapolate beyond the limits of the data or simply interpolate accurately within the range of the data.

In general, mathematical models cannot be produced on demand or according to a time schedule. It is difficult to do more than lay down a general procedure based on experience. Draper and Smith describe steps to take in formulating and developing a mathematical model, but actually they are describing steps to take to develop an acceptable multiple regression. They list four steps: planning, developing, verifying, and maintaining. Some modifications have been added.

In the *planning* step, define the problem and select the dependent (response) and independent variables. Then screen the variables without data. Run a test of the variables if data are available; if not, design a sample or experiment to collect relevant data. This is a better time to screen variables, especially in a new situation. Determine if a relationship with y is concentrated in one, two, or three major independent variables. It is suggested that variables be screened without data, but it may be much more effective to screen variables after a test is run with available data or with a designed sample test or experiment. The test will disclose intercorrelations between independent variables and any concentration of the relationship, as well as variables that contribute so little that they should be dropped.

In *developing* the model, it is suggested that y values be plotted against each x, that new x's be tested, that various models be run using different combinations of the x's, and that, finally, a regression analysis is run. It may be discovered in this stage that the relationship is definitely not linear, but curvilinear. The model will have to be changed to take this into consideration. There is no point in going ahead unless the model satisfactorily explains the data we already have. This is why it is so necessary to have adequate sample or experimental data right from the start, and not just some available data whose collection and quality we are not sure about.

In *verifying* the model, it is necessary to obtain an entirely new set of sample or experimental data or to use the split half method wherein half the data are used to establish the model and the other half are used to verify it. In verifying the model, it is necessary that the model explain existing data, that it gives adequate estimates within the range of the x's and y's to which it applies, and that it, hopefully, can be extrapolated within a short range if necessary. It is necessary to determine if the model is reasonable and that it does not contradict known facts, relationships, or principles. The deviations of the data from the model and the residuals are plotted and examined for trends, other patterns, and a normal distribution.

In *maintaining* the model, it is necessary to determine if the model continues to give satisfactory estimates or predictions. This requires comparisons with other independent sources of data or preferably with an entirely new sample or experiment.

A regression model of the form $y = a + bx - ct$ was used to predict very accurately the requirements one year in advance for bituminous coal in the United States for the purpose of wartime control.[7]

The dependent variable y was expressed in tons of 2,000 pounds; the independent variable x was the Federal Reserve Board index of industrial production; and the independent variable t was time in years.

The basic data to which the model was fitted were collected by the United States Bureau of Mines for the years 1923 through 1940. There was a positive upward trend due to the expansion of industrial production (increase in the Federal Reserve Board index), but a downward time trend due to the substitution of other fuels for bituminous coal, especially oil for coal in transportation and in industrial and domestic heating.

A graph showing a comparison of yearly actual versus calculated values gave an extremely close fit for the entire range of data. The model was so stable and predictive that it gave excellent estimates for two years in advance, 1941 and 1942.

Notes

[1]Leslie C. Edie. *Journal of the Operations Research Society of America,* vol. 2, (1954), p. 107. Charts and data used by permission of the author and ORSA.

[2]W. Leontief, *Input Output Economics,* Oxford University Press, New York, 1966.

[3]These figures were derived from data collected by the Internal Revenue Service.

[4]Adapted from K. J. Holzinger, *Statistical Methods, in Education,* Ginn, Boston, 1928, p. 325. The data are from a study by L. L. Thurstone.

[5]All data are from the official records of the War Production Board, Washington, D.C.

[6]N. R. Draper and H. Smith, *Applied Regression Analysis,* John Wiley, New York, 1966, p. 234.

[7]A. C. Rosander, *Elementary Principles of Statistics,* Van Nostrand, New York, 1951. Copyright by Wadsworth, Inc. Used by permission of Brooks Cole Publishing Co., pp. 359-361.

28

Control by Procedures

Nine aspects of control by procedures will be discussed:

1. Control by procedural manuals.
2. Control by time schedules.
3. Control by budgets.
4. Control by standards.
5. Control by specifications.
6. Control by written contracts and letters of agreement.
7. Control by techniques.
8. Control by training.
9. Control by graphics.

Control by procedures is basic to all operations. These consist of instructions, explanations, examples, guidelines, illustrations, and details. They describe processes, operations, tasks, methods, duties, division of responsibilities, and routines. They deal with specifics, with particulars, with the concrete, with connections, with sequences, and with relationships.

All of these methods and procedures are aimed at reducing human error and increasing efficiency through mastery and use of proper knowledge, skills, and abilities; by setting appropriate goals, guides, and directions to work behavior and performance; and by establishing and implementing methods, techniques, and processes which result in acceptable quality performance.

These procedures will, or should, make very clear the specific tasks that need to be performed, how they should be performed, what the specific methods and techniques to be followed are, and what the errors, defects and failures to be avoided are. If these procedures are carefully planned and implemented, they provide a foundation for acceptable quality performance at all levels and in all areas.

Control by procedural manuals. Procedural manuals contain detailed information and detailed instructions as to how a specific task or a specific job is to be performed. Often examples are described to guide the user in following instructions and resolving various problem situations. Indeed, a manual is incomplete without such detailed explanations. The purpose of these manuals is to explain how a task is to be performed so that the desired volume and quality are attained. Important decisions are explained in detail so standard practice will result. Procedural manuals are prepared for the following:

1. Statistical coding.
2. Commodity coding.
3. Industry coding.
4. Occupation coding.
5. Sample selection manual.
6. Sample receipts control.
7. Manual processing manual: reviewing, editing, verifying, transcribing (coding may be included).
8. Key punching.
9. Tabulating.
10. Computer processing manual.
11. Filing.
12. Writing, clearance, and distribution rules: memoranda, letters, reports, and instructions.
13. Preparing budget: personnel, travel, supplies, equipment, and training.
14. Travel guide.
15. Sample data processing, calculations, and data analysis (may be included in manual processing manual and/or computer processing manual).
16. Inspection and verification manual.
17. Quality control manual.

Control by time schedules. Time schedules are used in a wide variety of situations such as departure and arrival times in transportation, times for the completion of various components of a project, deadlines for the completion of a job, deadlines for the issuance of reports or publications, and time for the completion of a contract. These time limits may be expressed in terms of a day, week, month, quarter, or year.

Times of departure and arrival control various forms of transportation: airlines, railroads, intercity buses, city buses, subways, and ships. These schedules determine the work activities performed at certain times and determine the activities individuals must plan to take to meet the departure schedule as well as plans others must make to meet arrival schedules.

Adherence to these times is very significant because business, professional, and personal plans have to be made in terms of these schedules. A major measure of the quality of service is the amount of deviation from this printed schedule. This is because time very often means money and in some instances a considerable amount of money.

A time schedule of the work to be performed on a project organizes the work, determines how many and what kinds of employees are to be assigned to the project and when, and facilitates future planning by those paying for the project. Good practice calls for time scheduling, because time is directly related to cost and cost is very important.

Work sometimes is automatically time scheduled because significant data must be obtained and issued on a daily, weekly, monthly, quarterly, or annual basis. Examples are daily and weekly production and quality reports, monthly reports on national employment and unemployment and the consumer price index, and annual reports on Federal taxes, agricultural production, and carload freight shipments. Annual reports are very important because they show business receipts and income, business expenses and profit (or loss), total production, quality levels, and quality costs.

A time schedule should be accurate and reliable. Actual transportation times for arrivals, for example, should be published time, plus some tolerance, depending upon the distance and nature of transportation, say 90 percent of the time. On airplane flights, it might be published time plus 30 minutes 90 percent of the time. Also, a standard might be set stating that a minimum of 74 percent of the flights should be on time — that is, published time ± 5 minutes.

A time schedule should be realistic, reflecting human capabilities. Whenever possible it should be based on a learning curve study (described in an earlier chapter) that shows individual and group capabilities under all existing limiting conditions. In one instance, an official thought a sampling study could be finished in three months; actually it required 13 months. The three months was purely arbitrary, based on opinion and conjecture, not on knowledge of the job nor the capabilities of the existing staff. This practice is common because those in power lack know how. Setting arbitrary deadlines may only sacrifice acceptable quality performance because an enforced speed can only result in a sloppy job.

Random time sampling (RTS) described in Chapter 24 provides a very effective way of auditing time schedules. An example of auditing a suburban express bus morning time schedule was described in Chapter 19. There time arrivals were recorded over a period of 38 days to determine if the time was under control.

Control by budgets. Two kinds of budgets are in use for different but related purposes: the annual financial budget and a performance budget, which may be on a monthly or quarterly basis. The financial budget includes:

1. A description of the present functions, projects, and jobs, with job classifications and the number, wages, and salaries under each classification. A comparison of the workload covered by the past budget is made with the workload expected for the coming year, with workload directly related to staffing.

2. *Staffing.* A grouping of all employees by position title, number in each category, and total salaries and wages each group now receives with similar totals for the next year. Any increases in the number of employees in a job classification, together with salary or wage increases, should be clearly shown. Detailed justifications should be presented for these increases in terms of the additional workload expected and in terms of the increased efficiency and production that will result. If workload is not expected to increase, describe methods to be used to increase productivity.

3. *Space.* If additional employees or equipment, are requested, a request for additional space has to be included, unless present space is ample.

4. *Furniture and equipment.* The budget must include additional furniture and equipment needed not only for the present staff but for any additions. Equipment includes typewriters, desk calculators, hand-held calculators, computers, special testing and measuring instruments and apparatus, accounting machines, and duplicating machines.

5. *Computer.* If a computer installation is planned, this usually involves such a large investment that a special report involving the entire company or agency may be required. This report includes the following:

a. A detailed explanation of the need for a computer system.

b. The results from a feasibility study comparing competitive systems, giving advantages and disadvantages of each, together with comparative costs.

c. A recommendation of which system to buy, with original cost and annual cost.

d. Justification of the new system on the basis of cost, effectiveness in meeting present workload, and adequacy in meeting anticipated future workload.

e. Staff and space requirements, with anticipated new operating costs.

f. Future annual costs, compared with present costs.

6. *Supplies.* This includes the usual office supplies, plus any special supplies for the computer system, laboratory, field work, inspection, sample tests, sample surveys, and other operations.

7. *Travel.* The budget should contain travel expenses for any field work, professional conferences, seminars, conventions, training courses, research, visits to branch or field offices, and visits to vendors.

8. *Training and consultation.* Another item which should be in the budget is the amount of money required to upgrade employees, supervisors, professionals, and managers. These courses may be given in-house by a training division, outside, or by hiring one or more consultants in addition to training, technical advice, and managerial assistance.

9. *Other.* This item includes newspapers, books, magazines, technical papers, technical magazines, tapes, films, audio visual materials, projectors, and directories.

The *performance budget* lays out performance goals, usually on a quarterly basis, in terms of actual manhours versus planned or projected manhours by major programs or functions. The purpose of the budget is to measure progress toward these goals. These programs and functions have to be mutually exclusive so all manhours are accounted for without duplication or omission.

This budget requires considerable time to prepare because programs have to be identified, analyzed, employees allocated, manhours calculated after adjustments for leave, and quarterly divisions made. Then, actual manhours have to be recorded to match this format and accumulated to compare with the projected values.

The author has had to prepare such a budget and found it largely a waste of time since the demands on high cost professional time was extensive with no visible benefits to either professionals or top level management. The latter already had an annual budget plan and report and a monthly progress report by major programs so the performance of all personnel was accounted for. The performance budget was a lot of unnecessary work.

Control by standards. Standards exert considerable influence on the quality of performance and the quality of products and are beginning to exert considerable influence on the operations of service companies. Standards may be classified as formal and informal, official and unofficial, institutional and individual, or government and private. Standards are associated with government, especially the Federal government, with private and professional organizations, and with individual companies and agencies.

Federal government

1. National Bureau of Standards: measurements (metrology): length, weight, volume, time, and wave lengths.

2. Bureau of the Budget: standard commodity classification (STCC) and standard industry classification (SIC).

3. Department of Defense: standards for contractors, military standards such as MIL-STD-105-D, MIL-STD-414, and MIL-STD-9858-Q.

4. Environmental Protection Agency: standards for air pollution, water pollution, soil pollution, and environmental pollution.

5. Food and Drug Administration: standards for drugs, chemical additives in food, toxic minerals in food (e.g., mercury in swordfish), and GMP guidelines for "good manufacturing practice."

Private and professional organizations. Many professional and private organizations set standards which greatly influence the operations of industry, business, and government. Examples are the American Standards Association, the American Society for Testing Materials, the American Society for Quality Control, and the American National Standards Institute. Examples of private organizations are the Underwriters' Laboratories and the United States Pharmacoepia.

Individual companies and agencies. In the operations of a business or agency, working standards are used to distinguish acceptable quality performance from undesirable performance and an acceptable product from an unacceptable product. Sometimes these standards are quantified and known, but in some instances the rules for distinguishing good performance from bad exist only in the supervisor's or manager's mind. Examples of these operating or work standards are the following:

1. Number and percent of errors to be tolerated.
2. Size of the money error to be tolerated.
3. Number of idle workers to be tolerated.
4. Amount of idle time to be tolerated: employees.
5. Amount of down time on equipment and machines to be tolerated.
6. Delay time to be tolerated: frequency, percent, and amount.

Nature of standards. Standards take a wide variety of forms, but all aim to improve operations, to improve performance, to reduce costs, to reduce human error, to increase efficiency, to eliminate waste, to increase convenience, to foster safety, to facilitate communication, learning, and understanding, and to increase productivity. Examples are:

1. Standard sizes: money, 80 column punch card, $8^{1/2} \times 11$ paper, 2×4 lumber, football field, and baseball diamond.
2. Standard levels: 110 volts and 60 cycles.
3. Standard gauges: $56^{1/2}$ inches between railroad rails, nails, wires, bolts, and screws.
4. Standard plans: sample plans in MIL-STD-105-D and MIL-STD-414.
5. Standard symbols: mathematics, statistics, and highway signs.
6. Standard terminology: any subject matter field or specialty.
7. Standard formulas: mathematical formulas, recipes, chemicals, and drugs.
8. Standard codes: International Morse code in radio telegraphy, commodities, industries, and occupations.
9. Standard arrangement: typewriter keyboard and telephone dial.

In the service industries and operations, standards of some kind are usually set for each of the following: purchasing, employee performance, equipment performance, volume of production, quality of production (error rates and amount of error), delay time, service performance, and customer complaints.

Control by specifications. Specifications refer to a detailed description, presented in a written or printed document, of what is required in the product or service under consideration. The contents for a *purchased product specification* include the following:

1. Description of the product or products.
2. Characteristics of the product and their tolerances.
3. Performance requirements.
4. Quality standards to be met.
5. Reliability requirements.
6. Safety requirements.
7. Evidence that dimension, performance, reliability, quality, and safety requirements are met.
8. Steps to be taken in case the product does not meet all of the specified requirements.
9. Statement that the purpose of this specification and the goal of the company is to reduce, if not eliminate, a costly program of receiving inspection.

An actual *data processing operation* based on a nationwide probability sample used the following written specification:[1]

1. Purpose and uses of the data; the management problem to be solved.
2. Sample study specifications: the frame, sample plan, how to select the sample, sample receipts control, and quality controls.
3. Collection plans: questionnaire, data sheet, instructions, and field editing.
4. Processing the sample: quality control system, editing and reviewing, coding, key punching, closing out sample using computer and other counts, table formats, tabulations, and tabulation checks.
5. Analyzing the data: summary tables, exploratory tables, calculations, statistical analysis, plotting frequency tables, estimates, tests of significance, analysis of variance, regression analysis, graphs, final tables and charts, and conclusions.
6. The final report: purpose, major findings, graphic presentation, tabular presentation, comparisons with past data, limitations of data, recommendations of actions to be taken and changes, if any, to be made, and what is required to put the recommendations into effect.

An example of *operating specifications* would be the following:

1. Identification of the operation.
2. Components of the operation.
3. Key characteristics of each component.
4. How each component is produced with specific instructions.
5. How each component is controlled with specific instructions.
6. Quality control over products if any: product defects, quality, cost, and quality control techniques to be applied.
7. Quality control over services: service errors, error rates, delay times, customer satisfaction, customer complaints, cost, and quality control techniques to be applied.

Control by letters of agreement and written contracts. *Letters of agreement* are mutually agreed upon divisions of responsibilities, duties, and functions between a company or agency and a second party such as a consultant, service company, or vendor. It is less formal than a written contract but contains explanations and elaborations not found in the usual written contract.[2] Letters of agreement include:

1. Statement of the problem.
2. Description of the project and the nature of the assignment.

3. Responsibilities and duties of the service company.

4. Responsibilities and duties of the consultant or other second party.

5. Shared responsibilities and duties.

6. Areas of cooperation and their nature.

7. How the sample is to be selected.

8. How the sample is to be processed.

9. The controls to be put on sample selection and processing.

10. Nature and frequency of progress reports.

11. How problems which arise are to be resolved and by whom.

12. The analysis and interpretation of the data in terms of the problem.

13. The nature of the final report and recommendations.

A *written contract* is usually prepared when a firm submits a bid on a proposed job or project. Such a bid includes specific answers to the requirements specified in the proposal. These include:

1. Description of the major aspects of the job or project.

2. Statement of experience and competence to do the job and to meet the specifications in the time required and at the cost stipulated (if necessary).

3. Work schedule.

4. Time schedule: completion dates for various phrases.

5. Manning and staffing.

6. Financial accounting system.

7. Manhour accounting system.

8. Quality control system.

9. Inspection procedures.

10. Security system.

11. Cost estimates, by components, for total.

12. Performance standards to be met.

13. Certification of equal opportunity employer (on federal contracts).

There are contracts based on cost plus a percentage fee and contracts based on cost plus a fixed fee. Cost plus contracts are criticized because there is no incentive to be efficient, to do a high quality job, or to eliminate waste and unnecessary costs. Unless costs are carefully studied and estimated, taking into consideration various economic factors, an over-run is likely to occur. Over-runs on government contracts have been subject to considerable criticism. On large well-advanced government contracts, there has been nothing for the Federal government to do but re-negotiate and give the contractor additional money to complete the contract.

Control by techniques. The power of techniques is illustrated by the breakthrough due to Walter A. Shewhart in originating quality control charts. He used the principles and techniques from the new sciences of probability and mathematical statistics.

Neither engineering nor management knew what to do about variations in dimensions arising in manufacturing processes. That is why they took the problem to Shewhart, who used statistics to separate random variations which could be accepted from non-random variations due to assignable causes which could be eliminated.

Additional control techniques, not described by Shewhart, but applicable to service operations have already been described and illustrated in preceding chapters. More detailed applications of quality control to specific service industries and operations are described in Chapters 2 through 10.

Control by training. Training enters into quality control at four points: when hiring and promoting quality control technicians and specialists, when hiring and promoting any employee who has anything to do with quality performance, when on-the-job training of employees, including supervisors and managers, is necessary, and when special training or tutoring is required to reduce error rates that are too high or to meet a critical situation.

In the first and second cases, the hiring official or person looks for qualifications that relate to quality performance — abilities, attitudes, skills, technical knowledge, successful experience, and work attitudes. This care in hiring to meet quality requirements needs to be applied to all jobs, at all levels, and to all employees whether clerks or high-level managers.

In the third case, an on-the-job training program should be in effect so that every employee, including every supervisor and manager, is instructed with regard to the requirements of every job, every project, and every operation. The stress should be on error-free quality performance. This should be an on-going program, not a one-shot deal.

In the fourth case, special instruction is needed to meet an unusual situation such as a high error rate, some serious service or other failure, or an emergency situation that requires immediate attention. If only a few persons are involved, this can be accomplished by tutoring or by a small group session. Otherwise, a series of classes or seminars may be required at once. The goal is to reduce, if not eliminate, human error, poor decisions, inefficient decisions, or whatever the cause.

Control by graphic methods. Control by graphic methods includes use of various pictorial and graphic techniques to guide, direct, explain, and control individual performance. These include: blue prints, scale drawings, maps, sketches, pictures, graphs, charts, and plots. These graphic devices show such important ideas and concepts as dimensions, tolerances, distances, relationships, locations, distributions, differences, levels, and variations.

These are produced by a wide variety of specialists such as engineers, draftsmen, statisticians, map makers, photographers, artists, and architects. They may be prepared manually or some of them, such as graphs and charts, may be produced by a computer plotter.

Examples are:

1. Large plots of daily group error rates.

2. Posting of daily acceptable quality production in number of units.

3. Graphs of individual learning curves: trend of error rates, trend of errors, or trend of number of units produced or completed.

4. Graphs of group learning and production curves.

5. Graphs of frequency distributions.

6. Quality control charts.

7. Regression lines to show relationships.

Notes

[1]A. C. Rosander, *Case Studies in Sample Design,* Marcel Dekker, Inc., New York, 1977, pp. 92-93, 377-378, reprinted courtesy of Marcel Dekker, Inc.

[2]For a good example, see Dr. W. Edwards Deming's verified statement reproduced in full in Chapter 19 of A. C. Rosander, *Case Studies in Sample Design,* Marcel Dekker, Inc., New York, 1977.

29

Quality Management and Quality Costs

Quality management. Quality management is needed regardless of how large the company or agency is. In the former, there may be a formally organized division with a staff; in the latter, the functions will be performed by the regular staff possibly with the assistance of a consultant. Given below is a description of the seven steps necessary for developing a quality control and quality management program.

Steps in developing a quality management program.[1] Quality is so all-pervading in any company or agency that it requires, for its attainment, a quality management program if for no other reason than to assign responsibilities, set a policy, develop plans, and place a quality control program into operation. This can be done by any staff familiar with quality control techniques and practices even though no one is a quality engineer, quality technician, or statistician. The seven steps described below are not only necessary for a large organization, but also for a small business or agency. The seven steps not only describe a plan and a sequence of steps, but also a quality policy and some general notions as to how the plan is to be implemented.

1. *Define and document the service requirement.* This means determining buyer requirements, expectations, needs, and preferences. The methods used to do this include market surveys, opinion polls, market trends, customer preferences, customer complaints, sales figures, and customer surveys. Data from these sources are analyzed to determine what the buyer considers the most important quality characteristics.

One place to start is for the firm to survey both past and present customers to determine what specific characteristics and their levels are associated with acceptable quality service.

Once these characteristics are isolated and identified, they should be documented — that is, described in detail in a procedural manual. Service requirements include several characteristics which vary with the industry, but most of them are subject to waiting time, delay time, errors, prices, failures, warranties, courtesy, and meaningful communication.

Some standard or goal is set for every one of these service characteristics, so that the service is acceptable to the customer at the price charged. To the buyer, an appraisal of service may be based on one or more intangible facts such as appearance, treatment, politeness, attitude, and reliability since these may have a direct bearing on the kind and nature of the service received.

2. *Build the service requirement into each activity.* This means the service requirements have to be identified and analyzed into specific operations, and the various tasks and jobs associated therewith. Service tasks are assigned to jobs, and quality standards set so that the service quality goals are met. To meet these standards of acceptable quality services, it is necessary to:

1. Hire qualified personnel.
2. Institute the proper training courses.
3. Write the relevant tasks and standards into job descriptions.
4. Inform employees that the quality of their work will be observed and appraised.
5. Stress the importance, to both the employee and the company, *to have the job done right the first time.* This means that the ideal goal is Zero Errors and Zero Delay Time. Although they may be difficult to reach, these are the goals to aim at.

3. *Measure and report service performance.* If the jobs have been analyzed into tasks under item 2 above, the next step is to use the various sources of data to measure continuously the quality of job performance of every employee or group of employees. These data sources, depending upon the nature of the job, include:

1. Shewhartian control charts.
2. Sample audits.
3. Work sampling (random time sampling).
4. Error records.
5. Acceptance sampling.
6. Existing records.
7. Production records.
8. Operating failure records.
9. Ratings by supervisors.
10. Reports by supervisors.
11. Service department records.
12. Customer complaints.
13. Customer surveys.

The value of Shewhartian charts lies in the fact that they combine into one chart data collection, data analysis, the historical record, an indicator of possible trouble, and an indication as to whether the quality of the performance can be improved further.

In addition, it is necessary to measure the over-all quality of the system, as well as the quality of performance of individual employees and individual pieces of equipment.

This step implies that the necessary data are available. If not, a valid data collection system has to be designed using sound data collection procedures, such as:

1. Shewhartian charts.
2. Acceptance sampling.
3. 100 percent coverage and 100 percent inspection.
4. Sample audit.
5. Random time sampling.
6. Probability sample studies or surveys.

The merit of the Shewhartian charts is that the data were collected right off the production line, including paper processing, and plotted immediately. If a point fell outside the limits on the chart, immediate action could be taken to correct the situation. There were no long delay times between data collection and correction of the faulty situation or between data collection and corrective action.

4. *Analyze results and identify major problems.* One does not wait a week or a month to prepare a report and send it to management. If the condition needs top management attention and approval, then they should be informed immediately. Methods used to analyze the results, in addition to Shewhartian charts and acceptance sampling, are: close observation of the data, tabular analysis, graphic analysis, comparison with standards as needed, comparisons with past data, and comparisons with standards. Close observation of the data may reveal trouble without any detailed analysis.

The collection procedures are monitored while they are in progress to discover any deviations; the data are observed as they are coming in to detect anything unusual. If a serious problem is encountered under item 3, immediate feedback with corrective action is needed, e.g., the sample is not being selected correctly. *One does not wait for a formal analysis until after all the data are in.*

It is necessary to differentiate between two situations: those that need immediate attention (Zero Delay) and those that need attention but can wait. The goal is to eliminate any unacceptable quality performance as quickly as possible.

5. *Take corrective action.* Action is based upon the analysis of the data and a correct diagnosis of any trouble. The trouble discovered may be: too high an error rate, too long a delay time, too many customer complaints, and too much down time. As a result, standards are not being met. The steps to take are: find the trouble, decide how to correct it, consider various methods, and take corrective action. In Shewhartian charts out-of-control points require action by a specialist to determine if a real cause exists or just a random variation.

6. *Verify results.* Once corrective action has been taken, it is necessary by observation, sample audit, Shewhartian charts, or examination of appropriate records to determine if the "corrective action" really corrected the trouble. If the corrective action did not work, the data must be re-studied or new data must be collected and a different "corrective action" taken. It is to be noted that with Shewhartian charts once the corrective action has been taken and the assignable cause eliminated, succeeding points on the charts will stay within control limits.

7. *Modify or initiate a quality improvement program.* This assumes what is sound practice: conduct a small pilot study to isolate and resolve various problems and show that the quality program works before trying to put a large scale program into practice. The latter requires approval of top-level management as well as acceptance by lower-level employees and supervisors. Basically, a quality control program is a human relations problem requiring very careful attention to all human factors.

Quality costs.[2] Management needs to know what acceptable quality service costs and how it can be obtained at a minimum cost. Quality cost categories developed for manufacturing can be used as a guide for service industries in developing costs peculiar to service operations.[3] Quality costs incurred in manufacturing products in a factory are divided into four categories: prevention, appraisal, internal failures, and external failures.

Prevention costs. This category includes all those activities aimed at preventing defects, defective products, and errors, on the one hand, and building acceptable quality into products and services, on the other. It includes designing, implementing, maintaining, and auditing the quality system where such a system is actually identified. It includes all quality planning; all quality training, seminars, and courses; and it includes all customer surveys and consumer research aimed at improving quality or attaining the desired level of quality. It should be noted that "prevention" is a negative term aimed at stopping defects and errors from occurring. Stated positively this means "building acceptable quality" into products and services. The latter term obviously may include many more employees and operations than a specific group named "quality control" or "quality assurance" which aims at preventing defects and errors. This shows that a definition of terms and the identification and classification of operations should be objectified. Otherwise, quality costs will simply be a function of the person who calculates them.

Appraisal costs. Appraisal costs are those involved in attempting to measure the extent to which quality goals are being met. These activities include measuring, inspecting, testing, and auditing products and performance to determine conformance with acceptable quality levels, standards, and specifications. These include those activities associated with measuring and evaluating products, components, and purchased materials. It includes performance of employees, performance of equipment and machines, receiving inspection, laboratory testing, quality audits, field tests, life testing, quality data processing, failure testing, and random time sampling for appraisal.

Internal failures. These include defects, errors, rework, salvage, process failure, equipment failure, computer failure, and programming errors. These include costs associated with idle time, down time, correction and re-do, employee errors, performance failures, scrap and waste, service failure, delay time, and repair failure. These costs arise when personnel and equipment fail to perform in conformance with quality levels and standards. It is important to keep these failures at a low level because they directly affect the appraisal costs, on the one hand, and the external costs, on the other. They may also call for additional prevention costs.

External failures. These failures refer to those arising from products shipped and services rendered outside the company or agency. These include customer complaints, customer service related to poor quality products and service, field audits, defective merchandise, repair failure and other service failures, delay time cost to a customer, customer-discovered errors, and delay time loss to the company.

Three parties to service costs. In both manufacturing and service company quality costs, it is customary to ignore all costs except those to the manufacturing company or to the service company. This is because quality control specialists work for a company and the cost to the company is the only cost they are interested in. This is natural, but this restriction of interest ignores some very important costs that should not be overlooked. In services, costs should be developed not only for the service company which renders the service, but for the buyer who pays the bills. This means there are three parties to service costs, the service company and two buyers:

1. The company rendering the service.
2. The individual or household buyer.
3. The company buyer.

Service company costs. A service company may encounter serious problems in arriving at quality costs depending upon:

1. How it defines quality.
2. Whether it has a formal quality control department.
3. Whether it has quality specialists, as such.
4. Whether it can separate quality activities from non-quality activities and whether it can arrive at costs for the quality activities.
5. Whether quality is defined to permeate all tasks, jobs, functions, and operations.
6. Whether it has adequate cost accounting records.

The company need not be stopped from developing quality costs because it has no way of separating or decomposing joint costs or from separating quality costs from non-quality costs, where one or more employees perform both kinds of jobs. Random time sampling (RTS), described in detail in Chapter 24, is the answer; it is the only effective way known of decomposing joint costs, but it does require a proper sample plan and careful management to implement it. The company will need to develop this capability, if it does not already have this technical know how.

There are areas where quality costs may be easily measured from current records or from additional records that are easy to establish:

1. Correcting clerical errors.
2. Machine down time or idle time.
3. Equipment down time or idle time.
4. Employee idle time.
5. Correcting service errors and failures.
6. Correcting computer errors.
7. Auditing various operations for quality conformance.
8. Receiving inspection of purchased items.
9. Correcting other errors.
10. Some kinds of delay time.

Major causes of costs to the buyer, whether individual or firm. Quality of service creates costs in a wide variety of ways:

1. Service failure, such as *interrupted service,* e.g., telephone, electricity, water, or gas.
2. Service failure such as *unsatisfactory service,* e.g., failure to repair radio, television, automobile, or failure of prescribed medicine.
3. *Errors made by service company,* e.g., in billing, in installing, and in delivery.
4. *Delay time* in waiting to get service: pre-service delay or on-site delay.
5. *Excessive time* in performing service: this delay time can be very costly to a person who has to use an automobile to drive to and from work.
6. *Unnecessary service,* e.g., medical service; automobile, radio, and television repairs; and home insurance, where specific coverages are absolutely not applicable to the homeowner or renter. Specific examples are given later.
7. *Errors made by the buyer,* e.g., misuse of the product, failure to follow directions, failure to use due care, and lack of simple technical knowledge about automobiles, gasoline, carbon monoxide, electricity, gas, fire, poisons, drugs, etc.

A company as a buyer of services. A company as a buyer of services is in a much better position than the individual. As a general rule, it is the larger company and consumer that receives lower rates for public utilities and receives special prices and discounts. In purchasing items, such a company can prepare specifications and shop around until it finds a vendor willing to furnish acceptable quality products at the price the company is willing to pay. An individual, whether a householder or a small company, cannot do this; the individual has to pay the going price and may have little or no choice about what he or she is receiving.

Of course, a manufacturer has to buy many services such as water, electricity, gas, and telephone service. On special occasions, it may have to buy legal and accounting, or other expertise, if it lacks the capabilities needed.

Examples of service costs to individual buyers (householders). The following examples are actual cases which are cited to show that the costs of quality involve not only the service company but also the buyer — the person who pays the bills. We have ignored for far too long the costs of quality or, more accurately, the lack of quality to the individual householder or individual non-business buyer and consumer.

1. A family moving from one city to another has to pay an extra $200 in motel and restaurant bills because the moving van does not arrive on the day promised. Furthermore, the total bill is nearly $300 more than the original estimate.

2. A householder is forced to pay insurance on a "householder's insurance package" that includes three items which absolutely do not apply:

- Pays on "adjoining buildings", although they do not exist.
- Pays on possible unemployment, although the householder has a guaranteed income and will never be unemployed.
- Pays on household goods valued arbitrarily by the insurance company at 50 percent of the estimated sales price of the house also arrived at by the insurance company. This is absurd since once bought households goods have little resale value. Furthermore, many householders have older and better furniture than is now made and, hence, really cannot be replaced at any price. The 50 percent is much too high.

3. A mortgage company makes an error of $314 overcharge in connection with the monthly payment. It requires a letter, two telephone calls, a trip of 100 miles, and a personal investigation that required most of one day before the error was corrected after about four months. This shows how hard it is to correct a computer error.

4. A book seller repeatedly bills a customer for $33 worth of books that have already been paid for in full. It requires two letters and a duplication of both sides of a cancelled check to convince them of their error and of the fact that they need quality control in their accounting system.

5. A mortgage company makes an error of about $400 in handling a check of $3,400, issued solely to pay off the principal. It requires two letters, two telephone calls, and five months to correct this error.

6. In a city where monthly electric bills are only estimated twice a year, a new owner of a house is billed for

620 kwh, which is what the former owner used, although consumption is now about 400 kwh. It takes a letter and two telephone calls to shift the monthly billing to 400 kwh.

7. A customer orders drapes on sale from a well-known retail store 10 miles away. The customer is told it will take six weeks. After this time elapsed, an inquiry shows that the drapes are not in; the same is true after eight weeks. The customer goes to the manager of the drapery department with the sales slip showing details of order and that it was paid for. The manager rummages around in a store room and finds the drapes; they had been there for over two weeks. The store could not find their sales slip, so they had to use the customer's! Hardly quality performance.

8. Cases of where medical care was obviously unnecessary abound:

a. A man who is using eye medicine for glaucoma has a small part of his vision in one eye blurred. He goes to an ophthalmologist and explains. The doctor takes him off the medicine for three days; as a result spots appear in his vision, but the blur disappears. He complains to the doctor who tells him to go back to the eye medicine he has been using. The charge for this was $20. The doctor obviously paid no attention to the blur and never mentioned it until the man asked him directly about it. The doctor did not seem to know.

b. A rancher whose back was injured in an accident is placed in a nursing home to convalesce at a cost of $500 per month. He gets well, but cannot get out of the nursing home. He spends part of his time folding towels for the nurses. After considerable protests, he is finally released, but pays at least $1,000 for two unnecessary months in the nursing home.

c. This same man was discharged from the nursing home with the understanding that he is to see a doctor once a month ($20) and have his blood pressure taken every Tuesday ($5). Neither of these visits was necessary. After about a year and about $500 of unnecessary expenses, he was induced by a neighbor to drop both of them. Months later, when he was last heard from, he was doing fine.

d. A man sees his doctor about a recurring pain in the upper abdomen. The doctor thinks, to be on the safe side, that he should have an EKG (electrocardiogram) taken immediately, even though the man pointed out he had had one taken within a year. The EKG showed nothing, but cost $27.

Notes

[1]These steps are given and described by John T. Hagan and Ernest W. Karlin in a paper "Quality Management for Service Industries," *Annual Technical Conference Transactions,* American Society for Quality Control, Milwaukee, 1979, pp. 817-823.

[2]For background and other significant information on quality costs see the six papers in *Quality Progress,* April 1983.

[3]*Quality Costs — What and How,* Second edition, American Society for Quality Control, Milwaukee, 1971, p. 6.

30

Computers in Service Quality Control

Computer developments. Technical improvements are occurring so rapidly in the computer field that management and the professional have to be forever on the alert for new developments. This is just as true for the service industries as it is for the manufacturing industries. During the past five years the following significant changes have taken place:

1. Reduction in the *size* of computers.

This is due to the development of minicomputers, the programmable desk calculator, and the programmable and non-programmable hand-held calculator.

2. Reduction in the *price* of computers.

This is due to the reduction in size, in increased competition, and in improvements in the state of the art such as miniaturization.

3. Increased *capability* of small calculators and computers.

This is reflected in several ways. One is an increase in the number of operations involved in both calculations and programming, whether logical, mathematical, or statistical, which are built into the calculator. Also, a printing tape has been added to some small calculators. In addition, modules or peripherals, which can be attached to the calculator, have been designed. This greatly increases memory capacity, on the one hand, greatly increases the number of programming steps, on the other, and provides graphic recording and analysis.

4. Prepared programs or software have *increased* manyfold and become much *more specialized;* thus the amount of programming necessary for many standard operations has been reduced.

Examples of specialized fields related to service operations where programs are now available, in addition to programs for mathematics, probability, and statistics, include real estate, financial decisions, securities, clinical and nuclear medicine, lending and savings, marketing and forecasting, investment analysis, personal finance, quality control including $\bar{x}$ and R charts and operating characteristic curves, inventory control and waiting line analysis, and small business operations.

5. The above greatly *increase productivity* in data collection, data processing, data analysis, and data presentation.

These technological improvements make it possible to greatly increase the speed of obtaining basic information management needs for quality control, quantitative controls, problem-solving, and decision making. At the same time, it is possible to do this at a reduced cost. As far as the individual analyst is concerned, capability is increased by a magnitude of from five to 10 times. This means a person with present calculators can do the work formerly done by five to 10 persons using old-time desk calculators.

These advantages, however, cannot be attained without paying a price. The price is competent, qualified computer managers and programmers, careful feasibility studies, proper training in the use of the computers and calculators, careful matching of computers and calculators with the nature of the problems to be solved, and avoiding the kinds of errors and mistakes which are described below.

Four kinds of computers. Four classes of computers and calculators are now being manufactured and used:

1. Large scale computer systems.
2. Minicomputers.
3. Desk calculators, both programmable and non-programmable.
4. Hand-held calculators, both programmable and non-programmable.

All of these types can and are being used to varying degrees in the service industries. We discuss first large computers and large scale computer systems because they were the first to be introduced into service operations, and they remain today the dominant type of computer installation.

Problems with large computers. These problems are taken from actual computer practice; they diminish computer effectiveness and tend to increase costs. Therefore, they present a real challenge to management in service industries and operations.

1. Excessive data. It is very easy in computer work to become so involved with data elements, data banks, machine readable input, data analysis, data management, and computer software that the real purpose of the computer is lost sight of — using data to obtain information to help management solve its problems.

The place to start is with *problems* to be solved and questions to be answered which are bothering management and professional persons. The kinds and amounts of data arise from a detailed analysis of these problems. Data are always a means to an end, and the end is information and knowledge that will help solve important problems and answer significant questions.

A high-level official in a large corporation describes this situation as follows:

> "In today's industrial environment, cost effectiveness is 'first among equals.' This is especially true in military procurement where all systems are being clinically examined for austerity and ability to produce a hard commodity. Realistic, innovative data systems are in this category. Second and third generation computer capabilities have progressively simplified the data collection process. *Unfortunately with this capacity exists the temptation to collect all types and bits of information that 'might be needed by someone, somewhere, sometime, for some reason'.* Computer time is expensive. If a given system is not carefully planned in terms of minimum data requirements consistent with maximum use, a gross imbalance will exist between data acquisition costs and data utility.[1]" (Emphasis added)

What this means is that unless high-level officials and computer managers keep the computer operations based on management's need for important information, the computer memory can be filled with large amounts of useless but very costly data.

2. Separating the analyst from the data. Analysts, researchers, operators, and professionals generally are accustomed to collecting data themselves or being very familiar with how the data are collected. They examine and analyze the data first hand by careful examination, tabular analysis, and graphic analysis. In a large computer system, there is a danger that the data will be routed away from these professionals and they will not be able to do the careful job they have been accustomed to.

Snedecor and Cochran state the problem as follows:

"High-speed electronic computers are rapidly becoming available as a routine resource in centers where a substantial amount of data are analyzed. . . . We believe, however, that in the future it will be just as necessary that the investigator learn the standard techniques of analysis and understand their meaning as it was in the desk machine age. In one respect, computers may change the relation of the investigator to his data in an unfortunate way. When calculations are handed to a programmer who translates them into the language understood by the computer, the investigator, on seeing the printed results, may lack the self-assurance to query or detect errors that arose because the programmer did not fully understand what was wanted or because the program had not been correctly debugged. When data are being programmed it is often wise to include a similar example from this or another standard book as a check that the desired calculations are being done correctly".[2]

This is still a very real problem, especially where very large samples or very large amounts of data are being processed on a computer. Professional people, such as statisticians and quality control specialists, will have to lay down written specifications of the controls, tabulations, calculations, and analyses that are required and use test runs to see that the programming and operations are correctly done.

The situation, however, has improved greatly with the introduction of the powerful electronic desk calculator and hand-held calculator, and with a big expansion of the number of programs (software) available. The investigator, practitioner, or analyst can now perform, with a programmable calculator with plenty of program steps and memory capacity, fairly large size jobs without using a large computer, thereby keeping complete control over the processing and analysis of the data. Furthermore, if required, the necessary calculations and operations can be programmed on the desk calculator and discussed with the programmer who can program them in a language compatible with the large computer. In this way, management can be assured that the mathematical and statistical calculations are being programmed correctly.

3. Excessive analysis. Just as there is a temptation to collect all kinds of data simply because a computer has a tremendous memory capacity, so there is a temptation to engage in excessive calculations and analyses of data simply because the computer has a tremendous high-speed calculating capacity. Unnecessary statistical analyses are costly, waste computer time, and delay corrective action that may be needed.

An actual example is the case of a machine operator who complained of trouble. A lengthy series of samples were taken and sent to the computer where a straight line was fitted to the mean ($\bar{x}$) values and the value of the slope (b) was tested by means of a t test. This test indicated "out-of-control" which was reported to the operator who already knew something was wrong. This out-of-control could have been detected more quickly by using simple $\bar{x}$ and R charts derived from several samples taken at the work site. While there is nothing wrong in using a t test of the slope of the line fitted to the several $\bar{x}$'s, it was just too sophisticated statistically to apply to this situation. Interest seemed to center on doing a fancy statistical analysis rather than finding out quickly what was the matter with the machine. Statistics is a means to an end and should be so used.

4. Delay in detecting trouble. Conditions may exist which delay, rather than hasten, detection of trouble and corrective action. The computer may be too far away — it may be in another building or in another block. In these instances, the very remoteness of the computer introduces delay.

One could also find that the computer is loaded and there is a waiting line at the computer; the job has to wait its turn.

As described above, delay may be introduced by doing some time-consuming, but unnecessary, statistical analysis.

Finally, delay may be introduced because a sample audit is being used for quality control. This means days may elapse before the source of trouble is detected. A sample audit is a post mortem and no substitute for on-the-site quality control which allows continuous control of key characteristics relating to quality. A sample audit is for the purpose of assuring professionals and management that the quality control system installed is actually performing the way it is supposed to. It is a check on whether there is a deficiency in the system which has been overlooked or has developed due to unforeseen conditions.

5. Delay in detecting and correcting errors in the computer. This delay relates to the errors which are placed in the memory of the computer and have to be corrected. A significant weakness of large scale computer operations, as they now exist, is the extreme difficulty of an individual customer of getting an error corrected once it becomes a part of the computer system. It may take months, as the examples cited show. In the service industries, where these errors are likely to occur, there is a real need to introduce quality control at two points: first to eliminate the error arising in the first place. This requires better instructions and more care on the part of the employees involved. Then there should be more specific and efficient procedures developed to correct an error once it gets into the memory of the computer.

6. Lack of error control. Numerous examples can be cited to show a lack of proper control over the computer system in order to eliminate errors.

a. A class of ineligibles is included in a computer program in connection with a Federal financial aid program resulting in the payment of $18 million before the error was discovered. Obviously, the programmer was not instructed as to which classes of people were and which classes were not eligible for financial assistance.

b. A programmer did not know how to program the standard error of a ratio estimate of a sub-class aggregate and turned to a textbook to find the answer instead of conferring with a professional statistician who was avail-

500 miles away. To demonstrate the module, a few pairs of values were entered into the computer using the proper instructions entered by an on-line typewriter. The values a and b for a straight line $y = a + bx$ fitted by the method of least squares came back calculated to eight decimal places with the standard error of each! The computer cannot tell the difference between nonsensical data used for demonstration purposes and a real problem.

There are other warnings to be heeded relative to software. If they are purchased, the user should make sure that they are correct. This is done by testing either a real-world problem or a fictitious problem whose answers are known to be correct. Unless the canned program can calculate these answers correctly, it should be rejected. Wampler of the National Bureau of Standards showed many years ago that a program to solve polynomial equations was greatly in error; it did not even come close to the correct values.[5]

The software on the market may not fill one's needs or fill only one part of one's requirements. This means one is forced to lay out the entire set of calculations and either solve them using the keyboard or write a complete program. In doing this, the non-technical person may need a statistician or other specialist to assist in doing the calculations and programming.

There is no need to buy a program if its use is very limited. It would be more economical, in this situation, to work the problem directly from the keyboard; even a program is unnecessary. Programming pays off for kinds of problems that are met frequently or require so much programming or calculating time that it pays to once and for all program and record it on tape, magnetic cards, or other devices for future use. Furthermore, writing one's own program gives a flexibility and coverage that software simply does not have. It keeps not only the data but the calculations and the analysis under the direct control of the analyst, something which worried Snedecor and Cochran in the quotation given earlier in the chapter.

Software may be very inadequate because it deals with a very limited problem or even one aspect of a problem. Very often the analyst needs a much more complete treatment of the sample data than the software provides. In this case, the analyst writes, or has written, a program in several stages that produces all of the estimates, tests of significance, and standard errors that are required. For example, in one multi-stage program, the following calculations and results are obtained from a set of correlated variates; six problems are being solved in a single program:

1. Linear regression, a and b.
2. Residual variance.
3. Standard error of b.
4. t test of b.
5. Confidence limits, Y individual.
6. Confidence limits, Y average.

Large computer systems. The large computer systems representing the fourth generation or better are characterized by very fast memory access times (nanoseconds or 10^{-9}), miniature circuitry, mass memory, multiprogramming, and miniaturization using transistor chips. Examples of these computers are: CDC CYBER 205, IBM 3033, Burroughs 6800, Univac 1100-60 and 1100-80, and Honeywell.

Small Computers and Calculators. Small computers include those which require limited space, but are connected with a distant large computer, such as an on-line computer facility. Another form is the minicomputer. This term is used very loosely, but seems to refer to a computer facility which is much smaller than what passes as a large computer system.

There are also two kinds of programmable and non-programmable calculators: the desk type and the hand-held or pocket type which is called a hand calculator hereafter. Rapid technological advances in this area, as well as competition, are creating a number of revolutionary changes: miniaturization is making it possible to drastically reduce the size of computers and calculators, storage capacity and the number of memory steps are being greatly expanded, built-in mathematical and statistical calculations are drastically reducing the time required for numerous calculations, and programming and software are increasing the flexibility and power of the calculator.

These developments have a number of implications: what is considered an efficient and appropriate computer or calculator today may not be the best one tomorrow or next year. All one can write about today is what is on the market today, with a warning to those who are going to buy a computer or calculator in the future that they study carefully what is on the market then. We are moving toward a point when the desk programmable calculator with ample memory and programming capability, together with peripherals such as a printer and plotter, will be able to do what a large computer is doing today. The increased complexity and sophistication of the computer and calculator means that to obtain the most efficient use requires an understanding of the logic, mathematics, and statistics involved in the programs, in the data, and in the analysis. Technical proficiency is needed to make the best use of programs, plotters, and printers; otherwise, there will be a tremendous waste of time and resources "dabbling" with a highly mathematical instrument.

Examples of desk model computers. Some examples of desk programmable calculators or computer systems, which can include a CRT display, a printer, and a plotter, are the following:

HP 9800 Series System 45 addressable memory up to 449 K bytes.

For a larger data base this computer can be linked to the more powerful HP 1000 or HP 3000 systems.

Radio Shack TR80 System Models I and II.

With appropriate accessories can be used for business, engineering, and statistics, including inventory control, accounting and mailing lists. It has a printer but no plotter.

HP 85 — a personal computer for professionals.

Has CRT display, a printer, and a tape unit; a plotter can be added. A 16 K memory module can be plugged in to double the memory. Application pacs are available for statistics, mathematics, finance, and linear programming.

HP 97 — this is a desk programmable calculator with printing capabilities.

It is portable and can be run on batteries as well as AC. Fifteen programs come with the machine. Applications pacs include preprinted, prerecorded program cards; there are 20 in statistics, including one for the $\bar{x}$ and R control charts, and one for the OC curves.

Software issued by Hewlett Packard for the HP 67 and HP 97 include $\bar{x}$ and R charts, OC (operating characteristic) curves for sample plans, analysis of able. As a result, 26,000 estimates of the sampling variance printed in a report of a nationwide sample study for federal agency use were in error.

c. In another case, due to failure of the tape library system, four tapes were omitted from a tabulation accounting for about 17,000 sample documents or about eight percent of the total sample. This omission would have invalidated the entire study. The omission would not have been discovered unless a careful independent sample receipts control system had not been in use continuously for a number of years. This system showed that a shortage of eight percent had never occurred before in sample receipts. Indeed, an eight percent shortage in sample receipts would never have been accepted even for an individual company.

d. A mortgage company makes a $314 error in connection with a mortgage monthly payment due to the fact that the real estate company did not inform the mortgage company that the $314 was included in the settlement. As a result, it takes one letter, two telephone calls, an automobile trip of 100 miles and most of one day to reach the computer people so that the error could be corrected. This required about four months.

e. A mortgage company makes a $400 error in connection with a payment to reduce a loan. It requires two letters, two telephone calls, and about five months to have the error corrected in the computer to within $26 of what it should be. The error of $26 was allowed to stand because it was considered too costly by the customer to try to correct it any further. Finally, after another month, the $26 error was corrected by the company.

f. An individual familiar with the computer system transfers to his account $10 million because he knows the codes and how to communicate with the computer system. He then transfers it to a Swiss bank and uses the money to buy precious stones. This is an example of a computer crime which is becoming more common.

g. Another example of a computer crime was the case of a federal worker in Washington who transferred to his account $750,000 to be used for Federal revenue sharing. He was reported to the FBI by his neighbors because his buying of Cadillacs, yachts, and other luxuries made them suspicious. In this case, the agency in Washington did not discover it, nor did the office in Atlanta where the money was to be transferred. Quality controls were missing at every key spot.

7. Poor quality of the input data. The saying "garbage in, garbage out" shows that computer persons and others recognize the critical importance of the *quality of the input data.* Discussions of data elements, data banks, data management, and data analysis seldom, if ever, call attention to the critical role of the quality of the input data, to the various sources of error to which the data are subject, and to the need to obtain data that are relevant, clear, accurate, additive, and in the forms and amounts required by the problem.[3]

It should be emphasized that if the data are aimless, the computer will not give them purpose; if the data are ambiguous, the computer will not make them clear; if the data are meaningless, the computer will not give them significance; if the data are irrelevant, the computer will not make them pertinent; and if the data are invalid, the computer will not make them sound because it has no magic alchemy that will transform dross into gold.

8. Sampling and the computer. It is easy to conclude that with a high-speed computer at hand, the use of sampling is unnecessary. Nothing is further from the truth or sound efficient operations. Probability sampling is very effective in so many problems because of the redundant nature of numerical data. This means that adjacent values or measurements are so much alike that a 100 percent enumeration or tabulation is unnecessary; a properly designed sample will obtain the significant information. The same is true of a characteristic whose frequency distribution shows numerous values which differ very little. To obtain an estimate of male adult height, it is unnecessary to measure the heights of 50 million persons. A few hundred persons selected at random suffices because so many of the measurements are practically identical — the range is roughly from 60 to 75 inches. Sampling still applies to estimating the average income of individuals or families even though the range or variability is much greater; the sample will have to be larger.

An official of a large company expressed the very strong opinion that now that it had a large computer system, the use of sampling was no longer necessary. This official was carried away by the high speed of the computer, but he ignored its cost. Actually, this company had a large number of problems. A sample of no more than 10,000 documents was all that was necessary out of a total of over three million documents. Why waste a tremendous amount of resources on the latter, when the former contained all of the information needed? Indeed, while the three million documents were being processed by the computer, it would be possible to select an efficiently designed probability sample at the same time. Actually, this company had a type of problem that required expert judgments relative to each document included in the study. In this case, use of a sample was required; it would be impossible to do this study using three million documents.

The computer is an extremely effective way of selecting not only a simple sample but a stratified replicated sample where the frame is in the computer memory stored on tapes, disks, or drums. Examples are found in the reference.[4]

9. Dangers of canned programs (software). There are now available, from computer manufacturers and others, prepared programs or software for a wide variety of areas in business, science, and engineering where statistics, probability, and mathematics are applied to problems. These areas were listed at the beginning of this chapter. The techniques involved include statistical quality control, probability sampling, and statistical analysis of sample data.

These prepared programs stress calculations; they assume that the user knows how to use the programs properly. These programs do not explain what kinds of data the calculations apply to, how the data should be collected, what effect biases in the data may have upon the results, or what assumptions the collection of the data must meet. There may be no indication of the specific mathematical models which are implied in the calculations as must be the case with analysis of variance. This is why canned programs or software have to be used with caution by those without the necessary technical training. Preferably they should be used under the guidance of someone who understands how to test and use the programs.

A purely mechanical use of a computer can lead to nonsensical results. An actual example is the case of an on-line terminal which communicated with a computer variance, t test, chi-squared test, and multiple linear regression.

Examples of hand calculators.[6] A profusion of hand-held programmable and non-programmable calculators are being manufactured. The result is that highly efficient instruments are now being produced at relatively low prices. We discuss here only those that are most applicable to problems of quality control. We shall call them "hand" calculators and not "hand-held" or "pocket" calculators, "hand" in contrast to "desk".

Companies making this type of calculator include Canon, Casio, Hewlett Packard, Monroe, Sharp, and Texas Instruments. The models of all companies need to be examined and tested before making any final decision about which model to buy. A few models are described briefly below:

1. HP 32E. It is non-programmable but has built into it means with standard deviations, linear regression, correlation coefficient, normal distribution, and data accumulation for one or two variables; 15 addressable storage registers.
2. HP 67. It is fully programmable with 26 storage registers and 224 lines for programming. Statistical, mathematical, and quality control software available. Has magnetic card reader.
3. HP 41C. This is an alphanumeric programmable calculator with a continuous memory even when the calculator is turned off. Peripherals include magnetic cards, printer, card reader, memory modules to expand memory, application modules, and an optical wand for loading programs and data. A 14 program statistics application pac is available; also a 12 program pac on test statistics is available.
4. TI 59. This is a fully programmable calculator with 480 steps for programming and 60 storage registers. Modules are available for mathematics and statistics.

The HP 32E has programmed in it the normal probability distribution so that calculations can be made assuming the following:

The normal deviate: $z = \frac{x - u}{\sigma}$ or $z = \frac{\bar{x} - u}{\sigma_{\bar{x}}}$, and $P(z) = \int_{-\infty}^{z} p(x)dx$ (distribution function or cumulative probability distribution).

The computer solves for either of these two quantities:

1. Given the normal deviate z, it solves for P.
2. Given the probability P, it solves for z.

For example, if $z=3$ for three sigma, $P=0.9987$. If $P=0.975$, $z=1.96$. This program eliminates the use of the normal probability distribution tables.

Applications and software (prepared programs). Computer or calculator programs have several sources:

1. They may be built into the machine.
2. They may be modules that are plugged into the machine such as Stat Rom.
3. They may be printed, as in a book of programs or applications.
4. They may be published in a technical magazine.
5. They may be obtained from a user's exchange.
6. Users may write their own programs.

One of the major advantages of the computer or calculator is that so many important mathematical, statistical, and logical functions and operations are built into the machine. Mathematical functions include x^2, $1/x$, $x^{1/2}$, log x, ln x, e^x, 10^x, and y^x; statistical calculations include mean, standard deviation, variance, linear regression, correlation coefficient, and t test; logical operations such as store, recall, $x=y$, $x \neq y$; and moving operations such as up, down, in, and out. These features account for the tremendous saving in time and improvement in accuracy in both computational and non-computational operations since the use of many mathematical tables is eliminated, interpolations are eliminated, recording intermediate values is eliminated, answers to complex problems require only the pressing of a key, programming in multi-stages allows a whole series of related problems to be solved in one operation, and a program for an often-encountered problem can be saved and used as many times as necessary.

Modules may be purchased, which are ROMs (read only memory), containing a program covering several computations. An example is the Stat Rom for the HP 9810. This module provides calculation of the following: mean and variance for a single variate, and mean, variance, linear regression constants a and b fitted by least squares, and correlation coefficient squared. By using the key for $x^{1/2}$, the standard deviations of x and y could be calculated immediately, as well as the correlation coefficient r. The Stat Rom also handles three variables so a program can be easily written to solve the equation $Y = a + b_1X_1 + b_2X_2$. A 10-stage program was written to give A, b_1, b_2, r^2, r, residual variance, and the standard error of b_1 and b_2 in 449 programming steps.

Software in the form of printed programs. Application manuals furnished to buyers of calculators, or other printed matter, may contain the detailed steps for programming various types of problems. These may or may not meet the needs of the users because they are incomplete, cover only one part of a complex problem, and are inefficient because they require too many programming steps. Because of these and other deficiencies, the person in the market for software should be cautious; the programs may be inadequate and eventually require that the users write their own programs.

An example of inefficient programming, from the standpoint of the practitioner, is the 85 steps used in the applications book for the HP 25 to program permutations, combinations, and the factorial n (n!). Making use of the relation $P = Cn!$ and a three-stage program, the three quantities can be calculated in one operation instead of three by using only 30 programming steps. This eliminates over 60 percent of the programming steps used by the manufacturer. If m is the number of objects taken n at a time, the program features the following:

1. Enter $m - n$ in R_2 (register 2) before starting.
2. Enter m (also before starting).
3. Enter n at stop 1.
4. Record P at stop 1.
5. Record C at stop 2.
6. Record n! at stop 3.

Programs in technical magazines. An example is the computer programs which appear in every issue of the *Journal of Quality Technology.* An example applicable to the use of lot sampling for acceptance or rejection of work in the service industries is found in the October,

1980 issue of this journal. Actually, once a set of tables is derived from which sample sizes can be read for given values of p_1, p_2, α, and β, such as those given in the Appendix, there is not much need for any further calculations in this area, unless some special situation arises that calls for a new sample plan.

User's exchange. Programs are available for a wide variety of calculators by means of user's exchange which exist for both Hewlett Packard and Texas Instruments users. Some individuals are also interested in the interchange of programs for various programmable calculators.[7]

Writing one's own programs. There are many reasons why programs for sale may not be satisfactory and require users of computers and calculators to write their own programs.

1. No programs are sold to cover your particular problems and needs. Some specific areas are given below.

2. No programs are written for quality control problems such as $\bar{x}$ and sigma (σ) charts, p charts, sequential sampling, OC curves, API curves, AOQ curves, c charts, lot acceptance-rejection sample plans, regression quality control, and circular probability control.

3. No programs exist for selecting samples from computer memory (tapes, disks, or drums) such as:
   - Selecting simple random samples.
   - Selecting systematic samples with random starts.
   - Selecting replicated samples with random starts.
   - Selecting replicated samples with random starts in each zone.
   - Selecting stratified random samples.
   - Selecting stratified replicated samples with random starts.

4. No programs exist for selecting samples and then making estimates from them, such as selecting samples as above and then making estimates of means, standard deviation, coefficients of variation, aggregates, proportions, ratio estimates, standard errors, confidence limits, regression coefficients, and analysis of variance parameters.

5. The problem is too complex, such as large sample tabulation and calculation. An actual example is the calculation of 12 estimates of four types, including ratio estimates with absolute and relative sampling errors for each of about 6,000 classes and sub-classes. The calculations for one sub-class can be programmed on a desk programmable calculator, with 49 memory registers, using 500 programming steps and a Stat Rom with two variable linear regression. This is an example of where programming on a desk calculator can be used by a computer programmer to write a program for a large computer where the 6,000 cells involve about 500,000 punch cards.

How to get the most benefits from a programmable calculator has not received the attention it deserves. Experience with these calculators suggests the following.

1. To begin with, a machine should be purchased which meets actual needs determined beforehand by an actual survey of the calculation and analytical needs of the organizational unit or units involved. There is no point in buying a machine with trigonometric functions on the keyboard if these types of problems are never encountered. There is no point in buying a machine with large memory and programming capacity if these capacities are never used.

2. These calculators are sophisticated mathematical instruments which should never be put in the hands of untrained persons. Actually, one has to have a working knowledge of the basic principles of mathematics, probability, and statistics to use them intelligently and effectively, or at least have the assistance of such a person. Otherwise, utilization is very likely to be at a low level.

3. Many problems can be worked directly from the keyboard making programming, or the purchase of a program, unnecessary.

4. Programming pays off for a lengthy program used periodically or a short program used continuously.

5. A multi-stage program, combining a series of related calculations punctuated by the instruction "stop" for recording results, is a great time saver. It allows a whole series of related problems, problems dealt with separately in textbooks, to be solved simultaneously. Examples are a complete linear regression analysis from a sample of values, multiple regression, and analysis of variance involving interaction. In this type of programming, calculating the means, variances, and coefficients of variation is only the beginning.

6. There is no substitute for a thorough knowledge of how the computer operates, what each key on the keyboard means and how it is used, how an equation or a series of equations is programmed, how a sample of data is processed both from the keyboard and from a program, and how to select the most efficient sequence in calculating from the keyboard or in programming.

7. The importance of a computer lies in its use in assisting the user in getting meaning out of data, not only by calculation, but by analysis. Meaning does not flow out of numbers automatically; that is why knowledge of statistics and related fields is emphasized in item 2 above. Mere mechanical use of software (purchased programs), without understanding the basic principles and assumptions behind each program, can lead to unjustified applications and misleading inferences and conclusions. Calculations are but a part of the total process of getting meaning from numerical data. Calculations and data processing are not ends in themselves, although emphasis on computers and software seems to have led many to think so.

8. The computer can be used analytically to decompose mathematical and statistical equations and to show how various combinations of parameters affect the results, especially the dependent variable or variables of interest. This is the mathematical counterpart of conducting an experiment, except that computer parameters can be varied in any combination desired, including holding all parameters constant except one.

9. The computer can be used in seminars, demonstrations, and in-training classes to illustrate the basic principles of mathematics, probability, and statistics, to illustrate principles of sampling and statistical quality control, and to illustrate estimation, standard errors, confidence limits, and various tests of significance.

10. The programmable calculator can be used to show the programmer of a large computer the mathematical and statistical equations involved in a problem, how they would be programmed on the programmable calculator, and what intermediate and final results are required. This will be of great assistance to the programmer not familiar with mathematics and statistics. This is likely to be common, since programmers are not expected to be familiar with mathematics and statistics to the same degree as mathematicians and statisticians.

11. The most efficient methods of estimation can now be used because calculation limitations no longer exist. This means, for example, that the standard deviation as a measure of variability should be used where it is proper to use it. The range is still better for very small samples, but the standard deviation is better for larger samples.

12. One of the growing advantages of the desk calculator with additional memory and programming modules is that it can keep small and medium size jobs off of the large computer and thereby reduce costs and increase efficiency.

13. Mastering the keyboard saves considerable time in calculating from the keyboard as well as in writing programs. It means understanding what every key means, the logic of the machine, how it operates, and the rules governing keys in sequence. It includes a mastery of numerous short calculating and programming sequences, an understanding that some calculating sequences are better than others, and the importance of choosing the starting step and sequence and the use of sub-routines. It includes an understanding of how intermediate values can be created, stored, moved, and manipulated.

As learning continues from a study of instructions and solving problems in the owner's handbook and applications handbook, memory of key locations and sequences improves, unnecessary steps are omitted, a better starting point is selected, and shorter sequences are discovered. Working problems on the keyboard until it is mastered will make drastic reductions in the amounts of time required to solve a problem. Actual time tests show that calculating times can be reduced not only to one half but often to one third or even one quarter. When one arrives at this stage, it is very easy to translate calculations from the keyboard to the programming steps required.

A feasibility study. A total feasibility study for a computer installation of any service company or agency is for the purpose of selecting computer hardware and accessories (peripherals) which will be most efficient and economical in meeting the computer needs of the entire company or agency. This means a very careful survey and study must be made to determine what functions, projects, operations, and jobs can be most effectively handled by computers or calculators. A total feasibility study includes three parts:

A. A large computer system such as the CDC Cyber or the IBM 3033.

B. A desk computer system such as the HP Series 9800 System 45.

C. Hand calculators such as the TI 59 or the HP 41C.

The major steps in a feasibility study include the following:

1. A survey of the entire workload analyzed into functions, projects, operations, jobs, and tasks.

2. Analysis of the workload into problems, operations, sequences, amounts and kinds of data, listings, tabulations, calculations, graphics, and reports. Length, volume, and frequency are determined.

3. This analysis will reveal 1) the large jobs which require A above, 2) the small and medium jobs which can be done on one or more systems under B above, and 3) the jobs and tasks which need only the calculators listed under C above.

4. Write specifications for each type A, B, and C with regard to such needs as the following:
   a. Programming.
   b. Memory or storage capacity.
   c. Tabulations.
   d. Calculations.
   e. Printing requirements.
   f. Plotting and graphing requirements.
   g. Other analysis.

5. Determine how the various computer installations and operations are to be organized, managed, related, coordinated, and located.

6. Prepare a workload schedule for systems A, B, and C:
   a. Time schedule.
   b. Load distribution to A, B, and C.
   c. Time deadlines to meet.
   d. Length of each project and allocations to A, B, and C.
   e. Maximum load to be handled by each system A, B, and C.

7. Determine hardware required:
   a. System A: computer, storage, printer, plotter, other.
   b. System B: computer, storage, printer, plotter, other.
   c. System C: calculators, modules.

8. Determine software required, if any:
   a. For A, for B, for C.
   b. This depends upon whether the software (programs) required are available and can be purchased or whether all or most programs will have to be written to meet the requirements of the problems to be solved.

9. Study expandability of competing systems, including calculators, to meet future needs. It is very important to consider not only increased workload but the compatibility of new peripherals and new machines with the systems to be purchased. With the rapid rate of development of improved models and systems, it is well to keep this in mind to reduce possible losses due to obsolescence.

10. Survey of the personnel problems:
   a. Personnel needed if system A is installed.
   b. Personnel to be allotted to one or more systems of type B.
   c. Individuals who are to receive each type of C if more than one type is to be used.

11. Compare competing systems and calculators for A, B, and C using the foregoing requirements. Observe competing systems and calculators in operation, observe demonstrations, and obtain opinions of users.

12. Make final decisions about A, B, and C.

The objectives are to obtain a computer capability that 1) meets current needs relative to workload, time schedules, resources, and other limiting factors, 2) can be expanded without difficulty to meet future needs, 3) is adequate to meet current needs without any excessive equipment, and 4) is cost effective. This capability requires that highly competent personnel manage, program, and run the computer systems, and that technically trained personnel be hired and employed so as to obtain the maximum benefit from computer technology. This does not mean that all users need to be technically trained in proper mathematics, probability, and statistics, but it does mean that one or more experts in these areas should be available and consulted so that the maximum benefit can be obtained from the computer installation.

Experience shows that there are two extremes to be avoided:

1. Buying too large a computer, too much capacity, too many peripherals, and unnecessary software for the workload and personnel at hand.

2. Buying too small a computer for the workload. This is bound to place important projects and jobs in a waiting line, thereby losing one of the major advantages of the computer.

Experience also shows that it is best to make a feasibility study covering the entire firm or agency, rather than make computer decisions and installations on a piecemeal basis (such as a department-to-department approach).[8]

The computer as an alibi for human error. Nowhere is an error prevention program needed more than in all of the operations involving the computer. This is because the computer has become the Number One alibi to hide human error. We read and hear such expressions as the following:

- The computer made a mistake.
- The computer "blew up" (actual quote).
- We are having trouble with the new computer system.
- Our computer is antiquated.
- The computer was wrong.
- Something went wrong with the computer.
- The computer "translated 28,000 into 280,000" (actual quote).

For all of these human errors, the computer is to blame; it has become a convenient anthropomorphic alibi.

We are not going to receive full benefit from any computer or computer system, as a sophisticated mathematical instrument or as a simple storage or retrieval device, until we reject this myth.

We must see that individuals are properly trained, supervised, and motivated with regard to error prevention in:

- All entries and input.
- All programming.
- All tabulations.
- All read-outs and retrievals.
- Continuous updating as required.
- All console operations.
- Proper use of software and peripheral equipment.

Notes

[1]J. Y. McClure, *Quality Progress,* VIII, 1976, p. 14. Emphasis added.

[2]G. W. Snedecor and W. G. Cochran, *Statistical Methods,* 6th edition, Iowa University Press, Ames, 1967, p. vii. Used by permission.

[3]For examples of pre-designed probability sample studies that successfully solved management problems, see A. C. Rosander, *Case Studies in Sample Design,* Marcel Dekker, New York, 1977.

[4]Rosander, ibid., pp. 160-162, 363.

[5]R. H. Wampler, *Journal of the American Statistical Association,* vol. 65, no. 330, (June 1970), pp. 549-565.

[6]For more information, see R. W. Berger and M. Hale, "Using programmable calculators in quality control," *34th Annual Technical Conference Transactions,* 1980, American Society for Quality Control, Milwaukee, pp. 1-9.

[7]Both Hewlett Packard and Texas Instruments maintain a user's exchange. A specialist in this area to contact is Prof. Roger Berger, 4121 Dawes Drive, Ames, Iowa, 50010.

[8]The consequences of installing computers piece-meal is given in J. P. Wright, *On a Clear day you can see General Motors,* Avon, New York, 1978, pp. 122-124. How a new computer system was installed and what it accomplished is described on pp. 147-150.

Appendix

TABLE I

Factors for Computing Control Chartlines

	Chart for Averages			Chart for Standard Deviations						Chart for Ranges						
Number of Observations in Sample, n	Factors for Control Limits			Factors for Central Line		Factors for Control Limits				Factors for Central Line		Factors for Control Limits				
	A	A_1	A_2	c_2	$1/c_2$	B_1	B_2	B_3	B_4	d_2	$1/d_2$	d_3	D_1	D_2	D_3	D_4
2	2.121	3.760	1.880	0.5642	1.7725	0	1.843	0	3.267	1.128	0.8865	0.853	0	3.686	0	3.267
3	1.732	2.394	1.023	0.7236	1.3820	0	1.858	0	2.568	1.693	0.5907	0.888	0	4.358	0	2.575
4	1.500	1.880	0.729	0.7979	1.2533	0	1.808	0	2.266	2.059	0.4857	0.880	0	4.698	0	2.282
5	1.342	1.596	0.577	0.8407	1.1894	0	1.756	0	2.089	2.326	0.4299	0.864	0	4.918	0	2.115
6	1.225	1.410	0.483	0.8686	1.1512	0.026	1.711	0.030	1.970	2.534	0.3946	0.848	0	5.078	0	2.004
7	1.134	1.277	0.419	0.8882	1.1259	0.105	1.672	0.118	1.882	2.704	0.3698	0.833	0.205	5.203	0.076	1.924
8	1.061	1.175	0.373	0.9027	1.1078	0.167	1.638	0.185	1.815	2.847	0.3512	0.820	0.387	5.307	0.136	1.864
9	1.000	1.094	0.337	0.9139	1.0942	0.219	1.609	0.239	1.761	2.970	0.3367	0.808	0.546	5.394	0.184	1.816
10	0.949	1.028	0.308	0.9227	1.0837	0.262	1.584	0.284	1.716	3.078	0.3249	0.797	0.687	5.469	0.223	1.777
11	0.905	0.973	0.285	0.9300	1.0753	0.299	1.561	0.321	1.679	3.173	0.3152	0.787	0.812	5.534	0.256	1.744
12	0.866	0.925	0.266	0.9359	1.0684	0.331	1.541	0.354	1.646	3.258	0.3069	0.778	0.924	5.592	0.284	1.716
13	0.832	0.884	0.249	0.9410	1.0627	0.359	1.523	0.382	1.618	3.336	0.2998	0.770	1.026	5.646	0.308	1.692
14	0.802	0.848	0.235	0.9453	1.0579	0.384	1.507	0.406	1.594	3.407	0.2935	0.762	1.121	5.693	0.329	1.671
15	0.775	0.816	0.223	0.9490	1.0537	0.406	1.492	0.428	1.572	3.472	0.2880	0.755	1.207	5.737	0.348	1.652
16	0.750	0.788	0.212	0.9523	1.0501	0.427	1.478	0.448	1.552	3.532	0.2831	0.749	1.285	5.779	0.364	1.636
17	0.728	0.762	0.203	0.9551	1.0470	0.445	1.465	0.466	1.534	3.588	0.2787	0.743	1.359	5.817	0.379	1.621
18	0.707	0.738	0.194	0.9576	1.0442	0.461	1.454	0.482	1.518	3.640	0.2747	0.738	1.426	5.854	0.392	1.608
19	0.688	0.717	0.187	0.9599	1.0418	0.477	1.443	0.497	1.503	3.689	0.2711	0.733	1.490	5.888	0.404	1.596
20	0.671	0.697	0.180	0.9619	1.0396	0.491	1.433	0.510	1.490	3.735	0.2677	0.729	1.548	5.922	0.414	1.586
21	0.655	0.679	0.173	0.9638	1.0376	0.504	1.424	0.523	1.477	3.778	0.2647	0.724	1.606	5.950	0.425	1.575
22	0.640	0.662	0.167	0.9655	1.0358	0.516	1.415	0.534	1.466	3.819	0.2618	0.720	1.659	5.979	0.434	1.566
23	0.626	0.647	0.162	0.9670	1.0342	0.527	1.407	0.545	1.455	3.858	0.2592	0.716	1.710	6.006	0.443	1.557
24	0.612	0.632	0.157	0.9684	1.0327	0.538	1.399	0.555	1.445	3.895	0.2567	0.712	1.759	6.031	0.452	1.548
25	0.600	0.619	0.153	0.9696	1.0313	0.548	1.392	0.565	1.435	3.931	0.2544	0.709	1.804	6.058	0.459	1.541
Over 25	$\frac{3}{\sqrt{n}}$	$\frac{3}{\sqrt{n}}$	.....	.....	.....	*	**	*	**	.....	.....	.....	.....	.....	.....	.....

*$1 - \frac{3}{\sqrt{2n}}$

**$1 + \frac{3}{\sqrt{2n}}$

Note 1: For *No Standard Given* use A_1, A_2, c_2, B_3, B_4, d_2, D_3, and D_4

Note 2: For *Standard Given* use A, c_2, B_1, B_2, d_2, D_1, D_2

Source: *ASTM Manual on Quality Control of Materials,* Special Technical Publication 15-C, American Society for Testing Materials, Philadelphia, 1951, p. 115. Used by permission of ASTM.

TABLE II

Sample Size Code Letters for MIL-STD-105-D

Lot or batch size	Special inspection levels				General inspection levels		
	S-1	S-2	S-3	S-4	I	II	III
2 to 8	A	A	A	A	A	A	B
9 to 15	A	A	A	A	A	B	C
16 to .25	A	A	B	B	B	C	D
26 to 50	A	B	B	C	C	D	E
51 to 90	B	B	C	C	C	E	F
91 to 150	B	B	C	D	D	F	G
151 to 280	B	C	D	E	E	G	H
281 to 500	B	C	D	E	F	H	J
501 to 1200	C	C	E	F	G	J	K
1201 to 3200	C	D	E	G	H	K	L
3201 to 10000	C	D	F	G	J	L	M
10001 to 35000	C	D	F	H	K	M	N
35001 to 150000	D	E	G	J	L	N	P
150001 to 500000	D	E	G	J	M	P	Q
500001 and over	D	E	H	K	N	Q	R

Source: MIL-STD-105-D

TABLE II-A

Single Sampling Plans for Normal Inspection (Master Table)
(Use General Inspection Level II from Table II)

Sample size code letter	Sample size	Acceptable Quality Levels (normal inspection)																									
		0.010	0.015	0.025	0.040	0.065	0.10	0.15	0.25	0.40	0.65	1.0	1.5	2.5	4.0	6.5	10	15	25	40	65	100	150	250	400	650	1000
		Ac Re	Ac Re	Ac Re	Ac Re	Ac Re	Ac Re	Ac Re	Ac Re	Ac Re	Ac Re	Ac Re	Ac Re	Ac Re	Ac Re	Ac Re	Ac Re	Ac Re	Ac Re	Ac Re	Ac Re	Ac Re	Ac Re	Ac Re	Ac Re	Ac Re	Ac Re
A	2	↓	↓	↓	↓	↓	↓	↓	↓	↓	↓	↓	↓	↓	↓	0 1	↓	↓	1 2	2 3	3 4	5 6	7 8	10 11	14 15	21 22	30 31
B	3	↓	↓	↓	↓	↓	↓	↓	↓	↓	↓	↓	↓	↓	0 1	↑	↓	1 2	2 3	3 4	5 6	7 8	10 11	14 15	21 22	30 31	44 45
C	5	↓	↓	↓	↓	↓	↓	↓	↓	↓	↓	↓	↓	0 1	↑	↓	1 2	2 3	3 4	5 6	7 8	10 11	14 15	21 22	30 31	44 45	↑
D	8	↓	↓	↓	↓	↓	↓	↓	↓	↓	↓	↓	0 1	↑	↓	1 2	2 3	3 4	5 6	7 8	10 11	14 15	21 22	30 31	44 45	↑	↑
E	13	↓	↓	↓	↓	↓	↓	↓	↓	↓	↓	0 1	↑	↓	1 2	2 3	3 4	5 6	7 8	10 11	14 15	21 22	30 31	44 45	↑	↑	↑
F	20	↓	↓	↓	↓	↓	↓	↓	↓	↓	0 1	↑	↓	1 2	2 3	3 4	5 6	7 8	10 11	14 15	21 22	↑	↑	↑	↑	↑	↑
G	32	↓	↓	↓	↓	↓	↓	↓	↓	0 1	↑	↓	1 2	2 3	3 4	5 6	7 8	10 11	14 15	21 22	↑	↑	↑	↑	↑	↑	↑
H	50	↓	↓	↓	↓	↓	↓	↓	0 1	↑	↓	1 2	2 3	3 4	5 6	7 8	10 11	14 15	21 22	↑	↑	↑	↑	↑	↑	↑	↑
J	80	↓	↓	↓	↓	↓	↓	0 1	↑	↓	1 2	2 3	3 4	5 6	7 8	10 11	14 15	21 22	↑	↑	↑	↑	↑	↑	↑	↑	↑
K	125	↓	↓	↓	↓	↓	0 1	↑	↓	1 2	2 3	3 4	5 6	7 8	10 11	14 15	21 22	↑	↑	↑	↑	↑	↑	↑	↑	↑	↑
L	200	↓	↓	↓	↓	0 1	↑	↓	1 2	2 3	3 4	5 6	7 8	10 11	14 15	21 22	↑	↑	↑	↑	↑	↑	↑	↑	↑	↑	↑
M	315	↓	↓	↓	0 1	↑	↓	1 2	2 3	3 4	5 6	7 8	10 11	14 15	21 22	↑	↑	↑	↑	↑	↑	↑	↑	↑	↑	↑	↑
N	500	↓	↓	0 1	↑	↓	1 2	2 3	3 4	5 6	7 8	10 11	14 15	21 22	↑	↑	↑	↑	↑	↑	↑	↑	↑	↑	↑	↑	↑
P	800	↓	0 1	↑	↓	1 2	2 3	3 4	5 6	7 8	10 11	14 15	21 22	↑	↑	↑	↑	↑	↑	↑	↑	↑	↑	↑	↑	↑	↑
Q	1250	0 1	↑	↓	1 2	2 3	3 4	5 6	7 8	10 11	14 15	21 22	↑	↑	↑	↑	↑	↑	↑	↑	↑	↑	↑	↑	↑	↑	↑
R	2000	↑	↑	1 2	2 3	3 4	5 6	7 8	10 11	14 15	21 22	↑	↑	↑	↑	↑	↑	↑	↑	↑	↑	↑	↑	↑	↑	↑	↑

↓ = Use first sampling plan below arrow. If sample size equals, or exceeds, lot or batch size, do 100 percent inspection.

↑ = Use first sampling plan above arrow.

Ac = Acceptance number.

Re = Rejection number.

Source: MIL-STD-105-D

TABLE III

Values of np_1 and np_2 and Their Ratios for $\alpha = 0.05$ and $\beta = 0.10$ and for Values of c from 1 to 40

c	$m_1 = np_1$	$m_2 = np_2$	m_2/m_1	c	$m_1 = np_1$	$m_2 = np_2$	m_2/m_1
0	0.0513	2.303	44.893	21	14.894	28.184	1.892
1	.355	3.890	10.958	22	15.719	29.320	1.865
2	.818	5.322	6.506	23	16.548	30.453	1.840
3	1.366	6.681	4.891	24	17.382	31.584	1.817
4	1.970	7.994	4.058	25	18.218	32.711	1.796
5	2.613	9.275	3.550	26	19.058	33.836	1.775
6	3.286	10.532	3.205	27	19.900	34.959	1.757
7	3.981	11.771	2.957	28	20.746	36.080	1.739
8	4.695	12.995	2.768	29	21.594	37.198	1.723
9	5.426	14.206	2.618	30	22.444	38.315	1.707
10	6.169	15.407	2.497	31	23.298	39.430	1.692
11	6.924	16.598	2.397	32	24.152	40.543	1.679
12	7.690	17.782	2.312	33	25.010	41.654	1.665
13	8.464	18.958	2.240	34	25.870	42.704	1.651
14	9.246	20.128	2.177	35	26.731	43.872	1.641
15	10.035	21.292	2.118	36	27.594	44.978	1.630
16	10.831	22.452	2.073	37	28.460	46.083	1.619
17	11.633	23.606	2.029	38	29.327	47.187	1.609
18	12.442	24.756	1.990	39	30.196	48.289	1.599
19	13.254	25.932	1.954	40	31.066	49.390	1.590
20	14.072	27.845	1.922				

Source: J. M. Cameron, Tables for Constructing Single Sample Plans, *Industrial Quality Control,* (July 1952), pp. 38-39.

TABLE IV

Sample Sizes for Selected Values of p_1, p_2, c, with $\alpha=0.05$ and $\beta=0.10$*

p_1	p_2	c	n	p_1	p_2	c	n	p_1	p_2	c	n
0.25	1.0	4	800	1.75	5.5	6	192	3.0	7.0	12	257
	1.5	2	355		6.0	5	155		7.5	10	206
					7.0	4	115		8.0	9	181
0.50	1.0	18	2489		9.0	3	76		8.5	8	157
	1.5	7	797						9.0	7	133
	2.0	4	400	2.0	3.5	27	999		9.5	6	111
	2.5	3	268		4.0	18	623				
	3.0	2	178		4.5	13	424	3.5	6.0	30	642
	3.5	2	164		5.0	10	309		6.5	22	452
	5.5	1	71		5.5	8	237		7.0	18	356
					6.0	7	200		7.5	15	287
0.75	1.5	18	1659		6.5	6	165		8.0	12	223
	2.0	9	724		7.0	5	133		8.5	11	198
	2.5	6	439		8.0	4	100		9.0	9	158
	3.0	4	267		10.0	3	67		9.5	8	137
	4.0	3	176						10.0	7	118
	5.0	2	110	2.25	4.0	26	848				
	8.0	1	49		4.5	18	553	4.0	6.5	36	692
					5.0	13	380		7.0	27	500
1.0	2.0	18	1248		5.5	10	281		7.5	22	393
	2.5	10	617		6.0	9	242		8.0	18	312
	3.0	7	398		6.5	7	182		8.5	15	251
	3.5	5	265		7.0	6	155		9.0	13	212
	4.0	4	200		7.5	6	147		9.5	11	175
	5.0	3	134		8.0	5	117		10.0	10	155
	6.5	2	82		9.0	4	89				
	10.0	1	39					4.5	7.5	33	556
				2.5	4.0	39	1208		8.0	26	424
1.25	2.0	39	2416		4.5	25	729		8.5	21	332
	2.5	18	996		5.0	18	498		9.0	18	277
	3.0	11	554		5.5	14	370		9.5	15	225
	3.5	8	376		6.0	11	277		10.0	13	190
	4.0	6	264		6.5	9	219				
	4.5	5	210		7.0	8	188	5.0	8.0	39	604
	5.0	4	160		7.5	7	160		8.5	30	451
	6.5	3	106		8.0	6	132		9.0	25	365
	8.0	2	67		9.0	5	105		9.5	21	298
					10.0	4	80		10.0	18	249
1.50	2.5	33	1668								
	3.0	18	830	2.75	4.5	35	975				
	3.5	12	513		5.0	24	633				
	4.0	9	362		5.5	18	452				
	4.5	7	266		6.0	14	337				
	5.0	6	220		6.5	11	256				
	5.5	5	175		7.0	10	225				
	6.0	4	134		7.5	8	174				
	7.5	3	90		8.0	7	148				
	10.0	2	55		9.0	6	120				
					9.5	5	98				
1.75	3.0	30	1283								
	3.5	18	711	3.0	5.0	33	834				
	4.0	12	445		5.5	23	554				
	4.5	9	316		6.0	18	415				
	5.0	7	236		6.5	14	310				

*Where n_1 and n_2 are not equal, largest value is used; p_1 and p_2 are in percent.

TABLE V

Random Numbers

03 47 43 73 86	36 96 47 36 61	46 98 63 71 62	33 26 16 80 45	60 11 14 10 95
97 74 24 67 62	42 81 14 57 20	42 53 32 37 32	27 07 36 07 51	24 51 79 89 73
16 76 62 27 66	56 50 26 71 07	32 90 79 78 53	13 55 38 58 59	88 97 54 14 10
12 56 85 99 26	96 96 68 27 31	05 03 72 93 15	57 12 10 14 21	88 26 49 81 76
55 59 56 35 64	38 54 82 46 22	31 62 43 09 90	06 18 44 32 53	23 83 01 30 30
16 22 77 94 39	49 54 43 54 82	17 37 93 23 78	87 35 20 96 43	84 26 34 91 64
84 42 17 53 31	57 24 55 06 88	77 04 74 47 67	21 76 33 50 25	83 92 12 06 76
63 01 63 78 59	16 95 55 67 19	98 10 50 71 75	12 86 73 58 07	44 39 52 38 79
33 21 12 34 29	78 64 56 07 82	52 42 07 44 38	15 51 00 13 42	99 66 02 79 54
57 60 86 32 44	09 47 27 96 54	49 17 46 09 62	90 52 84 77 27	08 02 73 43 28
18 18 07 92 46	44 17 16 58 09	79 83 86 19 62	06 76 50 03 10	55 23 64 05 05
26 62 38 97 75	84 16 07 44 99	83 11 46 32 24	20 14 85 88 45	10 93 72 88 71
23 42 40 64 74	82 97 77 77 81	07 45 32 14 08	32 98 94 07 72	93 85 79 10 75
52 36 28 19 95	50 92 26 11 97	00 56 76 31 38	80 22 02 53 53	86 60 42 04 53
37 85 94 35 12	83 39 50 08 30	42 34 07 96 88	54 42 06 87 98	35 85 29 48 39
70 29 17 12 13	40 33 20 38 26	13 89 51 03 74	17 76 37 13 04	07 74 21 19 30
56 62 18 37 35	96 83 50 87 75	97 12 25 93 47	70 33 24 03 54	97 77 46 44 80
99 49 57 22 77	88 42 95 45 72	16 64 36 16 00	04 43 18 66 79	94 77 24 21 90
16 08 15 04 72	33 27 14 34 09	45 59 34 68 49	12 72 07 34 45	99 27 72 95 14
31 16 93 32 43	50 27 89 87 19	20 15 37 00 49	52 85 66 60 44	38 68 88 11 80
68 34 30 13 70	55 74 30 77 40	44 22 78 84 26	04 33 46 09 52	68 07 97 06 57
74 57 25 65 76	59 29 97 68 60	71 91 38 67 54	13 58 18 24 76	15 54 55 95 52
27 42 37 86 53	48 55 90 65 72	96 57 69 36 10	96 46 92 42 45	97 60 49 04 91
00 39 68 29 61	66 37 32 20 30	77 84 57 03 29	10 45 65 04 26	11 04 96 67 24
29 94 98 94 24	68 49 69 10 82	53 75 91 93 30	34 25 20 57 27	40 48 73 51 92
16 90 82 66 59	83 62 64 11 12	67 19 00 71 74	60 47 21 29 68	02 02 37 03 31
11 27 94 75 06	06 09 19 74 66	02 94 37 34 02	76 70 90 30 86	38 45 94 30 38
35 24 10 16 20	33 32 51 26 38	79 78 45 04 91	16 92 53 56 16	02 75 50 95 98
38 23 16 86 38	42 38 97 01 50	87 75 66 81 41	40 01 74 91 62	48 51 84 08 32
31 96 25 91 47	96 44 33 49 13	34 86 82 53 91	00 52 43 48 85	27 55 26 89 62
66 67 40 67 14	64 05 71 95 86	11 05 65 09 68	76 83 20 37 90	57 16 00 11 66
14 90 84 45 11	75 73 88 05 90	52 27 41 14 86	22 98 12 22 08	07 52 74 95 80
68 05 51 18 00	33 96 02 75 19	07 60 62 93 55	59 33 82 43 90	49 37 38 44 59
20 46 78 73 90	97 51 40 14 02	04 02 33 31 08	39 54 16 49 36	47 95 93 13 30
64 19 58 97 79	15 06 15 93 20	01 90 10 75 06	40 78 78 89 62	02 67 74 17 33
05 26 93 70 60	22 35 85 15 13	92 03 51 59 77	59 56 78 06 83	52 91 05 70 74
07 97 10 88 23	09 98 42 99 64	61 71 62 99 15	06 51 29 16 93	58 05 77 09 51
68 71 86 85 85	54 87 66 47 54	73 32 08 11 12	44 95 92 63 16	29 56 24 29 48
26 99 61 65 53	58 37 78 80 70	42 10 50 67 42	32 17 55 85 74	94 44 67 16 94
14 65 52 68 75	87 59 36 22 41	26 78 63 06 55	13 08 27 01 50	15 29 39 39 43
17 53 77 58 71	71 41 61 50 72	12 41 94 96 26	44 95 27 36 99	02 96 74 30 83
90 26 59 21 19	23 52 23 33 12	96 93 02 18 39	07 02 18 36 07	25 99 32 70 23
41 23 52 55 99	31 04 49 69 96	10 47 48 45 88	13 41 43 89 20	97 17 14 49 17
60 20 50 81 69	31 99 73 68 68	35 81 33 03 76	24 30 12 48 60	18 99 10 72 34
91 25 38 05 90	94 58 28 41 36	45 37 59 03 09	90 35 37 29 12	82 62 54 65 60
34 50 57 74 37	98 80 33 00 91	09 77 93 19 82	74 94 80 04 04	45 07 31 66 49
85 22 04 39 43	73 81 53 94 79	33 62 46 86 28	08 31 54 46 31	53 94 13 38 47
09 79 13 77 48	73 82 97 22 21	05 03 27 24 83	72 89 44 05 60	35 80 39 94 88
88 75 80 18 14	22 95 75 42 49	39 32 82 22 49	02 48 07 70 37	16 04 61 67 87
90 96 23 70 00	39 00 03 06 90	55 85 78 38 36	94 37 30 69 32	90 89 00 76 33

Source: Fisher and Yates, *Statistical Tables for Biological Agricultural and Medical Research,* 6th edition, 1963, Harlow, Longman Group, Ltd., p. 134. Used by permission of Longman Group, Ltd.

TABLE VI

Ordinates and Areas of the Normal Curve: $y = \frac{1}{\sqrt{2\pi}} e^{\frac{-z^2}{2}}$

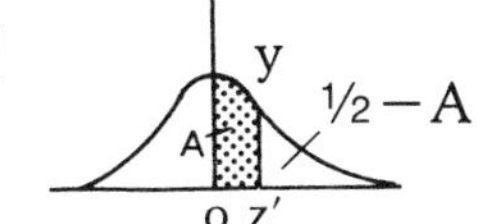

Normal Deviate z	Ordinate y	Area A	Area ½-A	z	y	A	½-A	z	y	A	½-A	z	y	A	½-A
.00	.3989	.0000	.5000	.36	.3739	.1406	.3594	.72	.3078	.2642	.2358	1.07	.2251	.3577	.1423
.01	.3989	.0040	.4960	.37	.3726	.1443	.3557	.73	.3056	.2673	.2327	1.08	.2226	.3599	.1401
.02	.3989	.0080	.4920	.38	.3712	.1480	.3520	.74	.3034	.2704	.2296	1.09	.2202	.3621	.1379
.03	.3988	.0120	.4880	.39	.3697	.1517	.3483	.75	.3011	.2734	.2266	1.10	.2178	.3643	.1357
.04	.3986	.0160	.4840	.40	.3683	.1554	.3446	.76	.2989	.2764	.2236	1.11	.2155	.3665	.1335
.05	.3984	.0199	.4801	.41	.3668	.1591	.3409	.77	.2966	.2794	.2206	1.12	.2131	.3686	.1314
.06	.3982	.0239	.4761	.42	.3653	.1628	.3372	.78	.2943	.2823	.2177	1.13	.2107	.3708	.1292
.07	.3980	.0279	.4721	.43	.3637	.1664	.3336	.79	.2920	.2852	.2148	1.14	.2083	.3729	.1271
.08	.3977	.0319	.4681	.44	.3621	.1700	.3300	.80	.2897	.2881	.2119	1.15	.2059	.3749	.1251
.09	.3973	.0359	.4641	.45	.3605	.1736	.3264	.81	.2874	.2910	.2090	1.16	.2036	.3770	.1230
.10	.3970	.0398	.4602	.46	.3589	.1772	.3228	.82	.2850	.2939	.2061	1.17	.2012	.3790	.1210
.11	.3965	.0438	.4562	.47	.3572	.1808	.3192	.83	.2827	.2967	.2033	1.18	.1989	.3810	.1190
.12	.3961	.0478	.4522	.48	.3555	.1844	.3156	.84	.2803	.2996	.2004	1.19	.1965	.3820	.1170
.13	.3956	.0517	.4483	.49	.3538	.1879	.3121	.85	.2780	.3023	.1977	1.20	.1942	.3849	.1151
.14	.3951	.0557	.4443	.50	.3521	.1915	.3085	.86	.2756	.3051	.1949	1.21	.1919	.3869	.1131
.15	.3945	.0596	.4404	.51	.3503	.1950	.3050	.87	.2732	.3078	.1922	1.22	.1895	.3888	.1112
.16	.3939	.0636	.4364	.52	.3485	.1985	.3015	.88	.2709	.3106	.1894	1.23	.1872	.3906	.1094
.17	.3932	.0675	.4325	.53	.3467	.2019	.2981	.89	.2685	.3133	.1867	1.24	.1849	.3925	.1075
.18	.3925	.0714	.4286	.54	.3448	.2054	.2946	.90	.2661	.3159	.1841	1.25	.1826	.3944	.1056
.19	.3918	.0754	.4246	.55	.3429	.2088	.2912	.91	.2637	.3186	.1814	1.26	.1804	.3962	.1038
.20	.3910	.0793	.4207	.56	.3410	.2123	.2877	.92	.2613	.3212	.1788	1.27	.1781	.3980	.1020
.21	.3902	.0832	.4168	.57	.3391	.2157	.2843	.93	.2589	.3238	.1762	1.28	.1758	.3997	.1003
.22	.3894	.0871	.4129	.58	.3372	.2190	.2810	.94	.2565	.3264	.1736	1.29	.1736	.4015	.0985
.23	.3885	.0910	.4090	.59	.3352	.2224	.2776	.95	.2541	.3289	.1711	1.30	.1714	.4032	.0968
.24	.3876	.0948	.4052	.60	.3332	.2258	.2742	.96	.2516	.3315	.1685	1.31	.1692	.4049	.0951
.25	.3867	.0987	.4013	.61	.3312	.2291	.2709	.97	.2492	.3340	.1660	1.32	.1669	.4066	.0934
.26	.3857	.1026	.3974	.62	.3292	.2324	.2676	.98	.2468	.3365	.1635	1.33	.1647	.4082	.0918
.27	.3847	.1064	.3936	.63	.3271	.2356	.2644	.99	.2444	.3389	.1611	1.34	.1626	.4099	.0901
.28	.3836	.1103	.3897	.64	.3251	.2389	.2611	1.00	.2420	.3413	.1587	1.35	.1604	.4115	.0885
.29	.3825	.1141	.3859	.65	.3230	.2422	.2578	1.01	.2396	.3438	.1562	1.36	.1582	.4131	.0869
.30	.3814	.1179	.3821	.66	.3209	.2454	.2546	1.02	.2371	.3461	.1539	1.37	.1561	.4147	.0853
.31	.3802	.1217	.3783	.67	.3187	.2486	.2514	1.03	.2347	.3485	.1515	1.38	.1540	.4162	.0838
.32	.3790	.1255	.3745	.68	.3166	.2518	.2482	1.04	.2323	.3508	.1492	1.39	.1518	.4177	.0823
.33	.3778	.1293	.3707	.69	.3144	.2549	.2451	1.05	.2299	.3531	.1469	1.40	.1497	.4192	.0808
.34	.3765	.1331	.3669	.70	.3122	.2580	.2420	1.06	.2275	.3554	.1446	1.41	.1476	.4207	.0793
.35	.3752	.1368	.3632	.71	.3101	.2612	.2388								

TABLE VI (continued)

Normal Deviate z	Ordinate y	Area A	Area $\frac{1}{2}$-A	z	y	A	$\frac{1}{2}$-A	z	y	A	$\frac{1}{2}$-A	z	y	A	$\frac{1}{2}$-A
1.42	.1456	.4222	.0778	1.83	.0748	.4664	.0336	2.24	.0325	.4875	.0125	2.65	.0119	.4960	.0040
1.43	.1435	.4236	.0764	1.84	.0734	.4671	.0329	2.25	.0317	.4878	.0122	2.66	.0116	.4961	.0039
1.44	.1415	.4251	.0749	1.85	.0721	.4678	.0322	2.26	.0310	.4881	.0119	2.67	.0113	.4962	.0038
1.45	.1394	.4265	.0735	1.86	.0707	.4686	.0314	2.27	.0303	.4884	.0116	2.68	.0110	.4963	.0037
1.46	.1374	.4279	.0721	1.87	.0694	.4693	.0307	2.28	.0296	.4887	.0113	2.69	.0107	.4964	.0036
1.47	.1354	.4292	.0708	1.88	.0681	.4699	.0301	2.29	.0290	.4890	.0110	2.70	.0104	.4965	.0035
1.48	.1334	.4306	.0694	1.89	.0669	.4706	.0294	2.30	.0283	.4893	.0107	2.71	.0101	.4966	.0034
1.49	.1315	.4319	.0681	1.90	.0656	.4713	.0287	2.31	.0277	.4896	.0104	2.72	.0099	.4967	.0033
1.50	.1295	.4332	.0668	1.91	.0644	.4719	.0281	2.32	.0270	.4898	.0102	2.73	.0096	.4968	.0032
1.51	.1276	.4345	.0655	1.92	.0632	.4726	.0274	2.33	.0264	.4901	.0099	2.74	.0094	.4969	.0031
1.52	.1257	.4357	.0643	1.93	.0620	.4732	.0268	2.34	.0258	.4904	.0096	2.75	.0091	.4970	.0030
1.53	.1238	.4370	.0630	1.94	.0608	.4738	.0262	2.35	.0252	.4906	.0094	2.76	.0088	.4971	.0029
1.54	.1319	.4382	.0618	1.95	.0596	.4744	.0256	2.36	.0246	.4909	.0091	2.77	.0086	.4972	.0028
1.55	.1200	.4394	.0606	1.96	.0584	.4750	.0250	2.37	.0241	.4911	.0089	2.78	.0084	.4973	.0027
1.56	.1182	.4406	.0594	1.97	.0573	.4756	.0244	2.38	.0235	.4913	.0087	2.79	.0081	.4974	.0026
1.57	.1163	.4418	.0582	1.98	.0562	.4761	.0239	2.39	.0229	.4916	.0084	2.80	.0079	.4974	.0026
1.58	.1145	.4429	.0571	1.99	.0551	.4767	.0233	2.40	.0224	.4918	.0082	2.81	.0077	.4975	.0025
1.59	.1127	.4441	.0559	2.00	.0540	.4772	.0228	2.41	.0219	.4920	.0080	2.82	.0075	.4976	.0024
1.60	.1109	.4452	.0548	2.01	.0529	.4778	.0222	2.42	.0213	.4922	.0078	2.83	.0073	.4977	.0023
1.61	.1092	.4463	.0537	2.02	.0519	.4783	.0217	2.43	.0208	.4925	.0075	2.84	.0071	.4977	.0023
1.62	.1074	.4474	.0526	2.03	.0508	.4788	.0212	2.44	.0203	.4927	.0073	2.85	.0069	.4978	.0022
1.63	.1057	.4484	.0516	2.04	.0498	.4793	.0207	2.45	.0198	.4929	.0071	2.86	.0067	.4979	.0021
1.64	.1040	.4495	.0505	2.05	.0488	.4798	.0202	2.46	.0194	.4931	.0069	2.87	.0065	.4979	.0021
1.65	.1023	.4405	.0495	2.06	.0478	.4803	.0197	2.47	.0189	.4932	.0068	2.88	.0063	.4980	.0020
1.66	.1006	.4515	.0485	2.07	.0468	.4808	.0192	2.48	.0184	.4934	.0066	2.89	.0061	.4981	.0019
1.67	.0989	.4525	.0475	2.08	.0459	.4812	.0188	2.49	.0180	.4936	.0064	2.90	.0060	.4981	.0019
1.68	.0973	.4535	.0465	2.09	.0449	.4817	.0183	2.50	.0175	.4938	.0062	2.91	.0058	.4982	.0018
1.69	.0957	.4545	.0455	2.10	.0440	.4821	.0179	2.51	.0171	.4940	.0060	2.92	.0056	.4982	.0018
1.70	.0940	.4554	.0446	2.11	.0431	.4826	.0174	2.52	.0167	.4941	.0059	2.93	.0054	.4983	.0017
1.71	.0925	.4564	.0436	2.12	.0422	.4830	.0170	2.53	.0162	.4943	.0057	2.94	.0053	.4984	.0016
1.72	.0909	.4573	.0427	2.13	.0413	.4834	.0166	2.54	.0158	.4945	.0055	2.95	.0051	.4984	.0016
1.73	.0893	.4582	.0418	2.14	.0404	.4838	.0162	2.55	.0154	.4946	.0054	2.96	.0050	.4985	.0015
1.74	.0878	.4591	.0409	2.15	.0396	.4842	.0158	2.56	.0151	.4948	.0052	2.97	.0048	.4985	.0015
1.75	.0863	.4599	.0401	2.16	.0387	.4846	.0154	2.57	.0147	.4949	.0051	2.98	.0047	.4986	.0014
1.76	.0848	.4608	.0392	2.17	.0379	.4850	.0150	2.58	.0143	.4951	.0049	2.99	.0046	.4986	.0014
1.77	.0833	.4616	.0384	2.18	.0371	.4854	.0146	2.59	.0139	.4952	.0048	3.00	.0044	.49865	.00135
1.78	.0818	.4625	.0375	2.19	.0363	.4857	.0143	2.60	.0136	.4953	.0047	3.09	.0034	.4990	.0010
1.79	.0804	.4633	.0367	2.20	.0355	.4861	.0139	2.61	.0132	.4955	.0045	3.29	.0018	.4995	.0005
1.80	.0790	.4641	.0359	2.21	.0347	.4864	.0136	2.62	.0129	.4956	.0044	3.72	.0004	.4999	.0001
1.81	.0775	.4648	.0352	2.22	.0339	.4868	.0132	2.63	.0126	.4957	.0043				
1.82	.0761	.4656	.0344	2.23	.0332	.4871	.0129	2.64	.0122	.4958	.0042				

Source: A. C. Rosander, *Elementary Principles of Statistics*, Van Nostrand, New York, 1951. Copyright by Wadsworth Inc. Used by permission of Brooks Cole Publishing Company.

TABLE VII

Distribution of t

n^1 = degrees of freedom

n^1	Probability .9	.8	.7	.6	.5	.4	.3	.2	.1	.05	.02	.01	.001
1	.158	.325	.510	.727	1.000	1.376	1.963	3.078	6.314	12.706	31.821	63.657	636.619
2	.142	.289	.445	.617	.816	1.061	1.386	1.886	2.920	4.303	6.965	9.925	31.598
3	.137	.277	.424	.584	.765	.978	1.250	1.638	2.353	3.182	4.541	5.841	12.924
4	.134	.271	.414	.569	.741	.941	1.190	1.533	2.132	2.776	3.747	4.604	8.610
5	.132	.267	.408	.559	.727	.920	1.156	1.476	2.015	2.571	3.365	4.032	6.869
6	.131	.265	.404	.553	.718	.906	1.134	1.440	1.943	2.447	3.143	3.707	5.959
7	.130	.263	.402	.549	.711	.896	1.119	1.415	1.895	2.365	2.998	3.499	5.408
8	.130	.262	.399	.546	.706	.889	1.108	1.397	1.860	2.306	2.896	3.355	5.041
9	.129	.261	.398	.543	.703	.883	1.100	1.383	1.833	2.262	2.821	3.250	4.781
10	.129	.260	.397	.542	.700	.879	1.093	1.372	1.812	2.228	2.764	3.169	4.587
11	.129	.260	.396	.540	.697	.876	1.088	1.363	1.796	2.201	2.718	3.106	4.437
12	.128	.259	.395	.539	.695	.873	1.083	1.356	1.782	2.179	2.681	3.055	4.318
13	.128	.259	.394	.538	.694	.870	1.079	1.350	1.771	2.160	2.650	3.012	4.221
14	.128	.258	.393	.537	.692	.868	1.076	1.345	1.761	2.145	2.624	2.977	4.140
15	.128	.258	.393	.536	.691	.866	1.074	1.341	1.753	2.131	2.602	2.947	4.073
16	.128	.258	.392	.535	.690	.865	1.071	1.337	1.746	2.120	2.583	2.921	4.015
17	.128	.257	.392	.534	.689	.863	1.069	1.333	1.740	2.110	2.567	2.898	3.965
18	.127	.257	.392	.534	.688	.862	1.067	1.330	1.734	2.101	2.552	2.878	3.922
19	.127	.257	.391	.533	.688	.861	1.066	1.328	1.729	2.093	2.539	2.861	3.883
20	.127	.257	.391	.533	.687	.860	1.064	1.325	1.725	2.086	2.528	2.845	3.850
21	.127	.257	.391	.532	.686	.859	1.063	1.323	1.721	2.080	2.518	2.831	3.819
22	.127	.256	.390	.532	.686	.858	1.061	1.321	1.717	2.074	2.508	2.819	3.792
23	.127	.256	.390	.532	.685	.858	1.060	1.319	1.714	2.069	2.500	2.807	3.767
24	.127	.256	.390	.531	.685	.857	1.059	1.318	1.711	2.064	2.492	2.797	3.745
25	.127	.256	.390	.531	.684	.856	1.058	1.316	1.708	2.060	2.485	2.787	3.725
26	.127	.256	.390	.531	.684	.856	1.058	1.315	1.706	2.056	2.479	2.779	3.707
27	.127	.256	.389	.531	.684	.855	1.057	1.314	1.703	2.052	2.473	2.771	3.690
28	.127	.256	.389	.530	.683	.855	1.056	1.313	1.701	2.048	2.467	2.763	3.674
29	.127	.256	.389	.530	.683	.854	1.055	1.311	1.699	2.045	2.462	2.756	3.659
30	.127	.256	.389	.530	.683	.854	1.055	1.310	1.697	2.042	2.457	2.750	3.646
40	.126	.255	.388	.529	.681	.851	1.050	1.303	1.684	2.021	2.423	2.704	3.551
60	.126	.254	.387	.527	.679	.848	1.046	1.296	1.671	2.000	2.390	2.660	3.460
120	.126	.254	.386	.526	.677	.845	1.041	1.289	1.658	1.980	2.358	2.617	3.373
∞	.126	.253	.385	.524	.674	.842	1.036	1.282	1.645	1.960	2.326	2.576	3.291

Source: Fisher and Yates, *Statistical Tables for Biological Agricultural and Medical Research,* 6th edition, 1963, Harlow: Longman Group Ltd., p. 46. Used by permission of Longman Group Ltd.

TABLE VIII

Distribution of χ^2

n^1 = degrees of freedom

Probability

n^1	.99	.98	.95	.90	.80	.70	.50	.30	.20	.10	.05	.02	.01	.001
1	$.0^3157$	$.0^3628$	.00393	.0158	.0642	.148	.455	1.074	1.642	2.706	3.841	5.412	6.635	10.827
2	.0201	.0404	.103	.211	.446	.713	1.386	2.408	3.219	4.605	5.991	7.824	9.210	13.815
3	.115	.185	.352	.584	1.005	1.424	2.366	3.665	4.642	6.251	7.815	9.837	11.345	16.266
4	.297	.429	.711	1.064	1.649	2.195	3.357	4.878	5.989	7.779	9.488	11.668	13.277	18.467
5	.554	.752	1.145	1.610	2.343	3.000	4.351	6.064	7.289	9.236	11.070	13.388	15.086	20.515
6	.872	1.134	1.635	2.204	3.070	3.828	5.348	7.231	8.558	10.645	12.592	15.033	16.812	22.457
7	1.239	1.564	2.167	2.833	3.822	4.671	6.346	8.383	9.803	12.017	14.067	16.622	18.475	24.322
8	1.646	2.032	2.733	3.490	4.594	5.527	7.344	9.524	11.030	13.362	15.507	18.168	20.090	26.125
9	2.088	2.532	3.325	4.168	5.380	6.393	8.343	10.656	12.242	14.684	16.919	19.679	21.666	27.877
10	2.558	3.059	3.940	4.865	6.179	7.267	9.342	11.781	13.442	15.987	18.307	21.161	23.209	29.588
11	3.053	3.609	4.575	5.578	6.989	8.148	10.341	12.899	14.631	17.275	19.675	22.618	24.725	31.264
12	3.571	4.178	5.226	6.304	7.807	9.034	11.340	14.011	15.812	18.549	21.026	24.054	26.217	32.909
13	4.107	4.765	5.892	7.042	8.634	9.926	12.340	15.119	16.985	19.812	22.362	25.472	27.688	34.528
14	4.660	5.368	6.571	7.790	9.467	10.821	13.339	16.222	18.151	21.064	23.685	26.873	29.141	36.123
15	5.229	5.985	7.261	8.547	10.307	11.721	14.339	17.322	19.311	22.307	24.996	28.259	30.578	37.697
16	5.812	6.614	7.962	9.312	11.152	12.624	15.338	18.418	20.465	23.542	26.296	29.633	32.000	39.252
17	6.408	7.255	8.672	10.085	12.002	13.531	16.338	19.511	21.615	24.769	27.587	30.995	33.409	40.790
18	7.015	7.906	9.390	10.865	12.857	14.440	17.338	20.601	22.760	25.989	28.869	32.346	34.805	42.312
19	7.633	8.567	10.117	11.651	13.716	15.352	18.338	21.689	23.900	27.204	30.144	33.687	36.191	43.820
20	8.260	9.237	10.851	12.443	14.578	16.266	19.337	22.775	25.038	28.412	31.410	35.020	37.566	45.315
21	8.897	9.915	11.591	13.240	15.445	17.182	20.337	23.858	26.171	29.615	32.671	36.343	38.932	46.797
22	9.542	10.600	12.338	14.041	16.314	18.101	21.337	24.939	27.301	30.813	33.924	37.659	40.289	48.268
23	10.196	11.293	13.091	14.848	17.187	19.021	22.337	26.018	28.429	32.007	35.172	38.968	41.638	49.728
24	10.856	11.992	13.848	15.659	18.062	19.943	23.337	27.096	29.553	33.196	36.415	40.270	42.980	51.179
25	11.524	12.697	14.611	16.473	18.940	20.867	24.337	28.172	30.675	34.382	37.652	41.566	44.314	52.620
26	12.198	13.409	15.379	17.292	19.820	21.792	25.336	29.246	31.795	35.563	38.885	42.856	45.642	54.052
27	12.879	14.125	16.151	18.114	20.703	22.719	26.336	30.319	32.912	36.741	40.113	44.140	46.963	55.476
28	13.565	14.847	16.928	18.939	21.588	23.647	27.336	31.391	34.027	37.916	41.337	45.419	48.278	56.893
29	14.256	15.574	17.708	19.768	22.475	24.577	28.336	32.461	35.139	39.087	42.557	46.693	49.588	58.302
30	14.953	16.306	18.493	20.599	23.364	25.508	29.336	33.530	36.250	40.256	43.773	47.962	50.892	59.703
32	16.362	17.783	20.072	22.271	25.148	27.373	31.336	35.665	38.466	42.585	46.194	50.487	53.486	62.487
34	17.789	19.275	21.664	23.952	26.938	29.242	33.336	37.795	40.676	44.903	48.602	52.995	56.061	65.247
36	19.233	20.783	23.269	25.643	28.735	31.115	35.336	39.922	42.879	47.212	50.999	55.489	58.619	67.985
38	20.691	22.304	24.884	27.343	30.537	32.992	37.335	42.045	45.076	49.513	53.384	57.969	61.162	70.703
40	22.164	23.838	26.509	29.051	32.345	34.872	39.335	44.165	47.269	51.805	55.759	60.436	63.691	73.402
42	23.650	25.383	28.144	30.765	34.157	36.755	41.335	46.282	49.456	54.090	58.124	62.892	66.206	76.084
44	25.148	26.939	29.787	32.487	35.974	38.641	43.335	48.396	51.639	56.369	60.481	65.337	68.710	78.750
46	26.657	28.504	31.439	34.215	37.795	40.529	45.335	50.507	53.818	58.641	62.830	67.771	71.201	81.400
48	28.177	30.080	33.098	35.949	39.621	42.420	47.335	52.616	55.993	60.907	65.171	70.197	73.683	84.037
50	29.707	31.664	34.764	37.689	41.449	44.313	49.335	54.723	58.164	63.167	67.505	72.613	76.154	86.661
52	31.246	33.256	36.437	39.433	43.281	46.209	51.335	56.827	60.332	65.422	69.832	75.021	78.616	89.272
54	32.793	34.856	38.116	41.183	45.117	48.106	53.335	58.930	62.496	67.673	72.153	77.422	81.069	91.872
56	34.350	36.464	39.801	42.937	46.955	50.005	55.335	61.031	64.658	69.919	74.468	79.815	83.513	94.461
58	35.913	38.078	41.492	44.696	48.797	51.906	57.335	63.129	66.816	72.160	76.778	82.201	85.950	97.039
60	37.485	39.699	43.188	46.459	50.641	53.809	59.335	65.227	68.972	74.397	79.082	84.580	88.379	99.607
62	39.063	41.327	44.889	48.226	52.487	55.714	61.335	67.322	71.125	76.630	81.381	86.953	90.802	102.166
64	40.649	42.960	46.595	49.996	54.336	57.620	63.335	69.416	73.276	78.860	83.675	89.320	93.217	104.716
66	42.240	44.599	48.305	51.770	56.188	59.527	65.335	71.508	75.424	81.085	85.965	91.681	95.626	107.258
68	43.838	46.244	50.020	53.548	58.042	61.436	67.335	73.600	77.571	83.308	88.250	94.037	98.028	109.791
70	45.442	47.893	51.739	55.329	59.898	63.346	69.334	75.689	79.715	85.527	90.531	96.388	100.425	112.317

For odd values of n between 30 and 70 the mean of the tabular values for $n-1$ and $n+1$ may be taken. For larger values of n, the expression $\sqrt{2\chi^2}-\sqrt{2m-1}$ may be used as a normal deviate with unit variance, remembering that the probability for χ^2 corresponds with that of a single tail of the normal curve. (n^1 = degrees of freedom.)

Source: Fisher and Yates, *Statistical Tables for Biological Agricultural and Medical Research,* 6th edition, 1963, Harlow: Longman Group, Ltd., p. 47. Used by permission of Longman Group Ltd.

TABLE IX

The Distribution of F*

ν_2†	ν_1 df in Numerator																								ν_2†
	1	2	3	4	5	6	7	8	9	10	11	12	14	16	20	24	30	40	50	75	100	200	500	∞	
1	161 **4,052**	200 **4,999**	216 **5,403**	225 **5,625**	230 **5,764**	234 **5,859**	237 **5,928**	239 **5,981**	241 **6,022**	242 **6,056**	243 **6,082**	244 **6,106**	245 **6,142**	246 **6,169**	248 **6,208**	249 **6,234**	250 **6,261**	251 **6,286**	252 **6,302**	253 **6,323**	253 **6,334**	254 **6,352**	254 **6,361**	254 **6,366**	1
2	18.51 **98.49**	19.00 **99.00**	19.16 **99.17**	19.25 **99.25**	19.30 **99.30**	19.33 **99.33**	19.36 **99.36**	19.37 **99.37**	19.38 **99.39**	19.39 **99.40**	19.40 **99.41**	19.41 **99.42**	19.42 **99.43**	19.43 **99.44**	19.44 **99.45**	19.45 **99.46**	19.46 **99.47**	19.47 **99.48**	19.47 **99.48**	19.48 **99.49**	19.49 **99.49**	19.49 **99.49**	19.50 **99.50**	19.50 **99.50**	2
3	10.13 **34.12**	9.55 **30.82**	9.28 **29.46**	9.12 **28.71**	9.01 **28.24**	8.94 **27.91**	8.88 **27.67**	8.84 **27.49**	8.81 **27.34**	8.78 **27.23**	8.76 **27.13**	8.74 **27.05**	8.71 **26.92**	8.69 **26.83**	8.66 **26.69**	8.64 **26.60**	8.62 **26.50**	8.60 **26.41**	8.58 **26.35**	8.57 **26.27**	8.56 **26.23**	8.54 **26.18**	8.54 **26.14**	8.53 **26.12**	3
4	7.71 **21.20**	6.94 **18.00**	6.59 **16.69**	6.39 **15.98**	6.26 **15.52**	6.16 **15.21**	6.09 **14.98**	6.04 **14.80**	6.00 **14.66**	5.96 **14.54**	5.93 **14.45**	5.91 **14.37**	5.87 **14.24**	5.84 **14.15**	5.80 **14.02**	5.77 **13.93**	5.74 **13.83**	5.71 **13.74**	5.70 **13.69**	5.68 **13.61**	5.66 **13.57**	5.65 **13.52**	5.64 **13.48**	5.63 **13.46**	4
5	6.61 **16.26**	5.79 **13.27**	5.41 **12.06**	5.19 **11.39**	5.05 **10.97**	4.95 **10.67**	4.88 **10.45**	4.82 **10.29**	4.78 **10.15**	4.74 **10.05**	4.70 **9.96**	4.68 **9.89**	4.64 **9.77**	4.60 **9.68**	4.56 **9.55**	4.53 **9.47**	4.50 **9.38**	4.46 **9.29**	4.44 **9.24**	4.42 **9.17**	4.40 **9.13**	4.38 **9.07**	4.37 **9.04**	4.36 **9.02**	5
6	5.99 **13.74**	5.14 **10.92**	4.76 **9.78**	4.53 **9.15**	4.39 **8.75**	4.28 **8.47**	4.21 **8.26**	4.15 **8.10**	4.10 **7.98**	4.06 **7.87**	4.03 **7.79**	4.00 **7.72**	3.96 **7.60**	3.92 **7.52**	3.87 **7.39**	3.84 **7.31**	3.81 **7.23**	3.77 **7.14**	3.75 **7.09**	3.72 **7.02**	3.71 **6.99**	3.69 **6.94**	3.68 **6.90**	3.67 **6.88**	6
7	5.59 **12.25**	4.74 **9.55**	4.35 **8.45**	4.12 **7.85**	3.97 **7.46**	3.87 **7.19**	3.79 **7.00**	3.73 **6.84**	3.68 **6.71**	3.63 **6.62**	3.60 **6.54**	3.57 **6.47**	3.52 **6.35**	3.49 **6.27**	3.44 **6.15**	3.41 **6.07**	3.38 **5.98**	3.34 **5.90**	3.32 **5.85**	3.29 **5.78**	3.28 **5.75**	3.25 **5.70**	3.24 **5.67**	3.23 **5.65**	7
8	5.32 **11.26**	4.46 **8.65**	4.07 **7.59**	3.84 **7.01**	3.69 **6.63**	3.58 **6.37**	3.50 **6.19**	3.44 **6.03**	3.39 **5.91**	3.34 **5.82**	3.31 **5.74**	3.28 **5.67**	3.23 **5.56**	3.20 **5.48**	3.15 **5.36**	3.12 **5.28**	3.08 **5.20**	3.05 **5.11**	3.03 **5.06**	3.00 **5.00**	2.98 **4.96**	2.96 **4.91**	2.94 **4.88**	2.93 **4.86**	8
9	5.12 **10.56**	4.26 **8.02**	3.86 **6.99**	3.63 **6.42**	3.48 **6.06**	3.37 **5.80**	3.29 **5.62**	3.23 **5.47**	3.18 **5.35**	3.13 **5.26**	3.10 **5.18**	3.07 **5.11**	3.02 **5.00**	2.98 **4.92**	2.93 **4.80**	2.90 **4.73**	2.86 **4.64**	2.82 **4.56**	2.80 **4.51**	2.77 **4.45**	2.76 **4.41**	2.73 **4.36**	2.72 **4.33**	2.71 **4.31**	9
10	4.96 **10.04**	4.10 **7.56**	3.71 **6.55**	3.48 **5.99**	3.33 **5.64**	3.22 **5.39**	3.14 **5.21**	3.07 **5.06**	3.02 **4.95**	2.97 **4.85**	2.94 **4.78**	2.91 **4.71**	2.86 **4.60**	2.82 **4.52**	2.77 **4.41**	2.74 **4.33**	2.70 **4.25**	2.67 **4.17**	2.64 **4.12**	2.61 **4.05**	2.59 **4.01**	2.56 **3.96**	2.55 **3.93**	2.54 **3.91**	10
11	4.84 **9.65**	3.98 **7.20**	3.59 **6.22**	3.36 **5.67**	3.20 **5.32**	3.09 **5.07**	3.01 **4.88**	2.95 **4.74**	2.90 **4.63**	2.86 **4.54**	2.82 **4.46**	2.79 **4.40**	2.74 **4.29**	2.70 **4.21**	2.65 **4.10**	2.61 **4.02**	2.57 **3.94**	2.53 **3.86**	2.50 **3.80**	2.47 **3.74**	2.45 **3.70**	2.42 **3.66**	2.41 **3.62**	2.40 **3.60**	11
12	4.75 **9.33**	3.88 **6.93**	3.49 **5.95**	3.26 **5.41**	3.11 **5.06**	3.00 **4.82**	2.92 **4.65**	2.85 **4.50**	2.80 **4.39**	2.76 **4.30**	2.72 **4.22**	2.69 **4.16**	2.64 **4.05**	2.60 **3.98**	2.54 **3.86**	2.50 **3.78**	2.46 **3.70**	2.42 **3.61**	2.40 **3.56**	2.36 **3.49**	2.35 **3.46**	2.32 **3.41**	2.31 **3.38**	2.30 **3.36**	12
13	4.67 **9.07**	3.80 **6.70**	3.41 **5.74**	3.18 **5.20**	3.02 **4.86**	2.92 **4.62**	2.84 **4.44**	2.77 **4.30**	2.72 **4.19**	2.67 **4.10**	2.63 **4.02**	2.60 **3.96**	2.55 **3.85**	2.51 **3.78**	2.46 **3.67**	2.42 **3.59**	2.38 **3.51**	2.34 **3.42**	2.32 **3.37**	2.28 **3.30**	2.26 **3.27**	2.24 **3.21**	2.22 **3.18**	2.21 **3.16**	13

*Lightface type = 5%; boldface type = 1%.

†df in denominator.

TABLE IX (continued)

ν_2†	1	2	3	4	5	6	7	8	9	10	11	12	14	16	20	24	30	40	50	75	100	200	500	∞	ν_2†
	ν_1 df in Numerator																								
14	4.60	3.74	3.34	3.11	2.96	2.85	2.77	2.70	2.65	2.60	2.56	2.53	2.48	2.44	2.39	2.35	2.31	2.27	2.24	2.21	2.19	2.16	2.14	2.13	14
	8.86	**6.51**	**5.56**	**5.03**	**4.69**	**4.46**	**4.28**	**4.14**	**4.03**	**3.94**	**3.86**	**3.80**	**3.70**	**3.62**	**3.51**	**3.43**	**3.34**	**3.26**	**3.21**	**3.14**	**3.11**	**3.06**	**3.02**	**3.00**	
15	4.54	3.68	3.29	3.06	2.90	2.79	2.70	2.64	2.59	2.55	2.51	2.48	2.43	2.39	2.33	2.29	2.25	2.21	2.18	2.15	2.12	2.10	2.08	2.07	15
	8.68	**6.36**	**5.42**	**4.89**	**4.56**	**4.32**	**4.14**	**4.00**	**3.89**	**3.80**	**3.73**	**3.67**	**3.56**	**3.48**	**3.36**	**3.29**	**3.20**	**3.12**	**3.07**	**3.00**	**2.97**	**2.92**	**2.89**	**2.87**	
16	4.49	3.63	3.24	3.01	2.85	2.74	2.66	2.59	2.54	2.49	2.45	2.42	2.37	2.33	2.28	2.24	2.20	2.16	2.13	2.09	2.07	2.04	2.02	2.01	16
	8.53	**6.23**	**5.29**	**4.77**	**4.44**	**4.20**	**4.03**	**3.89**	**3.78**	**3.69**	**3.61**	**3.55**	**3.45**	**3.37**	**3.25**	**3.18**	**3.10**	**3.01**	**2.96**	**2.98**	**2.86**	**2.80**	**2.77**	**2.75**	
17	4.45	3.59	3.20	2.96	2.81	2.70	2.62	2.55	2.50	2.45	2.41	2.38	2.33	2.29	2.23	2.19	2.15	2.11	2.08	2.04	2.02	1.99	1.97	1.96	17
	8.40	**6.11**	**5.18**	**4.67**	**4.34**	**4.10**	**3.93**	**3.79**	**3.68**	**3.59**	**3.52**	**3.45**	**3.35**	**3.27**	**3.16**	**3.08**	**3.00**	**2.92**	**2.86**	**2.79**	**2.76**	**2.70**	**2.67**	**2.65**	
18	4.41	3.55	3.16	2.93	2.77	3.66	2.58	2.51	2.46	2.41	2.37	2.34	2.29	2.25	2.19	2.15	2.11	2.07	2.04	2.00	1.98	1.95	1.93	1.92	18
	8.28	**6.01**	**5.09**	**4.58**	**4.25**	**4.01**	**3.85**	**3.71**	**3.60**	**3.51**	**3.44**	**3.37**	**3.27**	**3.19**	**3.07**	**3.00**	**2.91**	**2.83**	**2.78**	**2.71**	**2.68**	**2.62**	**2.59**	**2.57**	
19	4.38	3.52	3.13	2.90	2.74	2.63	2.55	2.48	2.43	2.38	2.34	2.31	2.26	2.21	2.15	2.11	2.07	2.02	2.00	1.96	1.94	1.91	1.90	1.88	19
	8.18	**5.93**	**5.01**	**4.50**	**4.17**	**3.94**	**3.77**	**3.63**	**3.52**	**3.43**	**3.36**	**3.30**	**3.19**	**3.12**	**3.00**	**2.92**	**2.84**	**2.76**	**2.70**	**2.63**	**2.60**	**2.54**	**2.51**	**2.49**	
20	4.35	3.49	3.10	2.87	2.71	2.60	2.52	2.45	2.40	2.35	2.31	2.28	2.23	2.18	2.12	2.08	2.04	1.99	1.96	1.92	1.90	1.87	1.85	1.84	20
	8.10	**5.85**	**4.94**	**4.43**	**4.10**	**3.87**	**3.71**	**3.56**	**3.45**	**3.37**	**3.30**	**3.23**	**3.13**	**3.05**	**2.94**	**2.86**	**2.77**	**2.69**	**2.63**	**2.56**	**2.53**	**2.47**	**2.44**	**2.42**	
21	4.32	3.47	3.07	2.84	2.68	2.57	2.49	2.42	2.37	2.32	2.28	2.25	2.20	2.15	2.09	2.05	2.00	1.96	1.93	1.89	1.87	1.84	1.82	1.81	21
	8.02	**5.78**	**4.87**	**4.37**	**4.04**	**3.81**	**3.65**	**3.51**	**3.40**	**3.31**	**3.24**	**3.17**	**3.07**	**2.99**	**2.88**	**2.80**	**2.72**	**2.63**	**2.58**	**2.51**	**2.47**	**2.42**	**2.38**	**2.36**	
22	4.30	3.44	3.05	2.82	2.66	2.55	2.47	2.40	2.35	2.30	2.26	2.23	2.18	2.13	2.07	2.03	1.98	1.93	1.91	1.87	1.84	1.81	1.80	1.78	22
	7.94	**5.72**	**4.82**	**4.31**	**3.99**	**3.76**	**3.59**	**3.45**	**3.35**	**3.26**	**3.18**	**3.12**	**3.02**	**2.94**	**2.83**	**2.75**	**2.67**	**2.58**	**2.53**	**2.46**	**2.42**	**2.37**	**2.33**	**2.31**	
23	4.28	3.42	3.03	2.80	2.64	2.53	2.45	2.38	2.32	2.28	2.24	2.20	2.14	2.10	2.04	2.00	1.96	1.91	1.88	1.84	1.82	1.79	1.77	1.76	23
	7.88	**5.66**	**4.76**	**4.26**	**3.94**	**3.71**	**3.54**	**3.41**	**3.30**	**3.21**	**3.14**	**3.07**	**2.97**	**2.89**	**2.78**	**2.70**	**2.62**	**2.53**	**2.48**	**2.41**	**2.37**	**2.32**	**2.28**	**2.26**	
24	4.26	3.40	3.01	2.78	2.62	2.51	2.43	2.36	2.30	2.26	2.22	2.18	2.13	2.09	2.02	1.98	1.94	1.89	1.86	1.82	1.80	1.76	1.74	1.73	24
	7.82	**5.61**	**4.72**	**4.22**	**3.90**	**3.67**	**3.50**	**3.36**	**3.25**	**3.17**	**3.09**	**3.03**	**2.93**	**2.85**	**2.74**	**2.66**	**2.58**	**2.49**	**2.44**	**2.36**	**2.33**	**2.27**	**2.23**	**2.21**	
25	4.24	3.38	2.99	2.76	2.60	2.49	2.41	2.34	2.28	2.24	2.20	2.16	2.11	2.06	2.00	1.96	1.92	1.87	1.84	1.80	1.77	1.74	1.72	1.71	25
	7.77	**5.57**	**4.68**	**4.18**	**3.86**	**3.63**	**3.46**	**3.32**	**3.21**	**3.13**	**3.05**	**2.99**	**2.89**	**2.81**	**2.70**	**2.62**	**2.54**	**2.45**	**2.40**	**2.32**	**2.29**	**2.23**	**2.19**	**2.17**	
26	4.22	3.37	2.98	2.74	2.59	2.47	2.39	2.32	2.27	2.22	2.18	2.15	2.10	2.05	1.99	1.95	1.90	1.85	1.82	1.78	1.76	1.72	1.70	1.69	26
	7.72	**5.53**	**4.64**	**4.14**	**3.82**	**3.59**	**3.42**	**3.29**	**3.17**	**3.09**	**3.02**	**2.96**	**2.86**	**2.77**	**2.66**	**2.58**	**2.50**	**2.41**	**2.36**	**2.28**	**2.25**	**2.19**	**2.15**	**2.13**	

†df in denominator.

TABLE IX (continued)

ν_2†	ν_1 df in Numerator																								ν_2†
	1	2	3	4	5	6	7	8	9	10	11	12	14	16	20	24	30	40	50	75	100	200	500	∞	
27	4.21 **7.68**	3.35 **5.49**	2.96 **4.60**	2.73 **4.11**	2.57 **3.79**	2.46 **3.56**	2.37 **3.39**	2.30 **3.26**	2.25 **3.14**	2.20 **3.06**	2.16 **2.98**	2.13 **2.93**	2.08 **2.83**	2.03 **2.74**	1.97 **2.63**	1.93 **2.55**	1.88 **2.47**	1.84 **2.38**	1.80 **2.33**	1.76 **2.25**	1.74 **2.21**	1.71 **2.16**	1.68 **2.12**	1.67 **2.10**	27
28	4.20 **7.64**	3.34 **5.45**	2.95 **4.57**	2.71 **4.07**	2.56 **3.76**	2.44 **3.53**	2.36 **3.36**	2.29 **3.23**	2.24 **3.11**	2.19 **3.03**	2.15 **2.95**	2.12 **2.90**	2.06 **2.80**	2.02 **2.71**	1.96 **2.60**	1.91 **2.52**	1.87 **2.44**	1.81 **2.35**	1.78 **2.30**	1.75 **2.22**	1.72 **2.18**	1.69 **2.13**	1.67 **2.09**	1.65 **2.06**	28
29	4.18 **7.60**	3.33 **5.42**	2.93 **4.54**	2.70 **4.04**	2.54 **3.73**	2.43 **3.50**	2.35 **3.33**	2.28 **3.20**	2.22 **3.08**	2.18 **3.00**	2.14 **2.92**	2.10 **2.87**	2.05 **2.77**	2.00 **2.68**	1.94 **2.57**	1.90 **2.49**	1.85 **2.41**	1.80 **2.32**	1.77 **2.27**	1.73 **2.19**	1.71 **2.15**	1.68 **2.10**	1.65 **2.06**	1.64 **2.03**	29
30	4.17 **7.56**	3.32 **5.39**	2.92 **4.51**	2.69 **4.02**	2.53 **3.70**	2.42 **3.47**	2.34 **3.30**	2.27 **3.17**	2.21 **3.06**	2.16 **2.98**	2.12 **2.90**	2.09 **2.84**	2.04 **2.74**	1.99 **2.66**	1.93 **2.55**	1.89 **2.47**	1.84 **2.38**	1.79 **2.29**	1.76 **2.24**	1.72 **2.16**	1.69 **2.13**	1.66 **2.07**	1.64 **2.03**	1.62 **2.01**	30
32	4.15 **7.50**	3.30 **5.34**	2.90 **4.46**	2.67 **3.97**	2.51 **3.66**	2.40 **3.42**	2.32 **3.25**	2.25 **3.12**	2.19 **3.01**	2.14 **2.94**	2.10 **2.86**	2.07 **2.80**	2.02 **2.70**	1.97 **2.62**	1.91 **2.51**	1.86 **2.42**	1.82 **2.34**	1.76 **2.25**	1.74 **2.20**	1.69 **2.12**	1.67 **2.08**	1.64 **2.02**	1.61 **1.98**	1.59 **1.96**	32
34	4.13 **7.44**	3.28 **5.29**	2.88 **4.42**	2.65 **3.93**	2.49 **3.61**	2.38 **3.38**	2.30 **3.21**	2.23 **3.08**	2.17 **2.97**	2.12 **2.89**	2.08 **2.82**	2.05 **2.76**	2.00 **2.66**	1.95 **2.58**	1.89 **2.47**	1.84 **2.38**	1.80 **2.30**	1.74 **2.21**	1.71 **2.15**	1.67 **2.08**	1.64 **2.04**	1.61 **1.98**	1.59 **1.94**	1.57 **1.91**	34
36	4.11 **7.39**	3.26 **5.25**	2.86 **4.38**	2.63 **3.89**	2.48 **3.58**	2.36 **3.35**	2.28 **3.18**	2.21 **3.04**	2.15 **2.94**	2.10 **2.86**	2.06 **2.78**	2.03 **2.72**	1.98 **2.62**	1.93 **2.54**	1.87 **2.43**	1.82 **2.35**	1.78 **2.26**	1.72 **2.17**	1.69 **2.12**	1.65 **2.04**	1.62 **2.00**	1.59 **1.94**	1.56 **1.90**	1.55 **1.87**	36
38	4.10 **7.35**	3.25 **5.21**	2.85 **4.34**	2.62 **3.86**	2.46 **3.54**	2.35 **3.32**	2.26 **3.15**	2.19 **3.02**	2.14 **2.91**	2.09 **2.82**	2.05 **2.75**	2.02 **2.69**	1.96 **2.59**	1.92 **2.51**	1.85 **2.40**	1.80 **2.32**	1.76 **2.22**	1.71 **2.14**	1.67 **2.08**	1.63 **2.00**	1.60 **1.97**	1.57 **1.90**	1.54 **1.86**	1.53 **1.84**	38
40	4.08 **7.31**	3.23 **5.18**	2.84 **4.31**	2.61 **3.83**	2.45 **3.51**	2.34 **3.29**	2.25 **3.12**	2.18 **2.99**	2.12 **2.88**	2.07 **2.80**	2.04 **2.73**	2.00 **2.66**	1.95 **2.56**	1.90 **2.49**	1.84 **2.37**	1.79 **2.29**	1.74 **2.20**	1.69 **2.11**	1.66 **2.05**	1.61 **1.97**	1.59 **1.94**	1.55 **1.88**	1.53 **1.84**	1.51 **1.81**	40
42	4.07 **7.27**	3.22 **5.15**	2.83 **4.29**	2.59 **3.80**	2.44 **3.49**	2.32 **3.26**	2.24 **3.10**	2.17 **2.96**	2.11 **2.86**	2.06 **2.77**	2.02 **2.70**	1.99 **2.64**	1.94 **2.54**	1.89 **2.46**	1.82 **2.35**	1.78 **2.26**	1.73 **2.17**	1.68 **2.08**	1.64 **2.02**	1.60 **1.94**	1.57 **1.91**	1.54 **1.85**	1.51 **1.80**	1.49 **1.78**	42
44	4.06 **7.24**	3.21 **5.12**	2.82 **4.26**	2.58 **3.78**	2.43 **3.46**	2.31 **3.24**	2.23 **3.07**	2.16 **2.94**	2.10 **2.84**	2.05 **2.75**	2.01 **2.68**	1.98 **2.62**	1.92 **2.52**	1.88 **2.44**	1.81 **2.32**	1.76 **2.24**	1.72 **2.15**	1.66 **2.06**	1.63 **2.00**	1.58 **1.92**	1.56 **1.88**	1.52 **1.82**	1.50 **1.78**	1.48 **1.75**	44
46	4.05 **7.21**	3.20 **5.10**	2.81 **4.24**	2.57 **3.76**	2.42 **3.44**	2.30 **3.22**	2.22 **3.05**	2.14 **2.92**	2.09 **2.82**	2.04 **2.73**	2.00 **2.66**	1.97 **2.60**	1.91 **2.50**	1.87 **2.42**	1.80 **2.30**	1.75 **2.22**	1.71 **2.13**	1.65 **2.04**	1.62 **1.98**	1.57 **1.90**	1.54 **1.86**	1.51 **1.80**	1.48 **1.76**	1.46 **1.72**	46
48	4.04 **7.19**	3.19 **5.08**	2.80 **4.22**	2.56 **3.74**	2.41 **3.42**	2.30 **3.20**	2.21 **3.04**	2.14 **2.90**	2.08 **2.80**	2.03 **2.71**	1.99 **2.64**	1.96 **2.58**	1.90 **2.48**	1.86 **2.40**	1.79 **2.28**	1.74 **2.20**	1.70 **2.11**	1.64 **2.02**	1.61 **1.96**	1.56 **1.88**	1.53 **1.84**	1.50 **1.78**	1.47 **1.73**	1.45 **1.70**	48

†df in denominator.

TABLE IX (continued)

ν_2†	ν_1 df in Numerator 1	2	3	4	5	6	7	8	9	10	11	12	14	16	20	24	30	40	50	75	100	200	500	∞	ν_2†
50	4.03	3.18	2.79	2.56	2.40	2.29	2.20	2.13	2.07	2.02	1.98	1.95	1.90	1.85	1.78	1.74	1.69	1.63	1.60	1.55	1.52	1.48	1.46	1.44	50
	7.17	**5.06**	**4.20**	**3.72**	**3.41**	**3.18**	**3.02**	**2.88**	**2.78**	**2.70**	**2.62**	**2.56**	**2.46**	**2.39**	**2.26**	**2.18**	**2.10**	**2.00**	**1.94**	**1.86**	**1.82**	**1.76**	**1.71**	**1.68**	
55	4.02	3.17	2.78	2.54	2.38	2.27	2.18	2.11	2.05	2.00	1.97	1.93	1.88	1.83	1.76	1.72	1.67	1.61	1.58	1.52	1.50	1.46	1.43	1.41	55
	7.12	**5.01**	**4.16**	**3.68**	**3.37**	**3.15**	**2.98**	**2.85**	**2.75**	**2.66**	**2.59**	**2.53**	**2.43**	**2.35**	**2.23**	**2.15**	**2.06**	**1.96**	**1.90**	**1.82**	**1.78**	**1.71**	**1.66**	**1.64**	
60	4.00	3.15	2.76	2.52	2.37	2.25	2.17	2.10	2.04	1.99	1.95	1.92	1.86	1.81	1.75	1.70	1.65	1.59	1.56	1.50	1.48	1.44	1.41	1.39	60
	7.08	**4.98**	**4.13**	**3.65**	**3.34**	**3.12**	**2.95**	**2.82**	**2.72**	**2.63**	**2.56**	**2.50**	**2.40**	**2.32**	**2.20**	**2.12**	**2.03**	**1.93**	**1.87**	**1.79**	**1.74**	**1.68**	**1.63**	**1.60**	
65	3.99	3.14	2.75	2.51	2.36	2.24	2.15	2.08	2.02	1.98	1.94	1.90	1.85	1.80	1.73	1.68	1.63	1.57	1.54	1.49	1.46	1.42	1.39	1.37	65
	7.04	**4.95**	**4.10**	**3.62**	**3.31**	**3.09**	**2.93**	**2.79**	**2.70**	**2.61**	**2.54**	**2.47**	**2.37**	**2.30**	**2.18**	**2.09**	**2.00**	**1.90**	**1.84**	**1.76**	**1.71**	**1.64**	**1.60**	**1.56**	
70	3.98	3.13	2.74	2.50	2.35	2.23	2.14	2.07	2.01	1.97	1.93	1.89	1.84	1.79	1.72	1.67	1.62	1.56	1.53	1.47	1.45	1.40	1.37	1.35	70
	7.01	**4.92**	**4.08**	**3.60**	**3.29**	**3.07**	**2.91**	**2.77**	**2.67**	**2.59**	**2.51**	**2.45**	**2.35**	**2.28**	**2.15**	**2.07**	**1.98**	**1.88**	**1.82**	**1.74**	**1.69**	**1.62**	**1.56**	**1.53**	
80	3.96	3.11	2.72	2.48	2.33	2.21	2.12	2.05	1.99	1.95	1.91	1.88	1.82	1.77	1.70	1.65	1.60	1.54	1.51	1.45	1.42	1.38	1.35	1.32	80
	6.96	**4.88**	**4.04**	**3.56**	**3.25**	**3.04**	**2.87**	**2.74**	**2.64**	**2.55**	**2.48**	**2.41**	**2.32**	**2.24**	**2.11**	**2.03**	**1.94**	**1.84**	**1.78**	**1.70**	**1.65**	**1.57**	**1.52**	**1.49**	
100	3.94	3.09	2.70	2.46	2.30	2.19	2.10	2.03	1.97	1.92	1.88	1.85	1.79	1.75	1.68	1.63	1.57	1.51	1.48	1.42	1.39	1.34	1.30	1.28	100
	6.90	**4.82**	**3.98**	**3.51**	**3.20**	**2.99**	**2.82**	**2.69**	**2.59**	**2.51**	**2.43**	**2.36**	**2.26**	**2.19**	**2.06**	**1.98**	**1.89**	**1.79**	**1.73**	**1.64**	**1.59**	**1.51**	**1.46**	**1.43**	
125	3.92	3.07	2.68	2.44	2.29	2.17	2.08	2.01	1.95	1.90	1.86	1.83	1.77	1.72	1.65	1.60	1.55	1.49	1.45	1.39	1.36	1.31	1.27	1.25	125
	6.84	**4.78**	**3.94**	**3.47**	**3.17**	**2.95**	**2.79**	**2.65**	**2.56**	**2.47**	**2.40**	**2.33**	**2.23**	**2.15**	**2.03**	**1.94**	**1.85**	**1.75**	**1.68**	**1.59**	**1.54**	**1.46**	**1.40**	**1.37**	
150	3.91	3.06	2.67	2.43	2.27	2.16	2.07	2.00	1.94	1.89	1.85	1.82	1.76	1.71	1.64	1.59	1.54	1.47	1.44	1.37	1.34	1.29	1.25	1.22	150
	6.81	**4.75**	**3.91**	**3.44**	**3.14**	**2.92**	**2.76**	**2.62**	**2.53**	**2.44**	**2.37**	**2.30**	**2.20**	**2.12**	**2.00**	**1.91**	**1.83**	**1.72**	**1.66**	**1.56**	**1.51**	**1.43**	**1.37**	**1.33**	
200	3.89	3.04	2.65	2.41	2.26	2.14	2.05	1.98	1.92	1.87	1.83	1.80	1.74	1.69	1.62	1.57	1.52	1.45	1.42	1.35	1.32	1.26	1.22	1.19	200
	6.76	**4.71**	**3.88**	**3.41**	**3.11**	**2.90**	**2.73**	**2.60**	**2.50**	**2.41**	**2.34**	**2.28**	**2.17**	**2.09**	**1.97**	**1.88**	**1.79**	**1.69**	**1.62**	**1.53**	**1.48**	**1.39**	**1.33**	**1.28**	
400	3.86	3.02	2.62	2.39	2.23	2.12	2.03	1.96	1.90	1.85	1.81	1.78	1.72	1.67	1.60	1.54	1.49	1.42	1.38	1.32	1.28	1.22	1.16	1.13	400
	6.70	**4.66**	**3.83**	**3.36**	**3.06**	**2.85**	**2.69**	**2.55**	**2.46**	**2.37**	**2.29**	**2.23**	**2.12**	**2.04**	**1.92**	**1.84**	**1.74**	**1.64**	**1.57**	**1.47**	**1.42**	**1.32**	**1.24**	**1.19**	
1000	3.85	3.00	2.61	2.38	2.22	2.10	2.02	1.95	1.89	1.84	1.80	1.76	1.70	1.65	1.58	1.53	1.47	1.41	1.36	1.30	1.26	1.19	1.13	1.08	1000
	6.66	**4.62**	**3.80**	**3.34**	**3.04**	**2.82**	**2.66**	**2.53**	**2.43**	**2.34**	**2.26**	**2.20**	**2.09**	**2.01**	**1.89**	**1.81**	**1.71**	**1.61**	**1.54**	**1.44**	**1.38**	**1.28**	**1.19**	**1.11**	
∞	3.84	2.99	2.60	2.37	2.21	2.09	2.01	1.94	1.88	1.83	1.79	1.75	1.69	1.64	1.57	1.52	1.46	1.40	1.35	1.28	1.24	1.17	1.11	1.00	∞
	6.63	**4.60**	**3.78**	**3.32**	**3.02**	**2.80**	**2.64**	**2.51**	**2.41**	**2.32**	**2.24**	**2.18**	**2.07**	**1.99**	**1.87**	**1.79**	**1.69**	**1.59**	**1.52**	**1.41**	**1.36**	**1.25**	**1.15**	**1.00**	

†df in denominator.

Source: Reprinted by permission from *Statistical Methods* by George W. Snedecor and William G. Cochran, 6th edition, © 1967 by Iowa State University Press, Ames, Iowa, Table A 14, Part 1.

TABLE X

Random Minutes and Days

Line Number	1	2	3	4	5	6	7	8	9	10
1	627	310	424	1200	443	949	1105	302	530	151
2	345	736	619	102	241	301	712	606	734	737
3	730	258	459	812	152	309	1139	556	809	553
4	912	402	658	807	851	441	1138	549	303	101
5	948	310	937	204	757	232	305	616	712	1229
6	133	533	657	642	1016	1208	647	250	1111	644
7	818	911	258	913	721	154	628	403	1239	945
8	109	635	534	1105	335	519	436	1207	423	805
9	330	231	336	431	235	931	451	309	136	851
10	805	630	355	437	1128	630	330	1121	1039	859
11	728	1027	704	335	1203	1238	1149	437	904	446
12	204	218	512	257	1043	203	342	131	924	342
13	217	1008	802	831	619	717	1116	1147	431	710
14	654	1154	923	556	148	656	1008	716	806	324
15	741	704	739	1039	858	1259	843	734	728	1039
16	1215	229	942	1120	636	755	131	414	346	1048
17	818	634	1005	1147	1249	524	309	159	634	1202
18	539	514	330	843	858	909	804	452	502	1152
19	834	1238	1245	824	653	1141	918	150	510	239
20	825	328	449	638	405	207	852	222	624	304
21	641	610	548	1056	645	924	1024	716	905	528
22	858	748	840	726	813	818	1129	943	941	801
23	1150	254	530	1130	939	243	747	1153	107	1139
24	547	744	336	1013	929	459	1152	329	334	1203
25	141	845	708	728	1015	726	855	123	1008	822
26	1254	823	748	1034	343	550	110	628	1007	1019
27	1158	111	658	1023	120	354	244	500	239	600
28	727	1117	1134	754	313	1115	1010	1101	101	506
29	554	924	1132	1030	830	925	207	125	709	404
30	123	131	1112	744	946	604	927	350	1225	131
31	854	1053	1101	842	617	504	231	517	559	1222
32	548	610	1207	348	1122	1043	351	1050	1219	307
33	637	706	752	248	135	1057	655	852	1037	510
34	554	1124	730	541	1003	1258	825	1028	322	237
35	531	754	1154	929	653	734	342	459	740	453
36	1215	834	513	1212	756	400	751	859	249	1126
37	1023	416	541	217	1223	1206	159	156	856	121
38	603	440	746	454	340	433	1041	331	1252	1155
39	1235	707	1028	1248	722	146	501	109	130	338
40	1049	317	423	925	540	124	839	441	1050	1200

TABLE X (continued)

Line Number	1	2	3	4	5	6	7	8	9	10
1	1022	247	231	827	1126	1258	515	1140	350	406
2	813	432	532	754	551	546	414	835	807	824
3	1027	1058	820	659	943	708	1242	639	1232	1019
4	759	1237	950	1201	812	1000	1047	131	737	312
5	845	640	1238	640	342	751	227	615	218	1120
6	1217	1226	333	949	620	617	258	1252	150	156
7	506	335	228	457	134	1142	645	727	659	844
8	102	221	342	904	111	858	412	1147	402	208
9	439	649	812	1026	738	754	1030	859	554	1128
10	842	504	101	1027	105	144	456	311	1126	314
11	544	610	649	128	639	1023	613	1023	944	335
12	413	256	701	642	327	1110	1216	448	530	906
13	931	412	1027	219	639	1241	653	458	829	459
14	345	327	1237	724	135	659	558	329	919	131
15	549	457	953	1141	1158	1046	956	655	521	523
16	941	533	829	1130	845	141	351	117	1225	926
17	1118	1059	739	437	911	232	147	510	503	1047
18	1224	934	1225	349	302	400	122	1020	516	1154
19	821	336	1106	704	314	1153	222	719	128	657
20	946	649	132	851	439	916	819	520	815	320
21	329	417	416	310	450	410	204	120	800	757
22	1221	538	112	723	1033	733	532	219	412	714
23	727	409	814	550	1254	103	254	604	1123	655
24	941	449	1127	424	1024	132	105	1035	237	948
25	1218	534	1118	318	846	512	1127	328	1021	1218
26	706	352	1004	428	343	1000	929	1044	1244	409
27	201	746	432	740	740	749	600	221	348	1051
28	859	107	1051	859	107	524	859	1125	938	454
29	641	137	716	1118	439	909	1120	544	905	812
30	530	1141	205	1100	1231	825	1137	742	312	742
31	312	742	908	332	438	249	553	458	219	336
32	327	815	1138	639	421	935	423	304	141	1253
33	400	510	421	117	943	933	1055	1224	851	519
34	1101	638	1005	1048	548	207	705	819	1027	119
35	1027	119	442	228	228	502	151	1247	128	935
36	357	1215	1118	233	902	450	210	947	719	856
37	624	1221	636	1248	716	245	107	833	820	744
38	320	648	840	736	113	322	650	1021	1024	1247
39	318	453	301	648	726	651	1200	224	629	333
40	1032	359	700	254	655	243	1239	134	1052	747

TABLE X (continued)

Line Number	1	2	3	4	5	6	7	8	9	10
1	334	233	1130	609	107	111	847	222	451	805
2	410	523	1034	353	907	827	440	1009	130	931
3	606	554	817	450	707	659	152	428	529	1030
4	145	645	451	818	754	656	140	516	724	1224
5	551	820	923	614	458	436	151	245	315	442
6	318	941	305	1205	838	1002	959	323	311	927
7	604	616	754	1208	1205	301	150	1014	210	853
8	308	1113	145	749	330	617	614	558	557	1138
9	1226	824	220	812	722	1113	859	116	547	527
10	739	450	453	1209	542	141	400	1016	1124	600
11	639	1130	203	907	818	625	907	1226	106	1233
12	715	953	526	1239	820	948	815	1215	1039	349
13	433	850	314	248	1127	345	1033	403	213	941
14	258	220	310	900	1147	448	706	513	623	204
15	808	730	356	410	1055	323	322	642	654	325
16	516	931	346	243	707	1209	1011	511	729	954
17	207	921	1116	312	1027	436	1207	212	1125	1133
18	756	252	917	116	747	927	715	748	1017	1111
19	243	707	1011	715	200	229	912	1153	1002	249
20	648	840	511	748	243	311	634	326	950	149
21	453	700	729	1017	1030	932	629	1130	836	126
22	359	514	954	1111	749	536	1113	844	240	941
23	522	123	207	1018	120	1101	100	734	649	523
24	107	431	921	918	233	551	224	910	834	526
25	907	456	1116	840	358	143	1003	808	557	625
26	707	302	312	924	1220	944	1037	1227	1045	552
27	754	1218	1027	347	738	124	831	1044	437	1234
28	458	943	436	913	647	157	206	827	909	841
29	838	757	1207	756	403	556	346	410	738	757
30	1205	816	212	1243	232	747	909	825	243	258
31	330	1237	1125	1126	928	306	314	450	1258	205
32	722	121	1133	1043	214	155	1159	552	1254	105
33	542	913	756	829	354	906	1024	949	541	429
34	818	511	252	856	731	1259	621	234	212	349
35	820	140	917	525	1118	618	239	948	1052	552
36	1127	452	116	733	1133	659	415	520	545	606
37	1147	939	747	332	427	244	1202	615	700	929
38	1055	1114	927	705	854	114	154	1108	1256	619
39	1116	443	107	839	921	351	215	231	856	1142
40	554	840	353	448	212	1030	353	1025	538	849

TABLE X (continued)

Line Number	1	2	3	4	5	6	7	8	9	10
1	721	650	802	753	416	554	802	749	816	342
2	705	145	706	1054	225	702	228	158	219	346
3	622	413	1229	918	700	1136	1242	338	832	647
4	849	145	930	132	1043	1159	805	210	256	817
5	223	1001	148	548	323	1053	518	110	934	322
6	147	949	1146	628	1228	1119	1234	834	743	1026
7	1251	855	237	408	117	205	234	546	745	129
8	810	135	600	1146	438	312	1213	958	507	157
9	558	511	1007	419	1034	540	1125	1020	135	844
10	1251	705	953	1247	942	452	1213	613	1241	318
11	222	1145	1225	450	530	453	1234	759	945	1022
12	548	441	458	753	750	1201	420	111	808	526
13	137	200	904	919	305	1226	504	1022	753	1100
14	800	236	559	157	859	609	656	134	1052	1223
15	935	1157	425	1147	713	136	631	755	720	1000
16	739	942	1200	1124	1258	1108	440	1115	714	659
17	906	408	1017	241	849	750	924	153	601	735
18	236	650	934	1241	531	844	535	1149	407	119
19	1034	358	635	1226	1103	840	801	509	519	245
20	157	712	1102	940	1254	157	1240	941	137	1149
21	1022	535	851	608	411	1218	1056	832	151	844
22	1131	1232	134	1200	250	1010	317	945	847	1052
23	925	107	618	702	735	907	713	203	233	940
24	734	356	955	902	729	126	608	820	742	921
25	1146	930	537	557	1044	139	1144	1232	235	222
26	927	146	357	1232	555	1104	538	802	548	224
27	1218	216	859	421	317	344	641	938	1130	115
28	236	457	942	1042	1034	1157	936	344	154	334
29	155	754	915	757	236	512	821	934	119	1209
30	1027	957	234	730	113	1119	747	1219	818	1037
31	800	303	205	406	316	1114	500	242	642	415
32	1252	143	734	159	543	219	311	118	538	220
33	1214	127	1204	1135	431	709	442	454	937	347
34	836	343	1150	817	600	701	843	807	1136	640
35	215	325	704	301	451	1140	125	436	1103	1120
36	317	845	1120	429	1034	111	943	357	758	1228
37	857	319	303	101	114	332	129	912	606	811
38	821	438	442	1237	1242	217	1136	755	155	348
39	1147	329	1007	135	522	729	835	714	601	303
40	103	943	651	225	228	1003	742	633	1108	751

TABLE X (continued)

Line Number	1	2	3	4	5	6	7	8	9	10
1	1219	813	247	1213	1030	334	1025	301	615	620
2	611	627	139	941	1159	939	1019	1222	1228	1147
3	810	846	1041	1051	1102	218	122	1004	1058	850
4	1230	242	424	713	1218	620	205	1238	645	519
5	628	653	348	259	244	120	1218	1145	346	151
6	1126	401	841	208	139	735	251	839	1056	714
7	512	1229	506	557	809	636	143	1227	1213	132
8	135	510	820	1217	1033	259	607	740	740	247
9	1227	716	611	105	942	326	822	810	512	743
10	1044	1243	246	734	104	1128	705	151	727	410
11	541	952	846	1257	105	405	718	659	217	314
12	551	700	307	519	507	1256	1052	124	920	1242
13	510	319	551	703	314	248	802	945	942	233
14	320	1153	204	559	901	247	326	723	1126	514
15	146	344	1007	133	120	654	152	535	735	641
16	1034	143	843	955	324	617	144	237	1109	659
17	1145	604	825	544	1248	150	1119	229	540	447
18	440	139	610	1105	216	436	658	338	630	226
19	840	1158	1011	259	230	129	328	1108	958	313
20	409	1001	426	207	948	203	933	1004	159	1216
21	915	153	716	125	516	1005	602	255	542	409
22	355	1042	939	650	556	737	652	121	1236	214
23	1121	225	652	450	1028	859	1127	403	1241	629
24	515	324	914	313	1225	209	1022	619	830	604
25	939	901	1236	446	706	1229	104	1054	602	516
26	647	224	310	624	1010	1043	858	1005	304	613
27	1007	503	1154	1032	746	809	255	311	657	506
28	752	809	956	1137	115	600	1213	922	639	1153
29	500	506	1239	935	1212	518	631	101	619	445
30	519	1211	716	354	356	138	1048	1112	439	440
31	1218	108	339	309	1212	807	328	137	735	101
32	115	1245	130	102	738	1218	554	735	623	722
33	320	1048	509	955	932	553	820	441	401	1027
34	719	910	1136	344	1044	942	154	1057	846	959
35	1204	120	648	807	1054	451	708	707	239	850
36	407	1154	836	351	253	926	220	921	1124	1040
37	617	538	501	215	901	453	823	306	528	628
38	823	940	218	138	154	100	1107	1120	714	1244
39	711	944	1105	1050	151	655	104	859	119	402
40	619	643	1017	718	1254	716	543	804	606	1035

Source: A. C. Rosander, *Case Studies in Sample Design*, Marcel Dekker, Inc., New York, 1977. Reprinted from appendix by courtesy of Marcel Dekker, Inc.

Selected Annotated Bibliography of Books and Pamphlets

Note: With a few exceptions, most of the books listed below deal with quality control applied to manufacturing. However, much of the text of these books deals with basic principles which are directly applicable to service operations or can be readily adapted to service operations. Annotations tell something about the book. Names are arranged alphabetically.

General References

Irwin D. J. Bross, *Scientific Strategies to Save Your Life: A Statistical Approach to Primary Prevention*, Marcel Dekker, Inc., New York, 1981.
A well known biostatistician reports on his experiences in low level radiation research, cancer research, and other public health areas over the last 25 years. Very readable, challenging, and controversial to many, but contains refreshing discussions of basic concepts and ideas in scientific research and the conflicts which arise therein. Highly recommended.

Philip Crosby, *Quality is Free*, McGraw Hill Book Company, New York, 1979.

W. Edwards Deming, *Quality Productivity and Competitive Position*, Massachusetts Institute of Technology, Cambridge, 1982. See especially Chapter 11, ''Quality and the Consumer'' and Chapter 12, ''Quality and Productivity in Service Organizations.''

G. B. Gori, ''The Regulation of Carcinogenic Hazards,'' *Science*, Vol. 208, 1980, pp. 256-261.

J. T. Hagan and E. W. Karlin, ''Quality Management For Service Industries,'' ASQC Technical Conference, 1979.

J. T. Hagan, E. W. Karlin and L. H. Arrington, ''Quality Costs For Service Industries,'' ASQC Technical Conference, 1979.

James F. Halpin, *Zero Defects*, McGraw Hill Book Company, New York, 1966.

Howard L. Jones, ''QC Techniques for Inspecting Clerical Work'', *Industrial Quality Control*, Vol. 11, No. 8 (May 1955), p. 42.

Howard L. Jones, ''Sampling Plans for Verifying Clerical Work'', *Industrial Quality Control*, Vol. 3, No. 4 (January 1947), p. 3.

Howard L. Jones, ''Some Problems in Sampling Accounting Records'', *Industrial Quality Control*, Vol. 10, No. 3 (November 1953), p. 6.

Roger G. Langevin, ''Quality Control in Bank Operations,'' *37th Annual Congress Transactions*, American Society for Quality Control, Boston, 1983, pp. 131–135.

W. J. Latzko, *The Application of Quality Control to Banking*, American Society for Quality Control, Milwaukee. In press.

The New Development of ZD Programs, Japan Management Association, 1978.

''QC Circle Koryo,'' *QC Circle Headquarter*, June 1980.

Frank Scanlon, ''Cost Reduction Through Quality Management,'' ASQC Technical Conference, 1980.

Zero Defects, Technical Report TR 9, Department of Defense, Washington, D.C., 1968.

Quality Control

ASTM Manual on Quality Control of Materials. Special Technical Publication 15-C, January 1951, American Society for Testing Materials, Philadelphia. Basic material prepared by an ASTM committee: distributions, control charts, four basic tables of factors to use in constructing control charts, sampling.

I. W. Burr, *Statistical Quality Control Methods*, Marcel Dekker, New York, 1976. Reproduces essential parts of MIL-STD-105-D for sampling inspection for attributes.

E. L. Grant and R. S. Leavenworth, *Statistical Quality Control*, McGraw-Hill, New York, 4th edition, 1974. A textbook that has withstood the test of time.

W. A. Shewhart, *Economic Control of Quality of Manufactured Product*, Van Nostrand, New York, 1931; reprinted by the American Society for Quality Control, 1980. The classical treatment of statistical quality control by the person who originated and developed its basic concepts.

W. A. Shewhart, *Statistical Method from the Viewpoint of Quality Control*, edited by W. Edwards Deming, U.S. Department of Agriculture Graduate School, Washington, 1939. Based on four lectures Shewhart gave in Washington, March 1938.

Management and Related Aspects

Philip B. Crosby, *Quality Is Free—The Art of Making Quality Certain*, McGraw-Hill, New York, 1979; also a Mentor paperback, 1980. The approach to quality of a former executive of the ITT Corporation. Describes 14 steps in the quality improvement program. Rejects quality assurance and quality control in favor of just ''quality''; obviously is critical of statistics and reliability. Stresses Zero Defects and defect prevention.

W. Edwards Deming, *Quality, Productivity, and Competitive Position*, MIT Press, Cambridge, 1983. The noted authority draws on his long American and Japanese experience to describe 14 points which management has to follow to attain quality and productivity. Grounded in Shewhart's logic and methods.

R. Likert, *The Human Organization, Its Management and Value*, McGraw-Hill, New York, 1967. A psychologist summarizes what his experience and studies showed about human organization, including business organization.

Early treatment of participative management and its superiority over a dictatorial type of work situation.

D. MacGregor, *The Human Side of Enterprise*, McGraw-Hill, New York, 1960. A psychological approach to management; description of Theory X, which is autocratic and dictatorial, and Theory Y, which is participative and cooperative.

W. G. Ouchi, *Theory Z*, Addison-Wesley, Reading, Mass., 1981; also an Avon paperback, 1982. A theory of management based on subtlety, trust, and intimacy. Describes 13 steps needed to make Theory Z, a participative theory, work. One of the most valuable parts of the book is the Hewlett Packard statement of corporate objectives in the appendix, a statement by David Packard and William Hewlett that represents practice in the real world, not just theory.

T. J. Peters and R. H. Waterman, Jr., *In Search of Excellence*, Harper and Row, New York, 1982. Describes eight characteristics which are associated with highly successful businesses.

J. P. Wright, *On a Clear Day You Can See General Motors*, Wright Enterprises, Grosse Point, Mich., 1979; also an Avon paperback, 1979. John DeLorean's story about his experience working for General Motors with a detailed description of mismanagement and successful management. Describes management practices found elsewhere in both private industry and government. A story of management in the real world.

Probability and Statistics

W. G. Cochran and G. W. Snedecor, *Statistical Methods*, 7th edition, The Iowa State University Press, Ames, 1980. One of the best elementary textbooks in statistics, but it does require knowledge of elementary algebra.

M. J. Monroney, *Facts from Figures*, Penguin, New York, 1956. A Penguin paperback. A very good elementary introduction to statistics.

F. Mosteller, R. E. K. Rourke, and G. B. Thomas, Jr., *Probability and Statistics*, Addison-Wesley, Reading, Mass., 1961. The official textbook of Continental Classroom which appeared on NBC-TV presented by Prof. Mosteller. One of the best elementary expositions of probability, including the normal and binomial distributions. Highly recommended.

Warren Weaver, *Lady Luck*, Doubleday and Co., Garden City, New York, 1963. An interesting but sound description of the theory of probability and its application to a wide variety of problems by a well-known mathematician. Written for the beginner.

Probability Sampling

W. G. Cochran, *Sampling Techniques*, 2nd edition, Wiley, New York, 1963.

W. Edwards Deming, *Sample Design in Business Research*, Wiley, New York, 1960.

W. Edwards Deming, *Some Theory of Sampling*, Wiley, New York, 1950.

Frank Yates, *Sampling Methods in Censuses and Surveys*, 4th edition, Macmillan, New York, 1981.

None of these books is simple or elementary. The minimum requirement to obtain the most value from them is an elementary course in statistics and probability.

Applications to Service Industries and Operations

Irwin D. J. Bross, *Scientific Strategies to Save Your Life—A Statistical Approach to Primary Prevention*, Marcel Dekker, New York, 1981. The author draws on his wide experience in applied biostatistics to discuss low-level radiation, scientific criticism, conflicts in interpreting scientific data, the need to emphasize quality control in data collection, and 10 commandments in speaking to the public about quantitative data in applied science. Refreshing and challenging; applicable to a wider field than public health or biostatistics.

W. Edwards Deming, *Sample Design in Business Research*, John Wiley, New York, 1960. Describes in detail how replicated sampling is applied to obtain acceptable quality data for management in making decisions about service operations, such as opinions of employees, motel reservations, accounts due, rail freight traffic, inventory volume, property deeds, and appraisal of physical property. Contains considerable sampling theory as well.

J. M. Juran, F. M. Gryna, Jr., and R. S. Bingham, Jr., *Quality Control Handbook*, third edition, McGraw-Hill, New York, 1974. See especially Section 46 on Support Operations and Section 47 on Service Industries which describe in detail medication errors, telephone service quality control, restaurant quality control, and transportation service in the form of household goods moving, mail service, parcel service, and airline service.

A. C. Rosander, *Case Studies in Sample Design*, Marcel Dekker, New York, 1977. Describes in detail cases in which significant problems facing management were resolved by applying quality performance at various stages: planning, sample designs to collect data, appropriate tabulations, effective analysis, and using this knowledge to improve operations. All cases are from service operations or industries and include quality control for processing insurance claims, measuring effectiveness of rules and regulations governing interstate household goods movements, reducing idle or down-time to increase productivity and reduce costs, and constructing delay time distributions in rail traffic to improve service to shippers in critical geographical areas.

A. C. Rosander, *Statistical Quality Control in Tax Operations*, U.S. Treasury Department, Internal Revenue Service, Washington, D.C., 1958. Purpose of this cooperatively prepared exposition was to expose tax administration people to quality control and its advantages, then practically unknown. Explains basic ideas underlying quality control, the advantages of control charts and other techniques such as the frequency distribution. Applications include clerical errors, key punch errors, tax assessments, deviations from a straight line calling for audit, and what is needed for a quality control program to succeed. Reflects current ideas on quality with the exception of error prevention and intensive training, although training is mentioned.

Index

C

D

J

K

L

M

R

S

T

U

V

W

XYZ